AF270229

Butterflies of
British Columbia

# Acknowledgments

This book would not have been possible without the financial support of several British Columbia government agencies over the last five years. The Research Branch of the Ministry of Forests, the Royal British Columbia Museum, and the Conservation Data Centre of the Ministry of Environment, Lands and Parks obtained and administered grants from Forest Renewal BC and the Habitat Conservation Trust Fund.

We are especially indebted to our wives, who have provided far more than the normal encouragement that spouses give. For 39 years Sigrid Shepard has helped in the field to gather the information for this book. She has provided original illustrations of butterfly genitalia, lent her considerable expertise in library cataloguing to the search for obscure bibliographic references, and prepared the final bibliography. Aud Fischer has helped in the field for the last seven years, and has provided the original illustrations of butterfly morphology and adult butterflies for the book.

Cris Guppy thanks his parents for encouraging him to study butterflies as a child, while putting up with escaped caterpillars and butterflies in the house. He and his mother, Peggy Guppy, went on many companionable camping trips together over the years, Cris looking for butterflies and Peggy looking for wildflowers. He and his father, Art Guppy, went on a number of expeditions throughout BC, the Arctic, and elsewhere, Cris single-mindedly looking for butterflies and Art photographing wildflowers and birds. Without his parents' encouragement and help, Cris could never have acquired the knowledge to write this book.

Jon Shepard would like to extend special thanks to two individuals who offered him very valuable help early in the process of gathering material for this book. Alan C. Jenkins of North Vancouver donated his comprehensive collection of Vancouver and Trail area butterflies and provided extensive notes on the rearing of British Columbia species. Peter Hughan of New Aiyansh donated a comprehensive collection of local butterflies collected by his daughter and son, and provided advice about interesting habitat on the north coast.

David L. Threatful of Vernon and Norbert G. Kondla of Castlegar have corresponded with the authors and contributed voucher specimens and data for the last 25 years.

We wish to thank the following, who have graciously allowed us to use photos they have taken of live adult and, especially, immature butterflies: Steven Ansell (Victoria, BC); Gregory Ballmer (Department of Entomology, University of California, Riverside, CA); Richard Beard (West Vancouver, BC); Arthur Guppy (Victoria, BC); Oleg Kosterin (Russia); Derrick Marven (Duncan, BC); David McCorkle (Monmouth, OR); Anna Roberts (Williams Lake, BC); and Stephen Walker (Williams Lake, BC).

The following provided computerized data for the maps: Ann Potter and Julie Stofel of the Washington State Department of Fish and Wildlife provided the data, 24,000 records compiled by John Hinchliff through 1990, for Washington species. Dr. Jeffrey C. Miller and Dana Ross provided the data, 12,000 records compiled by John Hinchliff through 1994, for Oregon species. Dr. J. Donald Lafontaine, Ross A. Layberry, and Larry Spears provided 9,000 records from the CNC for Alberta, Yukon, and the Northwest Territories. These comprise approximately half the total database used for mapping species distributions.

We thank the following museum curators for providing us with access to their collections: Philip Ackery, British Museum of Natural History (BMNH); George Ball, Strickland Entomology Collection, University of Alberta (UA); Robert Cannings, Royal BC Museum (RBCM); Harry Clench, Carnegie Museum of Natural History (CM); Julian Donahue, Los Angeles County Museum (LACM); William Field, Smithsonian Institution (NMNH); Lee Humble, Pacific Forest Research Centre (PFRC); J. Donald Lafontaine, Canadian National Collection of Insects, Arachnids and Nematodes (CNC); John Lattin, Oregon State University (OSU); Frank Merickle, Barr Entomological Collection, University of Idaho (UI); Norman Penny, California Academy of Sciences (CAS); Kenelm Philip, Alaska Lepidoptera Survey, University of Alaska (ALS); Naomi Pierce, Museum of Comparative Zoology, Harvard University (MCZ); Jerry Powell, University of California, Berkeley (UCB); Raymond Pupedis, Peabody Museum of Natural History, Yale University (PMNH); Frederick Rindge, American Museum of Natural History (AMNH); Geoffrey Scudder, Spencer Entomological Museum, University of British Columbia (UBC); and Richard Zack, James Entomology Collection, Washington State University (WSU). The abbreviations in parentheses following the names of the institutions were used to identify record sources in the database, and some are used in the text of this book.

Various people in British Columbia government ministries gave help and encouragement: David Blades, Robert Cannings, and Gerald Truscott, Royal BC Museum; Sydney Cannings, Adolf Ceska, and Andrew Harcombe, Conservation Data Centre; Janet Mason, Geographic Names, Ministry of

Northern Checkerspot
(*Charidryas palla*)

Environment, Lands and Parks, Victoria; Lisa Wilkinson, Ministry of Environment, Lands and Parks, Fort St. John; Randy Wright, Ministry of Environment, Lands and Parks, Williams Lake; and Hugh Philip, Ministry of Agriculture, Kelowna.

Many of our colleagues provided advice and help on various aspects of the species accounts or provided data for the database as indicated by the acronyms in parentheses: Paul Arnaud, California Academy of Sciences, San Francisco, CA; Arthur Bartsch, History Professor, Nelson University Centre, Nelson, BC; Judith Beattie, Hudson's Bay Archives, Winnipeg, MB; Alexei G. Belik, Saratov, Russia; Ann Carroll, Archivist, City of Vancouver; Nelson Curtis (NC), Moscow, ID; Judith Deon, Librarian, Selkirk College, Castlegar, BC; Julian Donahue, Los Angeles County Museum; Vladimir V. Dubatolov, Siberian Zoological Museum, Novosibirsk, Russia; David L. Eiler (DLE), North Manchester, IN; Thomas C. Emmel, University of Florida, Gainseville, FL; Clifford D. Ferris (CDF), Laramie, WY; Bruce Fladmark, Glacier National Park, West Glacier, MT; John Fleckenstein (JF), Washington State Department of Natural Resources, Olympia, WA; John Gordon (JG), Vancouver, BC; Paul Hammond (PH), Philomath, OR; Suzanne Haverkamp, Curator, Penticton Museum, Penticton, BC: Gerald Hilchie (GH), Edmonton, AB; John Hinchliff (JH), Portland, OR; Steve Kohler (SK), Missoula, MT; Norbert Kondla (NGK), Castlegar, BC; Oleg Kosterin, Novosibirsk, Russia; Kim Kratky, English Instructor, Selkirk College, Castlegar, BC; Vincent Lee, California Academy of Sciences, San Francisco, CA; Maria Leung (ML), Smithers, BC: David McCorkle (DVM) Monmouth, OR; Scott Muller (SM), Germansen Landing, BC; Dr. Judith H. Myers, Department of Zoology, University of British Columbia, Vancouver, BC; W. Dean Nicholson (WDN), Cranbrook, BC: Soren Nylin, Department of Zoology, Stockholm University, Sweden; Paul Opler, Loveland, CO; Jon Pelham (JPP), Seattle, WA; Ann Potter, Washington State Department of Fish and Wildlife, Olympia, WA; Robert M. Pyle (RMP), Greys River, WA; Carla Rickerson, Pacific Northwest Collection, University of Washington, Seattle, WA; Donald Rolfs (DAR), Wenatchee, WA; William (Bill) A. Sloan, History Instructor, Selkirk College, Castlegar, BC; Graham Smith, Geosense Consulting, Nelson, BC; Dmitry Sobanin, Voronezh, Russia; Ray Stanford (RES), Denver, CO; Julie Stofel, Washington State Department of Fish and Wildlife, Olympia, WA; Fay Tangemann, Yukon Archives, Whitehorse, YT; David Threatful (DLT), Vernon, BC; Vadim Tshikolovets, Zoological Museum, Kiev, Ukraine; Vasily K. Tuzov, Moscow, Russia; Niklas Wahlberg, Department of Ecology and Systematics, University of Helsinki, Finland; and Wayne Wehling (WW), Michigan State University, Lansing, MI.

© The Royal British Columbia Museum 2001

Publication date: March 1, 2001

Published by UBC Press in collaboration with the Royal British
Columbia Museum.



Printed in Canada on acid-free paper

ISBN 0-7748-0809-8

**National Library of Canada Cataloguing in Publication Data**

Guppy, Crispin Spencer, 1953-
  Butterflies of British Columbia

  Copublished by: Royal British Columbia Museum.
  ISBN 0-7748-0809-8

  1. Butterflies – British Columbia.  2. Butterflies – Northwest,
Pacific.  I. Shepard, Jon, 1941-  II. Royal British Columbia Museum. III.
Title.

QL552.G86 2001        595.78'9'09711        C00-911644-3

The Royal British Columbia Museum and UBC Press gratefully acknowledge
Forest Renewal BC for financial assistance in the production of this book.

UBC Press acknowledges the financial support of the Government of Canada
through the Book Publishing Industry Development Program (BPIDP) for our
publishing activities.

Canadä

We also gratefully acknowledge the support of the Canada Council for the
Arts for our publishing program, as well as the support of the British Columbia
Arts Council.

UBC Press
The University of British Columbia
2029 West Mall
Vancouver, BC V6T 1Z2
(604) 822-5959
Fax: (604) 822-6083
E-mail: info@ubcpress.ubc.ca
www.ubcpress.ca

Publishing Services
Royal British Columbia Museum
675 Belleville Street
Victoria, BC V8W 9W2
www.royalbcmuseum.bc.ca

*Illustrations*
page 1: Northwestern Fritillary (Williams Lake, Anna Roberts)
page 2: Mylitta Crescent (Camas Hill, A.G. Guppy)
pages 4-5: Western Sulphur (Shawnigan Lake, Derrick Marven)

Crispin S. Guppy and Jon H. Shepard

# Butterflies of British Columbia

Including

Western Alberta

Southern Yukon

The Alaska Panhandle

Washington

Northern Oregon

Northern Idaho

Northwestern Montana

A Royal British Columbia Museum book published by UBC Press

# Introduction

Butterflies have been likened to flowers that fly and are among the most beautiful and fascinating of animals in nature. They are found everywhere in British Columbia, from balcony planter boxes in the city to coastal bogs and wild ocean shorelines, alpine flower meadows of the coastal mountains, deserts of the southern interior, grasslands of the Chilcotin, and the vast unexplored expanses of boreal forest and mountains across the north. Although the butterfly species found in British Columbia represent only a small fraction of the world's approximately 15,000 species, they can still provide a lifetime of study. Observation of the complex behaviours and life histories of each butterfly species within its own unique habitat can provide years of enjoyment for the butterfly enthusiast.

This book has been written to provide both the naturalist and the professional biologist with an overview of the fascinating butterfly fauna of British Columbia. We hope that it will stimulate interest in BC butterflies and contribute to their conservation. We have widened the scope of the distribution maps beyond BC to show the distribution of BC butterfly species in the Alaska panhandle, southern Yukon, western Alberta, Washington, northern Oregon, northern Idaho, and northwestern Montana (Fig. 1). The inclusion of these adjacent areas provides a context in which to better understand the distribution of species and subspecies within BC.

One hundred and eighty-seven species are known from British Columbia, and an additional nine species that occur very near our borders are discussed and mapped. Three of the potential species are Arctic species that occur on Montana Mountain, south of Carcross, Yukon. The southern aspect, non-alpine slopes of Montana Mountain actually extend into British Columbia. There are similar mountains, but accessible only with difficulty, just south in British Columbia where these species will certainly be recorded for the province in the future. Two other potential BC species are rare migrants that have been recorded south of British Columbia, in Montana, Idaho, and/or extreme eastern Washington. With warmer climate due to global warming and with more observations in the East Kootenay, these migratory species should eventually be recorded in the province. Thus the total butterfly fauna of BC should eventually reach at least 192 species. The other four potential species are found adjacent to the eastern or southern borders of the province in Alberta, northwestern Montana, or Washington. There is a remote chance that they will eventually be recorded in British Columbia, bringing the potential fauna to 196 species.

The total known fauna of 187 species of butterflies is by far the largest in Canada. Alberta and Ontario, with about 160 species each, are the next most diverse (Table 1). If one ignores the migratory, non-breeding species, the greater diversity of butterflies in BC is even more obvious (Table 1).

In this book, the term *butterfly* is used to mean both true butterflies, in the superfamily Papilionoidea, and the other major group of day-flying Lepidoptera, the skippers, in the superfamily Hesperioidea. The old, bifold division into butterflies and moths is extremely artificial, but in the introductory chapters it is convenient to use the vernacular word *butterflies* to mean both Papilionoidea and Hesperioidea.

We follow the reassessment of the higher classification by Weller et al. (1996), which recognizes two superfamilies and six families. The Riodinidae are kept as a separate family, but monarchs (Danainae) and woodnymphs (Satyrinae) are subfamilies within the large family Nymphalidae.

The scientific classification of all animals began with the adoption of Carolus Linnaeus's binomial nomenclature in the 10th edition of his *Systema Naturae,* 1758. The binomial name (or scientific name) of an organism consists of a more inclusive generic name followed by a specific name. There are very specific rules set down by the International Code of Zoological Nomenclature to govern use of names. Essentially the "rule of priority" prevails. Changes of names in scientific and popular publications reflect changes in people's thinking regarding the boundaries of all levels of classification, for example, the amount of variation that defines a particular taxon, such as a species, genus, or family.

Many naturalists have found it frustrating that both higher classification and the binomial names of butterflies have varied greatly in butterfly books published during the last three decades. This reflects both the rapid increase in knowledge of classical morphological data and the use of chemical data from analysis of DNA or proteins. Even some scientists (Ehrlich and Murphy 1982) have not been tolerant of these changes, despite their being the inevitable result of increased scientific knowledge.

We hope that the additional species recognized and the additional subspecies described in this book will be met with equanimity. Our reasons for splitting several species into two

**1** Map of British Columbia and adjacent areas showing place names and ecoprovinces (BC)

| TABLE 1. COMPARISON OF BUTTERFLY SPECIES DIVERSITY IN THE PROVINCES AND TERRITORIES OF CANADA | | | | | | | | | | | | | | | |
|---|---|---|---|---|---|---|---|---|---|---|---|---|---|---|---|
| **FAMILY** | | YK | BC | AB | SK | MB | ON | PQ | NB | PE | NS | NF | NT | CAN |
| HESPERIIDAE | BREEDING | 6 | 26 | 24 | 33 | 36 | 40 | 26 | 15 | 10 | 14 | 5 | 4 | 58 |
| | MIGRANT | 0 | 1 | 1 | 1 | 2 | 10 | 0 | 0 | 0 | 0 | 0 | 0 | 10 |
| PAPILIONIDAE | BREEDING | 5 | 12 | 8 | 6 | 4 | 6 | 4 | 3 | 2 | 3 | 2 | 2 | 16 |
| | MIGRANT | 0 | 0 | 0 | 1 | 2 | 1 | 0 | 0 | 0 | 1 | 0 | 0 | 3 |
| PIERIDAE | BREEDING | 15 | 26 | 19 | 14 | 14 | 12 | 11 | 5 | 4 | 5 | 8 | 18 | 30 |
| | MIGRANT | 0 | 2 | 3 | 3 | 5 | 9 | 2 | 0 | 0 | 1 | 0 | 0 | 9 |
| LYCAENIDAE | BREEDING | 12 | 44 | 34 | 33 | 28 | 32 | 28 | 22 | 7 | 18 | 8 | 12 | 60 |
| | MIGRANT | 0 | 0 | 0 | 2 | 1 | 2 | 0 | 0 | 0 | 0 | 0 | 0 | 4 |
| RIODINIDAE | BREEDING | 0 | 1 | 0 | 1 | 0 | 0 | 0 | 0 | 0 | 0 | 0 | 0 | 1 |
| | MIGRANT | 0 | 0 | 0 | 0 | 0 | 0 | 0 | 0 | 0 | 0 | 0 | 0 | 0 |
| NYMPHALIDAE | BREEDING | 47 | 69 | 65 | 52 | 54 | 46 | 45 | 29 | 16 | 28 | 27 | 38 | 90 |
| | MIGRANT | 1 | 6 | 7 | 6 | 8 | 5 | 5 | 3 | 2 | 4 | 3 | 1 | 11 |
| TOTAL BREEDING | | 85 | 178 | 150 | 135 | 136 | 136 | 114 | 74 | 39 | 68 | 50 | 74 | 255 |
| % CANADIAN BREEDING FAUNA | | 33 | 70 | 59 | 53 | 54 | 54 | 45 | 30 | 15 | 27 | 20 | 30 | |
| TOTAL MIGRATORY | | 1 | 9 | 11 | 13 | 18 | 27 | 7 | 3 | 2 | 6 | 3 | 1 | 37 |
| TOTAL BREEDING + MIGRATORY | | 86 | 187 | 161 | 148 | 154 | 163 | 121 | 77 | 41 | 74 | 53 | 75 | 292 |
| % CANADIAN FAUNA | | 30 | 64 | 55 | 51 | 53 | 56 | 41 | 26 | 14 | 25 | 18 | 25 | |
| DUBIOUS | | 1 | 0 | 1 | 0 | 1 | 0 | 1 | 0 | 0 | 0 | 0 | 1 | 0 |
| OCCASIONAL* | | 0 | 0 | 2 | 0 | 0 | 1 | 0 | 0 | 0 | 0 | 0 | 0 | 0 |

* Occasional refers to species that have been recorded a few times, are not established species, and are not considered migratory.

separate species are given in the species accounts. This is largely the result of defining the limits of distribution for the allopatric species pairs and establishing that there appears to be little or no hybridization where they occur sympatrically. We have been able to find consistent morphological differences to differentiate each species pair.

Another problem is that some lepidopterists have arbitrarily chosen to ignore the International Code of Zoological Nomenclature rules regarding gender agreement between genus and species in binomial names. We follow these rules, so the general reader will find many differences in the spelling of species names between this book and the recently published *A Field Guide to the Western Butterflies* (Opler 1999). Opler ignored the International Code and reverted to original Latin endings for some species names and these names do not agree with the gender of the current generic assignment.

In this book, the historic common names of species, genera, and higher taxa are usually provided. These have been more stable than the scientific names. They first evolved in a haphazard manner, however, and do not reflect scientific concepts. Recent attempts by the North American Butterfly Association (NABA) (Cassie et al. 1995; Opler 1999) to make common names more closely reflect the scientific classification (based on the evolutionary origins of butterflies) will, if accepted, lead to as many changes in common names as in scientific names in the future.

The common names of most British Columbian and Canadian boreal butterflies were first coined by an early English naturalist, Phillip Gosse (1840, 1841a, 1841b, 1844, 1859), who lived in eastern Canada and then travelled to Alabama. The great American entomologist Samuel Scudder recognized the need to standardize the use of common names. Scudder (1874b, 1875) provided the first list of common names and recognized those of Gosse. Later Scudder used those names in his classic study *The Butterflies of the Eastern United States and Canada with Special Reference to New England* (Scudder 1889a, 1889b) and in several other popular books on butterflies. The reinvention of common names began early, when Scudder's rival, Charles Maynard (1886, 1891), published competing popular books and coined a different common name for virtually every butterfly species. Maynard was by no means Scudder's intellectual equal, however, and Scudder's usage has since prevailed.

In the last 15 years, there have been some drastic changes to common name usage, stemming from three major sources. Scott (1986b), in his *Butterflies of North America,* made many unwarranted changes in common names. NABA has produced a list of common names (Cassie et al. 1995) that are closer to historic usage (Miller 1992), but the organization has been very dogmatic and has suggested that only their names should be used. NABA has even criticized more recent scientific and popular works that increased or decreased the number of recognized species, on the grounds that they did not want "their" common names altered. Such arguments are like trying to decide how many angels can dance on the head of a pin and do not contribute to either scientific understanding or conservation of butterfly species. A third source of changes in common names is the tendency of government agencies to insist on using common names, often changing them to more politically correct names or to names reflecting regional bias. For example, in British Columbia and elsewhere, *Euphydryas editha taylori,* named after the province's first resident butterfly enthusiast, has always been known as Taylor's Checkerspot. In a list of Washington conservation management priority species of butterflies, however, Larsen et al. (1995) ignored the historic name and used "Whulge Checkerspot" instead!

For this book, we have kept closely to the historic common names of Gosse and Scudder and also to historic British Columbian usage. British Columbia was the first area in western North America to use common names in readily available publications (Anderson 1904; Harvey 1904; Anonymous 1906; Blackmore 1927). Other than Elrod (1906) and Comstock (1929), there were no books or provincial/state checklists in western North America using common names before the Second World War. Wright (1905) did not use common names. Elrod's usage was as unorthodox as Maynard's and has been ignored by subsequent authors. After the war, Brown et al. (1957) kept to historic usage for their book on Colorado butterflies. Leighton (1946) also kept to historic usage for Washington state butterflies, including the common name "Taylor's Checkerspot." In British Columbia, Jones (1951) and Harcombe and Underhill (1970) continued to follow historic usage; only in the last 15 years have authors or editors with particular viewpoints strayed from this. Another factor that influenced our usage of common names is the Holarctic nature of the British Columbia fauna for both species and genera. This is especially important for genera such as *Parnassius, Boloria,* and *Clossiana,* where we have often followed the British common names used in Europe.

In the species accounts, we provide the etymology of both Latin and common names. Hopefully, this will make the names more meaningful to the general public. In this we have been greatly aided by *The Scientific Names of the British Lepidoptera: Their History and Meaning,* by A. Maitland Emmet (1991), an extremely readable book that is at the same time very authoritative. For scientific names not in Emmet and not explained in the original publication, we have used *The Naturalist's Lexicon* (Woods 1944), *Room's Classical Dictionary* (Room 1983), *The Oxford Classical Dictionary* (Hornblower and Spawforth 1996), and *The Century Dictionary and Cyclopedia* (Smith 1902a, 1902b; Whitney 1902) to ferret out as many meanings of the Latin names as possible.

The introductory chapters emphasize general information about the biology and morphology of butterflies, and the special aspects of history, geological history, human impact, ecology, and conservation that are pertinent to the British Columbia fauna. For an excellent overall treatment of butterfly biology and morphology, we recommend *The Butterflies of North America* (Scott 1986b). For more in-depth reading, *The Lepidoptera: Form, Function and Diversity* (Scoble 1995) is recommended.

In the species accounts section, we have followed the general format of other books on butterfly faunas. There are, however, three ways in which this book is unique. First, we have made a special effort to examine the original scientific literature on the life histories and foodplants of the individual species. This is in contrast to most other North American books, which have copied each other and perpetuated many errors. We have provided citations to the original sources of these life histories and any subsequent references to observations made in areas closest to the provincial boundaries; referenced all papers that describe the life histories of butterfly populations observed within the province; and included information from our own and other unpublished sources in British Columbia, Alberta, and the United States Pacific Northwest region. Unpublished sources are identified by initials, as follows: A.G. Guppy (AGG), C.S. Guppy (CSG), G.A. Hardy (GAH), A.C. Jenkins (ACJ), J.P. Pelham (JPP), J.H. Shepard (JHS), and Forest Insect Survey (FIS).

For future research, it should be noted that besides the errors repeated in most butterfly books, we found many errors in other references to lepidopteran life histories. Foremost of these was the two-volume *Index to Life Histories* (Tietz 1972). We checked most of the references pertinent to the BC butterfly fauna and found many errors of both commission and omission, with Tietz listing larval foodplants that were not mentioned in the references cited, or providing the wrong volume, page, or sometimes even journal title for the references. In unravelling the literature, we were greatly aided by the extensive library and bibliographic index prepared by Jon Shepard, totalling more than 10,000 references, and by the bibliography published by Bridges (1993) covering all but the Nymphalidae; there is a 10% non-overlap between the two sources. Shepard's bibliography covering up to 1972 was also used by Scott (1986) for his book. Another major source of errors consisted of the larval foodplants recorded in Volume 1 of *Forest Lepidoptera of Canada* (McGugan 1958). Shepard examined the voucher specimens of the butterflies mentioned by McGugan (1958) to correct species misidentifications, and we have dismissed the recorded foodplants that appear to be data errors. *An Annotated Checklist of the Macrolepidoptera of British Columbia* (Jones 1951) contains many erroneous foodplant records derived from the literature, but the foodplant observations by Jones himself, which he had published earlier, are accurate.

We have used Brako et al. (1995), Douglas et al. (1998a, 1998b, 1999), Kartesz (1994a, 1994b), and Morin (1993, 1997) to standardize the names for larval foodplants of butterflies taken from the literature.

The museum specimen photographs in this book were taken by Cris Guppy using a custom-designed and handcrafted copy stand designed by Donald Rolfs of Wenatchee, Washington. The copy stand was an essential tool for producing high-quality photographs without background shadows and without the degradation of quality that results from digital replacement of the background. Most butterfly books use a white or off-white background colour for all photographs, and a few books use a black background. A white background obscures the white wing fringes characteristic of most hesperiids and lycaenids, and the wings of predominantly white butterflies of the family Pieridae. Similarly, dark butterflies disappear into a black background. Guppy chose a blue background for the hesperiids and lycaenids, which shows off the white wing fringes. Lighter backgrounds were used for the larger butterflies because most are dark in colour even along the wing fringes.

A second unique aspect of this book is the use of computer-generated distribution maps for each species and subspecies. The only other butterfly book to use computer-generated maps is *The Butterflies of Canada* (Layberry et al. 1998), which, however, does not show subspecies distributions. Shepard provided most of the British Columbia records for the Canadian book, and Guppy provided data from his collection. These totalled 13,827 of the more than 90,000 unique species/locality records for all of Canada available at the time. Shepard had not yet assigned mapping coordinates to the main database, resulting in errors in the assignment of coordinates for the maps in the Canadian book, which have been corrected for this book. Among the more conspicuous errors in the Canadian book are: (1) the records of the Monarch from northeastern BC, (2) the records of the West Coast Lady from northwestern BC, and (3) the records of the Red Admiral from the Queen Charlotte Islands and Atlin. These are all records of migratory species that were incorrectly extracted by Layberry et al. (1998) from the annual Season's Summary reports of the Lepidopterists' Society. Another major error was the mapping of an occurrence of *Parnassius clodius* at Copper Mountain near Terrace. There are three "Copper Mountains" in BC. The correct Copper Mountain for this and other species incorrectly attributed to near Terrace is the Copper Mountain near Nelson, in the Kootenays. Other, less obvious discrepancies

between the maps in this book and in the Canadian book are assumed to be due to similar errors in the Canadian book and we did not attempt to trace their sources.

The database for the mapping was almost completely compiled by Shepard, who personally examined most of the material except for some records of Guppy, Norbert Kondla, and David Threatful, who provided extensive records from their collections and sight record data. Otherwise, sight records were not used, and no sight records were used for species hard to distinguish in the field. Also, since 1975 Shepard has been receiving records in his capacity as Season's Summary Pacific Northwest coordinator for the Lepidopterists' Society. When there was any doubt about the correct identification of a species record from this source, the record was not included in the database (for example, some *Colias, Euphydryas,* and *Erynnis*).

The third unique aspect of this book is the use of computer-generated bar graphs of adult flight times. Each unit on the bar graphs represents a single species/locality/date combination. Thus if more than one specimen of a species was sighted or collected at the same site on the same day, the additional individuals were ignored in generating the bar graphs. Shepard experimented with plotting seasonal bar graphs using the total number of specimens of a species recorded, but the tendency of some collectors to collect large numbers of specimens greatly misrepresented seasonal flight, especially for species with limited data.

The total database consists of approximately 28,000 unique species/locality/date records for 196 species, including species of potential occurrence in BC. We estimate that we examined more than 100,000 individual specimens. About half were in private collections belonging to many people in British Columbia and the United States as well as Shepard and those mentioned earlier. The major public collections examined were those of the Royal British Columbia Museum, the University of British Columbia, the Pacific Forest Research Centre, the Alaska Lepidoptera Survey (the University of Alaska and the Smithsonian Institution), Washington State University, the California Academy of Sciences in San Francisco, the Canadian National Collection (CNC) of Insects in Ottawa, and the American Museum of Natural History in New York. Time did not permit a return visit to the Smithsonian Institution in Washington, DC, to record data other than those for *Parnassius, Boloria,* and *Clossiana.* The general BC material in the Los Angeles County Museum was not inventoried because of the museum's new policy of charging paid researchers for the privilege of examining material. No public institution should charge such fees, especially since most of the material in their holdings has been donated. Rather, they have an obligation to provide access or give their material to an institution that can.

Besides the institutions visited to obtain BC records, we examined the following institutional collections in order to resolve difficult taxonomic issues regarding BC species:

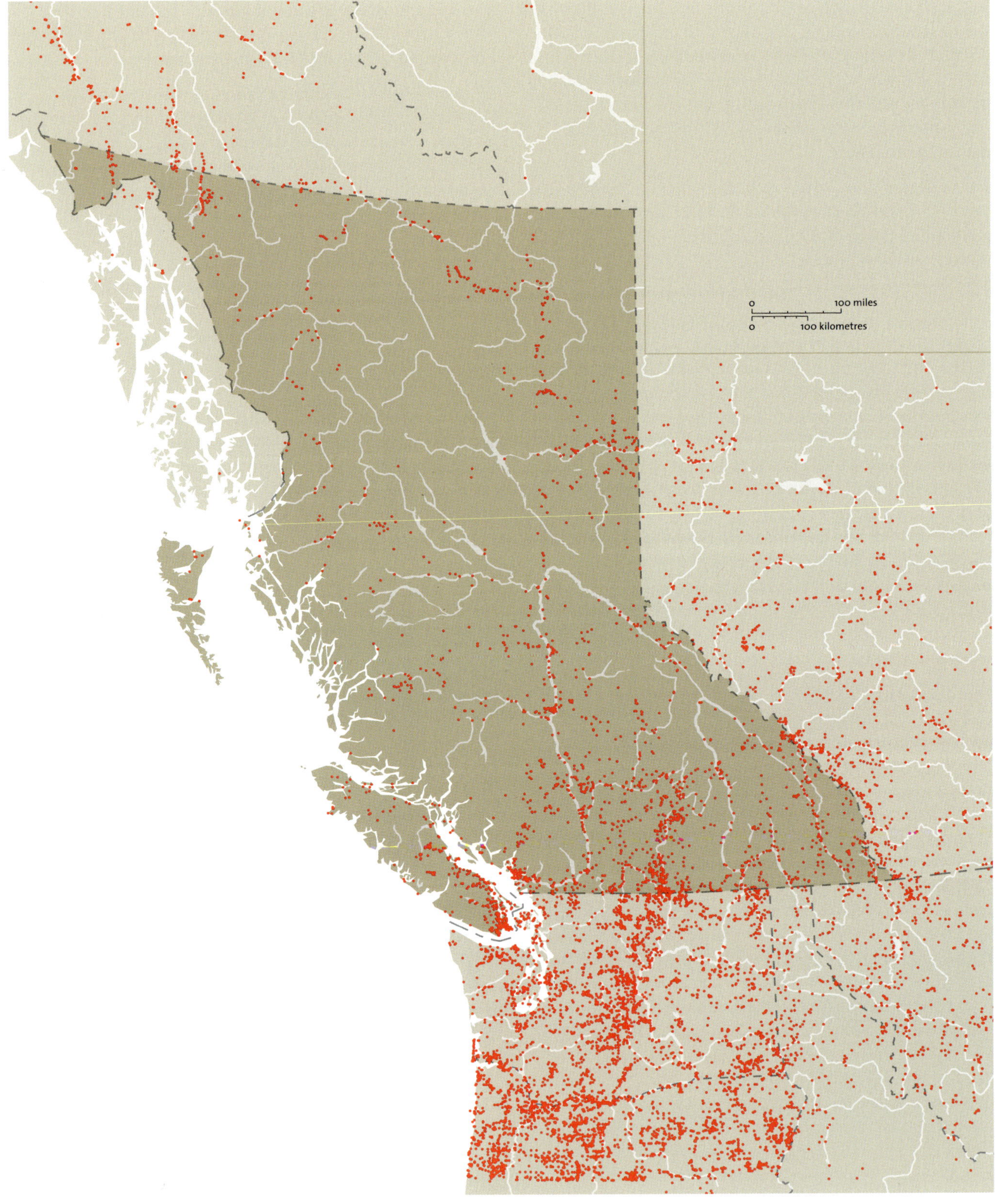

**2** Map of British Columbia and adjacent areas showing the localities where one or more butterfly species has been recorded

University of Alberta, University of Idaho, University of Washington, Oregon State University, University of California at Berkeley, Los Angeles County Museum, Harvard University, Yale University, Carnegie Museum of Natural History, Philadelphia Academy of Sciences, and the Smithsonian Institution. In addition, the curators at the British Museum lent material and examined specimens to answer our inquiries. For the genera *Boloria, Clossiana,* and *Parnassius,* records were also taken from all of these additional sources.

There are approximately 3,000 individual localities in British Columbia and the Alaska panhandle where at least one species has been recorded (Fig. 2). We also provide as many records of southern Yukon and western Alberta species as we were able to examine. Most of these are from the collections of the authors, Gerald Hilchie, Norbert Kondla, the University of Alberta, and the Canadian National Collection. Records from Washington, northern Idaho, and northwestern Montana were collected and compiled by Shepard, with additional personal records from Jon Pelham, Nelson Curtis, and Steve Kohler for their respective states. The database for United States records also includes records sent to Shepard for the Lepidopterists' Society Season Summary, especially records of David Eiler, Robert Pyle, and Clifford Ferris.

The sources of the unique species/locality/date records used in plotting the adult seasonal flight periods are summarized in Table 2. About 16,000 of these were from public collections, reflecting long periods of collecting in Victoria, the North Okanagan, and Cranbrook. About 12,000 records are from private collections.

The sources of the unique species/locality records used for mapping the distribution of each species are summarized in Table 3. There is almost no duplication of records between the locality records held in public collections and those in private collections, and both groups have contributed equally to the maps.

We have made a point of showing how much of our knowledge of BC butterflies is due to private efforts. Even the figures shown above are not completely accurate. Of the 16,000 voucher records in public collections, all but about 5,000 represent specimens donated by private individuals. Only the Canadian National Collection is based largely on material collected by employees of the institution. During the last 100 years, the major role of Canadian museums other than the CNC has been to preserve privately collected material for future generations to study. None of the regional collections, except the Halifax Institute of Science, has ever employed a lepidopterist in a professional capacity to conduct research on North American lepidopteran fauna and add significantly to the regional collection.

We would like to continue expanding the database, and we invite contributions. Voucher specimens or photographs to verify identification would be appreciated. To make contributions and obtain directions about the geographic information needed, please contact Jon H. Shepard, RR #2, S-22, C-44, Nelson, BC V1L 5P5 / (250) 352-3028 / shep.lep@netidea.com. For other information about butterflies, please contact either Shepard or Cris Guppy, 4627 Quesnel-Hydraulic Road, Quesnel, BC V2J 6P8 / (250) 747-1512 / cguppy@quesnelbc.com. Should our addresses change, contact us through the Royal BC Museum.

As co-authors, we worked jointly and cooperatively on all aspects of this book. Guppy was responsible for the photographs, while Shepard was responsible for maintaining and enlarging the database and producing the distribution maps. Guppy wrote the chapters on conservation, butterfly gardens, and biology of butterflies, while Shepard wrote those on the study of butterflies in BC, the origins of the BC fauna, and the morphology of adult and immature butterflies, as well as this introduction to the book. We shared the writing of the chapter on human impacts on butterflies. Shepard wrote the species accounts for the Hesperiidae, Lycaenidae, Riodinidae, Argynninae, and Melitaeinae, while Guppy wrote those for the Papilionidae, Pieridae, Nymphalinae, Limenitidinae, Satyrinae, and Danainae. We shared the determination of the origins of both scientific and common names for all the species. This division of efforts reflects our individual interests and strengths, but we share authorship of all parts of the book, except certain of the newly described subspecies as noted in the text.

# The Study of Butterflies in British Columbia

The study of butterflies in British Columbia began in 1858. None of the well-known British naturalists, such as David Douglas, Thomas Drummond, or Edward Ermatinger, who collected natural history specimens for their sponsors in Britain, collected butterflies or any other insects in British Columbia before that year (Shepard 1984). There is also no evidence that the early Russian naturalists, who collected in the Alaska panhandle and on the California coast near San Francisco, ever stopped in British Columbia (Tuzov 1997; Weiss 1998). An examination of the diaries and correspondence of the various people in charge of Hudson's Bay Company posts from 1800 to 1850 gives no indication that any company employee based in British Columbia, Washington, or Idaho or any casual visitor ever collected any insect material (Shepard 1984). A. McDonald, the company factor at Fort Langley, BC, and then at Fort Colville, Washington, from 1828 to 1844, wrote to the botanist Thomas Hooker: "I am extremely sorry to have to report that, with the single exception of our mutual friend Mr. Tolmie, the Gents, of the west side [BC and Washington] are very reluctant to dab in anything connected with the vegetable or animal kingdom" (Cole 1979). Mr. Tolmie was based at Nisqually, Washington, and was not interested in insects. In 1847 and 1849, Edward Doubleday described six species of butterflies from the "Rocky Mountains." For the next 50 years, the majority opinion was that they came from British Columbia (Elwes 1889; Strecker 1892), but we now know that they were collected near Jasper, Alberta (Shepard 1984).

The first naturalists to take an interest in British Columbia butterflies were associated with the boundary survey between British Columbia and Washington Territory, which began in 1858. John Kest Lord was appointed naturalist to the British North American Boundary Commission. He served from 1858 to 1862 and reported his findings in *The Naturalist in Vancouver Island and British Columbia* (Lord 1866). The appendix to this book listed 22 species of butterflies and three species of moths, two of which were described as new species. The descriptions of the new moths and apparently the list of butterflies were prepared by the curator of Lepidoptera at the British Museum of Natural History, Francis Walker. Lord travelled along the BC/United States border from the Gulf of Georgia to Osoyoos, BC. The survey parties first surveyed up to the Roche River in the Hope Mountains. Finding the mountains impassable, they returned to the coast and then travelled up the Columbia River to Colville, WA (Lyall 1863). From that supply base, they surveyed the border from the Ashnola River, near Keremeos, BC, to Alberta.

In addition, David Lyall, who was surgeon for *HMS Plumper,* a ship used by the boundary survey (Lyall 1863), collected 80 butterfly and moth specimens in the Victoria area. Both Lyall and Lord donated their material to the British Museum, where it can still be found. Lyall's specimens were collected in the spring of 1858. At the same time, the American Alexander Agassiz collected material that was labelled "Gulf of Georgia" or "Washington Territory." That material was donated to the Museum of Comparative Zoology (MCZ) at Harvard University, where Agassiz later became director. Hagen's comments (Hagen 1884) indicate that these were collected in 1859.

Samuel Scudder, the pre-eminent North American butterfly authority of this period, described the first British Columbia species of butterflies to be described from material collected in the province, *Pieris marginalis* in 1861 and *Colias occidentalis* in 1862 (Scudder 1861, 1862). These were part of the material collected by Agassiz. Victoria was the supply centre for both Canadian and American parties of the survey. They wintered there and returned for supplies in the summer of the first year of the survey. Agassiz's "Gulf of Georgia" specimens came from the Victoria region, the San Juan Islands, WA , and/or Boundary Bay, WA.

George Robert Crotch of Boston travelled to the interior of BC, reaching at least the Cariboo region on or before the summer of 1873. In 1874, William Henry Edwards described two varieties of fritillaries from butterflies collected by Crotch, *Speyeria hydaspe rhodope* and *S. mormonia opis*. In 1877, Henry Edwards described another fritillary from Crotch's material, *S. aphrodite columbia*. These specimens are preserved in the Museum of Comparative Zoology and the Carnegie Museum of Natural History.

George M. Dawson and his assistant, J. McEvoy, were the first to survey butterflies in the northern half of the province. In the summer of 1887, they collected sporadically between Telegraph Creek (17 May) and Dease Lake (17 June) while conducting geological surveys (Fletcher 1888). Fletcher (1888) also reported a single butterfly species taken at Chilkoot Pass in 1886. This material is now at the Canadian National Collection of Insects, Arachnids and Nematodes (CNC) in Ottawa.

The study of butterflies and other Lepidoptera in British Columbia began in earnest in 1882, when the first resident amateur entomologist, the Reverend George W. Taylor (Fig. 3), arrived in Victoria from Derby, England, at the age of 25

3 **Reverend George W. Taylor** (*Can. Ent.* 44 [1912]: pl. x, opposite p. 285)

(Fletcher 1904; F. 1912; Hanham and Fyles 1913; Taylor 1990).

In *Notes on the Entomology of Vancouver Island* (Taylor 1884), he noted observing "nearly forty species" of butterflies; his only other paper mentioning butterflies was published four years later (Taylor 1888). Taylor was content to collect and send butterfly specimens to many entomologists, including William Henry Edwards, who described Taylor's Checkerspot butterfly, *Euphydryas editha taylori,* in 1888. Taylor also contributed the British Columbia butterfly records to a list of Canadian insects published by Brodie and White (1883). Later, Taylor became the leading authority on North American geometrid moths (inchworms) after George Hulst died, a world authority on molluscs, and the first Government of Canada biologist for the Marine Biological Station at Departure Bay, Nanaimo. His collection of geometrid type specimens was purchased by Dr. William Barnes of Decatur, Illinois, and is now housed at the Smithsonian Institution in Washington, DC. Virtually every major natural history museum in the northern hemisphere has lepidopteran material collected by Taylor.

In 1888, two people who were very interested in butterflies arrived in the province. Charles de Blois Green came from England, where he had been trained as a land surveyor, and settled first in Nelson and then in Okanagan Landing. W.H. Danby moved to Victoria from New York City and had settled in Rossland by 1895. De Blois Green and Danby were very enthusiastic collectors and amassed the first significant collections of British Columbia butterflies. The de Blois Green collection is in the Canadian National Collection in Ottawa; Danby's collection was purchased by Cyril dos Passos and is now at the American Museum of Natural History in New York. In 1893 they published a checklist of 90 species of butterflies and 71 species of moths (Danby and de B. Green 1893). Overlooked by subsequent bibliographies, this checklist nevertheless contains the first illustrations of British Columbia Lepidoptera ever published. Danby (1894) republished an abbreviated form of this list. Together the two lists included most of the currently known butterfly fauna of southern British Columbia and 50% of the total butterfly fauna of the province.

From the 1880s, when railway travel became possible, eastern Canadian entomologists made regular summer trips to British Columbia. The most well known of these was James Fletcher, Dominion Entomologist. Two well-known California lepidopterists, Henry Edwards in the 1880s and William Greenwood Wright in 1891, also made trips to Vancouver Island. Fast on the heels of these early pioneers, the southern part of the province was filled with lepidopterists and local naturalists whose primary interest was butterflies and moths. Most of the early lepidopterists lived either on southeastern Vancouver Island, in Vancouver, or in New Westminster.

In the Kootenays, Danby was joined by J. William Cockle (Fig. 4) in the late 1890s. Cockle ran the Kaslo Hotel in Kaslo. He advertised the hotel in the *Canadian Entomologist* as a good place to collect from and spend the summer. Cockle quickly established himself as one of the most important lepidopterists in the province. He reared Lepidoptera, published observations on behaviour, and was the first resident lepidopterist to describe butterfly species and subspecies from British Columbia material (Cockle 1910). Besides his interest in natural history, Cockle was a good observer of all insects of medical and agricultural importance and corresponded with James Fletcher and Charles Bethune from 1904 to the 1930s. Fletcher regularly reported Cockle's observations in the *Canadian Entomologist*. In 1903, H.G. Dyar, curator of the Smithsonian Lepidoptera Collection, joined Cockle for the entire summer and later published a summary of his and Cockle's work (Dyar 1904). The Provincial Museum in Victoria could not find the funds to purchase Cockle's collection in the 1930s, and it was sold to the Canadian National Collection of Insects and Arthropods in Ottawa. Besides studying Lepidoptera, Cockle was the first to recognize that a local outbreak of paralysis in 1915 was "tick paralysis" and not spinal meningitis as both local and eastern Canadian doctors had thought. By the mid-1960s, doctors in the Kootenays had forgotten that "tick paralysis" was a significant disease and another lepidopterist, Jon Shepard, had to re-educate them.

Of the lepidopterists on Vancouver Island and in the Vancouver area at this time, only the most important can be mentioned in detail here.

4 **J. William Cockle (left) (BC Archives, HP 39042)**

E.M. Anderson was assistant curator of natural history at the Provincial Museum in Victoria from 1903 to 1916, and published the first comprehensive checklist of BC Lepidoptera in 1904. One hundred seventy-five species of butterflies were listed, 125 of which are recognized today. Only 108 were correctly identified, however, as Captain R.V. Harvey (1904) noted.

Capt. R.V. Harvey (1872–1915) arrived in Vancouver in 1900 from England, where he had received a Master of Arts degree from Cambridge University (Sherman 1918). He established the Queen's School for Boys in Vancouver and privately published *A Guide to Some of the Butterflies of British Columbia* (Harvey 1904). This rare work has been overlooked in all subsequent bibliographic compilations. It records 108 species of butterflies, or almost the entire fauna of southern British Columbia. Harvey stated that other publications listed many additional species that he did not credit, and time has proven him correct. His collection was donated to the Vancouver City Museum in 1915. Capt. Harvey's untimely death during the First World War dealt a serious blow to the study of the province's lepidopteran fauna.

Thomas Philip Oxenham Menzies was curator of the Vancouver City Museum from 1925 to 1953. He added much of his own material to the museum collection and distributed material to the museums in Ottawa and Victoria and to Ernest Henry Blackmore (see next page). The material from the Vancouver City Museum is now in the Spencer Entomological Museum at the University of British Columbia.

Abdiel William Hanham emigrated from England to Philadelphia in 1881, and then to Ontario, where he was bank manager for various branches of the Bank of British North America. He was based in Hamilton, Quebec City, and Winnipeg before moving to Victoria in 1901 and Quamichan by 1906, where he lived until his death in 1944 (Downes 1945). He had collected Lepidoptera extensively in eastern Canada and brought that material, all carefully curated and identified, with him to British Columbia. His collection is now at the Royal British Columbia Museum, Victoria.

From his home base at Quamichan, Hanham collected intensively in the Duncan area with George O. Day (1904–42), another important Vancouver Island lepidopterist of the time (Downes 1943). Day came directly to the Duncan area from England in 1905. He was a fellow of the Royal Entomological Society and brought with him an almost complete collection of British butterflies. He also amassed a large collection of macrolepidoptera from the Duncan area; both collections were later donated to the Shawnigan Lake School for educational purposes. The remains of the collections were recently given to the Royal BC Museum after Jon Shepard located them. Fortunately many duplicates had been given to Blackmore and the Royal BC Museum as well as to museums worldwide. Day published several papers on Vancouver Island Lepidoptera.

Other lepidopterists established in British Columbia by 1906 included: Arthur H. Bush (Vancouver), Theodore Bryant (Vancouver), William A. Dashwood-Jones (New Westminster) (Green 1947), R. Draper (Vancouver), Reverend G.H. Findlay (Ainsworth; Fort Steele), C. Livingston (Duncan), B. Marrion (Vancouver), R.S. Sherman (Vancouver), E.M. Skinner (Duncan), Jas. Towler (Vancouver), and E.P. Venables (Vernon) (Ross 1966). This group and those mentioned earlier, except George Day, represented 18 of the 22 original members of the Entomological Society of British Columbia (Downes 1943). The work of these early lepidopterists culminated in the publication of the second checklist of British Columbia Lepidoptera in 1906 by the membership of the Entomological Society of BC (Anonymous 1906). It contained fewer species of butterflies than Anderson's 1904 list and was more accurate.

While members of the first generation of lepidopterists in British Columbia were summarizing their work, William Greenwood Wright (1905) privately published his book *The Butterflies of the West Coast*. Wright had collected extensively on Vancouver Island (1891) and in eastern Washington (1889–1903), and he was the first to illustrate in colour most of the butterflies known from southern British Columbia, except alpine species. It is unfortunate that this otherwise laudable book contained many illustrations that were not identified correctly.

After 1906, the interests of amateur entomologists in BC became more varied, and the First World War took its toll on local lepidopterists, including Captain Harvey and Arthur Bush. There remained, however, a strong group of semi-professional and amateur lepidopterists, such as Cockle, Hanham, and Day, who continued their studies into the 1930s and who were reinforced by others.

Arthur William Armit (Artie) Phair was the best-known amateur entomologist during the second period of butterfly studies in the province, from roughly 1906 to 1930. He was also the first person born in BC to become well known for his discoveries of new and interesting butterflies. Also known as "Mr. Lillooet," Phair was born in the Lillooet area in 1880 and lived there his entire life, except for a few years as a student at Collegiate College in Victoria. He was a consummate collector of every conceivable natural, geological, and archeological artifact, which he housed in a family-owned warehouse in Lillooet. His invaluable collections were mostly destroyed by the Lillooet fire; fortunately, however, the Lepidoptera specimens he had exchanged with many other collectors are preserved in various museums. A published summary of his more significant findings of local Lepidoptera (Phair 1919) inspired first Hanham and then James McDunnough (1927), who would later become curator of Lepidoptera at the Canadian National Collection, to spend summers collecting in the Lillooet area.

Another exceptional amateur was Dr. W.R. Buckell of South Canoe, near Salmon Arm, who collected butterflies and moths in the Salmon Arm area between 1904 and 1929 and sent specimens to Blackmore and McDunnough for identification. His nephew, E.R. Buckell, published a *List of the Lepidoptera Collected in the Shuswap LakeDistrict of British Columbia by Dr. W.R. Buckell* (1947), containing 773 species. Dr. Buckell's material is now in the Spencer Entomological Museum.

Theodore (Ted) Albert Moilliet (1883–1935) of Vavenby was the first resident amateur lepidopterist to live north of Lillooet and Vernon (Moilliet 1947). He was responsible for all the Vavenby records in Blackmore's 1927 checklist, including several boreal species recorded for the first time in British Columbia. In 1992, Ted Moilliet's son located his collection in an unheated cabin and, encouraged by the authors of this volume, donated it to the Royal BC Museum. The collection was in remarkably good shape.

Foremost in studies of Lepidoptera during this period was Ernest Henry Blackmore (Fig. 5), who came to British Columbia from Shropshire, England, via Ontario and Alberta. By 1908 he was appointed to a post office position in Victoria and was a research associate at the Provincial Museum, thus beginning a very productive 20 years of work before dying unexpectedly on 2 March 1929, when he was only 47 years old (Anonymous 1929). Blackmore's main interests were the lycaenid butterflies and geometrid moths (he had studied the latter in England). He described several new species or subspecies in these groups, and was the third resident lepidopterist after George W. Taylor and William Cockle to describe new lepidopteran taxa. Blackmore also built up what was then the most geographically and faunistically comprehensive collection of British Columbia Lepidoptera, totalling 1,338 species and more than 12,000 specimens. This collection was most remarkable in that it also included microlepidoptera, a group that is still not well known in British Columbia or the adjacent United States. Blackmore was in constant correspondence with August Busck of the Smithsonian Institution, the leading North American

specialist for microlepidoptera at the time. Blackmore's work on macrolepidoptera, including butterflies, was synthesized in his *Checklist of the Macrolepidoptera of British Columbia (Butterflies and Moths)* (Blackmore 1927). One hundred thirty-nine species of butterflies, as recognized today, were documented, constituting 75% of the fauna now recognized.

Blackmore and E.M. Anderson initiated an annual

**5**  Ernest Henry Blackmore (*Ent. News* 40 [1929]: pl. vi, opposite p. 135)

summary of new and interesting lepidopteran species for the province, including black-and-white photographs of newly recorded species. The summary was published initially in the annual report of the Provincial Museum and then in the *Journal of the Entomological Society of British Columbia* until just after Blackmore's death. It was a prototype for the well-known annual *Season's Summary* of the Lepidopterists' Society, which first came out in 1948 and is still being published. Blackmore must have been able to work well with other amateur lepidopterists, as they all donated duplicate material to his collection and that of the Provincial Museum. At the peak of his career, the combined collections of Blackmore, Hanham, and the Provincial Museum totalled 47,000 specimens (12,000, 30,000, and 5,000, respectively). The importance of this research material was recognized by Jean Gunder (1929a) when he considered the BC Provincial Museum as one of the 16 outstanding collections of North American Lepidoptera. Only seven North American museums had significantly larger resources at the time, and only one west of Chicago, the California Academy of Sciences in San Francisco. Compared with the two other significant Canadian collections – approximately 60,000 specimens each at the CNC and at McGill University's Lyman Museum – this was a remarkable achievement. Today the collection of Lepidoptera at what is now the Royal BC Museum is still the largest in western Canada, covering all of western and central Canada better than any collection except the CNC. Unfortunately, when Blackmore died just before the Great Depression, the Provincial Museum could not afford to purchase his collection. George Spencer of the University of British Columbia had the foresight to purchase the bulk of it, and it is well preserved at the Spencer Entomological Museum. The remainder was given to J.F. Gates (Jack) Clarke and is at the Smithsonian Institution.

Clarke attended school in Victoria until the family moved to Bellingham, Washington. He developed his initial enthusiasm for and knowledge of microlepidoptera under Blackmore's tutelage from 1920 to 1929, and did his master's thesis on this subject at Washington State University. He was curator of entomology at the Provincial Museum during the first two years of the Great Depression, before obtaining his doctorate in Germany and moving on to an illustrious career at the Smithsonian.

The deaths of Blackmore and Cockle, the departure of Clarke, and the advancing age of Hanham and Day meant that there were no active lepidopterists in the province by 1930. The Great Depression also left the federal and provincial governments with no choice but to concentrate on the most important economic insect problems.

Between 1931 and 1951, the most important work on the Lepidoptera fauna of BC was that of James McDunnough, then curator of Lepidoptera at the Canadian National Collection. McDunnough collected extensively at Seton Lake, near Lillooet, in the 1930s, and later in the South Okanagan.

He was mainly concerned with moths, but also worked on butterflies and described several new subspecies of the latter based on his field studies in BC.

From 1930 to 1956, most other work on Lepidoptera in BC consisted of commercial collecting of butterflies for Americans whose fortunes had not been ruined by the depression. A.B. Garrett collected large amounts of butterfly material from Cranbrook and sold it to Jean Gunder, whose collection went to the American Museum of Natural History. Some material also went to the CNC. Other collectors during this period included John F. May for Gunder, John (Jack) D. Gregson for Cyril dos Passos of the American Museum of Natural History, and J.K. Jacob for the CNC. Harold Foxlee of Robson (near Castlegar) did commercial collecting of all groups of insects. He worked with curators at the CNC, and in 1945 published a more complete list of the Lepidoptera of the West Kootenay (Foxlee 1945). From the late 1940s to the early 1960s, Richard Guppy of Wellington and Thetis Island, a charter member of the Lepidopterists' Society, collected research material for several early members of the society and contributed to the annual Season's Summary. He also contributed several notes on Vancouver Island butterflies to the *Journal of the Lepidopterists' Society.*

J.R.J. Llewellyn Jones made significant contributions to knowledge of the lepidopteran fauna of BC from the 1930s to the 1950s. He immigrated from England in 1930 and settled at Mill Bay (Spencer 1956), where he built a house that now belongs to the headmaster of Brentwood School. Jones collected on southeastern Vancouver Island, but received materials from other parts of the province. He was mainly interested in rearing larvae and published a series of papers summarizing the known larval foodplants of BC lepidopteran species. In 1950, just before his death, he moved to Vancouver to have access to the Blackmore and other collections at UBC. He donated his collection and library to UBC and willed his estate to provide a permanent publication fund for the Entomological Society of British Columbia, of which he had been a president and longtime member. In 1951 Jones published an updated *Annotated Check List of the Macrolepidoptera of British Columbia,* following the updated taxonomy of the 1939 McDunnough checklist. It recorded 144 of the 187 species of butterflies now known for the province, five more than were found in the Blackmore checklist of 1927. Like the latter, it summarized the known geographic distribution of each species in the province, but it also included flight periods and larval foodplants and incorporated the work of W.R. Buckell and Harold Foxlee. It thus became a starting point for any further work on the butterfly fauna of BC.

In 1932, Josephine de Henry of Philadelphia collected butterflies in the alpine areas northwest of Fort St. John and recorded four of the five additional species that were recorded in the Jones checklist. When that checklist was published in 1951, the only parts of the province that were well known were Vancouver Island south of Nanaimo, Vancouver, Lillooet, Salmon Arm, Kaslo, and Cranbrook. The only information from north of these areas came from cursory collecting at Atlin, Telegraph Creek to Dease Lake, Smithers, Rolla, and near Pink Mountain.

The years from 1956 to 1965 saw four significant studies of the BC lepidopteran fauna. George A. Hardy (Fig. 6), curator of botany at the Provincial Museum, documented the butterfly and moth fauna of southeastern Vancouver Island just before the postwar development boom contributed to the local extinction of several butterfly species (Carl 1966). He also conducted the most extensive study to date of the larval stages and life histories of BC butterflies and moths, publishing many papers in Provincial Museum publications and in the *Proceedings of the Entomological Society of British Columbia* (see bibliography). James Edward (Ted) Underhill, naturalist at Manning Provincial Park, conducted an extensive survey of the Lepidoptera at the park, recording one new species of butterfly for the province, *Papilio indra* (Harcombe and Underhill 1970). The Forest Insect Survey collected larvae from trees and shrubs throughout the province and recorded significant new larval foodplants in BC, such as Rocky Mountain juniper for the Juniper Hairstreak, *Mitoura siva,* although this species was incorrectly identified. This was the first record of the Juniper Hairstreak in BC. The federal Northern Insect Survey after the Second World War surveyed extensively in northern BC, at Fort Nelson, Summit Lake, Lower Post, and Atlin, recording many new moth species for the province but only one new butterfly species, *Lycaena phlaeas.*

When we began our studies in 1965, 147 butterfly species were known from BC, or 78% of the fauna now known for the province. At that time, the provincial highway system was being greatly improved and extended, and mining exploration and logging in the northern half of the province had opened up many secondary roads. We have concentrated our efforts in areas that had not been well investigated previously, especially the Haines Highway, Atlin, Summit Lake, the BC Peace region, the Queen Charlotte Islands, the Skeena and Bulkley river drainages, and the Chilcotin and Bella Coola areas. Others have studied their own local areas, for example, David Threatful at Revelstoke (Threatful 1989) and Vernon, Hettie Miller and Helen Knight at Clearwater, Douglas Knight at Creston, Norbert Kondla (Kondla et al. 1994) in the Peace River and at McBride, and Gerald Straley in the Lower Fraser

6 George Austin Hardy, ca. 1964 (courtesy Province of British Columbia)

Valley. These efforts have added 40 species, bringing the provincial total to 187, which means that 22% of the BC fauna known today has been documented since 1965. Most of the additions are Beringian species recorded from the northern extremities of the province and Prairie species recorded from the Peace River. Three species were added from the extreme southeastern corner of the province.

Most of the historical collections made by resident lepidopterists are now housed at either the Royal BC Museum or the Spencer Entomological Museum. The only ones not in BC museums are the de Blois Green and Cockle collections at the Canadian National Collection, the Danby collection at the American Museum of Natural History, and the Taylor type specimens at the Smithsonian Institution. The Royal BC Museum has 50,000 lepidopteran voucher specimens, of which 12,000 are butterflies, and the Spencer Museum has 25,000 specimens, of which 5,000 are butterflies. Another 20,000 specimens, including 1,000 butterflies, are found at the Pacific Forestry Centre. It should be remembered that George Spencer single-handedly preserved many of the important historical collections of Lepidoptera at the University of British Columbia (Spencer 1957; Graham 1966). If not for this effort, we would not have nearly as good a

basis for managing currently endangered lepidopteran species.

The Shepard collection (50,000 specimens, 25,000 butterflies), the Guppy collection (17,500 specimens, 15,000 butterflies), and the efforts of David Threatful and Norbert Kondla have almost doubled the voucher material for Lepidoptera that is available in the province, and more than doubled the database for butterflies. The material, both historical and current, totals some 170,000 pinned, spread, and labelled specimens – the largest number of voucher specimens of Lepidoptera assembled in any western Canadian province or territory, and larger than that for any state in the United States west of the Mississippi except California and perhaps Colorado. Much information still needs to be filled in, however, concerning the limits of species distributions of butterflies in northern BC and the life histories of butterflies. The chemical systematics of the various east-west or Beringian-Cordilleran species pairs would provide material for several doctoral dissertations. For moths, the current state of knowledge is about what it was for butterflies in 1965. Naturalists can greatly increase our knowledge of moths in BC by making local observations, rearing larvae, and preparing voucher specimens.

# Postglacial Origins of the Butterfly Fauna of British Columbia

The maximum of the last, or Wisconsin, glaciation occurred around 13,500 BP in northern British Columbia and around 15,000 BP in southern British Columbia. At the glacial maximum, virtually the entire province was covered by ice (Fig. 7A). The only areas free of ice were portions of the Queen Charlotte Islands, the Brooks Peninsula on the northwest coast of Vancouver Island, and a high-elevation ice-free area straddling the British Columbia/Washington border near the Cathedral Lakes in the rain shadow of the northern Cascade Mountains.

This habitat was very inhospitable for butterflies, and there is no proof that any existed in the province at the glacial maximum. The unglaciated areas on the Queen Charlotte Islands and northern Vancouver Island could easily have been colonized from other refugia in the Alaska panhandle (MacDonald and Cook 1996) or on the west side of the Olympic Peninsula. Only one taxon, *Lycaena mariposa charlottensis,* is endemic to the entire area from the western lowlands of the Olympic Peninsula through the west coast of British Columbia to the Alaska panhandle. The area east of the north Cascades may have acted as a refugium for alpine species, but the area now contains only about half of the same taxa that occur in the Canadian Cordillera south to Glacier National Park. Thus the Alberta/Montana Rockies are the much more likely refugium for these northern Rocky Mountain alpine species and subspecies. At least one northern Rocky Mountain subspecies, *Oeneis melissa beanii,* is still present as far south as the Beartooth Plateau in northwestern Wyoming, which also indicates that the refugium for Canadian Rocky Mountain species was south, in the United States. Pike (1980) argued that the presence of endemic taxa in the Alberta and adjacent British Columbia Rocky Mountains proved that these species survived the Wisconsin glaciation in Alberta. It is more likely, however, that they developed in isolation at and after the hypsithermal, especially the high-altitude form *Clossiana eunomia nichollae,* which is derived from a boreal species, not an arctic/alpine species (Hilchie 1990). Hilchie (1990) showed that his new taxon, *Erebia magdalena saxicola,* was derived from populations formerly located in the Montana front ranges during the last glacial period, and refuted Pike's (1980) hypothesis.

Compared with other Canadian provinces and adjacent states, British Columbia had the most complex pattern of postglacial maximum melting of the glaciers, and the most complex sources for butterflies migrating into the area as it became free of ice. One must understand the pattern and the timing of the glacial retreat to realize why British Columbia has the most diverse butterfly fauna of any political entity in North America north of the 49th parallel.

By 12,500 BP, two major ice-free corridors running north to south had become available (Fig. 7B). The first corridor (Clague 1985; Luternauer et al. 1989; MacDonald and Cook 1996), on the west coast, allowed butterfly species that survived in small refugia in either the Alaska panhandle (MacDonald and Cook 1996), the Queen Charlotte Islands, the Brooks Peninsula, or the west side of the Olympic Peninsula to expand their range along the west coast. This west coast habitat is very inhospitable for butterflies, and only a few took advantage of the ice-free corridor. For example, only four butterfly taxa are resident on the Queen Charlotte Islands even today. Only one, *Lycaena mariposa charlottensis,* is derived from the western fauna; the other three are derived from the eastern, boreal fauna. The moth fauna of the Queen Charlotte Islands, some 200 species, is also primarily of eastern boreal origin (90%); only 10% are derived from the western fauna (J.H. Shepard, unpubl. data).

The second ice-free corridor (Fig. 7B) was much more significant for the makeup of the present-day British Columbia butterfly fauna. It ran parallel to the Rocky Mountains in the rain shadow on the east side of the Rockies in Alberta and extended north into British Columbia and south from the Yukon. This corridor allowed the eastern boreal forest butterfly fauna and the northern Rocky Mountain alpine fauna to reach British Columbia from Alberta. It also allowed minor penetration of the Beringian fauna – such as *Clossiana natazhati nabokovi, Erebia pawlowskii,* and *Erebia mackinleyensis* – into the northern extreme of the Rocky Mountains.

At the same time, southeastern Vancouver Island was ice-free but the current flora was not yet established. Heusser (1983) and Mathews and Rouse (1974) show that for both southeastern Vancouver Island and the Lower Fraser Valley around 12,000 BP, *Pinus contorta* and *Alnus rubra* were the dominant trees. They were replaced at the hypsithermal, 7,500 BP, by Douglas-fir and Garry oak. The development of the present-day western red cedar and western hemlock forest began as post-hypsithermal cooling progressed.

By 10,000 BP about half of the province was free of ice (Fig. 7C). The entire northeastern fourth was ice-free and available for the boreal fauna of butterflies to be established. Most of

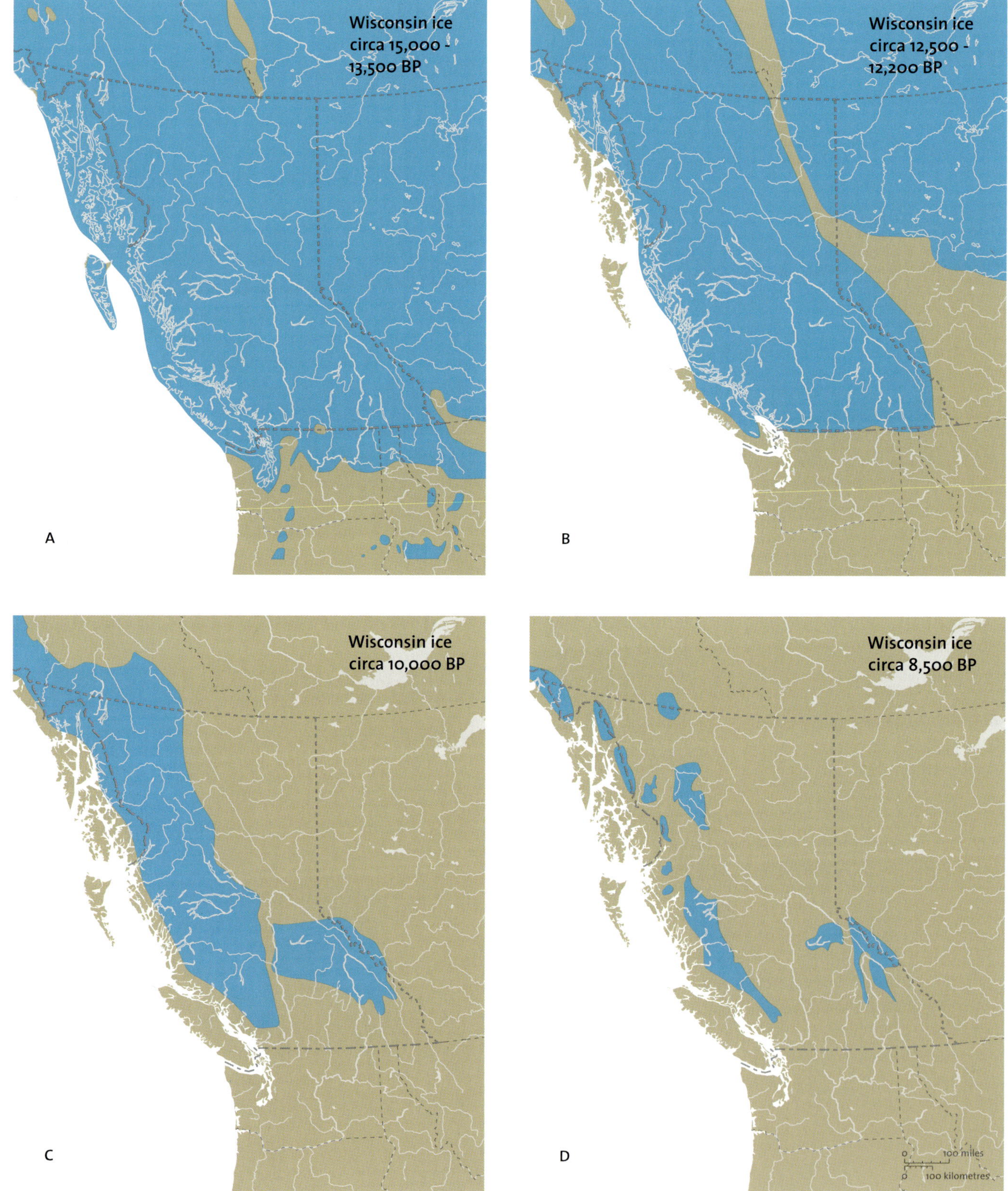



**7** Extent of late Wisconsin ice: (A) ca. 15,000–13,500 BP, (B) ca. 12,500–12,200 BP, (C) ca. 10,000 BP, (D) ca. 8,500 BP (redrawn from Prest 1969, Osburn and Luckman 1988, Pielou 1991)

the southern fourth of the province was also ice-free, enabling some of the different western faunal elements, such as Cascade Mountain and Rocky Mountain faunas, to become established. The most significant thing about this time period is that the eastern boreal fauna of northeastern British Columbia, which extends to Newfoundland, was not able to mix with the western faunas. The entire Coast Range was still glaciated, and southeast from Prince George the Cariboo, Columbia, and Rocky Mountains were also glaciated. Although there was a small, intermittent ice-free corridor along the Fraser River canyon between Prince George and Williams Lake, it was often plugged by pluvial lakes or Holocene glacial advances (Clague 1987, 1988; Osburn and Luckman 1988). Thus north-south dispersal of all but migratory species of butterflies was still blocked.

By 8,500 BP only about one-fifth of the province was glaciated (Fig. 7D), and the eastern and western floral and faunal elements began to overlap in distribution. At present, boreal flora and butterfly fauna have penetrated south to Williams Lake and Salmon Arm, and a few species of the western fauna, such as *Incisalia eryphon* and *Strymon sylvinum,* have managed to move north. Much of northwestern British Columbia was also then free of ice, which allowed the Beringian fauna to begin moving into the province.

Just after this, around 7,500 BP, there was a temperature maximum called the *hypsithermal.* This must have been the time that several montane Rocky Mountain species that are now found in disjunct populations between Atlin, BC, and Whitehorse, YT, became established there (e.g., *Polites draco, Euphydryas anicia,* and *Parnassius smintheus*). The hypsi-thermal must also have accounted for the Prairie species found in disjunct populations in the Peace River area of BC and Alberta (e.g., *Oeneis alberta, Oeneis uhleri, Satyrium titus,* and *Satyrium liparops*). When weather conditions cooled after the hypsithermal, these disjunct populations were separated from their parental, southern populations, and boreal flora and fauna again dominated. On the coast, the hypsithermal was concurrent with establishment of Garry oak meadow habitat (Heusser 1983). This allowed the modern butterfly fauna of southeastern Vancouver Island to become established. The colonization of Vancouver Island by *Phyciodes mylitta* showed that butterfly species can bridge the modern water barriers (Shepard 1977). This was also the period when the modern flora and butterfly fauna moved into the Columbia Basin from the Wisconsin refugium in the northern Great Basin (Nowak et al. 1994). At the peak of Wisconsin glaciation, the Columbia Basin had a very different arctic tundra flora, which persisted until nearly the hypsithermal.

At present the known butterfly fauna of British Columbia is 187 species and, with subspecies, is by far the most complex fauna in Canada. Eight species are migrants and 39 have a very restricted range within the province, however. Thus most of British Columbia is occupied by only 138 resident species, approximately the same breeding fauna as the other Canadian provinces east to Ontario. The Beringian element comprises 13% of the total non-migratory fauna, the boreal element 20%, and the western element 67%, of which 21% is Coastal/Cascade, 17% is Great Basin, 17% is Rocky Mountain, 5% is Prairie, and 7% is of undetermined western origin.

# Impact of Humans on the Butterfly Fauna of British Columbia

Although it is generally assumed that human immigration has drastically affected the butterfly populations of British Columbia, there is only limited concrete evidence for this.

The voucher specimens housed in the major public museum collections and information from the scientific literature give us a good idea of the butterfly fauna of southeastern Vancouver Island, the Lower Fraser Valley, the Okanagan Valley from Vernon to Osoyoos, and the West Kootenay before these areas experienced significant human impact. These were the areas first settled by Europeans, and large numbers of people have lived there since 1900. Most collecting and surveying was done before the Second World War. During the postwar period, the population of the province grew tremendously, and very few people made accurate observations of the butterfly fauna, especially in areas of greater population growth. The one exception was George Hardy's detailed study of Victoria and the surrounding Saanich Peninsula.

The most significant problem with this growth has been the alienation of natural habitat by increased housing, industrial sites, roads, hydroelectric dams, recreational sites, and farmland. This has been especially important on southeastern Vancouver Island, the Lower Fraser Valley, and the southern Okanagan Valley, where the only British Columbian or Canadian populations of many butterfly species are found. Intensive use by humans has completely altered most of the available natural habitat, and only butterfly species that were previously adapted to disturbed habitat have survived with viable, long-term populations.

The Garry oak meadow habitat on southeastern Vancouver Island, the sagebrush/antelope brush habitat of the southern Okanagan Valley, and the mature hemlock forests with associated mistletoe of the Lower Fraser Valley and Vancouver Island have been critically affected by the direct alienation of land for human use. In these areas, habitat has been so drastically reduced that some species have been extirpated or nearly so. In the West Kootenay, habitat alienation at the lowest elevations in the valley bottoms has caused many local extinctions of butterfly species, although the species still have healthy populations elsewhere in the province. The one exception is the lower Pend-d'Oreille River, which shares some of species of conservation concern with the Okanagan.

The second most important factor in the loss of butterfly species has been the introduction of weedy plants from other areas, primarily Europe. Introduced grasses have invaded native meadow and grassland habitat throughout the province, from the most urban areas in Victoria and Vancouver to the most remote areas on the Queen Charlotte Islands and at Atlin. They have replaced native vegetation and reduced the available habitat of butterfly species dependent on native grasses and sedges as larval foodplants. This has affected primarily skippers and woodnymphs. Another effect of the introduction of weedy grasses has been the elimination of non-grass species of plants that serve as either nectar sources or larval foodplants of butterfly species.

The most outstanding example of this is the spread of Scotch broom (*Cytisus scoparius*) on southwestern Vancouver Island. Broom is drought-tolerant and spreads by exploding seeds out of the seed pods. Much of the disturbed, peripheral lands that would support some butterfly species, as well as remnant native Garry oak meadows, has been taken over by Scotch broom. Compounding the natural spread of Scotch broom is hydroelectric line right-of-way management, which deliberately encourages growth of Scotch broom so that trees will not grow in the right-of-way. This has eliminated the penultimate population of the endangered Taylor's Checkerspot (*Euphydryas editha taylori*) in British Columbia. Elsewhere on Vancouver Island, Scotch broom has outcompeted virtually all native vegetation and will eventually prevent even oak seedling development if it is not controlled.

Introduced grasses have eliminated both nectar sources and larval foodplants. At Beacon Hill Park (Victoria), the classic habitat of Taylor's Checkerspot, the introduced and native plantain foodplants of the checkerspot's larvae are still abundant, but the adult nectar source, spring gold (*Lomatium* sp.), has been nearly eliminated by introduced grasses and other weeds. Taylor's Checkerspot has not been seen there since at least the 1950s. The physical impact of humans walking or driving over the open meadows is actually less today in Beacon Hill Park than in the late 1890s, when the open parts of the park served as a major recreational resource for Victoria (Fig. 8).

A third factor has been the suppression of fires, both naturally occurring and those used for Aboriginal landscape management. This has allowed meadows and early seral stages of climax forest regeneration to grow into mature forests (Turner 1994), an unsuitable habitat for most butterfly species. This is most significant on southeastern Vancouver Island. Lupines, *Lupinus* sp., are no longer found

**8** Holiday celebrations at the open west end of Beacon Hill Park (Victoria), ca. 1900. Prior to European settlement, Beacon Hill Park was habitat for the now endangered Taylor's Checkerspot (*Euphydryas editha taylori*) and the extirpated Island Large Marble (*Euchloe ausinides insulanus*). (BC Archives, A-02883)

in former lowland habitats and the three species of blues that utilized them as larval foodplants are now found only in meadows near timberline. Historically the lupines were common in low-elevation areas that were periodically burned.

The introduction of non-native butterfly species has also affected native species. The Cabbage White (*Pieris rapae*) has spread to all areas of the province, feeding on cultivated cabbage and native Cruciferae. This may have affected some native species of whites and marbles by eliminating native larval foodplants and by increasing the populations of larval parasites. The large numbers of the introduced white may also interfere with the normal breeding of native species, either by hybridization with or by disrupting the mating behaviour of the latter. Another, more recently introduced species, the European Skipper (*Thymelicus lineola*), has invaded grasslands, especially timothy hay fields. Because its peak flight is at the same time as the native Garita Skipperling (*Oarisma garita*), we expect that the native skipper will eventually decline.

Finally, pesticides and biological control agents can drastically reduce the populations of non-target species. The early use of pesticides in apple orchards in the Okanagan is considered to be the cause of the extinction of the Viceroy in BC and adjacent Washington (Hopfinger 1947). Recent research on the effects of *Bacillus thuringiensis* (*Bt*) spraying on non-target Lepidoptera shows that *Bt* drastically decreases the population density of all species of Lepidoptera that are in the larval stage of development at the time of spraying and for many weeks after the spraying (Miller and West 1987; Scriber and Gage 1995). Miller (1990, 1992) has shown that the species diversity and abundance of moths in Oregon were reduced for years after *Bt* application.

In the last two decades, there has been much debate and speculation about the effects of logging, mining, grazing, and global warming on the native flora and fauna of BC, including butterflies. There is, however, little evidence to support the existence of any specific effect on butterflies, and some data to indicate that logging and grazing have sometimes helped native butterfly species.

Mining is of no special significance except in restricted alpine habitat. So far, there is not enough mining in the province to affect any butterfly species. On the contrary, the development of roads for mining surveys and logging has given much greater access to naturalists inventorying butterfly populations in unexplored parts of the province.

Logging has positively affected some butterfly species in places where it has replaced fire as a means of creating open meadows and enabling butterfly larval foodplants to thrive. This has occurred on Vancouver Island, where low-elevation lupines have disappeared but where lupines are thriving in subalpine slopes that have been logged on Mt. Cokely and Mt. Arrowsmith. Here native species of blue butterflies no longer found at low elevations are also thriving on the lupines. On the other hand, the logging roads have also allowed Scotch broom to migrate into this habitat (Shepard 1995).

The grazing of native grasslands by excessive numbers of domestic or native mammals both destroys native plants and allows weed species to become more easily established. On Saltspring Island, however, where small numbers of feral and domestic sheep graze the southern slopes of Mt. Tuam, sheep grazing appears to have replaced fire as a means of maintaining the open meadow habitat with a fair number of native plants. This has enabled the only good population of Bremner's Fritillary, *Speyeria zerene bremnerii,* to continue to exist. Also, the only remaining Taylor's Checkerspot population is found in natural meadow habitat on Hornby Island that was, until the 1970s, spring sheep pasture. This presumably helped keep the habitat an open Garry oak meadow that has also not been invaded by a significant number of weeds (Shepard 1995).

Global warming is certainly a factor that will have to be taken into account in the future (Lester and Myers 1990), but there are no detectable effects on butterfly populations at this time. One paper attributed the extinction of populations of Taylor's Checkerspot on Vancouver Island to global warming (Parmesan 1996), but three of these populations – at Beacon Hill, Oak Bay, and Bright Angel parks – were already extinct well before global warming had significantly increased world temperature, and the other, on the Shawnigan Lake/Mill Bay Road became extinct because the habitat was no longer managed for a Christmas tree farm and was invaded by Scotch broom. Global warming was not a factor. In Colorado, drought and global warming were postulated as the reasons for the decline of the Uncompahgre Fritillary, *Clossiana improba acrocnema* (Britten et al. 1994), but the populations have since recovered.

The following sections discuss some of the critical areas of the province and human changes to their environment.

## Vancouver Island

The Victoria area originally had an abundance of butterflies, such that "nearly 40 species may be marked abundant. A patch of blossom in May, covered with blues and fritillaries, with an occasional sulphur and two or three magnificent species of swallowtails, later in the year hundreds of ladies, coppers, skippers and admirals make a very different but not less beautiful picture" (Taylor 1884) (Latin names have been replaced with common names). This description would have fit Beacon Hill Park, Uplands Park, and Thetis Lake Park equally well at that time. Over the past 100 years, however, many species of butterflies have nearly vanished from the Victoria area. Those still common are species that exploit

disturbed habitats. Nowhere on Vancouver Island can Taylor's observations can be repeated today. Most of the loss is due to habitat destruction, but other factors such as insecticide spraying have undoubtedly played a role.

The Garry oak forest and meadow habitat of southern Vancouver Island and the adjacent Gulf Islands in the Strait of Georgia are unique in Canada. This meadow habitat has been almost completely destroyed by human activity during the past 100 years. Garry oaks are still common, because they are prized by some as ornamental trees, but most of the associated open meadows and rocky knolls have been destroyed. The few surviving fragments of the original habitat have been greatly altered, partly through direct human development and park "improvements" and partly through the invasion of introduced weeds such as grasses and Scotch broom. The best examples of the original habitat were at Beacon Hill Park (Fig. 8) and Goldstream (Fig. 9), but only a few common and widespread butterfly species now remain. A major contributing factor to the loss of meadow habitat on southern Vancouver Island has been the control of fires over the past 100 years. Fire, both natural and those started by the Aboriginal people, maintained large areas of what is now Douglas-fir forest as grassland (Hebda and Aitkens 1993). The best remaining Garry oak meadow habitat is preserved in Helliwell Park on Hornby Island.

## Greater Vancouver

Habitat destruction has been almost complete at low elevations in the Greater Vancouver area. Most of the area was forest rather than the grassland favoured by most butterflies. The forest was logged, and the area either was converted to human settlement lacking in butterfly habitat or regrew into dense second growth forest lacking openings along streams that butterflies can utilize. The few grassland areas were seasonally flooded plains such as Sumas Prairie, and these have been converted entirely to farmland. At low elevations, fragments of habitat are left in parks, waste areas, floodplains, and vacant lots.

The Boundary Bay area of Delta is an example of an important remnant habitat. A portion of the dike along the bay supports a large population of Anise Swallowtails, apparently the last population west of Chilliwack in the Fraser Valley. The butterflies are a superb aesthetic resource, and people walking along the dike are treated to the sight of large numbers of Anise Swallowtails flying about the foodplants and nectaring at flowers. Breeding occurs on *Heracleum maximum* and *Angelica lucida,* which grow along both sides of the dike, with the *Angelica* mostly restricted to the seaward side. This side of the dike provides excellent opportunities for reproduction during the summer, but is inundated with winter tides and waves, making it impossible for pupae to survive over winter. Thus the landward side of the dike and the adjacent farmlands are the critical habitat

**9** Natural Garry oak meadow habitat on southern Vancouver Island, ca. 1900. Garry oak meadow habitats are now rare, as are many of the butterflies that inhabited them. (BC Archives, G-01564)

areas for the Anise Swallowtails. The sides of the dike have been mown since about 1997, which destroys eggs, larvae, and pupae on the cut vegetation. A population of Margined Whites (*Pieris marginalis*) is also present, one of a few remaining populations in the Vancouver area.

Another important remnant is Burns Bog. It has the only known low-elevation Fraser Valley population of Reakirt's Copper (*Lycaena mariposa*) as well as many other unique, non-butterfly species.

## Southern Interior and Kootenays

Hopfinger (1960, 1961) noted a major decline in Large Marbles (*Euchloe ausonides*) and other butterflies in central Washington state in 1959 and 1960, and attributed it to the destruction of their habitat by cattle and sheep. Drought conditions increased the effects of grazing by reducing the available forage for livestock. Similar livestock grazing was occurring in the grasslands of the BC interior at the same time. It is likely that the cattle simply ate all the larval foodplants, along with the eggs and caterpillars on them.

The grassland habitats of butterflies in the interior of BC have been greatly altered through a combination of livestock grazing, lack of wildfires, and introduced weeds. Livestock grazing and lack of wildfires both result in simplified plant communities dominated by grasses and introduced weeds. Many of the native larval foodplants have decreased in abundance to the levels that cannot sustain the butterfly populations that depend on them. For example, near Kootenay Lake in the early 1990s there were flourishing colonies of Boisduval's Blues along powerline right-of-ways. By the mid-1990s those colonies had vanished entirely as knapweed overwhelmed the native lupines used as larval foodplants (Chris Schmidt, pers. comm.).

Control of "undesirable" native species such as sagebrush, as well as introduced weeds such as thistles and knapweed, is done either by bulldozing or by applying herbicides to large areas of both private and Crown land and then replanting them with vigorous introduced grasses. Such areas are effectively devoid of native plant species, and all but the most common butterflies are lost. The Okanagan Valley is also undergoing rapid urbanization, which completes the destruction of the altered grasslands. Despite the changes to native grasslands, butterfly species other than those associated with antelope brush (*Purshia tridentata*) habitat are still reasonably abundant, although several are of conservation concern.

Much of the native parkland in the Peace River Lowlands has been irreversibly altered by agriculture, and the upper Peace River canyon has been modified or destroyed through hydroelectric dam building. Several butterfly species are found in the Peace River canyon of BC and Alberta as disjunct populations of Canadian Prairie species. In BC the range of 13 of these species and subspecies is confined to the canyon of the Peace River below Attachie and some tributary valleys, with only a few populations of some species found in adjacent upland areas. The only strong populations of most of these species are found only east of the mouth of the Beatton River. They are threatened in BC by the natural succession of grasslands to shrubs and trees in the absence of wildfire. Fortunately this habitat is considered prime wildlife habitat for game species and has been provided some protection. Management of the wildlife habitat should take into account the requirements of the butterflies found here that are unique in the province (Shepard 2000b).

# Conservation of Butterflies in British Columbia

Butterfly conservation in North America formally began in 1971 with the formation of the Xerces Society, which was initially dedicated to the conservation of butterflies but is now concerned with the conservation of all terrestrial arthropods. The Xerces Society initiated butterfly counts in North America, with the joint long-term objectives of monitoring changes in butterfly populations and of educating and involving the public with butterflies. It is also involved in public and political education programs to promote the conservation of butterflies and other invertebrates.

In British Columbia butterfly conservation has only recently begun to be formally considered. Guppy et al. (1994) and Scudder (1996) summarized the known status of butterflies of conservation concern in the province. As a result of these publications, the BC Conservation Data Centre commissioned Shepard (1995) to report on the status of the butterflies of conservation concern on southeastern Vancouver Island. Shepard has also written status reports for the Committee on the Status of Endangered Wildlife in Canada (COSEWIC) and the Conservation Data Centre on butterfly species of conservation concern on southeastern Vancouver Island and in the southern Okanagan Valley (Shepard 2000a).

The Nature Conservancy (United States) has completed global conservation status rankings for most North American butterflies, and the BC Ministry of Environment, Lands and Parks, in cooperation with The Nature Conservancy, has completed BC provincial rankings based on Guppy et al. (1994) and additional data from this book. These rankings are periodically updated as a result of new information. In this book we use the terms and criteria of both COSEWIC and The Nature Conservancy (TNC) for ranking the conservation status of the butterflies of BC. The classifications are ours alone, except where otherwise specifically stated, although we have tried to make the TNC ranks consistent with those established by the BC Conservation Data Centre. These conservation status rankings are based on the information available at the time this book was written, and many will change over time as new information is acquired.

The following COSEWIC terms are used here:

*Extinct:* any species or subspecies formerly indigenous to BC that no longer exists anywhere in the world

*Extirpated:* any indigenous species or subspecies no longer known to exist in BC but existing elsewhere

*Endangered:* any indigenous species or subspecies that is threatened with imminent extirpation or extinction throughout all or a significant portion of its range in BC

*Threatened:* any indigenous species or subspecies that is likely to become endangered in BC if the factors affecting its vulnerability are not reversed

*Special Concern:* any indigenous species or subspecies that is particularly at risk in BC because of low population numbers, occurrence at the fringe of its range or in restricted areas, or some other reason, but that is not a threatened species.

All our TNC rankings are preceded by **S**, indicating that the rank is for a species or subspecies within its British Columbia ("**S**ubnational") range, rather than a national or global rank. The following ranks are used:

1   critically imperilled because of extreme rarity (5 or fewer occurrences or very few remaining individuals) or because of some factor(s) making it especially vulnerable to extirpation or extinction

2   imperilled because of rarity (typically 6–20 occurrences or few remaining individuals) or because of some factor(s) making it vulnerable to extirpation or extinction

3   rare or uncommon (typically 21–100 occurrences); may be susceptible to large-scale disturbances (e.g., may have lost extensive peripheral populations)

4   frequent to common (more than 100 occurrences); apparently secure but may have a restricted distribution

5   common to very common; demonstrably secure and essentially ineradicable under present conditions

X   apparently extinct or extirpated, without the expectation that it will be rediscovered

H   historical records only; may be extinct or extirpated but with some expectation of rediscovery

B   breeding but not permanently resident

N   non-breeding and not permanently resident

E   exotic, non-native species

A range of ranks, such as S3S4, indicates that the exact status is uncertain. The number of *actual* occurrences and vulnerability were estimated from the known populations, historical population trends, and likely future population trends.

**Table 4.** **Summary of the provincial status of British Columbia butterflies and skippers, by predominant ecoprovince in the species' distribution (updated from Guppy et al. 1994)**

| | Ecoprovince of British Columbia | | | | | | |
|---|---|---|---|---|---|---|---|
| **Status** | **Georgia Depression** | **Coast and Mountains** | **Southern Interior** | **Southern Interior Mountains** | **Boreal Plains; Taiga Plains** | **Northern Boreal Mountains** | **Total taxa** |
| Extinct | – | – | – | – | – | – | 0 |
| Extirpated | 2 | – | 1 | – | – | – | 3 |
| Endangered | 3 | 1 | 2 | 2 | – | – | 8 |
| Threatened | 1 | – | 3 | 1 | 1 | – | 5 |
| Special Concern | 8 | 2 | 9 | 8 | 14 | 16 | 55 |
| **Total** | **14** | **3** | **15** | **11** | **14** | **16** | **75/73** |

*Note:* Two species of Special Concern were counted in each of two ecoprovinces, hence the difference in totals. Ecoprovinces are shown in Fig. 1.

There are 187 species and 77 additional subspecies of butterflies and skippers known from British Columbia, for a total of 264 resident or immigrant subspecies (see Species Checklist on page 362). Of these, 4 are non-breeding immigrants and 4 are immigrants that regularly breed within BC. Of the 264 subspecies, 73 (28%) are of conservation concern in BC (Tables 4 and 5). Thirty-three (13%) taxa are of concern for Canada as a whole, of which 2 (1%) are of concern globally. One subspecies is probably globally Extinct, and 1 species and 2 subspecies are Extirpated from British Columbia and Canada. Twelve species or subspecies are Endangered or Threatened (5%). One breeding immigrant, the Monarch, is ranked as of Special Concern in both BC and Canada, and another breeding immigrant, the coastal subspecies of the Silver-spotted Skipper (*Epargyreus clarus california*), is Extirpated.

There are 55 species and subspecies listed as of Special Concern, S3 or S3S4. It is critical to note that many of these species are given those ranks *only* because they are known from very few populations in BC. This is especially true for those in the Northern Boreal Mountains and Southern Interior Mountains ecoprovinces. Frequently the populations are apparently secure for the present, but significant natural and/or human-caused environmental changes could adversely affect them. The status of these species should therefore be monitored, and they have been assigned "of concern" ranks. It is likely that many of the apparently rare species are actually not rare in BC, and their status will eventually be upgraded to "not of concern" (usually S4). It is impossible to predict, however, *which* specific species are more abundant than present data indicate, and substantial additional inventory is required prior to reassessments. Other species ranked as of Special Concern are clearly restricted to small geographic areas of BC and/or are declining because of human alterations to their habitat.

These species will remain of conservation concern unless significant action is taken to appropriately manage their habitat.

Conservation efforts are probably too late for the one subspecies of Greenish Blue (*Plebeius saepiolus insulanus*) that is endangered and possibly extinct, although a concerted effort to find a surviving population should be made. Should apple orchard spraying with broad-spectrum insecticides (including *Bacillus thuringiensis* variety *kurstaki,* or *Btk*) ever cease in the southern interior, it may be possible to reintroduce the Viceroy (*Limenitis archippus*), or the Viceroy may recolonize the area naturally. It may be possible to reintroduce the Island Large Marble from San Juan Island, WA, if an area of suitable habitat can be found on southern Vancouver Island or the Gulf Islands. The *californicus* subspecies of the Silver-spotted Skipper was once a breeding immigrant, and unknown factors in the United States will determine whether it returns. The other species and subspecies of conservation concern can be retained as part of the fauna of the province. Hopefully this book will provide a basis for conservation efforts around the province.

## Legislation

It must be emphasized that legislation designed to protect butterflies from being collected, although fashionable and "feel-good," will not by itself protect any butterfly species from extirpation or extinction. The first butterfly to be legally protected from collection was the Apollo, *Parnassius apollo,* in Germany in 1913 (Winn 1914). That protection, and subsequent protection from collection of other species in Europe, the US, and elsewhere has only been a last desperate attempt to conserve butterflies that have been driven to the edge of extinction by habitat destruction. The Apollo is now nearly extirpated from Germany despite 87 years of

| Butterfly and skipper taxa (species and subspecies) | Suggested status (COSEWIC terms)[1] | | Provincial status (TNC ranks) |
|---|---|---|---|
| **Skippers (Family Hesperiidae)** | | | |
| Silver-spotted Skipper, *Epargyreus clarus californicus* | Extirpated | (BC, Can) | SXN |
| Propertius Duskywing, *Erynnis propertius* | Special Concern | (BC, Can) | S3 |
| Afranius Duskywing, *Erynnis afranius* | Endangered | (BC) | S1 |
| Checkered Skipper, *Pyrgus communis* | Special Concern | (BC) | S3S4 |
| Arctic Skipper, *Carterocephalus palaemon mandan* | Special Concern | (BC) | S3 |
| Common Branded Skipper, *Hesperia comma assiniboia* | Special Concern | (BC) | S3 |
| Common Branded Skipper, *Hesperia comma oregonia* | Special Concern | (BC, Can) | S3 |
| Sandhill Skipper, *Polites sabuleti* | Special Concern | (BC, Can) | S3 |
| Draco Skipper, *Polites draco* | Special Concern | (BC) | S3 |
| Sonora Skipper, *Polites sonora* | Threatened | (BC, Can) | S1 |
| Dun Skipper, *Euphyes vestris vestris* | Special Concern | (BC, Can) | S2 |
| **Swallowtails and Apollos (Family Papilionidae)** | | | |
| Clodius Apollo, *Parnassius clodius altaurus* | Special Concern | (BC, Can) | S3S4 |
| Phoebus Apollo, *Parnassius phoebus* | Special Concern | (BC) | S3 |
| Baird's Swallowtail, *Papilio bairdii pikei* | Special Concern | (BC) | S3 |
| Baird's Swallowtail, *Papilio bairdii dodi* | Special Concern | (BC) | S1S3 |
| Indra Swallowtail, *Papilio indra* | Special Concern | (BC, Can) | S2 |
| **Whites, Sulphurs, and Marbles (Family Pieridae)** | | | |
| Spring White, *Pontia sisymbrii beringiensis* | Special Concern | (BC) | S3 |
| Margined White, *Pieris marginalis guppyi* | Special Concern | (BC, Can) | S3 |
| Arctic White, *Pieris angelika* | Special Concern | (BC) | S3 |
| Large Marble, *Euchloe ausonides insulanus* | Extirpated | (BC, Can) | SX |
| Green Marble, *Euchloe naina* | Special Concern | (BC) | S3 |
| Western Sulphur, *Colias occidentalis* | Special Concern | (BC, Can) | S3S4 |
| Mead's Sulphur, *Colias meadii* | Special Concern | (BC) | S3 |
| Hecla Sulphur, *Colias hecla* | Special Concern | (BC) | S3 |
| Giant Sulphur, *Colias gigantea gigantea* | Special Concern | (BC) | S3 |
| **Gossamer Wings (Family Lycaenidae)** | | | |
| Bronze Copper, *Lycaena hyllus* | Special Concern | (BC) | S3 |
| Dione Copper, *Lycaena dione* | Endangered | (BC) | S1 |
| Lilac-bordered Copper, *Lycaena nivalis* | Special Concern | (BC, Can) | S3 |
| Coral Hairstreak, *Satyrium titus titus* | Special Concern | (BC) | S3 |
| Behr's Hairstreak, *Satyrium behrii* | Threatened | (BC, Can) | S2 |
| Sooty Hairstreak, *Satyrium fuliginosum* | Endangered | (BC, Can) | S1 |
| California Hairstreak, *Satyrium californicum* | Special Concern | (BC, Can) | S3 |
| Striped Hairstreak, *Satyrium liparops* | Special Concern | (BC) | S3 |
| Immaculate Green Hairstreak, *Callophrys affinis* | Special Concern | (BC, Can) | S3 |
| Johnson's Hairstreak, *Loranthomitoura johnsoni* | Endangered | (BC, Can) | S1S2 |
| Moss' Elfin, *Incisalia mossii mossii* | Special Concern | (BC, Can) | S3 |
| Eastern Pine Elfin, *Incisalia niphon* | Special Concern | (BC) | S1S3 |
| Eastern Tailed Blue, *Everes comyntas* | Special Concern | (BC) | S3 |

▶

| Butterfly and skipper taxa (species and subspecies) | Suggested status (COSEWIC terms)[1] | | Provincial status (TNC ranks) |
|---|---|---|---|
| **Gossamer Wings (Family Lycaenidae)** | | | |
| Greenish Blue, *Plebejus saepiolus insulanus* | Endangered | (BC, Global) | SH |
| Icarioides Blue, *Icaricia icarioides blackmorei* | Special Concern | (BC, Can) | S3 |
| Cranberry Blue, *Vacciniina optilete* | Special Concern | (BC) | S3 |
| Arctic Blue, *Agriades glandon lacustris* | Special Concern | (BC) | S3 |
| **Metalmarks (Family Riodinidae)** | | | |
| Mormon Metalmark, *Apodemia mormo* | Endangered | (BC, Can) | S1 |
| **Brushfoots (Family Nymphalidae)** | | | |
| Oreas Anglewing, *Polygonia oreas threatfuli* | Special Concern | (BC, Can) | S3 |
| Great Spangled Fritillary, *Speyeria cybele pseudocarpenterii* | Special Concern | (BC) | S3 |
| Aphrodite Fritillary, *Speyeria aphrodite whitehousei* | Special Concern | (BC, Can) | S3 |
| Aphrodite Fritillary, *Speyeria aphrodite manitoba* | Special Concern | (BC) | S3 |
| Bremner's Fritillary, *Speyeria zerene bremnerii* | Special Concern | (BC, Can) | S3 |
| Mormon Fritillary, *Speyeria mormonia erinna* | Threatened | (BC, Can) | S1S2 |
| Western Meadow Fritillary, *Clossiana epithore sigridae* | Special Concern | (BC, Can) | S3 |
| Albert's Fritillary, *Clossiana alberta* | Special Concern | (BC) | S3 |
| Beringian Fritillary, *Clossiana natazhati nabokovi* | Special Concern | (BC) | S3 |
| Boeber's Fritillary, *Clossiana tritonia distincta* | Special Concern | (BC) | S3 |
| Tawny Crescent, *Phyciodes batesii* | Special Concern | (BC) | S3 |
| Hoffman's Checkerspot, *Charidryas hoffmanni* | Special Concern | (BC, Can) | S3 |
| Taylor's Checkerspot, *Euphydryas editha taylori* | Endangered | (BC, Global) | S1 |
| Gillett's Checkerspot, *Euphydryas gillettii* | Threatened | (BC) | S2 |
| Viceroy, *Limenitis archippus idaho* | Extirpated | (BC, Can) | SX |
| Common Ringlet, *Coenonympha california insulana* | Special Concern | (BC, Can) | S2S3 |
| Common Ringlet, *Coenonympha california benjamini* | Special Concern | (BC) | S3 |
| Common Woodnymph, *Cercyonis pegala incana* | Special Concern | (BC, Can) | S3 |
| Common Woodnymph, *Cercyonis pegala ino* | Special Concern | (BC) | S3 |
| Small Woodnymph, *Cercyonis oetus charon* | Endangered | (BC) | S1 |
| Magdalena Alpine, *Erebia magdalena* | Special Concern | (BC) | S3 |
| Beringian Alpine, *Erebia mackinleyensis* | Special Concern | (BC) | S3 |
| Red-disked Alpine, *Erebia discoidalis* | Special Concern | (BC) | S3S4 |
| Mountain Alpine, *Erebia pawlowskii* | Special Concern | (BC) | S3 |
| Great Arctic, *Oeneis nevadensis* | Special Concern | (BC, Can) | S3 |
| Uhler's Arctic, *Oeneis uhleri* | Special Concern | (BC) | S3 |
| Alberta Arctic, *Oeneis alberta* | Special Concern | (BC) | S2S3 |
| White-veined Arctic, *Oeneis bore edwardsi* | Special Concern | (BC) | S3 |
| Jutta Arctic, *Oeneis jutta alaskensis* | Special Concern | (BC) | S3 |
| Polixenes Arctic, *Oeneis polixenes yukonensis* | Special Concern | (BC) | S3 |
| Rosov's Arctic, *Oeneis rosovi* | Special Concern | (BC) | S3 |
| Monarch, *Danaus plexippus* | Special Concern | (BC, Can) | S3B SZN |

[1] See text for definitions of the status ranks, and the species treatments for futher explanation. The status is for the noted subspecies only. If no subspecies is listed, then there is only one subspecies in BC and the status is for the species as a whole in BC.

protection from collection, because the real source of its population decline was, and is, lack of habitat protection. Only through protection and appropriate management of the habitat of butterflies, sometimes supported by management of collecting, will the goal of butterfly conservation be achieved. British Columbia requires legislation similar to the Flora and Fauna Guarantee Act 1988 of the state of Victoria in Australia, which aims to "guarantee that all taxa of flora and fauna and ecological communities in Victoria can survive, flourish, and retain their potential for evolutionary development in the wild" (Britton et al. 1995).

The Endangered Species Act of the United States, passed in 1973, provides for the establishment of conservation programs for endangered and threatened wildlife, including insects. There are now 28 insects, including 19 butterflies, on the US List of Endangered and Threatened Wildlife. Once listed as endangered or threatened, butterflies are protected from collection, commerce, and import/export. There are also requirements for recovery plans, including land (habitat) acquisition, and the species must be considered in all planning processes by federal agencies. Several states are also involved in the conservation of butterflies. Major development projects have been significantly modified in order to protect endangered or threatened butterflies in the US.

More than a quarter of a century after the passage of the Endangered Species Act, and 87 years after the first European butterfly was protected, Canada has no equivalent federal legislation. COSEWIC maintains a list of vertebrates and plants that are of conservation concern nationally, and recently began considering molluscs and Lepidoptera. It has designated three butterflies as being of conservation concern in Canada: the Karner Blue of Ontario (Extirpated), the Maritime Ringlet of Nova Scotia (Endangered), and the Monarch butterfly across all of southern Canada (Special Concern). Several butterflies in BC are being considered for listing as this book is written. The legislative proposals being considered by the federal government make it clear, however, that the designation and management of endangered species will probably be governed by political and economic factors rather than the requirements for adequate conservation of all species.

Currently, COSEWIC lists almost 200 endangered species of wildlife in Canada. If all invertebrates of conservation concern are included on the list, the number would certainly exceed 1,000. There is no legal requirement to protect a species or its habitat when COSEWIC has determined that it is Endangered, Threatened, or of Special Concern, but listing by COSEWIC can be used by the public as an argument for voluntary private or government conservation of a butterfly or its habitat. Such butterfly conservation efforts may be successful if they have no significant economic consequences.

The Convention on International Trade in Endangered Species of Wild Fauna and Flora (CITES) is an international agreement that regulates trade in endangered species of plants and animals, or their parts and derivatives. The government of Canada is a signatory of CITES and regulates the importation and exportation of species on the CITES list. Some butterflies are protected by CITES, but none of them occur naturally in Canada. Anyone interested in importing butterflies from outside Canada should obtain a list of species covered by the CITES agreement to avoid importing a regulated species. Some countries allow the export of CITES-listed butterflies that have been reared by licensed commercial farmers. This helps a local Third World community to earn a living while protecting adjacent natural habitats from further deterioration.

The US Fish and Wildlife Service has additional regulations that prohibit exportation or importation of any animals, including dead butterflies, unless a form is filled out at the border and customs agents are allowed to inspect the specimens both at the border crossing and, at any future date, in the public or private collections where they are kept. Many other countries require a permit for the exportation of any butterflies.

In BC the collection of butterflies and all other natural resources is prohibited in all national and provincial parks except with a permit. Permits are issued only for scientific research or as part of inventories of the wildlife resources of each park. The primary result of the requirement for permits is that parks have been poorly inventoried. Both amateurs and scientists prefer to collect elsewhere rather than undertake the paperwork to acquire a permit and then comply with its frequently highly restrictive conditions. Butterfly conservation is not aided by park restrictions on general collecting in their present form.

It should be noted that up to 2000, there has been a maximum period of six months after a butterfly was collected without a permit during which BC Parks officials could take legal action. A revised Park Act, which should be in force when this book is published, will remove this legal limitation; in the future, anyone collecting in a park without a permit may be held accountable at any time. BC Parks encourages anyone who has in the past collected butterflies in a BC provincial park to transfer them to the Royal British Columbia Museum in Victoria or to another public collection, so that they are available for public research and education.

In BC butterflies of conservation concern have no legal protection outside parks, even though the mandate of the Ministry of Environment, Lands and Parks (MELP) implicitly includes butterfly conservation. It must be noted, however, that many butterfly species on the red (endangered) and blue (species of special management concern) lists are rare in BC but not immediately endangered, as noted in the individual species accounts. The reason for listing them is to ensure that if major habitat changes occur naturally or through human activities, the effects of such changes on the rare species will be noted and considered. Collecting has little potential to adversely affect the majority of rare species, and is *essential* for documenting the discovery of additional populations.

## Habitat Loss

Habitat loss is the most critical threat to butterflies. Many species have a very limited natural range in BC because they require specialized habitats that are of limited extent in the province. The use of land for residential, industrial, forestry, agricultural, hydroelectric, and recreational purposes alters or destroys the natural habitats upon which butterflies depend. Unfortunately much of the land in BC considered most desirable and useful from the human perspective coincides with the small, specialized habitats that are the sole homes of many of our butterflies.

In the United Kingdom and continental Europe, roadsides are beginning to be seeded with native plants – sometimes including species that are butterfly foodplants and nectar sources – to provide a partial replacement for native habitats. At the same time, mowing and herbicide applications are being reduced for economic reasons, and strips of habitat for butterflies with open populations are being created (Thomas 1984). Something similar could be done in BC along roads, railways, powerline right-of-ways, and other linear developments. Garbage dumps could also be seeded with foodplants that thrive in disturbed habitats, such as rock cress (*Arabis*). Sources of native seeds would have to be developed, however, because the few seed sources available would not be able to supply the volume required. "Wildflower seed" from outside BC is of little value because few of the plant species are native to this area and using it may introduce more detrimental weed species.

## Pesticides

In 1947, in Washington state, just south of the BC border, "around the orchard districts, the use of DDT in spraying for coddling moth has just about wiped out all the local Lepidoptera, such as *Colias, Pieris rapae, P. beckerii,* as well as the housefly, a single one of which was seen, which is remarkable" (Hopfinger 1947). By this time, orchard spraying had extirpated the Viceroy from BC, as well as from northern and central Washington.

Spraying with herbicides to kill weeds, and insecticides to kill insects, is inevitably detrimental to butterfly populations. Most such spraying is now localized, and will affect only local populations of butterflies. The forest industry routinely applies broadcast treatments of herbicides over large areas, using Vision®, also sold as Roundup®, to control brush on regenerating plantations. This adversely affects the butterflies using the plantations at the time of spraying, but a required 10 m pesticide-free zone along all streams serves as a refuge for many species; in any case, most of the rest of the area will regrow into dense coniferous forest unsuitable for butterflies. The drier, more open forest types that support butterfly populations in mature forests are generally not sprayed with herbicide because brush is not a problem. In the long term, most butterfly populations will not be adversely affected by forest industry brush control. It must be noted, however, that there is a high diversity of moths in both coniferous and deciduous forests. Some of the moths are of conservation concern, and may be adversely affected by a combination of harvesting and herbicide brush control, through effects on rare populations similar to those discussed below for *Btk*.

The pesticide that poses the most serious threat to butterfly conservation is the otherwise environmentally neutral bacterial insecticide *Bacillus thuringiensis,* commonly known as *Bt. Bacillus thuringiensis* occurs naturally throughout the world in soils and on vegetation, and viable spores have even been found in Egyptian pyramids (Scriber and Gage 1995). Different varieties of *Bt* produce a crystal protein and spores that are toxic to specific groups of insects. *Bacillus thuringiensis* variety *kurstaki* (*Btk*) is especially toxic to butterfly and moth larvae. Other varieties of *Bt* include variety *tenebrionis* for control of Colorado potato beetle and elm leaf beetle larvae; variety *israelensis* for control of mosquito, black fly, and fungus gnat larvae; and variety *aizawai* for control of Wax Moth larvae and various other larvae, especially the Diamondback Moth caterpillar.

*Btk* kills the larvae of all butterflies and moths, although some species are partially resistant to its effects. When *Btk* is eaten by a caterpillar, the crystal protein and the spores damage the gut lining, leading to gut paralysis. Caterpillars stop eating, become limp and shrunken, die, and decompose. The crystal protein in commercial formulations of *Btk* is toxic only when eaten by lepidopteran larvae with an alkaline gut pH and gut membranes sensitive to the toxin. Any larvae feeding on the outside of leaves, which includes most butterfly larvae, will be affected when *Btk* is sprayed (Otvos and Vanderveen 1993). Caterpillars that eat doses of *Btk* too low to kill them may produce adults with reduced reproductive capability, sometimes resulting in additional population reductions the year after spraying (Otvos and Vanderveen 1993). Adult butterflies, such as Monarchs, are not affected by spraying with commercial preparations of *Btk*. The bacteria do enter the adult's hemolymph (insect "blood"), however, possibly contaminating the eggs and therefore the hatching larvae (Leong et al. 1992).

Swallowtail butterfly larvae feeding on foliage 30–38 days after normal field application of *Btk* suffer about 80% mortality, with very little difference whether the larvae were in the shade or in the sun. In contrast, the natural mortality of larvae on untreated leaves averaged only about 20% (Fig. 10). Similarly, toxic effects of *Btk* have been shown for Gypsy Moth larvae up to 64 days after spraying (Miller and West

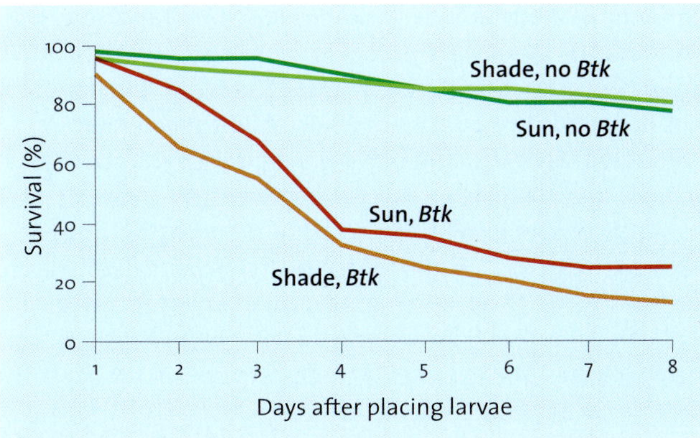


**10**  Survival of Eastern Tiger Swallowtail (*Papilio glaucus*) larvae when placed on leaves 30 days after *Btk* spraying, two years' data combined (modified from Scriber and Gage 1995).

1987). The conventional wisdom that *Btk* is rapidly inactivated after application – with sunlight breaking down the crystal protein and rain washing it off the foliage, so that within a few days to a week the *Btk* is no longer active (Otvos and Vanderveen 1993) – seems overly optimistic.

Garry oaks sprayed with *Btk* show significant reductions in both overall number of lepidopteran larvae (50% reduction) and in species diversity (38% reduction). Species diversity is still depressed in the *Btk*-treated area three years later, although overall caterpillar abundance has recovered. *Ceanothus* shrubs showed an 81% reduction in caterpillar abundance soon after spraying with *Btk,* and a 71% reduction during the same season the following year. Uncommon caterpillar species were the most adversely affected, and were completely lost from the area treated with *Btk* (Miller 1990, 1992).

Another repeated, but inadvertent, study involves the control of the Gypsy Moth, *Lymantria dispar* (L.), on the south coast of BC. Gypsy Moths are a rare species at any location in the province when they are detected. When *Btk* is applied to the area of occurrence, these small populations are effectively eliminated, demonstrating that a rare moth species will be consistently extirpated through application of the insecticide. Any other rare butterfly or moth species whose larvae are feeding on leaves when *Btk* is applied is also likely to be extirpated.

Gypsy Moths are excellent colonizers and soon return. Butterflies and moths of conservation concern are generally poor colonizers, however, and if there are no nearby populations from which recolonization can occur, they will have been extirpated forever. Controlling Gypsy Moths through *Btk* spray programs will therefore inevitably have a severe impact on, and likely extirpate, many of the Lepidoptera of conservation concern on southern Vancouver Island. The process may take decades, but it will probably be inexorable if *Btk* spraying continues to be the control method of choice for pest Lepidoptera species.

In addition to directly affecting non-target Lepidoptera, spraying with *Btk* may have adverse impacts on vertebrates that feed on butterflies or moths. For example, many rare and endangered songbirds, bats, and flammulated owls feed primarily on moths or their caterpillars during the nesting season. The number of adult moths of all species will be greatly depressed during the nesting season in the year of spraying (Miller 1990, 1992), which may result in reduced fledging success of nestlings. Of course, severe defoliation resulting from lack of control of a forest pest may also reduce fledging success, as the nests may become more visible and therefore vulnerable to predation (Scriber and Gage 1995).

Gypsy Moth control occurs regularly on southern Vancouver Island, in Greater Vancouver, and, less frequently, in the Okanagan Valley. These are the areas of BC most prone to accidental introductions of Gypsy Moth; unfortunately they are also the areas with the highest number of rare, Special Concern, Threatened, and Endangered butterflies and moths (Guppy et al. 1994).

These issues are not limited to the south coast. Large areas of the forests of BC are sprayed annually with *Btk* to control native forest pests such as spruce budworm. This control of native pest larvae will have the same effect on butterflies and moths as the control of introduced pests like the Gypsy Moth. Rare species will be extirpated throughout much of the province over time, and only the common species will be able to recolonize an area once a spray program is completed. The amount of forest area sprayed annually is to a large extent determined by budget: more spraying occurs when money is available than during periods of tight budgets. This indicates that much of the spraying for native forest pests is optional for adequate timber management and, given the effects on non-target lepidopteran species, should not be done.

A new threat from *Btk* has recently become apparent. The genes that produce the *Btk* toxin have been transferred from *Bacillus thuringiensis* to corn, so that cornfields do not need to be sprayed with pesticides to control corn earworm and other moth larvae. This is a wonderful idea in theory, because it would reduce pesticide spraying. However, the *Btk* toxin is also incorporated into corn pollen. The corn pollen drifts onto plants outside cornfields, and is toxic at least to Monarch butterfly larvae feeding on milkweed (Losey et al. 1999). There are also projects underway to incorporate the genes for the *Btk* toxin into forest trees such as trembling aspen (*Populus tremuloides*) to protect them from forest pests such as tent caterpillars. This would eliminate not only the pest species but also the numerous non-pest butterflies and moths whose larvae feed harmlessly on aspen. In addition, if the toxin is incorporated into aspen pollen, as in corn, the pollen rain would eliminate most of the very large number of butterfly and moth species that live in aspen forests.

## Predators, Parasites, and Competitors

Introduced predators such as birds, especially house sparrows and starlings, and parasites that are introduced either deliberately for pest control or accidentally must have a significant impact on butterfly populations, but there has been little research to document this.

The extirpation of the Island Large Marble (*Euchloe ausonides insulanus*) occurred so early in the history of European colonization of British Columbia that there is no documentation of its loss. One possibility is that parasitoids associated with the introduced Cabbage White built up large populations, resulting in incidental parasitism of Island Large Marble larvae at a rate high enough to push the subspecies to extinction. A more likely reason, however, was heavy grazing of the natural meadows by cattle and sheep, which would have severely depressed the populations of biannual and short-lived perennial rock cress, the presumed larval foodplant.

Mustard Whites are apparently displaced by Cabbage Whites that invade their habitat, resulting in greatly reduced distribution of Mustard Whites in eastern North America (Shull 1977). While documenting the spread of the Cabbage White, Scudder (1887, 1889b) noted that Checkered White and Mustard White populations disappeared as rapidly as Cabbage Whites spread over the continent. This could have been due to the direct effects of competition or to the arrival of Cabbage White parasites within two years after the Cabbage Whites had arrived in an area (Scudder 1887). Margined Whites in BC also tend to be extirpated from urban, suburban, and agricultural areas, possibly due in part to displacement by Cabbage Whites.

## Butterfly Collecting

Butterfly collecting is not extensive in BC, and never has been. There are fewer than 10 butterfly researchers residing in the province, and a smaller number of moth specialists. There are very few non-resident collectors, most of whom collect only along the Alaska Highway on the way to the Yukon and Alaska. Collecting is usually done by amateur (unpaid) scientific collectors except for occasional government or university inventory projects. Almost all the information on the butterflies of BC in this book originated from these amateur collections over the past 140 years. Sight observations of butterflies, without collection of voucher specimens and their placement in a permanent collection, are generally useless because of a high identification error rate.

Scientific collection of voucher specimens should have no adverse effect on the conservation status of the province's butterfly species, and is the basis for documenting the presence, distribution, and number of butterfly populations. It is very difficult to damage a population through collecting, even when one deliberately sets out to do so. Nearly all

extinctions of butterfly populations in the United Kingdom, which has many butterfly collectors, have occurred in the absence of collecting (Thomas 1984). The few extinctions that occurred during times of heavy collecting were of populations that were already on the brink of extinction due to habitat loss. Based on mark-release-recapture studies, it would take several collectors working every day for three weeks to extirpate a small (< 250 adults) lycaenid population (Thomas 1984). This is very unlikely to occur, especially if butterfly collectors follow guidelines similar to those of the Lepidopterists' Society (see Appendix 4) to ensure that their activities are conducted in an ethical manner. (The Lepidopterists' Society Web site, at <http://www.furman.edu/~snyder/snyder/lep/index.html>, includes links to other butterfly- and moth-related sites.)

## Butterfly Photography

Many people would like to inventory butterflies without killing them. Reasons include personal ethics, lack of desire to collect, or inability to maintain a collection. Non-lethal inventory is possible for some species, but identifications are much more difficult and errors cannot be corrected later. A sight observation without documentation is of limited value because it is too easy, even for an expert, to make a mistake. Clear photographs are the only adequate substitute for specimens, and still will not replace specimens for some species. Photographs should show both the upper and lower sides of the wings. It is frequently necessary to capture a butterfly to obtain proper photographs. Photographs that are intended to be permanent records of butterfly observations should be fully labelled with the same collection data as for specimens, and should be maintained and eventually transferred to a public museum for permanent record.

## Butterfly Counts

Butterfly enthusiasts in several areas of the province have started annual butterfly counts to track trends in population sizes of butterflies, but the data are often for too few years to be analyzed. Butterfly counts once or twice per year are useful for documentation of long-term general trends in populations averaged over a region, documentation of occurrences of migratory species, and location of new populations of rare species. Accurate identification is important but not critical, because any unusual sightings that lack supporting documentation are automatically dropped during data analysis or verified later through more rigorous methods.

Annual butterfly counts in North America were started by the Xerces Society but are now coordinated by the North American Butterfly Association (NABA). The recommended methodology should be followed as closely as possible to ensure that data are comparable between years and between areas. The standard date for the count range varies slightly between years, but centres around 4 July (13 June to 26 July in 1998). The count consists of recording the number of butterflies of each species seen within a circle 24 km in diameter in one day. It is critical to include the number of "person-hours" of observation so that yearly changes in number of butterflies counted can be adjusted to account for the number of hours in which they were counted. For details, see the NABA Web site at <http://www.NABA.org>.

Monthly butterfly counts can provide more valuable data than the single-day sampling, by including all species in an area and by accounting for some of the differences in flight periods between years. These counts can document year-to-year fluctuations in population size and, most importantly, the long-term trends in population numbers. Naturalists on southern Vancouver Island are the only group in BC doing these high-quality counts, the results of which are published annually in the *Victoria Naturalist*. An even more intense survey, consisting of weekly counts (Smith 1984), is desirable to track the population changes of butterflies of conservation concern.

Special methods of butterfly counts can provide detailed information on life span, population size, local movements, and dispersal of butterflies. These counts are readily done without elaborate equipment by capturing butterflies, marking and releasing them, and then recapturing them later the same day or after a day or two (mark-release-recapture, or MRR). MRR is time-consuming but it is the most informative method of estimating population sizes, and therefore of obtaining information on the conservation status of local populations. The marking instrument is a fine indelible felt-tip pen; either a coded arrangement of ink spots (Ehrlich and Davidson 1961) or a unique number on the wing is used to identify individuals. Very small butterflies require special marking techniques (Mattoni and Seiger 1963), and chilling in a camping cooler can take the place of anesthetizing with carbon dioxide. Care must be taken not to damage the butterflies during handling (Singer and Wedlake 1981; Orive and Baughman 1989). The process of capturing and marking can affect the subsequent behaviour of some butterflies (Morton 1982), so trauma must be minimized and the butterflies should be cooled in the shade for a few minutes before being released, to avoid a "panic" flight. Monarch butterflies have their own traditional marking method using paper tags (Urquhart 1960, 1976, 1987); this could also be used on other large, strong-flying butterflies such as Painted Ladies (*Vanessa cardui*) (Fig. 11). The very obvious white paper tags used on Monarchs can increase their predation rate, however (Sakai 1994). An excellent discussion of MRR methods, and associated simple to complex statistics for calculating population sizes, is provided by Gall (1985). MRR studies for butterflies in BC have been done only for Rocky Mountain Apollos (*Parnassius smintheus*) near Vernon by Cris Guppy and for Silver-bordered Fritillaries (*Clossiana selene*) in the Kootenays by Jon Shepard.

**11** Painted Lady (*Vanessa cardui*) with white paper tag

A method of determining relative butterfly population sizes, either over time or between populations, is that of "Pollard Walks." These intensive surveys consist of walking along a linear route at a constant pace, counting all butterflies of the chosen species within 5 m on each side. Pollard counts provide an index of the relative population size rather than an estimate of the total population size, and show seasonal changes, between-year changes, or differences between populations. In a long-term study of a butterfly species, Pollard counts and an MRR study can be done simultaneously one year to provide a "sightability index"; thereafter the Pollard counts can be used to provide estimates of population size. The details of the method are given by Pollard (1977), with modifications by Douwes (1977) and Thomas (1983).

## Larval Foodplants

The larval foodplants of butterflies in BC are very poorly known, which will make future butterfly conservation efforts difficult. Before a habitat can be managed to enhance the foodplant of a rare species of butterfly, that foodplant must be known. Frequently, foodplants for BC butterflies are recorded from faraway areas and may not even occur in the province. Local naturalists can add to our knowledge by providing detailed information about the foodplants associated with butterflies in the province.

For example, when a larva is reared, the foodplant should be documented with a prepared plant specimen. The adult that emerges from the pupa should be prepared as a museum voucher specimen and eventually deposited in a public museum. The nature of the foodplant should be specified:

*natural foodplant:* a plant upon or near which eggs are laid and on which larvae commence feeding and successfully mature, all under natural conditions

*laboratory foodplant:* a plant upon which a larva successfully matures under captive conditions

*natural oviposition substrate:* a plant or object on which oviposition occurs under natural conditions

*laboratory oviposition substrate:* a plant or object on which oviposition occurs under captive conditions

*foodplant by association:* the only plant present that is a likely larval foodplant, based on the foodplants known elsewhere

Details of reporting procedures for larval foodplants are provided by Shields et al. (1970).

## BUTTERFLIES IN SCHOOLS

Butterflies are useful tools for teaching children and adults about insects. If done properly, collecting butterflies and other insects can teach much more than the name of a butterfly. Observations can be made on the different habitats each species uses, the different behaviours, the seasons each butterfly flies in, and how butterflies interact with each other and their environment. Collecting butterflies can provide a complete course in butterfly behaviour and ecology, and give students an opportunity to determine what species have been observed and how they vary in wing pattern. A student collection of only a handful of butterfly species can be the basis for many discussions about what those species do in the wild.

Butterflies can be readily bred in captivity, and are frequently reared in schools to teach students about their life cycle.

Students find it fascinating to watch butterflies progress from eggs through caterpillars and pupae to adults that lay more eggs. Butterflies reared in classrooms should *never* be released into the wild, however, unless their wild parents came from the same area as the release. Reared butterflies can be killed painlessly by being placed in a deep freeze for 48 hours. They simply "fall asleep" as though it were a cold night, and never revive.

Monarchs are often reared in classrooms and then released into the wild. This may have adverse consequences. Monarchs from eastern North America released into western North America as adults or immatures could transmit diseases from one population to another. These diseases may be highly infectious if they are not already present in western populations or if they are a different strain. The eastern and western populations of Monarchs are also likely to be genetically different, and hybrids resulting from transfers between areas will not be adapted to either environment (Brower et al. 1995). Monarch butterflies in the Kootenays may be genetically adapted to migrate to eastern Mexico for hibernation, whereas those in the rest of southern BC probably hibernate in California. Even so, there is no assurance that Monarchs in the Okanagan are genetically the same as those rare individuals that migrate to coastal British Columbia.

Another commonly reared and released butterfly is the Painted Lady. Although butterflies from one area should not be released into another area, the risks associated with releasing Painted Ladies into the wild are much lower than in the case of the Monarch, because it is likely that all North American Painted Ladies share a common gene pool and a common set of diseases and parasites.

Cabbage Whites are also easily reared in classrooms. They are generally not commercially available but can be easily collected locally in September, at the start of a school year, in all areas except northern BC. The Cabbage White is an introduced "weed" species, and there is no concern with releasing small numbers back into the wild at the end of the school year. They will simply blend into the local population, and any detrimental effects will become part of the control of Cabbage White populations. Releasing large numbers may be unpopular with local gardeners and farmers, however.

# Butterfly Gardens

Butterfly gardening is the creation, enhancement, or maintenance of an area of land as butterfly habitat. Butterfly gardening is usually thought of in terms of individual suburban or city gardens, but it extends to the manipulation of large areas to enhance their habitat value for butterflies. The same basic concepts apply regardless of the size of area involved, because the principles of butterfly gardening are based on butterfly biology, behaviour, and ecology. In general urban and suburban butterfly gardens focus on butterflies with open populations, whereas those in parks, public lands, or large private tracts may focus on butterflies with closed populations.

Butterfly species have a wide range of biological characteristics, all of which determine how they will interact with a butterfly garden. A key characteristic is whether a species has "closed" or "open" populations. Adults of the former stay within discrete areas of suitable habitat; adults of the latter fly widely over the countryside without staying within discrete areas of suitable habitat. Butterfly gardens normally attract only butterflies with open populations, occasionally attract butterflies with semi-closed populations if such populations occur in the general vicinity, and only very rarely attract butterflies with closed populations. The larval foodplants chosen for a butterfly garden should be oriented primarily towards butterflies with open populations, unless one is fortunate enough to have existing semi-closed or closed butterfly populations very nearby.

## Gardening for Butterfly Species with Open Populations

Butterfly species with open populations will not establish a permanent, self-sustaining population in a butterfly garden; instead, the garden will be a patch of habitat in the area over which the species ranges. For the most part, a butterfly garden designed for butterflies with open populations just needs larval foodplants, so that passing females can lay eggs, and nectar sources so that both males and females will tarry within the garden for a while before moving on.

The garden should be arranged to provide sheltered warm, sunny areas with both the nectar sources and larval foodplants. Plant the taller and denser shrubs and trees on the north side of the garden to prevent shading. If they must be planted on the south side, use low, open-growing shrubs spaced to let sunshine into the garden. Some butterflies enjoy feeding on the sweetness of tree sap and the juice of overripe fruit, and also use a wide variety of other foods. Water in the form of puddles and mud should be present every sunny day, as butterflies dehydrate quickly on such days.

Native lupines, vetches, and wild peas provide not only beautiful flowers and nectar but also food for the larvae of many different blues, sulphurs, skippers, and moths. Hollyhocks provide food for the migratory West Coast Lady, and milkweed food for the famous Monarchs migrating north from their winter hibernation groves in California. The native mustards such as rock cress (*Arabis*) have small flowers that still produce abundant nectar, and the larvae of many whites such as Mustard Whites, orangetips, and marbles feed on the leaves. Many asters, as well as alfalfa, produce masses of flowers in late summer and fall, when few other flowers remain to produce nectar. Bumblebees and butterflies love the nectar of both plants; also the larvae of Field Crescents feed on aster leaves and those of Clouded Sulphurs and Orange Sulphurs feed on alfalfa leaves.

Flowers are the main food source for adult butterflies, but not all flowers produce nectar. Even if a flower produces nectar, it must still be accessible to butterflies; only swallowtails have "tongues" long enough to reach down the long tube flowers of daffodils and many rhododendrons to reach the nectar. Flowers that are large, brightly coloured, and sweetly scented advertise themselves to passing butterflies, who may then stop for a drink of nectar. In general large butterflies nectar on large flowers, and small butterflies nectar on small flowers or small flowers in large clusters. The best butterfly nectar sources are butterfly weed, lilac, butterfly bush, and (south coast only) lantana.

Vacant lots and other weedy areas can do their part as well. Alfalfa and red clover produce an abundance of scented flowers and nectar, and their leaves are eaten by the caterpillars of skippers, sulphurs and various moths. Unmown grass in sunny vacant lots is eaten by larvae of Woodland Skippers, whose golden-brown adults can be seen everywhere in late summer. Thistles, much reviled as "noxious weeds" in many areas, produce large amounts of nectar in their attractive flowers, and larvae of Painted Ladies and Crescents munch on the leaves. In moist areas, a patch of stinging nettle can provide spring greens for the table as well as food for the larvae of Red Admirals, West Coast Ladies, Satyr Anglewings, and Milbert's Tortoiseshells.

In the vegetable garden (unsprayed!), cabbage and its relatives support the Cabbage White, which, although its green larvae are unwelcome at times, still provides white butterflies that flit around the garden. In areas adjacent to the natural habitats of the native *Pieris* white butterflies, caterpillars of native species may also feed on garden cabbages. Parsley, anise, dill, parsnips, and carrots provide food for the yellow- or orange-spotted green larvae of Anise Swallowtails, as well as for your table. If excess larvae are present, such as a large mass of Mourning Cloak larvae on a small willow, some of them can be moved elsewhere rather than killed.

Native willows and the garden apple tree (unsprayed!) have attractive foliage that the larvae of Western Tiger Swallowtails, Mourning Cloaks, skippers, Eyed Hawkmoths, and many others feed on. On Vancouver Island, the tiny sluglike larvae of Brown Elfins feed unnoticed on the flowers of arbutus trees, and Garry oaks may have a scattering of larvae of the rare Propertius Skipper feeding on their leaves. Birches and aspens provide food for the larvae of swallowtails and Sphinx Moths along city boulevards. Even pine trees are eaten by larvae of Pine Elfins, and Pine Whites and many moths feed on pines and various other conifers.

Few butterflies or moths lay so many eggs on one plant that the larvae will damage it, although of course we notice the few that do! Most eggs produce caterpillars that never mature but instead become food for birds or are attacked by a fascinating array of spiders and insect predators and parasites. Of the hundreds of eggs laid by each female butterfly or moth, only two on average survive to produce the next generation. If more than two survived to replace their parents, there would be a population explosion of butterflies. Even those will of course succumb to pesticides used unwisely.

A network of wildlife gardens and parks throughout our cities is needed to maintain populations of butterflies and moths, for there is nowhere else for them to live. Individual gardens are needed, so that as butterflies and moths explore the city they constantly find new habitats in which to lay a few eggs before moving on. Each garden and park should provide both nectar and food for some larvae, with all the gardens and parks together providing habitat for all the species of butterflies and moths that once flew where our cities now stand.

## Gardening for Butterfly Species with Closed Populations

It is possible to establish a butterfly garden designed to support one or more species with closed populations, and then to artificially introduce eggs, larvae, or females to the garden to establish a population. The success rate will depend on the species of butterfly involved and how suitable the garden is as habitat. Establishing new habitat for butterflies with closed populations has never succeeded in BC, and *success is unlikely unless an expert on butterfly*

*ecology is extensively involved in planning and development.* Despite the difficulty, developing habitat for closed-population species is worth the risk and the effort. We should learn how to do it with common species before we are forced to do so with endangered species.

Habitat for closed-population species is much more than just foodplants for the larvae and nectar sources for adults, but the other critical habitat characteristics are unknown for most species. Wherever possible, examine the habitat used by several populations of the butterfly you are interested in, and think about what aspects might be important to them. Well-drained soils and leaf litter may be important for larvae and pupae that spend the winter on the ground. Sunshine is usually important, but some species prefer the broken shade of open forests.

As an example, the Mormon Metalmark is very rare in BC. There is only one small extant group of populations near Keremeos and one in the Okanagan, despite the fact that the larval food-plant is more widespread in southern BC. One key element of their habitat may be loose rocky slopes with abundant foodplant, so that the larvae can crawl down among the rocks to hibernate. There may be other critical habitat elements, but without this particular one, establishing a new colony of Mormon Metalmarks would be impossible.

Minimum sizes of suitable habitat for butterflies in BC are unknown, but range from perhaps a tenth of a hectare for some blues to most of western North America for the Monarch. As a general rule, the smaller the butterfly, the smaller the area required for a permanent population. Where possible, meadow habitat should be surrounded by forest, shrubs, or water, and shrub habitat by forest, meadow, or water. Surrounding the suitable habitat with clearly unsuitable habitat will reduce the tendency of butterflies to accidentally emigrate from the garden, and therefore increase the population size in the critical first few years after establishment. The size of the garden is effectively increased by including adjacent areas of suitable habitat, if they exist. To achieve the larger areas, it may be possible to combine several gardens in a neighbourhood, or gardens and adjacent park areas. The gardens or parks do not need to be in contact with each other, but should be within normal flight distance for the desired butterfly species.

Caterpillar foodplants and nectar sources do not need to be planted in solid stands; scattering them over the garden with other plants will work as well or better. Butterflies tend to lay eggs and nectar on the first plants encountered on the edge of a patch; interspersing the larval foodplants of several species as well as the adult nectar sources will spread out oviposition and nectaring and make it possible to cater to many different species. The overall appearance will be that of a natural wildflower meadow. The added bonus of this approach is that even if one or more species fails to establish a population, others may succeed, and of course open-population species will also be able to use the garden.

The minimum amount of larval foodplant within a small area that is required to maintain a population is not known for any butterfly in BC. Try spacing each kind of foodplant (individual plants or small patches) an average of 1–2 m apart from edge of plant to edge of plant, over the entire area of habitat.

It is also not known how many adult butterflies, pupae, larvae, or eggs need to be introduced so that there is a reasonable chance of successfully establishing a population. Try at least 20–30 females and the same number of males. The females will have mated but may be ready to mate a second time, so some males should be present. Too many males, however, may harass the females and drive them out of the garden.

Release the butterflies in the middle of the garden early in the morning, while they are too cold to fly, so that they start the day slowly and get used to their new surroundings before they become fully active. Take care not to remove more than 10% of the females from any wild population in any one year, or you might damage the population. If you can obtain only a few females, build a screened cage over some of the larval foodplants and nectar plants to protect and confine the females while they lay eggs.

In the United Kingdom, at least 10 butterfly species have been successfully introduced to new areas or reintroduced into areas from which populations had been extirpated and the habitat then rehabilitated (Thomas 1984). Most of the introductions lasted indefinitely, except when some were later lost because the habitat became unsuitable for lack of appropriate maintenance. A checkerspot (*Euphydryas aurinia*) population became extinct after introduction because the population exploded in the absence of parasites and all the foodplants were destroyed. In North America, Harrison (1989) found that when 100 postdiapause larvae (representing the offspring of a single female) of Edith's Checkerspot were transferred into each of 38 habitats, there was a 6.25% chance of establishing a new local population that lasted two years. When a single egg mass of Gillett's Checkerspot (*Euphydryas gillettii*) was introduced to each of eight areas of apparently suitable habitat, only one new population was established (Williams 1995). It lasted three years and then died out, apparently because its larval foodplant population was destroyed when a river altered course.

## Maintaining Butterfly Gardens

When planning a butterfly garden, whether for open- or closed-population butterflies, think about how the garden will be maintained. For open-population butterflies, normal gardening practices may need only slight modifications because the butterflies may not be present throughout the whole year. In closed populations, butterflies are present every day of the year, and care must be taken to maintain the

habitat for eggs, larvae, pupae, and adults at the appropriate times of year. If larvae or pupae spend part of the year on the ground, then litter cannot be removed at those times (and should never be entirely removed). The larval foodplants should not be pruned or manicured when eggs or larvae are on them, but rather at other times of the year. When adults are in flight, consider that they rest hidden in the vegetation when not in flight, and can be killed by lawn mowing, heavy hosing of leaves, or pruning of all plants, not just the larval foodplants. Any necessary maintenance that will kill eggs, larvae, pupae, or adults should be done on only part of the garden each year; this way, only part of the population is killed and the disturbed part of the garden can be recolonized the next year. If severe aphid or scale infestations require insecticide treatment, use soap sprays rather than chemical sprays. The soap will kill larvae if too much gets on them, but larvae (especially older ones) are less susceptible than aphids. Remember that most unsightly aphid infestations will not actually harm plants enough to damage the butterfly population, and that aphid control is likely to have a greater effect on the larvae. If particular plants are repeatedly attacked by garden pests, consider replacing them with other, less susceptible species or varieties.

Probably the biggest problem in most butterfly gardens is invasion by unwanted plants, especially weeds and grasses that might smother the larval foodplants and adult nectar sources. Weeds are a natural part of ecosystems, and should be controlled only if they are going to take over the habitat from the desired plants. For meadow habitats, mowing when the larval foodplant is dormant removes dead grasses and other vegetation that inhibit the growth of herbaceous plants, but the presence of eggs, pupae, or larvae should be taken into account. Herbicides may be required to control especially invasive weeds, but for obvious reasons only the bare minimum should be used, and only as a last resort. It may be desirable to apply herbicides before a butterfly garden is established, if one can foresee that existing weeds will compromise the garden.

An essential tool in reconstructing or rehabilitating large meadow habitats is fire (Reed 1997). Fire is a natural feature of grasslands, and frequently determines which plant species are present. It destroys shrubs and trees, and reduces the density of herbaceous vegetation and thatch. Fire also reduces the effects of competition between plant species, opens up bare soil to enable new young plants to become established, and creates a variety of soil conditions, which increases the diversity of plant species, including larval foodplants. Before fire is used to restore habitat, however, much of the accumulated fuel may need to be removed by cutting brush or trees or mowing grassland, otherwise the fire may be too hot and burn the soil and the roots of herbaceous vegetation. Burning can be more effective than mowing in maintaining habitat in large meadow areas, but it must be carefully planned.

Only part of the meadow should be mowed or burned each year; mowing and burning must be done at the right time of year; burns must not escape outside the planned area; and burns must be as cool as possible (with little fuel such as thatch) to avoid killing the roots of desirable herbaceous plants. These are short generalizations of complex management tools; the maintenance of each garden will have to be planned individually.

## Cabbage Whites in Your Garden

Cabbage Whites are usually the first butterflies to be seen in gardens in the spring, providing a welcome sign of the season as they flutter around nectaring and courting. After courting comes mating, followed by egg laying. The eggs are laid on a wide range of cole crops (garden vegetables related to cabbage) grown in home gardens, as well as nasturtiums. In fact, Cabbage Whites are probably the worst crop pest of any butterfly in North America (Fig. 12).

From 1914 to 1935, the method recommended by the BC Department of Agriculture for control of "cabbage-worms" (Cabbage White larvae) on cabbage crops was a heavy application of arsenate of lead (Treherne 1914; Ruhmann 1935), which was believed to have no effect on humans! It is now known, of course, that arsenate of lead and many other pesticides cause illnesses in humans.

For home gardens a good control method is to cover cabbage crops with Remay® cloth, to prevent eggs from being laid on the plants (Remay® is also good for controlling many other garden pests). Hand removal of large larvae, along with hosing the leaves with a jet of water to wash off small larvae and eggs, helps as well. Application of *Bacillus thuringiensis* (*Bt*) directly on the cabbages helps reduce the number of large larvae by killing them before they mature. *Bt* should not be used indiscriminately throughout the home garden, because it will kill many harmless butterfly and moth larvae

that supply food for nestling birds, spiders, and many predaceous and parasitic insects.

Books on organic gardening sometimes suggest interplanting various herbs with cabbage crops to discourage Cabbage Whites from laying eggs. Interplanting cabbage rows with marigold, nasturtium, pennyroyal, peppermint, garden sage, and thyme does not reduce the number of Cabbage White eggs and larvae compared with cabbage grown alone (Latheef and Irwin 1979). Interplanting collard rows with hyssop, southernwood, tansy, wormwood, catnip, or santolina actually *increases* the number of Cabbage White eggs on the collards compared with collards grown alone (Latheef and Ortiz 1983a, 1983b). Thus interplanting cabbage or collard crops with herbs commonly recommended as pest repellents is either ineffectual or detrimental (Latheef and Ortiz 1983a, 1983b).

## Raising Butterflies

Most butterflies are fairly easy to rear in captivity. Eggs, larvae, or pupae may be found in the wild and reared to adults. One of the best methods of obtaining immatures is to capture a female and induce her to lay eggs. Usually sufficient eggs for rearing will be laid in 1–3 days, and the female can be released again (but only in the same place where found!). A wild-caught female can be kept in a partly sunny cage placed over a planted or potted larval foodplant. Flowers can be included to supply nectar. This provides the most natural conditions for oviposition, but is frequently not possible. It may also be difficult to locate the eggs, especially if they are laid on debris rather than on the larval foodplant. Wild-caught females have almost always already mated.

Alternatively, a wild-caught female can be confined in a screen, cardboard, wooden, or plastic box with a top of cloth or screening. The size of the box depends on how many females are in the box (never more than 2–3) and how large they are, but should be a minimum of 25 cm on each side. Shine a reading lamp (40-watt) light on the top or place the container in sunlight. The cage should not be overheated, especially in the sun, lest the butterflies die. Include a piece of the foodplant in the container for eggs to be laid on. The objective is to raise the body temperature of the butterfly to 30–35°C so that she flutters around the cage without overheating. Position the foodplant relative to the light source so that when the butterfly lands, it comes into contact with the foodplant, to stimulate oviposition. If the container is too large, the butterfly will have trouble finding the foodplant. Observe the butterfly and adjust the position of the foodplant or cage so that she regularly lands on it while fluttering around, and also so that she does not wedge herself into a tight corner. Feed females at least once a day (unroll their proboscis with a pin and dip it in sugar water); they will often learn to feed themselves from a small container containing cotton soaked in sugar water. The

**12** Cabbage White (*Pieris rapae*) nectaring on garden buddlea

butterfly usually lives longer and produces more eggs if the warm temperatures are maintained for only a couple hours in the morning and afternoon to allow oviposition; the rest of the time the butterfly should be maintained at room temperature as more eggs mature in her body.

Caterpillars should normally be fed on the same larval foodplants as in the wild, either as individual leaves, cuttings, or potted plants. The stems of individual leaves or cuttings should be placed in water to keep them fresh, but the larvae must be prevented from crawling into the water and drowning. If the normal foodplant is not available, try related species of plants. When feeding Anise Swallowtails and relatives, carrot tops work well if organically grown (the tops of most store-bought carrots have high levels of pesticides). The more natural the rearing conditions, the better, although larvae should normally be sheltered from rain and provided with partial shade. Rearing can be done indoors or outdoors. Caterpillars should be caged with cloth or screening fine enough to keep them from wandering and to prevent

parasites and predators from attacking them. Cages may need frequent cleaning to prevent mould, and humidity must be kept low to prevent viral and bacterial diseases. The larvae of many species are cannibalistic, so this needs to be watched for. Sticks or other rough surfaces should be provided for species that hang themselves up to pupate, and fine leaf litter, loose soil, or peat moss for those that pupate on the ground.

Under appropriate rearing conditions, many species that are normally univoltine in the wild in BC will produce a second generation or a partial second generation. The high temperatures required for rearing may result in a high larval death rate, depending on the species. Alexandra's Sulphurs, Pink-edged Sulphurs, and Mead's Sulphurs will all produce partial second generations (some larvae go into diapause) when reared under conditions of 80°F (26.7°C) and 14 hours of light per day (Ae 1958a), as will White Admirals and Lorquin's Admirals under normal indoor temperature and light conditions.

Butterflies are a small group within the large phylum Arthropoda. Butterflies, other insects, and crustaceans, including crabs, shrimp, lobsters, crayfish, and sowbugs, are all arthropods. They all have a segmented body and an exoskeleton, or external skeleton, made up of layers of the complex polysaccharide, chitin.

Insects are the dominant life forms in the terrestrial environment, with at least 1 million described species. The primitive ancestor from which they evolved had a many-segmented body with one pair of walking legs on each segment, and looked much like present-day centipedes.

The most primitive living insects have the basic body pattern of all insects. The body is divided into the head, thorax, and abdomen. The head consists of several completely fused segments. The only evidence of the primitive segmentation is found in the paired mouthparts, the paired compound eyes, and the paired antennae. The most primitive mouthparts are chewing mouthparts made up of a fused labrum, paired mandibles, paired maxillae, and fused labium. The thorax consists of the next three body segments (prothorax, mesothorax, and metathorax), which are fused, thickened, and strengthened internally to provide muscle attachment for the locomotor structures, the functional legs and wings. The abdomen has 11 segments. Eight are large and unmodified; the others are highly reduced and the exoskeletal portions are modified as external reproductive structures.

Insects grow by periodically shedding the rigid exoskeleton (moulting) and replacing it with a larger exoskeleton that

can be filled with more body mass. Primitive insects such as dragonflies, stoneflies, mayflies, grasshoppers, crickets, and true bugs such as water striders and stink bugs grow into adults by simply increasing body size at each moult. At the last moult of the exoskeleton, both internal and external reproductive organs mature. This process is usually referred to as *incomplete metamorphosis*. A minority of insects undergo incomplete metamorphosis.

Most insects moult and mature into adults by complete metamorphosis. The egg hatches into a larva, and the larva undergoes a series of moults before going into a dormant pupal stage (Fig. 13). During the pupal stage, the internal and external body is completely transformed into the adult insect, still with the three basic body divisions of head, thorax, and abdomen.

The morphology of butterflies reflects their place in the evolutionary hierarchy. The larvae are usually external plant feeders, a primitive way of obtaining food. They feed on leaves, flowers, or, rarely, internally in legume pods or roots. None of the BC species feeds on roots. A few are predators. In Canada, only one species of butterfly has a larva that is predatory. It is the Harvester, *Feniseca traquinius* (Fabricius, 1793), found in southern Canada from Manitoba to Nova Scotia.

The eggs of butterflies are usually less than a millimetre in size. The outer surface is covered by a highly sculptured chorion (Fig. 14). Eggs are usually laid singly, but sometimes in clusters. In most species they are attached to the larval

**13** Transformation of a Sara's Orangetip (*Anthocharis sara flora*) caterpillar (left to right) into a pupa

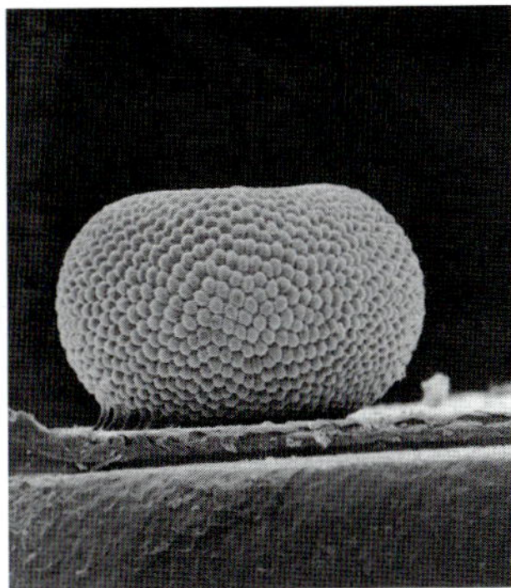

**14** Lateral view of the egg of a Rocky Mountain Apollo (*Parnassius smintheus*) viewed through a scanning electron microscope

**15** Dorsal view of the egg of a Rocky Mountain Apollo viewed through a scanning electron microscope, showing the micropyle

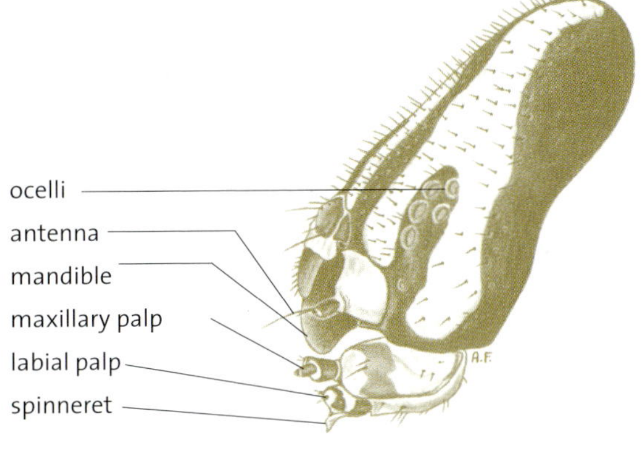


**17** Lateral view of an Anise Swallowtail larval head showing mandible, ocelli, and antenna

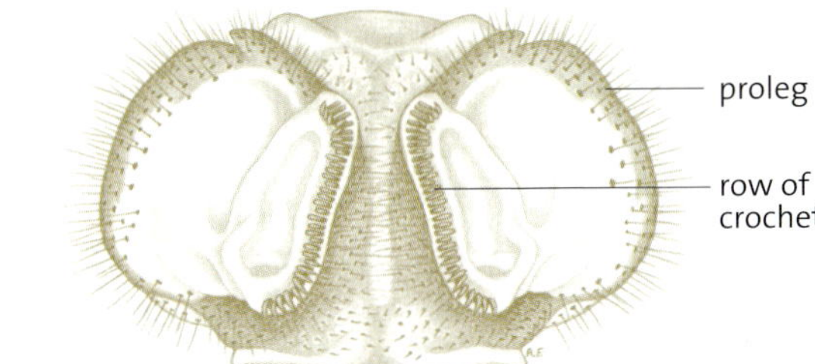


**18** Ventral view of one abdominal segment of an Anise Swallowtail larva showing prolegs and crochets

foodplants. Bog-inhabiting species often lay eggs randomly, whereas arctic/alpine species will even attach single eggs to rocks next to the larval foodplant. The developing eggs breathe through a dorsal opening called a *micropyle* (Fig. 15). The eggs do not feed, and development takes place by conversion of the stored energy into organized body structures.

When the larva first hatches from the egg, it is referred to as a *first instar larva*. The larva normally undergoes four moults; the last stage is usually the fifth instar. Butterfly larvae have a very primitive overall structure (Fig. 16). The head has chewing mouthparts with highly developed *mandibles* (Fig. 17). Viewed with the naked eye, the mandibles are dark because of black pigment in the chitin. There are light-sensitive structures called *ocelli* (Fig. 17), but no functional eyes. The ocelli occur in groups of six, one set on each side of the head above the mandibles. The *antennae* (Fig. 17) are very short. Each of the three *thoracic* segments (Fig. 16) has a pair of segmented legs that end with *tarsi* (claws). The abdomen (Fig. 16) has 10 segments, the first eight with *spiracle* openings for breathing. Abdominal segments 3 to 6 and 10 usually have a pair of prolegs (Fig. 16) each. The prolegs end with a group of small hooks called *crochets* (Fig. 18) arranged in various parallel rows. The arrangement is characteristic for each butterfly family. The larva in Fig. 16 has an additional structure found only on swallowtail and apollo larvae (family Papilionidae): a pair of hornlike structures called *osmeteria* that are everted from the upper

surface of the prothorax when the larva is disturbed. The only other insect larvae that can be confused with lepidopteran larvae are those of sawflies. There are usually more pairs of prolegs on sawfly larvae, and the prolegs lack crochets at their ends.

The main functions of butterfly larvae are to feed, moult, and accumulate body mass for conversion to adult structures. Thus their major internal organs are digestive organs (Fig. 19). The digestive system consists of external chewing mouthparts that break down food, and the internal digestive system. There is also an internal pair of salivary glands that empties into the mouth to aid digestion. The internal digestive system is one long tube running the length of the larval body. It ends in an external opening, the *anus*. The first part of the internal digestive system is the *crop*, a thin-walled tube that stores food. It is about one-eighth the length of the body. Most of the internal digestive system consists of the *midgut*, where food is digested and absorbed into the body. Much of the absorbed food is stored as fat. The *hindgut* and *rectum* compact the indigestible portions of the food, remove most of the water, and expel the remainder out the anus in small pellets called *frass*.

The other internal organs perform various necessary functions. Attached at the junction of the hindgut and rectum are variously shaped *Malpighian tubules*, which remove nitrogenous wastes from the larval body. They extend forward and backward from the point of attachment and reach all parts of the internal body cavity. Emptying into

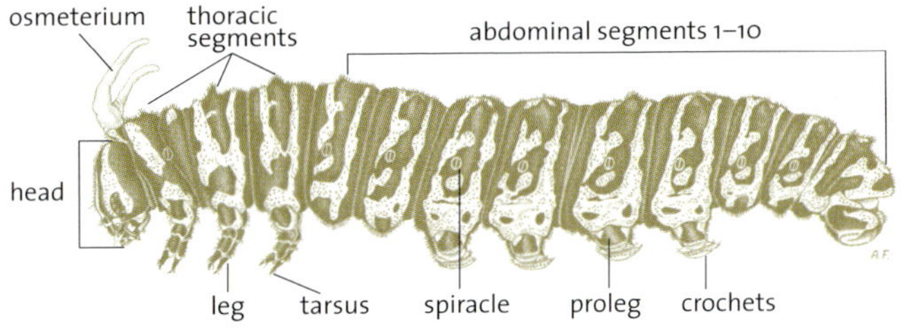


**16** Lateral view of an Anise Swallowtail (*Papilio zelicaon*) larva showing external anatomy

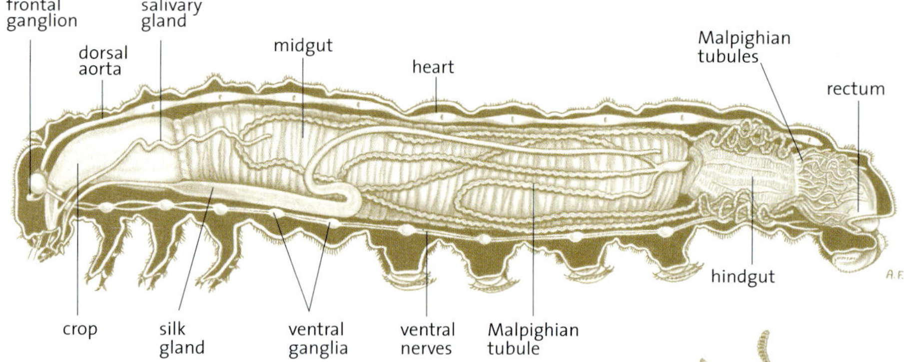


**19** Lateral view of an Anise Swallowtail larva showing internal anatomy

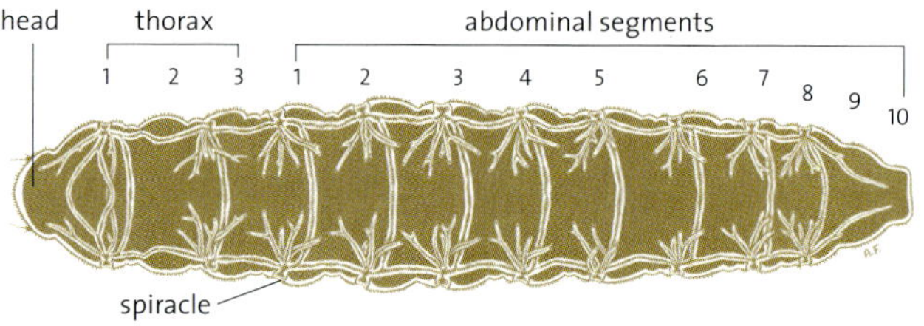


**20** Dorsal view of an Anise Swallowtail larva showing the tracheal system

the mouth are a pair of *silk glands*. The silk is used to attach the pupating larva to the substrate (such as a twig, a stem, or bark) or to form a loose wrapping of leaves and twigs around the pupating larvae of hesperiids and lycaenids. The circulatory system is a single dorsal tube, above the digestive system, called the *dorsal aorta,* with enlargements called "hearts" at most thoracic and abdominal segments. The nervous system is referred to as a *ventral nervous system* because it consists of a pair of *ventral nerves* running most of the length of the body, below the digestive system. At each thoracic and abdominal segment, there is a *ventral ganglion* joining the two nerves.

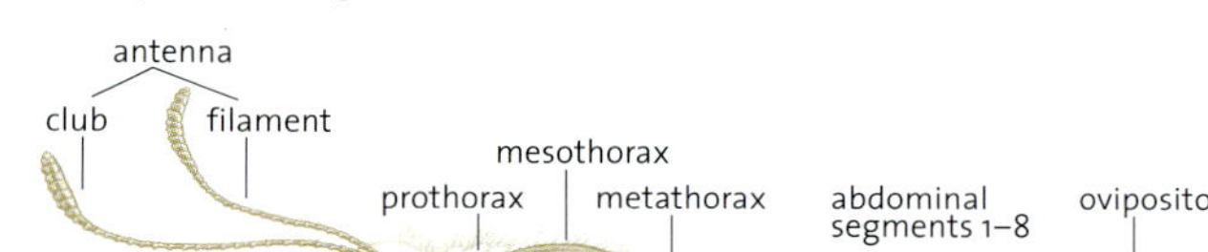

**22** Lateral view of an Anise Swallowtail adult showing external appearance and body scales. Wings are not shown.

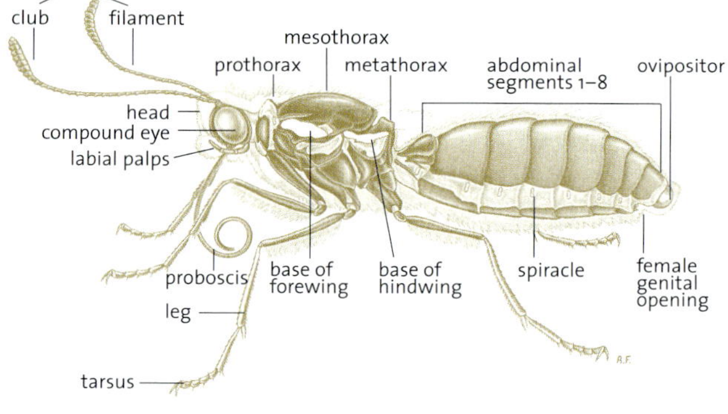


**23** Lateral view of an Anise Swallowtail adult showing details of the external anatomy

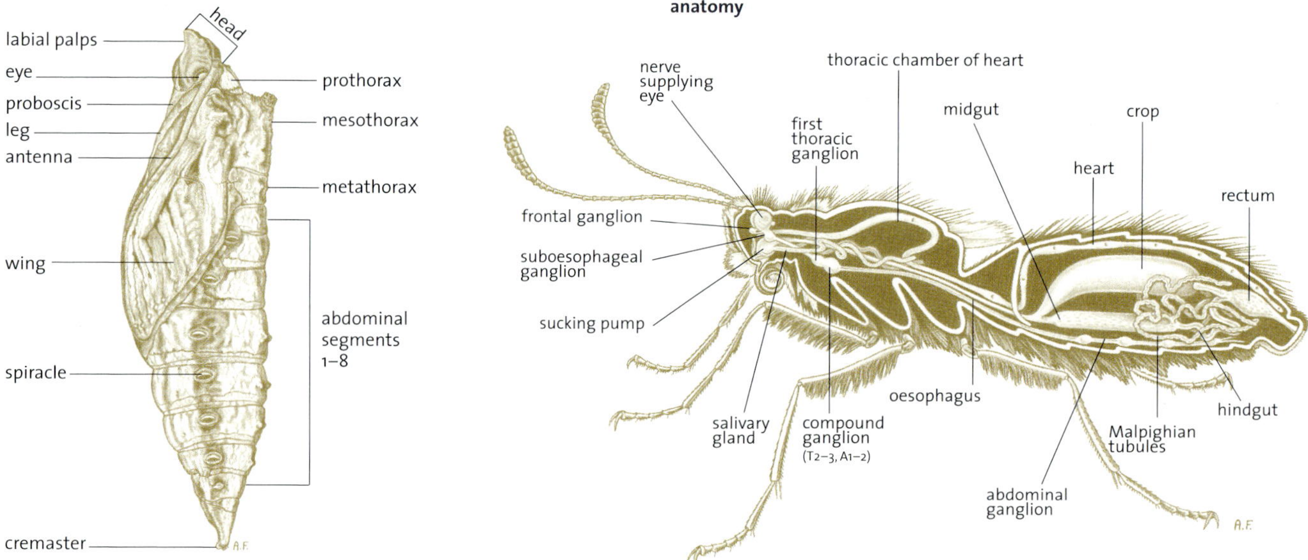


**21** Lateral view of an Anise Swallowtail pupa showing external anatomy

**24** Lateral view of an Anise Swallowtail adult showing details of the internal anatomy

The respiratory system of butterflies and other insects is very different from the lungs of vertebrates. At each side of the three thoracic and first eight abdominal segments is an external opening called a *spiracle*. Internally, each spiracle opens into an interconnecting system of tubes called *tracheae* (Fig. 20). The tracheae have fine side branches called *tracheoles*, which bring oxygen to all parts of the body and remove carbon dioxide.

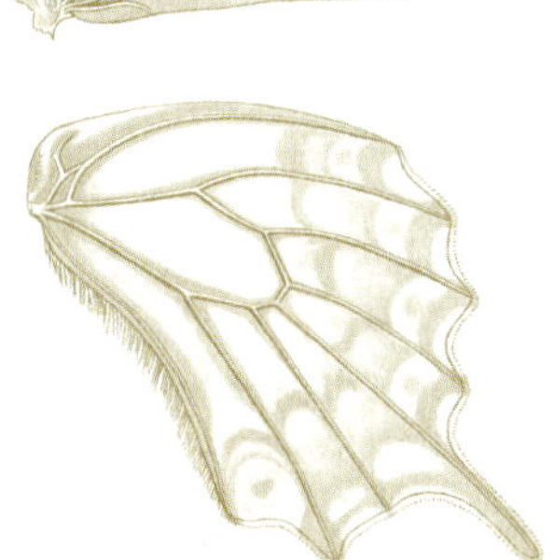

**25** View of Anise Swallowtail wing showing scale pattern

Once the larva has grown to full size, it moults and forms a dormant pupal stage (Fig. 21). The pupa is capable of only limited movement. Spiracle openings for breathing are still present. Internally there is an almost complete breakdown and rebuilding of body tissues. Externally the wings and antennae are visible on most pupae, but not on lycaenid pupae.

When the adult butterfly emerges from the pupa, its external and internal anatomy differ radically from those of the mature larva, but it still shows the basic division of the body into head, thorax, and abdomen (Fig. 22). Except for the tips of the antennae and the compound eyes, the wings and body are covered with scales, which are modified hairs. The colour patterns of these scales on the wings make butterflies one of the most beautiful groups of insects.

The morphology of the adult is so different from that of the larva because the adult has a very different primary function, namely, to produce eggs (females), fertilize eggs (males and females), and lay the eggs on or near the larval foodplants (females). This requires that males at least have an efficient means of moving around to locate receptive females. In butterflies, females are equally mobile and move around to find males and lay eggs. In some moth families, however, the females are wingless, requiring males to locate them for mating and to transport them to new oviposition sites.

Butterflies use wings to move around and to exploit the environment fully. Other changes in adult morphology (Fig. 23) compared with that of the larva also aid movement. The compound eyes provide visual acuity and colour perception, including the high end of ultraviolet light. The antennae and sometimes the legs are for chemoreception, and are used to recognize mating pheromones, larval foodplants, and adult nectar sources. The proboscis is used to suck up nectar from flowers, providing sugars for energy, and is used to obtain mineral salts from moist soil or dung. The thorax is larger, thicker, and harder (more sclerotized) to provide attachment and leverage for the massive flight muscles and stronger legs. The larger legs are important for gripping the substrate

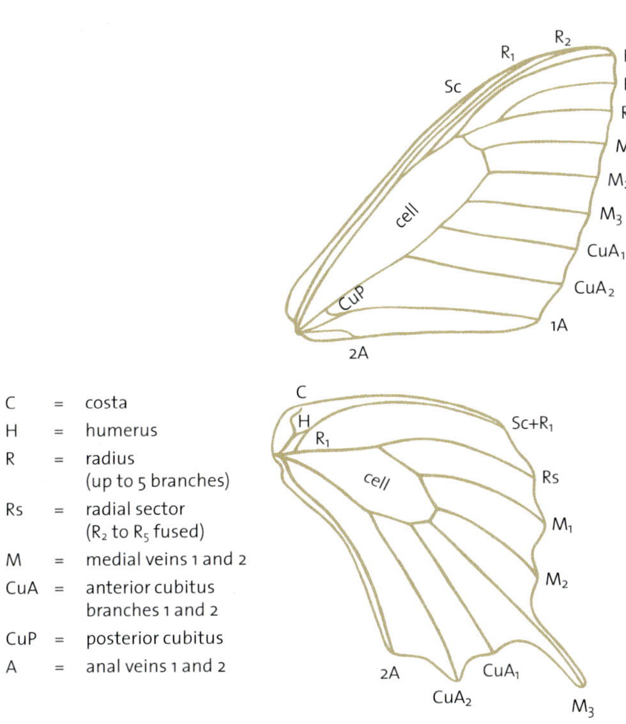

| | | |
|---|---|---|
| C | = | costa |
| H | = | humerus |
| R | = | radius (up to 5 branches) |
| Rs | = | radial sector (R₂ to R₅ fused) |
| M | = | medial veins 1 and 2 |
| CuA | = | anterior cubitus branches 1 and 2 |
| CuP | = | posterior cubitus |
| A | = | anal veins 1 and 2 |

**26** View of Anise Swallowtail wing showing venation

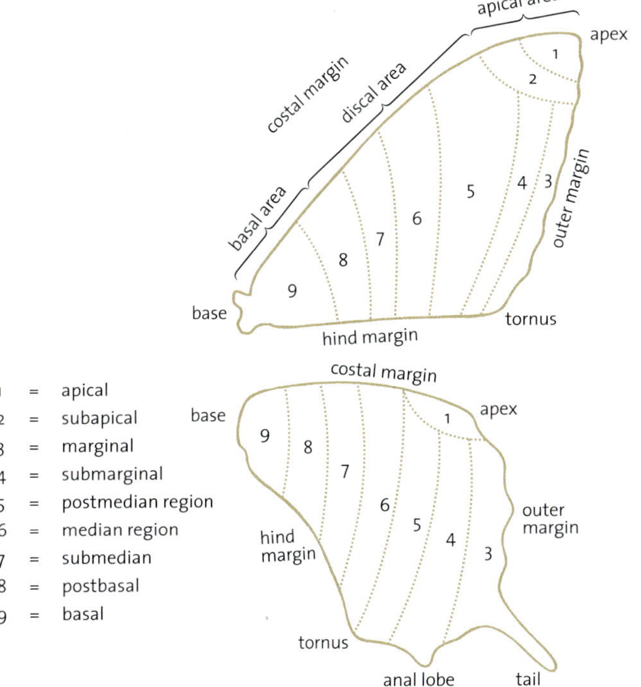

| | | |
|---|---|---|
| 1 | = | apical |
| 2 | = | subapical |
| 3 | = | marginal |
| 4 | = | submarginal |
| 5 | = | postmedian region |
| 6 | = | median region |
| 7 | = | submedian |
| 8 | = | postbasal |
| 9 | = | basal |

**27** View of Anise Swallowtail wing showing positional terminology and regions of the wing

and supporting the body without the prolegs. In males, the exoskeleton of abdominal segments 9 and 10 are highly modified into the uncus and a set of claspers called *valves* (Fig. 28), which hold the male and female together during mating.

Changes in the internal anatomy also reflect the different function of the adult (Fig. 24). The relatively large crop can store nectar gathered in short time periods in the morning and afternoon so that it can be digested and sugars absorbed as the adults fly. Adult males will also suck up mineral salts from moist, sandy soil and amino acids from decaying plant or animal tissues and dung. The second and third thoracic ganglia and first two abdominal ganglia have fused into a compound ganglion located in the thorax, providing the wings with a greater nerve supply. The circulatory, nervous, and respiratory systems are modified only so that they can fit around the wing muscles in the thorax and the internal reproductive organs in the abdomen (see Fig. 24). For a more detailed explanation of butterfly morphology, see Scoble (1995).

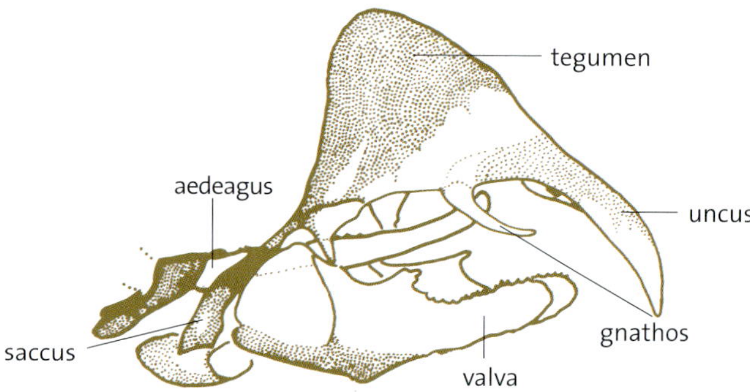


**28** Lateral view of the male genitalia of *Oeneis bore*

**29** Great Spangled Fritillary (*Speyeria cybele leto*) gynandromorph (6.8 cm). One side of the butterfly is male, the other side is female.

The preceding descriptions of larval and adult morphology are intended to give the reader a general overview. Special morphological terms are used to describe the adult butterfly in the species accounts that comprise most of the rest of this book. One must understand the position of the veins inside the wing, the terms used to describe various positions on the surface of the wings, and the morphology of external male genitalia to accurately distinguish between individual species.

The forewings and hindwings of a swallowtail butterfly with scale pattern are shown in Fig. 25. The terminology of wing venation is shown in Fig. 26. The presence and points of joining of the veins are important in defining family categories and higher classifications of Lepidoptera and other flying insects. Veins are derived from the tracheal system.

The terminology of positions and regions on the wing is shown in Fig. 27. Species descriptions in the text require an understanding of this terminology.

The terminology of the male external genitalia is shown in Fig. 28. This is a highly specialized terminology and is different in each order of insects.

# Biology of Butterflies

## Butterfly Reproduction

The reproductive process consists of mate location, courtship, and copulation. Mating is followed by oviposition (egg laying) by the female. Most female butterflies are able to mate during the first day after emergence from the cocoon, and some will mate immediately after emergence, while their wings are still expanding. A few females wait a day or two, and many mate more readily after a day or two even though they will also mate on the first day. Male butterflies usually need to mature for a few days before they mate, although some will mate on the first day after emergence (Scott 1973b).

### Mate Location

Male butterflies actively seek out and court females, while females either wait near the larval foodplant to be found by a male or seek out a general habitat where males are likely to occur. Mate location behaviour is primarily a male activity, although females sometimes actively seek out males if they have remained unmated for several days. The eyes of males are sometimes larger than those of females (e.g., *Parnassius*), possibly for better vision to locate females. Once females have mated they generally try to avoid being found by males, so as not to be harassed by them.

Males investigate flying or sitting objects of the approximate size, colour, and behaviour of females, although their visual discrimination is frequently very poor. The major patches of colour on the wings of females are often important in enabling males to recognize the right species, but small markings or complex patterns are not used in mate recognition. Rocky Mountain Apollo (*Parnassius smintheus*) males, for example, chase only flying objects that are light in colour, but they do not distinguish on the basis of size. They are equally likely to chase small blue lycaenids as they are to chase large white butterflies, but they ignore dark butterflies such as fritillaries (CSG).

Male butterflies have two primary methods of mate location: perching or patrolling. Unlike in many moths, pheromones usually come into play only after a female butterfly has been visually located by a male. Perching males investigate females by flying after them from a resting position, whereas patrolling males investigate resting or flying females after spotting them while flying. Perching and patrolling are not mutually exclusive behaviours in any species. A male of a perching species will sometimes detect a sitting or flying female as he flies between flowers to nectar, and a male of a patrolling species will sometimes detect a female in flight while he is nectaring or basking. In addition, males of some species may locate females by either patrolling or perching at different times or in different habitats (Scott 1973b, 1974, 1983). In most species, however, one behaviour or the other predominates.

Butterfly species with perching males are generally small and have thick, muscular thoraxes and short, "jet-plane"-shaped wings. This enables the males to rapidly accelerate and dart after approaching butterflies to see whether they are females willing to mate (CSG). Perch sites are characteristic of the species, and may be the ground itself, tree trunks, branches, and so on, at a characteristic height from the ground. Perches are used for only part of each day, often because the sun shines on them for a few hours; individual butterflies often move between perches. When another butterfly draws near, a perching male darts out towards it and determines its species. If it is a male of the same species, it is chased briefly, its sex is presumably determined, and usually the first male breaks off the chase and returns to his perch. This behaviour may be simple investigation or it may be aggressive defence of a territory against other males. If the approaching butterfly is a receptive female of the same species, the chase is prolonged until the female lands and courtship proceeds. It is likely that pheromones and/or flight behaviour tell the male whether the butterfly he is investigating is a receptive female.

Butterfly species with patrolling males are generally larger and have lightly built thoraxes and broad wings (CSG). The males are able to fly long distances with minimal energy expenditure, and they can therefore cover large areas in their search for females. Patrolling males sometimes locate females that are in flight, but frequently find resting females by investigating anything that is more or less the correct colour. Sometimes they locate females by their pheromones, once they are within a few metres of the females. Males fly continuously in search of females, stopping only to bask or feed. They generally stay within the area of the reproductive habitat, but in grasslands where there are no well-defined boundaries they may also patrol areas lacking females. When a male spots a butterfly or another object of approximately the correct colour, he investigates to determine whether it is a female. If it is another male or belongs to another species, there is a brief chase or circling flight, which is soon broken off. If it is a receptive female, she lands if she is not already sitting, and courtship and mating occur. A female that is highly receptive will sometimes fly towards a patrolling male and then drop to the ground to invite mating.

Patrolling males of Northern Blues recognize females resting head downward much more readily than females resting head upward. Visual recognition is based on the colour and shape of the underside of the resting female (wings closed), and resting males are not distinguished from females. The initial approach of the male is based entirely on visual recognition, with no apparent involvement of odour. Once the male has landed close to the female, however, phero-mones play a key role in sex recognition. This is demon-strated by the fact that long-dead females elicit no further courtship, with the males leaving; freshly dead females, on the other hand, continue to be courted, indicating that males recognize potential mates through pheromones when they come in close proximity (Pellmyr 1983).

Male butterflies investigate anything that vaguely resembles a female, and then use behavioural and odour cues to deter-mine whether it is in fact a female of the correct species. Wing colour is seldom important to a female's mating success, and female colours are frequently quite variable. Mimicry of distasteful species is often confined to females, with the male colour remaining constant. Females some-times use visual cues to identify male butterflies of the correct species, but behaviour and pheromones are usually the primary cues for mate acceptance.

Males are frequently brightly coloured, whereas females are more cryptic. Blues are excellent examples of this. Brightly coloured males may use their wings to "flash" signals at other males announcing their gender, thereby eliminating the need for prolonged or close encounters between males and reducing the amount of time wasted in such encounters. Males of territorial species defend their territories by flashing aggressive signals at intruding males rather than through physical contact. In species with brightly coloured males, signalling between males may be the primary source of selection for brilliant colour in males (Silberglied 1984). Apparently the depth perception of butterflies is poor, so they cannot distinguish between a small object nearby and a large object at a distance. Lorquin's Admirals (*Limenitis lorquini*) near Vancouver chase aircraft flying high overhead (W.G. Wellington, pers. comm.).

### Territoriality

Some male butterflies are territorial; they find an area with particular characteristics and occupy it for varying lengths of time. Individual butterflies remain within a territory from periods of a few hours (in forested areas where sunlit openings constantly change in location) to a few days. They may also leave a territory and return to it later the same day or another day. Defence of a perching or patrolling area also occurs, but it is generally difficult to distinguish simple investigative flights from defence of territory. Active territo-rial defence, with males grabbing each other with their

legs and battering with their wings while in flight, occurs in hilltopping Anise Swallowtails (*Papilio zelicaon*) (Shields 1968b; Pinheiro 1991) and Red Admirals (*Vanessa atalanta*) (Brown and Alcock 1991). Virgin females, or previously mated females seeking to mate again, search out habitats suitable for male territories in order to find mates, and then move away from the territory to find breeding areas. If there are few nectar sources in the territory, males may periodically abandon it to search for a nectar source. Strong winds some-times make it impossible for butterflies to remain in a hilltop or other open territory, and force them to use nearby areas of secondary preference that are sheltered from the wind.

Good territories typically have conspicuous features that make them easy for a butterfly to maintain as a territory and return to if it is blown off or if it moves off to feed or chase a potential mate. They are good sites for thermoregulation, with warm rocks, warm updrafts, and exposure to the sun early and late in the day. They are also near a larval feeding area, although "near" may be a kilometre or more. Most of all, a good territory for a male is one that has a high prob-ability of being visited by a female looking for a mate, which may be as difficult for the male butterfly to determine as it is for humans.

Lorquin's Admiral (*Limenitis lorquini*) males prefer a territory consisting of a bare or grassy patch on a south-facing slope or a opening in a riparian habitat, with dense shrubbery or trees at the upper end. The butterflies settle frequently on the shrubs or trees that are thermally optimal due to warm updrafts and/or local solar heating. The males dart out to investigate other butterflies, chase males and other species, and court females. Lorquin's Admiral males are frequently seen elsewhere, but the optimal territory types are always reoccupied rapidly after the resident is removed. Great Arctic (*Oeneis nevadensis*) males establish territories on bare patches of rock on level ground, low knolls, and open hilltops. They rest on these warm rocky or bare dirt areas, flying up to chase or court any intruding butterflies (Guppy 1970).

Male butterflies regularly aggregate on hilltops that are not the normal oviposition habitat, a behaviour called *hilltopping*. After many years of observing this phenomenon on Vancouver Island, Richard Guppy (1962) suggested that butterflies such as Anise Swallowtails (*Papilio zelicaon*) and Great Arctics (*Oeneis nevadensis*) hilltop to find mates (the "mating rendezvous hypothesis"). He suggested that males seek out hilltops and remain there waiting for females. Virgin females then seek out hilltops, encounter a waiting male, mate, and then leave the hilltop to lay eggs elsewhere. This hypothesis was supposedly confirmed for Anise Swal-lowtails by Shields (1968b), but his experimental design was badly flawed. He released male Anise Swallowtails at the base of a ridgeline, and recaptured about half of them along the top of the ridgeline. Shields interpreted this as evidence that the Anise Swallowtails had sought out the hilltop. However, this is actually the result to be expected from randomly directed flight away from the release point, rather than flight directed up the hill, because the ridgeline included about half of the possible flight directions. After further observations, Guppy pointed out weaknesses in his and Shields's assertions that hilltopping butterflies actively seek out hilltops. He suggested that males instead simply utilize hilltops as territories when hilltops encountered by chance are suitable as territories for a particular species (Guppy 1970). Under this hypothesis, therefore, hilltopping is an aspect of territorial behaviour where suitable territories are the tops of hills or ridges. Males patrol a hilltop territory looking for females that are in flight or sitting, or perch on the ground or vegetation and dart out to determine whether butterflies flying by are conspecific females.

This explanation for hilltopping may be correct for some butterflies. Shepard (1966), however, conducted a mark-release-recapture study of the most common hilltopping butterfly in BC, the Western White (*Pontia occidentalis*). His results clearly showed that males did not remain very long on a hilltop, and females never occurred on the hilltop. Thus the very obvious hilltopping behaviour of Western Whites cannot be part of their reproductive behaviour.

The more thinly dispersed a butterfly population is, the more difficult it is for males and females to locate the opposite sex in order to mate. Territorial behaviour makes it easier for thinly dispersed species to locate each other. Hilltops are especially well defined territories, and constitute only a tiny fraction of the landscape, so butterflies that use hilltops as their territories need to conduct a detailed search for mates within only a very small portion of the population's overall area. Rare and thinly dispersed butterfly species tend to use hilltops more exclusively as territories than species that are common or those that are rare but have dense populations concentrated in a small area (Shields 1968b).

### COURTSHIP

Once a male and a female have located each other visually, vision continues to play a role in courtship and recognition. However, pheromones (species-specific sex attractants) and/ or ritualized courtship behaviours are more important in helping butterflies determine whether they belong to the same species, whether they are of opposite sexes, and whether the female is receptive. The antennae are used to detect and identify pheromones.

Provided the female is receptive, courtship in most butterflies usually lasts less than a minute and is participated in by both sexes. Rocky Mountain Apollos (*Parnassius smintheus*) and Clodius Apollos (*P. clodius*) show no courtship at all: males simply dive on a female and force her to the ground (if she is not already there), and forcibly attempt to mate (Fig. 30). This is rapidly successful if she has not previously mated, but is seldom successful if she has already mated and possesses a sphragis (a structure that prevents multiple matings). Unsuccessful mating attempts may last an hour or

**30** Mating Rocky Mountain Apollos (*Parnassius smintheus magnus*)

more before the female manages to escape the grasp of the male (CSG).

The identification of correct mates is far from perfect: matings between species occur regularly in the wild, and the two species are occasionally very distantly related. Successful hybridization between closely related species is common, but distantly related species are unlikely to produce viable offspring. Canadian (*Papilio canadensis*) and Western (*P. rutulus*) Tiger Swallowtails, and White (*Limenitis arthemis*) and Lorquin's Admirals (*L. lorquini*), have bands of hybridization across southern and central BC. This regular hybridization does not fit a simplistic biological definition of species, which requires that species do not successfully interbreed. A more realistic species definition includes a statement that if two species interbreed, they still maintain their independent gene pools.

Besides courtship behaviours, females have rejection behaviours. Different butterfly species have different rejection behaviours, but typically a female in flight that is approached by a male will fly up and away to indicate that she is not receptive. A female resting on the ground or on vegetation may rapidly flap her wings (checkerspots, grass skippers), or may close her wings up tightly and then crawl or fly away. Unreceptive female Rocky Mountain Apollos and Clodius Apollos basking on the ground close their wings up tightly when a male approaches, apparently to avoid being seen (CSG). The "pierid rejection posture" is characteristic of all butterflies in the family Pieridae. A resting female approached by a courting male flattens her wings against the leaf or ground, with her forewings held back over her hindwings. She angles her abdomen upward into a nearly vertical position, at right angles to her thorax. The impression is that she is releasing "rejection pheromones," although this has not been demonstrated. The male hovers and flutters about her in courtship, but soon breaks off and flies away.

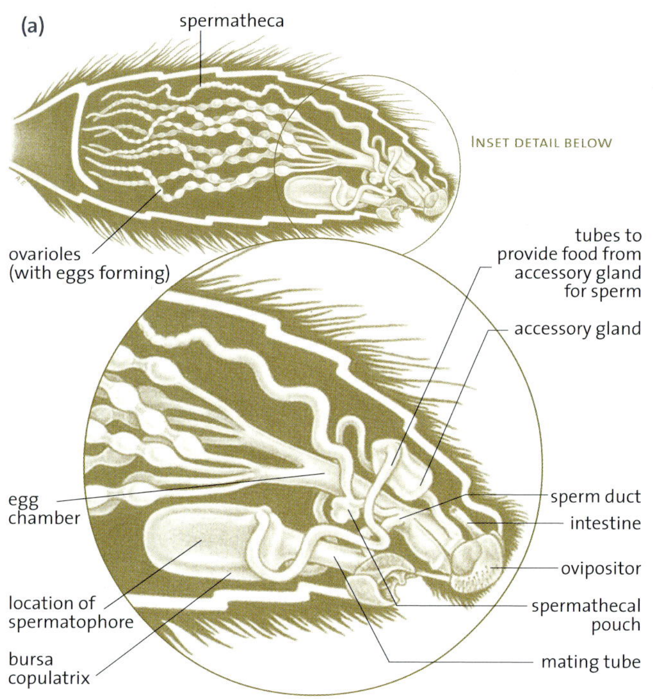



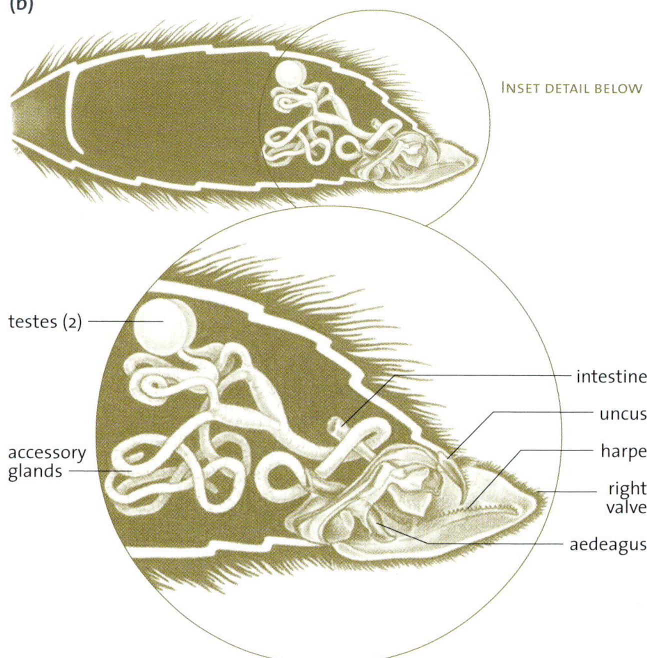



**31** Reproductive systems of (a) female and (b) male butterflies (*Papilio zelicaon*)

## COPULATION

Copulation is initiated by the male, who, while on top or beside the female, curves his abdomen around to grasp the underside of the tip of the female's abdomen with his valvae (Fig. 31). His uncus fits into a separate opening of the female reproductive anatomy under the female's ovipositor, and his tubelike aedeagus is inserted into the female's mating tube as far as the bursa copulatrix (Scott 1986b). Once copulation has been initiated, the male moves so that he is facing away from the female, with the tips of the abdomens joined. For a few minutes after they join, there is frequently a pumping movement of the male's abdomen, or his claspers may rhythmically squeeze the tip of the female's abdomen, after which the abdomens are still. Copulation generally lasts 1–3 hours but is highly variable, depending on weather conditions, time of day, and species. When copulation has been initiated late in the day, the butterflies may not separate until the next morning.

During copulation the male injects sperm and various materials produced by the male accessory glands into the female's bursa copulatrix. A granular material of unclear function is deposited in the inner end of the bursa copulatrix, and a large, tough white sac forming the spermatophore (sperm, proteins, fats, salts, and other materials) fills the rest of the bursa. The sperm mass is located in the base of the spermatophore, followed by a yellow granular secretion (Tschudi-Rein and Benz 1990). A hard, clear plug, whose function may be to keep the spermatophore in the bursa copulatrix, forms below the spermatophore. A count of the remnants of these structures indicates the number of times a female has copulated.

Males of some species deposit an internal genital plug during mating, which prevents the female from mating again and ensures that the male is the only one to produce progeny by that female. The sphragis of apollos (Fig. 32) is a large external shield that is particularly obvious on Clodius Apollo (*Parnassius clodius*) females. In Rocky Mountain Apollos (*P. smintheus*), a sphragis is occasionally not properly deposited, resulting in a female that lays fertile eggs but has no sphragis, or a male with the sphragis retained in his claspers (CSG).

About 10 hours after mating, muscular contractions of the female genital tract move the sperm from the spermatophore to the spermatheca (Tschudi-Rein and Benz 1990), where they are stored for periods ranging from a few minutes to many months. The sperm fertilizes each egg as it moves down the egg chamber just before oviposition. After

**32** A rare event: a Rocky Mountain Apollo female with two sphragis

the sperm have left the spermatophore, the female digests it and uses the materials as a source of nutrients (Boggs and Gilbert 1979). During oviposition, materials from the accessory glands are used to glue the egg to the larval foodplant or adjacent substrate. Stretch detectors in the bursa copulatrix inform the female when the spermatophore has been digested, and hence when it is time to mate again.

The spermatophore is very large, being 6–15% of the male's body weight, and consists of a cuticle-lined sac containing proteins, lipids, hydrocarbons, and water (Rutowski 1984; Boggs 1998). Included in it is a large amount of sodium salt, which males of many species replace by "mud-puddling." The production of the spermatophore may reduce the life span of the male by using up its stored nutrients (Shapiro 1982e), while at the same time extending the life span and egg production of the female.

If disturbed while mating, one of the two sexes takes flight and carries the other dangling from the end of its abdomen. The sex that carries the other is characteristic of the butterfly species. The male flies and carries the female in most species of Pieridae (whites, marbles, and sulphurs) and Danainae (the Monarch). The female flies and carries the male in most species of Papilionidae (apollos and swallowtails), Satyrinae (satyrs), Riodinidae (metalmarks), the tribe Theclini (hairstreaks) of the Lycaenidae, and Hesperiidae (skippers). Depending on the species, either sex may carry the other in the brushfoots, blues, and coppers. In some species either sex may carry the other at different times. Almost all butterflies fly while mating only if disturbed; the exceptions are Monarchs, which normally fly a short distance after starting to mate (Shields 1973).

Copulation is terminated at the initiative of the male butterfly. If the female dies during copulation, the male will eventually separate, but if the male dies the female does not separate from his body (Wickman 1986). Apparently the male clasps the female with his valvae, which do not relax even when he dies.

Males can mate more than once in their lives, and frequently can mate more than once in the same day. Checkered White (*Pontia protodice*) males can mate at least 13 times in their lives. Most females mate 1 to 3 times in their lives, but Monarch (*Danaus plexippus*) females may mate 2 to 4 times per day. Sperm from the last mating fertilizes all the eggs following that mating (Scott 1973b). Since there are usually roughly the same numbers of males and females in a population, this mating pattern presumably results in all females mating successfully, and in some males mating many times but others never being successful.

## OVIPOSITION

Eggs are usually laid (oviposited) on the leaves, flowers, or stems of the appropriate foodplants (Fig. 33). A butterfly may use only one plant species as a larval foodplant (monophagous), a few species (oligophagous), or many

**33** Mourning Cloak (left) and Rocky Mountain Apollo (right) ovipositing

species (polyphagous). When more than one plant species is used, females usually prefer one over the others, with individual females having different preferences. This variation in preference frequently results from genetic differences between females, but preferences also change as females encounter different plants (Thompson and Pellmyr 1991) and as an individual female ages and more "urgently" needs to oviposit (Chew and Robbins 1984). Caterpillars hatching from eggs accidentally laid on the wrong plants usually die. Oviposition alone is therefore not proof that a particular plant is a larval foodplant; the larvae must also be able to mature on the plant. Caterpillars from eggs laid on the "wrong" plant sometimes do mature successfully, and over time this may result in the evolution of the use of a new foodplant by a population (Chew and Robbins 1984).

Some butterflies prefer to lay their eggs some distance from the foodplant, under stones or twigs, and the newly hatched larvae have to find a suitable plant before they starve. The female may (as in *Speyeria* fritillaries) or may not (as in the Rocky Mountain Apollo, *Parnassius smintheus*) lay the egg adjacent to the larval foodplant. The reasons for laying eggs away from the larval foodplant are poorly understood, but in such species, either the eggs or unfed newly hatched larvae hibernate. The herbaceous foliage may die off in the fall (violets used by *Speyeria*, but not the stonecrop used by Rocky Mountain Apollos), and then start to re-grow in the spring before the larvae begin feeding. As a result, eggs placed on the foodplant would tend to be buried in decaying foliage, whereas a sheltered spot under a nearby rock or stick would enhance survival. The laying of eggs away from the larval foodplant could also be a strategy to reduce the chance of parasitoids or predators finding the eggs while the immature butterfly is hibernating (Wiklund 1984).

Female butterflies first locate a general habitat within which to search for larval foodplants, then locate potential foodplants from a distance by their general visual appearance. The butterfly's ability to recognize the correct plant may improve with practice (Rausher 1978), but not all butterflies have this learning ability (Renwick and Chew

1994). Overall plant shape, leaf shape, shade of green of a leaf, and water content of a leaf are all used by females to select the correct species of plant and the correct part of a plant (Chew and Robbins 1984). Preferences for plants with particular leaf shapes may change within and between seasons, depending on changes in foodplant quality (Chew and Robbins 1984). This is the case for Anise Swallowtails (*Papilio zelicaon*) on the Boundary Bay dike in Delta, BC (Roger Ashton, pers. comm.). Oviposition occurs more frequently on isolated plants or on plants at the edge of a patch than on plants in the middle of a patch, because a female flying from area to area encounters them first. When searching for eggs or larvae, it is best to search isolated plants, especially those that stand out from surrounding vegetation (CSG).

Once visual cues suggest that a plant is of the appropriate species, female butterflies approach and/or land on the plant. They confirm it as the larval foodplant through the particular combination of chemicals in the leaves or flowers (the "taste" and "smell" of the plant), and frequently through the texture and details of colour and leaf shape as well. Some plant chemicals stimulate oviposition, others inhibit oviposition, and the majority have no effect at all (Chew and Robbins 1984). The proportion of oviposition stimulants relative to deterrents is as important as the concentration of stimulants in determining whether oviposition will occur. The butterfly uses its antennae, legs (especially forelegs), ovipositor, and proboscis for "smelling" and "tasting" to determine the identity of the plant (Renwick and Chew 1994). Once the plant is recognized as the correct species, eggs are laid in specific locations on or near the plant, depending on the butterfly species: on or under leaves, in or on flowers, or adjacent to the plant. Different individual Edith's Checkerspot (*Euphydryas editha*) females from the same population prefer to oviposit on different plants, with the larvae from each female growing most rapidly on her preferred plant (Singer et al. 1988).

Some species of butterflies, notably some whites, marbles, and orangetips, have evolved the ability to recognize the presence of eggs on a foodplant. Females lay blue-green eggs on their larval foodplant, and the eggs turn bright orange within a day. Females lay fewer eggs on plants that already have orange eggs on them, presumably to avoid larval competition for food on small plants as well as cannibalism by older larvae (Shapiro 1981a). Reductions in seed production of larval foodplants resulting from pierid herbivory can be very high, with small plants being entirely consumed and plant populations sometimes suffering reductions in seed production of over 40% (Shapiro 1985c).

Sara's Orangetip (*Anthocharis sara*) females in California avoid ovipositing on small to medium-sized larval foodplants (rock cress, *Arabis*) that have orange eggs on them, but do not avoid ovipositing on large foodplants that already have orange eggs (Fig. 34). Females oviposit on foodplants of any size with eggs that have not yet turned orange; thus a female may return repeatedly to an isolated plant on the same day to lay additional eggs. Females fly between plants after laying each egg, ensuring that their eggs are dispersed throughout the available population of foodplant. Medium-sized plants are preferred for oviposition over large plants, despite the greater amount of food available from the latter (Shapiro 1981b).

**34** Sara's Orangetip egg (two days old)

Spring Whites (*Pontia sisymbrii*) in California commonly use *Arabis pulchra* as the larval foodplant. Females avoid laying eggs on plants with orange eggs when egg-free foodplants are easy to find, but not when most foodplants already have eggs on them. Spring Whites therefore prefer to lay eggs on plants that do not already have eggs, but accept plants with eggs when there is no choice (Kellogg 1986). The same is true for Large Marbles (*Euchloe ausonides*) in California (Shapiro 1985a). In BC, Large Marbles and Stella's Orangetips (*Anthocharis stella*) oviposit on flowerheads of *Arabis* early in the flight season, and on the flowers, leaves, and stems from mid-flight season onward (CSG); Sara's Orangetips usually oviposit on the leaves and stems during the entire flight season.

There is an ongoing evolutionary battle between butterflies and plants, with plants evolving their own defences. Spring Whites in serpentine soil areas of California commonly use *Streptanthus* species as larval foodplants. The plants are quite small, so one larva consumes one medium-sized or several small plants completely. *Streptanthus tortuosus* plants produce orange callosities, resembling pierid eggs, on their leaves. Plants with the callosities have a lower probability of having eggs laid on them than plants without the callosities. The plants have apparently evolved to produce the callosities as "egg mimics" to deter oviposition by pierids, thereby increasing their chances of surviving to successfully produce seeds (Shapiro 1981c). It must be noted, however, that heavy herbivory by pierids and beetles may not adversely affect seed production by *S. tortuosus* because "surplus" flowers are produced that would not produce mature seeds anyway (Karban and Courtney 1987). Tropical vines in the genus *Passiflora* have also developed egg mimics that reduce oviposition by *Heliconius* butterflies (Williams and Gilbert 1981).

Plant defences against oviposition by pierids go further. Some individual plants, but not all plants, of black mustard (*Brassica nigra*) kill Cabbage White eggs laid on them. A small circle of leaf underneath an egg dies and dries out, invariably killing the egg within 3–4 days. Apparently desiccation of the leaf results in desiccation of the egg (Shapiro and DeVay 1987). By killing a small piece of its own leaf, the plant saves the rest of the leaf from being eaten by the larva that would hatch from the egg.

Egg laying follows several patterns, including: a single egg laid on a leaf or plant before the female butterfly flies to the next plant (Pieridae); a small number of randomly arranged ovipositions on a leaf (Satyr Anglewing, *Polygonia satyrus*); a small number of eggs laid in a short row (Compton Tortoiseshell, *Roddia l-album*) or tight cluster; eggs laid in a large flat cluster (checkerspots in the genus *Euphydryas*); eggs laid in a cylinder around a twig (Mourning Cloak, *Nymphalis antiopa*); or eggs laid in large multilayered clusters (Milbert's Tortoiseshell, *Aglais milberti;* checkerspots in the genus *Charidryas*). The eggs in the bottom of the middle of a multilayered cluster may have a lower risk of parasitism because of the physical difficulty a parasitoid would have in reaching these eggs with its ovipositor.

When eggs are laid in clusters, the larvae are gregarious, at least in the early instars. The larvae are generally distasteful to many predators, so a predator may kill one larva in "testing" for palatability but leave its siblings alone ("kin selection"). The larvae tend to be brightly coloured to warn predators of their distastefulness, and are also frequently covered with spines or hairs. They generally have stereotypical behaviours when attacked; for example, all larvae may simultaneously rear the front part of their bodies and start twitching and swaying. The larvae of Mourning Cloak (*Nymphalis antiopa*) butterflies and the three BC species of tent caterpillar moths exhibit this behaviour. Gregarious larvae may also be more resistant to parasitism, and may be better able to deal with plant physical defences such as tough leaves and spines or hairs. They may require the presence of the other larvae to stimulate them to feed; for example, solitary larvae of *Charidryas* checkerspots normally die without feeding (Chew and Robbins 1984). There may, however, be a selective advantage for single-egg oviposition in some species, because hatching larvae frequently eat not only their own empty chorions but also adjacent unhatched eggs (Douglas 1986).

## Population Biology

Butterflies have complex behavioural and ecological interactions with their physical and biological environment. Each species has its unique patterns of behaviour and different environments in which it has evolved specialized interactions. Only a few aspects of this complexity are addressed in this section. Additional information is provided in the species descriptions, and much else can be learned through careful observation of butterflies.

### Population Structure
Butterfly species have different population structures and different minimum areas of suitable habitat required to maintain a population. These two factors are major determinants of both the conservation status of butterflies and the appropriate strategies for maintaining populations of each species. Population structure can be loosely divided into "open" and "closed" populations (Thomas 1984).

### Open Populations
Butterfly species with open populations consist of individuals that fly widely over large areas of suitable and unsuitable reproductive habitat, reproducing in both permanent and ephemeral patches of larval foodplants. Reproductive habitat and other key habitat features such as nectar sources may be widely separated. The reproductive habitat may consist of single, isolated larval foodplants or small, scattered patches of foodplants, or it may comprise large areas with many foodplants. Butterfly species with open populations are genetically identical over large geographic areas and have few, if any, subspecies because of constant genetic mixing. They are only rarely of conservation concern because they rapidly colonize areas of new habitat, rapidly recolonize areas from which they have been extirpated, and have large populations; their population sizes can recover rapidly if reduced by natural or human-caused events. Butterfly species with open populations are frequently medium-sized to large, and are strong fliers.

Butterfly species with open populations may be either permanently resident in BC or migrants that establish seasonal populations in BC. The resident species establish permanent populations in the larger patches of suitable habitat, and each year those populations provide the adults that roam over much larger areas. Migratory species all have open population structures, typically with permanent resident populations in a restricted portion of their habitat. In the United Kingdom about 15% of butterfly species form open populations (Thomas 1984). Western Whites (*Pontia occidentalis*) and Monarchs (*Danaus plexippus*) are typical species with open population structures in BC.

### Closed Populations
Butterfly species with closed populations tightly restrict their flight activity to areas of suitable habitat, and seldom cross areas of unsuitable habitat. Larval foodplants and adult nectar sources must be within the same area or very close by. Larval foodplants may be limited to a very small number of plants isolated within a large area otherwise lacking foodplants. Species with closed populations are genetically identical only within relatively small geographic areas and have many subspecies because of low rates of genetic mixing. Such species include most of those of conservation concern, because they are slow to colonize areas of new habitat, are slow to recolonize areas from which they have been extirpated, and frequently have small populations; their population sizes frequently cannot recover rapidly if reduced by natural or human-caused events. Butterfly

species with closed populations frequently have small to medium-sized adults, and are weak fliers. They are all permanent residents of BC and never migrate. In the United Kingdom about 77% of butterfly species form closed populations (Thomas 1984). Most lycaenids are species with closed population structures.

### Semi-closed Populations

Little in nature fits consistently into well-defined categories, and the structure of butterfly populations is no exception. A significant number of species have populations that are predominantly closed but have fairly high rates of emigration where adults cross expanses of unsuitable habitat. These populations can be called "semi-closed." Most butterflies in semi-closed populations tightly restrict their flight activity to areas of suitable habitat and do not cross areas of unsuitable habitat. Larval foodplants and adult nectar sources must be within the same area or close by. A significant number of adults, however, leave the area of suitable habitat and fly over unsuitable areas. Larval foodplants are generally scattered in frequent patches over a large area, so that dispersing adults have a significant probability of encountering them. Species with semi-closed populations are genetically similar over moderately large geographic areas and have a moderate number of subspecies because genetic mixing occurs within moderately large areas bounded by major geographic barriers. In the United Kingdom about 8% of butterfly species form semi-closed populations; Thomas (1984) treated them as closed populations. The Rocky Mountain Apollo (*Parnassius smintheus*) is a good example of a BC butterfly with semi-closed populations.

Even within a semi-closed or closed population there is tremendous variation. For example, many Rocky Mountain Apollos will spend their entire adult life (2–6 weeks) in a single meadow, whereas others fly many kilometres. The furthest documented distance by mark-release-recapture

methods (Guppy 1986b) is about 1.5 km across a forest barrier (Fig. 35). It is unclear how far adult Rocky Mountain Apollos will disperse. This is currently being studied by Jens Roland using radio-collars in the Rocky Mountains of Alberta (Struzik 1998).

### Metapopulations

Many butterfly species live as metapopulations, or groups of related closed populations spread over large areas containing patches of suitable habitat. All the closed populations of a butterfly species within a geographic area tend to increase and decrease in abundance at the same time, indicating that yearly fluctuations in weather probably cause the changes in abundance (Pollard 1984). When a species decreases sharply in abundance over much of its range, populations occupying less than perfect habitats may be extirpated. This will leave many patches of suitable habitat vacant, a condition that may last for many years. Eventually weather and other conditions will cause a population boom in one or more of the surviving populations, and extensive and unusual emigration will occur. Many of the vacant patches of suitable habitat will be recolonized, and separate closed populations will become established until the arrival of another especially adverse year.

### Migration

The most impressive migration of butterflies in North America is that of the Monarch. This phenomenon is now threatened as a result of clearing of forests in Mexico (eastern North American migrants) and the cutting of small stands and isolated trees in California (western North American migrants) that are used for hibernation roosts. In addition, milkweed, the larval foodplant of Monarchs, is considered a noxious weed in much of Canada and the US, and is subject to eradication. Nearly as impressive as the Monarch's migration is the mass northward migration of Painted Ladies (*Vanessa cardui*) through California in March, which creates a highway hazard in some years. The migration thins out as it moves north, and is barely noticeable when it arrives in southern BC in mid-April, with isolated individuals reaching the Yukon.

Other butterfly migrations in BC are generally less impressive or unnoticed. The other common migratory butterflies in the province are the Clouded Sulphur (*Colias philodice*), the Orange Sulphur (*C. eurytheme*), the West Coast Lady (*Vanessa annabella*), the Red Admiral (*V. atalanta*), the California Tortoiseshell (*Nymphalis californica*), and Milbert's Tortoiseshell (*Aglais milberti*). As noted in the species descriptions, some of these species also have resident populations. These and other species may also move vertically, from low elevations early in the summer to high elevations late in the summer, to follow nectar sources and cooler temperatures (Shapiro 1974a, 1975a). In the late summer of some years, huge numbers of Painted Ladies and West Coast Ladies gather to nectar in the alpine areas of the southern interior.

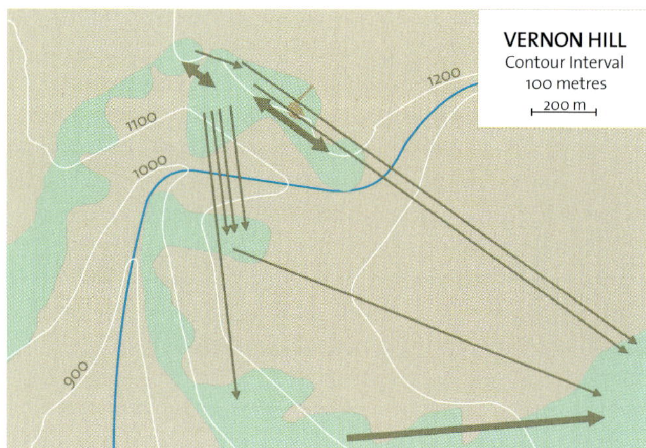



**35  Dispersal of male Rocky Mountain Apollos (*Parnassius smintheus magnus*) on part of Vernon Hill, near Vernon, BC (from Guppy 1986b). Thin lines indicate movements of a single male; thick lines indicate movements of more than four males. Brown areas are Douglas-fir forest; green areas are grassland.**

The term *migration* is used to refer to movement of butterflies in a specific direction away from their origin; the term *dispersal* is used for movement that is randomly directed away from their origin. Dispersal may be deliberate, in that the butterfly has "decided" to leave its place of origin, or it might be accidental, and result from a butterfly's leaving the population during escape flights from predators or during courtship flights, from wind, or from a butterfly simply getting lost. All butterfly populations have some individuals that leave through dispersal; only a few species are migratory.

Monarch butterflies west of the Rocky Mountains hibernate in California and then migrate northward in spring. A few, as well as more from later generations, migrate as far as southern BC, and breed in the southern interior as far north as Kamloops. It is not known whether the Monarchs that mature in BC migrate south to California. It is likely, but no Monarchs tagged in BC have been recovered in California. There have been no studies on the other migratory species in BC.

Northward migration ensures that the eggs for the next generation are laid in an area where foodplants will not dry out in the summer heat until the larvae have pupated. Butterflies migrate southward to escape the effects of winter and to be in a position to use the early spring plants in the south for the next generation. Dispersal and migration function to move butterflies between areas of suitable habitat, resulting in the genetic mixing of populations and enabling the colonization of newly created habitat. Migratory butterflies usually have few or no subspecies because constant genetic mixing prevents populations from diverging genetically.

## SOURCES OF BUTTERFLY MORTALITY

### WEATHER

Weather can cause high mortality of any stage of the life cycle. Heavy rain, wind, and hail can knock larvae off their foodplants, killing them, especially first instar larvae (Caldas 1995). Anise Swallowtail (*Papilio zelicaon*) larvae may be especially vulnerable to being knocked off the top of the broad leaves of cow parsnip by hail, which also shreds the leaves and probably reduces their nutritional value. Small larvae on the tops of leaves, such as those of swallowtails and Cabbage Whites, can drown in rainwater that pools on the leaves, or are knocked off the leaves and then die on the ground (Harcourt 1966; Blau 1981).

Unseasonable weather can extirpate populations by killing adults, immatures, and/or larval foodplants. Buckeyes (*Junonia coenia*) around San Francisco were completely eliminated by an unusually severe frost in December 1990. Recolonization occurred rapidly in 1991, because the adults have very high dispersal and colonization ability (Shapiro 1993). In contrast, a subalpine population of Silvery Blues (*Glaucopsyche lygdamus*) in Colorado was completely extirpated by a rare June snowstorm and freeze, and

recolonization had still not taken place several years later. Populations of several other butterflies in the same area were severely reduced and also had not recovered after several years (Ehrlich et al. 1972), apparently because all the species have poor recolonization abilities. Populations of butterflies whose larvae feed on the inflorescence or leaves of herbaceous plants, such as Silvery Blues (Breedlove and Ehrlich 1972) and checkerspots, are also adversely affected by droughts that cause the premature drying out of these above-ground plant parts, causing the larvae to die of starvation.

### PREDATORS

Predators of butterflies at all stages of their life cycle include both invertebrates and vertebrates. The intensity of predation on butterfly populations is poorly documented because it is very difficult to measure.

### PREDATORS OF EGGS, LARVAE, AND PUPAE

Predators of butterfly eggs are poorly known. Pentatomid bugs (Hemiptera) and snakeflies prey on Pine White (*Neophasia menapia*) eggs (Jennings and Toliver 1976). Predaceous mites destroyed 65% of all Alexandra's Sulphur (*Colias alexandra*) eggs in a Colorado population in 1981 (Hayes 1984). Birds also eat butterfly eggs when they find them.

Cabbage White (*Pieris rapae*) larvae are heavily preyed on by at least a dozen different invertebrates. About 45% of first instar larvae were preyed on by invertebrates in one study (Dempster 1967), with the predation level dropping steadily as the larvae matured, until few fifth instar larvae are killed by invertebrate predators. In contrast, birds kill few first and second instar Cabbage White larvae but kill a high proportion of the larger larvae. Large invertebrate predators such as yellow jacket wasps will prey on large Cabbage White larvae, and when a nest is nearby will kill a large proportion of the larvae. Gardeners should therefore retain wasp nests near their gardens where it is safe to do so. A similar pattern occurs for the swallowtail *Papilio xuthus,* whose young larvae suffer high mortality from predatory spiders, ants, and bugs in the first two instars, and from wasps and birds in the later instars (Watanabe 1976, 1981).

The mortality of larvae from invertebrate and vertebrate predation depends on the local populations of predators. Spraying with broad-spectrum chemical pesticides kills off the invertebrate predators, resulting in much higher survival rates of Cabbage White larvae in succeeding generations because the predator populations are slow to recover (Dempster 1968). Dense vegetation provides good habitat for a wide variety of invertebrates, both predators and prey. Since most invertebrate predators of butterfly larvae feed on a wide variety of prey, weedy habitats with a wide range of prey allow the predators to build up larger populations than they would if butterfly larvae were the only prey. The denser, taller, and more varied the vegetation of the larval habitat, the higher the rate of invertebrate predation. As a result, a weedy cabbage field has more invertebrate predators of Cabbage White larvae than a weed-free field (Dempster 1984). Butterfly larvae are also attacked by pentatomid bugs, reduviid bugs, *Polistes* wasps (Blau 1981), and ants. All of these can be major sources of mortality for larvae (Damman 1993).

Birds are apparently the most important vertebrate predators of larvae (Dempster 1984), although small mammals such as mice, chipmunks, and squirrels can be quite significant. Small birds feed on eggs and young larvae, and larger birds feed on larger larvae and pupae. Ground-feeding birds feed on larvae that have left the foodplant and are wandering to find a pupation site. Birds may ignore larvae if they are rare, but eat the same larvae when they are more common. Apparently birds have to learn that larvae are good to eat, but have no opportunity to do so when larvae are rare (Dempster 1984). Small mammals may be more important predators of pupae than birds, because pupae are generally relatively well hidden from sight but can be found by smell and touch. Pupae are also attacked by pentatomid bugs and ants (Blau 1981).

Mammalian herbivores can have large impacts on butterfly populations through the incidental predation of eggs laid on plants that they later eat. The subalpine population of Rocky Mountain Apollos (*Parnassius smintheus*) on Dividend Mountain near Penticton, BC, appears to be about one-half the size that the available habitat and foodplant are capable of supporting, because cattle graze the area after oviposition has occurred (CSG). Butterfly species whose larvae feed on plants unpalatable or toxic to herbivores, such as thistles and milkweed, are protected from large-mammal herbivory. Unfortunately these plants are frequently considered range pests and are killed with herbicides or by other means.



### Invertebrate predators of adult butterflies

There is a wide variety of invertebrate predators of adult butterflies, including dragonflies and damselflies (Order Odonata), robber flies (Order Diptera); ambush bugs (Order Hemiptera); wasps (Scudder 1889b; Price 1962; Fales 1976; Jennings and Toliver 1976; Neck 1977; Fales and Jennings 1977); crab spiders (Family Thomisidae); and orb-weaving spiders (Family Epeiridae). Ambush bugs (genus *Phymata*), which capture butterflies by their proboscis, and crab spiders (Fig. 36) lie in ambush on flowers, where they capture large numbers of insects, including butterflies and moths. Orb-weaving spiders (Fig. 37) are also significant predators. These three predators inject fast-acting venoms that subdue their prey within a minute or two, and successfully prey on the largest butterflies and moths. Dragonflies and robber flies are major predators of butterflies in flight, primarily medium-sized and small ones, although both will also chase swallowtails (CSG). Yellow jacket wasps are major predators of moths resting near lights; they are only minor predators of adult butterflies although they occasionally prey on hibernating Monarchs (*Danaus plexippus*) (Leong et al. 1990).

**36** Predation of a Large Marble (*Euchloe ausonides mayi*) by a crab spider

**37** Predation of a Rocky Mountain Apollo by an orb-weaving spider

Biting midges (female flies of the family Ceratopogonidae, known as punkies or no-see-ums) are well known to feed on the blood of birds and mammals, including humans. The subfamily Forcipomyiinae feed on the hemolymph ("blood") of invertebrates such as harvestmen, dragonflies, damselflies, stick-insects, grasshoppers, alderflies, lacewings, beetles, butterflies, moths, crane flies, mosquitoes, sawflies, and true bugs. Both adult and larval insects are attacked (Lane 1984). All the midges attacking adult butterflies and moths are in the subgenus *Trichohelea* of the genus *Forcipomyia*. The midges pierce the wing veins of moths and butterflies, and suck the hemolymph out of the body. Most midges attack the veins on the upper surface of the hindwings. *Trichohelea* midges have broad, round mandibles with blunt teeth capable of biting into tough wing veins, rather than the typical slender mandibles of most midges. The tarsal claws are also modified for gripping scaly wings (Lane 1984). Attack by midges usually results in the loss of scales in the small area of the wing being attacked. Ceratopogonid flies (*Forcipomyia baueri* Wirth) have been recorded attacking Spring Azures (*Celastrina echo*) in Arizona (Ehrlich 1962). Most other records are for moths or for butterflies not found in BC.

Ants are sometimes "partial predators" of adult butterflies such as Rocky Mountain Apollos (*Parnassius smintheus*) (CSG). On warm nights or warm cloudy days, when butterflies are resting, ants will cut away the margins of the wings and carry the pieces back to their nest as food. This "trimming" of the wings produces a characteristic scalloped pattern along the edge of the wings. The wings on both sides of the butterfly are trimmed equally by the ants when the butterfly has been resting with its wings closed over its back.



### VERTEBRATE PREDATORS OF ADULT BUTTERFLIES

Butterflies and moths have a spectrum of palatability to predators, rather than being either "palatable" or "not palatable." Butterfly palatability varies between species, between populations of a species, and between individuals within a population. Palatability is determined by factors such as butterfly age, sex, species and population of larval foodplant (which affects the secondary compounds present in individual larval foodplants), wing coloration, wing size to body size ratio, body size, and behaviour (which varies depending on environmental factors such as time of day/ night, temperature, and sun intensity). In turn, predator assessment of butterfly palatability depends on the predator's species, level of hunger, previous experience, number of partially palatable butterflies already eaten, and individual preferences. On average, butterflies are less palatable to birds than moths, and at best are only moderately palatable to birds. Larger butterflies tend to be preferred over smaller ones when there is a choice, butterflies with small wings and large bodies are preferred over those with large wings relative to body size, and brightly coloured butterflies appear to be preferred as much as cryptically coloured butterflies. Both the Viceroy (*Limenitis archippus*) and Monarch (*Danaus plexippus*) are moderately unpalatable to birds but are sometimes eaten (Sargent 1995).

Birds apparently determine whether a butterfly is palatable first from its visual appearance and then by its flavour, based on previous experience with eating butterflies (Brower 1969). Small mammals apparently determine palatability of butterflies first through odour and then possibly through taste. Batesian mimicry of distasteful butterflies, through the evolution of similar wing patterns by palatable butterflies, is therefore likely to result from selection pressure exerted by visually oriented predators such as birds and lizards. Chemical (taste and odour) mimicry (Brower 1969), which has been little studied, is likely to result from selection pressure exerted by odour-oriented predators such as rodents.

Birds produce distinctive patterns of damage on the wings of butterflies and moths that are attacked resting or in flight (Figs. 38 and 39). Beak marks form a triangular damage pattern on the wing, with a triangular chunk being torn out of the wing when the butterfly or moth tears itself out of the grasp of the bird. Sometimes the bird opens its beak and releases the butterfly or moth, leaving a triangular beak imprint and no other damage. The pattern of the wing damage depends on the manner in which the bird attacked the butterfly. Some moths rest with their wings held in positions that are never adopted by butterflies, hence not all possible damage patterns are seen in butterflies (Sargent 1973; Beck and Garnett 1984).

Butterflies attacked in flight, or at rest with the wings held flat and the forewings abnormally far forward, typically have damage restricted to one wing (type I damage). These are subdivided into type Ia (hindwing tear) and type Ib (forewing tear). When the butterfly is resting with its wings open, an attack from behind will result in damage to the margins of both the forewing and hindwing on the same side (type II damage, with only IIb damage normally occuring in butterflies). If a bird attacks a butterfly resting with its wings held over its back, either two or all four wings show matching damage patterns. Type IV damage results from attack from behind, and type V damage results from attack from in front. Type IVa damage results when the two forewings are grasped from behind, type IVb damage occurs when the two hindwings are grasped from behind, and type IVc damage occurs when all four wings are grasped from behind. Type V damage is usually only seen as beak imprints, because tearing of the forewing margin results in the butterfly no longer being able to fly. Type Va damage results from only the forewings being grasped by the bird beak, and type Vb damage results from all four wings being grasped.

Butterflies frequently escape bird attacks, and can be found flying around with beak-damaged wings. When birds successfully kill and eat butterflies, they usually clip off the wings and drop them to the ground, where they can be found.

Beak marks on the wings of butterflies confirm the occurrence of bird attacks on butterflies but do not reflect the actual rate of attack. Cabbage Whites (*Pieris rapae*) are attacked by captive blue jays five times more successfully when at rest (65% of attacked butterflies eaten) than when in flight (12% of attacked butterflies eaten). Beak marks are much more likely to be found on Cabbage Whites that survive attacks in flight than those that survive attacks when resting, but in general beak marks are very seldom (4% probability) left on Cabbage Whites that survive attacks. About 7% of a wild population of Cabbage Whites had beak marks from bird attacks, a typical proportion for various butterfly species. This suggests that, on average, surviving Cabbage Whites in the population had each been attacked 1.75 times by birds, excluding the attacks that resulted in the butterfly being eaten. The proportion of butterflies eaten by birds in the wild would depend on how many attacks were made on resting butterflies relative to attacks on flying butterflies. It is clear, however, that both attacks by birds and mortality due to bird predation are very high in butterfly populations (Wourms and Wasserman 1986).

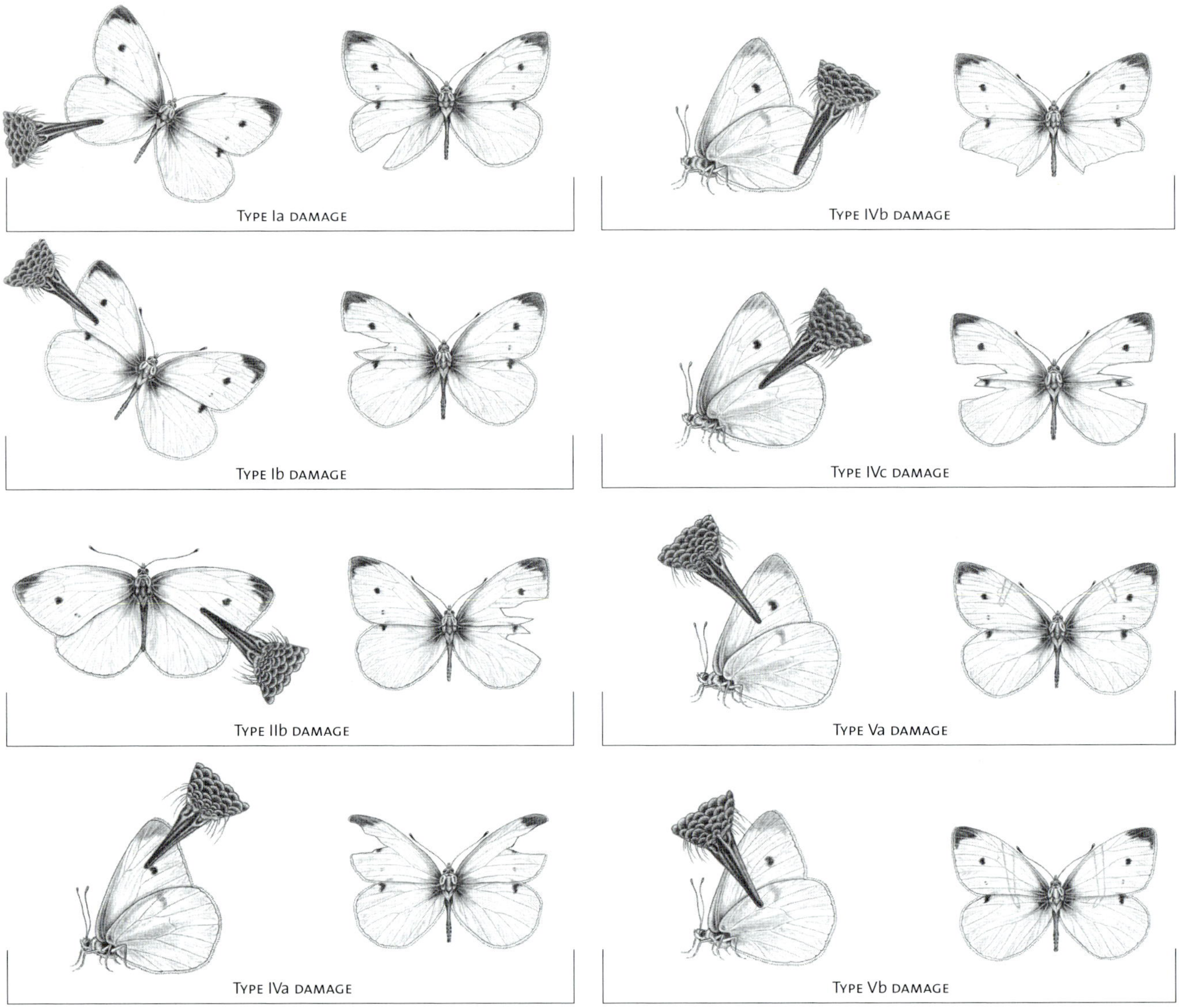

TYPE Ia DAMAGE

TYPE Ib DAMAGE

TYPE IIb DAMAGE

TYPE IVa DAMAGE

TYPE IVb DAMAGE

TYPE IVc DAMAGE

TYPE Va DAMAGE

TYPE Vb DAMAGE

**38** Common patterns of butterfly wing damage caused by bird attacks (adapted from Beck and Garnett 1984; Sargent 1973). The bird beak in the illustrations is that of an insectivorous bird commonly found in BC. See text for explanation of types of wing damage. Moths show damage pattern types not illustrated.

Small mammals are major predators of adult butterflies, and produce damage patterns on wings that are distinctly different from those of birds. The wings may show parallel scratches or tears from teeth and smudges from lips (Fig. 40).

Chipmunks eat more female than male Rocky Mountain Apollos (*Parnassius smintheus*) (Fig. 41). The chipmunk predator carried each butterfly back to a feeding station, on which it perched while clipping the wings off and eating the body (CSG). Rocky Mountain Apollos have a strong odour and their body fluid is a nasal irritant to humans. When attacked, they excrete from their anus drops of brown fluid that smells like their body. Their white wings with black and red markings are probably an initial warning of unpalatability to birds, their odour is an initial warning to rodents, and their taste is probably further warning should they be bitten by either

type of predator. Males presumably have more distasteful compounds in them than do females. Chalcedon Checkerspots (*Euphydryas chalcedona*) show the same predation pattern for birds, with females being attacked more frequently than males (Bowers et al. 1985).

**39** Rocky Mountain Apollo with minor type IVb and Ia bird damage

**40** Rocky Mountain Apollo with small-mammal damage



BIOLOGY OF BUTTERFLIES

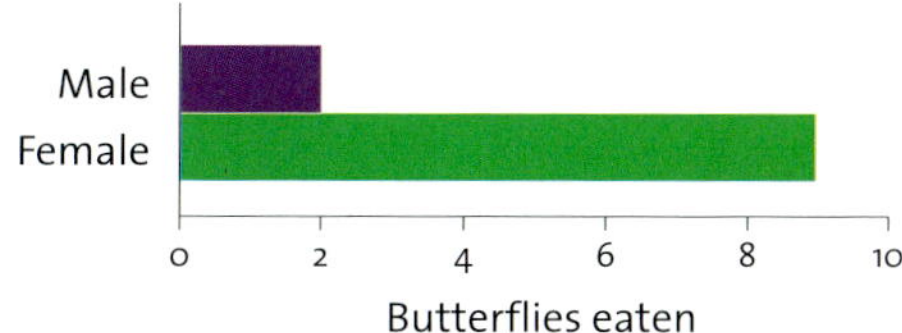



**41** Chipmunk predation on live male and female Rocky Mountain Apollos (*Parnassius smintheus magnus*). Shown are the number of males and females eaten by a chipmunk in one night after 20 males and 20 females were placed in tall grass in the evening, in an alternating grid pattern with 1 m spacing (CSG).

Monarchs (*Danaus plexippus*) hibernating in Mexico experience a 9% annual predation rate by small mammals, including three species of mice, and birds, including black-headed grosbeaks, Scott's oriole, and black-backed oriole (Calvert et al. 1979; Brower and Calvert 1985). Hibernating Monarchs at one Californian site suffered a 2% annual predation rate by chestnut-backed chickadees (Bell and Dayton 1986), and a single pair of rufous-sided towhees caused about 7% annual mortality at another hibernation site (Sakai 1994). Starlings and scrub jays are also known Monarch predators (Sakai 1994). In Mexico, male Monarchs are eaten more frequently than females because males are more palatable (Brower and Calvert 1985), but no difference in predation on male or female Monarchs occurs in California (Sakai 1994). Roosting aggregations of Monarchs in Mexico fall from their perches en masse in response to disturbance, with breath being the most effective stimulus causing the butterflies to drop. This mass behaviour may startle bird predators and make them less successful at catching the butterflies (Calvert 1994).

"FALSE HEADS" OF LYCAENIDAE AND PAPILIONIDAE: Many blues, coppers, and hairstreaks (Lycaenidae), as well as swallowtails (Papilionidae), have coloured spots on the rear edge of the hindwings, some of which strongly resemble eyes. Many also have long slender tails associated with the eyespots, and many of the tails of lycaenid species have a striking resemblance to antennae. Many lycaenids, whether or not they have tails, rub their hindwings together when resting on a leaf, flower, or stem or on the ground. Most butterflies, including the lycaenids and papilionids, rest facing head down after landing on a vertical or distinctly sloped surface.

These observations have resulted in the "false head hypothesis," which suggests that tails on the hindwings combine with spots at the bases of the tails to resemble a butterfly head (eyes and antennae) to a predator. The lycaenid subfamilies Theclinae, Polyommatinae, and Lycaeninae, and some species of metalmarks (Riodinidae), include butterflies with "false heads" (Robbins 1986). Many swallowtails also have false heads consisting of hindwing tails and eyespots, but they are relatively crude compared with those of the Lycaenidae.

The resemblance of the tails and eyespots to a head is enhanced by tails that are very long and slender and with the same colour pattern as antennae, by the lycaenid habit

of rubbing the hindwings together so as to move the tails like moving antennae, and by the butterfly's habit of resting head down on a vertical surface. It is suggested that a predator, such as a bird or lizard, will attack the false head rather than the real head. The butterfly will then tear itself away, losing only a bit of hindwing (Robbins 1980). Butterflies can fly well after having lost most or all of their hindwings. For example, an Eastern Tiger Swallowtail captured in Maine was flying well but had only its forewings developed; the hindwings had failed to develop (Fernald 1886).

The above is a very attractive hypothesis, and there is no doubt that many butterflies are found with pieces of their hindwings torn out near the tails by birds. What remains to be demonstrated is whether the survival of a butterfly species with tails and eyespots is greater than it would be if the same species lacked tails and eyespots.

### PARASITOIDS

"Parasites" of butterfly larvae are better called *parasitoids*. Parasitoids are insects whose larvae devour the host; they avoid eating the vital organs until they are ready to pupate. Parasitoids are essentially a special group of predators that eat their prey while keeping it alive as long as possible. In contrast, parasites, such as tapeworms in vertebrates, live within their host but generally do not kill it.

The parasitoids of most butterfly larvae are from four groups of insects: Tachinidae, Ichneumonidae, Braconidae, and Chalcidoidea. Tachinid flies (Tachinidae) lay eggs, either on a leaf of a larval hostplant or directly on the bodies of butterfly larvae. When a tachinid egg on a leaf is accidentally eaten by a caterpillar, the egg hatches within the caterpillar and burrows into the body. When the egg has been laid directly on the caterpillar, the hatching fly larva (maggot) penetrates the body of the butterfly larva through the base of the egg where it is attached to the host larva. Newly hatched fly larvae may also wait in ambush for a butterfly larva, and then crawl onto the surface of the caterpillar's body and penetrate to the inside. Regardless of how a tachinid fly larva gets inside the host butterfly caterpillar, the maggot consumes most of the inside of the caterpillar. The white maggot then emerges from the mature caterpillar and drops to the ground, where it develops into a smooth brown pupa. One or more tachinid larvae may parasitize a single caterpillar, and on rare occasions the host caterpillar will survive, pupate, and produce an adult (*Vanessa cardui*, CSG).

Ichneumonid wasps (Ichneumonidae) range in size from very small to very large. They lay eggs directly into eggs, caterpillars, or pupae; the eggs hatch and the parasitoid larvae develop, pupate, and emerge from the host as another adult parasitoid. Braconid wasps (Braconidae) are usually very small, but a few species are large. They lay many eggs on the back of caterpillars, and their larvae burrow inside the caterpillar to feed. When mature, the larvae sometimes pupate with the caterpillar, but usually emerge from the caterpillar and then pupate in white cocoons attached to the skin of the host caterpillar. Chalcid wasps (Chalcidoidea) range from very tiny, barely visible egg parasitoids to quite large species that emerge from pupae. The adults of many chalcids are a brilliant iridescent green or blue. Other egg parasitoids occur in other wasp families, including Trichogrammatidae, Encyrtidae, Eulophidae, and Scelionidae (Shapiro 1982a; Friedlander 1984; Garraway and Bailey 1992; Caldas 1995).

Butterfly larvae defend themselves against predation and parasitism in a variety of ways. When attacked they may drop off their foodplant to escape, but in doing so may risk starvation, desiccation, or ground predation. Dropping down a silk thread that they can climb back up reduces their risk of being separated from the larval foodplant, but ants can continue to attack by climbing down the thread. Rolling leaves or forming webs to hide in, and in some cases to feed from, helps protect the larvae from some predators and parasitoids, but other invertebrate predators and parasitoids will attack them within rolled leaves or webs and birds may use the rolled leaves as an indicator that a caterpillar is present within. Many larvae also vigorously defend themselves by thrashing their front end about to knock a predator or parasitoid away, regurgitating presumably distasteful material and wiping it on the predator or parasitoid, or biting the antennae and legs of the attacker. Swallowtails (*Papilio*) have osmeteria, and fritillaries (*Speyeria*) have glands on the ventral surface of the posterior of the head (Hammond and McCorkle 1988), that emit distasteful fluid that is wiped on a predator or parasitoid. The larger a caterpillar is compared with an insect predator, the more likely it is to succeed in repelling the attack (Stamp 1987). Swallowtail larvae will bite human fingers picking them up, although the skin on fingers is generally too tough for the bite to have any effect.

Caterpillars of checkerspot butterflies in the genus *Euphydryas* are usually attacked by one to three parasitoid species, often by a species of *Apanteles* wasp (Braconidae), a species of *Benjaminia* wasp (Ichneumonidae), and a tachinid fly (Stamp 1984). Parasitism rates vary from 0% to 67% of Edith's Checkerspot (*Euphydryas editha*) and Chalcedon Checkerspot (*Euphydryas chalcedona*) larvae, with the later instars having much higher rates of parasitism than early instars. The parasitoids of checkerspot larvae are in turn often parasitized by hyperparasitoids, which may kill half of the checkerspot parasitoids. Baltimore Checkerspot (*Euphydryas phaeton*) larvae in Virginia are parasitized by two specialist wasps, *Apanteles euphydryadis* (Braconidae) and *Benjaminia euphydryadis* (Ichneumonidae), both of which parasitize only larvae of the Baltimore Checkerspot (*Euphydryas phaeton*) and the closely related Harris' Checkerspot (*Charidryas harrissii*). First instar larvae are partly protected from the parasitoid by the silk tent they live in and by thrashing their heads. When young larvae are moulting, they are buried deep within the web and are out of reach of the parasitoids. The older, gregarious larvae

defend themselves by simultaneously thrashing the front half of their bodies to knock away the parasitoid, regurgitating on the parasitoid, and moving away from the parasitoid. These older larvae may also be protected by spines that are as long as the ovipositor of the parasitoid, making oviposition into the larvae difficult (Stamp 1984).

Butterfly pupae have a variety of defences against attack by parasitoids. Pupae are very vulnerable to parasitoids in the first day or two after pupation, when they are still soft and easily penetrated with an ovipositor. Ichneumonid parasitoids that attack pupae attack only pupae and not larvae. Once hardened, the pupae of *Aglais urticae* Linnaeus (the European relative of BC's *Aglais milberti*) and of Painted Ladies (*Vanessa cardui*) thrash violently to dislodge ichneumonid parasitoids, and then rapidly vibrate their abdomen. The thrashing makes it difficult for the ichneumonid to land on the pupa; if the wasp does manage to land, the vibrations and the smoothness of the pupa make it impossible for the ovipositor to be pressed against one spot long enough to penetrate. In general this behaviour occurs only in response to the touch of a parasitoid; artificial stimulation with antenna-like hairs does not stimulate the pupae. It is difficult for a parasitoid to grip satyrid pupae and then keep its ovipositor pressed strongly against one spot for penetration, because the pupae are smooth and rounded and have a thick cuticle. Ichneumonid parasitoids, however, have little difficulty ovipositing into Cabbage White pupae, which are rough and easily gripped and which do not move enough to inconvenience the parasitoid (Cole 1959).

### LIFE SPAN

Butterfly species vary greatly in their average life spans, both as immatures and as adults. If diapause (winter hibernation or summer aestivation) is ignored, the egg stage typically lasts from a few days to a week or two, the larval stage 4–6 weeks, the pupal stage 1–3 weeks, and the adult 1–6 weeks. The stage that hibernates has its time span increased by at least the winter months, and frequently the late summer and fall as well. In August one can sometimes see freshly emerged Mourning Cloaks and anglewings (*Polygonia*) flying with a few tattered oldsters from the previous winter. Non-hibernating adults typically live 1 or 2 weeks on the average, with a few tenacious individuals living up to 6 weeks. Some species take two years to develop from egg to adult, with young larvae hibernating the first year and nearly mature larvae or pupae hibernating the second year.

Species that hibernate as adults have long life spans, in addition to the hibernation period. A typical hibernator, the Mourning Cloak (*Nymphalis antiopa*), emerges as an adult in August and is active into October at low elevations on the south coast. The adults then hibernate, except for brief flights on abnormally warm winter days, and then are in flight from mid-March until they die sometime in June. Their "active" adult life span, excluding hibernation, is therefore 4–6 months.

## TEMPERATURE REGULATION

Butterflies are "cold-blooded" animals, unable to maintain a constant body temperature that is independent of their environment. They typically require a body temperature of 18–35°C before they are capable of flight. Preferred body temperatures range from 30° to 39°C (Kingsolver 1985), which are usually above the ambient air temperature. At the same time, their body temperature must not rise above about 42°C or they will die. Thus most butterflies need a body temperature within a few degrees of that of humans (37°C) in order to be fully active, and they cannot tolerate becoming very hot. At low temperatures wing muscles contract slowly, and at higher temperatures they can contract rapidly. Hence butterflies with slow wing beats can fly with lower body temperatures than those with fast wing beats, but even slow-flying butterflies appear to prefer a body temperature in the low to mid-30s.

All butterflies bask in the sun to absorb radiant heat when their body temperature is below the optimum, but the basking posture varies (Clench 1966) (Fig. 42). Many "dorsal bask" with their wings spread out flat so that the sun can warm their body and wings. Because they are very obvious to predators in this posture, they tend to be brightly coloured species that are unpalatable to most predators. Some species "acute bask," with their wings held at an angle above their body so that sunlight is reflected off the wings onto the body. These species also tend to be unpalatable to predators. Some skippers "double-acute bask," with the forewings and hindwings held at different angles over the body. A rarely observed basking behaviour is "body basking," where the wings are angled towards the sun and opened just enough (the width of the body) to allow the top of the body to be illuminated. Body basking is used by individuals of dorsal basking species such as Rocky Mountain Apollos (*Parnassius smintheus*) that are trying to bask and yet remain unseen at

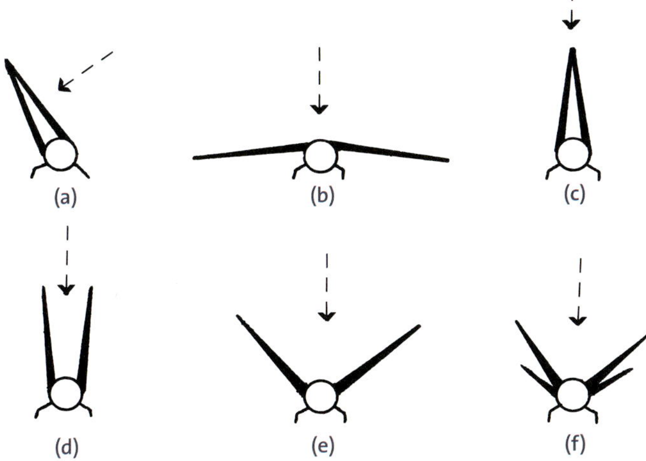

**42** Basking postures of butterflies: (a) lateral basking, (b) dorsal basking, (c) vertical basking, (d) body basking, (e) acute basking, (f) double-acute basking (CSG)

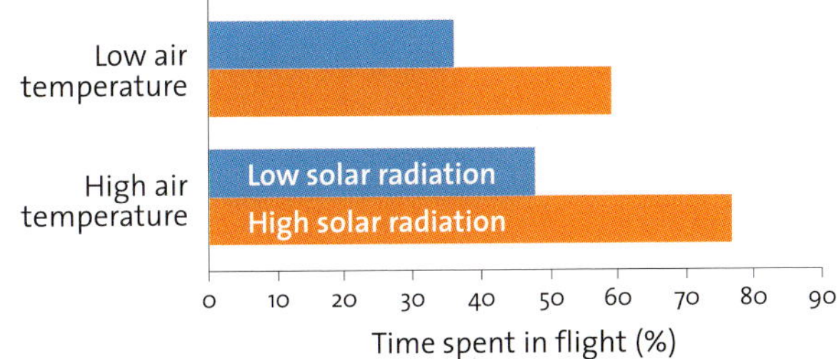


**43** Effect of solar radiation and air temperature on proportion of time spent in flight by male Rocky Mountain Apollos (redrawn from Guppy 1986b)

the same time, usually females that do not want to be seen by males. Many species "lateral bask" by closing their wings over their back and then angling them so that the sun illuminates the underside of one pair of wings and the side of the body. This is a comparatively cryptic posture, and is mostly used by relatively palatable butterflies. The last basking posture is "vertical basking," a heat avoidance posture in which the wings are closed over the back and pointed directly at the sun so as to shade the body.

The body temperature of a butterfly results from the combined effects of the air temperature and the heat available from the sun. The amount of time spent in flight, relative to the time spent basking to warm up, results from a combination of the ambient air temperature and the intensity of solar radiation (Fig. 43). At low air temperatures, heat absorption from the sun is critical to warming the butterfly enough so that it can fly. At high air temperatures, the intensity of solar radiation matters less, because the air alone warms the butterfly enough to fly (Guppy 1986b).

Butterflies with dark wings can absorb more heat from the sun than can light-coloured butterflies. Populations in cool environments (coastal, northern, or alpine) tend to have darker wings than populations of the same species in warm environments (interior, southern, or lowland). Apollos and sulphurs are good examples of this (Roland 1982; Guppy 1986a); the variation in wing melanism between populations results from both genetic differences and differences in the environment where the immature stages developed (Guppy 1989). Similarly butterfly species that have several generations each year generally produce light-coloured forms in summer and darker-coloured forms in spring and fall (Hoffman 1978).

When the air temperature is low, Rocky Mountain Apollos that have dark dorsal hindwings can remain in flight longer than those with light hindwings. When the air temperature is high, the colour of the dorsal hindwings has comparatively little effect on time spent in flight (Fig. 44).

Other adaptations enable butterflies to regulate their body temperature independently of their environment. Some butterflies, such as Monarchs (*Danaus plexippus*) and some swallowtails, generate metabolic heat by "shivering," contracting the thoracic muscles without moving their legs

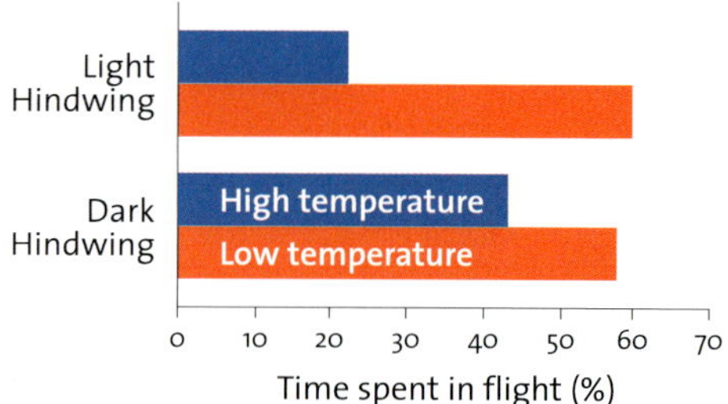


**44** Effect of hindwing colour on proportion of time spent in flight by male Rocky Mountain Apollos (redrawn from Guppy 1986b)

or wings. Many moths that have thick, muscular bodies (such as Sphingidae and many Noctuidae) also warm themselves by shivering. This is enhanced by dense modified scales that function like hair to insulate the body. Black Swallowtails (*Papilio polyxenes*) of eastern North America are dorsal basking butterflies. They raise their thoracic temperature by putting their abdomen above their wings to absorb solar radiation, and then pumping the warm abdominal hemolymph forward into their thorax. To cool their thorax, they drop their abdomen down into the shade of their hindwings, and then reverse the direction of their heart to pump warm hemolymph from thorax to abdomen; this acts as a radiator to cool the body (Rawlins 1980).

The coastal subspecies of the Zerene Fritillary (*Speyeria zerene hippolyta*) in Oregon has adapted to unusually cool, cloudy conditions. Butterflies are in full flight at 16°C in cloudy and foggy weather that provides no opportunity for solar warming. The base of the dorsal hindwings is very melanic compared with that of the subspecies inhabiting warmer environments (McCorkle and Hammond 1988). McCorkle and Hammond (1988) suggest that the subspecies *S. z. hippolyta* has physiologically adapted to flight at a lower body temperature than the warmer-environment subspecies. Flight in cloudy weather at an air temperature of 16°C suggests a body temperature of 16–18°C, which is comparable with the minimum body temperature of 17–18°C required for flight by the "sunny-environment" Rocky Mountain Apollo (Guppy 1986b). Rocky Mountain Apollos have a minimum solar intensity below which they will not fly, regardless of air temperature (Guppy 1986b); hence the *hippolyta* subspecies of the Zerene Fritillary may have adapted to flight in the absence of sunshine rather than at a lower body temperature.

A BC butterfly inhabiting a similar cool, cloudy, and foggy maritime environment is the coastal subspecies of the Mariposa Copper, *Lycaena mariposa charlottensis*. This subspecies has adapted to flight under cloudier and foggier conditions than interior populations, similar to the *hippolyta* subspecies of the Zerene Fritillary, and stays on the nectar source even when it rains (JHS). The Margined White (*Pieris marginalis*) also nectars in the rain on the coast of Alaska (Wright 1892b).

In the arctic many butterflies show great variation in wing melanism within populations. This may be due to the

frequent major changes in air temperature (20°C or more) over very short time periods (8 hours or less), resulting in "thermal shocks" to the pupa and hence the physiology of the developing butterfly (Ferris 1973b). In addition, the air temperature and intensity of solar radiation is more highly variable within each day, between days, and between years in the arctic than in more temperate areas further south. As a result, a wide range of variation in wing melanism in the arctic would help ensure that conditions at any time will be near the optimal for activity of a portion of any population.

### Hibernation

Butterfly species may hibernate at any stage in the life cycle. The stage at which hibernation occurs is specific to each species, with related species tending to hibernate at the same stage. The larval instar that hibernates is usually somewhat flexible, so that if most hibernate as third instars a few will be fourth instars and a few second instars. Butterfly species that have more than one brood during the summer and then hibernate as partly grown larvae frequently have one more larval instar in the hibernating generation than in the non-hibernating generation. Hibernation of immatures usually occurs in a sheltered, dry site such as under or on the side of rocks, logs, or debris. Adults may hibernate in hollow logs, stumps, loose bark, lumber piles, woodpiles, carports, or house crawl spaces.

Some species have an obligate diapause: they must pass through a hibernation stage before the life cycle can proceed. As a result they produce only one generation each year, even at low elevations in the south. Other species have a facultative diapause: their life cycle proceeds through two or more generations until the shortening of day length and cooling temperatures in the late summer triggers hibernation in the last generation of larvae, pupae, or adults.

Hibernation sites may not always be appropriate. Mourning Cloaks (*Nymphalis antiopa*) and California Tortoiseshells (*N. californica*) formerly hibernated in grain shocks (when they existed), with disastrous consequences at threshing time, especially on cold mornings when they were too torpid to fly (Bruggemann 1948). They, and other hibernating butterflies, also hibernate in debris piles left for burning, and are too cold to save themselves when the piles are burned in the late fall. This may be a major source of local mortality in logging areas in the BC interior, where logging debris is piled during the summer or previous winter and then burned in the late fall, when the risk of the fire escaping is low.

### Caterpillars and Ants

The larvae of many butterflies in the families Lycaenidae and Riodinidae live in association with ants. The associations range from simple coexistence through mutualism to parasitism. In many cases the association is obligatory; caterpillars have specialized behaviours and physiological adaptations and cannot survive without the ants. In other cases the association with ants increases the survival rate of

**45** Boisduval's Blue (*Icaricia icarioides*) larva tended by an ant

the caterpillars but is not obligatory. Caterpillars of the latter butterfly species are tended by ants when ants are present, but the presence of ants is not required to maintain a healthy butterfly population (Fig. 45).

Mutualism is the most common relationship between lycaenid larvae and ants. The butterfly larvae and pupae may have one or more organs that secrete chemicals for communication with ants or for feeding ants with a sugary solution (honeydew) containing amino acids and carbohydrates. They may also have organs that produce sound (Ballmer and Pratt 1992a; Schurian 1993; Fiedler et al. 1995). The organs found in BC species are:

*Lenticles.* Lenticles are also known as pore cupola organs. They are low cuticular structures located around the caterpillar spiracles (small openings for breathing) and apparently involved in chemical attraction of ants. Lenticle secretions appear to be detected by the ants through direct antennal contact. Lenticles also occur on lycaenids and hesperiids that are not attended by ants, but their function is unknown.

*Honey glands.* The honey gland is also known as the dorsal nectary organ. A single honey gland is located on the dorsal surface of the caterpillar's seventh abdominal segment. Honey glands secrete sugary solutions that ants feed on. The time spent feeding at the honey gland is relatively little compared with the total time ants spend attending caterpillars, which indicates the importance of the communication secretions from other organs.

*Eversible tubercles.* Eversible tubercles are located on the dorsal surface of the caterpillar's eighth abdominal segment. When everted, they secrete "ant alarm chemicals" that induce heightened activity and aggressive posturing in ants. The ants usually rush around with open mandibles, behaviour that presumably deters attack by predators.

*Dendritic setae.* Dendritic setae occur on both larvae and pupae, and tend to be concentrated around the honey gland and the lenticles. They are generally longer, stiffer, more branched, and lighter in colour than adjacent setae. They appear to secrete chemicals attractive to ants.

*Sound-producing organs.* Besides secretary organs, many lycaenid larvae and pupae also produce substrate-born vibrations that can be heard with a stethoscope (larvae) or by the unaided ear (pupae). A "file" and "stridulating plate" are located on the membrane between pupal segments 5 and 6 (Hoegh-Guldberg 1972). The pupal sounds can best be heard by shaking several pupae in a glass vial near one's ear (Downey and Allyn 1973). These vibrations appear to be involved in communication with ants, and may also have a defensive function against parasitoids and predators (Fiedler et al. 1995). Most pupae produce only a creaking sound, but European *Celastrina argiolus* Linnaeus pupae produce both a creaking sound and a faint buzzing, with some pupae producing much more sound than others when disturbed (Hoegh-Guldberg 1972).

The attractiveness of larvae to ants is highly variable, ranging from extremely attractive to no attraction depending on the species of butterfly and even individual populations of some species. As a general rule, the more types of ant-organs a caterpillar has, the more attractive it is to ants. Different colonies of the ant *Formica pilicornis* showed marked differences in their amount of attendance on larvae of the Californian lycaenid *Plebulina emigdionis* (Grinnell, 1905). Different populations of Blue Coppers (*Lycaena heteronea*) show marked differences in attractiveness to the ants. The ant-caterpillar interactions documented to date for BC butterfly species are for populations outside BC; it is possible that the interactions between ants and caterpillars may be different or absent in BC. Many butterfly habitats in BC have few ants, especially in the north and in alpine areas. Ballmer and Pratt (1989) summarize the available information for California lycaenids, many of which also occur in BC.

Ant species that do not attend lycaenid larvae normally ignore them after brief contact, but occasionally attack them. Mormon Metalmark larvae that are attacked thrash around and regurgitate a fluid that, when smeared on the ant, causes the ant to withdraw and preen itself (Ballmer and Pratt 1992a).

The most direct effect of caterpillar association with ants is protection of the caterpillars from parasitoids. The larvae of Silvery Blues (*Glaucopsyche lygdamus*) have much lower parasitism rates when tended by ants than when ants are absent. Silvery Blue pupae are also protected by ants from parasitism, but predation rates were unaffected (Pierce and Mead 1981; Pierce and Easteal 1986). The ants tend the larvae and pupae of the Silvery Blues in exchange for being allowed to feed on secretions produced by the honey gland on the backs of the larvae and pupae. The secretions contain sugar and amino acids that are valuable food for the ants.

Since food used to produce secretions for ants is diverted from larval and pupal growth and development, larvae sometimes grow more slowly and result in smaller adults. Slower larval growth increases the chance of death due to predation or accident, and smaller adults produce fewer eggs, but these disadvantages are outweighed by the reduction in parasitism. This metabolic cost is especially great in non-feeding prepupal larvae (Wagner 1993). In most cases, however, the nutrients secreted by the larvae are sugars that are abundant in the leaves that they eat, and there is no adverse effect on growth or weight at maturity. Caterpillars tended by ants often produce even heavier pupae than those of the same species not tended by ants (Fiedler and Hagemann 1995).

There are complex interactions between lycaenid larvae and some of their braconid wasp parasitoids. Lycaenid larvae may be parasitized by one or many braconid larvae, depending on the species of braconid. Some braconid larvae pupate within their host caterpillar; others leave their host and pupate externally in a silk cocoon attached to either the caterpillar or the foodplant. The caterpillar dies after the parasitoid larvae emerge from it.

Lycaenid larvae that are parasitized by braconid larvae belonging to the genus *Apanteles* (subfamily Microgasterinae) frequently show some signs of life for some time after the parasitoids have emerged and pupated next to them. Apparently the parasitoids feed on the internal organs of the caterpillar in a pattern that leaves the caterpillar's myrmeco-philous organs and vibratory ability functional so that the caterpillar remains attractive to ants for up to several days after the parasitoids emerge and pupate. The parasitoid larvae are also not attacked by the ants, despite being apparently vulnerable. In fact, they are protected by the ants as a result of this extended attractiveness of the caterpillars. Silver Blue caterpillars are the only BC species known to remain attractive to ants after the parasitoid larvae emerge, although it is not certain that the myrmecophilous organs remain active (Fiedler et al. 1995). Adult braconid parasitoids of a European lycaenid caterpillar fed on the secretions from the honey gland in addition to parasitizing the caterpillar (Schurian 1993).

### Adult Feeding

Adult butterflies require food containing sugar and water, and frequently require amino acids, proteins, salts, and minerals as well. Nectar provides water, sugar, and some-times amino acids. The sugar in nectar is a butterfly's primary source of energy for body maintenance, flight, and reproduction. The water is essential to prevent dehydration. Most small butterflies seldom last more than a day in hot weather without drinking water, either in the form of nectar or from puddles. All butterflies feed as adults.

The water in nectar may be more important to a butterfly than the sugar, at least in the short term. For example, after being deprived of both food and water for some time, sulphurs will refuse to drink nectar until after drinking pure water (Watt et al. 1974). Sulphur butterflies prefer dilute nectar solutions to sugar-rich nectar, apparently because their water requirements are greater than their sugar requirements. In contrast, the European Skipper (*Thymelicus lineola*) prefers nectar with high concentrations of sugar. When in flight, skippers use proportionately much more energy than sulphurs, and therefore need relatively more sugar than water (Pivnick and McNeil 1985).

Adult Chalcedon Checkerspots (*Euphydryas chalcedona*) tend to gather in areas with good nectar supplies and to lay more eggs on foodplants near good nectar sources (Murphy et al. 1984). Eastern Tiger Swallowtails (*Papilio glaucus*) oviposit more frequently on foodplants near patches of nectar sources than on foodplants in areas without nectar sources (Grossmueller and Lederhouse 1987). Edith's Checkerspots (*Euphydryas editha*) are much more sedentary when nectar sources are abundant within the breeding habitat than when nectar sources are mostly outside the breeding habitat (Gilbert and Singer 1973). Hence a cabbage patch with a nearby source of nectar is likely to have more eggs laid in it by Cabbage Whites (*Pieris rapae*) than would a cabbage patch without any flowers nearby. In the dry sagebrush areas of the southern interior of BC, butterflies such as blues, hairstreaks, fritillaries, and skippers concentrate in patches of nectar sources. Nectar sources are part of habitat require-ment for adult butterflies.

Butterflies are important flower pollinators because, like bees, they frequently visit flowers to nectar. They tend to visit the same species of flower several times in succession while nectaring, which increases the probability of success-ful pollination. Some tropical plants are pollinated only by butterflies (Webb and Bawa 1983), and various desert and tropical plants are pollinated only by moths. Some butterflies are extremely poor pollinators of some flowers, however, possibly because their smooth, slender proboscis picks up few pollen grains (Venables and Barrows 1986).

Woodnymphs (*Cercyonis*), Red Admirals (*Vanessa atalanta*), White Admirals (*Limenitis arthemis*), tortoiseshells (*Nymphalis*, *Aglais*, and *Roddia*), and anglewings (*Polygonia*) frequently obtain sugar and water from tree sap and rotting

**46** Compton Tortoiseshell (*Roddia l-album*) feeding on mountain alder sap.

**47** Butterflies feeding. *Left:* Anglewings (*Polygonia zephyrus* and *P. faunus*), feeding on animal dung. *Middle:* Grey Anglewing (*Polygonia progne*) feeding on rotten snowberry fruit. *Right:* Red Admiral (*Vanessa atalanta*) feeding on red plum juice.

fruit. Mourning Cloaks (*Nymphalis antiopa*), Compton Tortoiseshells (*Roddia l-album*) (Fig. 46), and anglewings emerge from hibernation on warm days in early spring and feed on sap from stumps of recently cut trees (Saunders 1869b; CSG; JHS). Various butterflies have been recorded feeding on an unknown substance (presumably a sugary secretion) on the inflorescences of two species of grasses in Texas (Neck 1980).

The most common food sources for butterflies consist of nectar, sap, rotting fruit, and other sugary solutions, but contrary to popular belief butterflies are not always fastidious epicures. They feed not only on nectar but on less savoury substances, including mud, carrion, dung (Fig. 47), and urine (Downes 1973; Adler 1982). Dung and urine of carnivores seems to be preferred over those of herbivores, but both are fed on. These items are used as sources of salts (sodium), minerals, amino acids, and proteins, which are frequently in short supply for herbivores such as larvae (and the adults they become) because they are generally in low concentrations in plants.

Amino acids and proteins are used for building the proteins of the butterfly, for both body maintenance and reproduction. Females require considerable amounts of protein for egg production, while the male butterflies use proteins in sperm production and in the production of accessory secretions to form a spermatophore (a package for the sperm). Without ingesting amino acids as adults, some butterflies can achieve only about 5% of their maximum potential reproductive level (Watt et al. 1974). Amino acid intake by the females also results in larger eggs and hence larger and healthier larvae (Murphy et al. 1983).

Butterflies may be attracted to the odours of decay in their search for sources of rotting flesh from which they can get amino acids, protein, and sodium (Young 1977). Dead fish appear to be especially attractive, and Underwing moths, Sphinx moths, American Ladies (*Vanessa virginiensis*),

Mourning Cloaks (*Nymphalis antiopa*), Compton Tortoiseshells (*Roddia l-album*), Hoary Anglewings (*Polygonia gracilis*), and Green Anglewings (*P. faunus*) have all been recorded feeding on rotten fish outside BC (Voss 1961; Nielsen 1977). In BC, blues, White Admirals (*Limenitis arthemis*), anglewings, Compton Tortoiseshells, and Mourning Cloaks are commonly seen at fishing areas feeding on fish gut piles, other carrion, and human and dog excrement. Various North American moths and butterflies (including Green Anglewings, Satyr Anglewings (*P. satyrus*), Common Woodnymphs (*Cercyonis pegala*), Silverspotted Skippers (*Epargyreus clarus*), and swallowtails) are attracted to pig carrion (Payne and King 1969) and dog carcasses (Reed 1958; Voss 1961). Clark (1932) recommends suspending dead snakes (road kills only!) from trees to attract woodland butterflies. Road kill snakes, toads, and frogs on BC backroads are fed on by butterflies, and the butterflies may then be driven over by vehicles while feeding. Bruce (1891) noted that damp sand above the grave of a mule was particularly attractive to swallowtails and other butterflies. All of these things have

**48** Green Anglewing (*Polygonia faunus*) feeding on moose bones.

been used in the past by ardent collectors as bait to collect both butterflies and moths.

One valuable source of protein for females is the spermatophore and accessory compounds that are transferred from male to female butterflies during mating. The nutrients acquired in this way make an important contribution to female egg production and probably body maintenance. In butterfly species with long life spans, such as Monarchs (*Danaus plexippus*), females mate repeatedly, and thus many of their nutrients are acquired from males (Boggs and Gilbert 1979).

Amino acids and proteins are obtained from other sources as well. The eastern North American subspecies of the White Admiral (*Limenitis arthemis astynax* Fabricius) is known to feed on fermented milk and on the spittle of spittlebugs (Perkins and Perkins 1975). Salts, water, and, in the case of the milk, alcohols are ingested. The alcohol is presumably not the reason for feeding on fermented milk, although butterflies sometimes get staggering drunk when feeding on fermenting fruit (CSG). Obtaining amino acids by feeding on pollen has been demonstrated in tropical *Heliconius* butterflies (Gilbert 1972), but not in any butterflies in BC.

### Mud-puddling

A familiar sight to many naturalists in BC is that of "mud-puddling" butterflies (Fig. 49). For most butterfly species, males greatly outnumber females at puddle parties, but some females may also be involved. Females rarely puddle, and when they do, they tend to puddle alone rather than in groups to avoid harassment by males (Berger and Lederhouse 1986). One reason for a butterfly to visit a puddle is to drink water to prevent dehydration, and both sexes require this.

Most mud-puddling is done by male butterflies, and occurs where water has been contaminated by feces or urine or where there are naturally occurring salts in soil. This behaviour apparently helps males replace sodium ions lost during the transfer of spermatophores and other secretions to females during mating. A female butterfly loses up to 75%

**49**  Mud-puddling Canadian Tiger Swallowtails (*Papilio canadensis*)

**50**  Western Spring Azure (*Celastrina echo*) feeding on ashes

of her body's supply of sodium by producing and laying eggs, but obtains as part of the spermatophore the sodium she needs in order to remain healthy and reproductively active. To replenish their own sodium, males need to ingest water containing high levels of sodium, resulting in the mud-puddling behaviour (Arms et al. 1974; Adler and Pearson 1982).

Other sources of salts are also available to butterflies. They may land on a person's arm and drink the sweat on it. Some skippers will exude a drop of fluid from their anus onto a stone or a finger and then suck it up again (MacNeill 1975). Presumably salts are dissolved in the fluid and are then taken up by the skipper. Western Spring Azures are frequently seen feeding on the ashes of old campfires (Fig. 50), where they may collect salts by exuding drops of moisture from the proboscis and then reimbibing them along with the dissolved minerals.

Sevastopulo (1974) suggests that butterflies in seashore areas where a lot of salt from sea spray is found naturally in the environment will have little need to mud-puddle for salt, because the larvae will have taken in large amounts with their food. Mass deaths of butterflies have occurred around cattle salt licks near Williams Lake, BC (Anna Roberts, pers. comm.), possibly because of an additive present in the salt blocks that is toxic to butterflies.

Many butterflies – such as Eastern Tiger Swallowtails (*Papilio glaucus*) (Clench 1958), Spring Azures (*Celastrina*) (Langston 1975), and Juniper Hairstreaks (*Mitoura siva*) near Keremeos, BC (CSG) – have been observed "pumping," or drinking water steadily while at the same time excreting a stream of water drops from their anus. This has also been observed in moths such as *Venusia cambrica* on Vancouver Island (Guppy 1952), and *Dyspteris abortivaria* and *Drepana arcuata* in Pennsylvania (Clench 1958). Similar observations for moths in Ohio and for tropical swallowtails are provided by Welling (1959). There is no obvious reason for this: gathering nutrients appears unlikely since the water is frequently very clean, and cooling down also appears unlikely because the conditions are not always hot (especially for the moths at night). The moth *Gluphisia septentrionis* has been shown to acquire salt through pumping (Boggs 1998), and it has been suggested that pumping by the Eastern Tiger Swallowtail flushes waste products from the body (Reinthal 1963).

In arid areas of the US, water and nectar are in short supply on the dry ridge breeding areas of some butterflies (Scott 1973a). Lycaenids such as Hedgerow Hairstreaks (*Satyrium saepium*) and Western Elfins (*Incisalia iroides*) move from the hillsides to the valley, and then down along the valley bottom to mud-puddle and nectar in the moister low-elevation areas. After feeding they fly directly up the hillside near their feeding area, and then move back up the ridgeline to their breeding areas. This behaviour is common in the drier areas of central and southern BC. Here lycaenids and other butterflies utilize introduced yellow (*Melilotus officinalis*) and white (*M. albus*) sweet clover for nectar.

### Larval Feeding

Lepidopteran larvae commence feeding after chewing their way through the egg shell (chorion). Some species simply eat their way out of the egg, rest awhile, and then commence eating leaves. Other species eat the egg chorion partially or completely before they start feeding on their foodplants.

Once larvae have commenced feeding on the foodplant, they spend only a portion of the time actually feeding, anywhere from 7% to 75% of the time, depending on species, instar, food quality, and the environment (Barbosa 1993). The feeding habits of lepidopteran larvae can be loosely divided into 13 groups, six of which are characteristic of some BC butterfly larvae. Most of the butterfly larvae in the province are *defoliators* (leaf feeders). Others (some lycaenids) are *flower feeders*, and some (mature larvae of marbles, orangetips, and some lycaenids) are *seed and pod feeders*. Butterfly caterpillars that feed on stinging nettle are *leaf tiers*, checkerspot larvae are *web makers*, and young defoliators are frequently *skeletonizers*.

Feeding requires a number of larval structures (Fig. 51), each with its own function. The serrated mandibles are moved from side to side while a leaf is chewed into pieces. A pair of maxillae with maxillary palpi lie below and behind the mandibles, and are used to taste the food while helping to push the pieces into the mouth. The maxillae are also used to clean the face. Behind and between the maxillae, a pair of fused labia form a "lower lip" and bears a pair of labial palpi with the spinneret between them. The labial palpi are also used to continuously "taste" the food and assist in moving the food into the mouth. The spinneret produces silk to anchor the larva to the leaf and to provide a lifeline for small larvae when they fall off the plant. The ocelli are very primitive eyes, sensing changes in light intensity and movement (= potential predators) without resolving images. Feeding involves a continuously repeated sequence of chewing and moving the food first into the mouth and then into the crop. Once the crop is full, the caterpillar stops feeding and rests while the food is digested and moved

further through the intestine. A typical caterpillar, such as the European Large White (*Pieris brassicae* Linnaeus) has about three bouts of feeding each hour, separated by much longer periods of rest. Other species spend a greater proportion of their time feeding, especially if the food is less nutritious.

Feeding is the primary objective of a caterpillar's life. The amount and quality of a caterpillar's food determines its growth rate and health, as well as adult size and reproductive success. The faster a caterpillar can eat, or the higher the nutritional quality of what it eats, the faster it can grow and mature. The less time required to mature, the less likely the caterpillar is to die from disease, parasitism, predation, or an accident. Rapid growth may also enable an individual to produce more than one generation in a year, thereby increasing the number of descendants.

Ideally all larvae should feed on the most nutritious and abundant food available. In practice, however, a large number of factors influence feeding behaviour. The nutritional value of plant tissues varies greatly, and depends on concentration and types of allelochemicals, fibre, toughness, plant tissue age, nitrogen content, and water content. Other variables include plant species, plant population, plant part (leaf, flower, and so on), the age of the plant or plant part, soil fertility, soil moisture, temperature, light, damage to the plant caused by herbivores, and time of day (Slansky 1993). In addition a caterpillar must eat while staying warm and avoiding disease, parasites, and predators. Tender young leaves and flowers, especially the ovaries and seeds, generally have higher nutritional value for larvae than leaves,

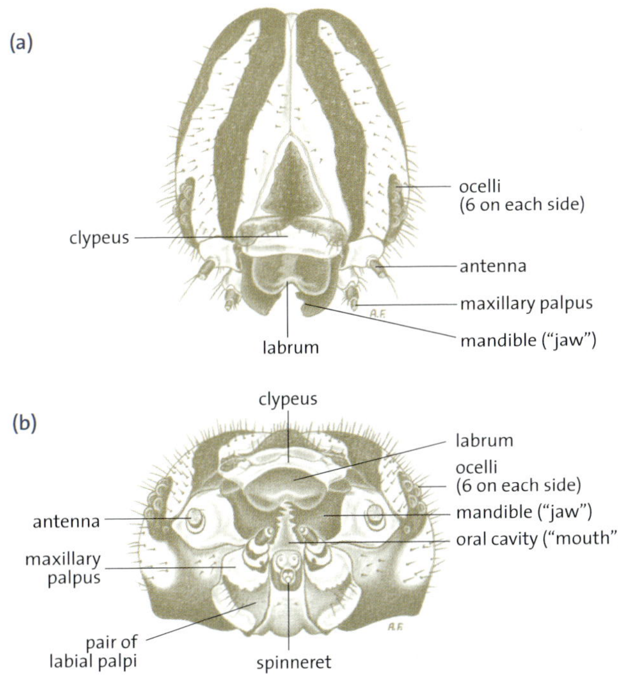


**51** Mouthparts of an Anise Swallowtail (*Papilio zelicaon*) caterpillar: (a) front view, (b) ventral view

especially tough, leathery mature leaves. Some lycaenid and riodinid larvae feed on other insects such as ant larvae.

### Conflicting Needs of Feeding, Thermoregulation, and Avoidance of Predation and Parasitism

Three major factors shape the feeding patterns of butterfly larvae in BC's temperate climate. Body temperature determines the metabolic rate of a caterpillar, which in turn determines growth rate and the ability to move around. Food quality determines the efficiency of nutrient intake, which is defined by the amount of food that has to be eaten to acquire a given amount of nutrients, and which includes the ability to effectively digest the food. Predation and parasitism rates determine the probability of a caterpillar living long enough to pupate and produce an adult.

Butterfly larvae are "cold-blooded" animals whose body temperature is determined entirely by the environment in which they live. They typically require a body temperature of 25–35°C before they are able to feed and grow at an optimal rate (Casey 1993). Such temperatures are usually above the ambient air temperature, especially in spring or fall, or at night. The lower the body temperature of larvae, the more slowly they grow and generally the smaller the resulting adult. Below 10°C larvae generally become inactive and stop growing altogether. Slow growth increases the chance of death through predation, parasitism, or accident, and smaller adult size results in lower reproductive success. In addition, high-quality food in the form of young leaves or flowers may be available for a limited period only, and rapid growth can help larvae optimize their use of this food (Stamp and Bowers 1990).

Most larvae are highly palatable to predators and hide from predators and parasites by remaining hidden under leaves while feeding (Fig. 52), seeking shelter when not feeding, or feeding at night. They have limited opportunity to raise their body temperature by basking in the sun, and therefore grow more slowly than they otherwise might. The cost of low growth rate is balanced by the benefit of avoiding predation and parasitism. An alternative approach by some larvae, such as those of checkerspots in the genus *Euphydryas,* is to mature at low air temperatures early in the spring. They maximize growth rate by basking in the sun and mature while the density of parasitoids is low. Checkerspot larvae are unpalatable to most predators (Porter 1982, 1983). Other larvae have

**52** Red Admiral (*Vanessa atalanta*) larval leaf shelter

solved the balancing act between feeding, thermoregulation, and avoidance of parasitism and predation in other ways.

The optimal body temperature for feeding is 20–26°C for Clouded Sulphur (*Colias philodice*) larvae and 25–29°C for Orange Sulphur (*Colias eurytheme*) larvae. Clouded Sulphur larvae cease feeding above 29°C, and Orange Sulphur larvae cease feeding and seek shade above 32°C; when reared at 35°C, the upper end of the adult optimal body temperature range, all Orange Sulphur larvae die before maturity. Sulphur larvae actively move around on the foodplant to optimize their body temperature. First to third instar larvae move from their normal position on the upper leaf surface to the lower leaf surface when they start overheating. Fourth and fifth instar larvae spend the night in the insulating litter on the ground at the base of their foodplant. In the morning they crawl up the foodplant and bask in the sun to warm up. When they begin overheating, they move downward into the shaded lower part of the plant, and if the environment is too hot, they may return to take shelter in the cool shaded litter on the ground. They spend most of the day alternately feeding on leaves until their crops are full and retreating to the stem of the plant to digest the food in a relatively sheltered location. In the evening or when the weather cools and the air temperature drops below 15°C, Orange Sulphur larvae move down the stem to spend the night on the ground (Sherman and Watt 1973).

Rocky Mountain Apollo (*Parnassius smintheus*) and Clodius Apollo (*P. clodius*) larvae also feed rapidly on their foodplants until their crops are filled, then cease feeding and move into a basking location (CSG). While basking they absorb heat from contact with the warm ground, heat re-radiated from the ground, and heat directly from the sun. Rocky Mountain Apollo larvae bask on the ground near their foodplant, stonecrop (*Sedum*), or on the basal foliage of the foodplant. Clodius Apollo (*Parnassius clodius*) larvae bask on open ground up to several metres away from their foodplant, Pacific bleeding heart (CSG), probably because Pacific bleeding heart typically grows in cool, shaded sites. The mostly black colour of the larvae likely optimizes absorption of heat from the sun and re-radiated heat from the ground. Solar heat absorption may be critical for apollo larvae because they are maturing early in the season, when air temperatures are low. Their colour, black with yellow markings, probably indicates that they are unpalatable to predators, so predation is probably low despite the high visibility of the larvae (they are easily found by humans). The basking behaviour and coloration of apollo caterpillars is very similar to that of desert populations of the White-lined Sphinx moth (*Hyles lineata*) (Casey 1976), suggesting convergent evolution of coloration and behaviour.

Caterpillars that form aggregations while feeding, such as those of checkerspots (*Euphydryas*) (Porter 1982) and Mourning Cloaks (*Nymphalis antiopa*), can retain warmth acquired through basking better than can single larvae. Heat loss through convection and re-radiation is reduced by the insulating effects of adjacent larvae. Aggregated larvae are effectively larger than single larvae, and therefore can achieve higher body temperatures (Stamp and Bowers 1990; Casey 1993). Butterfly larvae that form aggregations are usually black with red markings, to combine the most efficient colour for heat absorption with warning coloration indicating to predators that they are relatively unpalatable. Chickadees will not even peck at Mourning Cloak larvae or other similar black and spiny larvae (Heinrich and Collins 1983). Sheepmoth larvae are black and spiny, and feed and bask in aggregations similar to those of many butterflies (Stamp and Bowers 1990).

Swallowtail larvae have found another solution to the problem of how to thermoregulate while feeding without being preyed on. Swallowtail larvae of all ages feed and rest on the upper surface of leaves, where they can absorb heat from the sun. Young larvae are presumably palatable to predators, and are black with white markings, mimicking bird droppings. The black coloration maximizes solar heat gain, while their resemblance to feces protects the larvae from predation. As they mature, the larvae become too large to convincingly mimic bird droppings, and they become green to camouflage themselves against the green leaves. Higher temperatures due to basking increase the growth rate of larvae, but basking in an exposed location increases mortality from predation (Grossmueller and Lederhouse 1987). The green colour of the later instar larvae probably reflects a compromise between a lower body temperature on the one hand and reduced predation due to crypsis on the other.

### Finding a Plant to Eat
Most larvae do not have to search for a plant before eating after the egg hatches, because the egg has been laid in the appropriate place on their foodplant. Some eggs, however, are laid a significant distance from a larval foodplant, and the larvae have to find the plant before they starve to death. How they accomplish this has not been studied. It is likely that they follow a search pattern similar to that described below, but they may also be able to detect the "smell" of their foodplant from some distance and then home in on it. Greater fritillaries (*Speyeria*) larvae very likely use a slow, methodical search pattern because their eggs are normally laid within a few centimetres of their foodplant. Rocky Mountain Apollo (*Parnassius smintheus*) larvae have a fast, directional search pattern (CSG) because their eggs are generally laid a metre to many metres away from their foodplant.

Once larvae have started feeding on their foodplant, they normally remain on the plant until pupation, unless the plant dies or is defoliated. An attack by a parasite or predator, or wind, rain, hail, or various accidents, may cause them to fall off the plant. Alternatively, the caterpillar may completely defoliate a small foodplant, and have to leave it to find another to eat. Once on the ground, they must find a new plant within a day or two or perish; the amount of time depends on the species, the age of the larva, and the temperature. The details of their behaviour depend on the average spatial pattern of their foodplants (Jones 1977). If it is likely that the larvae will be able to locate the same or another foodplant near where they reached the ground, they move around slowly, turning frequently while lifting and waving their heads, and search a small area thoroughly. If foodplants are likely to be widely dispersed, they move rapidly in a relatively straight line to maximize the area searched. They may start by slowly and methodically searching a small area, and then switch to a fast, straight-line search as they get hungry and as the chances of finding a new plant nearby diminish.

### Plant Defences against Caterpillars
Plants have different defences against being eaten by larvae and other herbivores: chemical, structural (leaf toughness or hairiness), nutritional (minimizing the nutritional value of leaves), or phenological (the season and length of time the plant is available to be eaten). Because overcoming these defences requires special adaptations, larvae are frequently most effective at feeding if they specialize on only one or a few similar plants. Most butterfly species have only a few larval foodplants; either they will not feed on others or cannot mature on them even if they feed.

Different populations of plants, especially different subspecies, may have different chemical resistance to being eaten by butterfly larvae. A plant species used as a larval foodplant in one part of a butterfly's distribution may be toxic to the butterfly larvae elsewhere. In addition, different populations of a butterfly species may have evolved different levels of resistance to toxic chemicals in plants (Shields 1974).

Plants may also vary seasonally in their palatability to larvae, depending primarily on the age of the individual leaves. Young leaves are generally more palatable than older leaves, partly because in many plants they have fewer defensive chemicals and partly because they have a higher nutritional value and are easier to eat and digest. Stonecrop (*Sedum* sp.) has leaves all year, but in BC the leaves of all the commonly used *Sedum* species are toxic to the larvae of Rocky Mountain Apollos (*Parnassius smintheus*) from November to February (inclusive). The rest of the year the larvae feed and develop normally (CSG).

Milkweed leaves have a complex network of latex canals (following veins) containing a milky sticky latex with toxic cardenolides. It is under pressure and floods out when a canal is cut (Dussourd 1993); when exposed to air, it acts as a quick-setting glue. Milkweed is the foodplant of Monarch (*Danaus plexippus*) butterfly larvae, which have behavioural and physiological adaptations to avoid the latex and to use the toxic cardenolides in defence. Young Monarch larvae cut individual leaf veins before feeding, and mature larvae pinch and cut the leaf petiole. This prevents latex and sap from flowing to part of the leaf (young larvae) or the entire leaf (mature larvae), and the leaf droops from loss of latex and sap pressure. Pinching, rather than cutting leaf veins and petioles, crushes the latex canals internally, blocking the flow of latex while minimizing the amount of latex coming in contact with the caterpillar.

Leaves of stinging nettle are covered with urticating hairs, which may have both costs and benefits to larvae. Caterpillars that feed on nettle must avoid being "stung" by the hairs, but once they do this, the plant defences help protect them from parasites and predators. Red Admiral (*Vanessa atalanta*), West Coast Lady (*Vanessa annabella*), and Satyr Anglewing (*Polygonia satyrus*) larvae cut the midrib of a nettle leaf and tie the leaf into a "tent" as shelter from predators, parasites, and the weather. Milbert's Tortoiseshell (*Aglais milberti*) larvae notch the base of a leaf and then tie the leaf with silk into a shelter.

Plants in the parsley family (Apiaceae) have leaf veins containing toxic furanocoumarins, which young larvae can avoid by feeding between the veins (Dussourd 1993). Anise Swallowtail (*Papilio zelicaon*) larvae produce special enzymes in their guts that detoxify the furanocoumarins in the umbelliferous plants they are eating. They produce different intestinal enzymes depending on which species of Apiaceae they are feeding on, because each plant has different furanocoumarins. Thus Anise Swallowtail larvae can feed on Apiaceae that are highly toxic to vertebrates such as humans, including poison hemlock.

### Generalists and Specialists

The numerous ecological factors that affect larval feeding has given rise to two basic strategies for optimizing feeding, although many species fall somewhere between the extremes. Generalists use more than one plant family as larval foodplants, whereas specialists use only one family or part of a family (Scriber 1995). This is a somewhat artificial definition, because it assumes that the plant characteristics that are correlated with specialization are also correlated with plant family, which is only partly true. The larger number of species of Papilionidae in the tropics than in temperate areas is strongly correlated with increased specialization in larval foodplants (Fig. 54).

Specialists feed on one or a few closely related plants that share similar defensive compounds and frequently have similar structures and habitats. Specialists have adapted to detoxify the insecticidal chemicals of their hosts, to cope with physical barriers such as latex or hairs, to be closely synchronized with the seasonal growth pattern of the plants, and in all ways to deal with the defences of their larval foodplants. These species frequently feed on plants that are highly toxic to most insects, and may make use of the toxic compounds for their own purposes. For example, Monarchs (*Danaus plexippus*) use the cardenolides from

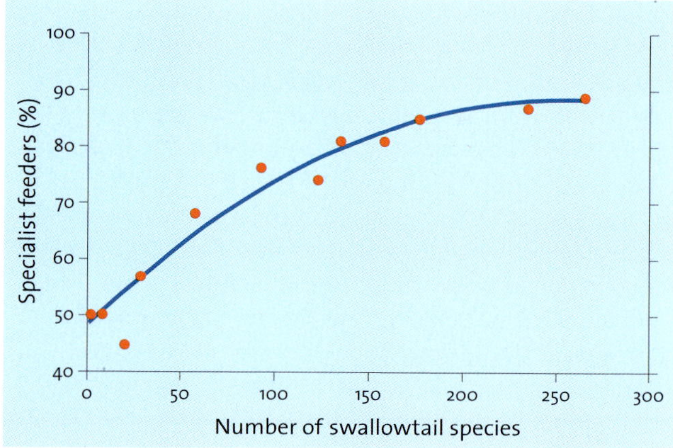

**53** Correlation between the number of swallowtail species and the proportion (%) of those species that are larval foodplant specialists (based on data in Scriber 1995)

their larval foodplant, milkweed, for their own chemical defence.

Generalists feed on a wide range of unrelated plants that may have a wide range of chemical and physical defences. Generalists have adapted to deal with a wide range of plant defences but cannot use plants that are highly toxic. They are "jacks of all trades, but masters of none."

### Patterns of Feeding

Anise Swallowtails (*Papilio zelicaon*) feeding on plants from the family Apiaceae with filiform and highly dissected leaves, such as fennel, feed from the tip of the leaf backward. When they have consumed the leaf from a tip back to a branching point in the leaf, they move out to the new leaf tip and eat backwards from it. This way they avoid wasting parts of the leaf by eating the base before the outer parts are consumed (Heinrich 1993). It is likely that Baird's Swallowtails (*Papilio bairdii*) feeding on tarragon have the same pattern of feeding.

Caterpillars feeding in aggregations, such as Mourning Cloaks (*Nymphalis antiopa*) and checkerspots (*Euphydryas*), are relatively unpalatable to predators and do not need to conceal their presence; they are therefore "messy" feeders that leave large areas of obviously damaged leaves. They focus on eating the tender and easily accessible parts of leaves and leave the tough veins and tips that are hard to feed on (Heinrich 1993). It is easy for predators, and naturalists, to spot plants fed on by these species.

Caterpillars that are relatively palatable, such as tiger swallowtails, avoid leaving large areas of obviously damaged leaves. They feed as individual larvae instead of forming aggregations that would produce many damaged leaves in a small area. They generally eat by paring a leaf down the side, so that the leaf simply looks smaller rather than damaged. Many palatable moth larvae feeding on trees and shrubs (not herbs) sever the leaf petiole and discard leaf remains after feeding, to eliminate evidence of their presence

(Heinrich 1993). These plants are more difficult to spot, although twigs completely stripped of all leaves and leaf remnants can be found when searching carefully. Palatable larvae feeding on low herbaceous plants frequently feed at night and retreat to hide in ground litter during the day; greater fritillaries (*Speyeria*) are an example of this (McCorkle and Hammond 1988). Other species, such as Orange Sulphurs (*Colias eurytheme*) feed during the day to benefit from solar warming but spend as much time as possible hiding in ground litter or on the plant away from the feeding area (Sherman and Watt 1973).

### Butterfly Larvae as Pests

Few butterfly species are major pest species, both through-out the world and in British Columbia. In BC, 1.1% (two species) of butterflies have larvae that are significant crop or forest pests. Both pests are in the family Pieridae: Cabbage Whites (*Pieris rapae*) are major pests on cole crops, and, rarely, Pine Whites are major pests on coniferous trees in the family Pinaceae (hemlock, pine, fir, and their relatives). About 1.6% of all temperate-climate pierids are pests, compared with 6% of pierids in BC for the two whites; the remaining lepidopteran families average about 0.4% of species for each family worldwide (Barbosa 1993).

The Orange Sulphur and the Common Sulphur can be pests on alfalfa and clover crops further south in the USA, where populations can build up because of higher temperatures and longer growing seasons, but they are not pests in BC. The European Skipper is a pest on timothy hay in eastern North America and Europe, but the populations in BC are not yet large enough for pest status. European Skippers are likely to become a hay pest in southern BC in the future. Many other butterfly caterpillars feed on a wide range of ornamental or food crops, but they are not significant pests within BC.

## Introduction of Cabbage Whites to British Columbia

The Cabbage White was one of the first, if not the first, pest species in North America that was recognized as having originated from abroad. Scudder (1887, 1889b) summarized what was known at that time about the introduction and spread of this pest butterfly; the scenario has been largely repeated for all pests introduced after the Cabbage White.

Cabbage Whites were introduced to North America via Quebec in the late 1850s, probably as larvae or pupae attached to cabbage from Europe. More than 70% of immigrants to Canada in the 1840s and 1850s were refugees from the Irish famine, most of whom were tenant farmers or agrarian labourers. Steerage passengers carried their own food with them on the transatlantic voyage, and cabbages and turnips (including leafy turnip tops) would have been common items because they store well. The quarantine station on Grosse Île, a tiny island near the port of Quebec City, was the primary Canadian entry point for immigrants.

A Cabbage White adult was recorded on a transatlantic steamer more than 1,600 km from land (Scudder 1889b), indicating that such transportation on cole plants to Quebec was possible. Further evidence that Ireland was the most likely source of the first introduction of the Cabbage White to North American is the bright yellow adult Cabbage Whites (almost always males) that appeared in Montreal by 1863. The yellow colour form was found in Quebec, Maine, and New Hampshire by 1869, with about 0.2% being yellow (Bowles 1872); they also occurred in Massachusetts by 1869 (Scudder 1872b). These may have been the form *flava*, which most frequently occurs in Ireland (Emmet 1990b).

William Couper, a taxidermist and general collector of natural history specimens (including Lepidoptera), first captured a few Cabbage Whites in the immediate vicinity of Quebec in 1860, when they were a rarity (Scudder 1887, 1889b). In 1863 G.J. Bowles, a lepidopterist new to Quebec, also captured some, and after talking with Couper, sent samples to the entomologists W. Saunders and S.H. Scudder, who identified the specimens as Cabbage Whites. By this time Cabbage Whites had become very common and destructive in the neighbourhood of Quebec and 24 km north at Laval, and had extended 48 km northwest along the shore of the St. Lawrence River. They had not yet been seen at Point Lévis to the south or at Ste. Anne 113 km downriver, where another lepidopterist lived. Given the extremely rapid rate of population increase, Cabbage Whites must have been introduced to Quebec in the preceding decade (Scudder 1887, 1889b), presumably first at Grosse Île and then spreading to the vicinity of Quebec City. By 1869 Cabbage Whites were widespread and abundant in southern Quebec (Saunders 1868), Maine, and Vermont (Sprague 1868). They were also on the north edge of New York City along the Hudson River (Mead 1869), apparently as the result of a second introduction at Hobomok, New York, in 1868, when adults emerging from pupae imported from Europe by a collector escaped (Scudder 1887). There were a number of point introductions from North American sources throughout the US that rapidly merged into the expanding general population. Cabbage Whites reached Nova Scotia by 1871 (Bethune 1873a), West Virginia and Nebraska by 1881 (Edwards 1882a; Dodge 1882), Alberta by 1882 (Geddes 1884), and Montana by 1884 (Scudder 1887). Their rapid spread was due to the transportation of cabbage, turnips, and mustard by train through the interior and by ship along the coast (Scudder 1887).

Yreka, California, may be the type locality of *Pieris yreka* Reakirt 1866, a spring form specimen of *Pieris rapae* (Shapiro 1978). Scudder (1887) rejected the suggestion of an early introduction on the west coast, apparently because he considered *P. yreka* to be a form of the Margined White, *Pieris marginalis*. An early introduction on the west coast is not unreasonable, given that cabbage or turnips (and therefore Cabbage White larvae or pupae) were likely to have been imported to California at that time of the California gold rush. Apparently this California introduction failed to become established, unlike the Quebec introduction, because there was insufficient farming of cabbage crops in California at the time to support a Cabbage White population. Cabbage Whites next reached California in 1883, but did not become established until 1890 or 1891 (Wright 1896), again perhaps because of lack of cabbage crops. By 1894 Cabbage Whites occurred from the Atlantic to the Rockies (Webster 1894), and to California in the southwest.

The rapid spread of Cabbage Whites suggests that they were transported south from Quebec into northern New England as part of ship and train shipments of cabbage and other cole crops. Once they were established in the northeastern US, trains would have carried them further south and west as part of the domestic trade in cole crops. Trains within Canada would have transported cole crops – and Cabbage Whites – from Quebec and Ontario east to the Maritimes. Westward movement of Cabbage Whites in Canada appeared to result initially from westward movement along rail lines in the northern United States, followed by movement into Canada from the south. Transport of cole crops and Cabbage Whites would have followed the westward extension of the transcontinental railway in Canada. In addition, weedy crucifers grew abundantly in the disturbed soils along railway right-of-ways, providing a linear strip of potentially suitable foodplant along which populations could spread (Dodge 1882). The spread of another introduced European butterfly, the European Skipper (*Thymelicus lineola*), may also have occurred through train shipments of timothy hay (Arthur 1966; Burns 1966; Irwin 1968) as well as through truck-transported waste from cleaning of timothy hay seed (Duchesne and McNeil 1978).

Curiously, there was a delay of more than a decade after the completion of the Canadian transcontinental railway in 1885 before Cabbage Whites appeared in BC. There may have been little importation of cole crops from eastern Canada because there was a thriving farm industry in BC and freight costs from eastern Canada would have been high. Cabbage Whites first appeared in Kaslo in 1899, reached Vancouver Island in 1900, and were a "very troublesome" pest throughout southern BC by 1901 (Treherne 1914). Once established, Cabbage Whites would have spread rapidly in cabbage crops transported within the province.

The first reported Cabbage White parasitoid was the introduced *Pteromalus puparum* L. (Ichneumonidae), a tiny wasp that attacks the pupa (Sprague 1871). By 1873 this parasitoid was apparently starting to control Cabbage White populations in the areas where the butterfly had first appeared (Bethune 1873a). By 1886 Cabbage Whites were still a pest, but "the worm of the Cabbage Butterfly, *Pieris rapae*, although still plentiful, is no longer the terror to cabbage growers it formerly was, its natural enemies having multiplied to an extent sufficient to keep it within some reasonable degree of subjection" (Saunders 1886). In general the parasitoids arrived in an area about two years after the butterfly, and rapidly reduced Cabbage White populations to bearable levels (Scudder 1887).

In the future the Cabbage White may be joined in BC by another European species, the Large White (*Pieris brassicae* L.). Two Large White larvae were found on nasturtium in Westmount, Quebec, on 4 September 1904. Both were parasitized, and fortunately the species did not become established (Fletcher 1905). As in the case of the Cabbage White, this unsuccessful introduction of the Large White probably stemmed from cabbages used as food by European immigrants during their transatlantic voyage. The Large White has been introduced to Chile, and was widespread in Chile by 1973 (Gardiner 1974). It is likely that over time it will spread throughout South America and then into North America. Large Whites are highly migratory, so it will be difficult to prevent their spread through the usual border controls on movements of agricultural products. They lay their eggs in masses on cabbage crops, resulting in complete defoliation of the affected plants.

# Seasonal Changes in Butterfly Fauna

In BC, adult butterflies can potentially be seen in flight during any month of the year. During a warm winter day in southern BC, hibernating adults of some species can awaken to fly around and delight the naturalist with a hint of spring to come. In central and northern BC, the same species are found hibernating in woodpiles, hopefully before the wood gets added to the fire. When spring finally arrives, a small number of species adapted to the cool conditions emerge from hibernation as adults and pupae to fly among the trees, shrubs, and herbs just sending forth their spring greenery. As summer approaches, the spring species begin to get worn and weary, and to be replaced by the summer species. These take advantage of the warm sun and still-green vegetation to fly about, mate, and lay their eggs. As the summer progresses the herbaceous vegetation dries out and the leaves of the trees and shrubs age and become less nutritious. A whole new set of butterflies, belonging to a fewer number of species, begins to fly. They had mostly hibernated as eggs or adults, and their larvae matured on the nutritious new growth in the spring and early summer. In the fall most butterflies are finished as adults for the year, except for those that hibernate as adults and a few multivoltine species that continue to fly until after the first hard frosts.

These seasonal changes in the adult butterfly fauna occur throughout BC, with remarkably similar assemblages at low elevations. The species or subspecies may change, but the general appearance remains the same. In arid areas most species fly in spring and early summer; in moister areas the species are fewer in number and more evenly spread throughout the year. At high elevations the seasons are shortened, so that the faunas of all the seasons are compressed within a three-month period. Although generally little seen, the assemblage of eggs, larvae, and pupae changes throughout the season as well. In BC the greatest abundance of butterflies occurs in midsummer (Fig. 54).

## Winter Butterflies (November to March)

On warm winter days on the south coast of BC, occasional members of a small group of butterflies take wing. They hibernate as adults, and all are in the family Nymphalidae, the brushfoots. The Zephyr Anglewing (*Polygonia zephyrus*), Satyr Anglewing (*P. satyrus*), Mourning Cloak (*Nymphalis antiopa*), and Milbert's Tortoiseshell (*Aglais milberti*) are all occasionally seen flying on warm and sunny winter days in Vancouver and Victoria. The same species, plus the Compton Tortoiseshell (*Roddia l-album*), are found in the interior of BC hibernating in woodpiles and other sheltered sites, but the winter weather is seldom warm enough for them to take flight. They are all widespread and commonly seen on the south coast during the spring, and continue to fly until they die in May or June (later at high elevations).

## Spring Butterflies (April and May)

Many of the butterflies that hibernate as pupae emerge in early spring just as the crocuses come into bloom and the first blades of grass push up through the brown winter lawns. This is a diverse group of butterflies from two families: Pieridae (whites and sulphurs) and Lycaenidae (blues, coppers, and hairstreaks). They join the hibernating brushfoots when they take wing in late March and April.

In forest openings formed by rock outcrops, streams, and logging roads, splashes of blue, white, orange, and yellow are seen cruising across brown landscapes. The first spring flowers are coming into bloom and the leaf buds of deciduous trees and shrubs are just opening. Tiny blue Western or

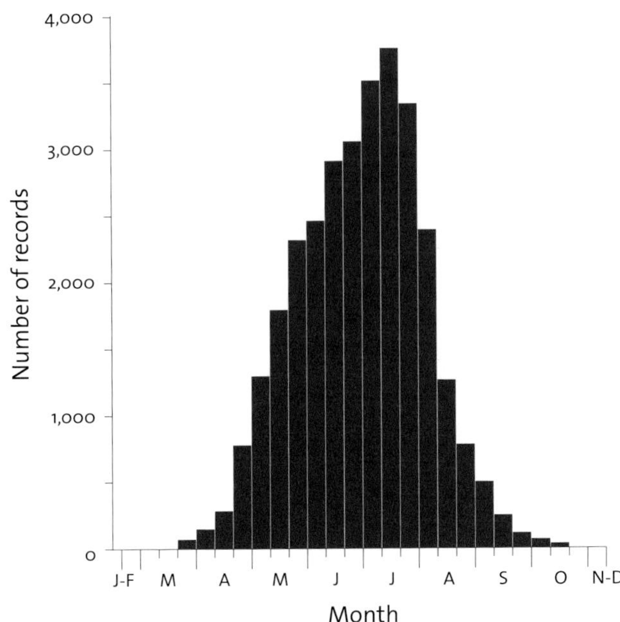

**54** Butterfly abundance in British Columbia through the year. Butterfly records are split into three groups per month; the first group in the graph includes all records from January and February, and the last group consists of all records from November and December.

Boreal Spring Azures (*Celastrina echo* or *C. ladon*) and Western Tailed Blues (*Everes amyntula*) visit muddy areas. Males of all three species have wings that are a startling bright blue above (females are more brown) and provide a beautiful sight when fluttering over new growth or drinking in mass groupings at mud puddles (see Adult Feeding in previous chapter). The Western Tailed Blue tends to be local in occurrence, because of the limited number of foodplants the larvae will accept. The larvae of the Spring Azure feed on many plants, including dogwood, blueberries, and hardhack, so in the spring the adults can be seen almost everywhere.

Frequently flying with the blues are orangetips. The males are white, the females range from white to bright yellow; both have orange wing tips. Their larvae feed on wild mustards that grow in open areas on rocky outcrops and stream banks on the coast, and the meadow areas of the interior of BC. Less colourful, but still providing welcome contrast to the remnants of winter brown, are the whites. Cabbage Whites (*Pieris rapae*), familiar to anyone with a vegetable garden, emerge from their pupae in early spring and fly out in search of the first seedling cabbages and broccoli. Native relatives of the Cabbage White are the Margined White (*P. marginalis*) and relatives, whose larvae feed on native mustards found in forest openings and moist meadows. The native Whites are usually found in relatively undisturbed habitats; Cabbage Whites are usually in disturbed areas.

Various elfins, hairstreaks, and skippers are among the first species to fly in spring, with different species being characteristic of different parts of the province. They can be locally common, but because of their small size, brown colour, and inconspicuous habits, they are seldom noticed. A sharp eye will spot adults flitting across rock outcrops and forest openings, or on hilltops. The most widespread of the spring lycaenids is the Western Elfin (*Incisalia iroides*). The larvae feed on various Ericaceae such as arbutus, salal, and bearberry, but apparently only one foodplant is used in any one area. Older books on butterflies claim that this species never drinks nectar as an adult. This may sometimes be true on the coast, where often no flowers are available and humidity is high enough so that the butterflies may not dehydrate rapidly. Western Elfins do nectar when flowers such as hawthorn are available, however, especially in the dry interior. Two-banded Checkered Skippers (*Pyrgus ruralis*) are a widespread and characteristic part of the spring butterfly fauna.

## Summer Butterflies (June to August)

Summer is the season of butterflies, with more species flying from June to August than at any other time. Swallowtails become common by June and dominate the early summer fauna. White Admirals (*Limenitis arthemis*) in the interior and Lorquin's Admirals (*L. lorquini*) on the coast fly in deciduous woodlands and riparian areas. Checkerspots, woodnymphs, and ringlets bounce saucily across grasslands and in open forests. A wide variety of brown, grey, and black skippers skip rapidly from flower to flower, with Common Branded Skippers buzzing around hill and ridgetops. Large white Rocky Mountain Apollos (*Parnassius smintheus*) drift lazily across sparse, dry grass hillsides, frequently with bright red and black checkerspots, various blues, and dull brown Chryxus Arctics (*Oeneis chryxus*) flying with them.

Woodland Skippers (*Ochlodes sylvanoides*) are the most abundant and widespread skipper in BC, and their appearance marks the arrival of late summer. At the same time Pine Whites (*Neophasia menapia*) appear, drifting like giant snowflakes around conifers. In rare "outbreak" years, the Pine Whites form a snowstorm in the air, and drifts of road kills accumulate along road edges.

Hilltops and ridgelines buzz with activity. Male Anise Swallowtails (*Papilio zelicaon*) and Baird's Swallowtails (*P. bairdii*) stake out their territories and chase all comers. Common Branded Skippers (*Hesperia comma*) buzz frenetically around summits, with chains of up to a dozen chasing each other around in the hope of finding a mate. Western White (*Pontia occidentalis*) males patrol up and down and around the hilltop, playing tag with each other. Checkerspot males bask in the sun on dirt and rocks, flying up to chase anything dark that comes flying by. A hiker's lunch stop on a prime hilltop in the summer results in many butterflies to see, and many behaviours to watch.

The multivoltine butterflies reach their peak populations in the summer. Clouded Sulphurs (*Colias philodice*) are everywhere in the interior, both around alfalfa fields and in the natural meadows and forest clearings. Cabbage Whites (*Pieris rapae*) hover around agricultural fields, gardens, and weedy disturbed areas. Monarchs (*Danais plexippus*), uncommon most years, are most likely to be seen in southern BC in midsummer.

In alpine areas summer is the time for all butterflies, with the early "spring" species appearing in June on exposed southern aspects from which the snow clears early. To see these butterflies, one must often hike across the snow still in the shade of forests lower down or on northern aspects. The "summer" butterflies are in full flight in July, fading by August. In August in the southern interior, alpine areas swarm with hordes of Painted Ladies (*Vanessa cardui*) and West Coast Ladies (*V. annabella*), with clouds of them flying up like locusts at every step. Other butterflies are still present but are difficult to see amid the distraction of ladies flying everywhere. By September the summer butterflies are mostly reduced to ragged leftovers.

## Autumn Butterflies (September and October)

The summer butterflies disappear through August, and are mostly replaced by the autumn species by September.

Species that hibernate as adults have passed their larval stage during the summer, and have emerged from their pupae in late July and August. They have nothing to do during autumn except feed on nectar, tree sap, decaying fruit, carrion, and dung. They are universally the colours of autumn – the golden browns of autumn leaves and the greys and browns on bark and dried leaves. The golden glow of anglewings is striking in the patches of sunlight on tree trunks or flitting between patches of sun on the ground when disturbed by a walker.

A few butterflies are left over from the summer, along with the anglewings waiting to go into hibernation. Cabbage Whites, Clouded Sulphurs, Western Whites, and, on the coast, Anise Swallowtails fly until the hard frosts or continuous heavy rains arrive. The females continue laying eggs until they die, with the last laid eggs having no hope of producing larvae or pupae that can successfully hibernate. When autumn ends, only the hibernating adults of the winter butterflies remain.

Species Accounts

Photo on previous page: Common Branded Skipper (*Hesperia comma harpalus*)

# ORGANIZATION OF THE SPECIES ACCOUNTS

The following species accounts for the skippers and butterflies of BC are presented in an order approximating the evolutionary relationships of the species. They summarize the important information available for each species, including all the information that has been gathered within British Columbia, most of the information that has been gathered in adjacent areas, and selected relevant information from other areas of North America and, occasionally, Europe and Asia.

For each species account, the information is arranged in the following categories:

**COMMON NAME:** This is the common name that we recommend for use. Most species have several common names, and we have chosen one based on appropriateness and historical usage.

**LATIN NAME:** The Latin name of the species consists of the genus (the first word) and the species (the second word), followed by the surname of the person who named the species and the year in which the species was named. If the author's name and the date are enclosed in parentheses, the species is now classified in a different genus from the original one. If the date is enclosed in square brackets, then it was not given correctly or was not stated in the original description of the species but has been deduced from other evidence.

**ETYMOLOGY:** This section discusses the origin and meaning of both the common name and the Latin name. It is hoped that knowledge of the meaning of the names, especially the common name, will result in greater stability of use. The earliest date for use of the common name is provided. In some cases, the name originally proposed has been emended somewhat over time, but we still acknowledge the original usage without tracking minor changes. A common name that refers to a modern person in honour of whom the butterfly was named is given in the possessive form with an apostrophe, such as "Christina's Sulphur." A number of names, especially those given by W.H. Edwards, appear to refer to unknown modern women. We have assumed that the butterflies were named in honour of specific women, and have therefore used an apostrophe in the common names, such as "Sara's Orangetip." When a name refers to a figure in classical Greek or Roman history or mythology, no apostrophe is used, such as "Clodius Apollo."

**ADULT:** This section summarizes the characteristics by which adults of the species can be distinguished from similar species.

**IMMATURE STAGES:** This section summarizes the appearance of the immature stages, with emphasis on the mature larva and pupa. In most cases the description is based on the original historical description, because that description has been repeated and distorted through much of the modern literature. Where there is a description of the immature stages in BC, we have used it.

**BIOLOGY:** This section summarizes what is known of the life history, ecology, and behaviour of the species. In most cases this is fragmentary, and much of the information originates from areas distant from BC.

**SUBSPECIES:** This section summarizes the distribution of each subspecies and their distinguishing characteristics. The third word in a Latin name is that of the subspecies. Subspecies are assumed to be able to interbreed freely with each other should they meet in the wild; hence a transition zone between subspecies can be expected. Such transition zones are remarkably few, however, because there is generally a significant topographic feature separating the distributions of subspecies. The type locality (TL) is given for each species and subspecies to indicate where they were originally collected. Eleven new subspecies are named.

**RANGE AND HABITAT:** This section summarizes the range and habitat in BC of the species as a whole.

**GENERAL DISTRIBUTION:** This section summarizes the general world distribution of the species. The official two-letter ZIP/postal code abbreviations are used for states and provinces. Canada, the United States, and Mexico are abbreviated CAN, USA, and MEX, respectively. Pacific Northwest is abbreviated PNW.

**CONSERVATION STATUS:** This section summarizes the conservation status of the species or subspecies. The terminology of both The Nature Conservancy (TNC) and the Committee on the Status of Endangered Wildlife in Canada (COSEWIC) are used (see p. 31). The classifications are ours alone except where otherwise specifically stated. These conservation status ranks are based on the information available as this book was being written, and many will change over time as new information is acquired.

FLIGHT SEASON: The bar graph shows the relative abundance of each species throughout the known flight season. It shows the number of times a species has been observed in each of the three 10-day intervals forming a month. For months with 31 days, the third period is 11 days long. A single species/locality/date observation record consisted of one to many individuals. The first bar in the graph includes all records from January and February and the last bar consists of all records from November and December. When there was limited or no data from British Columbia, data from adjacent areas was used.

MUSEUM SPECIMEN PHOTOGRAPHS: The photographs of museum specimens of each species are the key to identification. They show the characteristics that separate species, which, unfortunately, often cannot be easily seen in live butterflies. The identifying characteristics sometimes differ only slightly between species. Some very similar species can be identified with certainty only through dissection and examination of the genitalia, although there is usually also a slight difference in wing pattern that an expert can use to tentatively separate the species. The captions for the photographs include the subspecies name where there is more than one subspecies in BC, a symbol indicating the sex of the specimen (male, ♂; female, ♀), and a letter indicating whether the upperside (dorsal, D) or underside (ventral, V) is shown. The size of the specimen is given as the "wingspan," i.e., the distance in centimetres across the specimen from the outside edge of one forewing to the outside edge of the other forewing.

The full collection data for each specimen is given in Appendix 3.

LIVE SPECIMEN PHOTOGRAPHS: Where possible, photographs of live adults, eggs, larvae, and pupae are included. Their number was limited by the availability of photos of some species and by space constraints. The captions indicate the subspecies each specimen represents and the photographer's name. The full data for each photograph is given in Appendix 3. Some of the immature stages are for subspecies that do not occur in BC, as noted in Appendix 3, and therefore may look slightly different from those in BC.

ILLUSTRATIONS OF ADULTS: Illustrations of certain species pairs or complexes are provided to show the important characteristics for distinguishing the species.

ILLUSTRATIONS OF GENITALIA: Illustrations of the genitalia are provided for species pairs or complexes that cannot be reliably distinguished by any consistent wing pattern character. For such species, dissection of the male external anatomical structures is necessary. There are reliable characteristics of the female reproductive anatomy that have been extensively used to identify moth species, but these have not been used for most North American butterfly groups. Dissection of the female reproductive structures requires a more delicate touch due to the minimal amount of chitin and is beyond the skill of even most keen lepidopterists. Full data for each genitalia drawing is given in Appendix 3.

DISTRIBUTION MAPS: The distribution maps show the locations of all known observations of each species in BC, and most observations in adjoining areas. The latter are shown in order to provide context for the distributions in BC and to indicate areas along the border where a search should be made to extend the known BC distributions. An additional 20 maps are included in Appendix 1. These maps show the distribution of species not found in BC but occurring in WA, northern ID, northwestern MT, and western AB. Thus all species known to occur within the limits of the distribution maps are covered, even when it is unlikely that these species will be found in BC.

wingspan

♂ D (3.9 CM)

wingspan

♂ V (3.9 CM)

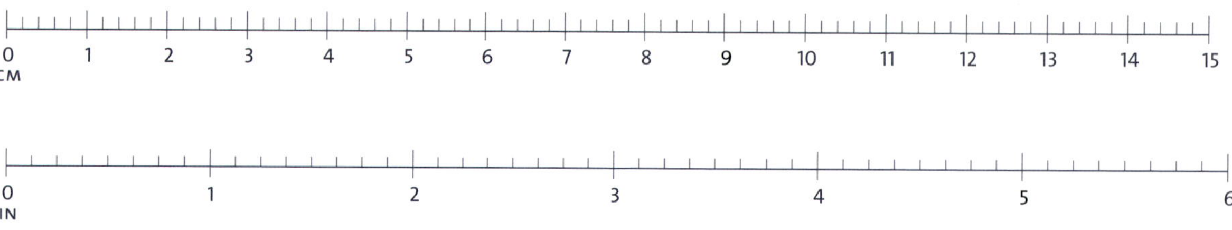




ORGANIZATION OF THE SPECIES ACCOUNTS

# Superfamily Hesperioidea Latreille, 1809

## Skippers

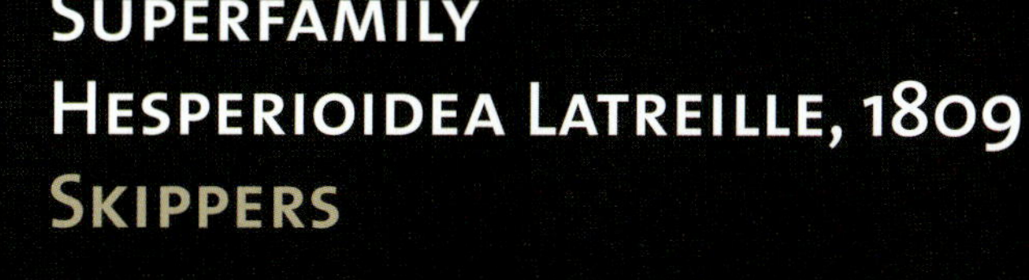

In North America this superfamily is characterized by the lack of a frenulum and retinaculum to hold forewings and hindwings together, a simple antenna with expanded tip, three pairs of legs that are all functional for walking, and forewings with 12 veins unbranched from base to outer margin of the wing. The last two characteristics in combination separate the Hesperioidea (skippers) from the Papilionoidea (butterflies). In some skippers the expanded tip of the antenna is bent back to form a hook, but this is not a definitive trait for all skippers.

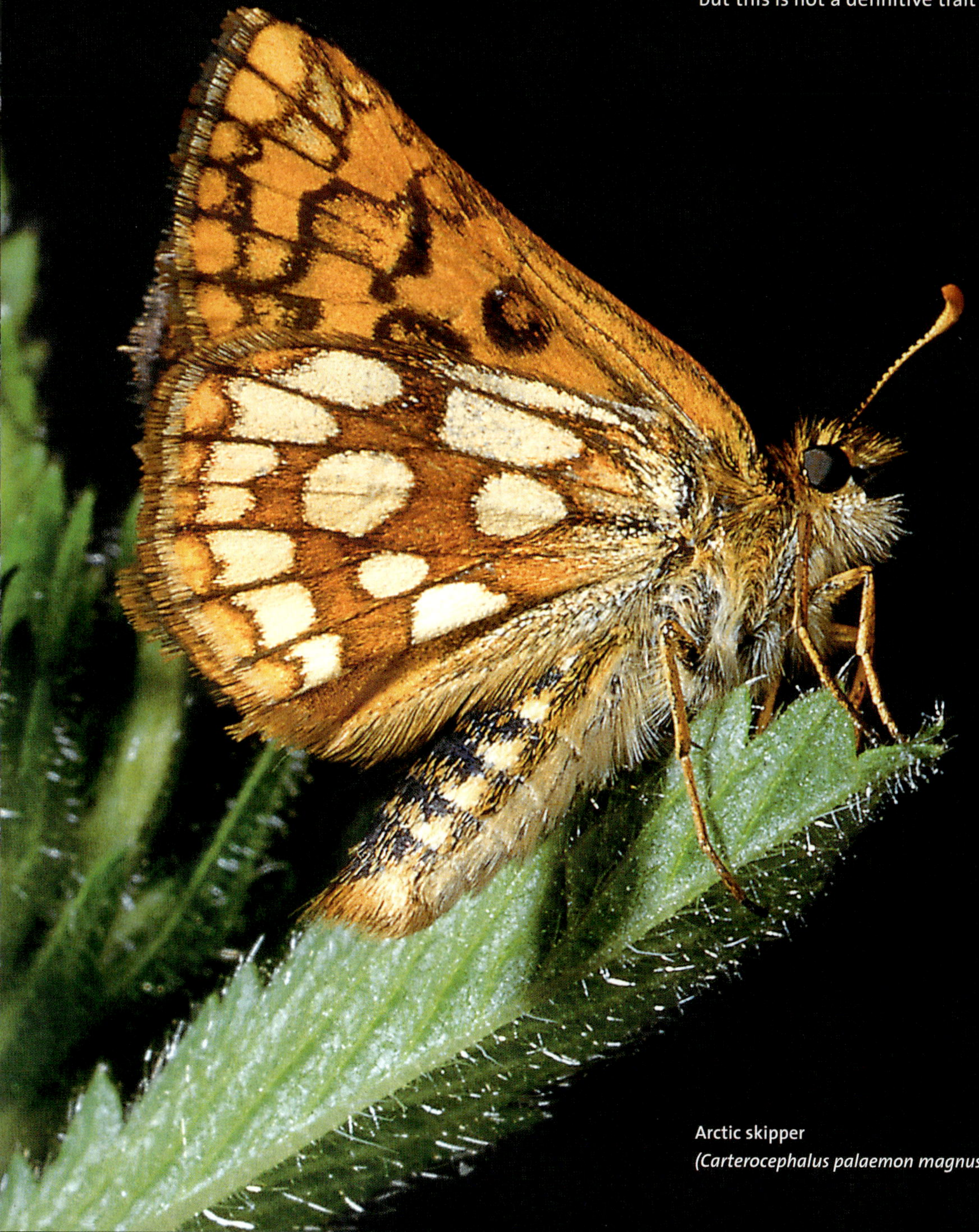

Arctic skipper
*(Carterocephalus palaemon magnus)*

# FAMILY HESPERIIDAE LATREILLE, 1809  SKIPPERS

The common name "skippers" is derived from the insects' characteristic rapid, darting flight, in contrast to the generally slower, flapping flight of butterflies.

There is only one family of Hesperioidea in British Columbia. The characteristics of the superfamily are sufficient to separate the family Hesperiidae from families of Papilionoidea. In addition, adult skippers are characterized by a wide, low head, stout thorax and abdomen, and palpi with long, bushy hairlike scales. In some species the expanded tip of the antenna is bent back to form a hook. The larvae have a wide head and a narrow first thoracic segment. Larvae and pupae are found in a loose silk-bound cell of leaves or grass. Two of the four North American subfamilies are found in British Columbia. We follow Evans (1955) in not placing *Carterocephalus* Lederer, 1852 in a separate subfamily. Evans (1951, 1952, 1953, 1955) provides the only comprehensive review of the family in the Americas. An earlier work by Lindsey, Bell, and Williams (1931) reviewed the North American species.

**Propertius Duskywings (*Erynnis propertius*) mating**

The common name "spread-wing skippers" refers to the typical basking posture, with the wings spread out flat rather than partly angled over the back as in the grass skippers (subfamily Hesperiinae).

Evans (1952) characterizes the adults of this subfamily as having the $M_2$ vein of the forewing midway between the $M_1$ and $M_3$ veins or closer to the $M_3$ vein, the antennal club with a bend, and the male genitalia with a well-developed gnathos. Sometimes males have a costal fold on the upper forewing or hair pencils on the hind tibia. Eggs are ribbed. The larvae feed on various dicotyledonous plants. Two BC genera, *Pyrgus* Hübner, [1819] and *Erynnis* Schrank, 1801 are Holarctic; the others are restricted to the New World. Eleven species of Pyrginae occur in the province. They utilize Malvaceae, Fabaceae, *Salix*, *Populus*, *Quercus*, *Ceanothus*, *Chenopodium*, and *Amaranthus* as larval foodplants.

### GENUS *EPARGYREUS* HÜBNER, [1819]

The name *Epargyreus* is derived from the Greek *argyros* (silvered), referring to the white spot on the ventral hindwing. There is no common name for the genus.

Evans (1952) recognized 15 species in this Neotropical genus. The larvae feed on various genera of Fabaceae.

## SILVER-SPOTTED SKIPPER
*Epargyreus clarus* (Cramer, [1775])

**ETYMOLOGY:** The species name *clarus* means bright and, like the common name, refers to the white spot on the ventral hindwing. Gosse (1859) first used the name "White-spotted Skipper." Maynard (1886) changed it to "Silver-spotted Skipper," and this name has been in continuous use since, one of the very few Maynard common names in use today.

Ssp. *clarus* ♀ D  (4.7 CM)

Ssp. *clarus* ♂ V  (4.4 CM)

Ssp. *californicus* ♂ V  (4.7 CM)

**ADULT:** The Silver-spotted Skipper is easily distinguished from all other BC skippers by the presence of a large white area in the centre of the ventral hindwing. Another key characteristic is the median row of yellow spots on the central fourth of the dorsal forewing. The ground colour of the wings is brown. This is the largest skipper in the province, with a forewing length of 2.2–2.7 cm for BC and WA specimens.

**IMMATURE STAGES:** This species has not been reared in BC but the common and conspicuous larvae have often been described elsewhere. Scudder (1889b) gives the best account. The egg has 16–18 ribs and is green with a red spot at the top in eastern North American populations. Eggs from the East Kootenay are white with pink around the micropyle and a band of pink around the middle of the egg (CSG). The mature larval head is dark brownish red, with a bright orange spot

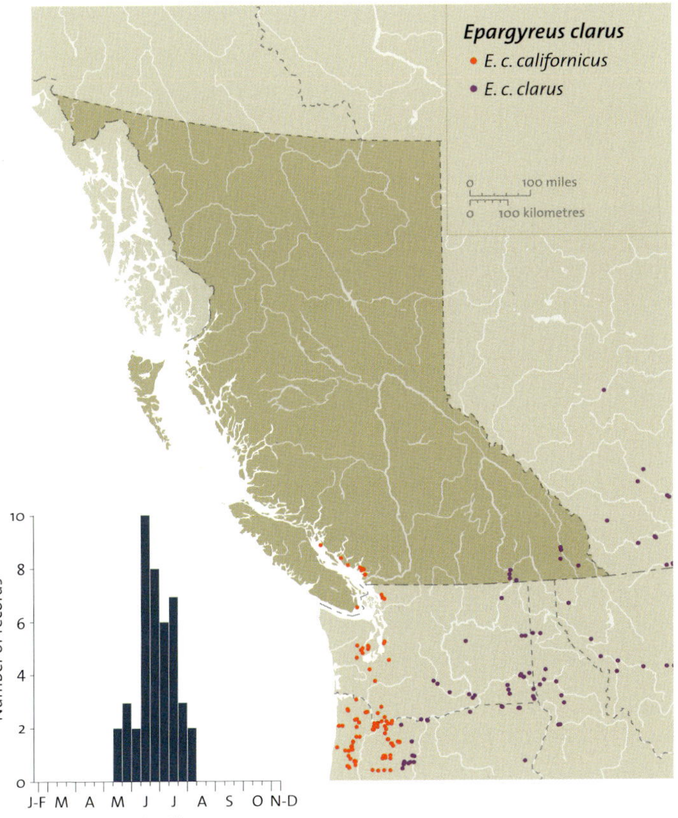

between labrum and ocellus on each side of the head. The body is yellow green, with darker green transverse lines.

**BIOLOGY:** Adults are on the wing from late May to early August, with peak flight occurring in the first week of July. Eggs are laid singly on the foodplant, hatch within the week, develop into mature larvae by late September, and pupate before winter (Scudder 1889b). There is one brood per year in BC. Further south, the Silver-spotted Skipper is multivoltine. Elsewhere the larvae have been found on a number of woody Fabaceae. J. Pelham (pers. comm.) reared subspecies *E. c. californicus* on *Lotus crassifolius* in western Washington. CSG has observed oviposition by subspecies *E. c. clarus* at the leaf bases of *Glycyrrhiza lepidota* (wild licorice) at Bummer's Flats, north of Cranbrook. This plant is one of the two major larval foodplants for this species in the southern Rocky Mountains (Ferris and Brown 1981). In our region, the introduced black locust, *Robinia pseudoacacia,* is also a likely foodplant for west Kootenay populations.

**SUBSPECIES:** Specimens from southeastern BC belong to the nominate subspecies, *E. c. clarus* (TL: Dayton, VA), which is widely distributed in temperate North America. Hinchliff (1996) also recognized *E. c. californicus* MacNeill, 1975 (TL: China Flat, El Dorado Co., CA). This name applies to the rare individuals that were historically seen on Vancouver Island and the Lower Mainland. They utilize a different larval foodplant, as noted above, and have a slightly different pattern on the ventral hindwing.

**RANGE AND HABITAT:** The Silver-spotted Skipper has been recorded from the Gulf Islands, the Lower Mainland, and the southern Kootenays. In the Trail area, it is found in recently established populations in disturbed areas, associated with black locust trees. The Wasa Lake and Bummer's Flats populations are the only known native, permanently breeding populations.

**GENERAL DISTRIBUTION:** The Silver-spotted Skipper is found throughout temperate North America. In adjacent PNW states it is recorded from the Puget Sound Trough, the Columbia Basin, and the Rocky Mountains, but not from the Olympic Peninsula, the Cascades, or the Okanogan Highlands, except for one record from near Kettle Falls, WA.

**CONSERVATION STATUS:** Subspecies *californicus* records in southwestern BC are of historical, probably non-breeding, immigrants (SXN). Subspecies *clarus* populations in southeastern BC appear to be expanding, through both increased migration and new colonization, and are not of concern (S4).

---


## GENUS *THORYBES* SCUDDER, 1872  CLOUDYWINGS

The name *Thorybes* may be derived from the Greek *thorybos* (uproar or clamour), but this has no apparent relevance to the butterfly. The name "cloudywings" refers to the indefinite pattern of the ventral hindwing, and was first coined by Scudder (1889b).

Evans (1952) recognized six species in this Neotropical genus. The larvae feed on various genera of Fabaceae.

## NORTHERN CLOUDYWING
*Thorybes pylades* (Scudder, 1870)

**Northern Cloudywing (*Thorybes pylades*)**

♂  D  (3.2 CM)

♂  V  (3.2 CM)

**ETYMOLOGY:** The species name *pylades* is from Pylades, a figure in Greek legend who was married to Electra and was the friend of her brother Orestes. All three names were used as species names of related skippers. The common name "Northern Cloudywing" was first used by Scudder (1889b).

**ADULT:** The Northern Cloudywing is similar to the Silver-spotted Skipper except that it lacks the silver spot on the ventral hindwing and the median row of spots on the dorsal forewing is much smaller. The ground colour of the wings is milk-chocolate brown. The sharply bent back antennal club, pointed forewing, and long discal cell separate this species from the other Pyrginae. Forewing length is 1.8–2.1 cm.

**IMMATURE STAGES:** The Northern Cloudywing has not been reared in the PNW. Comstock and Dammers (1933), for California populations, and Scudder (1889b), for eastern populations, give good descriptions of the immature stages. The egg has 15 ribs and is pale green. The head of the mature larva is black with dense hairs. The first thoracic segment is black. Scudder describes the remaining body as dark green with a narrow, brownish green dorsal stripe, a dull salmon lateral stripe, and a similar infrastigmatal band. Comstock

and Dammers describe the larva as buff orange, with stripes similar to the eastern ones but maroon in colour. It is not clear whether these colour variations are geographically based or represent small, non-typical samples from the respective areas.

**BIOLOGY:** Adults are on the wing from late April to early August, with peak flight occurring in the last third of June. Eggs are laid singly on the under surface of leaves. In California the larvae mature by fall, overwinter as larvae, pupate in March, and emerge as adults in May. In the east the larvae pupate from late July to September, and there is a partial second brood. The pupae overwinter. There is one brood in British Columbia. The larvae have been recorded on a number of plants: in the east on *Trifolium pratense, T. repens, Lespedeza capitata,* and *L. hirta;* in California on *Amorpha californica.* In Colorado the Northern Cloudywing has been observed ovipositing on *Lathyrus* sp. (Scott 1992). *Trifolium* sp. and *Lathyrus* sp. occur in BC.

**SUBSPECIES:** The BC populations belong to the nominate subspecies (TL: "Massachusetts").

**RANGE AND HABITAT:** The Northern Cloudywing is found across the southern third of the province and in the northeast. The habitat is aspen parkland, mesic meadows, and edges of streams below 1,000 m elevation. Usually only single individuals are seen at any given locality.

**GENERAL DISTRIBUTION:** The Northern Cloudywing is generally distributed across temperate North America.

**CONSERVATION STATUS:** Not of concern (S5).



---

## GENUS *ERYNNIS* SCHRANK, 1801  DUSKYWINGS

The name *Erynnis* is derived from the Erinnyes or Furies who harried wrongdoers (Emmet 1991). Schrank used the generic name for all skippers, and used it to describe the erratic flight characteristic of skippers, as though they were avoiding the Furies. The Erinnyes sprang from the dark, thus the common name "duskywings" in reference to the dark wings. Scudder (1889b) was the first to use the name "duskywings" for the genus.

This and the remaining genera of the Pyrginae have rounded forewing tips, short discal cells, inconspicuous antennal tips, and porrect palpi. The genus *Erynnis* is distinguished by the mottled black background of the wings, hence the common name "duskywings"

(Fig. 55). This Holarctic genus contains 17 Nearctic and 5 Palearctic species. Five species are found in BC. Closely related genera are all Neotropical. The larvae of this genus have been recorded as feeding on *Quercus, Salix, Populus, Ceanothus,* and Fabaceae. The various BC species cover this wide range of foodplant preferences, but individual species are restricted to one or a few closely related foodplants. Burns (1964) wrote the definitive work on the genus. Reference to Lindsey et al. (1931) is necessary for drawings of male genitalia, the only reliable structures for distinguishing most species. The shape of the left valve is the critical characteristic. Fig. 56 shows the left valve of BC species.

**55** Right forewing of *Erynnis* species: (a) *E. pacuvius,* (b) *E. persius,* (c) *E. icelus*

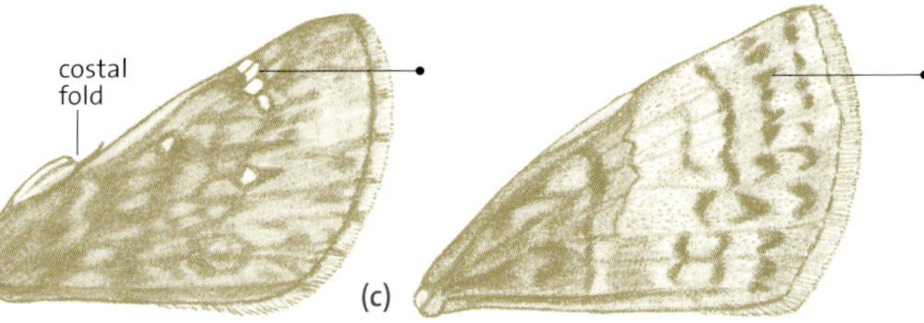

 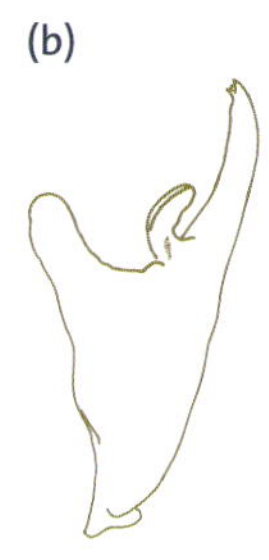 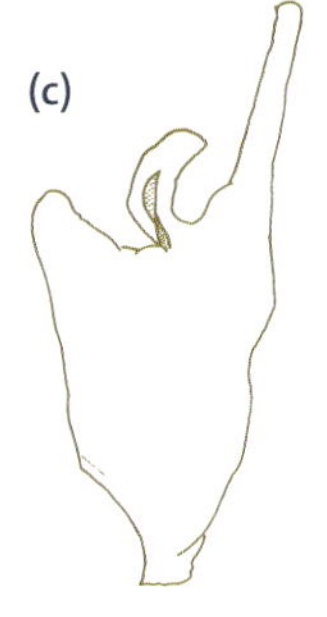 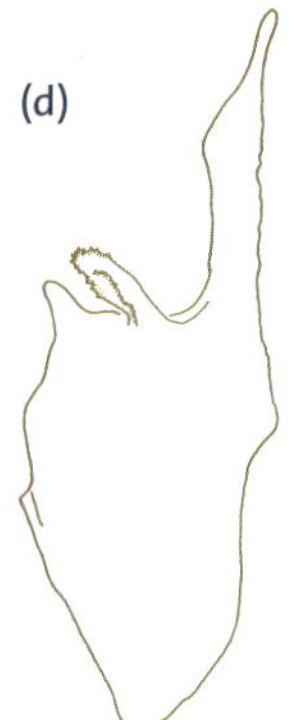 

**56** Left valve of *Erynnis* species:
(a) *E. icelus*
(b) *E. afranius*
(c) *E. persius*
(d) *E. pacuvius*
(e) *E. propertius*

## DREAMY DUSKYWING
*Erynnis icelus* (Scudder & Burgess, 1870)

**ETYMOLOGY:** The species name *icelus* is derived from Ikelos, son of Hypnos, the Greek winged god of sleep (Bird et al. 1995); hence the common name "Dreamy Duskywing," first used by Scudder (1889b).

**ADULT:** *E. icelus* is easily distinguished from other *Erynnis* in BC by the lack of a submarginal group of clear white spots on the forewing (Fig. 55c). The male valve has a short ventral process and a small, rounded middle lobe (Fig. 56a). Although the wings are variable in pattern, there is no clear way to distinguish the sexes except that the male has a costal fold. The length of the forewing is 1.4–1.6 cm.

**IMMATURE STAGES:** Scudder (1889b) described the life cycle. The egg is very pale green. The larval head is uniform light red brown; the ocelli are red brown on a black stripe. The first thoracic segment is yellow and the rest of the body is grey-green, caused by fine whitish granulations on a pale green surface. The larva is marked by a dark median stripe and a pale lateral stripe.

**BIOLOGY:** Adults fly from early May to late June, with peak flight usually occurring in early June, depending on spring weather. Eggs are laid singly on the upper surface of leaves of the foodplant, hatch within the week, and develop into mature larvae by late July. Mature larvae overwinter and pupate the following spring (Scudder 1889b). In BC there is a single brood per year. Cockle (Harvey 1908) recorded *Salix* as the foodplant at Kaslo. The Forest Insect Survey has two additional records of larvae collected from *Salix*. Elsewhere *Salix* and *Populus* have often been recorded as foodplants (Burns 1964).

**SUBSPECIES:** None. The type locality of the species is New England.

**RANGE AND HABITAT:** The Dreamy Duskywing is found across the southern fourth of BC and throughout the rest of BC east of the Coast Ranges. The species should be looked for near willows and aspens in meadows and along streams.

**GENERAL DISTRIBUTION:**
Across CAN from Vancouver Island to NS and south in the Sierras, the Rockies, and the Appalachians. In the PNW it is almost universally distributed below 1,000 m except in dry, sage- or grass-dominated areas.

**CONSERVATION STATUS:**
Not of concern (S5).

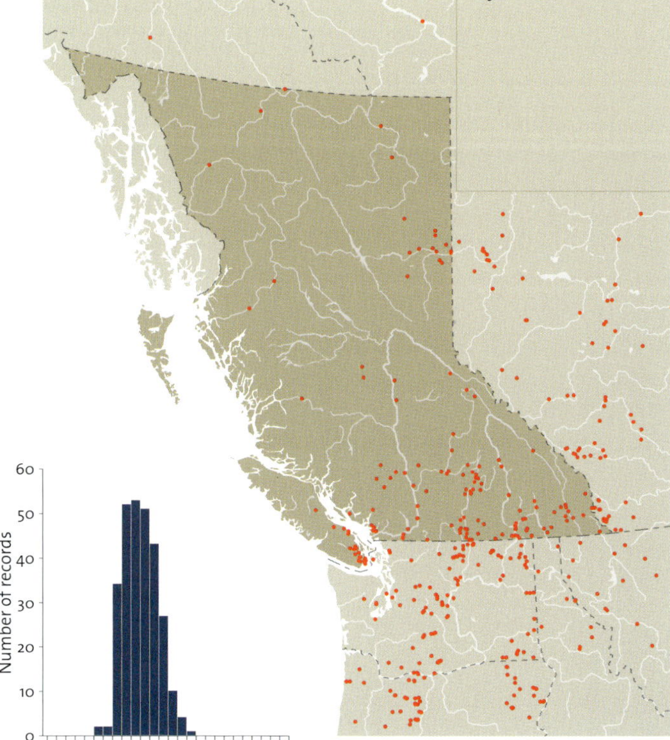

♂ V  (2.8 CM)

♂ D  (2.8 CM)

♀ D  (2.9 CM)

# Propertius Duskywing
*Erynnis propertius* (Scudder & Burgess, 1870)

**ETYMOLOGY:** The name *propertius* refers to a Latin elegiac poet, with no obvious connection to the butterfly. The common name was first used by Comstock (1927).

**ADULT:** This is the largest species of *Erynnis* in BC, with a forewing length of 1.8–2.1 cm, and it generally flies in close association with Garry oak. It is best, however, to examine the male genitalia to ensure proper identification. The ventral process of the left valve is extremely long; the middle lobe is elongated and strongly curved towards the ventral process (Fig. 56e). The sexes are dimorphic, with the female wings being lighter. In particular the spots of the post-median row of both wings are larger in the female. The male possesses the costal fold characteristic of the genus.

**IMMATURE STAGES:** Hardy (1958b) described the immature stages of the Vancouver Island populations in detail from larvae collected on the foodplant. The mature larval head is flesh-coloured with orange spots on each side and covered with fine hair. The body is sage green with pale yellow subdorsal lines and pale yellow intersegmental rings. The spiracles are white.

**BIOLOGY:** Adults are on the wing from late April to early July, with a single record from 27 July and peak flight occurring in early June. A second instar larva was collected on 29 June (Hardy 1958b). The larva matured by late August and over-

♂ D  (3.9 CM)

♀ D  (3.9 CM)

♂ V  (3.9 CM)

wintered in a nest of food-plant leaves. In spring it spun a light cocoon, pupated on 29 April, and emerged as an adult on 28 May. The species is univoltine, with an occasional possible second-brood specimen. In BC the foodplant is Garry oak (*Quercus garryana*). Besides Hardy's observations, there are two records from the Forest Insect Survey, also on Garry oak. In California, it has been reared from *Quercus agrifolia* (Burns 1964).

**SUBSPECIES:** None. The type locality of the species is California.

**RANGE AND HABITAT:** In BC *E. propertius* has historically been collected only on the southeastern tip of Vancouver Island and the adjacent Gulf Islands where Garry oak is known to occur. There is a single record from the 1930s of the species at the disjunct Garry oak population on Sumas Mountain, and numerous specimens from "Vedder Crossing," presumably the Sumas Mountain site. There are three records of the species from Lower Mainland localities, where botanists have so far not recorded Garry oak (Ross Lake, Hope, and 24 km north of Mt. Currie). Until either Garry oak can be found in these areas or an alternative foodplant can be shown for *E. propertius,* these specimens must be considered strays. All of the disjunct specimens were recorded late in the flight period, indicating that they were strays. Both Garry oak as the larval foodplant and nectar sources to sustain the long adult flight are equally important for maintaining the species. In addition, leaf litter must be left at the base of horticultural trees to protect the larvae during hibernation.

**GENERAL DISTRIBUTION:** The species is found along the entire Pacific coast of North America from southern Vancouver Island to northern Baja California, in association with oak species.

**CONSERVATION STATUS:** The Propertius Duskywing is of Special Concern in BC (S3).

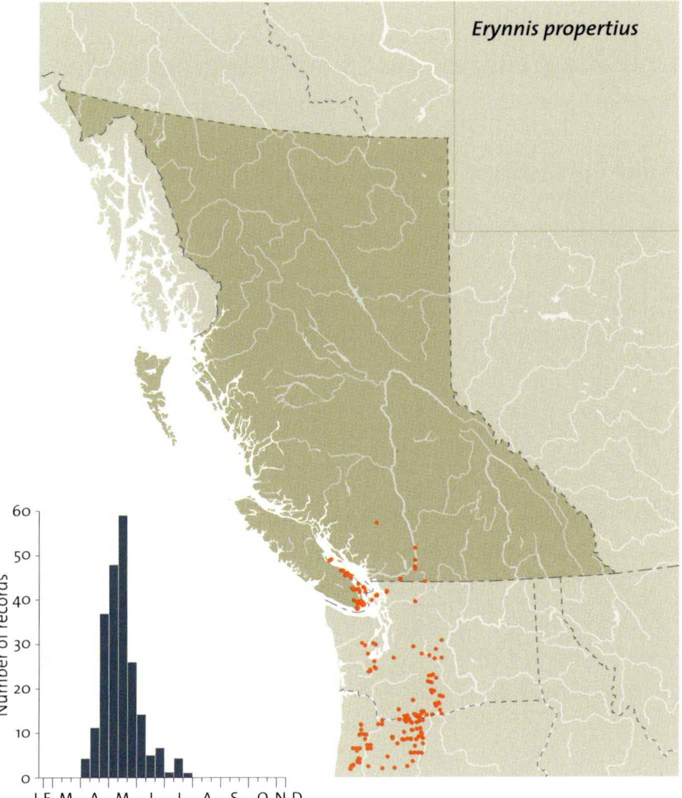

# Pacuvius Duskywing

*Erynnis pacuvius* (Lintner, 1878)

**Etymology:** The name *pacuvius* refers to a Latin tragic poet. Several other *Erynnis* species were named for classical poets. The common name was first used by Holland (1898).

**Adult:** This species and the next two are nearly impossible to separate by wing pattern characteristics. Once a series of *E. pacuvius* is separated by examination of the male genitalia, the forewing submedian dark spots appear to be in an irregular line, whereas in both *E. persius* and *E. afranius* the spots are in a more regular line (Fig. 55a). This is not a reliable characteristic, however. The shape of the left valve of the male is the only reliable characteristic for separating the three species. In *E. pacuvius* the ventral process is long and the middle lobe is curved away from the ventral process (Fig. 56d). *E. persius* and *E. afranius* have a similar ventral

♂  D  (3.4 cm)          ♂  V  (3.4 cm)

process but the median lobe is curved towards the ventral process.

**Immature stages:** Undescribed.

**Biology:** Adults fly from early May to mid-July, with peak flight occurring in late June. The species is univoltine. Elsewhere the larvae have been collected from various *Ceanothus* species (Burns 1964). In our area the Pacuvius Duskywing is associated with *Ceanothus velutinus*.

**Subspecies:** BC populations are *E. pacuvius lilius* (Dyar, 1904) (TL: Kaslo, BC). The subspecies ranges over the entire PNW and south in the California Sierras. In contrast to the California subspecies, *lilius* has dark markings that contrast highly with the ground colour. This characteristic is shared with the southern Rocky Mountain subspecies, which has an even sharper contrast of dark markings and ground colour.

**Range and habitat:** The Pacuvius Duskywing is found in the southern fourth of BC from east of the Cascades to the Rocky Mountain Trench. It is found in open pine forest areas and on dry hillsides in association with *Ceanothus* bushes. In the past this species and the next have been unrecognized by BC lepidopterists; thus previous collection records are not a good indication of its range and density prior to recent human disturbance.

**General distribution:** The species ranges from southern BC to northern Baja California and AZ in mountains where *Ceanothus* occurs.

**Conservation status:** Not of concern (S4).

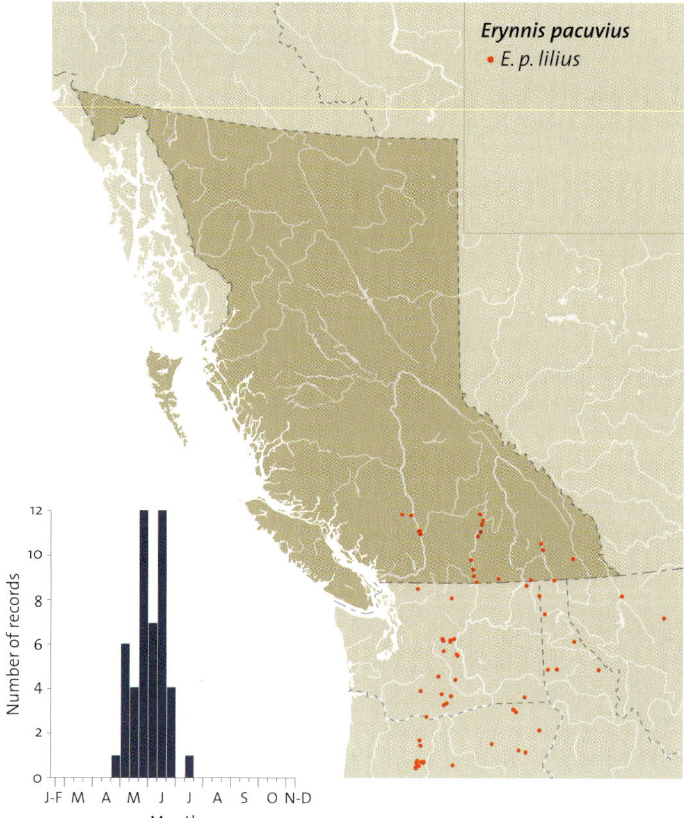



# Afranius Duskywing

*Erynnis afranius* (Lintner, 1878)

**Etymology:** Given the derivation of other species names in this genus, *afranius* likely refers to Afranius, a Latin comic, not the name of a Roman family as suggested by Bird et al. (1995). The common name was first used by Holland (1898).

**Adult:** Adults of *E. afranius* and *E. persius* can sometimes be distinguished from *E. pacuvius* as noted earlier (Fig. 55a, b).

Without examination of male genitalia, *E. afranius* and *E. persius* are impossible to differentiate in BC. Books on the USA fauna state that males of *E. persius* have long grey hairs on the dorsal surface of the forewings that are lacking on *E. afranius,* and that *E. afranius* has a wider light band on the outer side of the fringe that is visible with the naked eye.

These characters are not reliable in BC. Females of *E. afranius* and *E. persius* are impossible to distinguish except by association with males. Males are distinguished by the shape of the valve (Fig. 56). In *E. afranius* (Fig. 56b), the anterior lobe of the left valve is more pointed and the central lobe narrower than in *E. persius* (Fig. 56c). Two BC specimens with the genitalia of *E. afranius* have been identified. They have a wing pattern closer to that of *E. pacuvius* in that the forewing spots contrast strongly with the ground colour.

**IMMATURE STAGES:** Comstock and Dammers (1932) described the immature stages. The egg is white when laid, changing

♂ D  (2.9 CM)        ♂ V  (2.9 CM)

to cream or light yellow with 15 vertical ridges. The head of the mature larva is black with minute white points. The body is pale green with small white spots. The mid-dorsal line is dark and the lateral line is yellow. The pupa is vivid green, with a prominent black spiracle on the thorax.

**BIOLOGY:** The eggs are laid on the tips of new growth of the foodplant. Elsewhere the species is reported to have two or three generations per year, but in BC it is univoltine. All of the *afranius/persius* complex specimens examined from BC indicate only one brood. The larvae have been recorded elsewhere from a variety of Fabaceae, including *Lupinus* (Lindsey 1927), *Lotus* (Comstock and Dammers 1932), and *Medicago sativa* (Emmel and Emmel 1973). Our populations must be reared to thoroughly understand their relationship to *E. persius* and other *E. afranius* populations.

**SUBSPECIES:** None. The type locality of the species is Colorado.

**RANGE AND HABITAT:** So far recorded in BC only from New Aiyansh, on the edge of a pasture.

**GENERAL DISTRIBUTION:** The species occurs from AB to NM in the Rockies, and across AZ to southern CA. It has not yet been recorded elsewhere in the PNW, except ID, northeastern OR, and extreme southeastern WA.

**CONSERVATION STATUS:** The Afranius Duskywing is presumed to be Endangered in BC (S1).

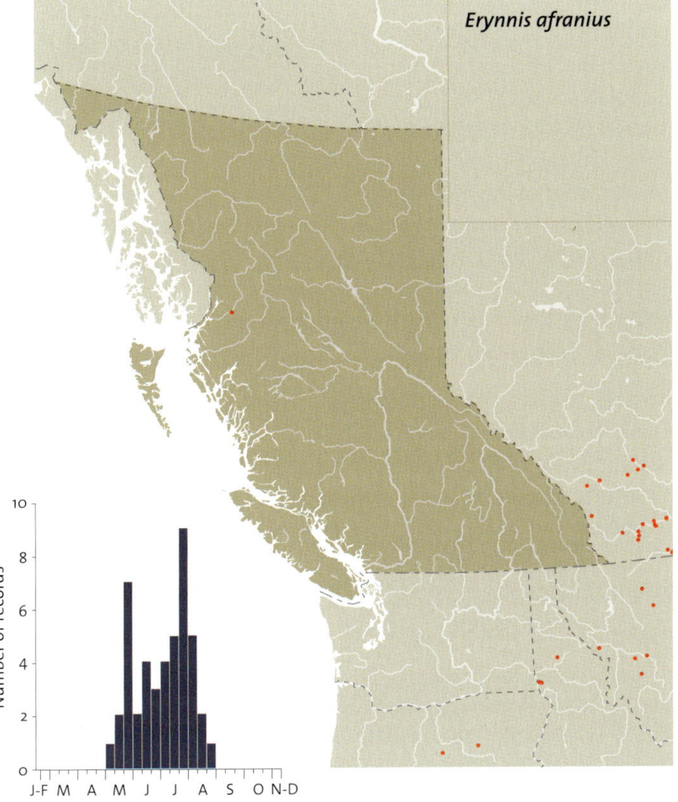



# PERSIUS DUSKYWING
*Erynnis persius* (Scudder, 1863)

**ETYMOLOGY:** The species is named after the Roman satirist Persius. The name of subspecies *borealis* is derived from the Latin *boreus* (northern). Subspecies *fredericki* is named after R.C. Frederick, who, with his wife, collected most of the type specimens. The common name was first used by Scudder (1889b).

**ADULT:** The Persius Duskywing is difficult to distinguish from the preceding species, *E. afranius*. Refer to the previous discussion for details.

**IMMATURE STAGES:** Scudder (1889b) gives the only description. The egg has 10–14 ribs, with usually 11 or 12, and is pale yellowish green. The mature larval head is variable in colour

but is usually reddish brown with pale, inconspicuous vertical streaks. The body is pale green with numerous white spots.

**BIOLOGY:** Adults fly from early May to late June, with peak flight depending on elevation and latitude. Eggs are laid singly on the upper surface of foodplant leaves, hatch within the week, and develop into mature larvae by late July. Mature larvae overwinter and pupate the following spring (Scudder 1889b). Scudder reported a partial second brood in some areas of New England. Burns (1964) reports that the species is univoltine everywhere in its range except California and Oregon. In BC there is only one brood per year. Scudder (1889b) recorded various *Salix* and *Populus* species

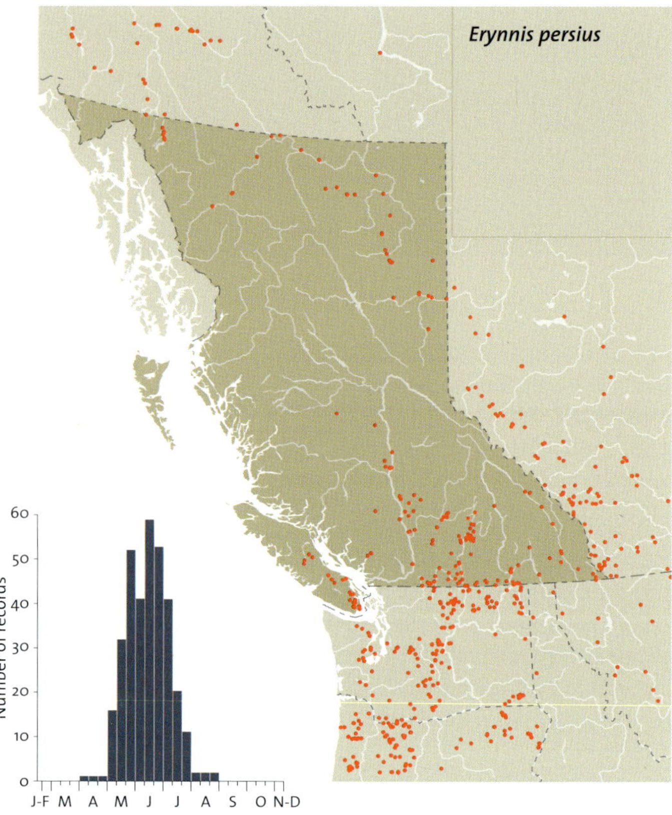



Number of records

60
50
40
30
20
10
0

J-F M A M J J A S O N-D
Month

♂ D (3.1 cm)

♀ D (3.0 cm)

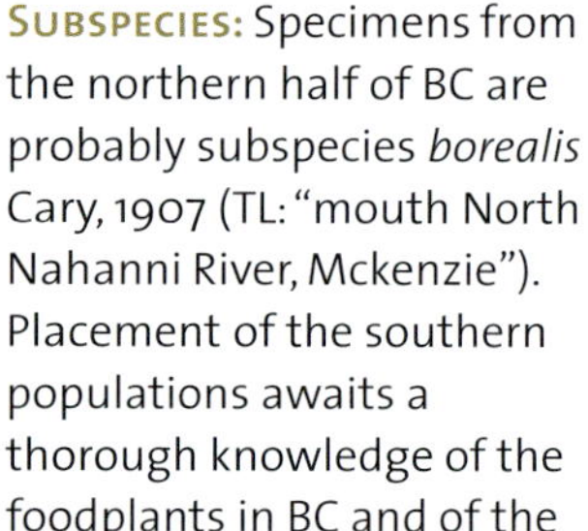

♂ V (3.1 cm)

as foodplants. In Arizona, Burns (1964) recorded several ovipositions and partial rearings of *"persius"* on *Thermopsis rhombifolia* var. *montana*. In Saskatchewan, Hooper (1973) found adults associated with *Thermopsis rhombifolia* and *Oxytropis* sp., and seldom near *Salix*. Observations of oviposition by J. Pelham (pers. comm.) indicate that *Lupinus arcticus* is the foodplant in most of Washington. If southern Rocky Mountain *persius* are monophagous on *Thermopsis* sp., as maintained by Ferris and Brown (1981), then there must be a sharp change to another foodplant in BC, as *Thermopsis* is found only in southeastern BC. *Lupinus* species appear to be the most likely foodplants in BC.

**SUBSPECIES:** Specimens from the northern half of BC are probably subspecies *borealis* Cary, 1907 (TL: "mouth North Nahanni River, Mckenzie"). Placement of the southern populations awaits a thorough knowledge of the foodplants in BC and of the relationship of southeastern BC populations to the *Thermopsis*-feeding subspecies, *fredericki* Freeman, 1943, described from SD. In southern BC specimens, the wing pattern contrasts more with the ground colour than in northern populations. We did not attempt to map the putative subspecies ranges.

**RANGE AND HABITAT:** As presently understood, the Persius Duskywing is the most widely distributed species of *Erynnis* in BC, having been recorded in every area but the Queen Charlotte Islands. In the southern part of the province, it is usually found in mid- and high-elevation meadows, but is also taken in lower meadows and along low-elevation streams in association with *E. icelus*. It is usually associated with *Lupinus* species.

**GENERAL DISTRIBUTION:** The Persius Duskywing is found across North America from AK to New England. In CAN the species is not found east of MB. South in the Sierras/ Cascades and the Rockies it is found in mountainous regions.

**CONSERVATION STATUS:** Not of concern (S5).

---

### GENUS *PYRGUS* HÜBNER, [1819]  CHECKERED SKIPPERS

The name *Pyrgus* is derived from the Greek *pyrgos,* meaning a tower on a wall, a battlement. This presumably refers to the checkered terminal cilia on the edges of the wings (Emmet 1991). The common name "checkered skippers" refers to the "checkerboard" black and white pattern of the wings. Holland (1898) is responsible for the common name of the genus.

This genus is structurally similar to *Erynnis* and *Pholisora,* with rounded wing tips, short discal cell, inconspicuous antennal tips, and porrect palpi. The genus *Pyrgus* is distinguished by the checkered black and white wing pattern, as mentioned above. In England they are

referred to as "grizzled skippers." A closely related genus, *Heliopetes,* has been recorded from just south of the BC border, in the Washington Okanogan. It has the same black and white colours but they are not arranged in as obvious a checkered pattern. The genus *Pyrgus* is Holarctic and Neotropical, with at least 19 Palearctic species, 1 Holarctic species, 3 Nearctic species, and 8 Neotropical species. Three species are found in British Columbia. Larval feeding and oviposition have been observed on *Potentilla* species and various Malvaceae. Evans (1953) provides the only comprehensive review of American species. In our area, examination of genitalia is not necessary to determine species.

# Grizzled Skipper

*Pyrgus centaureae* (Rambur, 1842)

**Etymology:** In 1839 Rambur named six species of *Pyrgus*: *P. centaureae, P. serratulae, P. carlinae, P. cirsii, P. onopordi,* and *P. cinarae.* All but the last of these species names are also names of genera of composites. The subspecies name *freija* is derived from Freya, the Norse goddess of love and fertility. The subspecies name *loki* refers to Loki, a Norse god who was noted for being a troublemaker. The common name, Grizzled, refers to the black and white wing pattern. It was first used by Scudder (1889b) in the form "Grizzled Tessellate" (mosaic), which Holland (1898) changed to the modern usage.

**Adult:** The dorsal forewing and hindwing have two rows of white spots. There is a basal spot on the dorsal hindwing that is much larger than in *P. ruralis.* Although this species is similar to *P. ruralis,* it is easily distinguished by its larger size and very different range and habitat.

**Immature stages:** Undescribed.

**Biology:** The adults fly from late June to early August, depending on elevation and latitude. There is no evidence from adult flight period data to indicate that the species

Ssp. *loki* ♂ D (2.7 cm)

Ssp. *freija* ♂ D (2.9 cm)

Ssp. *loki* ♂ V (2.7 cm)

Ssp. *freija* ♂ V (2.9 cm)

takes two years to develop, as suggested by Ferris and Brown (1981). Nielsen (1985) reported the Grizzled Skipper ovipositing on *Fragaria virginiana* in mid-May in Michigan. He found middle instar larvae in nests on *Fragaria.* They fed on the *Fragaria,* pupated by late summer, and overwintered as pupae. In Europe the species has been reared on *Rubus chamaemorus.* In BC both *Rubus* and *Fragaria* are potential foodplants.

**Subspecies:** Southern populations are referable to subspecies *loki* Evans, 1953 (TL: Long Peak trail, CO). Northern populations are subspecies *freija* (Warren, [1924]) (TL: Labrador). *P. c. loki* has slightly larger white spots on both wings.

**Range and habitat:** The Grizzled Skipper ranges throughout BC, except for Vancouver Island and the Queen Charlotte Islands, above timberline. In the north it also occurs below timberline. The lack of records from the central coast is probably an artifact of scant collecting in the area and the scarcity of the species everywhere in its range.

**General distribution:** The Grizzled Skipper is Holarctic. It is found across arctic North America from AK to Labrador, except the high arctic, and south in the Rockies to NM. There is an isolated subspecies in the Appalachians.

**Conservation status:** Not of concern, with subspecies *loki* S5 and subspecies *freija* S4S5.

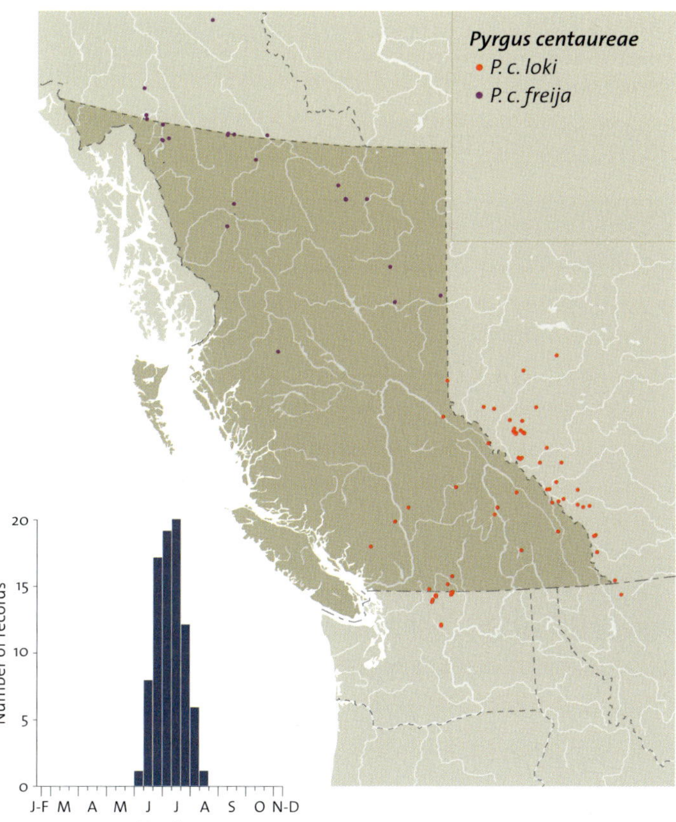

# TWO-BANDED CHECKERED SKIPPER
*Pyrgus ruralis* (Boisduval, 1852)

**ETYMOLOGY:** The name *ruralis* is Latin for "belonging to the countryside." The common name refers to the two rows of white spots on the dorsal wings, which many other *Pyrgus* species also have. The common name was first used by Holland (1898).

**ADULT:** Similar to *P. centaureae,* with differences as noted earlier.

**IMMATURE STAGES:** Undescribed except the egg (Coolidge 1909), which is light green when first laid, changing to lemon yellow.

**BIOLOGY:** Adults fly from late April to mid-June; there are records in mid-July from high-elevation populations. The species is univoltine. Nothing is known of the hibernating stage but it is likely the mature larva. Oviposition was observed on native *Fragaria* sp. near Victoria. The resulting larvae were successfully reared to adults on the *Fragaria* by J. Tatum (pers. comm.). Although Tietz (1972) lists *Potentilla*

♂ D  (2.6 CM)  ♂ V  (2.6 CM)

*douglasii, Horkelia tenuiloba,* and *Sidalcea* sp. as foodplants, they are not mentioned in his only reference, Coolidge (1909).

**SUBSPECIES:** BC populations are the nominate subspecies. The TL of the nominate subspecies was restricted to Murphy Creek, Plumas Co., CA (Emmel et al. 1998a).

**RANGE AND HABITAT:** The Two-banded Checkered Skipper is found across the southern fourth of BC. It appears in moist open parts of the forested areas generally below 1,000 m, but can be found up to 1,700 m.

**GENERAL DISTRIBUTION:** From southern BC and southern AB south in the mountains to the Sierras of CA and to UT and CO in the Rockies.

**CONSERVATION STATUS:** Not of concern (S5).

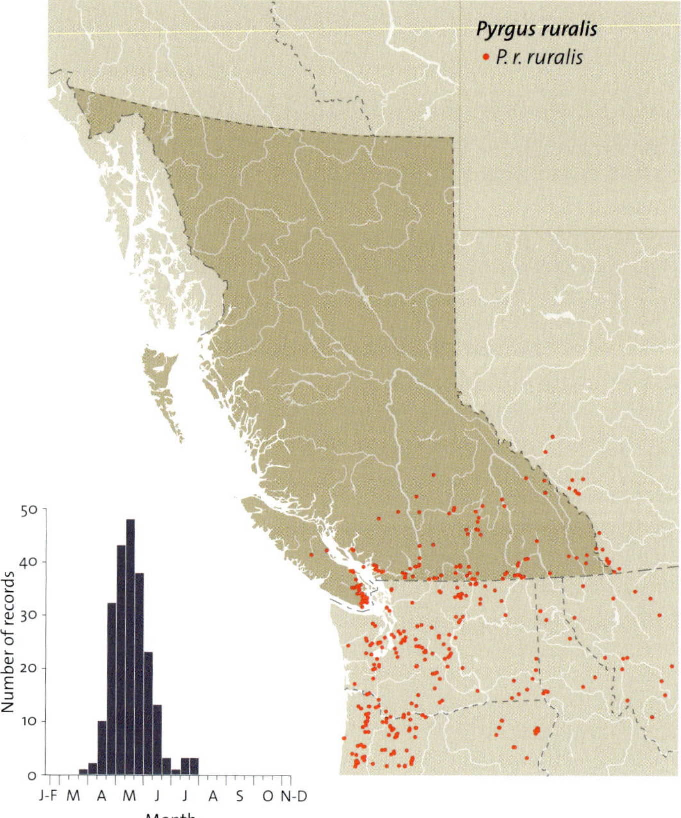

Two-banded Checkered Skipper (*Pyrgus ruralis*)

# CHECKERED SKIPPER
*Pyrgus communis* (Grote, 1872)

**ETYMOLOGY:** The name *communis* is Latin for common. The common name refers to the strongly checkered pattern of black and white on the dorsal wings, first used by Grote as "checkered Hesperia" (Scudder 1889b).

**ADULT:** The wing pattern is similar to that of other species in the genus, except that the white spots are noticeably larger, especially the median row of the dorsal and ventral hindwing. Males lack the tibial tufts present in other members of the genus, causing several authors to speculate that the species should be in a separate genus. The Checkered Skipper is the same size as the Grizzled Skipper. The Checkered Skipper may be confused with *Heliopetes ericetorum* (Boisduval, 1852), which should eventually be recorded from the southern Okanagan. *H. ericetorum* is considerably larger.

**IMMATURE STAGES:** Scudder (1889b) described the immatures. The egg is white with 24 ribs. The mature larval head is shiny black with an overlay of short, brownish hairs. The first thoracic segment is slightly lighter than the head. The rest of the body is green with short, knobbed hairs. The dorsal stripe is white; the lateral stripes are indistinct. The pupa is dark

♂ D (3.0 cm)

♀ D (3.0 cm)

♂ V (3.0 cm)

green and heavily marked with dark reddish brown areas.

**BIOLOGY:** In BC the adults are on the wing from mid-May to mid-June, and again in August. There are two broods per year. Further south this species can be trivoltine. Eggs are laid on the upper surface of leaves. It is not known for certain which stage hibernates, but it is likely the mature larva. There are foodplant records from many members of the family Malvaceae, including garden hollyhocks (Scudder 1889b).

**SUBSPECIES:** BC populations are the nominate subspecies. The TL of the species is central Alabama.

**RANGE AND HABITAT:** The Checkered Skipper is found in the Okanagan and the Kootenays in xeric areas favouring the foodplants.

**GENERAL DISTRIBUTION:** The species is found from southeastern BC east to MB in CAN, east to MA in the USA, and south to the southern USA, where it is replaced by the related species *P. albescens* Plotz. The Checkered Skipper has been recorded in the Peace River region of AB but not in the Peace River region of BC.

**CONSERVATION STATUS:** The Checkered Skipper is of Special Concern in BC (S3S4).

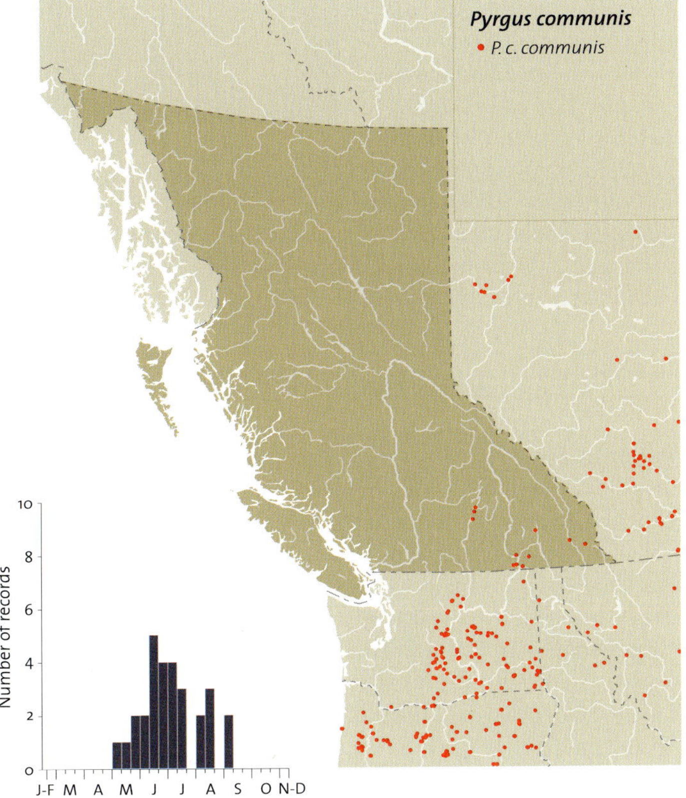

*Pyrgus communis*
• *P. c. communis*

Number of records

10
8
6
4
2
0

J-F M A M J J A S O N-D

Month

---

## GENUS *HELIOPETES* BILLBERG, 1820  WHITE SKIPPERS

The generic name *Heliopetes* is derived from the Greek *heliacus* (of the sun) and *petes* (flying); thus it means "flying in the sun." The common name "white skippers" refers to the checkered black and white pattern being predominately white (Comstock 1927).

This genus is structurally very close to *Pyrgus,* but the species are all larger than any *Pyrgus*. The wing pattern is similar, but white predominates and black only outlines the white areas. The genus is primarily Neotropical, with at least 12 species ranging from Argentina to the southwestern United States, where several species occur as strays from Mexico.

# LARGE WHITE SKIPPER
*Heliopetes ericetorum* (Boisduval, 1852)

**ETYMOLOGY:** The derivation of *ericetorum* is unknown. The two Latin meanings of *erice* are "hedgehog" and "beam with iron nails"; the word can also mean "heath." None of these appear to apply to the butterfly. The common name, first used by Comstock (1927), was meant to show that this skipper is similar to Checkered Skippers, but larger.

**ADULT:** The Large White Skipper is very similar to the Checkered Skipper in wing pattern, but has proportionately more white and is considerably larger.

**IMMATURE STAGES:** Coolidge (1923) described the immatures in detail. The egg is white and spherical. Where the horizontal and vertical reticulations cross on the surface, blunt spines project upward from the egg surface. The mature larva is pale yellow green with a pale green dorsal line and two lateral lines. The pupa is yellow brown; the cremaster

♂ D (3.5 CM)

♀ D (3.8 CM)

♂ V (3.5 CM)

and a pair of lateral lines on the thorax are black.

**BIOLOGY:** In Washington the Large White Skipper is bivoltine, with peak flights in late June and early September. It is a resident, breeding species in the state, contrary to Opler (1999), which shows the Pacific Northwest range as strays. The published literature says that the larvae feed on a variety of plants in the family Malvaceae. The larval foodplant is unknown in the Pacific Northwest.

**SUBSPECIES:** None. The type locality of the species was restricted to 2 air miles E. Magalia, Butte Co., CA (Emmel et al. 1998a).

**RANGE AND HABITAT:** In WA, the Large White Skipper occurs very close to the BC border, in the Okanogan River valley. It should eventually be found in BC. If it is found, at least one voucher specimen should be taken and forwarded to us or to the Royal BC Museum for verification of the identification. The maps of Opler (1999) and Hinchliff (1996) showing the Large White Skipper range extending to Pend Oreille Co. in northeastern WA are errors that are based on specimens labelled Ruby, WA. They were actually collected at Ruby in Okanogan Co., not Ruby in Pend Oreille Co.

**GENERAL DISTRIBUTION:** The Large White Skipper ranges from extreme southern AZ and NM and northern Baja California north to the Columbia Basin of WA and just into adjacent ID.

**CONSERVATION STATUS:** Not yet known from BC.

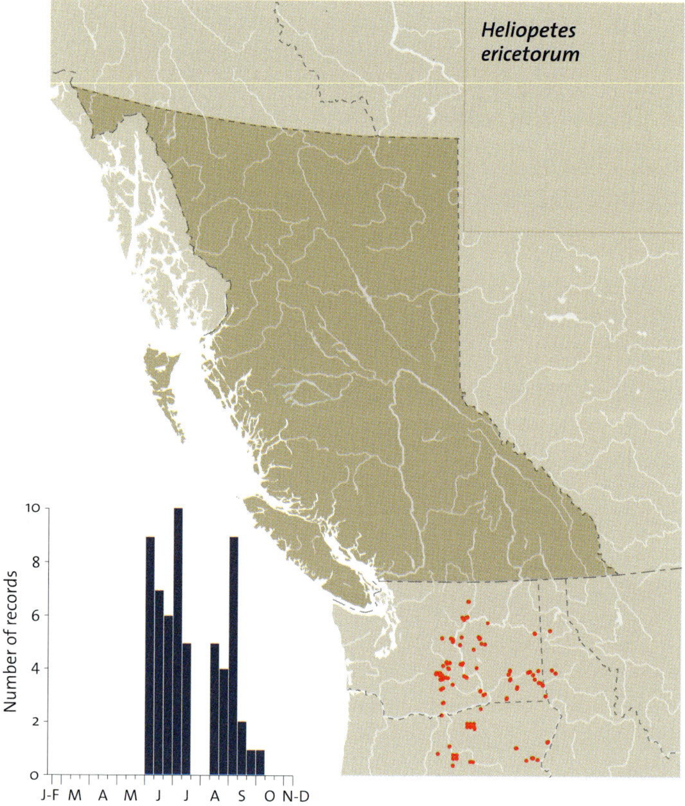

## Genus *Pholisora* Scudder, 1872  Sootywings

The name *Pholisora* may be derived from the Greek *pholis,* meaning lurking in a hole, and *ora,* meaning season, especially spring (Bird et al. 1995). This may refer to the wings being the dark colour of a hole, and the spring flight period. The common name for the genus was first used by Scudder (1889b).

This genus is structurally similar to *Erynnis, Pyrgus,* and *Heliopetes.* The only species in our area has a uniform black ground colour with several groups of small white spots arranged in rows. The genus is Nearctic and centred in the US southwest. There are either two or five species in the genus, depending on whether *Hesperopsis* Dyar is given generic rank. Regardless of whether *Hesperopsis* is recognized, the BC species falls in the genus *Pholisora. Pholisora* larvae feed on *Chenopodium* (Chenopodiaceae) and *Amaranthus* (Amaranthaceae). *Hesperopsis* larvae feed on *Atriplex* (Chenopodiaceae).

## COMMON SOOTYWING
*Pholisora catullus* (Fabricius, 1793)

**ETYMOLOGY:** The name *catullus* refers to Catullus, a Roman lyric poet, not the adversary of Pompeius (Bird et al. 1995). The latter's name was spelled with one l (Catulus). The common name was first used by Gosse (1859).

**ADULT:** The ground colour of the wings is black. There are white spots in the median area of the forewing and a submarginal row of very small white spots on both wings. The females have more median spots than the males. This species can be confused with *Amblyscirtes vialis,* but comparison of the ventral hindwings will show that *P. catullus* is uniform black with white spots, whereas *A. vialis* is mottled and overlaid with grey on the outer half.

**IMMATURE STAGES:** Edwards (1885d), Scudder (1889b), and Comstock (1927) all describe the immatures. The egg is white

♂ D  (2.6 cm)

♀ D  (2.6 cm)

♂ V  (2.6 cm)

to pale yellow brown with 15–18 ribs. The head of the mature larva is blackish brown with a thin pile of fulvous hairs mixed with black. The body is pale yellow green and is overlaid by small white tubercles with short hairs. There is a pair of faint yellow lateral stripes near the dorsal surface.

**BIOLOGY:** In BC the adults are on the wing from late May to mid-June, and again from mid-July to late August, with two broods per year. Eggs are laid on the upper surface of the leaves (Scudder 1889b). The larvae apparently hibernate in various instars. There is no record of a BC foodplant. Elsewhere the species feeds on *Chenopodium* and *Amaranthus* (Edwards 1885d; Scudder 1889b; Comstock 1927).

**SUBSPECIES:** None. The type locality of the species is presumably Georgia.

**RANGE AND HABITAT:** In BC the Common Sootywing is known only from Lillooet, the lower Thompson River, the Okanagan, and Grand Forks, always in very xeric areas.

**GENERAL DISTRIBUTION:** Found across the southern fringe of CAN from BC to PQ, and south to MEX.

**CONSERVATION STATUS:** Not of concern (S4).

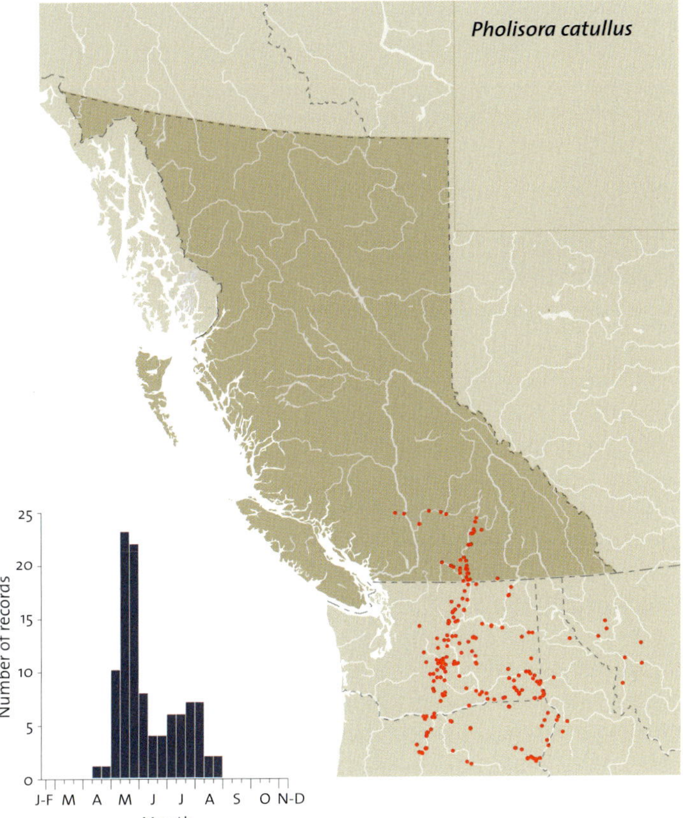

The common name "grass skippers" was first used by Scott (1986b) in reference to the skippers' use of grasses as larval foodplants, in contrast to the dicotyledons utilized by the Pyrginae. This has replaced the earlier American usage of "branded skippers" (Klots 1951), thus emphasizing foodplants rather than morphology.

In adults of this subfamily, the $M_2$ vein of the forewing is closer to the $M_1$ vein than to the $M_3$ vein. In males of most species, the dorsal forewing has a stigmal area of dark black scales. The males never have a costal fold on the dorsal forewing or hair pencils on the hind tibia, as do some males of the subfamily Pyrginae. The larvae of most species feed on grasses. Three BC genera, *Hesperia* Fabricius, 1793, *Ochlodes* Scudder, 1872, and *Carterocephalus* Lederer, 1852, are Holarctic. One genus, *Thymelicus* Hübner, [1819], was introduced from the Palearctic. The other five genera are restricted to the New World. Sixteen species of Hesperiinae occur in BC; they utilize grasses as larval foodplants.

### GENUS *CARTEROCEPHALUS* LEDERER, 1852

The name *Carterocephalus* is derived from the Greek *karteros* (strong) and *kephale* (head), referring to the wide head (Emmet 1991). There does not seem to be a common name for the genus.

This genus is easily recognized by the unique tan and black checkered pattern of the wings. There are four species in this genus, one Holarctic and the others Palearctic.

### ARCTIC SKIPPER
*Carterocephalus palaemon* (Pallas, 1771)

**ETYMOLOGY:** The species name *palaemon* is derived from Palaemon, a Greek god of the sea (Emmet 1991). The subspecies name *mandan* refers to the Mandan Indians, who once lived on the northeastern Great Plains. The subspecies

Arctic Skipper (*Carterocephalus palaemon magnus*)

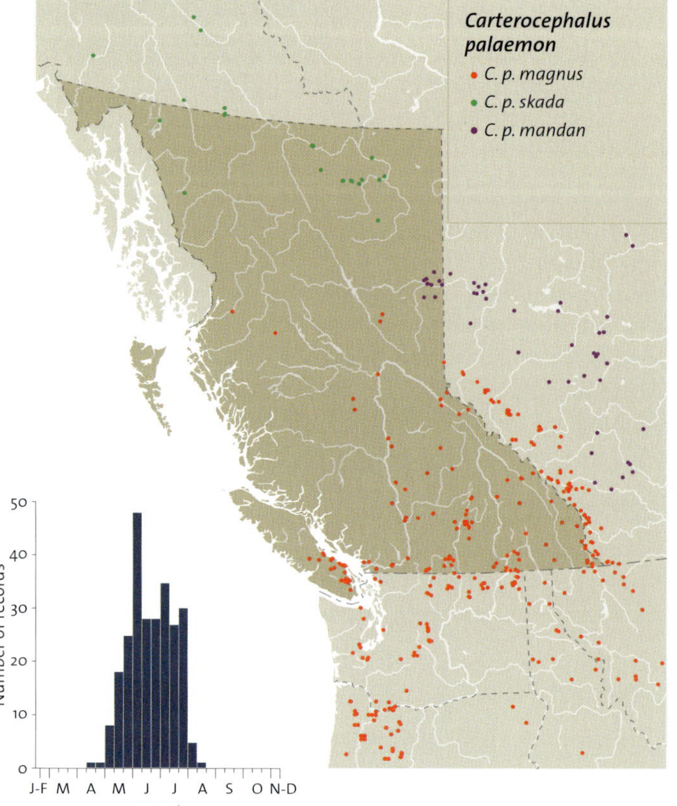


name *magnus* refers to it being the largest subspecies. Subspecies *skada* is presumably named for a local Native tribe, like other Edwards and Scudder names for skippers. It cannot be derived from Skadi, of Norse mythology, because she was a giant, and Edwards characterized *skada* as smaller than *mandan,* then the only related taxon. In England the species is called the "Checkered Skipper" because it has a checkered orange and black pattern on the wings. The common name "Arctic Skipper" used in North America is poor usage because this species is not found primarily in arctic or alpine habitat, as Scudder (1889b) thought when he coined the name. American use of "checkered skippers" for the genus *Pyrgus* precludes its use for this genus, however.

**ADULT:** The Arctic Skipper is easy to identify by the tan and black checkered wing pattern.

**IMMATURE STAGES:** Fletcher (1889b) reared the species to mature larvae. The eggs were pale green and round. The mature larvae were 1.12 inches (2.8 cm) in length. The colour was pale green, with white lateral stripes on the abdomen. The prothorax is not heavily sclerotized as in some other skipper larvae.

**BIOLOGY:** Adults are found flying from early May to early August, depending on elevation and latitude. There is no evidence of a second brood. The eggs hatch within two weeks of being laid and develop to mature larvae by fall. Presumably the mature larva is the overwintering stage. This species has not been reared in BC or elsewhere in the Pacific Northwest. Garth and Tilden (1986) report purple reed grass (*Calamagrostis purpurascens*) as the foodplant. In Europe it has been reared on *Bromus* sp. (Higgins and Riley 1970).

**SUBSPECIES:** Previous authorities have treated BC populations as part of the widely distributed North American subspecies *C. p. mandan* (W.H. Edwards, 1863) (TL: Pine Ridge, MB). The only BC populations that appear to be of this subspecies are those in the Peace River canyon. Other northern populations have a darker ground colour on the ventral hindwing and are subspecies *C. m. skada* (W.H. Edwards, 1870) (TL: Kodiak, AK). This subspecies is found in Alaska, the Yukon, and northern BC. Mattoon and Tilden (1998) incorrectly apply this name to the rest of BC and south to Colorado. The central and

Arctic Skipper (*Carterocephalus palaemon magnus*)

Ssp. *magnus* ♂ D (2.8 CM)

Ssp. *magnus* ♂ V (2.8 CM)

Ssp. *mandan* ♂ V (2.5 CM)

Ssp. *skada* ♂ V (2.8 CM)

southern BC populations are larger, with relatively larger tan spots in the checkered pattern. The first two spots of the submarginal row of spots are also elongated, a character that separates them from all other subspecies. The newly described subspecies *C. m. magnus* Mattoon & Tilden, 1998 is the correct name for the subspecies from northern California north to central BC, and the subspecies is not restricted to northern California as the authors thought (Mattoon and Tilden 1998).

**RANGE AND HABITAT:** The Arctic Skipper is known from all parts of BC that have been surveyed, but it is never an abundant species. Usually it is found in moist, open meadows along streams in the south, often in association with *Clossiana epithore* or *C. selene*.

**GENERAL DISTRIBUTION:** The Arctic Skipper is Holarctic and is found across boreal North America from central AK to NF and south to northern CA, northwestern WY, and New England. Despite the common name, it is primarily a boreal species, at least in BC.

**CONSERVATION STATUS:** Subspecies *mandan* is of Special Concern in BC (S3), whereas subspecies *skada* and *magnus* are not of concern (both S5).

---

### GENUS *OARISMA* SCUDDER, 1872  SKIPPERLINGS

The name *Oarisma* is the Greek word meaning "familiar discourse" or "loving conversation." The common name "skipperlings," first used by Holland (1931), refers to the small size of these skippers.

This Neotropical genus contains six species, of which three occur from BC or Alberta south to northern Mexico along the Rockies and the prairies east of the Rockies. One species occurs in BC.

# Garita Skipperling

*Oarisma garita* (Reakirt, 1866)

**Etymology:** The name *garita* may be derived from La Garita Mountains of southwestern Colorado, since the type locality is "Rocky Mtns., Colorado Terr." The common name was first used in the simple form "Garita" by Holland (1898) and in the current form by Holland (1931).

**Adult:** The upperside of both wings is a uniform dark greenish brown. In addition the males have a black patch in the centre of the forewings and often a tan flush to the costal area of the upper forewing. Underneath, the species is very similar to the introduced European Skipper, *Thymelicus lineola* (Ochsenheimer, 1808).

**Immature stages:** Gibson (1910) reared the species from a single egg received from Regina, Saskatchewan. The egg was creamy white and round. The fourth instar larva was 8 mm long and pale green with seven white stripes on each side of the thorax and abdomen. The seventh, subspiracular stripe

♂ D (2.2 cm)

♀ D (2.4 cm)

♂ V (2.2 cm)

was the most noticeable and was pure white. Both the body and head had short black bristles.

**Biology:** The adults fly from mid-June to mid-July in one generation. The single egg that Gibson studied hatched on 17 July, and by mid-September was a mature fourth instar larva. The fourth instar larva is the presumed overwintering stage. Gibson reared the larva on the grass *Poa pratensis*.

**Subspecies:** None. The type locality is "Rocky Mtns., Colorado Terr."

**Range and habitat:** The Garita Skipperling is found in extreme southern BC from Okanagan Falls east to the AB border and in the Peace River area as a disjunct distribution. The habitat is undisturbed grassy meadows in the Peace (Kondla et al. 1994). In the south the species is found in undisturbed moist meadows, especially at lake outlets, often in association with *C. selene*. The species has been restricted to moist meadows for years, but expanded greatly in the late 1990s. The introduced European Skipper occupies a similar habitat and may affect future numbers of the Garita Skipperling.

**General distribution:** The Garita Skipperling occurs in disjunct populations in the Peace River area of BC and AB, then from southern BC to MB south to AZ and NM and extending into adjacent MEX.

**Conservation status:** Not of concern (S4).

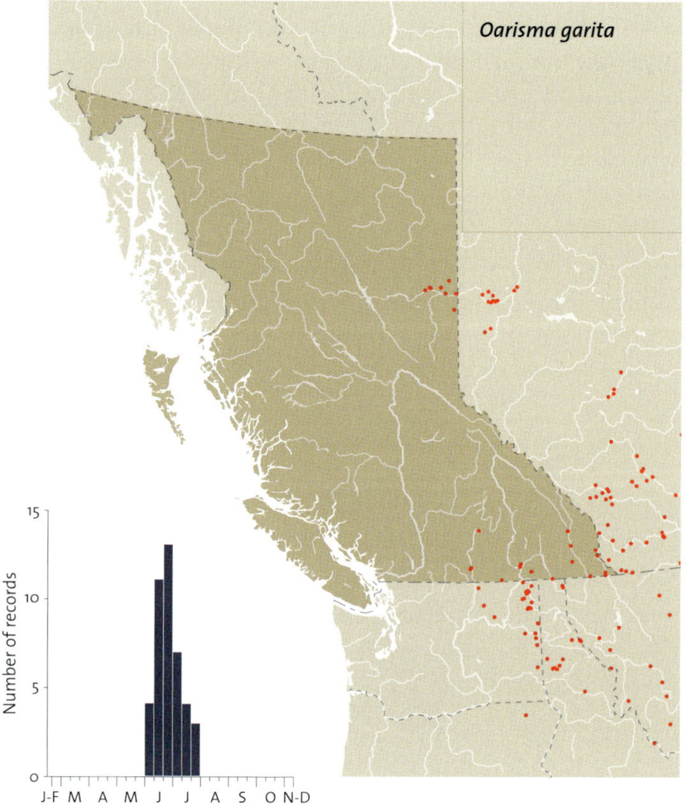

## GENUS *THYMELICUS* HÜBNER, [1819]

The name *Thymelicus* is derived from the Greek *thumelikos* (a member of the chorus in Greek drama) (Emmet 1991).

This genus was recently introduced from Europe. There are nine Palearctic species. Superficially the genus looks like *Oarisma*, except that the upperside of the wings is predominantly tan.

## EUROPEAN SKIPPER
*Thymelicus lineola* (Ochsenheimer, 1808)

**ETYMOLOGY:** The name *lineola* is Latin for small line, which refers to the thin, black stigma or sex-brand of the male (Emmet 1991). The name "European Skipper" originated with Saunders (1916), when he first recorded the species for North America. The historic British name "Essex Skipper" (Higgins and Riley 1970) would have been an unmeaningful name in North America.

**ADULT:** The European Skipper is similar to the Garita Skipperling except that the ground colour of the upperside of the wings is dark tan, not greenish brown. No other skipper except the Garita Skipperling has such a plain appearance. From the underside, the two species cannot be easily distinguished in the field.

**IMMATURE STAGES:** The egg is white and hemispherical. The mature larva has a dark dorsal stripe that distinguishes it from the larva of the Garita Skipperling.

**BIOLOGY:** Adults are seen from late June until mid-July in the south and from mid-July to mid-August in the Skeena River

♂  D  (2.4 CM)

♀  D  (2.5 CM)

♂  V  (2.4 CM)

drainage. Males are patrollers. Eggs are laid in a row at the base of grass leaves (Burns 1966). The first instar larvae overwinter inside the egg and hatch the following spring. The larvae of the European Skipper are a minor pest of timothy (*Phleum pratense*), the preferred larval foodplant.

**SUBSPECIES:** None. This species was originally described from Germany.

**RANGE AND HABITAT:** This introduced species was first recorded from BC at Terrace in 1960 (Burns 1966). Up to 1987, it was not known to have spread from Terrace. In 1987 it was recorded at Cedarvale by J. Shepard. In 1980 a separate introduction was noted at Sicamous by D. Threatful (Shepard 1983); the species was already well established. From that introduction it has spread east to Revelstoke and south to Nelson. In the mid-1990s the European Skipper appeared on Vancouver Island but has not yet dispersed from Victoria. It was recorded at Burnaby in 1991 but may not yet be established in the Fraser Valley. It is associated with hay fields and roadside ditches, never with undisturbed areas.

**GENERAL DISTRIBUTION:** This species was first noted at London, ON, in 1910 (Saunders 1916). Until the separate 1960 introduction to Terrace, BC, it had only spread east to NB and south to MD. Since the second BC introduction at Sicamous in 1980, the European Skipper has spread to much of temperate North America.

**CONSERVATION STATUS:** Not of concern; an introduced exotic species (SE).

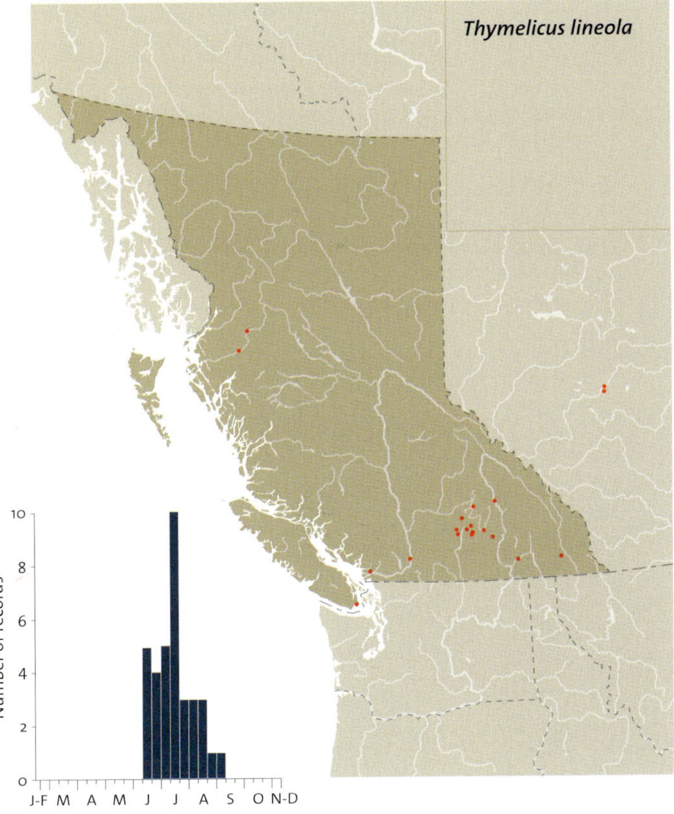

## GENUS *HESPERIA* FABRICIUS, 1793

*Hesperia* refers to the Hesperides, the nymphs who guarded the apples of Hera (Emmet 1991). The genus originally included all the skippers and lycaenids but was later restricted to the skippers and now to just this small number of species.

The genus *Hesperia* contains 1 Palearctic species, 1 Holarctic species, and 17 Nearctic species. The definitive revision is by MacNeill (1964). The species can often be determined only by dissection of the male genitalia. Three species are known for BC (Figs. 57 to 60). Fig. 57a shows the left valve of an additional species, *Hesperia uncas* W.H. Edwards, 1863, which may eventually be found in BC.

The genus *Hesperia* and the genera *Polites* and *Ochlodes* are very similar in general appearance. *Hesperia* males can be distinguished by the black stigma on the dorsal forewing: in *Hesperia* it is one long black line, whereas in *Polites* and *Ochlodes* it is two separate black areas that are narrowly separated from each other. In the field these three closely related genera can often be separated by their flight period and the habitat where they are found. *Polites* flies generally in June and occupies moist grassy areas, except for *P. sabuleti,* which is bivoltine in May and late July/early August and is found only in mesic meadows and lawns in the Okanagan. *Hesperia* is found in dry mesic to xeric areas, the males establishing territories on the tops of ridges. The only species that occurs in June is found in the Okanagan, usually on xeric ridgetops. The other two species fly in spring or after early July. *Hesperia* larvae appear to be associated with bunchgrasses. *Ochlodes* flies from late July to mid-September, and the one species is very common. It is not associated with bunchgrasses.

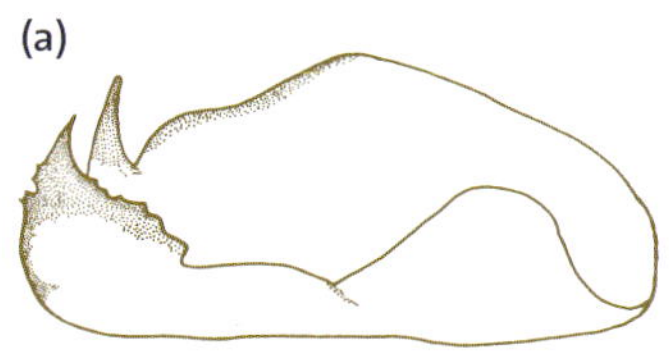  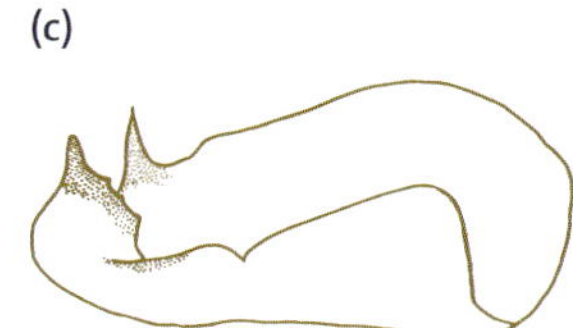 

**57** Left valve of *Hesperia* species: (a) *H. uncas*, (b) *H. nevada*, (c) *H. juba*, (d) *H. comma*.

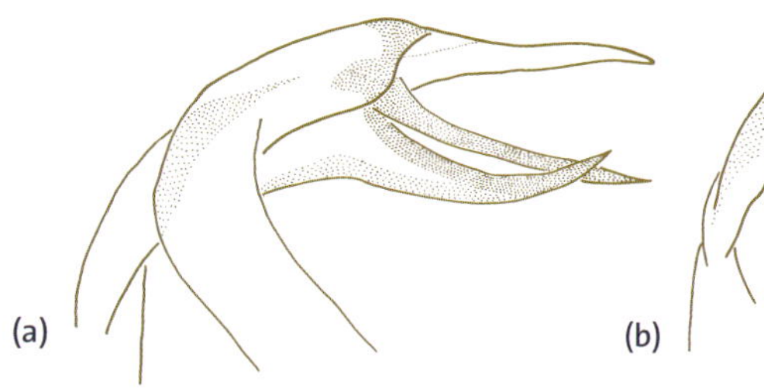 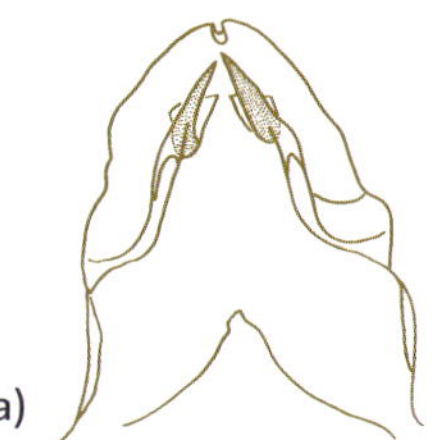 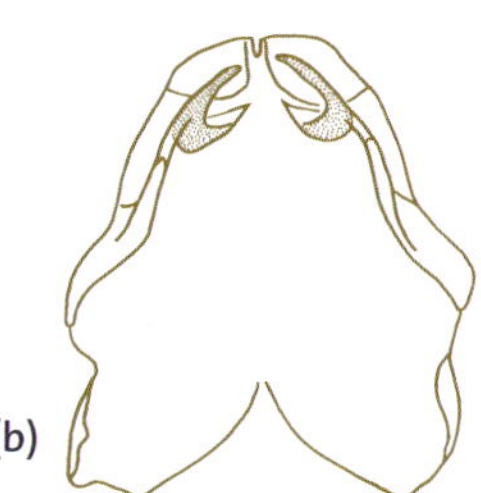

**58** Lateral view of the uncus and tegumen of *Hesperia* species: (a) *H. nevada*, (b) *H. comma*.

**59** Ventral view of the uncus of *Hesperia* species: (a) *H. comma*, (b) *H. juba*.

   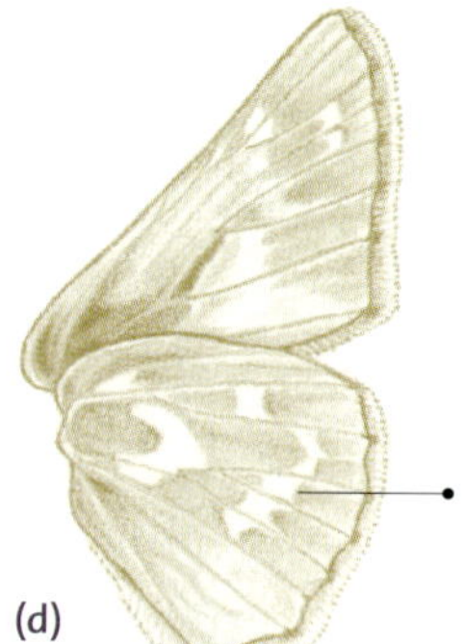

**60** Left wing underside of *Hesperia* species: (a) *H. nevada*, (b) *H. juba*, (c) *H. comma* (northern), (d) *H. comma* (southern interior).

# JUBA SKIPPER
*Hesperia juba* (Scudder, 1874)

Juba Skipper (*Hesperia juba*)

♂ D  (3.1 cm)    ♀ D  (3.3 cm)

♂ V  (3.1 cm)

**ETYMOLOGY:** The supposed type locality of the Juba Skipper was California and the source of most California material at the time was the gold fields near the Yuba River. Also, Boisduval (1852) referred to the locality Juba in California. Scudder (1874) described the species *juba* because he did not accept Boisduval's opinion that the specimens in front of Boisduval were *comma*. The spelling *juba* is the original Spanish spelling of Yuba. Holland (1931) first employed the common name.

**ADULT:** The Juba Skipper is the largest *Hesperia* species in BC, but Common Branded Skippers from the Southern Interior, with which the Juba Skipper is most likely to be confused, are almost as large. The white spots in the postmedian line on the ventral hindwings are larger in the Juba Skipper, and the spots are not in a regular line (Fig. 60b). The adult males can be reliably separated from the other species in the genus only by examination of the genitalia (Figs. 57c, 59b). Females can be reliably identified only by association with males. In the fall this species can be confused with *H. comma;* in the spring, with *H. nevada*. Habitat preference also separates the Juba Skipper from the other *Hesperia*.

**IMMATURE STAGES:** The egg is white (Lindsey 1923). Mature larvae are dark brown and have four transverse ridges at the back of each body segment (Scott 1992).

**BIOLOGY:** The adults are found in early spring and again in the fall. Scott (1992) believed that the fall flight represents a second brood. Berkhousen and Shapiro (1994), however, showed that fall-blooming plant pollen was present on spring adults, proving conclusively that fall-flying adults hibernate and that this species has only one generation each year. Scott (1992) demonstrated that *Poa* spp. are the natural foodplants.

**SUBSPECIES:** None. The TL of the species is Utah, collected by Mead on 2 October (Scudder 1874a), not the California and Nevada quoted by previous authors.

**RANGE AND HABITAT:** The Juba Skipper is found in the Southern Interior from Clinton southeast to Castlegar and south to the border. It is found only in the most xeric lowland areas and is not known to congregate on ridgetops like the other two species in the genus.

**GENERAL DISTRIBUTION:** The Juba Skipper occurs from southern BC south to Baja California and NM.

**CONSERVATION STATUS:** Not of concern (S4).

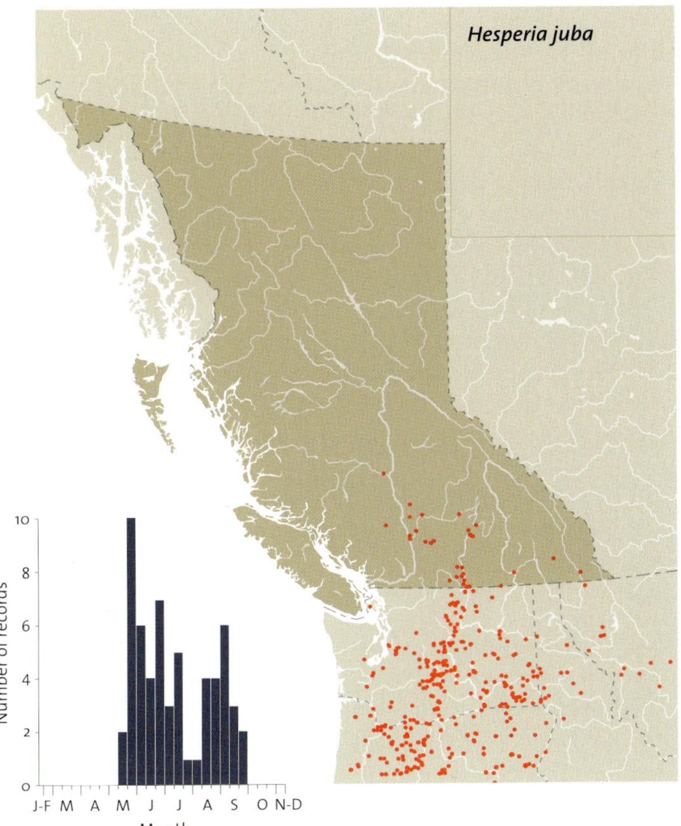

# COMMON BRANDED SKIPPER
*Hesperia comma* (Linnaeus, 1758)

**ETYMOLOGY:** The name *comma* refers to the sex-brand stigma on the dorsal forewing (Emmet 1991). The subspecies names *oregonia, assiniboia,* and *manitoba* are each derived from the area where they were originally collected. The meaning of the subspecies name *harpalus* is not known. It cannot refer to the genitalia, as Edwards does not include them in the original description and was never known to look at genitalic structures, unlike his more scientific contemporary Samuel Scudder. The name "Common Branded Skipper" refers to the same black stigma on the forewing of the male as well as its wide distribution. The common name originated with Klots (1951) as the common name of the subfamily. Pyle (1981) adapted the name to use it for this specific species.

**ADULT:** In the Common Branded Skipper, the postmedian line of white spots on the ventral hindwing is more regular than in the other *Hesperia* species, and the spots are smaller in proportion to total wing size (Fig. 60c, d). The adult males can be reliably separated from the other species in the genus only by examination of the genitalia (Figs. 57d, 58b, 59a). In the fall this species can be confused with *H. juba.* At low elevations the two come in close contact, but at higher elevations in the south Okanagan and in the rest of BC, only *H. comma* occurs. The Nevada Skipper (*H. nevada*) occurs as adults earlier than the Common Branded Skipper.

**IMMATURE STAGES:** The egg is white and similar to that of the Juba Skipper (Scott 1992). The mature larva is brownish

Ssp. *oregonia* ♂ D (3.0 cm)

Ssp. *oregonia* ♀ D (3.7 cm)

Ssp. *oregonia* ♂ V (3.0 cm)

Ssp. *harpalus* ♂ V (3.0 cm)

Ssp. *manitoba* ♂ V (2.8 cm)

Ssp. *assiniboia* ♂ V (2.7 cm)

purple, with black spiracles and an obscure dark dorsal line (Hardy 1954).

**BIOLOGY:** In southern BC, the Harpalus Skipper flies from early July to early September. The southern, mountain populations of the Manitoba Skipper fly from mid-July to mid-August; in northern BC, this subspecies flies from mid-June to mid-August. The Assiniboian Skipper flies from late July to late August. The Oregon Skipper flies in August but on occasional early warm summers, it can be seen by mid-July. Hardy (1954) reared a single egg of the Oregon Skipper to adult. It hibernated as an egg laid on "lawn" grass on 10 September, and the larva emerged the following 24 April. The larva fed on the grass genera *Lolium* and *Bromus.* Throughout development, the larva stayed near the silken cell it had made. If the foodplant became scarce and it had to move, it built another cell. The mature larva pupated on 24 August and the adult emerged on 19 September. J. Pelham (pers. comm.) has reared the Manitoba Skipper at Slate Peak, WA, on *Festuca* sp. Scott (1992) reported that *Carex* sp. was the preferred native foodplant in Colorado, and confirmed that the Assiniboian Skipper also overwinters as an egg.

**SUBSPECIES:** The Vancouver Island populations are the distinctive subspecies *H. c. oregonia* (W.H. Edwards, 1883) (TL: Trinity Co., CA), which is found west of the Cascade Mountains from northern California to Vancouver Island. The Oregon Skipper is characterized by a richer brown

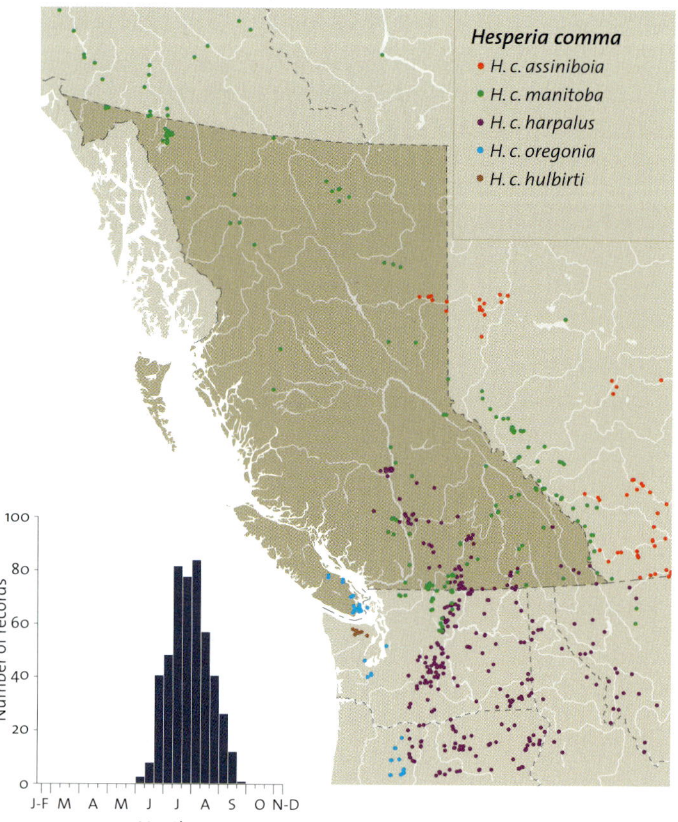

ground colour on the ventral hindwing. Populations on southern Vancouver Island are now rare because of land alienation. Peace River populations are a disjunct distribution of the Canadian Prairie subspecies *H. c. assiniboia* (Lyman, 1892) (TL: Regina, SK). The Assiniboian Skipper is distinguished by the very lightly coloured ventral hindwing and the almost complete lack of a contrasting medial area. The boreal populations are subspecies *H. c. manitoba* (Scudder, 1874) (TL: Lac la Hache, BC). The populations from the mountains in the southern third of BC are also assigned to *H. c. manitoba*. The underside of the Manitoba Skipper hindwing has a dark greenish ground colour. The low-elevation populations from the Southern Interior and the southern Kootenays are the Great Basin subspecies *H. c. harpalus* (W.H. Edwards, 1881) (TL: Carson City, NV). The ventral hindwing is lightly coloured as in *assiniboia,* but the median area is contrasting.

**RANGE AND HABITAT:** The Common Branded Skipper is found throughout BC east of the Coast Ranges, and also on Vancouver Island. The habitat requirements are as varied as the range of the species, but the species always occurs in open, grassy areas where the larval foodplant occurs. South from Williams Lake there are low-elevation populations below 1,000 m and others above 1,700 m. From 1,000 to 1,700 m elevation there is usually no suitable habitat, except for a few south-facing mountain slopes that are bare of

**Common Branded Skipper (*Hesperia comma oregonia*)**

trees. Males of the Assiniboian Skipper congregate at the tops of the south-facing banks of the Peace River.

**GENERAL DISTRIBUTION:** The Common Branded Skipper occurs across boreal North America from central AK to NF and south to southern CA, NM, MN, and ME. In the Palearctic it is found in the entire boreal area from Scandinavia to the Chersgoga Mountains in eastern Siberia, but not Kamchatka or extreme eastern Siberia.

**CONSERVATION STATUS:** Subspecies *assiniboia* and *oregonia* are of Special Concern in BC (S3). Subspecies *manitoba* (S5) and subspecies *harpalus* (S4) are not of concern.

## NEVADA SKIPPER
*Hesperia nevada* (Scudder, 1874)

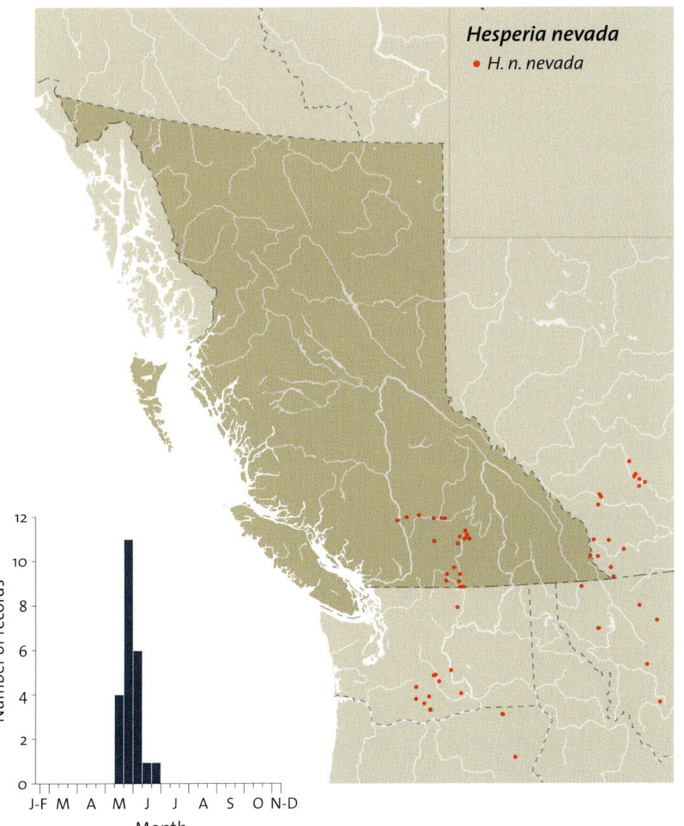

♂ D (3.0 CM)          ♀ D (3.2 CM)

♂ V (3.0 CM)

**ETYMOLOGY:** The name *nevada* means "snowy," and refers to the mountains of the type locality around South Park, Colorado. The common name was first used by Comstock (1927).

**ADULT:** The Nevada Skipper is hard to distinguish from other *Hesperia* species. The most reliable character is the postmedian band in the ventral hindwing, which appears to be a connected series of irregular rectangles (Fig. 60a). There is a slight overlap with the spring flight of the Juba Skipper, but no overlap with the flight of the Common Branded

Skipper. Again, genitalia dissection is the only reliable way to distinguish species of this genus (Figs. 57b, 58a).

**IMMATURE STAGES:** Eggs are cream-coloured when first laid, changing to pale green within 24 hours (MacNeill 1964). By contrast, Scott (1992) states that the egg is white. The mature larva is similar in colour pattern to that of *H. comma.*

**BIOLOGY:** In the Southern Interior, adults appear from mid-May to mid-June. MacNeill (1964) found that the eggs hatched shortly after being laid, and assumed that the overwintering stage was a partially mature larva. J. Pelham (pers. comm.) has found larval shelters and empty pupal cases at the base of *Stipa* sp. grasses in central Washington. Scott (1992) found that Colorado populations prefer *Festuca* sp.

**SUBSPECIES:** None. The type locality of the species is Colorado.

**RANGE AND HABITAT:** The Nevada Skipper is found in the Similkameen and Okanagan valleys of the Southern Interior. It is associated with very xeric ridgetops where the larval foodplant is found.

**GENERAL DISTRIBUTION:** From southeastern BC to NB, and south to CA and NM.

**CONSERVATION STATUS:** Not of concern (S4).

## GENUS *POLITES* SCUDDER, 1872

The name *Polites* is from Polites, a son of Priam of Troy and Hecuba. Polites was a swift runner, and perhaps Scudder meant that members of the genus were swift fliers.

The genus *Polites* is Nearctic, occurring from northern Mexico to southern Canada. The 12 species (MacNeill 1993) are well represented in British Columbia, where six species occur. A seventh species, *Polites mardon* (W.H. Edwards, 1881), occurs just south, in the Puget Trough of Washington, and is one of the few possible species that did not colonize Vancouver Island.

## PECK'S SKIPPER
*Polites peckius* (W. Kirby, 1837)

**ETYMOLOGY:** Peck's Skipper was named by Kirby in honour of the late Professor William Dandridge Peck of Cambridge, Massachusetts (Kirby 1837), an early North American

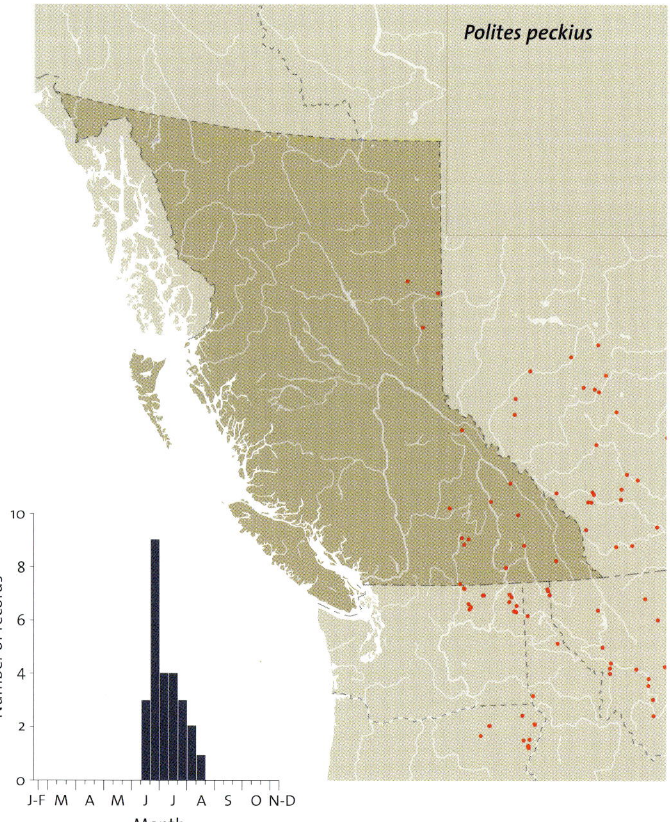

♂ D (2.5 CM)

♀ D (2.7 CM)

♂ V (2.5 CM)

economic entomologist whose first publication on economic entomology was "The description and history of the canker-worm" (Peck 1796). Kirby (1837) used the common name Peck's Hesperia. Harris (Scudder 1889b) modified the name to Peck's Skipper, and Holland (1898) reintroduced it to the popular literature.

**ADULT:** Peck's Skipper is easily distinguished by the large yellow spots on the ventral hindwing.

**IMMATURE STAGES:** The egg is pale green and round (Saunders 1869b). Scudder (1889b) described the body of the third instar larva as pale brown with minute black spots and a narrow, black dorsal line. Dethier (1940) described the remaining immature stages. The head of the mature larva is distinctive within the genus; the body is dark maroon with light brown mottling and is covered with fine, dark hairs. The mature larva pupated inside a loose cocoon constructed from a bent grass blade. The pupa is dull reddish purple.

**BIOLOGY:** In BC, Peck's Skipper flies from late June to early August and has only one brood per year. Elsewhere at the southern limits of its distribution, it has two or three broods per year. The mature larva is the presumed overwintering stage. Shapiro (1974b) records the grass *Leersia oryzoides* as the larval foodplant in New York. This grass is found throughout the range of Peck's Skipper, including BC, in habitat where the skipper is found.

**SUBSPECIES:** None. The type locality is not defined.

**RANGE AND HABITAT:** Peck's Skipper is found only in the southeastern corner of BC, from the eastern Okanagan to the AB border, and in northeastern BC. In southeastern BC it is found in moist meadows and the edges of marshes, often with *Clossiana selene* (Silver-bordered Fritillary).

**GENERAL DISTRIBUTION:** Peck's Skipper ranges from southeastern BC to central ID in the PNW. It ranges from northeastern BC to Labrador and NF, and from there south to central CO, AR, and GA, with disjunct populations in AZ and TX.

**CONSERVATION STATUS:** Not of concern (S4).

## SANDHILL SKIPPER
*Polites sabuleti* (Boisduval, 1852)

**ETYMOLOGY:** The etymology of the name *sabuleti* is unknown. Holland (1898) first coined the common name for California populations living in sand dune habitat.

**ADULT:** The Sandhill Skipper is a very distinctive skipper that is identified by the transverse white lines that run perpendicular to the median spots on the ventral hindwing. For BC this is a unique character. If the Uncus Skipper (*Hesperia uncas* W.H. Edwards, 1863) is ever found in the province, it could be confused with the Sandhill Skipper. The two are not likely to occur in the same area, however.

♂ D  (2.5 CM)

♀ D  (2.8 CM)

♂ V  (2.5 CM)

**IMMATURE STAGES:** Dethier (1944) described the immatures in detail. The egg is light green to blue green when first laid. The mature larva has a dark head covered with fine, colourless hairs. The pattern of the head is distinctive for the species. The body of the larva is dull green and heavily mottled with chocolate brown. The mid-dorsal line is dark chocolate brown. The pupa is initially pale rusty brown with a dark brown mid-dorsal line; it darkens eventually.

**BIOLOGY:** The few specimens recorded in BC indicate that the Sandhill Skipper is bivoltine even at the northern limit of its distribution. Comstock (1929) reared this skipper on Bermuda grass (*Cynodon dactylon*), which is often used in lawn grass mixes. Scott (1986a) found eggs of the Sandhill Skipper on *Eragrostis trichodes*. This genus of grasses includes several introduced species as well as native ones, but it has not yet been noted in BC (Taylor and MacBride 1977).

**SUBSPECIES:** BC populations are the nominate subspecies, *P. s. sabuleti*; TL: San Francisco, CA (Emmet et al. 1998a).

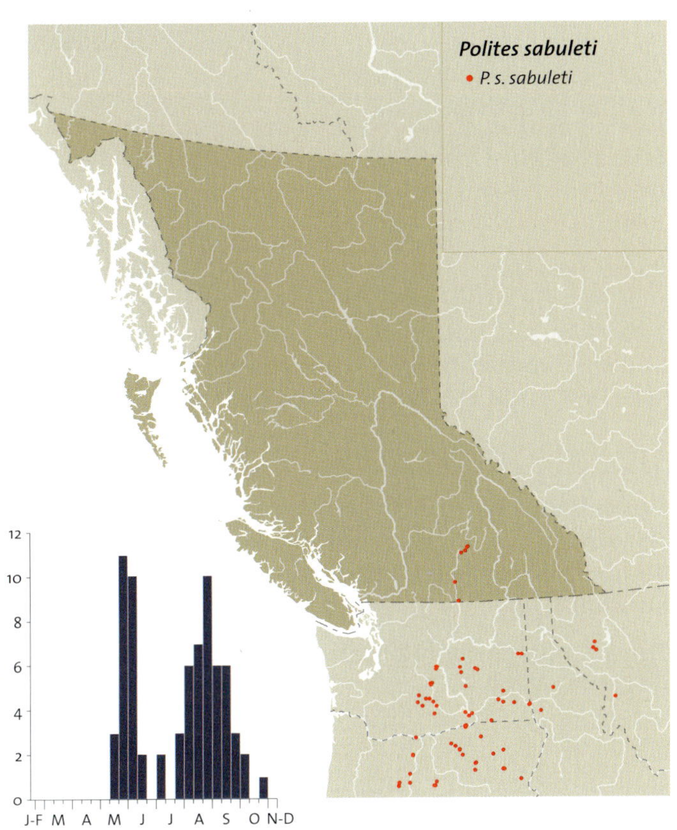

**RANGE AND HABITAT:** The Sandhill Skipper is known in BC by four populations ranging from Vernon to Osoyoos. It has been found both in natural mesic meadows at mid elevations and in lawns in suburban areas. In WA, it was not mentioned by Leighton (1946). By the mid-1950s it was common in the parts of eastern WA serviced by various irrigation projects (JHS). The WA populations are assumed to have been established after the Second World War. The Sandhill Skipper was not recorded in BC until the 1970s, and is assumed to be recently established. It may be much more widespread in suburban lawns of the South Okanagan than has been reported.

**GENERAL DISTRIBUTION:** The Sandhill Skipper is found from the Okanagan Valley in BC south through eastern WA to CA, AZ, and NM.

**CONSERVATION STATUS:** The Sandhill Skipper is of Special Concern in BC (S3).

## DRACO SKIPPER
*Polites draco* (W.H. Edwards, 1871)

**ETYMOLOGY:** Draco was an Athenian legislator who prescribed death for very minor crimes, hence the term "draconian" or penalty of death. The original description gives no clue to the derivation, which is, however, consistent with that of *P. themistocles* (see below). The common name was first used by Holland (1931).

**ADULT:** The median row of light spots on the ventral hindwing is arranged in an irregular line, as in the Sandhill Skipper. The Draco Skipper, however, lacks the transverse lines overlying this median row that are present on the wings of the Sandhill Skipper. Also, the ranges of the Draco Skipper and the Sandhill Skipper do not come close to overlapping in BC.

**IMMATURE STAGES:** Undescribed.

**BIOLOGY:** The Draco Skipper flies from late June to early July in northeastern BC. The foodplant is unknown.

**SUBSPECIES:** None. The type locality of the species has been restricted to Twin Lakes, Lake Co., CO (Brown and Miller 1980).

**RANGE AND HABITAT:** In BC, the Draco Skipper is known only from north of Atlin and the Stikine River. There are several populations just north of BC, near Whitehorse, YT. The Draco Skipper is found in the dry grassland habitat in the rain shadow of the St. Elias Mountains and Coast Ranges.

**GENERAL DISTRIBUTION:** The Draco Skipper is found in a series of disjunct populations from north of Whitehorse to the Stikine River of BC and then from Hinton, AB, south to AZ and NM in the Rocky Mountains. In BC it is found in dry pine areas with extensive grass, flying with *Parnassius smintheus yukonensis*.

**CONSERVATION STATUS:** The Draco Skipper is of Special Concern in BC (S3).

♂ D  (2.6 CM)

♀ D  (2.7 CM)

♂ V  (2.6 CM)

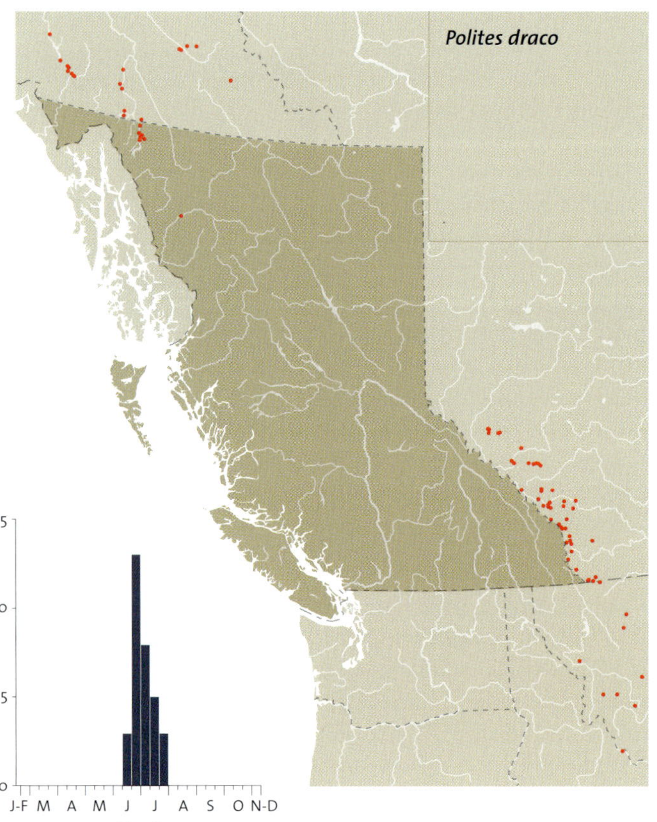

# Tawny-edged Skipper
*Polites themistocles* (Latreille, [1824])

**Etymology:** Themistocles was an Athenian statesman and commander. The name *turneri* honours J.R. Turner, an authority on the megathymid skippers when *turneri* was described by H.A. Freeman, another megathymid specialist. The common name refers to the "tawny," or brownish yellow, front edge of the forewing, and was first used by Gosse (1859).

**Adult:** The Tawny-edged Skipper is easy to identify in BC; it is the only skipper in which the dorsal hindwing is a solid dark khaki colour in both sexes. This, combined with the golden tan of the basal half of the dorsal forewing of all males and some females, distinguishes this species. Females can be confused with *Euphyes vestris* in BC.

**Immature stages:** Scudder (1889b) described the egg and larva. The egg is pale, pea green. The mature larval head is chocolate brown with a dark median line and paler lateral lines. The body of the larva is pale green mottled with dark green areas and with a dark green dorsal line.

**Biology:** The Tawny-edged Skipper flies from early June to late July in a single generation. The literature is contradictory

♂ D (2.7 cm)

♀ D (2.8 cm)

♂ V (2.7 cm)

regarding its larval foodplant. Shapiro (1966) states that the foodplant is the grass genus *Panicum,* but Shapiro (1974b) says that the foodplant is unknown. If *Panicum* is the correct foodplant, *Panicum occidentale* is probably the species used in BC.

**Subspecies:** American writers have either ignored (Miller and Brown 1981) or misplaced (Ferris 1989) the subspecies of the Tawny-edged Skipper described from BC: *P. t. turneri* Freeman, 1944 (TL: Jesmond, BC). Most BC and adjacent northeastern WA populations are disjunct from the rest of the range of the Tawny-edged Skipper, and Freeman characterized subspecies *turneri* as darker than the nominate subspecies. Populations in the Rocky Mountain Trench are tentatively assigned to the nominate subspecies (TL: "Amerique meridionale").

**Range and habitat:** The Tawny-edged Skipper is known from Prince George south and east across the Southern Interior and the Kootenays. It is usually found in moist grass areas around small lakes or springs in dry grassland habitat, which are frequently impacted by cattle grazing.

**General distribution:** Known from the disjunct BC and WA populations and then east from AB to NS and south to AZ and FL.

**Conservation status:** Not of concern (S4).

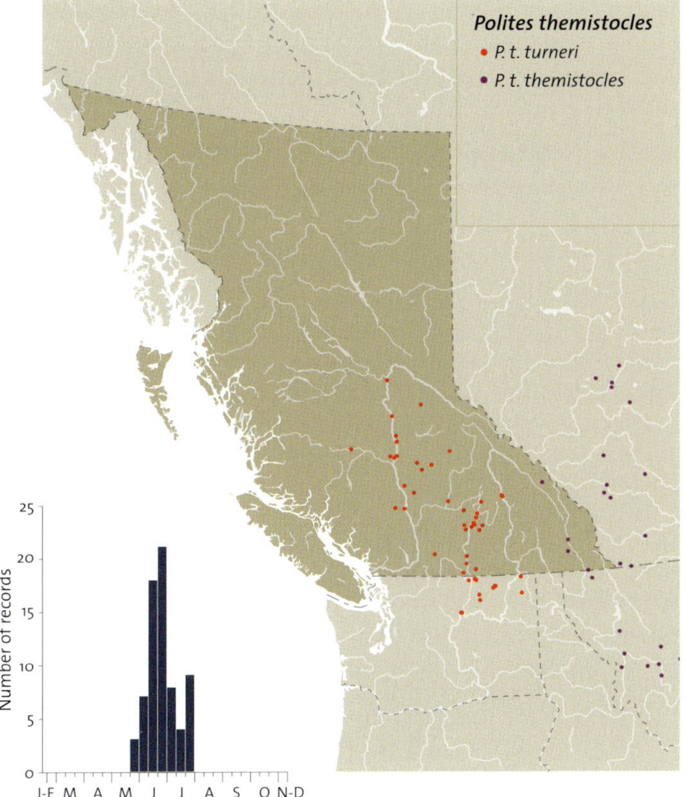

Egg

# Long Dash Skipper
*Polites mystic* (W.H. Edwards, 1863)

**Etymology:** The species name *mystic* refers to Mystic, Connecticut. Edwards mentions Connecticut among the known states in his description of this species. The common name refers to the long black stigma or "dash" on the dorsal forewing of the male (Scudder 1889b).

**Adult:** Adults of the Long Dash Skipper can be distinguished from the preceding *Polites* species by the characteristics already discussed, and from *Hesperia* species by generic characters. It can be separated from the remaining *Polites* species, *Polites sonora,* by its larger size and very different habitat choice. In *P. mystic* the median pale band of the ventral hindwing is indistinct, whereas in *P. sonora* this band is crisply distinct. Another species that can be confused with the Long Dash Skipper is the Woodland Skipper (*Ochlodes sylvanoides*), but the flight periods of the two species hardly ever overlap. When they do overlap, worn females of the Long Dash Skipper fly concurrently with fresh males of the Woodland Skipper.

**Immature stages:** Saunders (1869b) described the egg as pale green and slightly taller than broad. The mature larval

♂ D (2.9 cm)

♀ D (3.0 cm)

♂ V (2.9 cm)

head is reddish brown and edged posteriorly with white hairs. The body is brownish green and has white hairs like the head. There is also a dorsal line and lateral spots of a darker shade. The first thoracic segment of the larva is pale white, with light brown sclerotization across the dorsal surface.

**Biology:** Adult Long Dash Skippers fly from mid-June to mid-July, and females are present well after the male flight period is over. The species is univoltine in BC. Eggs are laid in mid-July and pupation occurs by mid-August (Saunders 1869b). Shapiro (1966) records *Poa* species as the natural foodplant, but the larvae will feed on several grasses in the lab (Scudder 1889b).

**Subspecies:** MacNeill (1975) states that the PNW populations represent an undescribed subspecies. This is a situation analogous to that of the Tawny-edged Skipper. The type locality of the species is Hunters, NY (Brown and Miller 1980).

**Range and habitat:** In BC the Long Dash Skipper is known from the Kootenays and the Peace River region. In south-eastern BC it is found in wet meadows and near bog habitat where *C. selene* occurs. This same habitat preference is found in WA, OR, ID, and northwest MT populations. By contrast, the Peace River populations are not associated with as moist a habitat.

**General distribution:** The Long Dash Skipper is found in the Columbia River drainage of BC and the PNW and otherwise from the Peace River region of BC and AB east of the Rocky Mountains to NS, and south to central CO and VA, with disjunct populations in southwestern CO and AZ.

**Conservation status:** Not of concern (S4).

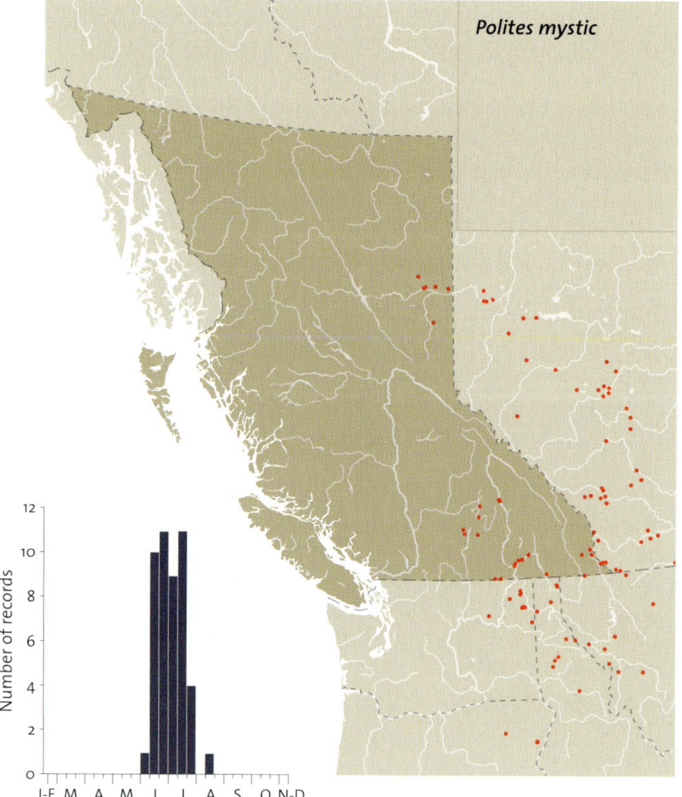

# Sonora Skipper

*Polites sonora* (Scudder, 1872)

**ETYMOLOGY:** The name *sonora* may refer to the Sonoran region of eastern California, Arizona, and Mexico, with the type locality being "Sierra Nevada, California." The common name was first used by Comstock (1927).

**ADULT:** The Sonora Skipper can be distinguished from other *Polites* species by the crisp appearance of the pale median line on the ventral hindwing and the uniform darker ground colour of the rest of the ventral hindwing.

♂ D (2.6 CM)

♂ V (2.6 CM)

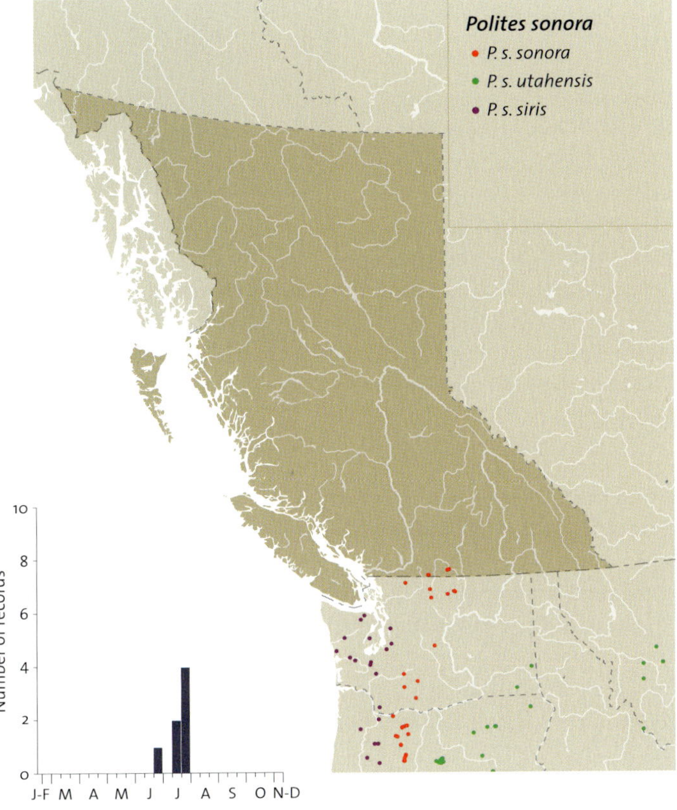


**IMMATURE STAGES:** Newcomer (1967b) described some of the immature stages. The egg is round and light green. The third instar larva is 5 mm long, grey green, with many fine black scales.

**BIOLOGY:** The few adults that have been observed were all flying in July. The Sonora Skipper presumably overwinters as an immature larva. The foodplant is suspected to be *Festuca idahoensis,* and Newcomer (1967b) reared it in captivity on this grass.

**SUBSPECIES:** The BC populations are the nominate subspecies, *P. s. sonora* (TL: Sierra Nevada, CA). The Rocky Mountain subspecies has not been recorded near BC.

**RANGE AND HABITAT:** The Sonora Skipper reaches its extreme northern limit in the Hope Mountains, the only part of the Cascade/Sierra Mountains to extend into BC. It is found on dry grassy slopes along the Similkameen River drainage, especially near Keremeos.

**GENERAL DISTRIBUTION:** The Sonora Skipper occurs from the Hope Mountains of BC south to southern CA, northern AZ, and CO.

**CONSERVATION STATUS:** The Sonora Skipper is Threatened in BC (S1).

---

## GENUS *ATALOPEDES* SCUDDER, 1872

*Atalopedes* may be derived from the Greek words *atalus* (delicate) and *pedes* (on foot) – delicate walking. When the genus was first described, however, it contained only the species *campestris,* which could not be characterized as delicate. There does not appear to be a common name for the genus.

*Atalopedes* is a small Neotropical genus with only three species. The species recorded from British Columbia is a rare migrant.

# Sachem

*Atalopedes campestris* (Boisduval, 1852)

**Etymology:** The species name *campestris* is Latin for "of the plain," presumably referring to coastal California. The common name "Sachem" was the name used for the peace chiefs of the aboriginal Iroquoian League by Scudder (1889b).

**Adult:** Sachem males can be separated from similar species by the presence of a large black spot surrounding the stigma. This makes the centre of the dorsal forewing appear to be a large, square black area. Females have a translucent spot in the centre of the forewing that is not found in other BC skippers. In both sexes, the ground colour of the ventral hindwing is a tweedy green. The Sachem is most likely to be seen nectaring on garden composites in association with the Woodland Skipper, and is easy to overlook.

**Immature stages:** In New England the egg is white and, according to Scudder (1889b), broad and short. The head of the mature larva is black, and the body is dark fuscous green, covered with dark tubercles, each with a short, fulvous hair. In California larvae, the hairs are black and the larvae have a mid-dorsal dark greenish brown line. The pupa has a white spot near the eye (Comstock 1929).

**Biology:** In its breeding range, the Sachem has three generations per year. In BC adults have been seen from late July to late September. Of the known native grass foodplants (Opler and Krizek 1984), *Digitaria* (crabgrass), *Cynodon dactylon* (Bermuda grass), and *Eleusine indica* (goosegrass)

♂ D (2.9 cm)

♀ D (3.5 cm)

♂ V (2.9 cm)

occur in BC. So far there is no evidence that the species breeds in the province.

**Subspecies:** The three stray BC specimens presumably belong to the nominate subspecies, *A. c. campestris;* TL: Sacramento, CA (Emmel et al. 1998a). Other migrant species that breed only east of the Rocky Mountains are known to occasionally reach southeastern BC, however, such as the Variegated Fritillary, *Euptoieta claudia* (Cramer, [1775]). Thus the stray BC specimen records could be the eastern subspecies.

**Range and habitat:** The Sachem has been recorded three times in BC: in 1937 at Jesmond by J.K. Jacob; in 1953 at Blackwall, Manning Provincial Park, by J.E.R. Martin; and in 1960 north of Creston by D.F. Hardwick. These records represent rare migrant individuals. The species had not been known in WA up to the mid-1980s (Leighton 1946), when it began turning up in extreme southern WA, especially the Tri-Cities area at the confluence of the Columbia and Snake rivers. The Tri-Cities area had been collected extensively in

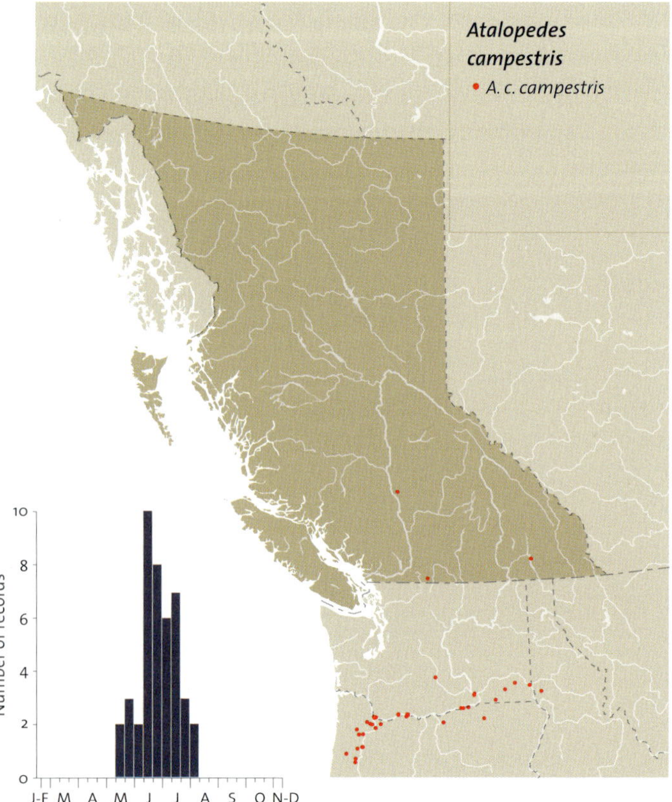

**Egg hatching**

**First instar larva**

**Mature larva**

**Pupa**

the 1950s and 1960s (JHS) without this species being seen. In eastern WA the habitat consists of disturbed areas and lawns.

**GENERAL DISTRIBUTION:** The Sachem occurs in continuous breeding populations from central CA east to FL and north to southern IL and VA. It has been recorded north as temporary populations or as strays in all US states except ID and MT. The BC specimens represent the only authentic Canadian records except for a temporary breeding colony established in southern ON in 1968 (Riotte 1992).

**CONSERVATION STATUS:** The Sachem is a rare non-breeding migrant (SAN).

Sachem (*Atalopedes campestris*)

## GENUS *OCHLODES* SCUDDER, 1872

The name *Ochlodes* is Greek for turbulent or unruly, from the swift, erratic flight of the members of this genus (Emmet 1991).

The genus is Holarctic, with five Nearctic species and four Palearctic species. Only one species has been recorded in BC, but habitat for a second species, *O. yuma* (W.H. Edwards, 1873), was historically present in the southern Okanagan (JHS).

## WOODLAND SKIPPER
*Ochlodes sylvanoides* (Boisduval, 1852)

**ETYMOLOGY:** The name *sylvanoides* is derived from the Latin *silva* (woods or forest); the common name is therefore a translation of the Latin name (Holland 1898).

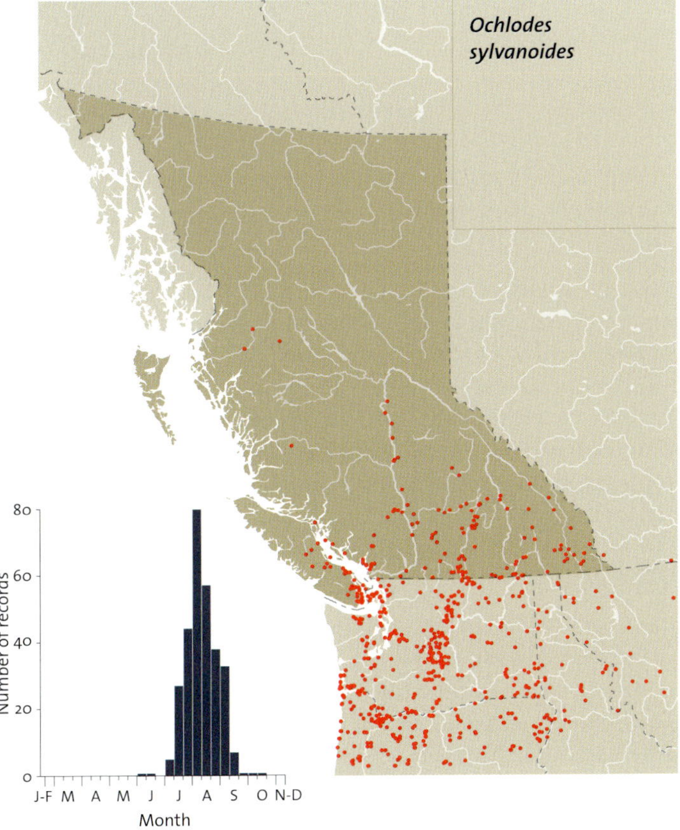

♂ D  (2.7 CM)

♀ D  (2.9 CM)

♂ V  (2.7 CM)

**ADULT:** The Woodland Skipper is most easily confused with the Long Dash Skipper, *Polites mystic*. The flight times of the two barely overlap, however, and their habitat choices are very different.

**IMMATURE STAGES:** Comstock (1929) described the mature larva. The head is black and the body buff yellow with seven black longitudinal stripes. Otherwise, the immatures of this very common species have not been described in the literature. G.A. Hardy (GAH) described the first instar larva as having a jet-black head and a pale, putty-coloured body.

**BIOLOGY:** The Woodland Skipper is by far the most abundant skipper throughout southern BC from late July until frost kills the adults. It is found nectaring at the flowers of many garden composites, especially marigolds. There is only one

**Woodland Skipper (*Ochlodes sylvanoides*)**

brood. Larvae are known to feed on grass but no specific species has ever been identified. GAH observed eggs being laid on 8 September. They hatched on 3 October and the larvae ate most of the chorion. The larvae did not feed on grasses. Presumably the Woodland Skipper hibernates as a first instar larva.

**SUBSPECIES:** The Woodland Skipper shows remarkable variation across BC and south in the United States. Scott (1981) attempted to double the number of recognized subspecies from three to six to accommodate this variation. The huge, nearly continuous breeding populations indicate that there is little if any barrier to gene flow, however, and that the variation in colour pattern represents only phenotypic responses to different climatic conditions, not genetically different subspecies. The type locality was restricted to near Beldon, Plumas Co., CA, by Emmel et al. (1998a).

**RANGE AND HABITAT:** The Woodland Skipper is ubiquitous and very abundant across the southern fourth of BC below 1,000 m elevation. There are also populations in the lower Skeena drainage. It probably occurs along the coast between the Skeena drainage and the Lower Mainland, but that area is very poorly surveyed.

**GENERAL DISTRIBUTION:** The Woodland Skipper is found from southern BC south to Baja California and NM.

**CONSERVATION STATUS:** Not of concern (S5).

## GENUS *EUPHYES* SCUDDER, 1872  SEDGE SKIPPERS

The name *Euphyes* is derived from Euphues, an Athenian youth who was elegant and handsome. The common name for the genus was coined by Scott (1986) to reflect the fact that many species in the genus feed on sedges.

The genus *Euphyes* is primarily Neotropical, with 20 species (Shuey 1994). A few species occur in temperate North America, and one is found in British Columbia.

## DUN SKIPPER
*Euphyes vestris* (Boisduval, 1852)

**ETYMOLOGY:** The name *vestris* may be derived from the Latin *vestis* (a covering, or clothing). When the species was first described, it was placed in the then large genus *Hesperia*, most species of which had a distinct pattern. *Euphyes vestris* has a uniform, milk-chocolate ground colour, obscuring the typical hesperid pattern. Scudder (1889b) was the first to use the common name, replacing his earlier name "Immaculate Skipper."

**ADULT:** The dorsal forewing of the male Dun Skipper is uniformly chocolate brown, with a black stigmal patch. At most, there is a slight tan flush around the stigma. The Dun Skipper can be confused only with the male Tawny-edged Skipper, in which the tan flush usually extends to the costal (front) edge of the forewing. The median spots on the dorsal forewing of the female Dun Skipper are very inconspicuous compared with those on the female Tawny-edged Skipper. The ranges of the two species overlap in the Chilcotin area of BC between Lillooet and Riske Creek.

**IMMATURE STAGES:** Heitzman (1965) described the immatures of the eastern populations. The egg is green. The body of the mature larva is pale green, with overlying fine white lines. The front of the head is caramel brown with two vertical light bands. The back is black with a dorsal oval spot.

**Dun Skipper (*Euphyes vestris*)**

**Biology:** In BC the Dun Skipper flies from late June to mid-August in one brood. Time of flight is keyed to the time when summer begins on the coast, a highly variable event. Further south, there are two generations per year. Elsewhere the only known foodplant is the sedge *Cyperus esculentus* (Heitzman

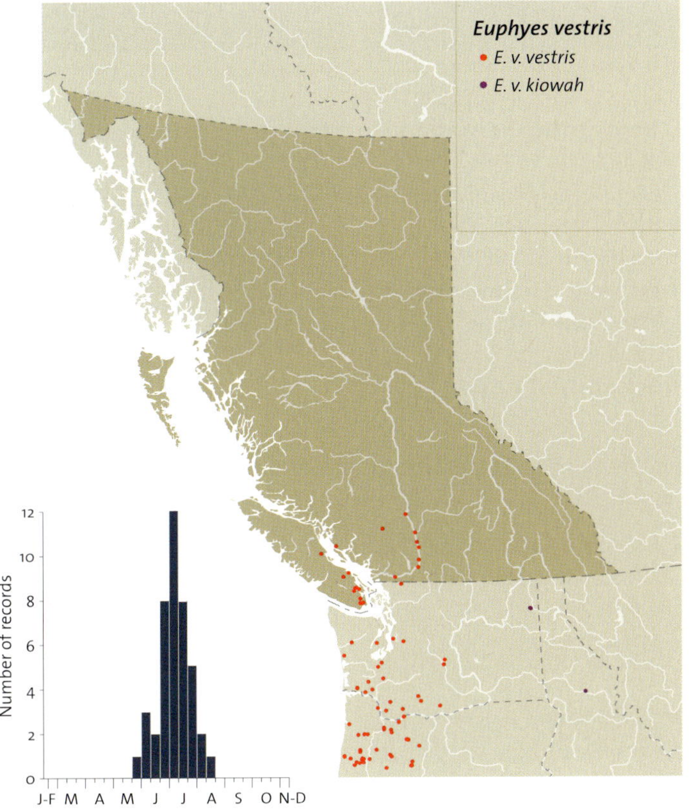

♂ D (2.9 см)

♂ V (2.9 см)

1965). This plant and other *Cyperus* occur in BC, and the Dun Skipper should be looked for on this and other sedges.

**Subspecies:** BC populations are the nominate subspecies, *E. v. vestris*; TL: Spanish Ranch Rd., Meadow Valley, Plumas Co., CA (Emmel et al. 1998a). The eastern subspecies may eventually be found in southeastern BC.

**Range and habitat:** The Dun Skipper is known from southern Vancouver Island, the Lower Mainland, and the Fraser River canyon upstream to Lillooet. Except for the Lillooet population, the species has been encountered only as single individuals in mesic grassy areas, often along old railway right-of-ways. The single Lillooet observation was at a spring surrounded by an extremely xeric area. The spring area had enough sedge to support a population of the Dun Skipper.

**General distribution:** The western subspecies is found from southwestern BC through western WA to CA. The eastern form occurs from southern SK to NS, and south to AZ and FL.

**Conservation status:** The Dun Skipper is of Special Concern in BC (S2).

---

### Genus *Amblyscirtes* Scudder, 1872   Roadside Skippers

The name *Amblyscirtes* may be derived from *ambl* (blunt) and *skirtao* (leap), perhaps referring to the short, hopping flights of members of this genus (Bird et al. 1995). The generic common name "roadside skippers" (Scudder 1889b) is derived from the most common species in the genus, "the" Roadside Skipper, *A. vialis*.

The genus *Amblyscirtes* is Nearctic, with 20 species found mostly from northern Mexico through the eastern United States. Only one species occurs in British Columbia.

## Roadside Skipper
*Amblyscirtes vialis* (W.H. Edwards, 1862)

**Etymology:** The Latin *vialis* means "highway," thus the origin of the common name "Roadside Skipper" used by Scudder (1889b). At the time Edwards described this species from Lake Winnipeg, Manitoba, and Rock Island, Illinois, there were few roads in either area and it is unlikely that he had the Latin meaning in mind; the original description does not help explain the intended meaning. It is unfortunate that the common name was coined because, at least in the west, this species is not common along roadsides.

**Adult:** The Roadside Skipper is a uniform dark chocolate brown on the uppersides of the wings. There is a faint row of median light spots on the dorsal forewing. On the underside, the same brown ground colour appears grey due to the presence of white scales on the outer half of both forewing and hindwing.

**Immature stages:** Scudder (1889b) described the immatures. The egg is pale green and wider than high. The mature larval

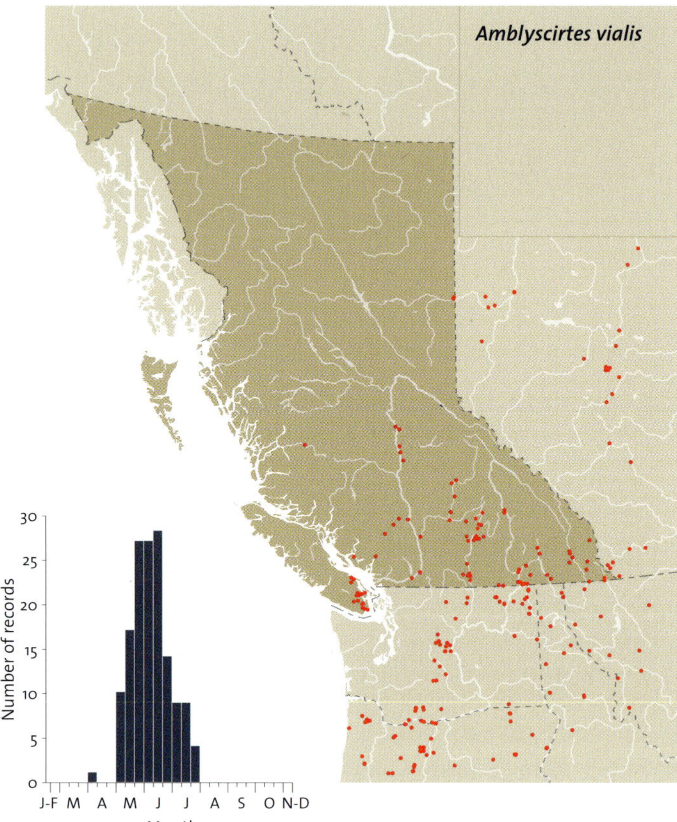

♂ D (2.4 cm)

♀ D (2.4 cm)

♂ V (2.4 cm)

**BIOLOGY:** In BC the Roadside Skipper is univoltine. It flies from mid-May to mid-June, with occasional females seen until mid-July. Elsewhere in the southern limits of the species' range, it is double-brooded. Scudder mentions the larval foodplant as "grasses."

**SUBSPECIES:** None. The type locality of the species is western IL.

**RANGE AND HABITAT:** The Roadside Skipper is found across southern BC and in the Peace River region. It is found in mesic grassy meadows, especially in aspen forest and riparian habitat.

**GENERAL DISTRIBUTION:** The Roadside Skipper ranges from BC to NS and south in the western USA to central CA and northern NM in mesic habitats, especially aspen forests. In the east it occurs south to TX and FL.

**CONSERVATION STATUS:** Not of concern (S5).

head appears white due to a covering of short white hairs. There are also several reddish brown vertical stripes on the head. The body of the larva is pallid green and has many pale green spots, each with a short hair. The prolegs are pale green.

# Superfamily
# Papilionoidea Latreille, [1802]
## The Butterflies

The superfamily Papilionoidea is characterized by antennae
that are straight and never hooked, forewings either with
some R veins branched or (in some Hairstreaks) with only
three R veins present, and hindwings with the $M_2$ vein
present (Scott 1986b). There are six butterfly families in
the world, and all occur in British Columbia except the family
Libytheidae, the snout butterflies.

Painted Lady
(*Vanessa Cardui*)

# Family Papilionidae Latreille, [1802]
## Swallowtails and Apollos

The family Papilionidae contains the largest and most brightly coloured BC butterflies. All our swallowtails and apollos are white or yellow with black markings. Our swallowtails have tails on the hindwings, whereas our apollos do not. Many swallowtails elsewhere in the world lack tails, and some Eurasian apollos have tails.

All the Papilionidae have smooth, hairless eyes, and short antennae with (in BC species) well-developed elongated clubs. They have three fully developed pairs of walking legs. The middle leg tibia has no spurs, each foot has a single pair of claws and lacks paronychia and pulvilli, and the foreleg tibia has an epiphysis. The forewing has 11 or 12 veins; veins 2A and 1A are separated. The discal cell is closed in both fore- and hindwing. The hindwing has one anal vein; hindwing vein 3A is absent in two subfamilies (Papilioninae and Parnassiinae) but is present in a third Eurasian subfamily (Baroniinae). The wings do not have androconial scales.

There are three subfamilies within the family Papilionidae: Papilioninae (27 genera, 476+ species), Parnassiinae (8 genera, 40+ species), and Baroniinae (11 genera, 247+ species) (Scriber 1995). Most swallowtail species are tropical or subtropical, but most apollos are temperate Eurasian. In BC there are only 8 species of swallowtails, all in the genus *Papilio,* and 4 species of apollos, all in the genus *Parnassius.*

Eggs are hemispherical, and have a slightly to strongly granular surface. Larvae have Y-shaped, eversible defensive secretory glands called osmeteria on the top of the thorax. The osmeteria are fully functional in swallowtails but are vestigial (present but without producing defensive chemicals) in BC apollos. Swallowtail pupae have a silk girdle around the middle; BC apollos have a loose silk cocoon. Swallowtails and apollos clean their antennae with their forelegs.

Males seek females by patrolling hilltops, meadows, forest openings, streamsides, or roadsides. They generally have slow wing beats except when disturbed, but because of their large size can travel remarkably fast with little effort. Males fly faster than females, probably because of their lighter body weight and smaller wings, and also because they are searching for females.

Two-tailed Swallowtail (*Papilio multicaudatus*)

The Latin and common names for the subfamily are derived from the type species of the genus *Parnassius, P. appollo* Linnaeus.

The subfamily Parnassiinae has only 11 veins in the forewing; vein Cu-v is missing. The forewing radius is four-branched in our species. The male, and sometimes the female, has asymmetric tarsal claws. The antennae are separated by less than one-half the width of the base of the antennae, and the labial palpi are longer than the head.

There are two tribes in the subfamily: the completely Eurasian Zerynthiini (5 genera, 11 species) and the primarily Eurasian Parnassiini (3 genera, 31 species) (Scriber 1995; Shepard and Manley 1998). Five species occur in North America, all in the genus *Parnassius,* and four of them occur in BC. Of the four, the Clodius Apollo and the Rocky Mountain Apollo are frequently seen, but Eversmann's Apollo and the Phoebus Apollo are seldom seen in their northern mountain habitats.

### GENUS *PARNASSIUS* LATREILLE, 1804   APOLLOS

The name *Parnassius* is derived from the Parnassus mountain range near Delphi in Greece, in reference to the alpine habitats of most species (Emmet 1991). Linnaeus divided butterflies into several groups, the second of which was the Heliconii, which took their name from the Muses and Graces that lived on Mt. Helicon, the highest peak in the Parnassus range. Apollo was the patron god of the Muses and Graces, and the first species of Linnaeus's Heliconii was *Papilio apollo* (Emmet 1991), now known as *Parnassius apollo.* The common name "apollo" was first applied in Britain by British lepidopterists to the one species *P. apollo* (Bretherton 1990a), and was later extended to apply to the genus as a whole.

Apollos are medium-sized to large white or yellow butterflies with black wing markings. Red eyespots are usually present on the hindwings and, in two species, on the forewings. The outer borders of the wings are semi-transparent due to lack of scales. Two hooks on the forewing base help in the emergence of the adult from the pupal cocoon (Scott 1986b). Females have a brown or white sphragis, a hard structure deposited in the female mating tube by the male during mating to prevent further matings.

The abdomens of the males are very hairy, possibly to reduce heat loss during their long flights searching for females. In contrast, the abdomens of the females are naked or sparsely haired, possibly enabling them to reduce overheating on the hot ground, where they spend most of their time.

Unlike most butterflies, the eyes of males are much larger than those of females. It is unlikely that the only reason for this is that males locate females visually at a distance, because that is true for most butterflies. It may be correlated with the lack of courtship prior to mating: a male simply grapples with a female as soon as he spots her, and attempts copulation. If the female has already mated, the male attempts to grasp the sphragis with his claspers and remove it (CSG).

Eggs are round with a pebbled or pitted surface, and are white to tan in colour. They are laid singly under the edges of objects in the general vicinity of the larval foodplants. Phoebus Apollos may lay eggs directly on the larval foodplant (Shepard and Manley 1998). The embryo develops into a larva within the egg chorion within a few weeks of oviposition, but the egg does not hatch until the following spring (Edwards 1868–72).

The larvae have small, vestigial osmeteria (Y-shaped, eversible defensive secretory glands) on the top of the thorax; these are frequently not everted when a larva is "attacked" with forceps, and do not produce any chemical secretion. Pupation occurs in weak cocoons in loose soil or debris on the ground.

# Eversmann's Apollo

*Parnassius eversmanni* Ménétriés, [1851]

**ETYMOLOGY:** The species *eversmanni* was named for the prominent early entomologist Eduard Eversmann, who was born in Germany but lived and worked in Russia. The subspecies name *thor* refers to the Norse god Thor. The common name was first used by Holland (1931), and has been used by most books since then.

**ADULT:** Eversmann's Apollos have yellow males and white to light yellow females, with dark markings all the same shade of grey. The ventral hindwing veins are light brown at the base. There are red spots on the hindwings but never on the forewings. In females the red spot in the centre of the hindwing is connected to the other spots with a black bar, forming a black band containing a series of red spots. The antennae are black. The sphragis of the female is moderately large, and is white to light brown.

**IMMATURE STAGES:** Eversmann's Apollos have not been reared in North America. The descriptions of the egg, larvae, and pupae given by Scott (1986b) and Layberry et al. (1998) are based on descriptions in the literature of Asian populations. Eggs are white. Larvae are black, with short black hair and a lateral row of white to yellowish dashes separated by dots of the same colour; the head is black. Pupae are dark reddish brown, in a thin silk cocoon (Scott 1986b).

**BIOLOGY:** Eversmann's Apollos are univoltine, flying in June and July. The life cycle is probably the same as for the Asian

♂ D (5.2 cm)

♀ D (4.8 cm)

populations, with the eggs laid singly in the general vicinity of the larval foodplant, under objects such as leaves, sticks, and rocks. In Asia, eggs containing fully developed larvae hibernate the first year and pupae hibernate a second year, resulting in a two-year life cycle (Tuzov 1997). Males patrol large areas and are easily spotted, while females spend little time in flight and are more difficult to find.

The larval foodplant of Eversmann's Apollos is unknown in North America. Known larval foodplants elsewhere are in the Fumariaceae: *Corydalis arctica, C. gorodkovi, C. paeonipholia,* and *C. pauciflora* in Asia (Tuzov 1997), and *Dicentra peregrina* in Japan (Scott 1986a). The most likely larval foodplant in North America appears to be *Corydalis pauciflora,* although the individual plants are so small that one larva would have to eat many plants to reach maturity.

**SUBSPECIES:** Eversmann's Apollo was originally described from Kansk, in the Krasnoyarsk region of Russia. The actual collection site for the type specimens was probably in spurs of the East Sayan Mountains closest to the town of Kansk (Tuzov 1997). The North American subspecies is *P. eversmanni thor* Hy. Edwards, 1881 (TL: Yukon River, about 160 km from the river mouth). Eversmann's Apollos from Pink Mountain in northeastern BC were named subspecies *meridionalis* Eisner, 1978, and then renamed *pinkensis* Gauthier, 1984 because the name *meridionalis* had previously been used for another *Parnassius*. Both names are synonyms of subspecies *thor*; Eversmann's Apollos from Pink Mountain are the same as in the rest of Alaska, Yukon, and BC.

**RANGE AND HABITAT:** Eversmann's Apollos are known from scattered locations in montane northern BC near Atlin, Pink Mountain, Smithers, and Terrace. The species is likely more widely distributed in the mountains of BC north of Highway 16, but has not been found there due to limited access. Most populations are associated with the shrub zone between timberline and alpine tundra, although individuals are also found in high alpine tundra and in lower forest clearings.

**GENERAL DISTRIBUTION:** Eversmann's Apollos occur across northern Asia. In North America the distribution is restricted to AK, YT, northern BC, and extreme western NT.

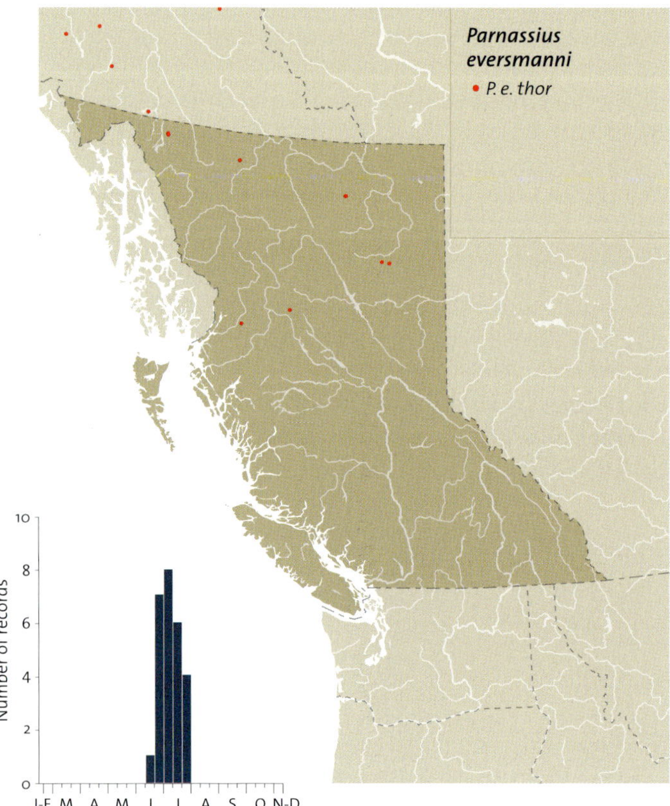

# Clodius Apollo

*Parnassius clodius* Ménétriés, 1855

**Etymology:** *Clodius* is an alternative spelling of Claudius. It and the subspecies name *claudianus* refer to Claudius I, emperor of Rome from 41 to 54 AD, so in naming the subspecies Stichel was using *claudianus* as a form of *clodius*. The subspecies name *pseudogallatinus* means "a false subspecies *gallatinus*." The subspecies name *altaurus* refers to the type locality, Alturas Lake, ID. The common name was first proposed by Holland (1898) and has since been used by most other authors.

**Adult:** Clodius Apollos are white with dark markings all the same shade of grey. The ventral hindwing veins are light brown at the base, and there are red spots on the hindwings but never on the forewings. The antennae are black. The sphragis of the female is large and white.

**Immature stages:** Eggs are white to pale coffee brown, and round with flattened top and bottom; they have close-set rounded pits over the surface. Mature larvae are normally black with a stripe formed of elongated yellow spots down each side. They are smooth aside from fine black hairs. An alpine form of the larva, unknown in BC, is grey-brown to pink-grey, with cream yellow lateral spots and dorsal rows of narrow chevron markings (McCorkle and Hammond 1986). The pupae are dark red brown, oval, and smooth, and are formed within a sturdy cocoon.

**Biology:** Clodius Apollos are univoltine. Adults fly from late May at low elevations to early August at high elevations,

Ssp. *claudianus* ♂ D (6.0 cm)

Ssp. *claudianus* ♀ D (6.0 cm)

Ssp. *altaurus* ♂ D (6.1 cm)

Ssp. *altaurus* ♀ D (5.9 cm)

Ssp. *pseudogallatinus* ♂ D (5.9 cm)

Ssp. *pseudogallatinus* ♀ D (5.6 cm)

with each population having a four- to six-week flight period. Mating consists of a male visually locating a female in the air or on the ground, forcing her to the ground, and then copulating with her. Eggs are laid singly on or in the general vicinity of the larval foodplant, with females possibly being stimulated to oviposit by the odour of the larval foodplant. The embryos develop into fully formed larvae within a month of oviposition, but remain dormant within the egg chorion until early the following spring. Larvae feed during the day (contrary to Layberry et al. 1998), but spend much of their time hidden under debris on the ground or basking in the sun on the ground up to several metres from

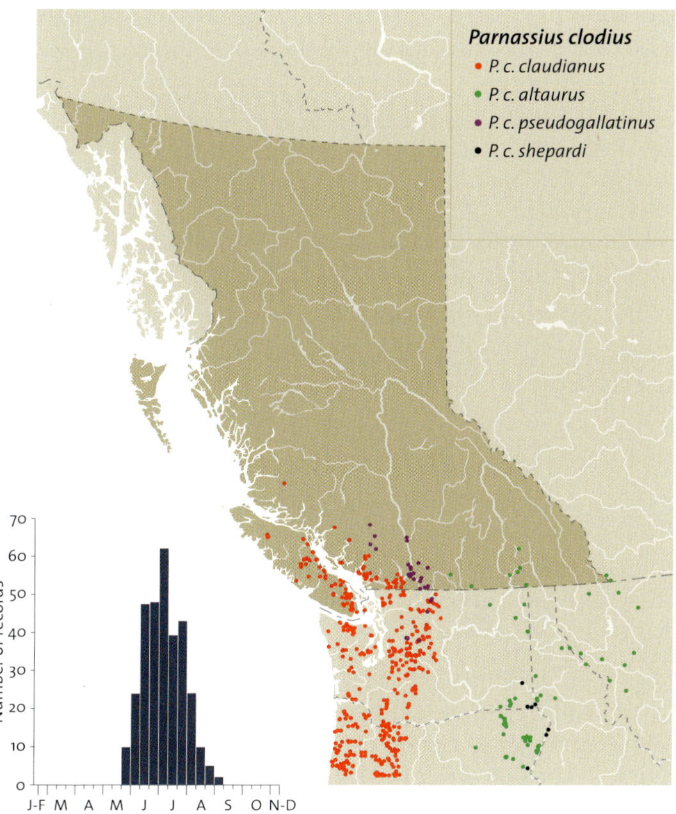

Eggs

Mature larva

**Clodius Apollo (*Parnassius clodius claudianus*)**

the nearest larval foodplant. How they relocate the larval foodplant is unknown. Ten to 12 weeks after the egg hatches, the pupa is formed within a strong cocoon placed in loose debris on the ground. The pupal period is about three weeks (McCorkle and Hammond 1986).

On the BC coast the larval foodplant is Pacific bleeding heart, *Dicentra formosa* (Fumariaceae). Steer's head, *Dicentra uniflora,* is the larval foodplant in the Sierra Nevada Mountains of California, with oviposition occurring in the vicinity of the plants after they have dried up for the season (Emmel and Emmel 1974). It may be the larval foodplant for the high-elevation populations that occur in the east Cascade Mountains of Oregon, Washington (McCorkle and Hammond 1986), and interior BC (JHS).

**Subspecies:** Subspecies *claudianus* Stichel, 1907 (TL: "Washington, British Columbia, Vancouver Island") occurs on the south coast of BC, including Vancouver Island, from sea level to subalpine areas without change in wing coloration. Adults are large and white. The band of grey chevrons on the outer part of the hindwings is usually absent or only poorly developed in males but is well developed in females. The anal spot band on the central hindwings is distinct in both sexes. Subspecies *pseudogallatinus* Bryk, 1913 (TL: Yale, BC) occurs in the Coast Range from elevations of 100 m at the type locality to subalpine habitats. Adults are generally small and grey, but are highly variable. The unusual range of subspecies *pseudogallatinus* is apparently the result of the subspecies having evolved from subspecies *altaurus* rather than *claudianus*. Populations in the Kootenays are the subspecies *altaurus* Dyar, 1903 (TL: Alturas Lake, ID). Males lack the hindwing chevron band and anal spot band, but these are present in females. Subspecies *gallatinus* Stichel, 1907 (TL: Gallatin River, MT) is a synonym of *altaurus* (Shepard and Shepard 1975), because the only distinguishing characteristic between the two subspecies' wing pattern is a higher proportion of yellow (rather than red) ocelli at the type locality of *altaurus*. There is a cline with declining numbers of yellow ocelli away from Alturas Lake, with no identifiable boundary between the two supposed subspecies (Nelson Curtis collection).

**Range and habitat:** Clodius Apollos occur from Vancouver Island across the southern Coast Range and northern Cascades to the southern Okanagan Valley, with a few populations in subalpine habitats of the southern Selkirk Mountains. Populations on the coast occur in moist riparian habitats along low-elevation streams. Logging frequently temporarily increases the population size by increasing breeding habitat, but it may adversely affect populations if the second-growth forest lacks the riparian openings present in old-growth forests (Layberry et al. 1998), or if herbicides or broadcast burning after harvesting eliminate foodplants or isolated populations of adults. Wet subalpine meadows and subalpine riparian habitats are used at higher elevations.

**General distribution:** Clodius Apollos range from southwestern BC to central CA in the coastal mountains and Cascades/Sierra Nevada. Inland, they occur in the Rockies from southeastern BC and adjacent AB south to UT. The known northern limit of the species is just south of Bella Coola, BC, which is also the northern limit of the coastal larval foodplant, *Dicentra formosa.* Subspecies *incredibilis* Bryk, 1932 (TL: Mt. St. Elias, AK) must be based on specimens bearing incorrect locality labels, and we consider it a synonym of subspecies *claudianus.*

**Conservation status:** Subspecies *altaurus* is of Special Concern in BC (S3S4). Subspecies *claudianus* and *pseudogallatinus* are not of concern (both S4).

## Rocky Mountain Apollo
*Parnassius smintheus* Doubleday, [1847]

**Etymology:** The species name *smintheus* is an alternative name of the Greek god Apollo that refers to the Sminthia district, where Apollo was worshipped. The subspecies name *olympiannus* refers to the Olympic Mountains, the type locality of the subspecies. The subspecies name *magnus* is Latin for great, large, big, referring to the large size of adults. The subspecies name *yukonensis* refers to the Yukon Territory.

The common name was used for the first time by Layberry et al. (1998), and refers to the distribution being centred on the Rocky Mountains.

**Adult:** Rocky Mountain Apollos are translucent yellowish white with black, grey, and red markings. On the dorsal forewing the marginal grey band is variably developed, and the submarginal band is weakly developed and pale grey.

**Rocky Mountain Apollo (*Parnassius smintheus magnus*)**

The forewing margin usually has small triangles of black at each vein. The dorsal or ventral hindwings usually have at least faint marginal and submarginal grey markings. The wing fringes are mostly distinctly black at the ends of the veins. The hairs and scales on the head, legs, and ventral abdomen of both sexes are yellowish. Both sexes usually have red spots on the hindwings, and frequently also on the forewings, that are much smaller on average than those of Phoebus Apollos. High-elevation populations have darker females, and both males and females are smaller than at low elevations. The antennae are ringed with black and white bands. The sphragis of the female is small and dark brown. Shepard and Manley (1998) have shown that this and the following species are separate species.

**IMMATURE STAGES:** Eggs are white, spherical but flattened at the top and bottom, and with a pebbled surface. The micropylar area is brownish and sunken. First instar larvae have black bodies, are quite hairy, and have dull black heads. Mature larvae are black, with many short, fine black hairs and with two lateral and two dorsal rows of bright yellow spots. The vestigial osmeteria are small and pale yellow. Pupae are dark yellow brown to red brown, and are formed in a weak cocoon in litter on the ground.

**BIOLOGY:** Rocky Mountain Apollos are univoltine, and fly from the first week of June at low elevations to late September in alpine tundra. Eggs are laid singly on the underside of flower heads, leaves, sticks, stones, moss, clumps of dirt, and even on the larval foodplant. The embryo develops into a first instar larva within a month of oviposition, but the egg does not

Ssp. *magnus* ♂ D (6.1 CM)

Ssp. *magnus* ♀ D (6.4 CM)

Ssp. *olympiannus* ♂ D (5.6 CM)

Ssp. *olympiannus* ♀ D (5.8 CM)

Ssp. *yukonensis* ♂ D (5.6 CM)

Ssp. *yukonensis* ♀ D (5.8 CM)

Ssp. *smintheus* ♂ D (5.4 CM)

Ssp. *smintheus* ♀ D (5.9 CM)

hatch until the snow melts the next spring. Larvae spend much of their time off their foodplant, basking in the sun on the ground. They prefer to feed on the leaves and flowers of the inflorescence, but also eat the basal leaves. The larval period is about 10–12 weeks, with pupation occurring within a feeble cocoon in loose debris on the ground.

Larval foodplants are stonecrop species (Harvey 1908), including *S. divergens, S. oreganum* (both mostly for coastal BC populations), *S. lanceolatum, S. stenopetalum* (interior BC populations), and *S. integrifolium* (occasional alpine interior populations) (CSG). Many garden stonecrops are toxic to the larvae; and at least the first four stonecrop species listed are toxic to larvae from November to February, inclusive (CSG). The lowest elevation for successful reproduction of Rocky Mountain Apollos is that above where the snow melts before March. If the snow melts before March, the eggs hatch while the larval foodplant is still toxic and the larvae die. In years when snow remains on the ground at low elevations into March in the Okanagan Valley, adult Rocky Mountain Apollos are seen at unusually low elevations. They apparently stem from eggs laid the previous year by females flying down from higher elevations (CSG).

**SUBSPECIES:** Shepard and Shepard (1975) reduced the number of subspecies to approximately the modern concept. Rocky Mountain populations are the nominate subspecies, *smintheus;* TL: Rock Lake, near Jasper, AB (Shepard 1984). Adults are moderate in size and quite light in colour. Coast Range, Southern Interior, and Central Interior populations are subspecies *magnus* W.G. Wright, 1905 (TL: Enderby, BC) (= *xanthus* Ehrmann, 1918; TL: Moron [sic; Moscow], ID). Adults are generally large and are highly variable in colour, ranging from white with most black markings reduced or absent, to a unique population in Silver Star Mountain Provincial Park in which females are almost completely slate grey. Females are frequently yellowish in low-elevation populations of both *smintheus* and *magnus,* especially when fresh. Tom Manley (in prep.) suggests that within the Southern Interior distribution of *magnus, smintheus* occurs at very high elevation (2,500 m) in Cathedral Lakes Provincial Park. This hypothesis is not as unreasonable as it may at first

Egg

Pupa

**Mature larva**

appear, once late glacial and early postglacial history is considered. Vancouver Island populations are subspecies *olympiannus* Burdick, 1941 (TL: Hurricane Ridge, Olympic Peninsula, WA). Adults are small and the females are very dark. Subspecies *guppyi* Wyatt, 1969; TL: Mt. Cokely, Vancouver Island, BC (Guppy 1998), is a synonym of *olympiannus.* Northwestern BC populations are subspecies *yukonensis* Eisner, 1969 (TL: 5 miles southwest of Haines Junction, YT). The type series was collected by Arthur M. Pearson, and supplied to Eisner by James Ebner (Ebner, pers. comm.; Pearson, pers. comm.). Adults are small and very white, with greatly reduced black and grey markings and with prominent red markings.

**RANGE AND HABITAT:** Rocky Mountain Apollos are found in most areas of BC. The habitat is that of their larval foodplants: thinly vegetated, dry, open areas that are generally south-facing slopes or short-grass alpine tundra.

**GENERAL DISTRIBUTION:** Rocky Mountain Apollos are found in western North America from southwestern YT south to northern CA, northeastern NV, and northern NM.

**CONSERVATION STATUS:** Not of concern, with subspecies *smintheus, olympiannus,* and *yukonensis* all S4, and sub-species *magnus* S5.


## PHOEBUS APOLLO
*Parnassius phoebus* (Fabricius, 1793)

**ETYMOLOGY:** The name *phoebus* is an epithet of Apollo in his capacity as sun god, indicating the affinity of *P. phoebus* with the European *P. apollo* (Emmet 1991). The subspecies name *apricatus* is Latin for "to warm in the sun," in this case in the warmth of the sun god, Apollo. The common name is derived directly from the Latin species name. When coined by Pyle (1974), it applied to the combined species of *P. phoebus, P. smintheus, P. behrii,* and the European *P. sacerdos* Stichel, 1906.

**ADULT:** Males are a clean, opaque white with sharply defined black, grey, and red markings. On the dorsal forewing, both the marginal and the submarginal grey bands are strongly developed and dark grey; the forewing margin usually lacks small triangles of black at each vein. The dorsal and ventral hindwings lack marginal and submarginal grey markings. Both sexes usually have red spots on the hindwings, and frequently also on the forewings, that are much larger on



FAMILY PAPILIONIDAE (SWALLOWTAILS AND APOLLOS)

average than those of Rocky Mountain Apollos. The wing fringes are mixed black and white at the ends of the veins; the black is not distinct. The hairs and scales on the head, legs, and ventral abdomen of both sexes are greyish white. The antennae are ringed with black and white bands. The sphragis of the female is small and dark brown.

**IMMATURE STAGES:** Eggs are similar to those of Rocky Mountain Apollos. A circle of 6–8 pie-wedge–shaped divisions surrounds the micropyle of a Rocky Mountain Apollo egg, whereas cuboidal divisions surround Phoebus Apollo micropyles (Shepard and Manley 1998). The mature larva is black with two small orange spots and an orange dot between them on either side of each segment except for segments 1 and 2; there is one spot on either side of segment 1 and two spots on either side of segment 2. Each segment bears three warts, one on the back and one on either side. Some individuals also have two rows of small red spots along the back. Another colour form of the larva resembles that of *P. smintheus* (see illustrations). A teneral (recently formed and still soft) pupa is of pale sandy colour but later darkens to chocolate (Korshunov and Gorbunov 1995).

**BIOLOGY:** Phoebus Apollos are univoltine, and fly in July and August. The single BC record is from mid-July. Adults are found in mesic tundra areas, frequently in association with solufluction terraces. Solufluction terraces are wavelike ripples of the surface of a slope, with the flatter top of the

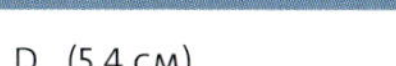

♂ D (5.4 CM)

♀ D (5.6 CM)

wave being moister and supporting more roseroot than the drier, steeper slopes. They are apparently caused by soil moving downhill on top of permafrost during spring thaw. The larva feeds in warm weather both during the day and at night; in cold weather it is usually rolled up on the foodplant (Korshunov and Gorbunov 1995).

The larval foodplant appears to be roseroot (*Sedum integrifolium*). Oviposition on roseroot has been observed in the northern Yukon by Sigrid Shepard (Shepard and Manley 1998), and it is the larval foodplant in Asia (Tuzov 1997).

**SUBSPECIES:** The only BC record is a female collected near the Haines Highway at Chilkat Pass by Oscar Dorfmann of Utah. The BC subspecies is *apricatus* Stichel, 1906 (TL: Kodiak Island, AK), with *elias* Bryk, 1934 (TL: Mt. St. Elias, AK) and *alaskensis* Eisner, 1956 (TL: Mt. McKinley National Park, AK) being synonyms (Shepard and Manley 1998).

**RANGE AND HABITAT:** Phoebus Apollos are known in BC only at Chilkat Pass on the Haines Highway in extreme northwestern BC. They inhabit alpine tundra.

**GENERAL DISTRIBUTION:** Phoebus Apollos occur in Siberia, east across AK to western YT and northwestern BC.

**CONSERVATION STATUS:** Phoebus Apollos are of Special Concern in BC (S3).

Mature larva, ssp. *phoebus*

Mature larva, ssp. *phoebus*

The Latin and common names of this subfamily are derived from the type genus, *Papilio* (see below).

Swallowtails include some of the largest and most spectacular butterflies in the world. There are only seven species in BC, all of which are large black and yellow, or black and white, butterflies with tails on their hindwings at the end of $V_4$. Swallowtails elsewhere in the world are highly variable in colour, and sometimes lack tails on the hindwing.

The forewing of swallowtails has 12 veins. Vein Cu-v is present, and the forewing radius is five-branched. The tarsal claws are symmetric in both males and females.

There are 3 tribes within the subfamily Papilioninae: Triodini (8 genera, 121 species), Graphini (8 genera, 128+ species), and Papilionini (11 genera, 247+ species) (Scriber 1995). Most swallowtail species are tropical or subtropical. Only 8 species, all in the genus *Papilio,* occur in BC.

### GENUS *PAPILIO* LINNAEUS, 1758  SWALLOWTAILS

Linnaeus divided butterflies into several groups. The first group was the swallowtails, which were called equites or knights. Those with red on the thorax were Greek heroes, those with no red on the thorax were Roman heroes (Emmet 1991). *Papilio,* which is Latin for butterfly, was the original generic name that Linnaeus used for all butterflies. The common name was first used in Britain in 1766 for "The Swallowtail," *P. machaon* (Bretherton 1990b), in reference to the resemblance of the tails on the hindwings to the tails of swallows. The name was later extended to include the entire genus. Gosse (1840) was the first to use the common name "swallowtails" in North America.

Swallowtails found in North America are large, brightly coloured butterflies with tails on their hindwings. Six of the eight species in BC are yellow with black stripes. In addition, Pale Swallowtails are white to very pale yellow with black stripes, and Indra Swallowtails are mostly black. Swallowtails also have an orange eyespot at the base of each hindwing tail, and orange and blue spots on the ventral hindwings.

Eggs are smooth and hemispherical, and are cream, yellow, yellow green, or green when laid. The egg colour darkens, and a red ring develops around the top before hatching. Young larvae are black with a white saddle, and resemble bird droppings. Larvae of all ages have well-developed osmeteria, extrusible Y-shaped glands on the top of the thorax that produce defensive chemicals in response to attack. Pupae have two small horns on the head and a point at the top of the thorax. A silk girdle holds them head up against a stem.

The eggs are laid on the leaves of the larval foodplants. On plants with large flat leaves, the eggs are laid on the top or occasionally just under the leaf edge. Both the top and bottom of small leaves are used. The pupae overwinter. In BC two species have more than one generation each year in some populations; the other six are univoltine.

Hancock (1983) split the genus *Papilio* into six genera, two of which (*Papilio* and *Pterourus*) are in BC. We treat Hancock's genera as subgenera of a single large genus, *Papilio,* as do most recent authors.

Higgins (1975) suggested that the North American populations of Old World Swallowtails may not be the same species as *Papilio machaon*. Eitschberger (1993) found that the rings of plates surrounding the egg micropyle are significantly different between one European *machaon* subspecies and subspecies *aliaska*, the North American subspecies closest to European *machaon*. European *P. machaon* has 3 rings with about 112 plates around the micropyle, while *aliaska* has 5 rings with about 142 plates. A second character used by Eitschberger to separate *P. machaon* from *aliaska*, the

*Papilio eurymedon* × *canadensis*
♂ D  (8.9 CM)

*Papilio eurymedon* × *canadensis*
♂ V  (8.9 CM)

*Papilio rutulus* × *multicaudatus*
♂ D  (8.5 CM)

*Papilio rutulus* × *multicaudatus*
♂ V  (8.5 CM)

**61** Examples of hybrid swallowtails, subgenus *Pterourus*

number of teeth on the harpe of the male genitalia, is not useful in separating species in North America. The difference in egg structure is insufficient to split the species without additional data, hence we continue to treat the North American populations as subspecies of *Papilio machaon*.

All the tiger swallowtails (subgenus *Pterourus*) hybridize in the wild to some extent (Fig. 61). In southern BC there is a broad zone of hybridization between Canadian Tiger Swallowtails and Western Tiger Swallowtails from Manning Provincial Park east to Creston. In the areas where their ranges overlap, Western Tiger Swallowtails prefer low-elevation deciduous forest habitats whereas Canadian Tiger Swallowtails prefer higher-elevation boreal forest habitats. Hybridization between Pale Swallowtails and Western Tiger Swallowtails is rare, but Wagner (1978) collected a perfectly intermediate male hybrid in the wild in Idaho. Jon and Sigrid Shepard found a male hybrid of the Pale Swallowtail and the Canadian Swallowtail 10 km south of Galloway, BC. It is intermediate in appearance between the two species.

Similarly all the Old World swallowtails (subgenus *Papilio*) occasionally hybridize in the wild. The Old World swallowtail species are most easily distinguished by the overall coloration of the hindwing and by the colour of the eyespot at the base of the tail on the hindwing.

## BAIRD'S SWALLOWTAIL
*Papilio bairdii* W. H. Edwards, 1869

**ETYMOLOGY:** The name *bairdii* honours Spencer Baird of the Smithsonian Institution. Subspecies *pikei* was named for Alberta entomologist E.M. (Ted) Pike, who discovered the subspecies with Felix Sperling in 1980 (Sperling 1987). The subspecies name *oregonius* refers to the type locality being in Oregon. Subspecies *dodi* was named for F.H. Wolley-Dod, the first significant lepidopterist in Alberta (Bird et al. 1995). The common name was first proposed by Holland (1898).

**ADULT:** Baird's Swallowtails can be distinguished from Old World Swallowtails by the black pupil of the dorsal hindwing eyespot, which forms a thick black line, usually club-shaped,

Ssp. *oregonius* ♂ D (6.2 cm)

Ssp. *oregonius* ♂ V (6.2 cm)

Ssp. *pikei* ♂ D (6.9 cm)

Ssp. *dodi* ♀ V (6.2 cm)

at the bottom of the red area (Fig. 61). Baird's Swallowtails are also larger and a brighter yellow.

**IMMATURE STAGES:** Eggs are white (soon turning yellow), spherical, flattened at the base, and with a finely pitted surface. The first to third instar larvae resemble bird droppings: they are black with a white splotch in the middle of the back and, in the second and third instars, orange spots down the sides. Mature larvae are green above and bluish green below. Around the middle of each segment is a black ring with yellow spots, except that 68% of the larvae of subspecies *pikei* have orange spots (Sperling 1987). Pupae are long and cylindrical, tapering towards the back. There is a projection on each side of the thorax and in the middle of the top of the thorax. The front of the head is elongated forward as two projections. Pupae are usually brown but sometimes yellow green.

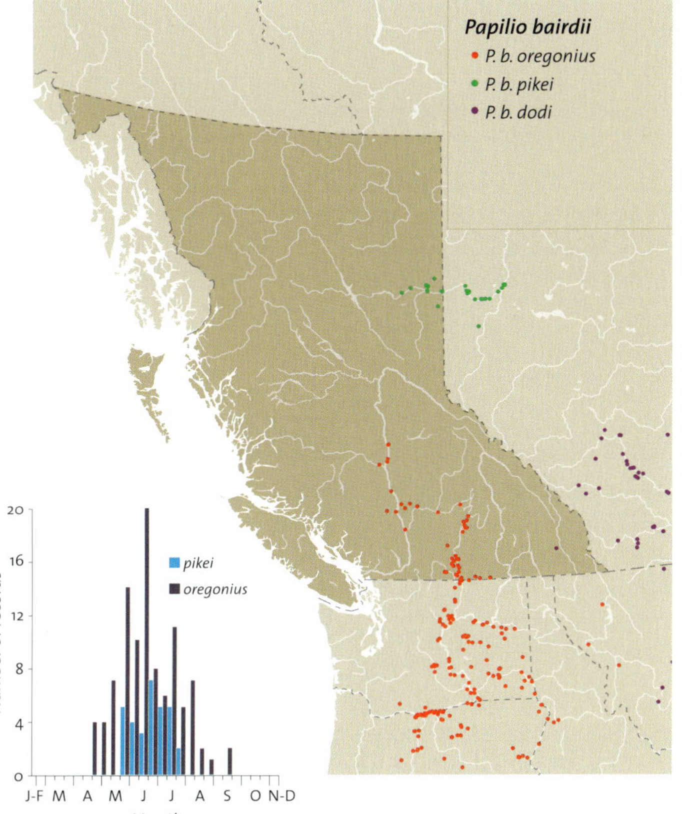

**BIOLOGY:** Subspecies *pikei* is univoltine, and flies from late May through early July. Subspecies *oregonius* is bivoltine, with flight periods from April to June and July to September. Eggs are laid singly on the leaves of the foodplant. First instar larvae produce silk to anchor themselves; when knocked off a leaf, they hang from a silk thread. Larvae spin silk pads on which to rest. They are readily visible on the plants as they rest on the leaves or stems. Adult males commonly "hilltop" on hill or cliff tops.

Baird's Swallowtails use tarragon (*Artemisia dracunculus*) as the only natural larval foodplant. When subspecies *pikei* from Dunvegan, Alberta was reared from wild-collected and captive-laid eggs, however, larvae fed on leaves of cow parsnip (*Heracleum maximum* [Apiaceae]) grew at roughly twice the rate of those fed on tarragon, with no mortality (CSG).

**SUBSPECIES:** Subspecies *oregonius* W.H. Edwards, 1876 (TL: Columbia River, near The Dalles, OR), the Oregon Swallowtail,

**Eggs**

**First instar larva**

**Mature larva**

**Pupa**

**Baird's Swallowtail (*Papilio bairdii oregonius*)**

inhabits the dry sagebrush areas of the Southern Interior and Chilcotin, north to Soda Creek. Adults are large and yellow, and the black pupil of the dorsal hindwing eyespot forms a thick black line, usually club-shaped, at the bottom of the red area. On the ventral hindwing there is usually a substantial amount of orange in addition to the eyespot. It is the state insect of Oregon (Dornfield 1980). Subspecies *pikei* Sperling, 1987 (TL: Dunvegan, AB) inhabits the dry grassland and clay bank slopes along the Peace River near the Alberta border. Adults are smaller and darker than *oregonius,* and on the ventral hindwing there is little or no orange other than the eyespot. Subspecies *dodi* is known in BC from one specimen collected near Cranbrook by CSG. The adults have the base of the ventral forewing solid black rather than mixed black and yellow, and the black of the eyespot is clubbed.

**RANGE AND HABITAT:** Baird's Swallowtails inhabit open, dry grass slopes along the Peace River canyon, the dry grasslands of the Southern and Central Interior, and the east Kootenay.

**GENERAL DISTRIBUTION:** Baird's Swallowtails occur in the Peace River area of AB and BC, and from central BC, central AB, and southern SK south to northern CA, AZ, NM, and KS.

**CONSERVATION STATUS:** Subspecies *pikei* is of Special Concern in BC (S3). Subspecies *oregonius* is not of concern (S4). Subspecies *dodi* is of uncertain status, but rarely seen (S1S3), and is of conservation concern.

## OLD WORLD SWALLOWTAIL
*Papilio machaon* Linnaeus, 1758

**ETYMOLOGY:** Machaon was a physician who served on the Greek side during the Trojan War. He was the son of Asclepius, the god of healing, and Epione (Emmet 1991). This was one of the names of several Greek heroes used by Linnaeus for swallowtails with red marks on the thorax. The subspecies name *aliaska* refers to the type locality being in Alaska. The common name was first used by Klots (1951), referring to this swallowtail being the only one known from Europe and Asia (the Old World) as well as North America.

**ADULT:** Old World Swallowtails can be separated from Anise Swallowtails and Baird's Swallowtails by the black pupil of

the red eyespot on the dorsal hindwing, which forms a thin black line attached to the bottom of the red area (Fig. 61). They are also generally smaller and darker than Baird's Swallowtails.

♂ D (5.8 CM)

In BC and Alberta, *P. machaon* consists of subspecies *aliaska* and *hudsonianus,* and *P. bairdii* of subspecies *pikei, oregonius,* and *dodi*



FAMILY PAPILIONIDAE (SWALLOWTAILS AND APOLLOS)

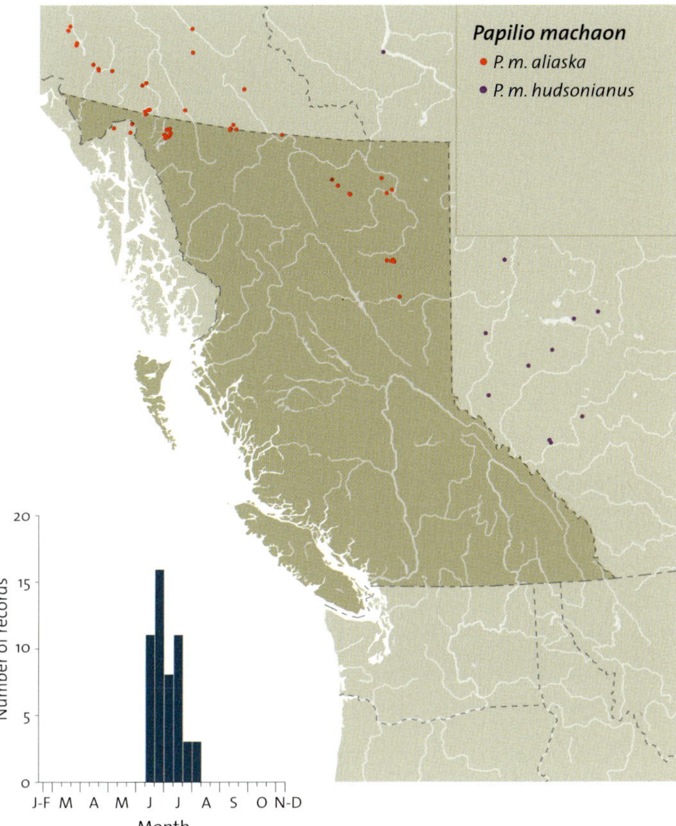


of some *aliaska* wing pattern and protein characters into the *oregonius* genotype, leading to the *pikei* genotype. The genotype of *pikei* is now stable, lacking hybridization with nearby *aliaska* populations, which indicates reproductive isolation and hence speciation. The close relationship of *P. machaon* and *P. bairdii* is indicated by the high similarity in their mitochondrial DNA (Sperling 1993).

**IMMATURE STAGES:** Eggs are white, soon turning yellow, spherical, flattened at the base, and with a finely pitted surface. The first to third instar larvae resemble bird droppings, similar to those of Baird's Swallowtails. Mature larvae are green above and bluish green below; around the middle of each segment is a black ring with yellow spots (Sperling 1987). Pupae are similar to those of Baird's Swallowtails.

**BIOLOGY:** Old World Swallowtails are univoltine, and are in flight in June and July. Eggs are laid singly on the leaves of the foodplant. Larvae are readily visible on the plants as they rest on the leaves or stems. Adult males commonly "hilltop" on the top of mountains.

In BC the larval foodplant is mountain sagewort, *Artemisia norvegica*. Coltsfoot (*Petasites frigidus* var. *palmatus*) has been reported for subspecies *hudsonianus* outside BC (Sperling 1987).

**SUBSPECIES:** Subspecies *aliaska* Scudder, 1869 (TL: Nulato, AK) occurs in BC. Subspecies *hudsonianus* A.H. Clarke, 1932 may occur in the unsampled boreal forests of northeastern BC.

**RANGE AND HABITAT:** Old World Swallowtails inhabit openings in the northern boreal forests, northern subalpine willow-shrub habitats, and alpine tundra.

**GENERAL DISTRIBUTION:** Old World Swallowtails occur from Britain, Europe, and north Africa across temperate Asia to North America. In North America the species occurs from AK across boreal Canada to PQ.

**CONSERVATION STATUS:** Not of concern (S5).

(McCorkle and Hammond 1989), contrary to Sperling (1987). Subspecies *hudsonianus* and possibly *aliaska* freely hybridize with *P. zelicaon* and the related eastern *Papilio polyxenes* Fabricius, 1775. In contrast, the *P. machaon* subspecies rarely hybridize with the *P. bairdii* subspecies *dodi* and *pikei,* even when they are sympatric or near sympatric, indicating reproductive isolation. The Peace River lowlands were probably colonized by *P. bairdii oregonius* (along dry riverbanks through Pine Pass) during the warm hypsithermal period, at which time a small amount of hybridization with *P. machaon aliaska* occurred. This resulted in the introgression

## ANISE SWALLOWTAIL
*Papilio zelicaon* Lucas, 1852

**ETYMOLOGY:** The name *zelicaon* is derived from the Greek *zelos* (emulation). The suffix "caon" likely refers to the species' similarity to *Papilio machaon,* thus the species "emulates *machaon*" (Bird et al. 1995). The common name was first used by Comstock (1927), and reflects the frequent use of anise as a larval foodplant in California.

**ADULT:** Anise Swallowtails can be distinguished from Old World and Baird's swallowtails by the shape and position of the black pupil of the red eyespot on the dorsal hindwing. The black pupil forms a black spot centred in the red eyespot, and is completely surrounded by red (Fig. 61). Adults of interior populations tend to be smaller and darker than

those of coastal populations. Occasional hybrids with Old World and Baird's swallowtails occur, and they are intermediate in appearance.

**IMMATURE STAGES:** Eggs are spherical and pale yellow. The micropylar area turns red, and a red ring develops around the middle of the

♂ D (6.6 CM)

egg as it matures. The first to third instar larvae resemble bird droppings, being black with a white splotch in the middle of the back and, in instars 2 and 3, orange spots down the sides. In mature larvae, black and green rings alternate around the body, with yellow or orange spots in the black rings. The spots are usually yellow on the coast and orange in the Peace River and eastern interior, and yellow and orange in equal frequency in the Southern Interior (Sperling 1987). Pupae are long and cylindrical, tapering posteriorly. Each side of the thorax has a lateral protuberance, and down each side of the back is a row of small protuberances. Pupal colour is highly variable, ranging from very dark brown through green to green yellow.

**BIOLOGY:** Anise Swallowtails are univoltine east of the Coast Range. Adults are in flight in April and May at low elevations, and until August and September above the timberline. In subalpine and alpine habitats west of the Coast Range, they are also univoltine. At low elevations on the coast, they are univoltine (April and May) in dry areas around Victoria, and trivoltine (April to September) along the wetter coastal areas. In dry summers occasional adults occur in the "univoltine" areas of southeastern Vancouver Island throughout the summer, due to either immigration from the west coast of Vancouver Island or a partial second and/or third brood. In years with a wet May and June, there is definitely a partial second brood. In general, populations using *Lomatium* species as their primary larval foodplant (such as those on southeastern Vancouver Island) are univoltine, because the foodplants dry up by midsummer. Females from different populations prefer to lay eggs on different foodplant species among those available. The differences in oviposition preference do not result

Second to fifth instar larvae, coastal BC

Mature larva, interior BC

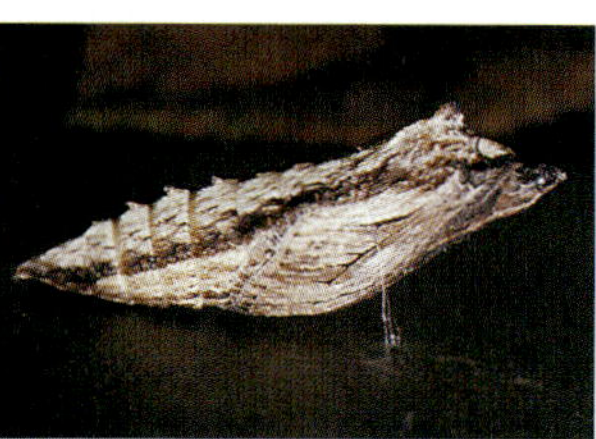
Brown pupa

from genetic differences, and change between seasons in some populations (Wehling 1994), including the one at Boundary Bay, Delta (R. Ashton, pers. comm.).

Green pupa

Many different plants in the family Apiaceae are used as larval foodplants, including garden parsley, parsnips, and carrots. Wehling (1994) recorded 69 foodplant species in 39 genera of Apiaceae and Rutaceae, with 70% of the host species in the genera *Lomatium, Angelica, Foeniculum,* and *Heracleum.* In BC *Lomatium* species (spring gold, chocolate tip) are commonly used in dry areas, both on southern Vancouver Island and in the Southern Interior. Cow parsnip (*Heracleum maximum*) is used in wet meadows and ditches at all elevations. Cow parsnip and angelica are used along beach tops. Females straying into urban and suburban areas frequently lay eggs in vegetable gardens on carrots, parsley, anise, parsnips, and dill (Jones 1933; R.A. Cannings, pers. comm.; CSG). Most of the resulting larvae are eliminated as pests, unless the gardener recognizes them as Anise Swallowtail larvae.

**SUBSPECIES:** The Anise Swallowtail type locality has been restricted to San Francisco, CA (Emmel et al. 1998b), and there are no recognized subspecies. Within California there is no significant genetic differentiation (Tong and Shapiro 1989), but the unpublished electrophoretic data of Wehling (1994) suggest the possibility of a California subspecies separate from the rest of western North America. There are also some isolated, genetically distinct populations along the

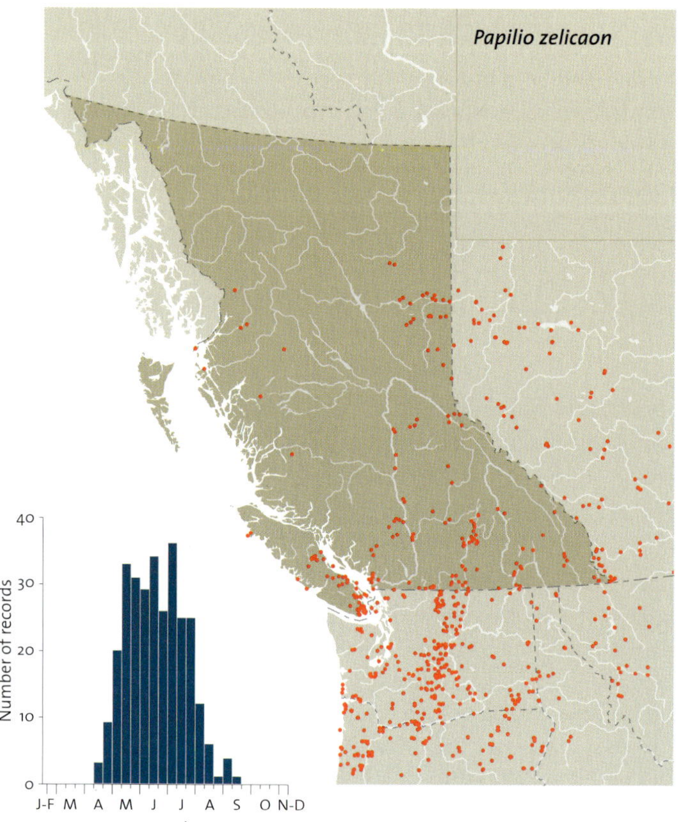

coast of Washington and Oregon that may have resulted from postglacial colonization patterns as the sea level rose and climate changed.

**RANGE AND HABITAT:** Anise Swallowtails inhabit forest edges, stream banks, beach tops, hilltops, open rocky knolls, moderately dry grasslands, and subalpine meadows both on the coast and in the interior. Males are frequently found hilltopping.

**GENERAL DISTRIBUTION:** Anise Swallowtails occur only in North America, from the central coast of BC east to SK and south to CA, AZ, and NM.

**CONSERVATION STATUS:** Not of concern (S5).

## INDRA SWALLOWTAIL
*Papilio indra* Reakirt, 1866

**ETYMOLOGY:** The species name *indra* refers to the Hindu god Indra, the head deity of the deities of the air. This in reference to the type locality of Pike's Peak, CO, a high, prominent peak that sticks up into the air. The common name was first used by Holland (1898).

**ADULT:** Indra Swallowtails in BC are small and almost entirely black except for a yellow band across the middle of the wings. They have very short tails and cannot be confused with any other swallowtail.

**IMMATURE STAGES:** Eggs are spherical and creamy with a slight greenish tinge; the surface is very finely pitted. Within two days a brownish ring develops around the circumference as well as a brownish area at the micropyle. Before hatching, the egg becomes totally black. Mature larvae are velvety black, with a pinkish transverse stripe across each segment. There is an ochre yellow spot on each side of the dorsal line on each segment, and a smaller yellow spot on each side of

the back. There is another series of yellow spots along each side, with the last spot being white. Pupae are similar in shape to those of Anise Swallowtails, but are much smoother. The two projections in front of the head are absent, and the projections on the thorax are

♂ D (6.9 CM)

reduced. The wing cases and the thorax are initially dull green, then become olive green. The abdomen is light creamy brown, and mottled with darker brown in Yakima Co., WA (Newcomer 1964b).

**BIOLOGY:** In BC Indra Swallowtails have a single generation that is in flight in June and July. Eggs hatch after 6 days at room temperature. The larval period is 18 days, and the pupal period is about 11 months. Males frequently mud-puddle, while females are rarely seen except occasionally nectaring or near the larval foodplant. The newly hatched larva eats the egg chorion, and after each moult the larva eats all the exuviae except for the head capsule (Newcomer 1964b).

The larval foodplants have not been determined for BC. In Yakima Co., WA, the larval foodplant is *Lomatium grayi* (Newcomer 1964b). Other populations of Indra Swallowtails use *Aletes acaulis, Harbouria trachypleura, Lomatium lucida, L. parryi, L. eastwoodiae, Pteryxia petreae, P. terebinthina, Tauschia parishii,* and *T. arguta* (Emmel and Emmel 1962, 1963, 1964, 1967, 1968, 1973, 1974; Emmel 1982; Scott 1992).

**SUBSPECIES:** The nominate subspecies, *P. i. indra* (TL: Pike's Peak, CO), occurs in BC.

**RANGE AND HABITAT:** Indra Swallowtails are known in BC only from the subalpine areas of Gibson Pass and Allison Pass in Manning Provincial Park.

**GENERAL DISTRIBUTION:** Indra Swallowtails occur from the Cascade Mountains of BC south to Baja California. They extend east to the Black Hills, SD, and south through the Rockies to northern NM.

**CONSERVATION STATUS:** Indra Swallowtails are of Special Concern in BC (S2).

# Canadian Tiger Swallowtail

*Papilio canadensis* Rothschild & Jordan, 1906

**Etymology:** The name *canadensis* refers to the Canadian type locality of the species. The common name, Tiger Swallowtail, was first used by Gosse (1841a) for the Eastern Tiger Swallowtail, *Papilio glaucus,* which included *canadensis* before the Canadian Tiger Swallowtail was recognized as a separate species. The name "Canadian Tiger Swallowtail" was first used by Macy and Shepard (1941).

**Adult:** Canadian Tiger Swallowtails are yellow with black bands running from front to back across the wings. The spots in the black margins on the dorsal hindwings are mostly yellow, but the first spot is always orange (in contrast to the Western Tiger Swallowtail), except that sometimes it is missing entirely. These marginal spots are all partly or entirely orange on the underside of the hindwing. The blue markings on the underside of the hindwings are much more developed in the Canadian Tiger Swallowtail than in the Western Tiger Swallowtail. There is a zone of hybridization with the Western Tiger Swallowtail where their ranges overlap.

For many years the Canadian Tiger Swallowtail was considered to be a subspecies of the Eastern Tiger Swallowtail. Recently Hagen and Scriber (1991) and Hagen et al. (1992) demonstrated that they are separate species that fly together in southern Ontario. Eastern Tiger Swallowtails are bivoltine and can mature more successfully on different larval foodplants than Canadian Tiger Swallowtails (with

♂ D (8.2 cm)       ♂ V (8.2 cm)

some overlap). Many pupae of hybrids die rather than produce adults.

**Immature stages:** The eggs are green, approximately hemispherical, and smooth. Mature larvae are velvet green with eyespots on their middle thoracic segment and a yellow and black stripe between that segment and the next. The eyespot on each side of the thorax of a mature larva is yellow, outlined in black and bisected by a black line. A black line encloses a blue centre spot, and the black transverse band is narrower than the anterior yellow band and does not extend to the spiracular line (Sugden and Ross 1963). The osmeteria are orange (CSG) and the prepupal larvae turn dark reddish brown (Saunders 1869c). Pupae are light brown with a darker brown lateral stripe, two short horns on the head, and a peak at the top of the thorax.

**Biology:** Canadian Tiger Swallowtails are univoltine. They are in flight from May to July at low elevations, and until August at higher elevations. Eggs are laid singly on the upperside of the leaves of larval foodplants, and hatch in 7–10 days. Pupation occurs in another 6–8 weeks, and pupae diapause over the fall and winter. The males patrol forest openings and stream banks looking for females, and are frequently seen mud-puddling.

In BC larval foodplants include alder, birch, black cottonwood, trembling aspen, and willow (Sugden and Ross 1963; CSG; FIS). Outside BC other larval foodplants include ash, cherry, cultivated apple, and cultivated crab apple (Fletcher 1889c; Ferguson 1954; Brower 1959; Bird et al. 1995).

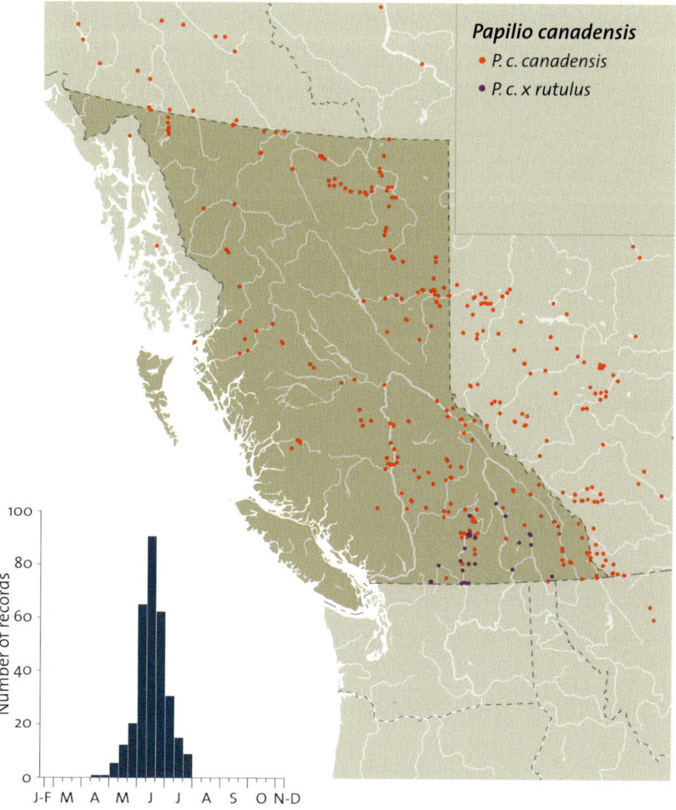

**Canadian Tiger Swallowtail (*Papilio canadensis*)**

**Egg**

**Mature larva**

**Prepupal larva**

**Pupa**

**SUBSPECIES:** Canadian Tiger Swallowtails were originally described from Newfoundland. There are no recognized subspecies. *Papilio rutulus arcticus* Skinner was published in December 1906 and is a synonym of *P. canadensis*, which was published in August 1906 (Martin Honey, pers. comm.).

**RANGE AND HABITAT:** Canadian Tiger Swallowtails are found throughout southeastern, central, and northern BC in predominantly boreal forest habitats. They occur in coniferous forest and aspen grove habitats.

**GENERAL DISTRIBUTION:** Canadian Tiger Swallowtails are found from AK south to southeastern BC, and east across to NF. East of the Rocky Mountains, the species extends south into the northern USA.

**CONSERVATION STATUS:** Not of concern (S5).

# WESTERN TIGER SWALLOWTAIL
*Papilio rutulus* Lucas, 1852

**ETYMOLOGY:** The name *rutulus* refers to the people called Rutuli in Roman legendary history. Their king was Turnus, the Latin name formerly used for the larger Eastern Tiger Swallowtail; thus *rutulus* is closely related to *turnus*. The common name "Western Tiger Swallowtail" was first used by Comstock (1927) in reference to its western North American distribution.

♂ D (7.6 CM)

♂ V (7.6 CM)

**ADULT:** Western Tiger Swallowtails are large yellow butterflies with black stripes running from front to back across the wings. All the marginal spots on the dorsal hindwings are yellow, including the first one. The ventral surface of the hindwings frequently has extensive blue and orange spots or areas, especially in females, but much less so than in the Canadian Tiger Swallowtail.

**IMMATURE STAGES:** The eggs are green, hemispherical in shape, and smooth. Mature larvae are smooth, hairless, and green, with two sky blue dots on each side of each segment. The head is orange, with a yellow "collar" behind it. The eyespots of mature larvae consist of a large main eyespot with a smaller satellite spot below. The larvae have orange osmeteria. The prepupal larvae turn brown, the blue spots fade, and a prominent eyespot (orange with a black centre) replaces the eyespots on the third segment. The pupae are brown and have a girdle, and are fastened head up on the side or underside of logs, branches, fences, or house siding (CSG).

**BIOLOGY:** Western Tiger Swallowtails are univoltine in BC, but populations further south have two or three generations

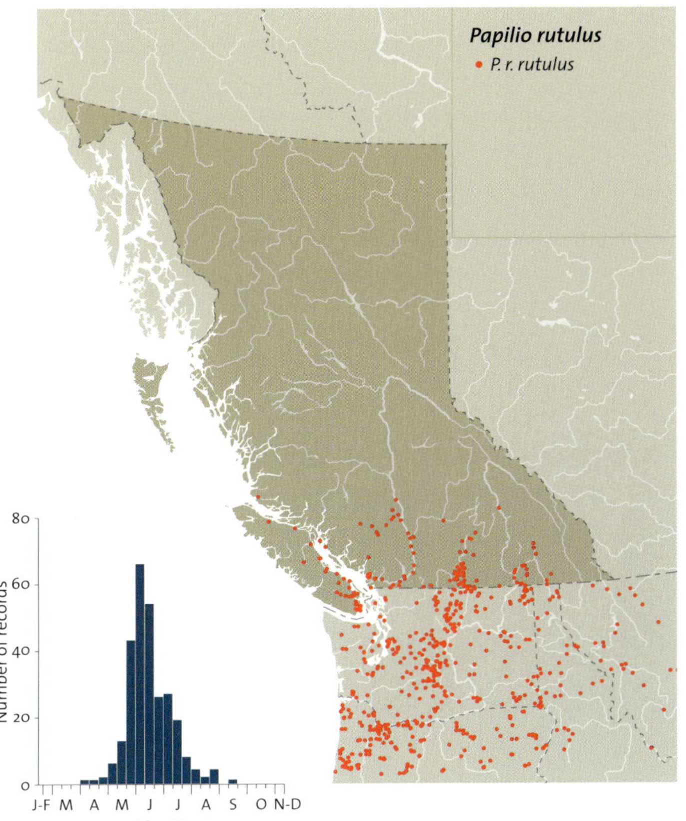

in a summer. They are in flight from late May to July, and until August at mid elevations. Eggs are laid singly on the upperside of the leaves of foodplants, with a female laying only one or a few eggs on a plant before flying on to find another. Eggs hatch in 7–10 days, and pupation occurs about 6–8 weeks later. There are five larval instars. The brown prepupal larvae, which spend a day or two wandering before pupating, are frequently found crossing roads and paths in August before finding a sheltered spot in which to pupate. Pupae hibernate, and adults emerge early the next summer. Males patrol forest openings and stream banks looking for females, and are frequently seen mud-puddling. Vehicles on roads frequently kill both adults and prepupal larvae (CSG). An individual Western Tiger Swallowtail has lived at least 39 days in the wild (Smith 1982).

Larval foodplants in BC are alder, cultivated apple, birch, bitter cherry, poplar, and the willows *Salix hookeriana* and

Egg

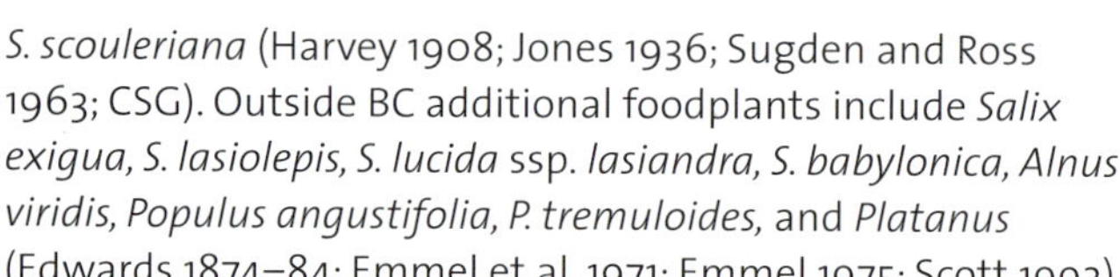
Mature larva

*S. scouleriana* (Harvey 1908; Jones 1936; Sugden and Ross 1963; CSG). Outside BC additional foodplants include *Salix exigua, S. lasiolepis, S. lucida* ssp. *lasiandra, S. babylonica, Alnus viridis, Populus angustifolia, P. tremuloides,* and *Platanus* (Edwards 1874–84; Emmel et al. 1971; Emmel 1975; Scott 1992).

SUBSPECIES: The nominate subspecies of the Western Tiger Swallowtail (TL: restricted to near Belden, Plumas Co., CA [Emmel et al. 1998b]) occurs in BC. The subspecies name *arcticus* Skinner, 1906, which has sometimes been considered to apply to a northern subspecies of Western Tiger Swallowtail, is a synonym of the Canadian Tiger Swallowtail (Hagen et al. 1992).

RANGE AND HABITAT: Western Tiger Swallowtails are common along the southern edge of BC from Bella Coola and Vancouver Island east to Creston. Populations are free of hybridization with the Canadian Tiger Swallowtail west of the Cascade and Coast mountains. They are found wherever their larval foodplants occur, but are particularly common in or near riparian habitats along streams and rivers, where their larval foodplants are most abundant. They are often quite common in residential areas due to the use of birches, willows, and poplars as boulevard and garden trees.

GENERAL DISTRIBUTION: Western Tiger Swallowtails occur from southern BC south to CA and northern MEX, and east to the Rockies and NM.

CONSERVATION STATUS: Not of concern (S5).

**Western Tiger Swallowtail (*Papilio rutulus*)**

## TWO-TAILED SWALLOWTAIL
*Papilio multicaudatus* W. Kirby, 1884

ETYMOLOGY: The name *multicaudatus* is derived from the Latin *multi* (many) and *cauda* (tail). The subspecies name *pusillus* is Latin for very little, paltry, petty, in reference to the relatively small size of the subspecies (Austin and Emmel 1998b). The common name was first used by Holland (1925).

ADULT: Two-tailed Swallowtails, which actually appear to have three tails due to a pronounced "caudal lobe" on the hindwings, are the largest butterflies in BC. The wing colour is yellow with narrow black bands. No other swallowtail in BC has more than one tail on each hindwing.

♂ D (7.8 CM)

♂ V (7.8 CM)



FAMILY PAPILIONIDAE (SWALLOWTAILS AND APOLLOS)

**Immature stages:** The eggs are pale green yellow, smooth, and hemispherical (CSG). The eyespots of mature larvae consist of a large main eyespot with a smaller satellite spot below. The main spot is yellow green to yellow and is surrounded by a very fine black line; it has a pale blue centre spot that is surrounded by yellow and then by a black line. A black line below the blue centre spot bisects it. The satellite spot is yellow green to yellow and is enclosed by a very fine black line. The black transverse band, three to four times as wide as the anterior yellow band, extends below the spiracular line. A narrow black line frequently occurs on the top of the front margin of some of the abdominal segments (Sugden and Ross 1963). Pupae are mottled green brown to yellow brown, with a darker brown lateral stripe.

**Biology:** Two-tailed Swallowtails are univoltine in BC and multivoltine in the southern USA. They are in flight in May and June, rarely as late as August at mid elevations. The adult males patrol forest edges and openings, lake margins, and stream banks looking for females, and are frequently seen mud-puddling. Females do not start to lay eggs for two or three days after fertilization (except old females), during which time she nectars. The female lays 6–7 eggs per day,

Two-tailed Swallowtail (*Papilio multicaudatus*)

over a 10-day period, usually producing about 55 eggs. Eggs are laid singly on the upperside of the leaves of larval foodplants, and tend to be laid in partial shade. Eggs hatch within 7–10 days; immediately after hatching, the larva eats the egg chorion. It makes a silk mat on the surface of the leaf upon which to rest. The silk mat leads to the edge of the leaf, where the larva feeds. All instars make the silk mat (Pronin 1955). Larvae pupate within 5–6 weeks, and pupae diapause over winter.

The larval foodplant normally used in BC, in Yakima County, WA, and in California is chokecherry (*Prunus virginiana*) (Pronin 1955; Newcomer 1964a; CSG), but saskatoon (*Amelanchier alnifolia*) is also used (McDunnough 1927). The introduced hop tree (*Ptelea trifoliata* [Rutaceae]) is used when available in gardens (Grant 1963). Outside BC additional foodplants include ash and tulip tree (Brower 1959; Kendall 1964; Scott 1992).

**Subspecies:** There are three subspecies of Two-tailed Swallowtails in North America. Subspecies *pusillus* Austin and Emmel, 1998 (TL: Independence Mountains, Elko Co., NV) occurs in BC (Austin and Emmel 1998), and is smaller and less richly coloured than the other two.

**Range and habitat:** Two-tailed Swallowtails inhabit the arid grasslands and associated riparian area of southern BC, at low to mid elevations. They range as far north as Soda Creek in the Chilcotin.

**General distribution:** Two-tailed Swallowtails occur from southern BC and AB south to Guatemala and east to the central Great Plains and TX.

**Conservation status:** Not of concern (S4).

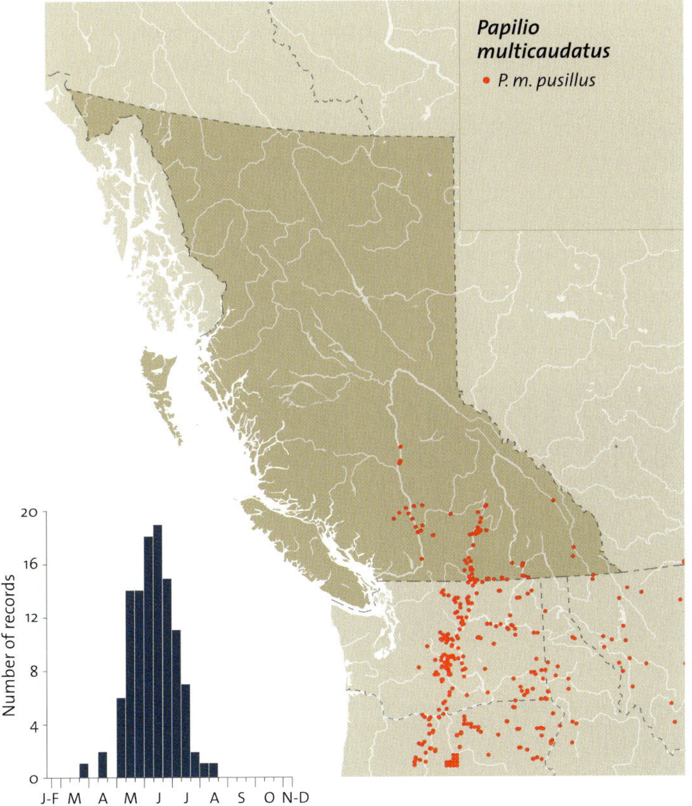

# Pale Swallowtail

*Papilio eurymedon* Lucas, 1852

**Etymology:** The name *eurymedon* is derived from the Greek *eurys* (broad) and *medon* (guardian) (Bird et al. 1995), in reference to the broad black border on the wings. The common name was first used by Comstock (1927) in reference to the adult's whitish ground colour compared with the yellow colour of most other species.

**Adult:** Pale Swallowtails are the only swallowtails in BC that are white (males) or very pale yellow (females) with black stripes, rather than yellow and black. Pale Swallowtails have a small orange crescent, sometimes missing, at the base of the dorsal hindwing tail. All other tiger swallowtails have a large yellow crescent instead.

The report of *Papilio ajax* var. *marcellus* (= *Eurytides marcellus*) from the Cowichan Valley on Vancouver Island (Fletcher 1899), based on a painting sent to Fletcher by C. de Blois Green, may have been either an accidental introduction or a misidentification based on an aberration of *P. eurymedon*.

**Immature stages:** Eggs are yellow green, smooth, and hemispherical in shape. A pink tinge develops around the side of the egg within 1–2 days after oviposition. The day before hatching, the egg turns green brown, with the black head capsule showing through the top. Mature larvae are cylindrical, tapering towards the back. They are apple green, except for the whitish underside. They have eyespots similar to those of the Western Tiger Swallowtail, but the blue central spot is smaller. The blue centre of the Western Tiger

Swallowtail eyespot is about 1 mm, that of the Pale Swallowtail is 0.5 mm. Behind the eyespots, the two segments are separated by a broad yellow band followed by a black band. Osmeteria are bright orange. Prepupal larvae turn brown, with bright orange eyespots with black centres. The pupae are cylindrical, with their greatest diameter near the middle, tapering slightly towards the head and rapidly backwards. The pupae are usually brown, streaked with black and brown, with a dark brown band along each side. Some pupae are green (Edwards 1874–84; Sugden and Ross 1963; CSG).

♂ D (9.0 cm)

**Biology:** Pale Swallowtails are univoltine in BC but multivoltine further south. They are in flight in June and July in southern BC. Eggs are laid singly on the upper surface of the leaves of larval foodplants. Eggs hatch 9–10 days after oviposition, and the larvae eat the egg chorion. The first instar lasts about 5 days, and larvae mature in 5–6 weeks. Pupae diapause over winter. Males patrol along forest edges and openings and stream banks looking for females, and are frequently seen mud-puddling. They hilltop on Vancouver Island (Guppy 1970), but this may be because the tops of the hills are the best forest openings available for patrolling. Larvae from the eggs of an interior population using *Ceanothus* grew abnormally slowly when fed red alder (R. Ashton, pers. comm.). In captivity, females from the Pend-d'Oreille River in the West Kootenay refused to lay eggs on

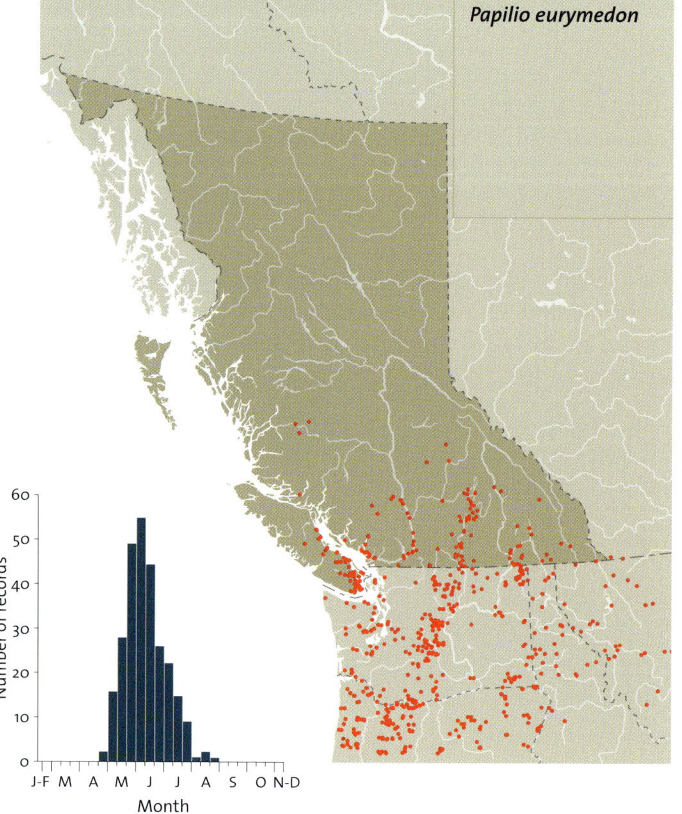



Number of records

60
50
40
30
20
10
0

J-F M A M J J A S O N-D

Month

Egg

Mature larva

Prepupal larva

Pupa



Family Papilionidae (Swallowtails and Apollos)

Western Tiger Swallowtail (*Papilio rutulus*) and Pale Swallowtails (*Papilio eurymedon*) mud-puddling

*Ceanothus,* and then laid eggs only on red alder when a mixture of the two plants was offered the next day (CSG).

The known larval foodplants in BC are cultivated apple, *Alnus rubra, Amelanchier alnifolia, Betula, Ceanothus sanguineus, Holodiscus discolor,* and *Prunus emarginata* (Harvey 1908; Dyar 1904b; McDunnough 1927; Jones 1936, 1939, 1942; Sugden and Ross 1963; ACJ; FIS). Red alder is the primary larval foodplant on the coast and *Ceanothus* the primary larval foodplant in the dry interior (CSG). Outside BC additional foodplants include garden crab apple, *Ceanothus fendleri, Crataegus rivularis, Frangula californica, Prunus emarginata, P. ilicifolia,* and *Rhamnus crocea* (Edwards 1874–84; Remington 1952; Emmel 1975; Bird et al. 1995; Scott 1992; ACJ).

SUBSPECIES: None. The type locality has been restricted to near Belden, Plumas Co., CA (Emmel et al. 1998b).

RANGE AND HABITAT: Pale Swallowtails occur throughout southern BC, in low- to mid-elevation forest openings and riparian habitats as well as on dry *Ceanothus* slopes in the interior.

GENERAL DISTRIBUTION: Pale Swallowtails occur from southern BC and AB east to the central Great Plains and south to Baja California and NM.

CONSERVATION STATUS: Not of concern (S5).

# Family Pieridae Duponchel, [1832]
## Whites, Marbles, and Sulphurs

The family name Pieridae is derived from the type genus, *Pieris*. The common name for the family is derived from the common names of the three subfamilies.

The Pieridae are medium-sized butterflies with white, yellow, orange, or (one BC species) yellow green wings. All species have black markings in addition to their basic colour on the upperside. They are generally sexually dimorphic, with females frequently very different in appearance from the males. None of the species in BC have tails on the hindwings.

Their eyes are smooth, hairless, round, and not indented next to the antennae. The antennae are of medium length, slender, and with a tapering club. The palpi are longer than the head, and the face is as tall as it is wide between the eyes. The forewing has 10, 11, or 12 veins, and the discal cell is closed in both wings. The hindwing has two anal veins, and forewing veins 1A and 2A are joined. Androconial scales are usually present on the dorsal forewing. All three pairs of legs are fully functional for walking. All legs are about the same size, and the foreleg has no projection (epiphysis) on the upper part of the tibia. There are two claws on the end of each leg, each of which is forked in two, and paronychia and pulvilli are sometimes present.

Eggs are tall and conical, 2–3 times as tall as they are wide, and with vertical ribs. Larvae are slender, cylindrical, covered with a fine layer of short hairs, and with longitudinal stripes. Young larvae frequently have glandular hairs that secrete drops of an unknown fluid. Pupae are roughly cylindrical and smooth, and have a pyramidal projection on the back of the head. The pupae are held head up against a stem with a girdle as well as a cremaster. Pupae or larvae hibernate.

There are about 1,000 species of Pieridae in the world, with about 75 species in North America. British Columbia has 28 confirmed species.

**Spring White (*Pontia sisymbrii flavitincta*)**

The name of the subfamily Pierinae is derived from the type genus, *Pieris* (below). The common name is derived from the typically white ground colour.

Whites are medium-sized butterflies that are typically white with black markings, but many species have females or ventral wings that are pale yellow. They typically have a moderately fast, straight, and fluttering or bobbing flight pattern.

Antennae are moderately long, with a well-defined club. The forewing has 10 or 11 veins, and the hindwing precostal vein is present. The prothorax is without patagia, which are two hardened lobelike subdorsal lumps present in the subfamily Coliadinae. All the whites in BC are in the tribe Pierini. The dorsal forewing has androconial scales.

All whites in BC except for the Pine White feed on plants in the family Brassicaceae. Whites may be distasteful to vertebrate predators because of the mustard oil glycosides that they acquire from their larval foodplants. Both chipmunks and ground squirrels find Cabbage Whites tasty (CSG), however, and birds apparently eat all species of whites, suggesting that any chemical protection is not widely effective. It may decrease the rate of predation rather than eliminate it.

### GENUS *NEOPHASIA* BEHR, 1869  PINE WHITES

The name *Neophasia* is from the Latin *neo* (new) and *phasis* (phase or appearance). "New" refers to the genus *Leucophasia* Stephen, 1827, to which Behr compared the new genus. The common name "pine whites" refers to the larval foodplants being pines and other trees in the family Pinaceae.

There are only two species of pine whites in the world. The Pine White occurs throughout western North America, including BC. The other species, the Chiricahua Pine White, occurs only in the southwestern USA and northwestern Mexico. The larvae of both species feed exclusively on trees in the family Pinaceae (pines, firs, hemlocks).

The genus *Neophasia* is characterized by a straight hindwing humeral vein, and by the cross-veins at the outer ends of the hindwing and forewing discal cells being straight and not curved.

The genus appears to be quite primitive, and may have evolved before flowering plants replaced conifers and related plants as the dominant vegetation.

## PINE WHITE
*Neophasia menapia* (C. & R. Felder, 1859)

**Pine White (*Neophasia menapia*)**

**ETYMOLOGY:** *Menapia* is Latin for the people of Gallia Belgica, or modern Belgium and Netherlands. Alternatively, the name may be derived from either *meno* (to stay or remain) or *menos* (force or strength). None of the three possibilities has an obvious connection to the butterfly. The subspecies name *tau* is of unknown derivation. The common name was first used by Holland (1898) in reference to the larval foodplants being pines and other trees in the family Pinaceae.

**ADULT:** Pine Whites are easily recognized white butterflies. The outer margins of the forewings and hindwings are black with white spots. The front edge of the forewing is black, and the wing veins on the ventral hindwing are outlined in black. Orange markings are frequently present on the ventral hindwings, especially in females.

**IMMATURE STAGES:** Eggs are flask-shaped and emerald green, with vertical ribbing and a circle of white beadlike bumps below the narrow upper end. Mature larvae are dark green with a narrow dorsal white stripe and a broad white stripe along each side; they have two short anal tails. The head is yellow green, sometimes with blackish patches. The legs are black, and the prolegs are green yellow. Pupae are slender and dark green, with a white dorsal line and two white lateral lines (Edwards 1887–97; Holland 1931).

**BIOLOGY:** Pine Whites are univoltine throughout their range. The adults are on the wing from late July at low elevations to late September at high elevations and on the west coast of Vancouver Island. The females spend most of their time resting in the upper branches of conifers, making short flights for oviposition or to forage for nectar. Males spend most of their time patrolling the surface of the conifers (where females rest), but spend considerable time nectaring at flowers, especially in the morning and evening. Their flight is very slow and weak, making it possible to capture them in midair with bare hands.

Pine Whites undergo occasional population outbreaks that result in severe defoliation of conifers, which is sometimes controlled by aerial spraying. Fletcher (1902) noted that "towards the end of the season, in August, the dead butterflies may be seen in vast numbers floating on the sea around Vancouver Island, or thrown up along the beach in windrows sometimes an inch or two in depth." The most recent population outbreak in BC was in 1961 on Vancouver Island, when there were "snowdrifts" of Pine Whites along the highway through Cathedral Grove (CSG). Population outbreaks are rare now, possibly because most of the old forests of southeastern Vancouver Island have been eliminated.

♂ D  (4.4 cm)

♂ V  (4.4 cm)

♀ V  (4.7 cm)

High population densities in Yakima County, WA, have damaged ponderosa pines (Newcomer 1964a), but very large numbers of adults in flight do not necessarily result in noticeable defoliation (Young 1987).

Eggs are laid in groups at the base of conifer needles, where they overwinter. The eggs hatch in the spring, and the larvae feed on the new needles. Pupation may occur on the tree branches, or the larvae may drop to the ground on silk threads to pupate on tree bases or shrubs (Edwards 1887–97).

In BC Pine Whites use a wide range of conifers in the family Pinaceae as foodplants, including amabilis fir, Douglas-fir, lodgepole pine, ponderosa pine, western hemlock, western white pine, and the introduced Scotch pine (Harvey 1908; CSG; FIS). Within a given area, one coniferous species is generally preferred over the others that are present. Outside BC additional larval foodplants include Jeffrey pine, pinyon pine, and subalpine fir (Howe 1975; Scott 1992).

**SUBSPECIES:** Subspecies *tau* (Scudder, 1861), described from the Gulf of Georgia, is the Pacific Northwest subspecies that occurs in BC (Austin 1998a). Compared with the nominate subspecies (TL: restricted to vicinity of Davis Creek Park, Washoe Co., NV [Emmel et al. 1998d]), male *tau* lack red ventral markings and females have more heavily marked ventral hindwing veins and orange, rather than red, markings.

**RANGE AND HABITAT:** Pine Whites occur throughout southern BC in low-elevation (below 1,500 m) coniferous forests, and occasionally in subalpine forests up to timberline above 2,000 m. Partially open forest or forest margins appear to be preferred, possibly because of access to nectar sources.

**GENERAL DISTRIBUTION:** Pine Whites occur from southern BC and southwestern AB south to central CA, NM, and AZ, and east to SD and NE.

**CONSERVATION STATUS:** Not of concern (S5).

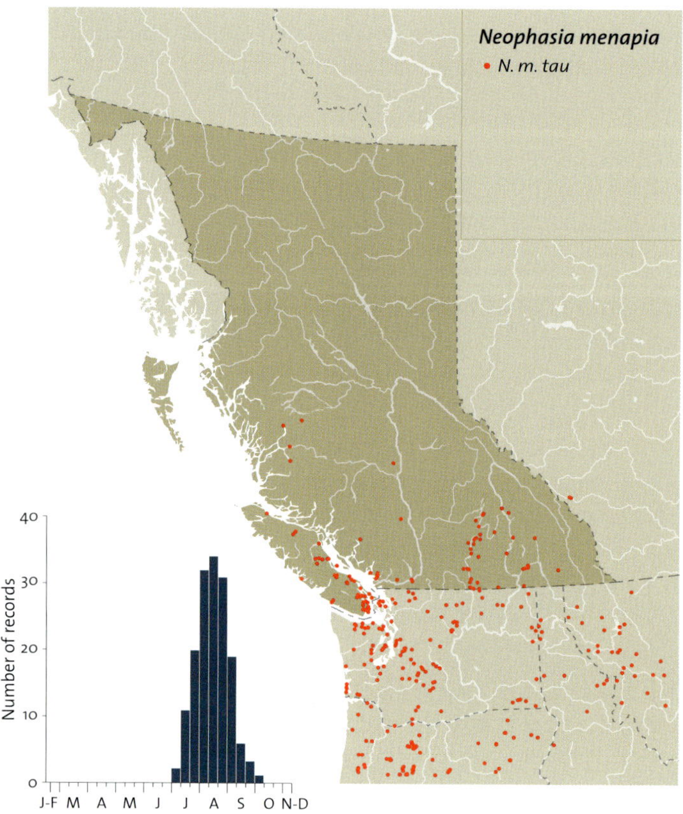




FAMILY PIERIDAE (WHITES, MARBLES, AND SULPHURS)

The name *Pontia* is from the Greek *pontios,* especially in reference to Aphrodite (Venus), the goddess of beauty, who was born from the sea. Pontia and Colias were both names associated with Aphrodite. Fabricius divided what we now call the family Pieridae into those that are yellow (*Colias*) and those that are white (*Pontia*), two related aspects of beauty. Fabricius was probably also referring to Linnaeus having originally included all white and yellow/orange butterflies in the group he called the Danai, with many species named after the 50 daughters of Danaus. The common name is shared with the genus *Pieris* and refers to the basic white colour of the wings. The common name for the genus was first used in North America by Scudder (1875), but apparently had been in use in Britain for the white Pieridae since at least 1717 (Warren 1990).

Whites in the genus *Pontia* are all medium-sized butterflies with white or pale yellow wings with black markings. A given species may look quite different at different elevations, latitudes, or seasons. They are usually smaller and darker in the spring, at high elevations, and in the north.

The eggs of whites are conical, with vertical ribs down the sides and numerous small horizontal ridges between the vertical ribs. The eggs are pale yellow when laid, but within a day or two turn bright orange. Eggs are laid singly on the leaves or flowers of plants in the mustard family (Brassicaceae), with the most commonly used native plants being in the genus *Arabis.* Mature larvae are smooth-skinned with a thin coat of fine hairs.

In the genus *Pontia* the cross-vein at the end of the forewing discal cell is strongly curved inward towards the wing base. This vein is white but is surrounded by a black "discal cell spot" that is lacking in the genus *Pieris.* There are grey green or grey markings following the venation on the ventral hindwings.

There are four species in the genus in North America, all of which occur in BC, and another six in Europe, Asia, and Africa.

All whites were included in the genus *Pieris* until relatively recently, but the differences between the groups of species appear to be great enough to separate them into two genera, *Pontia* and *Pieris,* based on morphological characters (Higgins 1975) and electrophoretic data (Geiger 1990). A third generic name sometimes applied to whites, *Artogeia* Verity, 1947, is a synonym of *Pieris* (Geiger 1990).

# BECKER'S WHITE
*Pontia beckerii* (W.H. Edwards, 1871)

Becker's White (*Pontia beckerii*)

**ETYMOLOGY:** The species *beckerii* is named after Dr. Ludwig Becker, "who laid down his noble life in the cause of science in Australia ... He died of fatigue and privation at Cooper's Creek, New South Wales in 1861." Henry Edwards had been thinking of Becker in the few moments before he collected the type specimens, which he gave to W.H. Edwards with the request to name a species after Becker (Edwards 1868–72). The common name was first used by Holland (1898).

**ADULT:** Becker's Whites are white with black markings above, and yellow veins broadly outlined in green on the ventral hindwing and the ventral forewing apex. The black discal cell spot is large and roughly square, and curves inward to follow the shape of the cross-vein at the end of the cell. Males have androconial scales in the discal cell spot. Spring and fall forms of Becker's Whites are smaller and darker than the summer form.

**IMMATURE STAGES:** Eggs are columnar with vertical ribbing. Mature larvae are green white, thickly marbled or sprinkled with grey, and with orange yellow rings between each segment. Each segment has 16–18 black tubercles capped with bristles. The head is tinged

Pupa

with yellow. Pupae are roughly cylindrical and smooth, and are held against a stem with a girdle. They are grey, with the wings nearly white, and the top of the thorax is dark grey brown. There is a pale line along each side of the abdomen, and four black dots in a row across the back between thorax and abdomen (Mead 1878).

**BIOLOGY:** Becker's Whites are multivoltine in BC, with broods emerging from hibernating pupae in late April and additional generations emerging from pupae in early July, mid-August, and late September. There is considerable overlap between the later broods. Becker's Whites lay eggs on the flowerheads of *Arabis,* with the larvae feeding on the flowers and fruit of the inflorescence. Pupation occurs on the stem of the larval foodplant.

Outside BC foodplants are Brassicaceae such as *Arabis, Brassica nigra, Descurainia sophia, Lepidium perfoliatum, Sisymbrium loeselii, Stanleya pinnata, Streptanthus,* and

♂ summer form D (4.4 CM)

♀ summer form D (4.7 CM)

♂ spring form V (3.9 CM)

♀ spring form V (4.0 CM)

♀ summer form V (4.7 CM)

*Thlaspi* (Emmel et al. 1971; Howe 1975; Ferris and Brown 1981). In California *Cleome isomeris* (Capparidaceae) is also a major foodplant (Hovanitz 1962). The *Cleome* remains green even under very arid conditions, and may permit the multiple broods of the Becker's White (Hovanitz 1969).

**SUBSPECIES:** The type locality of *Pontia beckerii* is Virginia City, NV. There are no recognized subspecies. The spring form of *P. beckerii* was incorrectly named subspecies *pseudochlorodice* (McDunnough, 1928) (TL: Oliver, BC).

**RANGE AND HABITAT:** Becker's Whites inhabit the arid lowland grasslands of the Southern Interior of BC.

**GENERAL DISTRIBUTION:** Becker's Whites occur from the Southern Interior of BC south to Baja California east of the coastal mountains and in south coastal CA. They extend east to MT, WY, CO, and NM.

**CONSERVATION STATUS:** Not of concern (S5).

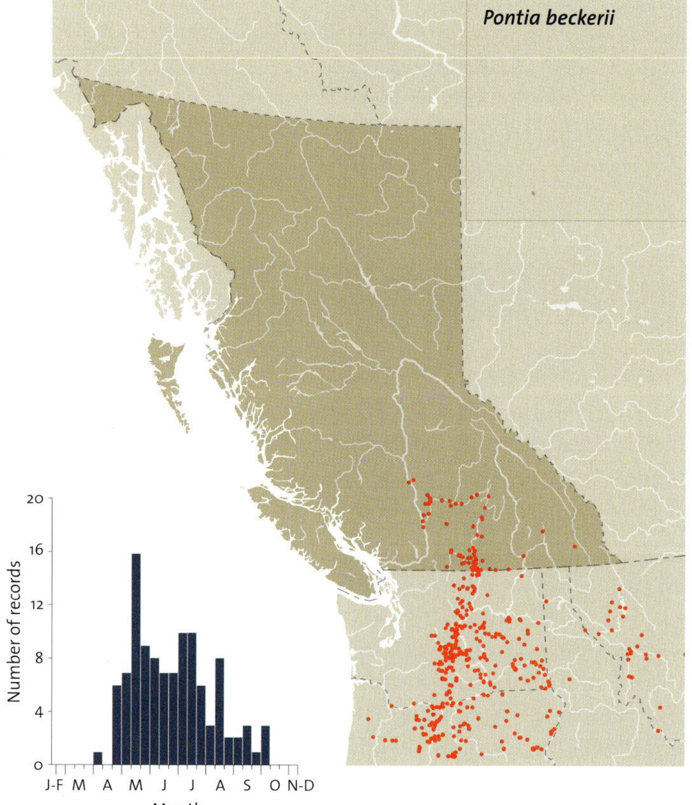




FAMILY PIERIDAE (WHITES, MARBLES, AND SULPHURS)

# Spring White
*Pontia sisymbrii* (Boisduval, 1852)

**Etymology:** The name *sisymbrii* refers to weedy mustards in the genus *Sisymbrium* (Brassicaceae), which are among the larval foodplants. The subspecies name *flavitincta* is derived from the Latin *flava* (yellow) and *tinctus* (coloured), in reference to the yellow colour of the females. Subspecies *beringiensis* is named for its northern distribution, which is associated with "Beringia," the area of Siberia, Alaska, and Yukon that was ice-free during the last ice age. The common name "Spring White" was first used by Pyle (1981) in reference to the early spring flight period of this white.

**Adult:** Spring Whites are white (males) or pale yellow (southern females) with thin black veining on the dorsal forewings. The veins on the ventral hindwings are yellow and are bordered with grey or brown grey. The black spot at the end of the forewing discal cell is small, narrow, and completely black, and the centre curves slightly inward towards the base of the wing. Males have androconial scales in the discal cell spot, and females have a medial spot on the dorsal forewing. Forewing vein $R_3$ is much longer than in the Western White or the Checkered White (Chang 1963).

**Immature stages:** Eggs are columnar, with 14 vertical ribs. When newly laid they are yellow (Edwards 1874–84) or blue green (Scott 1986b), soon turning to orange. Mature larvae are cylindrical, tapering slightly from the middle to the rear end. They are light yellow with two wavy black lines, separated by a white or pale yellow line, on each segment.

Ssp. *flavitincta* ♂ D  (3.8 cm)

Ssp. *flavitincta* ♂ V  (3.8 cm)

Ssp. *flavitincta* ♀ D  (4.0 cm)

Ssp. *flavitincta* ♀ V  (4.0 cm)

Ssp. *beringiensis* ♂ D  Holotype
(3.7 cm)

Ssp. *beringiensis* ♂ V  Holotype
(3.7 cm)

Ssp. *beringiensis* ♀ D  (3.4 cm)

Ssp. *beringiensis* ♀ V  (3.4 cm)

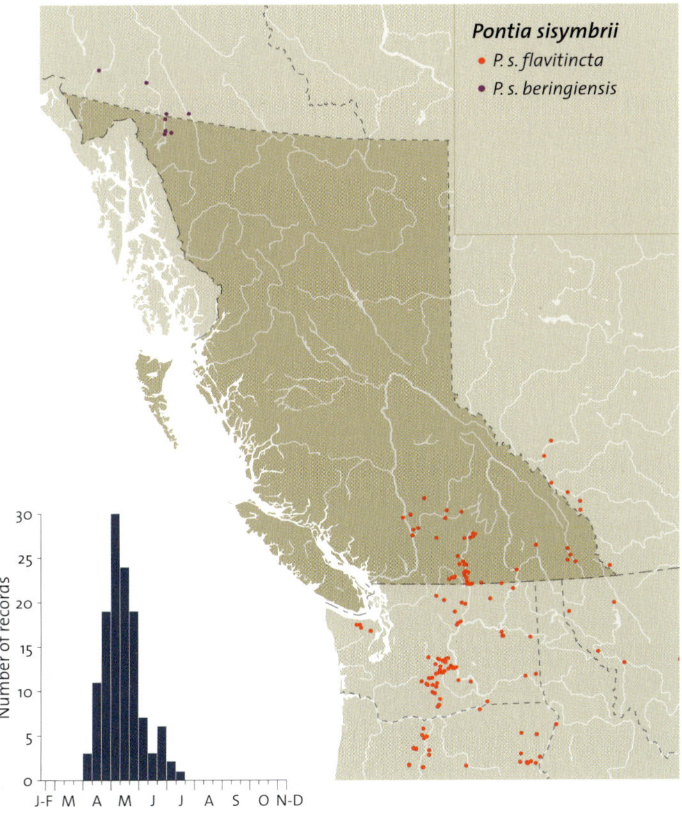



There are several ridges across the top of each segment, each with small yellow tubercles of irregular size. There is also a black line along each side. The head is black with a granular surface and darker spots. Pupae are dark brown with a granular surface and darker spots (Edwards 1874–84; Scott 1986b). The mature larvae and pupae of subspecies *beringiensis* have the same coloration as subspecies *flavitincta* (CSG).

**Biology:** Spring Whites are univoltine. They fly in April and May at low elevations in the south, and in early July above timberline in the north. Eggs are laid on the basal rosette leaves of flowering *Arabis* by subspecies *beringiensis* near Atlin, and on stem leaves near Keremeos (CSG). Small larvae feed on leaves, but larger larvae prefer to feed on flowers and fruit. Shapiro (1981a) found six eggs on an isolated rosette of

Fourth instar larva (ssp. *beringiensis*)

*Streptanthus glandulosus,* but this large number of eggs on a plant is unusual and resulted from a scarcity of foodplants. Feeding larvae routinely cannibalize any eggs they encounter. Pupae in California may hibernate through more than one winter (Shapiro 1981a).

The larval foodplants in BC are mostly unknown, but include *Arabis* species for both subspecies (CSG). Outside BC larval foodplants include *Arabis glabra, A. pulchra, Caulanthus coulteri, Draba, Streptanthus breweri,* and *S. glandulosus* (Kellogg 1986; Shapiro 1981a; Emmel et al. 1971).

**SUBSPECIES:** Subspecies *flavitincta* (J.A. Comstock, 1924) occurs in the arid Southern Interior of BC (TL: Cranbrook, BC), and is characterized by yellow females. The populations of northern BC, Yukon, and NT are a previously undescribed subspecies.

***Pontia sisymbrii beringiensis* Guppy & Norbert G. Kondla, new subspecies.** *Pontia sisymbrii beringiensis* has strongly developed dorsal black markings, and the borders of the ventral hindwing veins are wider when compared with those of all other subspecies of *P. sisymbrii*. The borders of the ventral hindwing veins and the apex of the ventral forewing are dark grey rather than the brown to brown

grey of *flavitincta.* The ground colour of females is white, or white with a partial very pale yellow tint, rather than the continuous darker pale yellow of subspecies *flavitincta.*
**Types.** Holotype: male, BC, Atlin, 9 km N on Atlin Rd at Ruffner Mine Rd, 21 June 1999, C.S. Guppy; a label "HOLOTYPE / *Pontia sisymbrii / beringiensis* Guppy & Kondla" is attached. The holotype is deposited in the Royal British Columbia Museum, Victoria, BC, CAN. Paratypes: 1 male, same data as holotype (CSG); 1 male, YT, Tarfu Lake, 2 km south on Atlin Road, 23 June 1999, C.S. Guppy (CSG); 1 female, BC, Atlin, Hitchcock Creek, 2.9 km north of creek, 24 June 1999, C.S. Guppy (CSG); YT, Atlin Road, 6 km N of BC border, 17 June 1999, Norbert G. Kondla (NGK); YT, Atlin Rd at Tarfu Lake Rd, 15 June 1999, Norbert G. Kondla (NGK); BC, Atlin, 7 km E on Surprise Lake Rd, 16 June 1999, Norbert G. Kondla (NGK); 1 male, BC, Atlin, near airport, 21 June 1999, Norbert G. Kondla (NGK); 1 female, BC, Atlin, near airport, 23 June 1999, Norbert G. Kondla (NGK).

**RANGE AND HABITAT:** Spring Whites inhabit the arid low-elevation areas of the Southern Interior of BC, and also the dry grass slopes and dry, open pine forests near Atlin. We have confirmed that the only record for Vancouver Island (Guppy 1970) was a misidentified Western White. Habitats are typically open, exposed rocky or gravelly areas with cool (spring) temperatures but lots of sunlight. Subspecies *beringiensis* has seldom been seen because few lepidopterists have visited the north during the normal May flight season. The type series was collected at the end of an unusually late flight season.

**GENERAL DISTRIBUTION:** Spring Whites occur in the southwest YT, southwest NT, and northern BC and AB. They also occur from the Southern Interior of BC and southern AB, south to Baja California and NM. There are populations in the mountains of the Olympic Peninsula. There is a gap of more than 700 km in the range of *P. sisymbrii* between subspecies *flavitincta* and subspecies *beringiensis.*

**CONSERVATION STATUS:** Subspecies *beringiensis* is of Special Concern in BC (S3). Subspecies *flavitincta* is not of concern (S4).

## CHECKERED WHITE
*Pontia protodice* (Boisduval & LeConte, 1829)

**ETYMOLOGY:** The species name *protodice* is from the Latin *proto* (first) and *dice* (season), in reference to the early flight of the first brood. The common name was first used by Scudder (1875), and refers to the somewhat square black wing markings forming a "checkered" pattern.

**ADULT:** Checkered Whites are white with black markings above, and are very similar to Western Whites. Compared with Western Whites, the black markings of the dorsal forewing are browner, and in males the black markings on the outer edge of the dorsal forewing apex are reduced and broken into spots. Male Checkered Whites have a large, black,

square spot on the ventral forewing, near the back margin. This spot is usually much smaller and rounder on male Western Whites. In Checkered Whites, vein $R_{3+4+5}$ branches from $M_2$ first, then vein $R_2$ branches from $R_{3+4+5}$. In Western Whites, forewing vein $R_2$ branches from vein $M_2$ before vein $R_{3+4+5}$ (Chang 1963). This character is the most reliable for confirming the identity of Checkered Whites. If photographing a live specimen, a clear close-up of the ventral forewing should show the veining.

**IMMATURE STAGES:** Eggs are orange when first laid (Shapiro 1981a), or cream that soon turns orange (Scott 1992). Mature larvae are blue green to blue grey with black dots. There is a dorsal line on the front half of the body, and yellow sub-dorsal and lateral stripes. The underside is light olive green with a yellow dash on the side of each proleg base, and the head is light olive green with a yellow spot on each side. Pupae are light blue grey, grey cream, or green cream, and speckled with black. They have a narrow dorsal line that is orange brown on the thorax and yellow cream on the abdomen, and sometimes have yellowish or orangish sub-dorsal and lateral stripes or markings (Scott 1992). Scudder (1889b) described the colours inaccurately, because he used larvae preserved in glycerine.

**BIOLOGY:** Checkered Whites are multivoltine in California, and BC specimens are migrants from further south. They are extremely rare in BC, but based on the flight period in Alberta (Bird et al. 1995), they should occur in BC from midsummer to late fall. They cannot survive through winter north of the extreme southern USA (Scott 1986b).

Eggs are laid on the flower buds (preferred), stems, and leaves of native and introduced Brassicaceae, with the female flying to a new plant after laying each egg. Females prefer some *Arabis* plants to others; in part, the preference is for young plants over older plants but it is also due to factors other than plant age. If eggs are already present on a larval foodplant, additional eggs are seldom laid. This egg-load assessment is visual, and pheromones are not involved. The eggs are just as visible in the ultraviolet spectrum (which butterflies can see) as in the colours visible to humans. Larvae feed on leaves, buds, flowers, and seeds, and eat the

♂ D  (4.3 cm)  ♀ D  (4.7 cm)

♂ V  (4.3 cm)  ♀ V  (4.7 cm)

*Arabis* plants from the top down. Cannibalism is frequent (Shapiro 1981a).

Outside BC larval foodplants are native and introduced Brassicaceae, including turnips, cabbage, *Arabis drummondii*, *Barbarea orthoceras*, *Berteroa incana*, *Capsella bursa-pastoris*, *Cardaria draba*, *C. pubescens*, *C. latifolia*, *Chorispora tenella*, *Descurainia sophia*, *D. pinnata*, *D. incana*, *Lepidium virginicum*, *L. densiflorum*, *L. campestre*, *Lobularia maritima*, *Rorippa curvisiliqua*, *R. nasturtium-aquaticum*, *R. sinuata*, *R. teres*, *Schoenocrambe linearifolia*, *Selenia aurea*, *Sisymbrium altissimum*, *Thelypodium wrightii*, *Thelypodiopsis elegans*, *Thlaspi arvense*, and *Wislizenia refracta*. Also used is *Cleome serrulata* in the family Capparidaceae (Scudder 1889; Shapiro and Shapiro 1974; Shapiro et al. 1981; Howe 1975; Scott 1986a, 1992).

**SUBSPECIES:** There are no subspecies. The type locality is probably Screven Co., GA. As recently as the 1960s, the Western White and the Checkered White were considered to be one species (Hovanitz 1962).

**RANGE AND HABITAT:** Checkered Whites migrate into BC from the south, and hence could potentially occur anywhere in the south, east of the Coast Range in dry meadows, forest clearings, and roadsides. They are most likely to be seen in the Kootenays.

**GENERAL DISTRIBUTION:** Checkered Whites have permanent populations across extreme southern USA and MEX, and summer populations resulting from migrations that extend north to southern CAN, including central AB and SK.

**CONSERVATION STATUS:** Checkered Whites are rare non-breeding immigrants in BC (SAN).

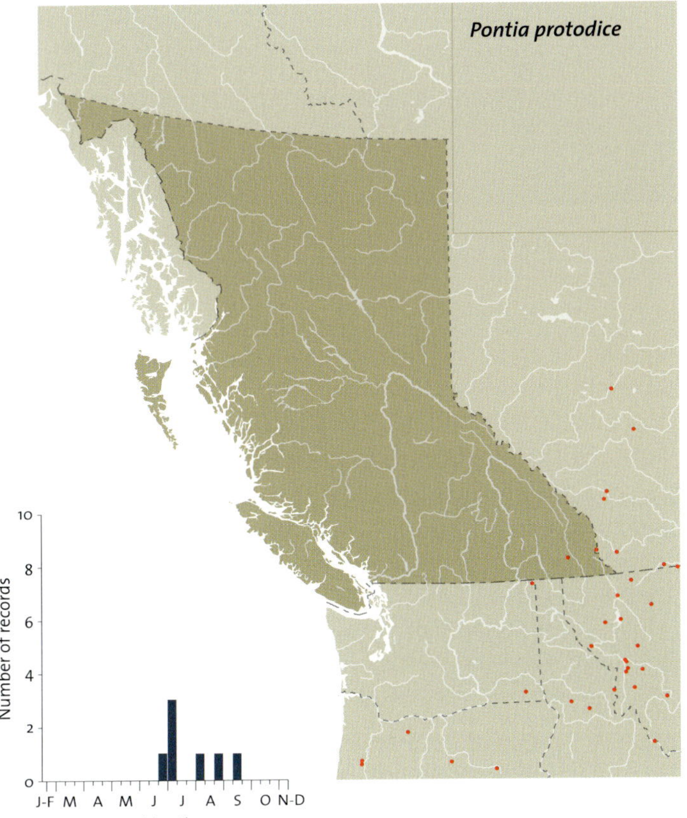

# Western White
*Pontia occidentalis* (Reakirt, 1866)

**Western Whites (*Pontia occidentalis occidentalis*) mating**

**ETYMOLOGY:** The name *occidentalis* (residing in the west), like the common name, refers to the western North American distribution of the species. Subspecies *nelsoni* was named for J.W. Nelson, the collector of the type specimen (Edwards 1874–84).

**ADULT:** Western Whites are white with black markings above, and veins broadly outlined with yellow green on the ventral hindwing and the ventral forewing apex. The forewing discal cell spot is rectangular and curved outward, opposite to the curvature of the cross-vein at the end of the cell. Males do

SSP. *occidentalis* ♂ spring form D (3.7 cm)

SSP. *occidentalis* ♂ spring form V (3.7 cm)

SSP. *occidentalis* ♀ summer form D (4.4 cm)

SSP. *occidentalis* ♂ fall form V (3.6 cm)

SSP. *nelsoni* ♂ D (4.1 cm)

SSP. *occidentalis* ♀ summer form V (4.4 cm)

SSP. *nelsoni* ♀ D (3.8 cm)

SSP. *nelsoni* ♂ V (4.1 cm)

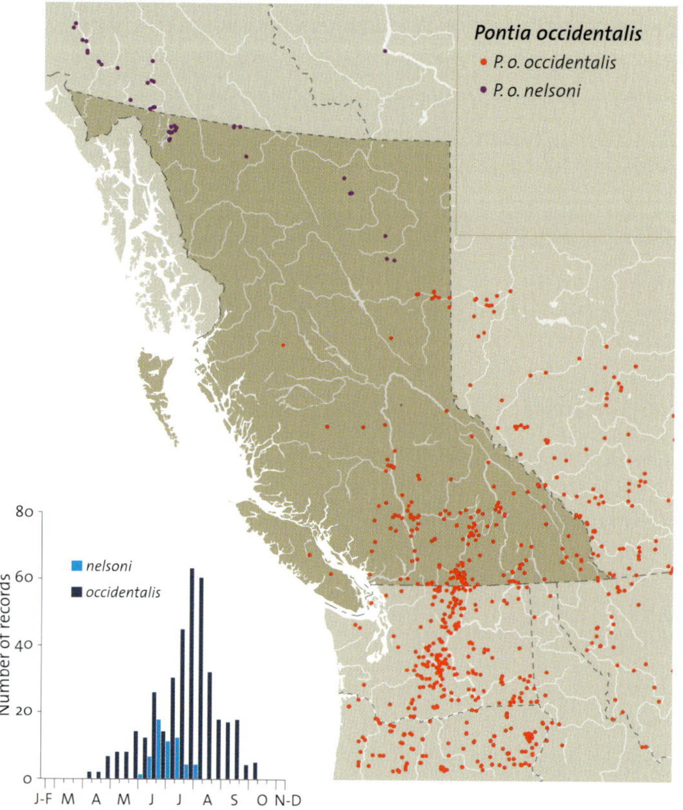



not have androconial scales in the discal cell spot. Spring and fall forms of Western Whites are smaller and darker than the summer form, and northern montane populations are darker than those in the south. Additional characters are given under the Checkered White description.

**IMMATURE STAGES:** Mature larvae of subspecies *occidentalis* are blue grey with small black spots. There is a yellowish stripe on each side of the back, and another along each side. There may be orange spots towards the front end of the yellow stripes on the back. Pupae are light grey, with a few small black spots on the wing veins and body. The ridges on the pupae are red-tinted (Shapiro 1980a; Scott 1986b).

**BIOLOGY:** Subspecies *occidentalis* is trivoltine at low elevations in the Southern Interior and probably the Peace River



FAMILY PIERIDAE (WHITES, MARBLES, AND SULPHURS)

lowlands (June to September), and univoltine in alpine areas (July to September). Eggs are laid on the leaves and inflorescences of native Brassicaceae, with larvae feeding on leaves, buds, flowers, and seeds. Pupation occurs on the stem of the larval foodplant. Males hilltop and are commonly seen circling open mountain summits. The spring and fall broods are much smaller and darker coloured than the summer brood. The darker colour is triggered by the maturation of larvae under either short daylight hours at warm temperatures, or the subjection of pupae to 10 or more days of low temperature with long daylight hours (Shapiro 1982c). The large, light-coloured summer generation matures at warm temperatures under long daylight hours. Subspecies *nelsoni* is at least partially bivoltine at low elevations as far north as Fairbanks, AK (Shapiro 1980a). In BC it is mostly univoltine (June, July), but is bivoltine at low elevations. The dark wing colour of females is present in both generations. Subspecies *nelsoni* develops from egg to adult in 20–25 days at 25°C (Shapiro 1980a).

All recorded larval foodplants are in the family Brassicaceae. Outside BC subspecies *nelsoni* feeds on *Lepidium densiflorum* at Fairbanks, AK (Shapiro 1976a, 1980a). Subspecies *occidentalis* uses *Cleome serrulata, Draba cuneifolia, Lepidium virginicum, Sisymbrium altissimum, Thlaspi arvense,* and *T. montanum* (Shields et al. 1970; Emmel et al. 1971; Howe 1975; Shapiro 1977).

**Subspecies:** Subspecies *occidentalis* (Reakirt, 1866) (TL: vicinity Empire, Clear Creek Co., CO) occurs throughout most of BC, including the Peace River lowlands. Subspecies *nelsoni* (W.H. Edwards, 1883) (TL: St. Michael's, AK) occurs in the mountains across northern BC, including the lower-elevation boreal forest along the Alaska Highway in the northeast. The black markings are smaller on male *nelsoni* than on male *occidentalis,* but are more prominent on females. Both sexes of *nelsoni* are smaller than *occidentalis.*

Some authors have considered the Western White to be a subspecies of the European *P. callidice.* However, electrophoretic data indicate that *P. o. occidentalis* is a separate species from *P. callidice* (Shapiro and Geiger 1986); hybridizing *P. o. nelsoni* and *P. o. occidentalis* results in full genetic

Eggs

Mature larva

Pupa

compatibility (Shapiro 1976b); and hybridizing *P. o. nelsoni* with *P. callidice* results in almost complete mortality of eggs and larvae (Shapiro 1980a). *P. occidentalis* and *P. callidice* are therefore biologically distinct species.

**Range and habitat:** Western Whites occur throughout BC, although they are rare on Vancouver Island. They inhabit dry meadows, forest clearings, roadsides, and alpine areas up to at least 2,500 m elevation.

**General distribution:** Western Whites occur throughout AK, YT, and boreal western and central NT, south to northern CA and NM. The distribution extends east to ON.

**Conservation status:** Not of concern. Subspecies *nelsoni* and *occidentalis* are both S5.

Pieris was one of the Muses (Pierides) who lived on Mt. Pierus, close to Mt. Olympus. Pieris was one of the five families into which Schrank divided the butterflies. Originally Schrank applied the name to all the swallowtails and whites, with the first species name being *apollo* (Apollo was the patron of the Muses). Latreille later separated the swallowtails and whites, applying the name Pieris to what we now call the family Pieridae (Emmet 1991). The common name "whites" is shared with *Pontia* and refers to the predominantly white colour of the wings.

Whites in the genus *Pieris* are all medium-sized white butterflies with black markings and, especially when newly emerged, with pale yellow ventral hindwings. Some females of all the species are entirely yellow. The genus *Pontia,* discussed earlier, includes other species of whites.

The eggs of whites are conical, with vertical ribs down the sides and numerous small horizontal ridges between the vertical ribs. The eggs are pale yellow or yellow green. Eggs are laid singly on the leaves or flowers of plants in the mustard family (Brassicaceae), both cultivated species and native mustards. Mature larvae are green and smooth-skinned with a thin coat of fine hairs. Whites hibernate as pupae, which are roughly cylindrical and smooth except for dorsal and dorsolateral ridges; the pupae are held against a stem or other vertical surface with a girdle.

In the genus *Pieris,* the cross-vein at the end of the forewing discal cell curves inward towards the wing base. Unlike in the genus *Pontia,* this vein is not surrounded by a black spot. There are dark borders to the veins on the ventral hindwing and ventral forewing apex, except in the Cabbage White and summer broods of some other species. Geiger and Scholl (1985) and Robbins and Hensen (1986) showed that *Artogeia* Verity, 1947, into which some authors have placed the BC species of *Pieris,* is a synonym of the genus *Pieris.*

Until recently only two species of the genus *Pieris* were thought to occur in BC, the introduced Cabbage White and the Holarctic *Pieris napi.* Instead, there are four species of *Pieris* in BC, and it is now clear that *Pieris napi* does not occur in North America (Geiger and Shapiro 1992).

Eitschberger (1983) proposed four BC species: *P. rapae, P. marginalis* (four subspecies), *P. oleracea,* and *P. angelika,* which are distinguished by small differences in wing pattern characteristics (Fig. 62). The fact that Eitschberger's book was in German, its high cost, and unreasonably critical reviews of this book by Kudrna and Geiger (1985), Ferris (1989), and Shapiro (1985b) combined to make Eitschberger (1983) ignored by most North American authors. Eitschberger's revision of *Pieris* is substantially supported by the electrophoretic data of Geiger and Shapiro (1992), however, which has led to a grudging acceptance of parts of his work.

During the preparation of this book, we examined several series of northern *Pieris* collected by Norbert Kondla and CSG, and various museum specimens. CSG conducted further sampling and rearing of northern BC *Pieris.* Two *Pieris* are sympatric in various northern localities, from Valemont to the southern Yukon: *P. oleracea* and *P. marginalis tremblayi.* Near the Yukon border, they are joined by a third species, *P. angelika.* Eitschberger's *P. marginalis guppyi* is electrophoretically quite distinct from *P. marginalis marginalis* (Geiger and Shapiro 1992), and the wing pattern of *tremblayi* indicates that it is closely related to *guppyi.* This suggests that *guppyi* and *tremblayi* may be a separate species from *P. marginalis.* We do not raise *Pieris guppyi* to species status, with *tremblayi* as a subspecies of *guppyi,* because the electrophoretic data are, by themselves, insufficient to separate the taxa into two species.

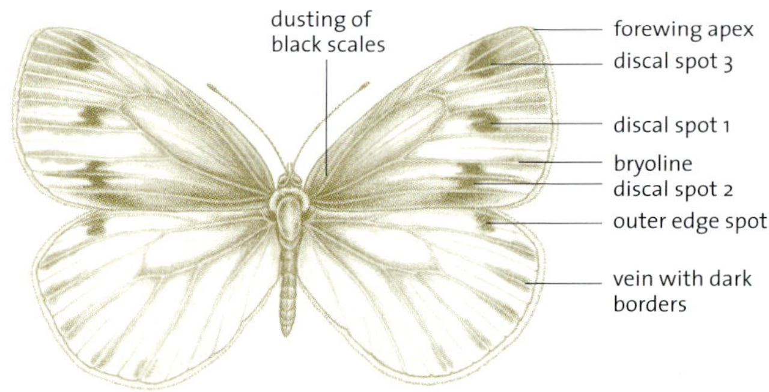


**62** **Wing pattern characters of *Pieris* species**

# Margined White
*Pieris marginalis* Scudder, 1861

**Margined White (*Pieris marginalis marginalis*)**

**ETYMOLOGY:** The name *marginalis* refers to the markings on the upperside of the two type specimens, which are small lines and patches of grey and black around the margin of the forewings and hindwings. Subspecies *reicheli* was named for Johann (John) Reichel of Revelstoke, BC, subspecies *guppyi* was named for Crispin S. Guppy (given in error as Cyril S. Guppy in the original description), and subspecies *tremblayi* was named for Norman Tremblay of Ontario. All three supplied specimens to Eitschberger (1983). The common name "Veined White," sometimes used for this species, should be reserved for the Californian species *Pieris venosa* Scudder, 1861.

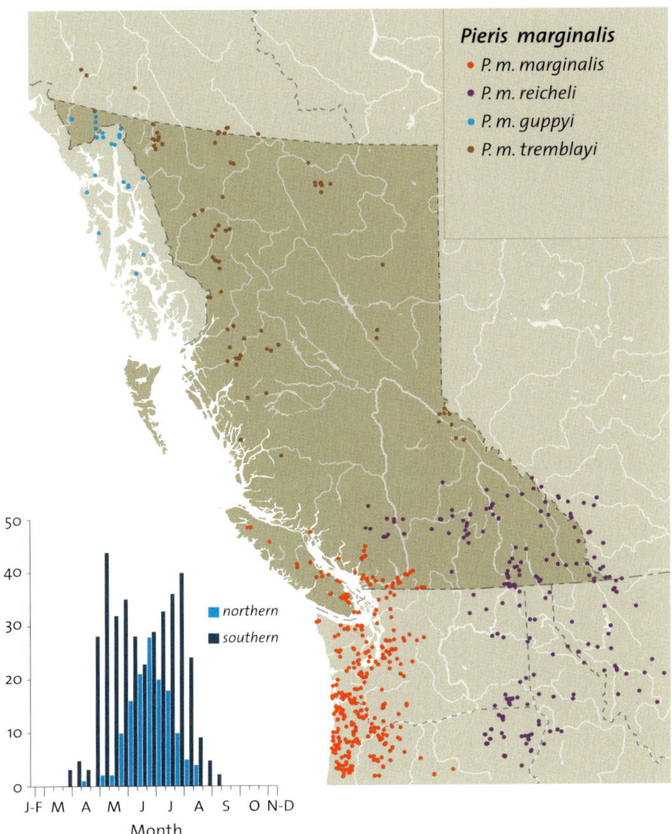

Ssp. *marginalis* ♂ spring form
D (4.2 CM)

Ssp. *marginalis* ♂ spring form
V (4.2 CM)

Ssp. *marginalis* ♀ spring form
D (4.2 CM)

Ssp. *marginalis* ♀ summer form
V (4.3 CM)

Ssp. *marginalis* ♀ summer form
D (4.3 CM)

Ssp. *reicheli* ♂ spring form
D (4.1 CM)

Ssp. *reicheli* ♀ spring form
D (4.3 CM)

Ssp. *reicheli* ♀ spring form
V (4.3 CM)



**ADULT:** Margined Whites are extremely variable in wing pattern between subspecies, between sexes, and between seasonal forms. They are best identified by elimination of other species. In southern BC there are no other *Pieris* except *P. rapae*, which has a black spot in the middle of the ventral forewing that is lacking in all other BC *Pieris*. In central and northern BC, Margined Whites occur with Mustard Whites, which have strongly defined narrow green vein borders on the ventral hindwings in the spring form. In northern BC, Margined Whites also occur with Arctic Whites, which have strongly defined dark grey vein borders on the ventral hindwings, with the vein borders narrowing strongly towards the wing margin. In contrast, in central and

Ssp. *tremblayi* ♀ D  (4.1 cm)

Ssp. *guppyi* ♀ D  (4.0 cm)

Ssp. *tremblayi* ♀ V  (4.1 cm)

Ssp. *guppyi* ♀ V  (4.0 cm)

Ssp. *tremblayi* ♂ V  (4.1 cm)

Ssp. *guppyi* ♂ V  (4.1 cm)

northern BC, the hindwing vein borders in Margined Whites are wide and show little or no narrowing towards the wing margin, and in southern BC populations the vein borders are not sharply defined and at most narrow only weakly towards the wing margin.

**IMMATURE STAGES:** Eggs are pale yellow and conical, and have 12 vertical ribs. First instar larvae are yellow when newly hatched, turning to pale green. The mature larvae of subspecies *tremblayi* from McBride, BC, and *marginalis* from Duncan, BC, are an even green without any markings. The pale lateral line that appears in the photographs of subspecies *reicheli* larvae from Revelstoke, BC (Eitschberger 1983) is actually a lateral band of stiff, shiny hairs rather than colour formed by pigment. Pupae have well-developed rounded apical and dorsal projections, and are pale tan to dark green with dark markings (Eitschberger 1983; CSG).

**BIOLOGY:** Margined Whites are one of the first butterflies in flight in the spring after emerging from hibernating pupae. They are bivoltine at low to mid elevations, and univoltine at high elevations (July, August). A partial third generation occurs in September near Vancouver, and possibly elsewhere at low elevations. Females oviposit on the leaves of foodplants, and larvae feed on leaves, flowers, and fruit. At room temperature, subspecies *tremblayi* matures from egg to pupa in 19 days, with the adult emerging about 10 days later. Some pupae of each generation go into diapause, and

adults emerge the next spring. Adults of subspecies *guppyi* may feed "on honey-dew, which is found on the leaves and twigs of Alder bushes," and may occur in very high densities (Wright 1892b).

Margined Whites normally utilize native Brassicaceae such as *Arabis* and *Dentaria* species that grow in open forest or subalpine meadows, but cultivated Brassicaceae are used when grown in the correct habitat. Oviposition by subspecies *tremblayi* on *Cardamine pennsylvanica* (det. A. Ceska) was observed near McBride, and the larvae were reared on that plant (CSG). Subspecies *guppyi* near the Tatshenshini River lays eggs on annual Brassicaceae growing deep under the surface of lush subalpine meadows (CSG). *Arabis* species are commonly used as the larval foodplant along the Alaska Highway in northern BC (CSG). Outside BC, foodplants include *Arabis glabra, Barbarea verna, Cardamine cordifolia, Dentana, Rorippa nasturtium-aquaticum,* and *Thlaspi arvense* (Shields et al. 1970; Emmel et al. 1971; Shapiro 1976d, 1980a; Chew 1980). Subspecies *guppyi* and/or Arctic Whites in Sitka, AK, use garden cabbage (Wright 1892b), unless modern competition with Cabbage Whites prevents it. In captivity Margined Whites are easily reared on common watercress *(Rorippa nasturtium-aquaticum).*

**SUBSPECIES:** On the ventral hindwing of the nominate subspecies, *P. m. marginalis* Scudder, 1861, the vein borders narrow only slightly in width from wing base to outer wing margin in the spring generation. Females are commonly yellow, and the upperside of the forewing has relatively undeveloped dark pattern elements. *Pieris pallida* Scudder, 1861 is a synonym of nominate *Pieris marginalis,* being its summer form. In subspecies *reicheli* Eitschberger, 1983, the vein borders on the ventral hindwing are narrower towards the outer wing margin in the spring generation. Females are almost always white, rarely yellow, and have strongly developed dark pattern elements. The summer low-elevation generation of both sexes of subspecies *marginalis* and *reicheli* have, at most, only traces of ventral vein borders. *P. m. reicheli* is a weakly defined subspecies, with no detected electrophoretic difference from Oregon *P. m. marginalis* (Geiger and Shapiro 1992). In subspecies *marginalis* and *reicheli,* the dark vein borders on the dorsal forewing of females are wider near the wing apex than near the discal cell. The vein borders on the dorsal hindwing are narrower and more faded towards the outer edge of the wings. The discal cell spots, when present, usually contrast strongly with the ground colour. The dorsal cell is usually less than half-filled with dark grey scale dusting. Females are white to clear pale yellow.

Subspecies *guppyi* and *tremblayi* usually have the vein borders on the ventral hindwing equally wide from wing base to outer margin, and the vein borders are well defined in the summer generation. The forewing outer margin is relatively straight, and the hindwing is less round than in subspecies *marginalis* and *reicheli.* Females of these two northern subspecies have dark vein borders on the dorsal forewing, and usually on the dorsal hindwing, all of even width from wing base to outer margin. The discal cell spots are usually present, faded and merging into the ground colour. The dorsal cell is usually mostly filled with dark grey scale dusting. Females are white to brownish yellow. Subspecies *guppyi* (TL: Skagway, AK) occurs in the wet coastal forests of the Alaska panhandle, extending into BC only on the wet coastal side of the height of land on the Haines Highway and the lower end of the Tatshenshini River. The ventral hindwing vein borders are dark grey to brown grey, and the female dorsal vein borders and dusting between veins are brown grey. Subspecies *tremblayi* Eitschberger, 1983 (TL: Pink Mountain, BC) occurs throughout northern BC, from Bella Coola northward in the west and McBride northward in the east. The ventral hindwing vein borders are light grey to medium grey, and the female dorsal vein borders and dusting between veins are grey.

During the last glaciation, there was an ice-free refugium in the Alexandra Archipelago of the Alaska panhandle (Pielou 1991), where subspecies *guppyi* may have survived, with postglacial inland dispersal prevented by continued glaciation along crest of the Coast Range. There was another ice-free refugium on the Queen Charlotte Islands (Pielou 1991), where subspecies *tremblayi* may have survived before dispersing inland postglacially through the Nass River and Skeena River gap in the Coast Range. Alternatively *tremblayi* may have evolved from *guppyi* during postglacial colonization inland.

**RANGE AND HABITAT:** Margined Whites occur locally throughout southern and central BC, north to Atlin. The habitat at low elevations is damp deciduous forest areas with partial shade and cool temperatures; at mid elevations, willow/alder scrub river floodplains or avalanche chutes; and at high elevations, cool, damp subalpine meadows. Their habitats are cool and moist, and usually have regularly occuring low to

Ssp. *tremblayi* mature larva

Ssp. *tremblayi* pupa

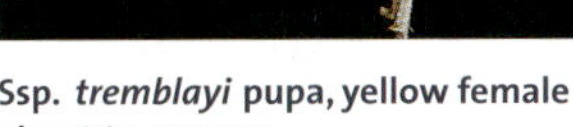

Ssp. *tremblayi* pupa, yellow female about to emerge

moderate disturbance levels that expose patches of soil to allow their short-lived (annual, biennial, or short-lived perennial) cruciferous foodplants to constantly produce new plants from seed.

**GENERAL DISTRIBUTION:** Margined Whites occur from southwestern YT east to the AB Rockies, and south to CA on the coast and in the Rockies to MEX.

**CONSERVATION STATUS:** Subspecies *guppyi* is of Special Concern in BC (S3). The remaining subspecies are not of conservation concern, with subspecies *marginalis* S4 and *reicheli* and *tremblayi* S5.

## MUSTARD WHITE
*Pieris oleracea* Harris, 1829

**Mustard White (*Pieris oleracea*)**

**ETYMOLOGY:** The name *oleracea* refers to the cultivated Brassicaceae (mustard family), including *Brassica oleracea* (cabbage), on which the species was originally discovered feeding as larvae (Harris 1829). The common name "Mustard White" refers to the use of plants in the mustard family as larval foodplants. Gosse (1840) called the species the "Grey-veined White," a name whose use was discontinued.

**ADULT:** Mustard Whites are chalky white on the dorsal wing surface. They have black dusting at the wing bases, and sometimes grey triangles at the dorsal forewing apex pointing inward along the veins. In the spring generation, there are strongly defined narrow green borders to the veins on the ventral hindwing and ventral forewing apex that show through to the upper surface. These distinctive vein borders are absent in the summer generation, resulting in a featureless pale yellow ventral hindwing without any dusting of grey scales. Females sometimes have grey "Cabbage White" markings on the dorsal forewing and hindwing, but seldom as extensively as in Margined Whites.

IMMATURE STAGES: The immature stages were described from New Hampshire and Massachusetts by Harris (1829) and Ontario by Eitschberger (1983), and are similar in BC (Quesnel) (CSG). Eggs are pale green when laid, soon turning pale yellow to pale orange (CSG); they are conical and have 16 vertical ribs. First instar larvae are pale translucent tan to green tan, with glandular hairs. Mature larvae are green with a darker dorsal stripe, fine hairs over the body, numerous black speckles, and a few small white spots on each segment. Pupae have strongly developed and pointed apical and dorsal projections; they are white, pale green, or brown with black spots.

BIOLOGY: Mustard Whites are displaced by Cabbage Whites where Cabbage Whites invade their habitat, resulting in a greatly reduced distribution of Mustard Whites in eastern North America (Scudder 1889b). Mustard Whites were common in Nova Scotia in 1864 (before the arrival of Cabbage Whites), but were uncommon and local by 1954 (Ferguson 1954). The mechanism of the displacement is unknown, but it may be similar to the extirpation of Viceroys from BC and northern Washington. Since cultivated cabbage crops are used where available, they may become the

dominant oviposition sites for Mustard White populations near human habitation. Pest control for Cabbage Whites may result in the extirpation of the bivoltine to trivoltine Mustard Whites, while the multivoltine Cabbage Whites survive. The distribution and abundance of Mustard Whites probably increased greatly early in European settlement of eastern North America, before the appearance of Cabbage Whites, because of the introduction of crops of cabbage, turnips, radishes, and other Brassicaceae.

Mustard White biology in eastern North America has been summarized by Shull (1977). Females have a much greater

♂ D  (4.1 см)

♀ D  (4.3 см)

♂ V  (4.1 см)

♀ V  (4.3 см)

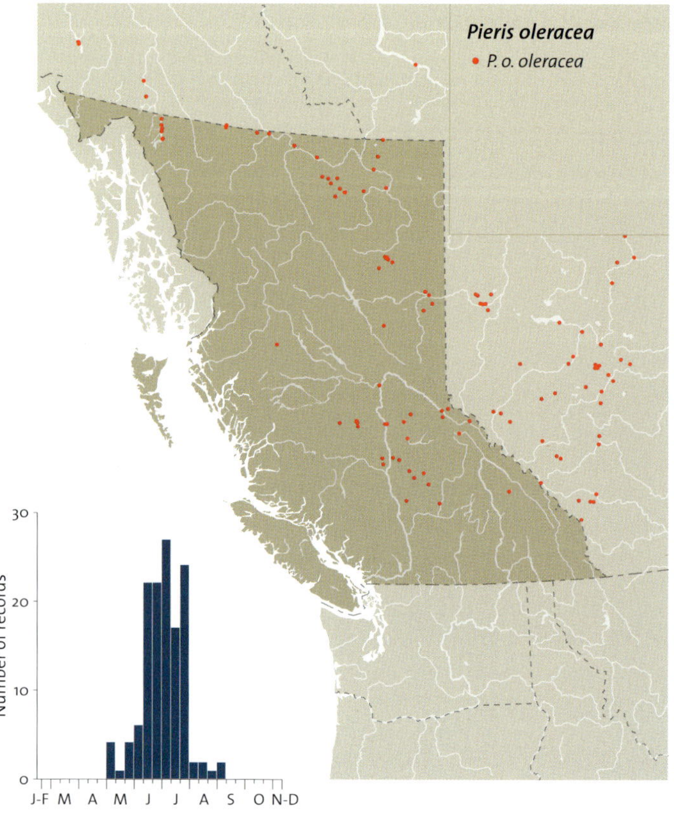

**Eggs**

**Mature larva**

**Pupa with adult about to emerge**

probability of dispersing from the main colony area than males, and dispersing females frequently lay eggs on larval foodplants not normally used. Females can apparently detect the odour of the larval foodplants when flying nearby, even when the plant is hidden under other vegetation. When larval foodplants are common, a female lays only one egg on a plant before moving on; when the plants are rare, she may lay more than one egg on the same plant. Mustard Whites are trivoltine across temperate North America, according to Scudder (1889b), and in BC they are at least bivoltine.

The larval foodplants are all in the family Brassicaceae. Mustard Whites normally utilize native species such as *Arabis* and *Dentaria* species that grow in open forest, but cultivated Brassicaceae were utilized in the absence of competition with Cabbage Whites. The larvae that produced the type series of the species were found on turnip, cabbage, and radish (Harris 1829), and Mustard Whites were a pest on cabbage and radish in gardens in Newfoundland in the early 1830s (Gosse 1883). Harris (1862) reported that "about the last of May, and the beginning of June, it [the Mustard White] is seen fluttering over cabbage, radish, and turnip beds, and patches of mustard, for the purpose of depositing its eggs ... I have seen these butterflies in great abundance during the latter part of July and the beginning of August, in pairs, or laying their eggs for a second brood of the caterpillars." Thus for most of the 19th century Mustard Whites were an economic pest in eastern North America, only to be displaced by the Cabbage White.

Near Quesnel, BC, Mustard Whites use *Arabis* species. Outside BC other larval foodplants include *Arabis drummondii, A. perfoliata, Barbarea vulgaris, Cardamine pratensis, C. flexuosa, Dentaria, Nasturtium armoracia, Rorippa nasturtium-aquaticum,* and *Sinapis* (Scudder 1889b; Shull 1977; Chew 1980). Eggs laid on introduced garlic mustard in New England result in larvae that sometimes successfully mature and sometimes die (Courant et al. 1994).

**SUBSPECIES:** Mustard Whites were originally described from northern and western Massachusetts. The nominate subspecies occurs in BC; the only other subspecies occurs in Newfoundland.

**RANGE AND HABITAT:** Mustard Whites occur locally in central and northern BC. The habitat is cool, damp meadows and wet shrubby forest openings.

**GENERAL DISTRIBUTION:** Mustard Whites occur from AK to central BC, east across boreal CAN to Labrador and NF. East of the Rockies they extend south roughly to the USA border, except south to the southern Appalachians south and east of the Great Lakes.

**CONSERVATION STATUS:** Not of concern (S5).

## ARCTIC WHITE
*Pieris angelika* Eitschberger, 1983

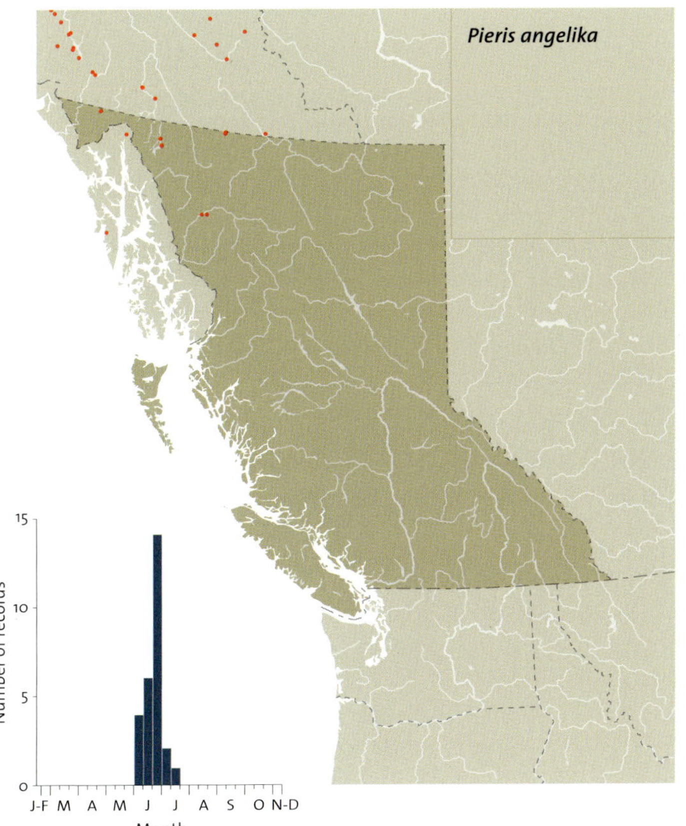


**ETYMOLOGY:** The species *angelika* was named for Angelika, wife of Ulf Eitschberger (Eitschberger 1983). The common name was first used by Layberry et al. (1998), and refers to the species' arctic distribution.

**ADULT:** Male Arctic Whites are usually completely white on the dorsal wing surface. There are usually strongly defined dark grey vein borders on the ventral hindwings, with the vein generally the same dark colour as the borders. The vein borders on the ventral hindwing narrow strongly towards the wing margin. The ventral hindwing ground colour is frequently bright yellow. The veins on the dorsal wing surface form fine, contrasting black lines. Females have dark borders to the dorsal wing veins that widen towards the wing margin. The other dark markings are well developed, and the ground colour varies from white to yellow.

The name *Pieris angelika* was first used by Eitschberger (1981), but, as noted by Ferris et al. (1983), it was not accompanied by a description, definition, or indication differentiating the taxon, and hence is a *nomen nudum*. Eitschberger (1983) redescribed *Pieris angelika,* and fulfilled all the requirements for describing a new species, hence the date for the original description of *Pieris angelika* is 1983 rather than 1981.

**IMMATURE STAGES:** The immature stages were described and illustrated from Dempster Highway, YT, by Eitschberger (1983), from specimens and photographs supplied by CSG. Eggs are pale green when laid, soon turning pale yellow; they are conical and have 14–16 vertical ribs. Mature larvae are dark green with fine hairs, numerous black speckles, and a few small white spots on each segment. Each spiracle is highlighted by a yellow ring, the only *Pieris* larva in BC that has this character. Pupae have reduced projections compared with related species, with short and round apical and dorsal projections. There are no black spots down the back of the abdomen.

**BIOLOGY:** Arctic Whites are univoltine, with adults in flight in June and July. Shapiro (1985c) reared larvae (as *P. marginalis*) from eggs obtained from Dawson, YT. He reared them under continuous light and obtained only diapause pupae, indicating that the univoltine life cycle is genetically based rather than a result of the developmental environment.

The larval foodplants are presumably various Brassicaceae. Oviposition of two eggs has been observed on northern

♂ D  (3.6 cm)          ♀ D  (4.0 cm)

♂ V  (3.6 cm)          ♀ V  (4.0 cm)

parrya, *Parrya nudicaulis,* in the Ogilvie Mountains of the Yukon (N.G. Kondla, pers. comm.).

**SUBSPECIES:** No subspecies are recognized. The type locality of the species is Elsa and Keno, YT.

**RANGE AND HABITAT:** Arctic Whites occur in northwestern BC. The habitat is forest openings and wet tundra.

**GENERAL DISTRIBUTION:** Arctic Whites occur in AK, YT, western NT, and northern BC.

**CONSERVATION STATUS:** The Arctic White is of Special Concern in BC (S3).

Egg

Mature larva

## CABBAGE WHITE
*Pieris rapae* Linnaeus, 1758

**ETYMOLOGY:** The name *rapae* is derived from *Brassica rapa* (field mustard, now known as *B. campestris*), a foodplant of the species. At the same time, Linnaeus named the related European species *P. napi* after *Brassica napus* (turnip) (Emmet 1991). The common name "Cabbage White" was first used by Scudder (1875), referring to the larvae being a pest on cabbage and other plants in the mustard family. In Britain the species is called the Small White.

**ADULT:** Cabbage Whites are white, with the apex of the dorsal forewings black, a black spot near the centre of the dorsal forewings, and a black spot on the front margin of the hindwing. There is a black spot in the middle of the ventral forewing. The ventral hindwing and the apex of the ventral forewing are pale yellow in fresh specimens, and never have the dark veins of other whites. Females are cream-coloured rather than white, and have a black spot and/or streak along the hind margin of the forewings. In the spring brood of Cabbage Whites, the dark markings are strongly reduced and less distinct than in the summer broods (Bowles 1872).

**IMMATURE STAGES:** Cabbage White eggs are yellow, conical, and ribbed down the sides. Newly hatched larvae are initially yellowish, and turn green after they commence feeding. Mature larvae are green, with a pale yellow stripe down the back. Pupae have a dorsal ridge and a lateral projection on each side. Pupal colour is usually similar to that of their substrate, ranging from brown to green.

**BIOLOGY:** Cabbage Whites are multivoltine throughout most of their range. In BC adults emerge in the spring (March on the coast, April in the interior) from hibernating pupae. Females usually begin to lay eggs on the second or third day after emergence (Jones 1977). Eggs are laid singly, usually on the underside of the edges of leaves, and hatch in 3–6 days. A generation is completed in 4–8 weeks (depending on the temperature), with 3–4 generations per year in southern BC and probably 3 in the Peace River lowlands. The larvae are very resistant to cold, wet autumn weather, and in mild years occur on cabbage crops well into December on the south coast.

**Mature larva, defecating**

In BC larval foodplants include many plants of European origin in the family Brassicaceae, such as cabbage, broccoli, cauliflower, mustards, rape, canola, and weedy Brassicaceae (Hovanitz 1962; CSG). Wild radish (*Raphanus raphanistrum*) growing as a roadside weed can support large populations (CSG). They also feed on garden nasturtium (*Tropaeolum majus*), but it is not as suitable a larval foodplant as cabbage crops (Hovanitz and Chang 1962). Native species of Brassicaceae are used, but Cabbage Whites apparently never establish permanent populations on them (Hovanitz and

♂ D (4.6 cm)

♀ D (4.9 cm)

♂ V (4.6 cm)

Chang 1962). The multivoltine Cabbage White may find it difficult to adapt to the seasonal nature of most native Brassicaceae. Outside BC other foodplants include common watercress, *Arabis glabra, Barbarea vulgaris, B. orthoceras, Brassica carinata, B. kaber, B. nigra, Cardaria pubescens, C. latifolia, Cleome serrulata, Eruca vesicaria* ssp. *sativa, Hirschfeldia incana, Lunaria annua, Lepidium virginicum, L. campestre, Rorippa teres, Sisymbrium officinale, S. altissimum,* and *Thlaspi arvense* (Remington 1952; Shapiro and Shapiro 1974; Shapiro 1976d; Scott 1992).

**SUBSPECIES:** Only the nominate subspecies, described from Sweden, occurs in North America.

**RANGE AND HABITAT:** Cabbage Whites commonly occur across southern and central BC, and in the Peace River lowlands in agricultural areas. The Queen Charlotte Islands and Atlin records represent temporary introductions that resulted from larvae or pupae imported on fresh produce. The northernmost permanent population is associated with a farm on the Alaska Highway west of Racing River. The range of Cabbage Whites in BC is probably determined by lack of agricultural habitats rather than climatic factors. Typical habitats are fields of cruciferous crops, urban and suburban gardens, and vacant lots containing weedy introduced species of Brassicaceae. Fresh specimens are occasionally seen far from agricultural areas, suggesting sporadic breeding on native species of Brassicaceae.

**GENERAL DISTRIBUTION:** Cabbage Whites are native to northern Africa and Europe, and are found across Asia to Japan. They have also been introduced to Australia, Hawaii, and many other places, and are now found throughout most parts of the world in which cabbage and other cole crops are grown.

**CONSERVATION STATUS:** A widespread introduced exotic species (SE).

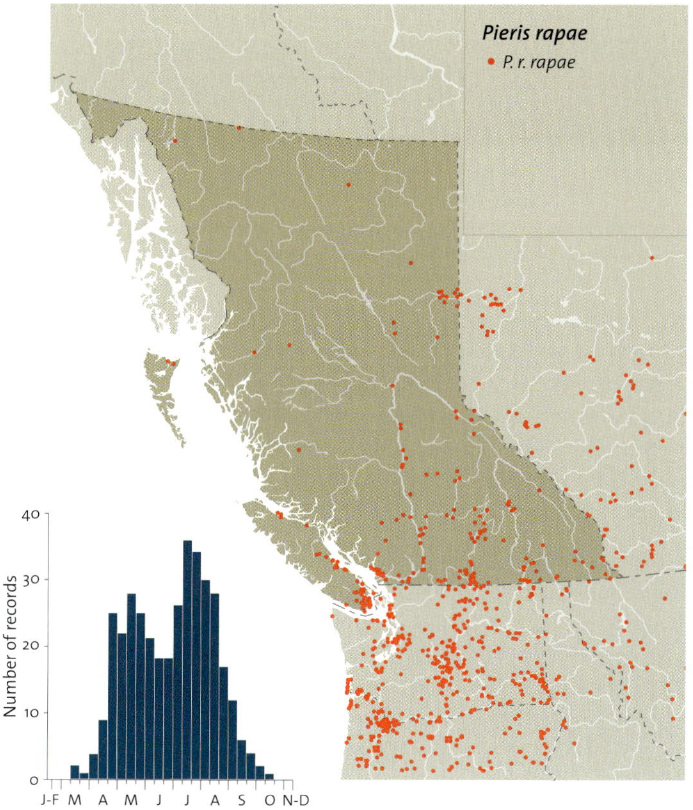

The subfamily name is derived from the type genus, *Anthocharis*. The common name "marbles" is derived from the green marbled pattern of the ventral hindwings of the genus *Euchloe*. The common name "orangetips" is derived from the orange or yellow tips of the dorsal forewings of most species of the genus *Anthocharis*.

Marbles and orangetips are small butterflies that are typically white with black markings on the upperside and green marbling on the ventral hindwings. Orangetips may have yellow females, and marbles may have cream dorsal hindwings. They typically have a slow to moderately fast bobbing flight pattern.

Antennae are less than half the length of the forewing, with a well-defined oval club. The forewing has 12 veins, and the hindwing precostal vein is present. The prothorax is without patagia (two hardened lobelike subdorsal lumps occurring in subfamily Coliadinae). The dorsal forewing has androconial scales.

The larvae of all the species of marbles and orangetips feed on plants in the family Brassicaceae.

**Rock cress (*Arabis* sp.), typical foodplant of Anthocharinae**

### GENUS *EUCHLOE* HÜBNER, [1819]   MARBLES

The name *Euchloe* was written as *Euchloë* until the taxonomic code eliminated the use of diacritical marks. The name is apparently composed of the Greek *eu* (good) and *khloë* (light green colour of spring vegetation), referring to the green marbling on the ventral hindwings (Emmet 1991). The common name refers to the green marbling on the ventral hindwings.

The three species of marbles in BC are all small white butterflies with black markings on the upperside of the wings and a pattern of greenish marbling on the ventral hindwings. Marbles are generally found east of the Coast Range in BC. The most recent review of the genus *Euchloe* was by Opler (1967b, 1967c, 1967d, 1967e, 1970, 1971, 1975).

Eggs are greenish or cream white when laid, and then soon turn bright orange. Eggs are columnar, with the micropylar area broadly rounded. There are 15–20 prominent vertical ridges, connected by less prominent horizontal ridges. Larvae are green with longitudinal stripes, becoming purplish when prepupal. Pupae are variable shades of brown; they are long and slender, and the head end tapers to a narrow tip. A girdle holds the pupa head up and tightly against a stem.

Marbles lay eggs and feed on the leaves, stems, or flowerheads of Brassicaceae, especially *Arabis*, that are bolting into flower. Preferred plants grow in full sunlight and are more than 12 cm in height with an erect configuration. Young larvae feed on leaves or within flower buds or inflorescences of flower buds. Older larvae feed externally on seed pods (Opler 1975). Marbles hibernate as pupae, and all have only a single spring brood.

# LARGE MARBLE
*Euchloe ausonides* (Lucas, 1852)

**Large Marble (*Euchloe ausonides mayi*)**

Ssp. *mayi* ♂ D  (4.1 cm)

Ssp. *mayi* ♀ D  (4.3 cm)

Ssp. *mayi* ♂ V  (4.1 cm)

Ssp. *ogilvia* ♂ V  (4.2 cm)

Ssp. *ogilvia* ♀ V  (4.1 cm)

**ETYMOLOGY:** The name *ausonides* refers to the resemblance of the species to the European species *E. ausonia* (*ides* = to resemble). In turn, *ausonia* is derived from the name Ausonius, a part of the Italian peninsula. Subspecies *mayi* was named for J.F. May, who supplied the Chermocks with many butterflies from Manitoba (Chermock and Chermock 1940). The subspecies name *insulanus* is Latin for islander, in reference to the island locations of all historical and modern populations. The common name "Large Marble" refers to the species being the largest of the marbles in North America.

The common name "Island Large Marble" is recommended for subspecies *insulanus* because all known populations are found on islands.

**ADULT:** The ventral hindwing marbling of Large Marbles is green tinged with yellow. The marbling occurs in large patches with equally large white areas between them. In females the white of the hindwings is frequently tinged with yellow, while the forewings remain pure white. The black bar at the end of the forewing discal cell always has some white scales mixed in with the black scales (Opler 1970).

**IMMATURE STAGES:** Eggs are light bluish green when laid, turning light orange within a day and then becoming vermilion. Before hatching they become yellowish brown (Opler 1975). Young larvae have black heads, with orange yellow bodies in the first instar, changing to greenish in the second instar. Third to fifth instar larvae are grey green above the spiracles and green below the spiracles, with small black spots all over. There is a yellow green stripe down each side of the back, and a yellow (instars 3 and 4) or white (instar 5) stripe along the side at the spiracles. The head of instar 3 is green black, that of instars 4 and 5 is green grey. Prepupal larvae turn a striking dark purple (Opler 1975). In contrast, in BC the stripe along the sides at the spiracles is white in instars 3 to 5, and the prepupal larva turns brown (CSG). The pupae have a straight elongated head projection (Opler 1975), and are grey brown and covered with fine longitudinal darker streaks (Edwards 1874–84).

**BIOLOGY:** Large Marbles have only one brood in BC, which is in flight in May and June at low elevations, and as late as

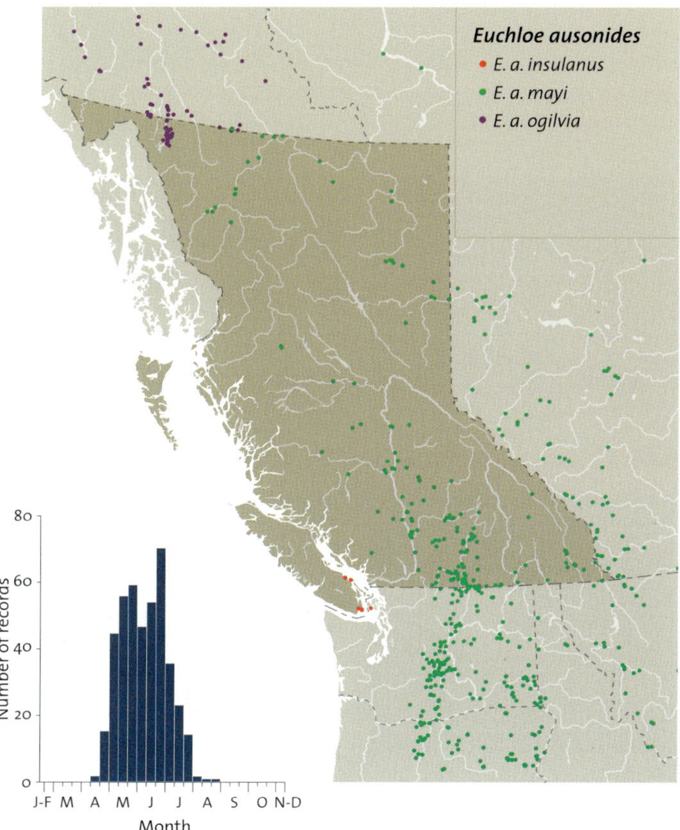

Ssp. *insulanus* ♂ D (4.1 cm)

Ssp. *insulanus* ♀ D Holotype (4.4 cm)

Ssp. *insulanus* ♂ V (4.1 cm)

Ssp. *insulanus* ♀ V Holotype (4.4 cm)

early August in alpine habitats. In California eggs are laid exclusively on the flower buds of Brassicaceae (Opler 1975). This exclusive use of flower buds is so complete that Shapiro (1985a) explicitly noted how unusual was one egg laid on a small leaf adjacent to an inflorescence. In California the eggs are seldom laid on plants less than 1 m high (Shapiro 1981a), and first instar larvae die if fresh unopened flower buds are not available for them to feed on when they hatch (Opler 1975). Young larvae position themselves vertically between flower buds and cover the inflorescence with loosely spun silk (Opler 1975). In contrast, in BC oviposition by Large Marbles is normally on leaves in various positions along the stem of *Arabis,* from near the ground to near the top of the stem. Early in the flight season, subspecies *ogilvia* lays eggs primarily on *Arabis* flower buds, but later in the season eggs are laid on the leaves and stems. Stems at the time of oviposition may all be less than 0.5 m in height, or near 1 m, depending on the population. Larvae feed on the leaves when young, and move to the seed pods as they mature (CSG).

All larval foodplants are in the family Brassicaceae, primarily *Arabis* in BC (CSG). Outside BC foodplants include *Arabis drummondii, A. fendleri, A. glabra, A. hirsuta, A. lyallii, A. suffrutescens, Barbarea vulgaris, B. orthoceras, Berteroa incana, Brassica campestris, B. kaber, B. nigra, B. vulgaris, Descurainia incana, D. californica, Draba nemorosa, Erysimum asperum, Hirschfeldia incana, Isatis tinctoria, Lepidium, Raphanus sativus, Sisymbrium altissimum,* and *S. officinale* (Shields et al. 1970; Shapiro 1981a, 1974b; Howe 1975; Opler 1975; Shapiro et al. 1981; Scott 1992; Bird et al. 1995).

**Subspecies:** Subspecies *mayi* F. & R. Chermock, 1940 (TL: Riding Mountains, MB) occurs in the Southern Interior, Central Interior, Kootenays, and northeast. The marbling on the ventral hindwings is yellow green with yellow veins. The nominate subspecies (TL: restricted to San Francisco, CA

[Opler 1967e]), to which the BC populations have in the past been assigned, occurs only in California. The populations of northwestern BC are subspecies *ogilvia* Back, 1990 (TL: Ogilvie Mountains, YT). The ventral green marbling is more extensive, darker, and less yellowish than in the nominate subspecies. An undescribed subspecies formerly occurred on southern Vancouver Island, the Gulf Islands, and San Juan Island, WA, for which a name is provided here.

***Euchloe ausonides insulanus*** Guppy & Shepard, new subspecies. *Euchloe ausonides insulanus* is larger and differs in wing pattern from subspecies *mayi* on the adjacent mainland. The most easily recognized trait is the greatly expanded marbling of the ventral hindwings, which is frequently strongly suffused with yellow scales and hairs. This expansion and yellowing of markings also occurs on the ventral apical and subapical forewing. The dark markings of the dorsal forewing are expanded and the wing bases are heavily suffused with black scaling. **Types.** Holotype: female, BC, W[ellington]., 3 June [19]04, Rev. [George W.] Taylor; a label "Holotype / *Euchloe ausonides / insulanus* Guppy & Shepard" is attached. The holotype is in the Canadian National Collection of Insects and Arthropods, Ottawa, ON, CAN. Paratypes: 1 male, BC, B[eacon]. H[ill]. P[ark]., Vict[oria]., 28 May [18]99, E. A[nderson]. (RBCM); 1 male, BC, J[ames]. Bay Vict[oria]., 17 May [18]98, E. A[nderson]. (RBCM); 1 female, BC, Victoria, 26 May [18]82, (CNC); 1 male, BC, S[outh]. Gabriola I[sland]., 30 May 1908, B.R. Elliott (JHS); 1 male, BC, Vancouver [Island], [1859] 59.7., 25/252, Dr. Lyall. (BMNH); 1 male, BC, Victoria, 25 April [18]85 (CNC); 1 female, BC, Langfords, 27 May [18]98, E. A.[nderson] (RBCM); 1 female, BC, [Beacon Hill Park] Vict[oria]., 17 May [18]98, E. A.[nderson]. (RBCM); 1 female, BC, Victoria, 24 May 1905, J. Fletcher. Barnes Collection (NMNH); 1 male, BC, Victoria, 18 May [18]87 (NMNH); 1 male, [BC], Vic[toria]., 11 May [19]04 (CNC).

**Egg**

**Mature larva, ssp. *mayi***

**Prepupal larva, ssp. *mayi***

**Pupa, ssp. *mayi***

**RANGE AND HABITAT:** Large Marbles occur in meadows at all elevations east of the Coast Range, with occasional populations in the coastal mountains (Garibaldi, Hope). Island Large Marbles formerly occurred on Vancouver Island and on Gabriola Island. Victoria was mentioned as a locality in early provincial checklists and in Harvey (1904), but Vancouver Island has not been recognized in the more recent publications as part of the natural range of the species. They were first found in 1858 or 1859 in Victoria, and last recorded from Gabriola Island in 1908. In 1997 John Fleckenstein discovered an extant population on San Juan Island, WA.

**GENERAL DISTRIBUTION:** Large Marbles occur from central AK, YT, and NT south to the Great Lakes in the east and to central CA and NM in the west.

**CONSERVATION STATUS:** Subspecies *insulanus* is Extirpated from BC (SX). Subspecies *mayi* and *ogilvia* are not of concern (both S5).

## NORTHERN MARBLE
*Euchloe creusa* (Doubleday, [1847])

**ETYMOLOGY:** In classical legends Creusa was the daughter of Priam of Troy and wife of Aeneas. The common name refers to the northern distribution of the species, relative to most of the species in the genus.

**ADULT:** In Northern Marbles the green marbling on the ventral hindwings is never tinged with yellow, although the veins within the marbling are yellow. The marbling occurs in many small patches, with equally small areas of white

♂ D  (3.5 CM)          ♂ V  (3.5 CM)

between them. The green and white patches are elongated to give a banded appearance to the marbling. The forewings and hindwings of both sexes are always completely white. The black discal cell spot on the forewing has at most very few white scales in it, and it is narrow and angular.

**IMMATURE STAGES:** Undescribed.

**BIOLOGY:** Northern Marbles fly in June and July, with only one brood each year. Oviposition occurs on the flower buds of *Draba* species. The only known larval foodplant is lance-leafed draba, *Draba cana,* at Moraine Lake, Banff National Park, AB (Opler 1975; Bird et al. 1995).

**SUBSPECIES:** None. The species was named from Rock Lake, Jasper, AB (Shepard 1984).

**RANGE AND HABITAT:** Northern Marbles inhabit dry, thinly vegetated forest openings, river flats, and alpine tundra across northern BC and down through the Rocky Mountains.

**GENERAL DISTRIBUTION:** Northern Marbles occur from western AK across northern YT to southwestern NT, south through northern BC and the Rockies to Glacier National Park, MT, and east to northwestern SK.

**CONSERVATION STATUS:** Not of concern (S5).

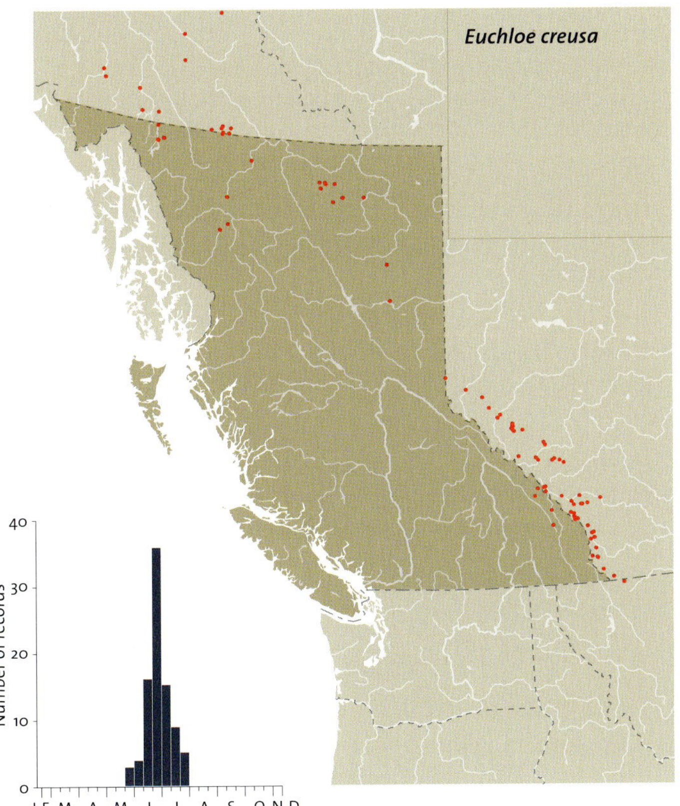

# Green Marble

*Euchloe naina* Kozhantshikov, 1923

**ETYMOLOGY:** The derivation of the name *naina* is unknown. The common name was first used by Kondla and Pelham (1995) in reference to the dull green marbling of the ventral hindwing.

**ADULT:** The ventral hindwing marbling of Green Marbles is 80–90% dull grey green, with very few, small white patches present. The costal margin and the base of the dorsal forewing are heavily covered with black scales. The dorsal hindwing has a wide streak of black scaling extending from the wing base to the middle of the wing. Females are frequently quite grey over the entire upper wing surface. The white of the dorsal surface is pearly, and the white of the ventral hindwing is silvered. The forewing apex is more rounded than in other *Euchloe*.

♂ D (3.6 cm)

♀ D (3.2 cm)

♂ V (3.6 cm)

♀ V (3.2 cm)

**IMMATURE STAGES:** Undescribed.

**BIOLOGY:** Green Marbles fly in June in BC, with only one brood each year. Oviposition occurs on *Draba* and *Braya humilis* in the northern YT (Kondla and Pelham 1995). *Sisymbrium* and *Descurainia* are used as larval foodplants in Siberia (Tuzov 1997).

**SUBSPECIES:** The taxonomy of this group has yet to be fully worked out. The subspecies from the Yukon and BC may be undescribed.

**RANGE AND HABITAT:** Green Marbles inhabit dry, barren scree mountain slopes in the northern Yukon (Kondla and Pelham 1995). In BC they inhabit dry, gravelly habitats at low elevations.

**GENERAL DISTRIBUTION:** Green Marbles occur in Siberia, the Kenai Peninsula, AK, and the Ogilvie Mountains in northern YT, and in northwestern BC along the YT border.

**CONSERVATION STATUS:** The Green Marble is of Special Concern in BC (S3).

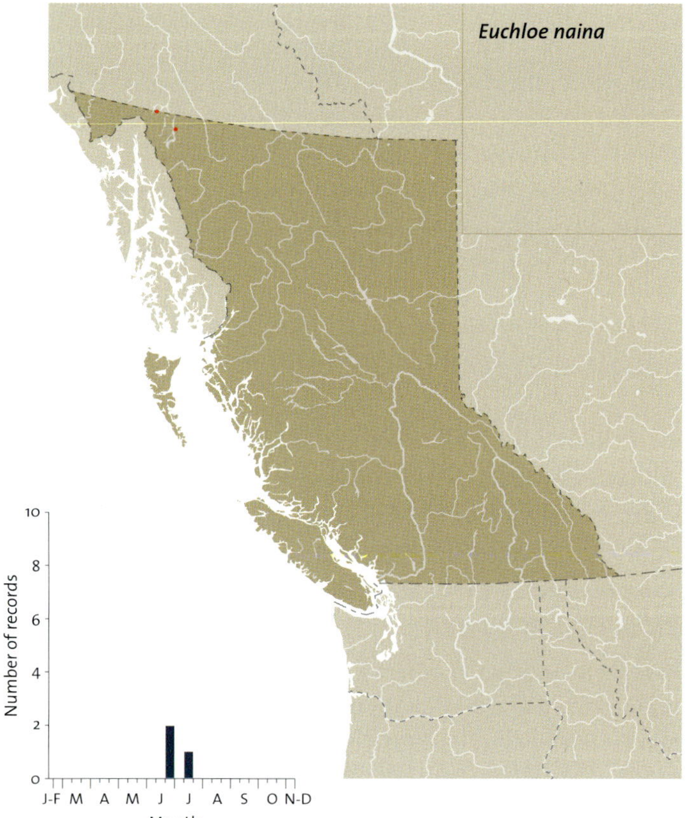



# Desert Marble

*Euchloe lotta* Beutenmüller, 1898

**ETYMOLOGY:** The name *lotta* is of unknown derivation, but perhaps refers to a woman named Charlotte. The common name refers to the arid sagebrush grasslands inhabited by the species, and was first used by Layberry et al. (1998).

**ADULT:** In Desert Marbles the green marbling on the ventral hindwings is never tinged with yellow. The marbling occurs in large patches with substantially smaller, pearly white areas between them. The forewings and hindwings of both sexes are always completely white. The black discal cell spot on the forewing does not have any white scales in it, and is large and square (Opler 1970).

**IMMATURE STAGES:** Eggs and young larvae are similar to those of the Large Marble. Third to fifth instar larvae are green above and below the spiracles, with few or no black spots except on the head. There is a narrow purplish stripe above the spiracles, and sometimes down the back. There is a white stripe along the side at the spiracles, and the head is green. Larval colour is geographically variable, with the basic green colour sometimes being greyish or yellowish green in some populations. The prepupal larva is dark purple with a prominent white stripe down each side. The elongated head end of the pupa is hooked (Opler 1975).

**BIOLOGY:** Desert Marbles are on the wing in April and May, with only one brood each year. Oviposition is on flower buds, leaves, or stems of Brassicaceae. The first instar larvae move immediately to the flower buds to feed, either by walking over the plant surface or by boring through the leaves clasping the inflorescence. The larvae then bore through the side of the flower petals, enter, and feed on the inside of the flower (Opler 1975). Where they occur close to each other in the US, Desert Marbles fly earlier than California Marbles, *Euchloe hyantis*. Desert Marbles lay eggs on, and larvae feed on, a wide range of crucifers, whereas California Marbles use species of *Streptanthus* as larval foodplants almost exclusively (Shapiro 1982b). Desert Marbles in Benton County, WA, use *Descurainia pinnata* and *Sisymbrium altissimum* as larval foodplants (Opler 1975).

**SUBSPECIES:** None. The species was named from "Colorado, Arizona, Utah, southern California." Until recently the Desert Marble was treated as a subspecies of the California Marble,

♂ D  (3.7 CM)     ♀ D  (3.7 CM)

♂ V  (3.7 CM)     ♀ V  (3.7 CM)

*Euchloe hyantis* (W.H. Edwards, 1871), but is now recognized as a full species. Desert Marbles are pure white above, and have a wider forewing discal black bar and a somewhat square forewing. The white areas of the ventral hindwing are distinctly pearly iridescent. In contrast, California Marbles are off-white above with a narrow forewing discal black bar, and have a pointed forewing; the white area of the ventral hindwing is only slightly pearly iridescent (Opler 1999).

**RANGE AND HABITAT:** Desert Marbles occur in the arid lowland grasslands of the Southern Interior and Kootenays of BC.

**GENERAL DISTRIBUTION:** Desert Marbles occur from southern BC south through the transmontane Columbian drainage, Great Basin, and Colorado plateau to north-central MEX (Opler 1999).

**CONSERVATION STATUS:** Not of concern (S4).

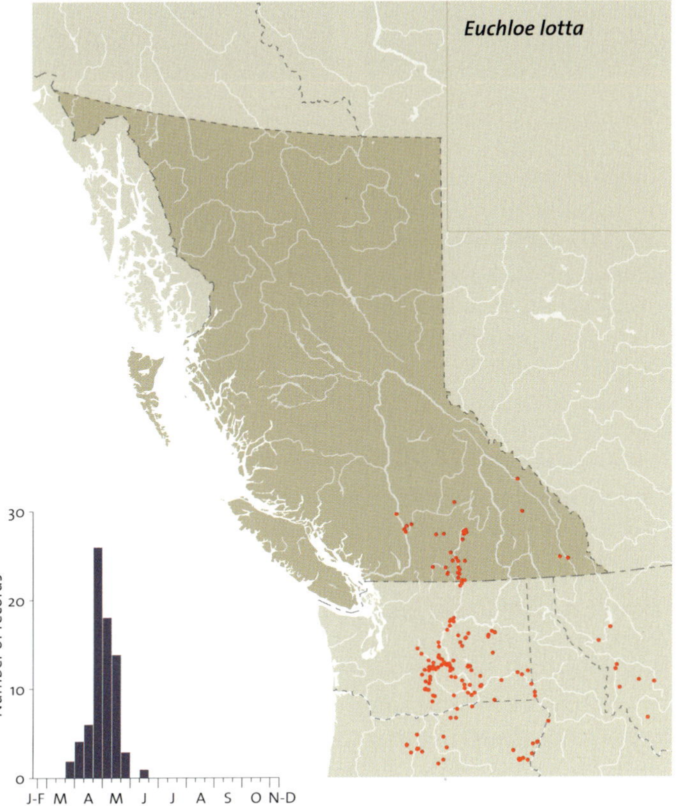

**Desert Marble (*Euchloe lotta*)**

The name *Anthocharis* is derived from the Greek *anthos* (flower) and *kharis* (grace), either because the butterflies have the grace of a flower or because they lend grace to the flowers they frequent (Emmet 1991). The common name "The Orange-tip" was first used for *A. cardamines* (Linnaeus) in Britain in 1747–49 (Emmet 1990a). The name was later extended to the entire genus, and was first used for the genus in North America by Scudder (1875). The common name refers to the orange or yellow apex of the dorsal forewings of most species in the genus.

Orangetips are similar to marbles, but the tips of the forewings are orange in most species. Male orangetips are usually white, occasionally pale yellow, with black and red orange markings. Females range from white to rich yellow, with black and pale orange markings. There are about 18 species in the genus.

Two species of orangetips occur in BC, as suggested by Layberry et al. (1998). Research in California (Geiger and Shapiro 1986) showed that what has traditionally been treated as the species *Anthocharis sara* is actually a group of closely related species. Their electrophoretic data indicated that three "subspecies" of Sara's Orangetip are all separate species, *Anthocharis sara* Lucas, 1852; *A. stella* W.H. Edwards, 1879; and *A. julia* W.H. Edwards, 1872. Opler (1999) recognizes a fourth species in this group, *Anthocharis thoosa* (Scudder, 1878).

Two of these four species occur in BC: Sara's Orangetip (*Anthocharis sara*) on the coast and Stella's Orangetip (*Anthocharis stella*) in the interior. They occur nearly sympatrically on the west side of the Cascade Mountains on Mount Cheam, west of Hope. Sara's Orangetip occurs at the bottom of the mountain in the coastal forest, and Stella's Orangetip occurs in the alpine. More sampling is required to determine whether they occur together in any habitat.

## Sara's Orangetip

*Anthocharis sara* Lucas, 1852

**Etymology:** The derivations of both the species name *sara* and the subspecies name *flora* are unknown. Both appear to be named after unknown women. Flora was the Roman goddess of flowers, but there is no hint of a relationship to flowers in the original description of subspecies *flora* (Wright 1892a). Subspecies *alaskensis* was named for the fact that the type locality was in Alaska. The common name was first used by Holland (1898).

**Adult:** Sara's Orangetips are small white (males, some females) or yellow (most females) butterflies with orange wing tips. In males, the black bar across the base of the

Egg

Mature larva

orange wing tip is straight and usually strongly black across the entire wing. The ventral hindwing marbling is grey or green, and is usually found in crisply defined patches. Populations near the ocean have many white females; further from the ocean most of the females

Pupa

are yellow and only a few are white. Males or females usually have a row of black spots along the margin of the dorsal hindwing.

**Immature stages:** Eggs are pale green when laid, but turn bright orange after a day. Third to fifth instar larvae of subspecies *flora* are medium green, shading to lighter green down the sides to white lateral stripes. Below the white lateral stripe there is a dark green line that is obvious only when the larva is newly moulted; below that they are dark

Sara's Orangetip (*Anthocharis sara flora*)

green on the underside. The spiracles are grey green and centred in the white lateral stripe. There is a sparse covering of hairs, with a black spot at the base of each hair. Most hairs are short and non-secretory; a few are longer, blacker, and thicker, have a larger black spot at the base, and are secretory. Pupae are long and thin, and are light brown to dark green (CSG).

**BIOLOGY:** Sara's Orangetips have one generation each year, and fly from March to May at low elevations in the south. In the north they fly in June, and in the mountains of Vancouver Island in July and August. Eggs are laid on the leaves, stems, and buds of *Arabis* in California (Shapiro 1981a, 1981b) and BC. Young larvae feed on leaves, but older larvae prefer flowers and fruits in both California and BC. Many pupae in some Californian populations diapause over two or more winters (Edwards 1887–97), apparently to provide insurance against population extinction caused by a single catastrophic drought season (Shapiro 1981b). Pupae from coastal BC seldom, if ever, diapause more than one winter (CSG), possibly because the environment is more predictably moist than in California. In California the nominate subspecies has a partial second generation when there has been a wet spring (Evans 1975), but subspecies *flora* is always univoltine.

Larval foodplants include *Arabis glabra* in the area of Victoria, BC (Jones 1935; J.B. Tatum, pers. comm.; CSG; GAH) and *Arabis drummondii* at Silverhope Creek near Hope (ACJ). Various

Ssp. *flora* ♂ D  (4.0 cm)

Ssp. *flora* ♀ D  (4.0 cm)

Ssp. *flora* ♂ V  (4.0 cm)

Ssp. *alaskensis* ♂ V  (3.5 cm)

Ssp. *alaskensis* ♂ D  (3.5 cm)

Ssp. *alaskensis* ♀ D  (3.7 cm)

Brassicaceae are used elsewhere, including *Arabis sparsiflora, Barbarea vulgaris, Brassica kaber, Descurainia,* and *Sisymbrium officinale* (Opler 1967a).

**SUBSPECIES:** Subspecies *flora* W.G. Wright, 1892 nec 1905 authors (TL: Tenino, WA [Tilden 1975]) occurs in southern coastal BC. The ventral hindwing veins are orange yellow; the marbling on the underside of the wings is dark green or occasionally grey (lighter further from the ocean); and females are white, partly yellow, or completely yellow. Subspecies *alaskensis* Gunder, 1932 (TL: Skagway, AK) occurs in the Coast Range of northwestern BC, east to Atlin. The ventral hindwing veins are orange, the marbling is grey green, and females are usually white but occasionally yellow.

**RANGE AND HABITAT:** Sara's Orangetips inhabit forest openings, rock slopes, and cliffs from sea level to subalpine habitats. They prefer partly wooded or brushy areas to completely open meadows. They occur along coastal BC, north to the Yukon in the Coast Range.

**GENERAL DISTRIBUTION:** From YT south to Baja California west of the coastal mountain ranges of BC and the Cascade and Sierra Mountains of the USA.

**CONSERVATION STATUS:** Not of concern. Subspecies *flora* is S4 and subspecies *alaskensis* is S5.

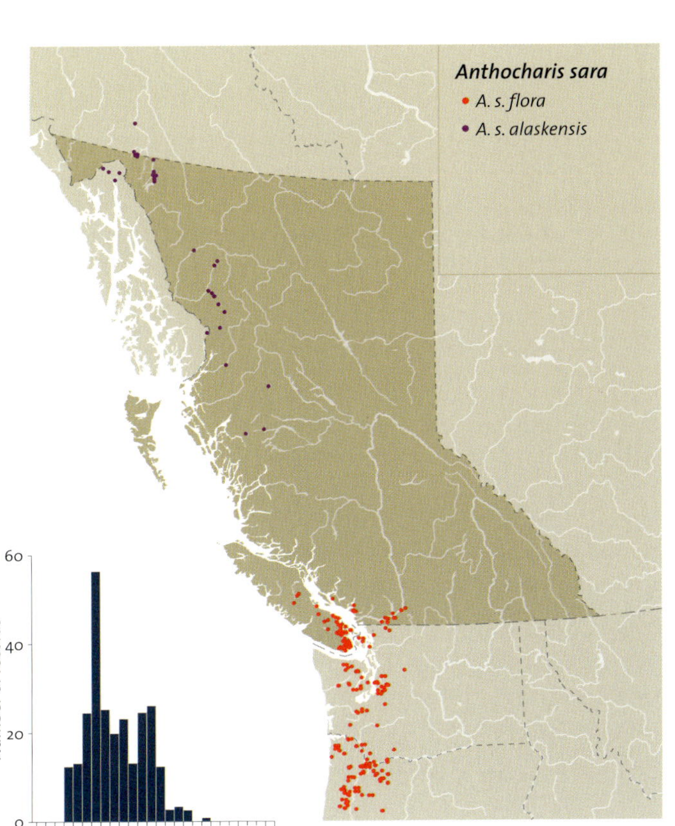

# STELLA'S ORANGETIP
*Anthocharis stella* W.H. Edwards, 1879

Stella's Orangetip (*Anthocharis stella*)

♂ D (3.6 cm)

♀ D (3.9 cm)

♂ V (3.6 cm)

**ETYMOLOGY:** The derivation of the name *stella* is unknown. Edwards named many species using women's names of unknown origin. The common name was first used by Holland (1898).

**ADULT:** Stella's Orangetips are small white (males) or yellow (almost all females) butterflies with orange wing tips. In males, the middle of the black bar across the base of the orange wing tip is displaced towards the wing tip, rather than connecting smoothly with the black rectangle at the end of the discal cell. This black bar usually fades to grey in the middle. The ventral hindwing marbling is yellow green, and usually appears smudged. In populations along the eastern side of the Cascades, the black bar is more strongly developed, at first looking similar to that in Sara's Orangetip. The bend in the bar is still present, however, although less obvious because of its increased width, and the ventral markings are still typical of Stella's Orangetip. Occasional males or females have a row of black spots along the margin of the upperside of the hindwing.

**IMMATURE STAGES:** Unknown.

**BIOLOGY:** Stella's Orangetips have one generation each year, and fly from April to June at low elevations. At high elevations they fly in July and August. The larvae feed on *Arabis* species in the Southern Interior and Pine Pass, BC (CSG).

**SUBSPECIES:** The nominate subspecies occurs in BC (TL: Marlette Peak, Washoe Co., NV).

**RANGE AND HABITAT:** Stella's Orangetips inhabit forest openings and meadows from valley bottoms to subalpine habitats. They prefer partly wooded or brushy areas to completely open meadows. They occur across the Southern Interior of BC, and north to near Chetwynd. The only population on the west side of the Cascade Mountains is near the summit of Mount Cheam, west of Hope.

**GENERAL DISTRIBUTION:** Stella's Orangetips occur almost entirely east of the coastal mountain ranges of BC, from north-central BC and western AB south to WY and the Sierra Nevada Mountains of CA and NV.

**CONSERVATION STATUS:** Not of concern (S5).

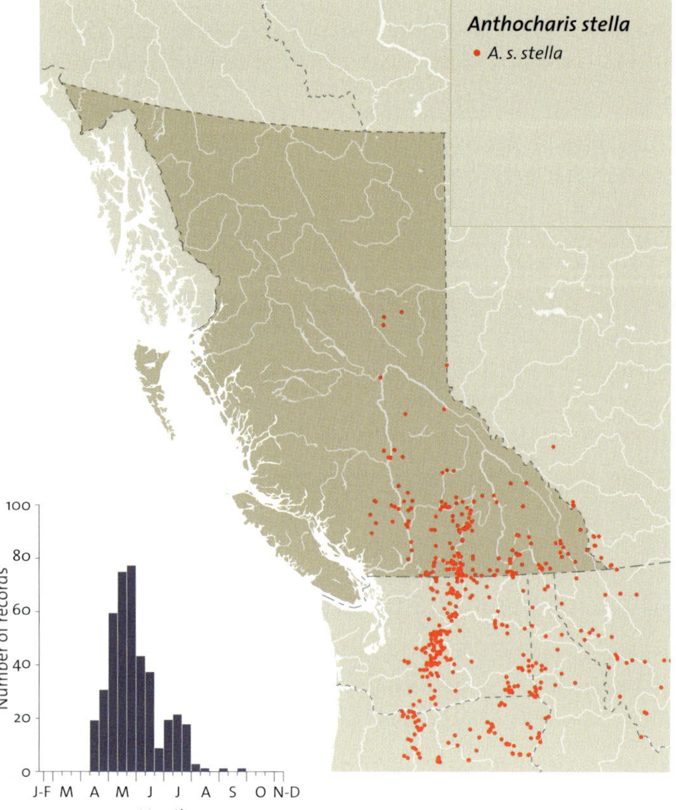

The Latin and common names for the subfamily are derived from the type genus, *Colias* (below). The generic common name "sulphurs" was first used by Scudder (1875). The British use the generic common name "yellows" for *Colias*.

Sulphurs are medium-sized butterflies that are yellow, orange, or greenish white with black markings. They are sexually dimorphic, and frequently have white females lacking the black border on the upperside of the wings. They have a moderately fast, bobbing flight pattern.

Antennae are short, with a thick shaft that gradually expands into a slender club. The forewing has 11 veins, and the hindwing precostal vein is absent. The prothorax has patagia, two hardened, lobelike subdorsal lumps. The dorsal forewing has androconial scales.

The sulphurs are represented in BC by only one genus, *Colias*, except for the possible migratory occurrence of the Dainty Sulphur (genus *Nathalis*) and the Sleepy Orange (genus *Eurema*).

### GENUS *COLIAS* FABRICIUS, 1807 SULPHURS

Colias is the name of a promontory on the coast of Attica where there was a temple of Aphrodite. There is no obvious relationship to the butterfly, but the name may be a pun (Emmet 1991). An alternative explanation is suggested under *Pontia*. The common name "sulphurs" is derived from the yellow "sulphur" colour of most species.

Sulphurs in BC are generally medium-sized butterflies that are yellow, orange, white, or (one species) yellow green with black markings. The wings of males always have a solid black border, with the exception of the Arctic Sulphur. The black borders of females contain extensive pale areas, or may be greatly reduced or absent. There are several multivoltine species that show considerable seasonal variation in wing colour.

There are about 70 species of *Colias* in the world. The centre of distribution in North America is BC, with more species (13) than any other province or state. *Colias* species may all be inter-fertile, with natural hybrids known for most species combinations where they occur together in the wild. The species have behavioural, ecological, and physiological differences that maintain separation of the species in the wild (Hovanitz 1963).

Eggs are laid singly on the leaves of the foodplants, and are pale yellow green to cream, later turning orange. Young larvae are slender, yellowish or green, and smooth-skinned with a thin coat of fine hairs. Mature larvae are yellow green or green with fine black dots all over, and stripes of various colours running along the back and sides. Sulphurs hibernate as second to fourth instar larvae (except Canadian Sulphurs, which hibernate as fifth instar larvae), and then complete development in the spring. There are five larval instars in all *Colias* (Ae 1958a). Pupae are fastened head up with a girdle around the middle.

Members of the genus utilize a wide range of foodplants, although each species specializes to a greater or lesser extent. Larvae of sulphurs feed on plants in three groups: legumes (Fabaceae), *Vaccinium* (Ericaceae), and *Salix* (Salicaceae). Sulphurs occur in a wide range of habitats, including arid sagebrush areas, alfalfa fields, meadows, alpine tundra, and forest bogs.

Sulphurs always rest with their wings folded over their backs, and bask in the sun by leaning to the side to allow the sun to warm the underside of their wings. It has been demonstrated for several species (*C. meadii, C. nastes, C. philodice,* and *C. eurytheme*) that the darker the pigmentation on the underside of the wings, the more heat can be absorbed from the sun while basking, permitting greater flight activity in cold environments (Kingsolver 1985).

There is relatively little variation in wing pattern between many species, making identification difficult. The key characters mentioned in the species discussions are shown in Figure 63.

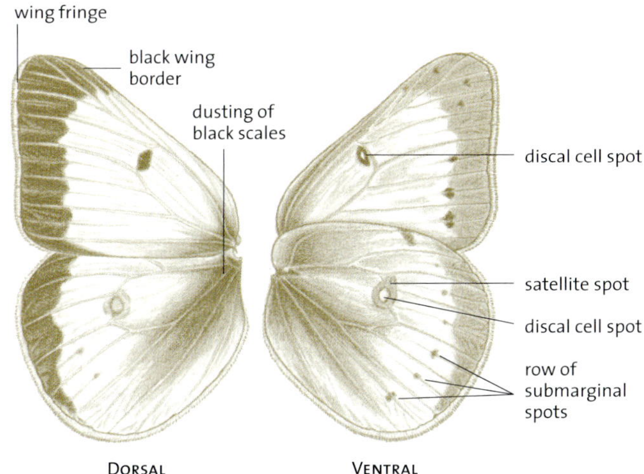


**63  Wing pattern of the Orange Sulphur (*Colias eurytheme*)**

# Clouded Sulphur

*Colias philodice* Godart, [1819]

**Clouded Sulphur (*Colias philodice eryphyle*)**

**ETYMOLOGY:** The name *philodice* is derived from the Greek *philos* (friend) and *dice* (one of the hours or seasons) (Reed 1870), hence "friend of the seasons." The subspecies names *eriphyle* and *vitabunda* are of unknown origin. The common name was first used by Scudder (1875).

**ADULT:** Clouded Sulphurs are predominantly yellow with a black wing border. Most northern females are white. The discal cell spot is a small black dot on the dorsal forewing, and yellow or orange on the dorsal hindwing. Females have yellow (or white) spots within a wide black forewing border. The discal cell spot on the ventral hindwing is white and

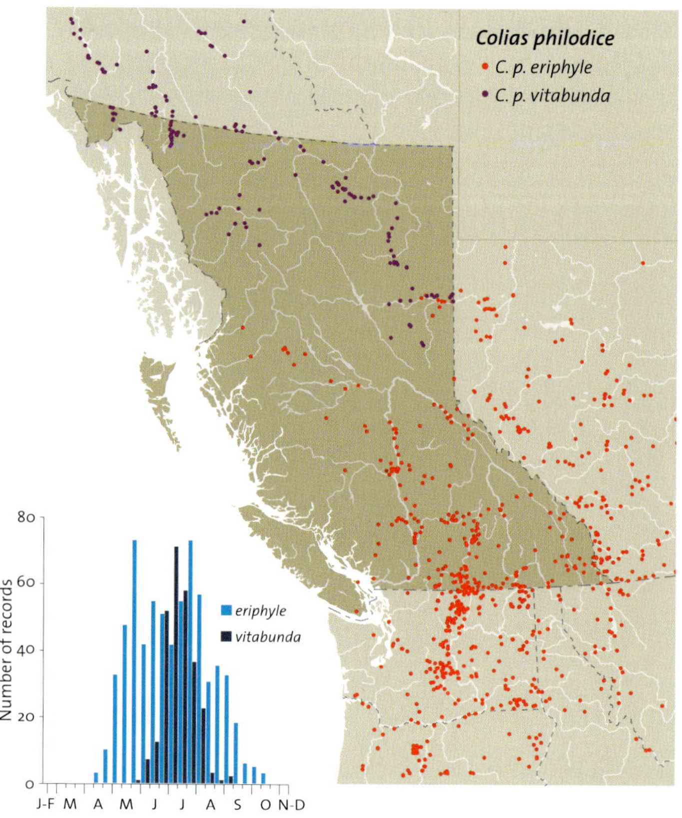

Ssp. *eriphyle* ♂ spring form D (4.3 CM)

Ssp. *eriphyle* ♀ spring form D (4.0 CM)

Ssp. *eriphyle* ♂ spring form V (4.3 CM)

Ssp. *eriphyle* ♂ summer form V (4.6 CM)

Ssp. *eriphyle* ♀ white form D (4.1 CM)

Ssp. *vitabunda* ♂ V (4.3 CM)

Ssp. *vitabunda* ♀ D (4.7 CM)

Ssp. *vitabunda* ♀ V (4.7 CM)

surrounded by two red or pink rings, and it may have a satellite spot. There is usually a ventral row of up to six small, brown submarginal spots on each wing, but they may be reduced or absent, especially in northern populations. Females are sometimes white, especially in northern BC, with the usual dark markings. The upperside of the wings of males does not reflect ultraviolet light, in contrast to the Orange Sulphur.

**IMMATURE STAGES:** Eggs are cylindrical and tapered at both ends, with 12–15 longitudinal ribs down the sides. The eggs are pale yellow when laid, changing to red in a few days



FAMILY PIERIDAE (WHITES, SULPHURS, AND MARBLES)

(both subspecies). Mature larvae of subspecies *philodice* are dark green, with slightly paler raised points. Many very short hairs give it a downy appearance. There is a fainter green dorsal line and a yellowish-white lateral band with a rosy line through it; the head is green (Saunders 1869a; Bethune 1873b). Mature larvae of subspecies *eriphyle* from Quesnel, BC, are dark green, with numerous black slightly raised points, with a short black hair arising from each black point, a darker green dorsal line, and a narrow white lateral line with pink red ("rosy") dashes in the lower half of the white line (CSG). The dashes are broken with white between each segment. The pupae of subspecies *philodice* are pale green, with a yellowish tinge and a ventral darker yellow line down each side formed of tiny spots. The abdomen has a blackish brown line low down on each side, and there is a dorsal dark green line (Saunders 1869a; Bethune 1873b). There is a red dot below the yellow lateral line on each abdominal segment (Edwards 1874–84). Pupae of subspecies *eriphyle* from Quesnel, BC, have three bright red dashes on each side of the abdomen (CSG).

**BIOLOGY:** Clouded Sulphurs are multivoltine at low elevations in BC, and are in flight from early May to October. Southern subalpine and alpine populations are univoltine, and are in flight from late July to September. Subspecies *vitabunda* is univoltine, and flies in June and July. Eggs are laid singly on the foodplant starting 2–4 days after emergence; the larvae feed on the leaves. Subspecies *philodice* is multivoltine, with larvae present in the fall and hibernating in the third instar or occasionally in the fourth instar. In the spring these larvae break diapause and feed on the new growth of their larval foodplants, and then pupate (Ae 1958a). This is also the case with subspecies *eriphyle,* but subspecies *vitabunda* is univoltine, with the larvae hibernating in third instar (CSG).

Larval foodplants in BC include introduced red clover, white clover, and alfalfa (Harvey 1908; CSG). Presumably native clovers are also used. Outside BC additional larval foodplants include garden peas, lupines, *Astragalus carodurpus, A. bisulcatus, Hedysarum boreale, Lathyrus lanzwetii* var. *leucantus, Trifolium fragiferum,* and *Vicia americana* (Edwards 1874–84; Emmel et al. 1971; Shapiro and Shapiro 1974; Hayes 1981; Scott 1992).

**SUBSPECIES:** Subspecies in BC are poorly defined. Subspecies *eriphyle* W.H. Edwards, 1876 (TL: Lac la Hache, BC) definitely occurs in southern and central BC and the Peace River lowlands. Both males and females are warm yellow and frequently have an orange tint to the basal area of the fore- and hindwings. White females are uncommon but occur regularly. There are many populations that do not appear to be subspecies *eriphyle,* because the adults are a colder yellow and lack the orange tint, and females are very rarely white. This phenotype is most often associated with alfalfa, suggesting that the nominate subspecies has followed alfalfa into the province. Subspecies *vitabunda* Hovanitz, 1943 (TL: Mt. McKinley National Park, AK) occurs throughout

**Eggs (ssp. *eriphyle*)**

**Mature larva (ssp. *eriphyle*)**

**Pupa (ssp. *eriphyle*)**

montane northern BC and much of the Peace River. The black wing borders are narrow, the females are frequently white, and the submarginal brown spots on the ventral hindwings are frequently absent. Subspecies *eriphyle* and *vitabunda* fly together at the Clayhurst crossing of the Peace River, in closely adjacent habitats, suggesting they may be separate species. More observations are needed to determine whether their contact is the result of the recent introduction of alfalfa (used by *eriphyle* larvae), and whether they may simply have not yet interbred enough to merge into a single population.

**RANGE AND HABITAT:** Clouded Sulphurs occur throughout BC east of the Coast Range. Clouded Sulphurs in the Vancouver area appear to be migrants or to have emerged from larvae or pupae imported from the interior of BC with hay, rather than resident populations. Clouded Sulphurs occur in all grassland and open forest areas that support alfalfa, clover, or vetches, from low elevations to alpine meadows. The largest population densities are now found in alfalfa fields, but it is also a common and widespread species in natural meadows and grasslands.

**GENERAL DISTRIBUTION:** Clouded Sulphurs inhabit almost all of subarctic, temperate, and subtropical North America from AK and YT east to NF and south to southern MEX. They do not occur in Nunavut or western CA.

**CONSERVATION STATUS:** Not of concern, with both subspecies S5.

# Orange Sulphur

*Colias eurytheme* Boisduval, 1852

**Etymology:** The name *eurytheme* is in reference to the similar European Lesser Clouded Yellow, *C. chrysotheme* Esper. The Greek *eurys* (broad) in the first part of the word refers to the black wing borders of *eurytheme* being wider than those of *chrysotheme*. The common name was first used by Scudder (1875) and refers to the colour of the wings.

**Adult:** Orange Sulphurs are very similar to Clouded Sulphurs, but are always predominantly orange on the upperside of males and most females. The black borders of the upperside tend to be broader than in Clouded Sulphurs, and the row of submarginal spots on the underside of the wings are more pronounced. The summer generation has completely orange wings with black borders, and the spring and fall broods have yellow wings with an orange patch restricted to the base of the wings. Most females are orange, or yellow and orange, with black borders. White female Orange Sulphurs have wider black borders, and the spots within the black border are larger, than those of white female Clouded Sulphurs (Hovanitz 1948).

**Immature stages:** Eggs are initially white, translucent, and pointed at the upper end; they also have longitudinal ribs. They turn red after a day or two, then black before hatching. Mature larvae have a dark velvety green back, with each segment finely creased. On each side, there is a narrow white lateral line, through the middle of which is a line of vermilion red to orange yellow dashes. The underside is green, and the

♂ D (4.5 cm)

♀ D (5.0 cm)

♂ V (4.5 cm)

♀ V (5.0 cm)

♀ white form D (4.9 cm)

head is green and translucent. Pupae are light green with a lateral yellow line, above which is a brown point on each segment. A subdorsal brown patch commences at the edge of the wing covers and occupies two or three segments. Another colour form has a whitish lateral line along the abdomen, with a black stripe above the line for the first two or three abdominal segments. Compared with Clouded Sulphurs, the mature larvae of the Orange Sulphur are larger, their lateral spots are a brighter scarlet, and they lack semi-circular black spots under the lateral band. The pupae of Orange Sulphurs have a more attenuated head case, the mesonotal process is less prominent, and some abdominal markings are absent (Edwards 1868–72).

**Biology:** Orange Sulphurs are on the wing from April to October, with at least three broods. In BC they are seldom seen before midsummer, when they migrate into the province. Orange Sulphur larvae do not go into hibernation in the fall, unlike those of Clouded Sulphurs (Ae 1958a), so they cannot survive through the winter in BC. Hybridization with Clouded Sulphurs is minimized because females normally mate only with males that have ultraviolet reflective wings and the correct male pheromone. Females start laying eggs when they are several days old, and in captivity lay an average of 700 eggs in their lifetime. Eggs are laid singly on leaves, and young larvae eat holes in the top of the leaves. Older larvae feed on the tip of the leaves, and the largest larvae eat each half of the leaf separately. If the

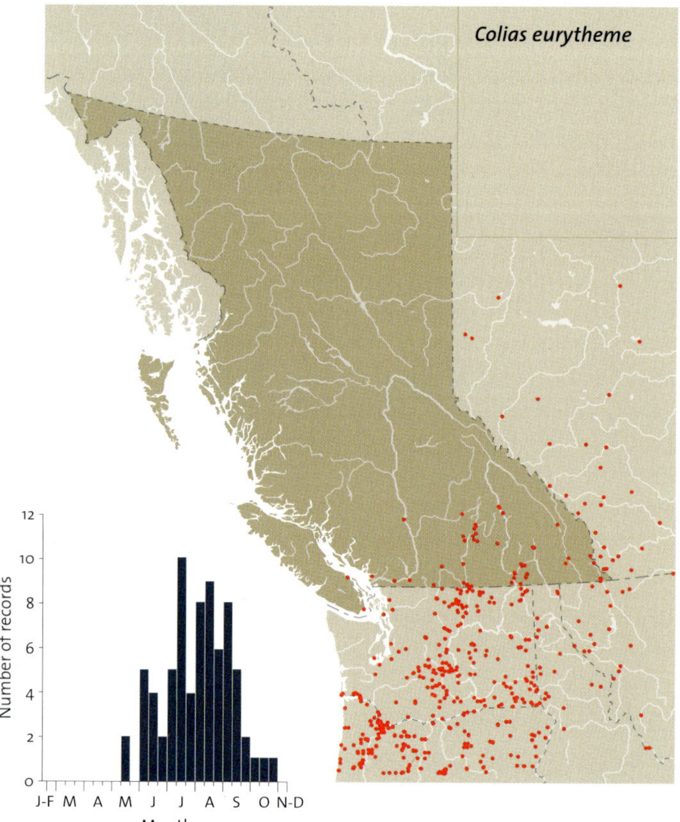

weather is too hot, resulting in larval body temperatures over 32°C, or too cold (body temperature below 15°C), the larger larvae move to the base of the larval foodplant (Sherman and Watt 1973). Orange Sulphurs can be a crop pest on alfalfa in warm areas of the southern USA (Allen and Smith 1958), but in the cool climate of BC their popula tions never get large enough to be a serious problem.

Orange Sulphurs commonly use alfalfa as a larval food-plant in BC, and it appears to be the preferred foodplant. Outside BC other foodplants include red clover, white clover, *Astragalus flexuosus, A. agrestis, A. bisulcatus, A. whitneyi, A. drummondii, Lupinus argenteus, Medicago lupulina, Melilotus officinalis, Trifolium fragiferum, T. hybridum, T. longipes, T. nanum, T. wormskholdii, Vicia americana,* and *V. sativa* (Emmel and Emmel 1962; Emmel et al. 1971; Shapiro and Shapiro 1974; Shapiro 1975b; Scott 1992).

**Subspecies:** None. The type locality has been restricted to Sacramento, CA (Emmel et al. 1998a).

## Alexandra's Sulphur
*Colias alexandra* W.H. Edwards, 1863

**Etymology:** The name *alexandra* refers to an unknown woman, as was often the case with names given by Edwards. Pyle (1981) was incorrect when he suggested that it referred to Queen (then Princess) Alexandra of Great Britain. The preprint of the original description of *C. alexandra* is dated March 1863, the month Alexandra married the future

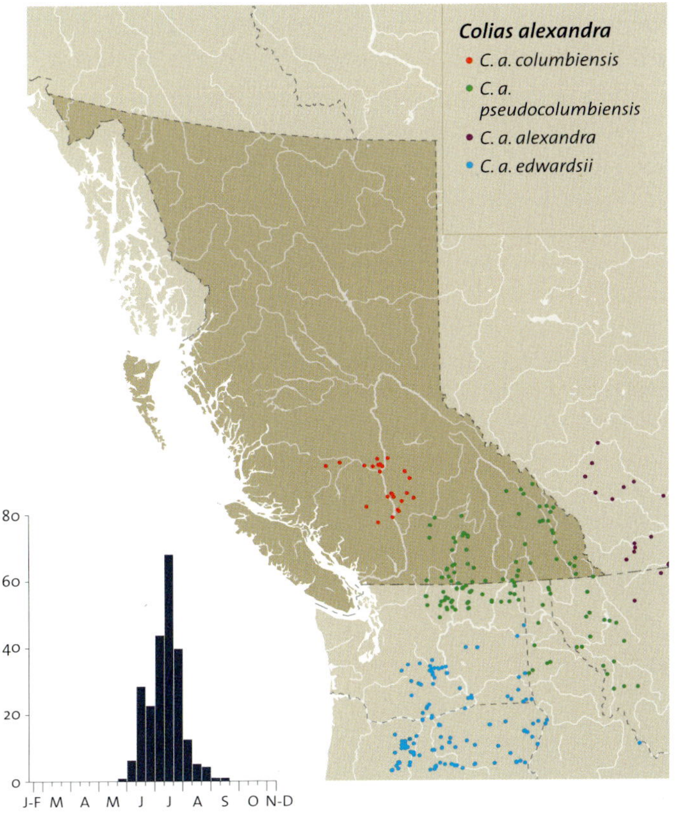


**Range and habitat:** BC is at the northern edge of the range of Orange Sulphurs. They migrate north from WA, ID, and MT in the late summer, resulting in their widespread occurrence across southern BC in warm summers. They rarely occur north of the Thompson River valley. Orange Sulphurs are sometimes encountered in subalpine and alpine areas during migratory flights, but they primarily inhabit low-elevation alfalfa fields and nearby natural or weedy habitats containing naturalized alfalfa.

**General distribution:** Orange Sulphurs occur from BC to NF as breeding migrants, and south throughout the USA and MEX, from coast to coast. Migrants extend north to NT, east of the Rockies.

**Conservation status:** Orange Sulphurs are breeding immigrants (SZB).

Ssp. *columbiensis* ♂ D (4.9 cm)    Ssp. *columbiensis* ♀ D (5.4 cm)

Ssp. *columbiensis* ♂ V (4.9 cm)    Ssp. *columbiensis* ♀ V (5.4 cm)

Edward VII, 40 years before he became king and she became widely known. In 1870 Edwards named *C. edwardsii* for Henry Edwards, not Edward VII, as would have been expected if he had named *C. alexandra* for Princess Alexandra. In addition, the American Civil War was raging in 1863. Britain supported the South, and it would have been very politically incorrect to honour a British prince or princess at that time. Subspecies *columbiensis* is named for the type locality of British Columbia. The subspecies name *pseudocolumbiensis* (false subspecies *columbiensis*) refers to the name *columbiensis* having been incorrectly applied to two subspecies by the author of *columbiensis* (Ferris 1973a). The correct common

Ssp. *pseudocolumbiensis* ♂ D
Holotype (4.9 cm)

Ssp. *pseudocolumbiensis* ♀ D
(4.7 cm)

Ssp. *pseudocolumbiensis* ♂ V
Holotype (4.9 cm)

Ssp. *pseudocolumbiensis* ♀ V
(4.7 cm)

name of the species is "Alexandra's Sulphur," which was first used by Holland (1898).

**Adult:** Male Alexandra's Sulphurs are yellow, occasionally with orange patches, and have narrow to moderately wide black wing margins. Males reflect ultraviolet light from the outer half of the dorsal hindwing. In normal light this area usually has a deeper yellow tint, occasionally deepening to orange, than the base of the hindwings. Females are white to yellow, with no black border or only faded remnants of the black border. The ventral hindwing is dark greenish. Both sexes usually have a pink ring around the ventral hindwing discal cell spot, and the dorsal hindwing discal cell spot is orange. In 5–10% of individuals, the discal cell spot on the dorsal hindwing is yellow and barely distinguishable from the ground colour, and on the ventral hindwing it is white and lacks a red ring. This is similar to subspecies *edwardsii* W.H. Edwards, 1870, which occurs in southern Washington.

**Immature stages:** Eggs of subspecies *alexandra* (Colorado) are conical and have 18–20 vertical ribs. They are pale yellow green, turning pink within 48 hours if fertile. First instar larvae are whitish, turning green with a black head once feeding commences. Mature larvae are yellow green, with small black tubercles, a white lateral band with a line of red orange dashes through it, and a green head. Pupae are yellow green with lighter lines that mimic leaf highlights, and three small reddish spots on the ventral side of the abdomen next to the wing (Edwards 1868–72, 1887).

**Biology:** In BC Alexandra's Sulphurs are in flight from June to August, depending on elevation. Males emerge earlier than females (Watt et al. 1977), as with most butterflies. Females can lay up to 600 eggs, which are laid singly on the upperside of foodplant leaves. A reduction in the number of eggs laid due to inclement weather or adult mortality can strongly affect the population size the next year. Larvae feed

on the foodplant leaves, and in subspecies *alexandra* hibernate in the third instar (Edwards 1887; Ae 1958a), possibly because the larval foodplants dry up before larvae can mature (Ellis 1974). Late third instar larvae stop feeding after about seven days, and move to an untouched leaf or to a neighbouring plant. The larvae are sluggish, somewhat swollen, and lighter green in colour with a relatively small dark green head. The diapause larvae either fall to the ground with the dead plant foliage or crawl into the litter on the ground, where they hibernate. In spring, diapause is broken when the new growth of the foodplant is available for food (Hayes 1981). When more than one egg is laid on a plant, only one larva successfully matures; the rest are dislodged during encounters between larvae or are eliminated through cannibalism (Hayes 1981). Most populations are univoltine, but isolated bivoltine populations occur in Colorado (Ellis 1974; Hayes 1981) and they are bivoltine in Alberta (Bird et al. 1995). Adults are highly mobile and fly long distances, resulting in populations being thinly dispersed over large areas. In Colorado they move an average of 1.3 km, with a maximum recorded dispersal of 8 km, and average about 2 adults per hectare (Watt et al. 1977). The larval foodplants tend to be weedy "pioneer" species, and disturbance such as selection logging or ground fires may increase the populations of larval foodplants and therefore of the butterfly. Population size and dispersal rates may also be affected by the availability of nectar sources in their dry environment (Ellis 1974).

Outside BC recorded foodplants are in the family Fabaceae, including *Astragalus bisulcatus*, *A. canadensis*, *A. eremiticus* complex, *A. lentiginosus*, *A. miser*, *A. robbinsii*, *Lathyrus lanzwetii* var. *leucantus*, *Oxytropis lambertii*, *Thermopsis pinetorum*, and *Thermopsis rhombifolia* var. *divaricarpa* (Shields et al. 1970; Ellis 1974; Watt et al. 1977; Hayes 1981).

**Subspecies:** Subspecies *columbiensis* Ferris, 1973 (TL: Anderson Lake, D'Arcy, BC) occurs only in the Chilcotin District of BC, from D'Arcy in the south to Riske Creek in the north. Males are lemon yellow, with the ultraviolet-reflective areas of the outer hindwing deep yellow. Females are always cream white, with only a hint of a grey wing border. Across the remainder of the Southern Interior and the Kootenays is a new subspecies, found also in northeastern Washington and northern Idaho, which is characterized by a geographic difference in female coloration. This difference in female coloration within BC was noted as long ago as 1950 by Hovanitz, but was missed by Ferris (1973a) because he included in error two yellow female *C. gigantea* as paratypes for *columbiensis* (N.G. Kondla, pers. comm.; Fig. 6c of Ferris [1973a]).

***Colias alexandra pseudocolumbiensis* Guppy & Shepard, new subspecies.** *C. a. pseudocolumbiensis* males are similar to those of *columbiensis*. On the average they are slightly darker yellow and have a slightly wider black wing border than *columbiensis*. The outer part of the hindwing is sometimes orange-tinted, which never occurs in *columbiensis*. Females are distinctly different from those of *columbiensis*, ranging from nearly as yellow as the males to pale yellow, but almost never the cream white of *columbiensis*. Females have grey wing borders that vary from barely present to well developed. **Types.** Holotype: male, BC, Hall Creek at Hwy. 33, 19 July 1986, J. and S. Shepard; a label "HOLOTYPE / *Colias alexandra* / *pseudocolumbiensis* / Guppy & Shepard" is attached. The holotype is deposited in the Royal British Columbia Museum, Victoria, BC, CAN. Paratypes: 11 males, 2 females, BC, Hall Creek at Hwy. 33, 19 July 1986, J. and S. Shepard (JHS); 6 males, BC, Christian Valley, 13 mi. south, 22 June 1966, J. and S. Shepard (JHS); 2 males, 1 female, BC, Joe Rich Creek, 27 km E of Kelowna on Hwy 33, 13 June and 17 June 1983, C.S. Guppy (CSG).

## CHRISTINA'S SULPHUR
*Colias christina* W.H. Edwards, 1863

**ETYMOLOGY:** The species *christina* was named for Christina Ross, the wife of the chief factor of Fort Simpson, NT. She collected the type specimens of the species (Edwards 1863). Subspecies *kluanensis* was named for the type locality of Kluane Lake, YT. The common name was first used by Holland (1931).

**ADULT:** Male Christina's Sulphurs are highly variable in colour, ranging from predominantly orange to mostly yellow with an orange tint. Females are even more variable, with the same range of colours as the males, plus various pale shades

♂ D  (4.6 CM)      ♀ D  (4.6 CM)

♂ V  (4.6 CM)      ♀ V  (4.6 CM)

**RANGE AND HABITAT:** Alexandra's Sulphurs occur in the dry Southern and Central Interior of BC, east of the Cascade and Coast ranges at low and mid elevations. They inhabit dry open pine and fir forests, occurring in grasslands only when dry open forest areas are nearby.

**GENERAL DISTRIBUTION:** Alexandra's Sulphurs occur throughout the dry interior areas of montane western North America, from central BC and southern AB south to CA and NM.

**CONSERVATION STATUS:** Not of concern. Subspecies *columbiensis* is S4, and subspecies *pseudocolumbiensis* is S5.

through to cream white. Females may have well-developed dark wing borders, or they may lack any dark markings. Both sexes always have a pink ring around the ventral hindwing discal cell spot, and the dorsal hindwing discal cell spot is orange. There are sometimes submarginal spots on the underside of the wings.

**IMMATURE STAGES:** Mature larvae from Laggan (Lake Louise), AB, are dark yellow green. There is a white basal stripe, with a short red dash through it behind each spiracle. Pupae are yellow green, with a broad yellow stripe from the wings to the underside of the end of the abdomen. Below the yellow stripe, a dash of red brown crosses the three abdominal segments behind the wings (Edwards 1887–97).

**BIOLOGY:** Adult Christina's Sulphurs are in flight in June and July. They occur in open, dry pine and spruce forests that have legumes such as *Vicia* growing in forest openings.

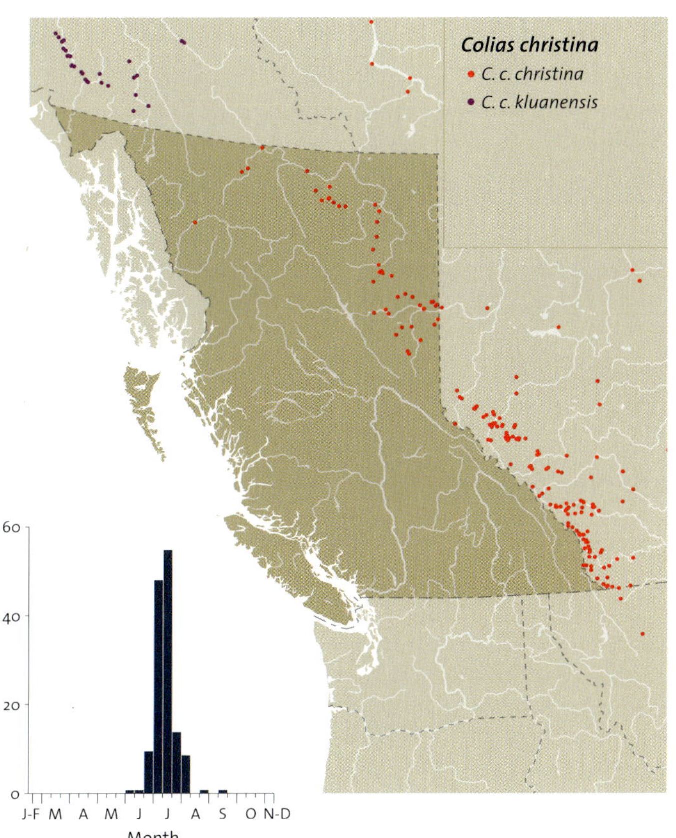

**Christina's Sulphurs (*Colias christina kluanensis*) mating**

Outside BC larval foodplants are in the family Fabaceae, including *Hedysarum sulphurescens, Lupinus,* and *Thermopsis rhombifolia* (McDunnough 1922; Klassen et al. 1989; Bird et al. 1995; Layberry et al. 1998).

SUBSPECIES: The nominate subspecies (TL: Mountain Rapids, Slave River [near Fitzgerald], AB [Kondla 1995]) occurs in northeastern BC and the East Kootenay. The ventral hindwing is yellow with a heavy dusting of grey scales.

The upperside of females is either yellow with prominent pale orange flushes on the wings and a broken black wing border, or cream with faint orange flushes and only a fragmentary grey wing border. *Colias alberta* K. Bowman, 1942 (TL: Wembley, AB) is a synonym of the nominate subspecies (Kondla 1986), with the type series being all Christina's Sulphurs except one paratype from the Kootenays of BC, which is *C. eurytheme* (N.G. Kondla, pers. comm.). Subspecies *kluanensis* Ferris, 1981 (TL: Kluane Lake, YT) may occur in northwestern BC; we have been unable to confirm the report from the Haines Highway of Layberry et al. (1998). This subspecies has a dark green ventral hindwing, and the dark dorsal wing borders are darker and wider than in *christina,* especially in females. There may be a long cline between subspecies *kluanensis* and subspecies *christina* in the southern Yukon (Layberry et al. 1998).

RANGE AND HABITAT: Christina's Sulphurs commonly occur at low elevations in northeastern BC, and are rare at low to mid elevations in the Kootenays.

GENERAL DISTRIBUTION: Christina's Sulphurs occur from southern YT through northern BC and southwestern NT to MB, and south to MT and SD.

CONSERVATION STATUS: Not of concern (S5).

## WESTERN SULPHUR
*Colias occidentalis* Scudder, 1862

ETYMOLOGY: The name *occidentalis* is Latin for "residing in the west," referring to the western North American distribution of the species. Scudder considered that there were three geographically defined boreal *Colias* species, which he described in the same paper: *Colias labradorensis* in the east, *C. interior* in the interior, and *C. occidentalis* in the west. The common name "Western Sulphur" was first proposed by Holland (1931) in reference to the western distribution of the species.

ADULT: Western Sulphurs are always yellow with black wing borders, and show little variation in BC. They usually lack all submarginal spots, although one or two may be faintly visible on the ventral hindwing. The ventral hindwing discal cell spot has only a single red ring around it, usually has a pink centre, and seldom has a satellite spot. Unlike those of Alexandra's and Christina's Sulphurs, the dorsal hindwings of male Western Sulphurs are an even, clear yellow except for the black wing border and a characteristic extensive dusting of black scales near the dorsal wing bases. The females are pale yellow, with a poorly defined and reduced black wing border.

IMMATURE STAGES: Eggs are shiny and whitish at first, becoming rosy red with an ivory white tip as they mature. Mature larvae are dark velvety green, with many black dots surrounded by green and with white hairs thickly distributed

♂ D (4.8 CM)    ♀ D (5.1 CM)

♂ V (4.8 CM)    ♀ V (5.1 CM)

on the sides of the body. The spiracular line is thin but conspicuously white; the white ring around the black dots is replaced with green. There is a faint pink suffusion along the spiracular line. The pupal head has a projecting beak, and there is a decided hump on the top of the thorax. The pupae are smooth, emerald green at first but becoming darker and assuming a yellowish colour towards maturity. The beak is dark green above and yellow below. The three abdominal segments beyond the tip of the wing cases each have two small black dots. The spiracular line is distinctly yellowish (Hardy 1960).

**BIOLOGY:** Western Sulphurs are univoltine, and fly from June, at low elevations on Vancouver Island and the mainland, to September in subalpine habitats on Vancouver Island. Eggs hatch a week or so after oviposition, and the egg chorions are partially or entirely eaten by the larvae. Third instar larvae commence aestivation in the fold of a shrivelled leaf

about two weeks after the eggs hatch, and remain quiescent until early April the next year. Pupation occurs about a month after feeding recommences in the spring. Pupation occurs with the head up and a girdle around the thorax. The adult emerges about 17 days after pupation. When disturbed, fifth instar larvae raised their thoracic segments in a sphinxlike position (Hardy 1960).

The larval foodplants are unknown in BC, but *Lathyrus nevadensis* var. *nuttallii* is a likely one on Vancouver Island (Hardy 1960). Oviposition has been recorded on sweet white clover, lupines, and *Vicia sativa* in Yakima Co., WA (Newcomer 1964a).

**SUBSPECIES:** Only the nominate subspecies occurs in BC. Western Sulphurs were originally described by Scudder (1862) from "Gulf of Georgia (A. Agassiz), Fort Simpson, British America (W.H. Edwards)." There was a Fort Simpson on the coast near Prince Rupert, BC, at the time Scudder described *C. occidentalis.* It is clear, however, that the Fort Simpson referred to is the one on the Mackenzie River, NT, because Edwards (1868–72), who supplied the Fort Simpson specimens, states that the type series included specimens from "Mackenzies River." The female types may have included the one illustrated by Edwards (1868–72) as the female for *C. occidentalis,* which appears to be a *Colias philodice vitabunda.*

**RANGE AND HABITAT:** Western Sulphurs inhabit low-elevation dry grassy slopes and forest edges on the lee side of the Cascade and Coast ranges on the mainland of BC. On Vancouver Island they occur in sea-level forest openings and edges, and upward into subalpine and alpine meadows.

**GENERAL DISTRIBUTION:** Western Sulphurs are restricted to Vancouver Island and the Cascade Mountains of BC, western WA, western OR, and northwestern CA.

**CONSERVATION STATUS:** Western Sulphurs are of Special Concern in BC (S3S4).



## MEAD'S SULPHUR
*Colias meadii* W.H. Edwards, 1871

**ETYMOLOGY:** The species *meadii* was named for Theodore L. Mead, who collected the type specimens in Colorado for Edwards (Edwards 1868–72), along with many other Rocky Mountain butterflies (Brown 1965). Mead was also Edwards's son-in-law, the husband of his only daughter, Edith. The subspecies name *elis* is the name of the plain of northwestern Peloponnese, famed for horse breeding; the town of Elis was built in 471, BC. The common name was first used by Holland (1898).

**ADULT:** Mead's Sulphurs are a beautiful deep orange over the entire dorsal wing surface. The orange has a purple sheen,

especially visible in males, that is correlated with the presence of strong ultraviolet reflectance. Females are paler orange than males, but the dorsal wing surface is still entirely orange. White females are rare in BC populations, but are common in the nominate subspecies. White males are extremely rare, and the only one ever seen may be the one described by Gall (1983). Males have a sex patch on the front edge of the dorsal hindwing; the sex patch is usually covered by the forewing and hence not visible.

**IMMATURE STAGES:** Immature stages of subspecies *elis* are identical to those of subspecies *meadii* (Edwards 1892). Eggs are conical, tapered towards each end, and yellow green. Mature larvae are dark yellow green with short black hairs. There is a pale yellow stripe along each side of the back, with a black spot on each segment under the line. There is a white line along each side. Pupae are yellow green, with a dark line down the centre of the back and a faint line down each side of the back. The abdomen is dotted and mottled with whitish markings (Edwards 1889).

**Biology:** Mead's Sulphurs are univoltine, and fly in July and August. Males emerge earlier than females (Watt et al. 1977), as with most butterflies. Eggs hatch in early August; the larvae hibernate in the second or third instar starting in mid-August and emerge from hibernation the following May. Reared specimens are larger and more brightly coloured than wild-caught specimens (Bean 1890; Edwards 1892). According to Ferris (1972) the nominate subspecies in Colorado and Wyoming flies only at high elevations above timberline and is fast-flying, wary, and difficult to capture. Other populations in Colorado, like those of subspecies *elis,* fly in subalpine forest meadows and only the lower edge of the alpine (Watt et al. 1977). Populations range from low densities over large areas to high densities (120 adults per hectare) in small areas, with up to 2,500 adults in one 8-hectare island of habitat in Colorado. Different populations differ in the proportion of the population that disperses, and average

dispersal distances vary between populations. Average dispersal distances range from 0.3 to 0.7 km, with 1.3 km being the greatest recorded dispersal distance (Watt et al. 1977).

Larval foodplants for subspecies *meadii* in Wyoming and Colorado are in the family Fabaceae, including *Astragalus alpinus, Trifolium dasyphyllum, T. nanum,* and *T. parryi* (Ae 1958a; Emmel et al. 1971; Watt et al. 1977; Scott and Scott 1980; Scott 1992).

**Subspecies:** The subspecies of Mead's Sulphur found in BC is *Colias meadii elis* Strecker, 1885. The type locality is hereby re-restricted to Kicking Horse Pass railway station, elevation 1,628 m, Yoho National Park, BC. This is contrary to Kondla (1996), who restricted the type locality to the Alberta side of the BC/AB border, apparently because he was unaware that the railway station existed. As Kondla (1996) states, the purported elevation of 10,000 feet is impossible. This change of type locality by at most a few kilometres makes no difference to taxonomy but clarifies how the type locality should be cited.

**Range and habitat:** Mead's Sulphurs are known from a few localities in the central and southern Rocky Mountains of BC. They inhabit steep, dry, south-facing alpine slopes in BC, and nearby in Alberta also occur in subalpine and alpine meadows.

**General distribution:** Mead's Sulphurs occur in disjunct populations in subalpine and alpine areas of the Rocky Mountains from central BC and AB south to NM.

**Conservation status:** Mead's Sulphurs are of Special Concern in BC (S3).

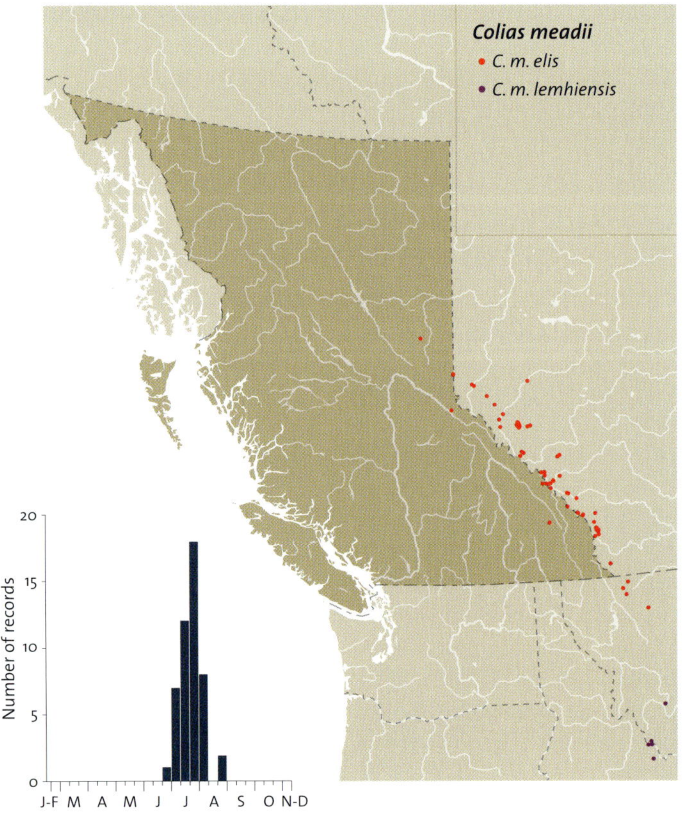

# Hecla Sulphur

*Colias hecla* Lefèbvre, 1836

**Etymology:** Wolff (1964) suggests that the name *hecla* refers to a volcano in Iceland, the result of Lefèbvre's confusing Iceland with Greenland (the actual type locality). On the other hand, *Hecla* is also old Icelandic for a fell or hill in general, which suggests that the name *hecla* refers to the fell habitat in Greenland. The common name was first used by Holland (1931).

**Adult:** Male Hecla Sulphurs are dark orange and have wide black wing borders. Females are either orange or white. The ventral hindwings are dark grey green and the red ring of the discal cell spot is smeared towards the outer margin of the wing. There are no ventral submarginal spots.

**Immature stages:** Mature larvae are slender and green, with lateral lines and tiny black spots (Layberry et al. 1998).

**Biology:** Hecla Sulphurs are univoltine, and fly from late June to early August. Klots (1975) suggested that they have a two-year life cycle. They fly rapidly across alpine tundra, and blend in perfectly with the tundra vegetation when they land.

♂ D (4.3 cm)

♀ D (4.6 cm)

♂ V (4.3 cm)

♀ V (4.6 cm)

♀ white form D (4.4 cm)

Larval foodplants include members of the Fabaceae, with *Astragalus alpinus* being the most common foodplant in North America (Klots 1975; Layberry et al. 1998) and Europe (Higgins and Riley 1970). *Hedysarum alpinum* var. *americanum* is also used (Petersen 1967).

**Subspecies:** Hecla Sulphurs in BC are the nominate subspecies, originally described from Greenland. One other subspecies is presently recognized, *hela* Strecker, 1880, which occurs on the west side of Hudson Bay (Ferris 1982).

**Range and habitat:** Hecla Sulphurs occur in subalpine and alpine tundra areas in northwestern BC.

**General distribution:** Hecla Sulphurs occur from northern AK south to northwestern BC, and east across the arctic to Greenland.

**Conservation status:** The Hecla Sulphur is of Special Concern in BC (S3).

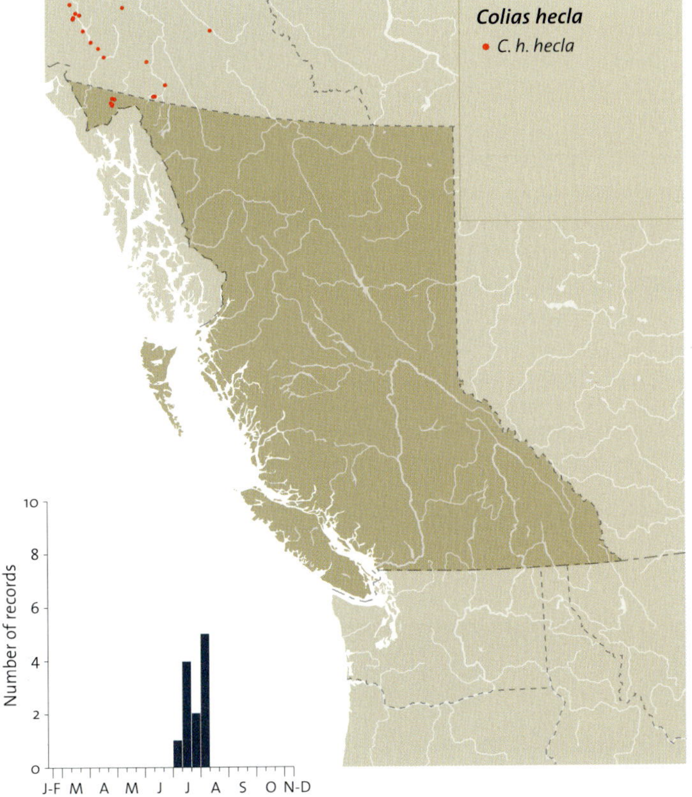

# Canadian Sulphur

*Colias canadensis* Ferris, 1982

**Etymology:** The name *canadensis* refers to the type locality and to much of the species' distribution being in northwestern Canada. The common name was first proposed by Bird et al. (1995).

**Adult:** Canadian Sulphur males are orange or yellow orange, with a black wing border of variable width. Females have white or pale orange dorsal wings. The ventral hindwings are yellow green and the red ring of the ventral discal cell spot is seldom smeared towards the outer margin of the wing. There are no ventral submarginal spots. Canadian Sulphurs were recently shown to be a distinct species from Hecla Sulphurs by Ferris (1988). The two species fly together in some areas of Alaska, and in northwestern BC along the Haines Highway.

**Immature stages:** Mature larvae are dark green. They have numerous black slightly raised points, with a short black hair arising from each black point, and a darker green dorsal line. The upper half of the lateral line is white and the lower half pink red (CSG).

**Biology:** Canadian Sulphurs are univoltine, and fly in June and July. Mature larvae hibernate (CSG). Larval foodplants are probably Fabaceae such as *Hedysarum* (Ferris 1988). In captivity a Canadian Sulphur from Atlin oviposited on *Hedysarum* and larvae were reared on red clover (CSG).

**Subspecies:** None. The type locality of the species is Mile 209, Alaska Highway, BC.

♂ D (4.2 cm)

♂ V (4.2 cm)

♀ orange form D (4.2 cm)

♀ D (4.2 cm)

♀ D (4.0 cm)

**Range and habitat:** Canadian Sulphurs occur in boreal forest openings in northern BC, with most known populations occurring along the Alaska Highway. Populations commonly occur at Second World War emergency airstrips, now overgrown with low vegetation (Ferris 1982). Both Canadian and Hecla Sulphurs occur along the Haines Highway in extreme northwestern BC.

**General distribution:** Canadian Sulphurs occur from northern AK and YT south to northern BC and AB.

**Conservation status:** Not of concern (S5).

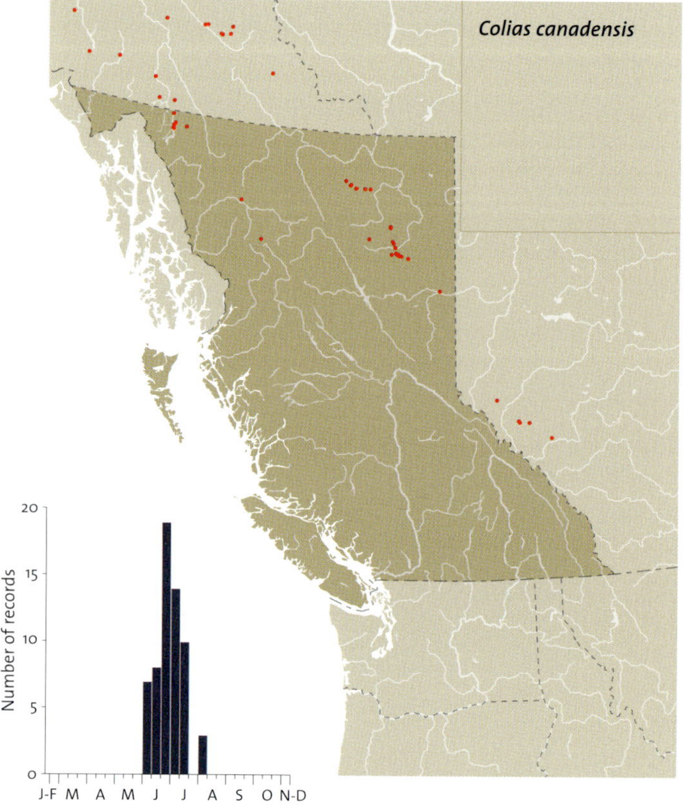

**Mature larva**



Family Pieridae (Whites, Sulphurs, and Marbles)

# Arctic Sulphur

*Colias nastes* Boisduval, [1834]

Arctic Sulphur (*Colias nastes streckeri*)

Ssp. *streckeri* ♂  D  (3.7 cm)

Ssp. *streckeri* ♀  D  (4.4 cm)

Ssp. *streckeri* ♂  V  (3.7 cm)

Ssp. *aliaska* ♂  V  (3.7 cm)

Ssp. *aliaska* ♂  D  (3.7 cm)

Ssp. *aliaska* ♀  D  (4.0 cm)

**ETYMOLOGY:** The name *nastes* may be derived from the Greek *nastes* (solid or firm), with no obvious connection to the butterfly. Subspecies *streckeri* was named for Herman Strecker, a prominent lepidopterist of late 1800s (Holland 1931). Subspecies *aliaska* was so named because the type locality was in Alaska. The common name was first used by Holland (1898) in reference to the alpine and arctic distribution of the species.

**ADULT:** Arctic Sulphurs are slate grey, grey green, green, or green yellow with extensive black markings. This is the only sulphur in which males as well as females have light spots within the black wing border. The forewing colour ranges from entirely slate grey through green to nearly white, with the variation occurring within as well as between populations. Jon and Sigrid Shepard found an aberrant Arctic Sulphur with orange dorsal wings on Sweeney Mountain in the central Coast Range, as part of a population of normal Arctic Sulphurs.

**IMMATURE STAGES:** Mature larvae are dark green and have two pink-edged stripes down each side (Bird et al. 1995).

**BIOLOGY:** Arctic Sulphurs are univoltine, and fly in July and August. They fly low across dry alpine tundra, and when they land, they blend in perfectly with the short tundra vegetation. They are alert and difficult to approach, flying up and letting the wind whisk them out of sight when approached.

Larval foodplants are unknown for BC. In Manitoba and Alberta larval foodplants include *Astragalus alpinus, Oxytropis campestris,* and *O. splendens* (Klots 1975; Bird et al. 1995). In Europe *Astragalus alpinus* and *A. deflexus* are used (Petersen 1967; Higgins and Riley 1970).

**SUBSPECIES:** The subspecies that occurs in most of BC is *Colias nastes streckeri* Grum-Grschimailo, 1895 (TL: Laggan [vicinity

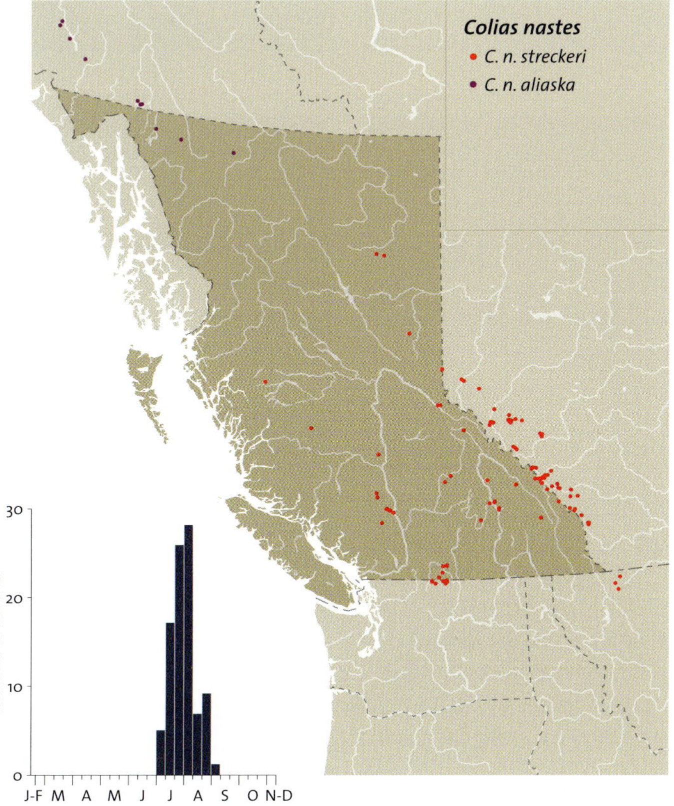

**First instar larva**

**Mature larvae in diapause, ssp.** *streckeri*

of Lake Louise], AB). The dorsal wing colours are highly variable, but are generally dull yellow green, with poorly defined dark wing borders that include poorly defined pale spots. The submarginal dark spots on the ventral forewings are usually only faintly present. Subspecies *aliaska* Bang-Haas, 1927 (TL: Rampart, AK) occurs in extreme northwestern BC. The dorsal wing colours are highly variable, but are generally bright yellow green with dark wing borders containing well-defined pale spots. The submarginal dark spots on the ventral forewing are strongly developed.

**RANGE AND HABITAT:** Arctic Sulphurs occur in alpine tundra across the mainland of southern BC. They are also known to occur on a few scattered mountains in northwestern BC, but are probably much more common there than our current inventory indicates.

**GENERAL DISTRIBUTION:** Arctic Sulphurs occur across northern Eurasia and arctic North America. In western North America they occur through the mountains of BC and AB, south to extreme northern WA and MT.

**CONSERVATION STATUS:** Not of concern. Subspecies *aliaska* is S4 and subspecies *streckeri* is S5.

## CHIPPEWA SULPHUR
*Colias chippewa* W.H. Edwards, 1872

**ETYMOLOGY:** The name *chippewa* refers to the Chippewa Indians of eastern Canada. Edwards apparently liked this Native group, as he also used the name *pembina* for a species of blue in reference to a group of Chippewa near Lake Winnipeg. The common name was first used by Holland (1931).

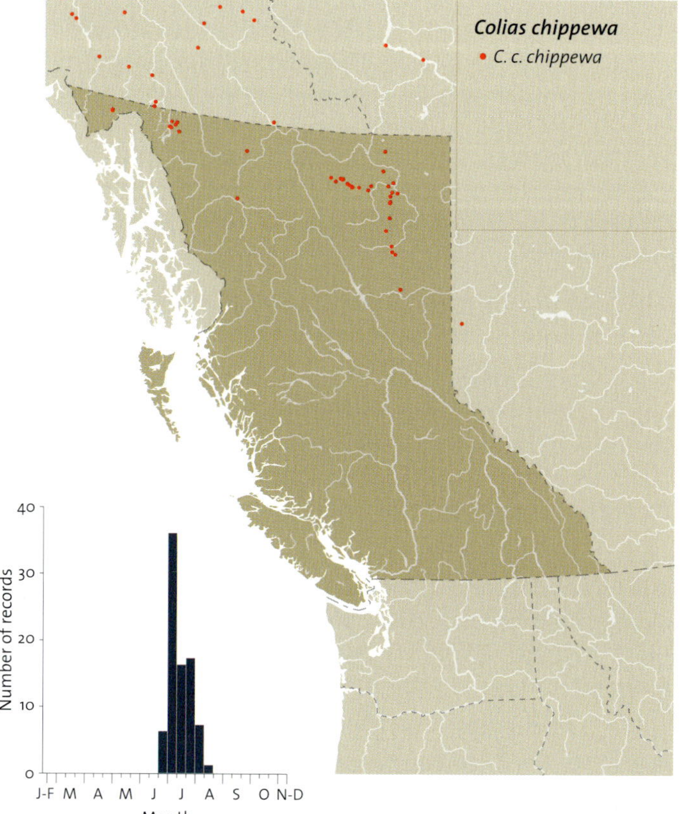

♂ D (4.2 CM)

♀ D (4.0 CM)

♂ V (4.2 CM)

♀ V (4.4 CM)

♀ yellow form D (4.4 CM)

**ADULT:** Chippewa Sulphurs have yellow males and white, occasionally pale yellow, females. There are no submarginal spots on the ventral hindwings. The hindwing discal cell spot is white or pale yellow on the upperside, nearly the same colour as the surrounding wing area, and normally lacks a red ring on the ventral side.

In recent decades, *Colias chippewa* has been considered a subspecies of the Eurasian *Colias palaeno* (Linnaeus, 1761). Compared with *C. palaeno*, *C. chippewa* is notably smaller and the black border on the wings is reduced. Tuzov (1997) treated *C. palaeno* and *C. chippewa* as separate species because in the Magadan Region of Siberia, *Colias palaeno orientalis* Staudinger, 1892 occurs only in low-elevation forested swamps, whereas *Colias chippewa* occurs only along streams in dry alpine tundra (V.K. Tuzov, pers. comm.).

**IMMATURE STAGES:** Undescribed.

**BIOLOGY:** Chippewa Sulphurs are univoltine, and fly from late June through July. The larval foodplants are apparently various species of blueberry. *Vaccinium caespitosum* and perhaps *V. uliginosum* are used at Churchill, MB (Klots 1975; Oosting and Parshall 1980).

**SUBSPECIES:** The nominate subspecies occurs in BC (TL: west end of Great Slave Lake, NT).

**RANGE AND HABITAT:** Chippewa Sulphurs inhabit high-elevation bogs and wet tundra in northern BC.

**GENERAL DISTRIBUTION:** Chippewa Sulphurs occur from Siberia across AK, across arctic and northern boreal North America to NU and northern ON and PQ, southward into the Rocky Mountains of northern BC.

**CONSERVATION STATUS:** Not of concern (S5).

## PINK-EDGED SULPHUR
*Colias interior* Scudder, 1862

**ETYMOLOGY:** The name *interior* is Latin for internal (Opler and Krizek 1984), referring to the occurrence of the species in the interior of North America. This is in contrast to two other species named by Scudder at the same time, from the east and west coasts. The common name "Pink-edged Sulphur" was first used by Scudder (1875) in reference to the prominent pink fringes of the wings in populations east of the Rocky Mountains.

**ADULT:** Pink-edged Sulphurs are small to medium-sized yellow sulphurs. They have traditionally been identified by conspicuous pink wing fringes, hence the common name,

♂ D  (4.3 CM)     ♀ D  (4.5 CM)

♂ V  (4.3 CM)     ♀ V  (4.5 CM)

and a uniformly yellow ventral hindwing with little dark scaling. Unfortunately, this description holds true only for populations from northeastern BC and elsewhere east of the Rocky Mountains. In central and southern BC, the pink wing fringes are similar to those in many other sulphurs, and the ventral hindwing is darkly scaled. The dorsal hindwing discal cell spot is orange, and ventrally the spot has a large silver centre surrounded by one well-defined ring and a diffuse outer ring; and there is never a satellite spot. There are no submarginal spots on the underside of the wings, and the black wing borders are narrow in males and almost absent in females. The outer margin of the forewing is usually more rounded than in other sulphurs. Females are very rarely pale yellow, nearly white.

**IMMATURE STAGES:** In eastern North America eggs are initially white, tinged with greenish yellow, and soon turn reddish orange. Mature larvae are dark yellow green with a dark

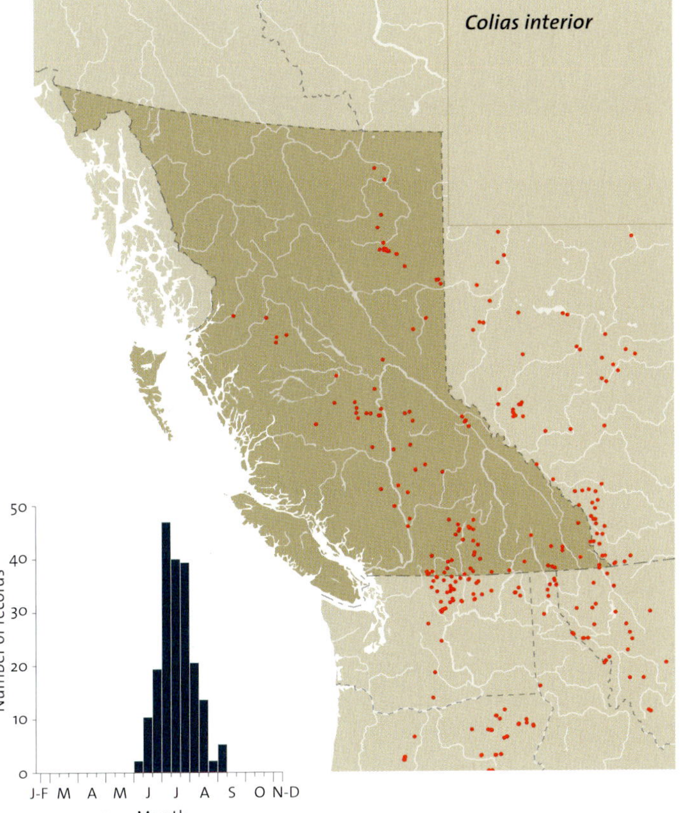

**Pink-edged Sulphur (*Colias interior*)**

stripe bordered with bluish green along the back, and a white stripe with a bright red line through it along each side. Pupae are green with yellowish white markings, and a green line runs down the back. Initially there is a white and red stripe down each side, but it soon becomes faded yellow. There are three red dashes on the side of the abdomen (Lyman 1897).

## LABRADOR SULPHUR
*Colias pelidne* Boisduval & Leconte, 1829

**ETYMOLOGY:** The name *pelidne* is from the Greek word *pelidnos* (livid), probably in reference to the prominent pink wing fringes. The subspecies name *minisni* is of unknown

**BIOLOGY:** Pink-edged Sulphurs are univoltine, and fly in June and July. Larvae hibernate in either the first instar (Klots 1951), the second instar (Lyman 1897), or (most likely) the third instar (Ae 1958a). Larval foodplants are *Vaccinium* species (Edwards 1892; Lyman 1897; Ferris 1988), such as *V. myrtilloides* and *V. caespitosum* (Klots 1975).

**SUBSPECIES:** None. The species was named from near Grand Rapids, MB (Ferris 1988).

**RANGE AND HABITAT:** Pink-edged Sulphurs occur east of the immediate coast through most of BC, but are most abundant in the northern and eastern two-thirds of mainland BC. They inhabit moist coniferous forest openings in which blueberries grow.

**GENERAL DISTRIBUTION:** Pink-edged Sulphurs occur from BC east to NF, with one record in southwestern NT, and south to OR, ID, and MT in the west and the Great Lakes and the Appalachians in the eastern USA.

**CONSERVATION STATUS:** Not of concern (S5).

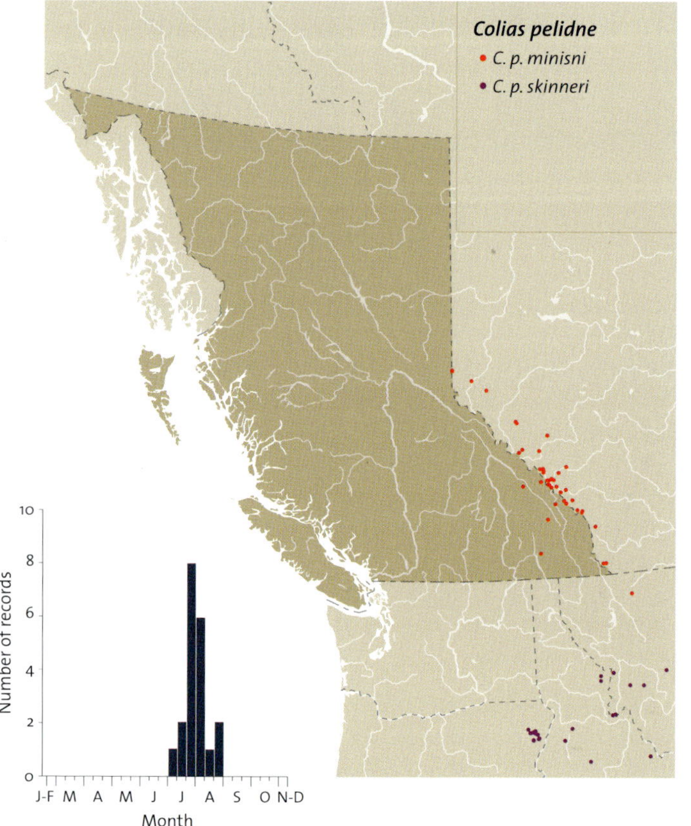

♂  D  (3.7 CM)

♀  white form  D  (4.4 CM)

♂  V  (3.7 CM)

♀  V  (4.4 CM)

derivation. The common name refers to the type locality of the species.

**ADULT:** Labrador Sulphurs in BC have yellow males and white, occasionally pale yellow, females. The dorsal hindwing discal cell spot is

♀  yellow form  D  (4.4 CM)



FAMILY PIERIDAE (WHITES, SULPHURS, AND MARBLES)

yellow or missing in males, and is small and pale orange in females. The ventral hindwing discal cell spot is small and surrounded by one thin red ring that is occasionally missing. There are never submarginal spots on the underside of the wings. The ventral forewing and hindwing have extensive black scaling, and there is black scaling at the base of the upperside of the wings. The black marginal band on the forewings is well developed in both males and females.

**IMMATURE STAGES:** Undescribed.

**BIOLOGY:** Labrador Sulphurs are univoltine, and are in flight from mid-July to August. Larval foodplants are *Gaultheria humifusa* and *Vaccinium* (Klots 1975; Ferris 1988).

**SUBSPECIES:** The subspecies in BC is *Colias pelidne minisni* Bean, 1895. Ferris (1988, 1989) incorrectly states that *Colias pelidne minisni* is a *nomen nudum* because the name is not accompanied by a description. Bean, however, does provide an indication of a bibliographic reference, which is to the Strecker (1885) description of white females of *Colias elis*. The name *minisni* Bean, 1895 is therefore available, and is hereby restricted to Kicking Horse Pass railway station, elevation 1,628 m, Yoho National Park, BC (see also *Colias meadii elis*).

**RANGE AND HABITAT:** Labrador Sulphurs occur at timberline near Paradise Mines in the Columbia Mountains, and at a few BC Rocky Mountain localities.

**GENERAL DISTRIBUTION:** Labrador Sulphurs occur in three well-separated areas: in YT and eastern AK, southeastern BC and the AB Rockies south to OR and WY, and Hudson Bay east to NF.

**CONSERVATION STATUS:** Not of Concern (S4).

## GIANT SULPHUR
*Colias gigantea* Strecker, 1900

**ETYMOLOGY:** The name *gigantea* refers to the often large size of this sulphur that is also reflected in the common name. Subspecies *mayi* was named for J.F. May, who supplied the Chermocks with many butterflies from Manitoba (Chermock and Chermock 1940). The common name was first used by Klots (1951).

**ADULT:** Giant Sulphurs, the largest sulphur in BC, have yellow males and pale yellow to white females. The hindwing discal

Ssp. *mayi* ♂ D  (4.6 CM)

Ssp. *mayi* ♀ D  (5.5 CM)

Ssp. *mayi* ♂ V  (4.6 CM)

Ssp. *gigantea* ♂ V  (4.3 CM)

Ssp. *gigantea* ♂ D  (4.3 CM)

Ssp. *gigantea* ♀ D  (4.5 CM)

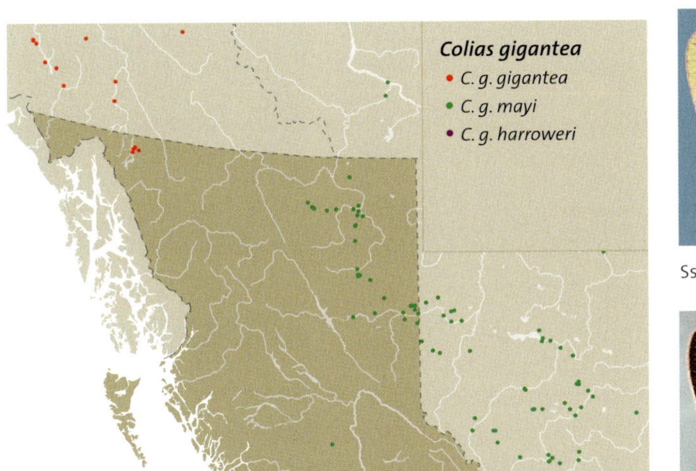



cell spot is very large on both the upper and lower hindwing, and usually has a satellite spot on the underside. There are no submarginal spots on the underside of the wings. Unlike in most sulphurs, on the upperside of the wings there is very little dark scaling at the wing bases.

**IMMATURE STAGES:** Undescribed.

**BIOLOGY:** Giant Sulphurs are univoltine, and fly from late June through July. Near Nordegg, AB, oviposition occurred on several varieties of dwarf willows on hillocks in muskeg. The eggs hatched after 10 days and larval growth was slow, suggesting that the larvae hibernate (McDunnough 1922). In the Peace River region of Alberta, they are universally restricted to wet willow fens and catchment marshes with willows (Kondla et al. 1994).

The larval foodplants are various species of willow (McDunnough 1922). Oviposition on *Salix reticulata* at Churchill, MB, was observed by Klots (1975), who reared the resulting larvae to the third instar.

**SUBSPECIES:** The northwestern BC populations are the nominate subspecies *C. g. gigantea* Strecker, 1900 (TL: "west coast of Hudson Bay above Fort York, [Manitoba]"). Females are usually white (occasionally yellow) and males are a cold yellow with a narrow black wing border. Atlin populations have been called *C. pelidne* or *C. interior* by some authors. The northeastern BC populations are subspecies *mayi* F. & R. Chermock, 1940 (TL: Riding Mountains, MB), with females usually yellow (occasionally white) and males a warmer yellow than the nominate subspecies. Adults of subspecies *mayi* are much larger than those of subspecies *gigantea*.

**RANGE AND HABITAT:** Giant Sulphurs inhabit boreal willow fens in northern BC and in the Cariboo Region.

**GENERAL DISTRIBUTION:** Giant Sulphurs occur from northern AK and YT, across southwestern NT and central and northern BC, to Hudson Bay and ON.

**CONSERVATION STATUS:** Subspecies *gigantea* is of Special Concern in BC (S3). Subspecies *mayi* is not of concern (S5).

---

## GENUS *EUREMA* HÜBNER, [1819]  SULPHURS

The name *Eurema* is from the Latin *heurema* (discovery) (Whitney 1902), perhaps because the author was excited about discovering this new genus.

Small sulphurs are small to medium-sized, bright yellow or orange butterflies with black wing margins. They fly quite rapidly, as do sulphurs in the genus *Colias*. One species may occur in BC as a rare migrant.

## SLEEPY ORANGE
*Eurema nicippe* (Cramer, [1779])

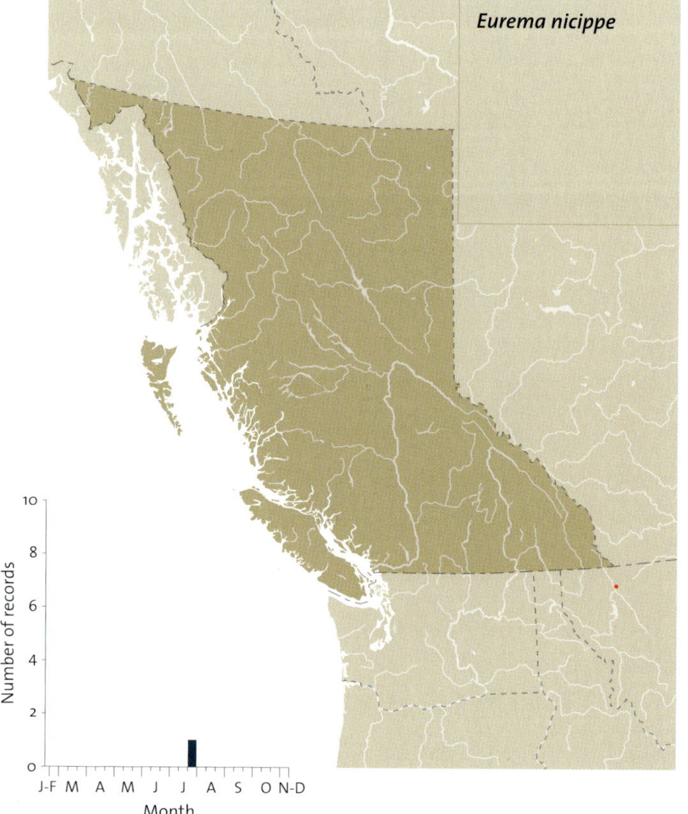

♂ D  (4.9 CM)

♀ D  (4.9 CM)

♂ V  (4.9 CM)

**ETYMOLOGY:** Nicippe was the wife of Sthenelusm, the son of Perseus and Andromeda, and she was the daughter of Pelops. Her name meant "conquering mare," from *nice* (victory) and *hippos* (mare) (Room 1983). The common name "Sleepy Orange" was first used by Klots (1951); before that, the name "Sleepy Yellow" was used (Macy and Shepard 1941). The name may have originated from the butterfly's habit of sleeping through cool periods in the southern USA (Pyle 1981), in combination with the orange wing colour.

**Adult:** Adults are medium-sized butterflies that are orange with wide black wing borders. The forewings and hindwings are both quite rounded, and all wings are similar in size.

**Immature stages:** Eggs are elongate and spindle-shaped, tapering at both ends and with longitudinal ribs. They are pale green yellow when laid, darkening with time and turning reddish before hatching. Mature larvae are velvety green, usually with a fine white yellow lateral band that is sometimes white with orange spots. There are tiny black spots on the body, and the head is yellow green. Pupae are long and slender except for a very deep ventral keel. They are olive green to black, with white and brown markings (Edwards 1881; Comstock 1927; Scott 1986b).

**Biology:** The Sleepy Orange is a tropical and subtropical butterfly that is highly migratory. It breeds all year in the southern USA and Mexico, and each summer migrates north through the USA. Larval foodplants outside BC are senna and clover (Comstock 1927; Pyle 1981; Scott 1986a).

**Subspecies:** None. The type locality of the species is Virginia.

**Range and habitat:** A Sleepy Orange has been found in northwestern Montana, indicating that they are likely to be seen eventually in the eastern Kootenay region of BC. Sleepy Oranges inhabit dry, open areas, and will most likely be found in late summer in BC.

**General distribution:** Tropical America, migrating north each summer throughout the southern and central USA and straying north to the Canadian border.

**Conservation status:** Not yet known from BC.

---

## Genus *Nathalis* Boisduval, 1836  Dwarf Sulphurs

The derivation of *Nathalis* is unknown. The common name refers to the very small size of the species in the genus, compared with other sulphurs.

Dwarf sulphurs are small, bright yellow butterflies with black wing margins. They are delicately built and slow-flying. One species may occur in BC as a rare migrant.

## Dainty Sulphur
*Nathalis iole* Boisduval, 1836

**Etymology:** The name *iole* is from the Greek *iov* (violet), referring to the purple flush that is sometimes present.

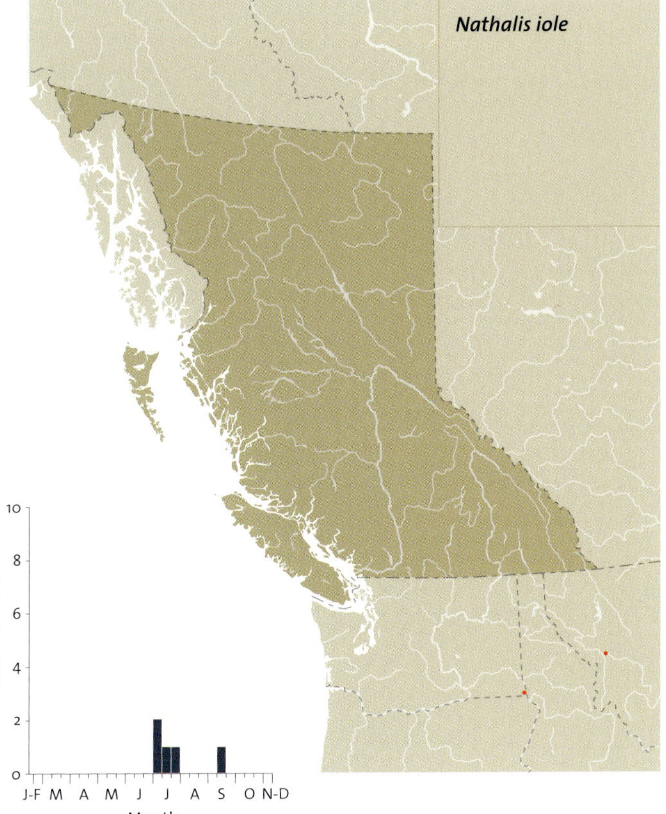

♂ D  (2.5 cm)          ♂ V  (2.5 cm)

The common name "Dainty Sulphur" was first used by Klots (1951); before that, the name "Dwarf Yellow" was used (Holland 1898). The common name refers to the fact that it is a very small, delicate butterfly.

**Adult:** Adults are unmistakable, tiny yellow butterflies with black wing borders. They cannot be confused with any other butterfly.

**Immature stages:** Eggs are orange yellow when laid and do not change colour to orange. Mature larvae are dark green with a broad purple dorsal stripe bordered on each side by a narrow yellow stripe, and there is a yellow lateral stripe on each side with an equally wide purple band above. The body is covered with numerous white hairs, and the head is green with white hairs. Pupae are green with fine whitish markings. The underside of the abdomen is white green. There is a purple dorsal line on the thorax, and a purple green dorsal line from the thorax back along the abdomen that is bordered in yellow. There is a yellow lateral band in the middle area of the abdomen, and spiracles have purple spots above them (Scott 1992).

**BIOLOGY:** Dainty Sulphurs are highly migratory, and each summer migrate north throughout the USA. Outside BC larval foodplants include *Bidens pilosa, B. frondosa, Dyssodia papposa, Helenium autumnale, Palafoxia arida* var. *gigantea, Stellaria media, Thelesperma filifolium,* and *T. megapotamicum* (Comstock 1927; Pyle 1981; Scott 1992).

**SUBSPECIES:** None. The type locality of the species is Mexico.

**RANGE AND HABITAT:** Dainty Sulphurs have been recorded from northwestern MT and eastern WA, indicating that they will eventually be seen in the eastern Kootenay region of BC. They inhabit dry, open areas.

**GENERAL DISTRIBUTION:** Tropical America, migrating north each summer throughout the southern and central USA and straying north into CAN.

**CONSERVATION STATUS:** Not yet known from BC.

# Family Lycaenidae Leach, 1815    Gossamer Wings

The common name "gossamer wings" refers to the flimsy, delicate (gossamer) wings of many species.

Gossamer wing males are usually bright blue or copper/golden, especially the blues and coppers but also the Theclini hairstreaks, which are not represented in our area but are very diverse in northwestern China, Korea, Japan, and the Amur region of Russia (Shirôzu 1956). The blue and copper colours are structural, not pigmental.

Adult lycaenids have forelegs that are not usable for walking. In the males, the fused foreleg tarsi are reduced from five articulated segments to one long segment. The tibial spurs are reduced to one pair that are often absent on the foreleg tibia. The eyes are indented next to the antennae, and the face is taller than the width between the eyes.

The larvae of lycaenids have reduced true legs and prolegs, and the overall appearance is more that of a grub than a normal lepidopteran larva. The dorsal surface is covered with short hairs. Except those of coppers, the larvae have a honey gland on the seventh abdominal segment that secretes a fluid attractive to ants. Many lycaenid larvae live part of their life cycle in symbiosis with ants in ant nests or on plants. The larvae of coppers are also associated with ants. Most lycaenids have four larval instars, one less than other butterflies. Ballmer and Pratt (1989) surveyed the last instar larvae of 69 California lycaenids, including 32 species that occur in BC.

This book takes the conservative approach of recognizing the Riodinidae as a separate family (Weller et al. 1996), and recognizes three subfamilies of Lycaenidae in the British Columbia fauna.

Purplish Copper (*Lycaena helloides*) showing type IVa bird damage (see p. 62)

The common name of this subfamily is derived from the copper colour of the wings of many species in the subfamily. It is presumably of British origin, but in North America the usage goes back to Emmons (1854).

In most species of coppers, the upper surfaces of the wings have a bright to dull copper sheen. For one BC species, however, the males are blue, not copper. Males of most species are copper, blue, or at least some metallic pattern. For several of the species, the female has little or no structural colour, and appears melanic compared with males of the same species.

The valves of the male genitalia are the most significant morphological characteristic of the subfamily. The valves are not joined at the base like those of most other lycaenids except Theclini. They are also weakly attached to the rest of the genitalia. The eyes are bare. Males lack androconial scales on the dorsal forewings. Larvae lack a honey gland or eversible tubercles. The larval prolegs have lateroseries crochets not normally found on Theclinae or Polyommatinae. Most larvae lack any obvious colour pattern and are a pale green. They also have "mushroom" setae over the dorsal and lateral body, a character, at least in North America, unique to the subfamily (Ballmer and Pratt 1989).

The generic classification of the subfamily is not at all stable. Palearctic workers recognize four to seven genera and do not agree on placement of species in the genera. Recent North American workers, except Miller and Brown (1979), have ignored structural differences and lump all species in one genus. This would mean, however, including more than 40 species worldwide, with three from New Zealand, two from New Guinea, and African species and several temperate Asian species with limited distribution (Sibatani 1974). Clearly this is geologically an ancient group with multiple genera (Shields 1977). At least two of the species Opler (1999) and others put in *Lycaena*, *Tharsalea arota* (Boisduval, 1852) and *Hermelycaena hermes* (W.H. Edwards, 1870), belong in distinct genera. Freeman (1937) showed how different the male genitalia of *T. arota* were from other North American coppers. Ballmer and Pratt (1989) did not find distinctive larval characters for *T. arota* but support Miller and Brown (1979) by showing that the larva of *H. hermes* is morphologically distinct. What is most remarkable is that Freeman (1937), Miller and Brown (1979), and Ballmer and Pratt (1989) are in almost complete agreement as to the various groups of North American coppers, despite the fact that neither of the more recent papers referenced Freeman (1937). The major difference with Miller and Brown (1979) is that both Freeman (1937) and Ballmer and Pratt (1989) place *L. rubida* with *L. xanthoides* (Boisduval, 1852) and *L. gorgon* (Boisduval, 1852) with *L. heteronea*. In this book we include all nine BC species in the above restricted genus *Lycaena*, and recognize all but one of the genera recognized by Miller and Brown (1979) as subgenera. The subgenus name *Chalceria* Scudder, 1876 has page priority over the subgenus *Gaeides* Scudder, 1876, and we synonymize the latter under *Chalceria*. When American species are more carefully compared to Old World species, however, the other genera as proposed by Miller and Brown (1979) may prove valid.

## GENUS *LYCAENA* FABRICIUS, 1807   COPPERS

The name *Lycaena* is most likely derived from the Greek *Lukaios* (Arcadian), as several of the species names are those of Arcadian shepherds (Emmet 1991). The common name refers to the copper-coloured wings of most species. It was first used in North America by Emmons (1854).

The characteristics given above for the subfamily also define the genus as used in BC. The larvae of northern Palearctic species all feed on plants of the family Polygonaceae, such as *Rumex* (dock/sorrel) and *Polygonum* (knotweed). Most North American species also feed on these genera, but some feed on *Eriogonum* or *Oxyria* (Polygonaceae), *Potentilla* (Rosaceae), and *Vaccinium* (Ericaceae). There are 15 North American species, of which nine occur in BC.

## SMALL COPPER

*Lycaena (Lycaena) phlaeas* (Linnaeus, 1761)

**ETYMOLOGY:** The name *phlaeas* is derived from either the Greek *phlego* (to blaze up) or the Latin *floreo* (flourish), both in reference to the firelike copper colour of the dorsal forewings (Emmet 1991). The subspecies name *arethusa* refers to Mt. Arethusa, AB, near the type locality of the subspecies (Kondla 1996). The English common name "Small Copper" refers to the butterfly's small size in contrast to a similar British species, the Large Copper (*Lycaena dispar*). Gosse (1840) called the species the Small Copper but Scudder (1875) treated the subspecies *americana* as a full species and used Harris's common name, "American Copper" (Scudder 1975). As a result, what is now the regional common name "American Copper" is frequently used in North American butterfly books for the entire species.

**ADULT:** The Small Copper and the Lustrous Copper are the only BC species that have the copper colour on the upperside of the wings. In the Small Copper, only the dorsal forewings are copper-coloured. The underside of the wings is typical of the genus. It has a tawny ground colour with a submarginal row of orange spots. In extreme northeastern BC, individual females of the Mariposa Copper can look very similar to the Small Copper but have very different ventral hindwings.

**IMMATURE STAGES:** A mature larva from a California population is figured by Ballmer and Pratt (1989), and CSG photographed a mature larva on *Rumex* at Mile 96 of the Dempster Highway, YT. Otherwise, the immatures of native populations are undescribed for North America. The mature larva is green with a pink dorsal stripe. The sides from the spiracles down are also pink. The larva from the Yukon is more brightly coloured than the one from California.

**BIOLOGY:** Adult Small Coppers fly from late July to mid-August. The native North American subspecies have been associated with *Oxyria digyna* (Polygonaceae) by several observers (Shields and Montgomery 1967). European and Asian populations (Shields 1968a) and the introduced subspecies from the northeastern USA feed on *Rumex* sp. (Polygonaceae). BC populations have one generation each year. Since the adults emerge late in the season, the species

♂ D  (3.1 CM)          ♀ D  (3.0 CM)

♂ V  (3.1 CM)

must hibernate as an egg or first instar larva and mature the following year.

**SUBSPECIES:** Northern Rocky Mountain populations are subspecies *L. p. arethusa* (Wolley-Dod, 1907), described from near Calgary, AB. Populations from extreme northwestern BC, near Atlin, may be the undescribed central Alaskan subspecies characterized by Ferris (1974). Populations from the Lillooet/Lytton area are a separate segregate left over from the ice-free area that straddled the BC/Washington border at Cathedral Lakes Provincial Park.

**RANGE AND HABITAT:** The Small Copper is found above timberline in the mountains of BC east of the Coast Ranges. It was been recorded only infrequently because adults emerge in late July and early August after the emergence of all other alpine butterflies in the province. In AB, the Small Copper can sometimes be found well below timberline, on the gravel bars of braided streams where the foodplant occurs. It should occur in BC in similar situations on the east side of the Rocky Mountains.

**GENERAL DISTRIBUTION:** The Small Copper is Holarctic. In North America it ranges from northern and central AK to Baffin Island and Greenland. It ranges south through BC to Beartooth Plateau and the Tetons of WY, with disjunct populations in the central Sierras of CA. There are other populations in CAN from ON to NS and adjacent USA that were introduced from Europe.

**CONSERVATION STATUS:** Not of concern (S4).

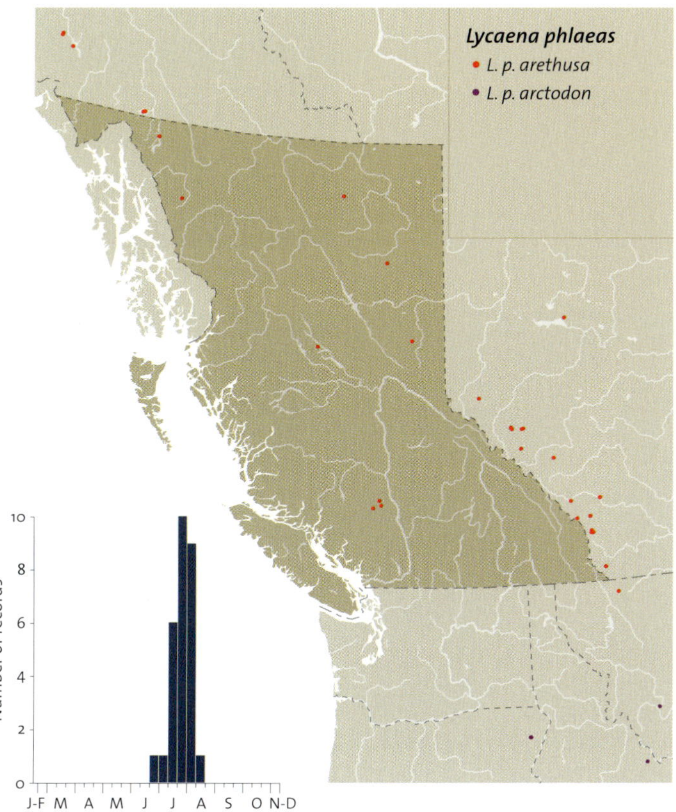

**Mature larva, Dempster Hwy, YT**

# Lustrous Copper

*Lycaena (Lycaena) cuprea* (W.H. Edwards, 1870)

**ETYMOLOGY:** The species name *cuprea* is Latin for copper (Woods 1944). The subspecies name *henryae* honours the collector of the type specimens, Miss Josephine de Henry. The common name Lustrous Copper (Comstock 1927) refers to the lustrous shine of the dorsal wing colour.

**ADULT:** The Lustrous Copper is the only BC butterfly in which the entire upperside of both wings is a bright copper colour. The ventral hindwing is a dull chalky white. The Lustrous and Small Coppers are the only alpine coppers, and occur together in the BC Coast Range on Big Dog Mountain.

**IMMATURE STAGES:** Scott (1992) partially described the immatures. The egg is white, changing to dirty white, which matches the colour of the rocks on which the egg is often laid, next to the larval foodplant. The mature larva is green with only faint markings. The pupa is a typical lycaenid brown.

**BIOLOGY:** The adult Lustrous Copper flies from mid-July to mid-August. Where it occurs with the Small Copper in Alberta, it always emerges before the Small Copper, but they fly together in the BC Coast Range. In Colorado, Scott (1992) has observed females ovipositing on *Oxyria digyna* (Polygonaceae); this plant occurs in the BC range of the Lustrous Copper. Elsewhere the species is associated with *Rumex* sp. (Scott 1992).

**SUBSPECIES:** The BC and Alberta subspecies, *L. c. henryae* (Cadbury, 1937), was described from "Caribou Pass," near Pink

♂ D (2.6 cm)     ♀ V (2.8 cm)

♂ V (2.6 cm)

Mountain. This subspecies has the smallest spots on the underside of both wings, especially the hindwing, giving the ventral hindwing a washed-out appearance. It ranges south to Okanogan Co., WA, and Glacier National Park, MT. This pattern is similar to that of Colorado populations, but geographically intermediate populations from ID, WY, and southern MT have larger spots. Thus *L. c. henryae* must be considered a valid subspecies.

**RANGE AND HABITAT:** The Lustrous Copper is found across mainland southern BC at and above timberline, and north in the Rockies to near Pink Mountain.

**GENERAL DISTRIBUTION:** The Lustrous Copper ranges from near Pink Mountain in the northern Rockies south to CO with disjunct populations in southwest OR and CA.

**CONSERVATION STATUS:** Not of concern (S5).

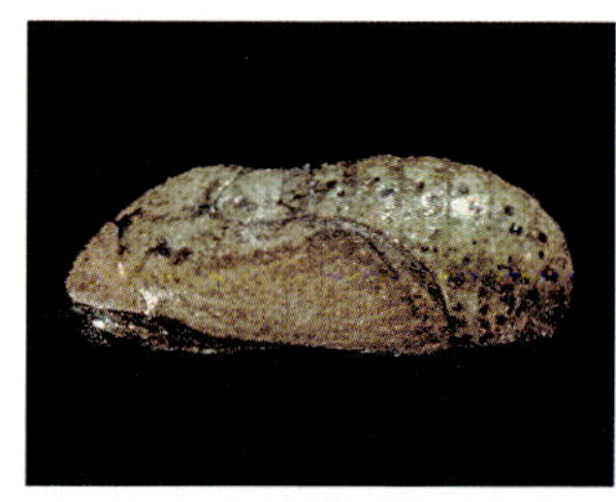

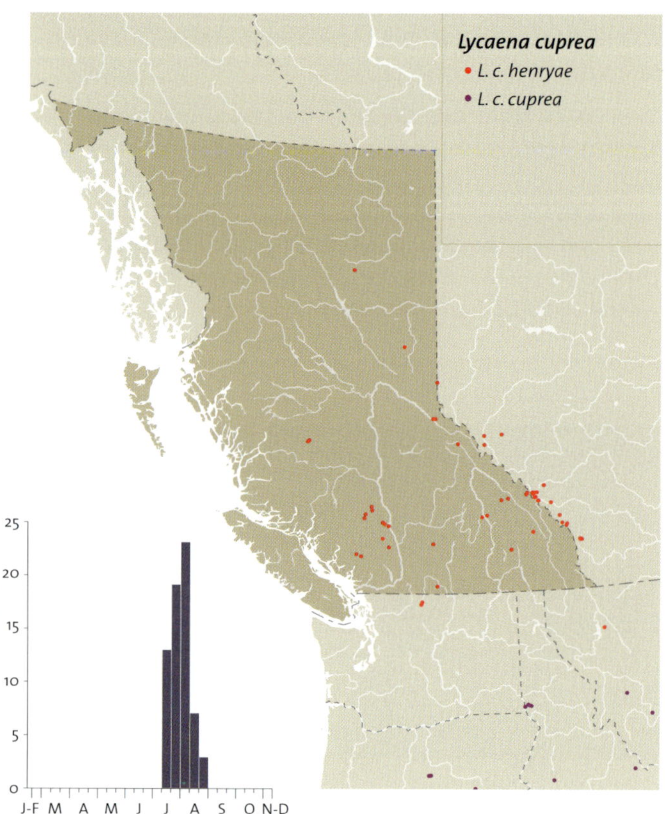



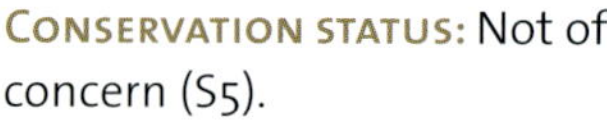

Pupa

Mature larva

# Bronze Copper
*Lycaena (Hyllolycaena) hyllus* (Cramer, [1775])

**Etymology:** The species name most likely means "of the woods" (cf. Emmet [1991] for *Hylaea, Hyles,* and *Hylephila*). The common name (Scudder 1875) refers to the dull copper colour.

**Adult:** The Bronze Copper is easily distinguished from the other BC coppers by the uniform, white ground colour and very wide orange submarginal band on the ventral hindwings.

**Immature stages:** Scudder (1889b) partially described the immatures. The egg is pale green with faint reticulation. The first instar larval head is fuscous, with the ocelli region black. The body is dusky yellow with pale hairs. The pupa is light yellow brown.

♂ D  (3.3 cm)

♀ D  (3.7 cm)

♂ V  (3.3 cm)

**Biology:** In the east, Scudder (1889b) found that the species had two broods, one in mid-June and the second in late August. The latter laid eggs that hibernated. The only record for BC is fresh specimens in early August in extreme northeastern BC. Based on the few additional records from the Peace River region of Alberta, the species is presumably univoltine in BC. BC specimens were found associated with *Rumex* species, presumably *R. acetosa*, but the plant has not been positively identified.

**Subspecies:** None. The type locality of the species is not fixed.

**Range and habitat:** The Bronze Copper was found by Shepard in 1999 along the Liard River Road. It was on the edge of wet, marshy habitat that also supported *Lycaena dorcas*. The species should eventually be found throughout northeastern BC in small, scattered colonies.

**General distribution:** The Bronze Copper is found from northeastern BC and southeastern NT east and south to NS, CO, AR, and MD.

**Conservation status:** The Bronze Copper is of Special Concern in BC (S3).

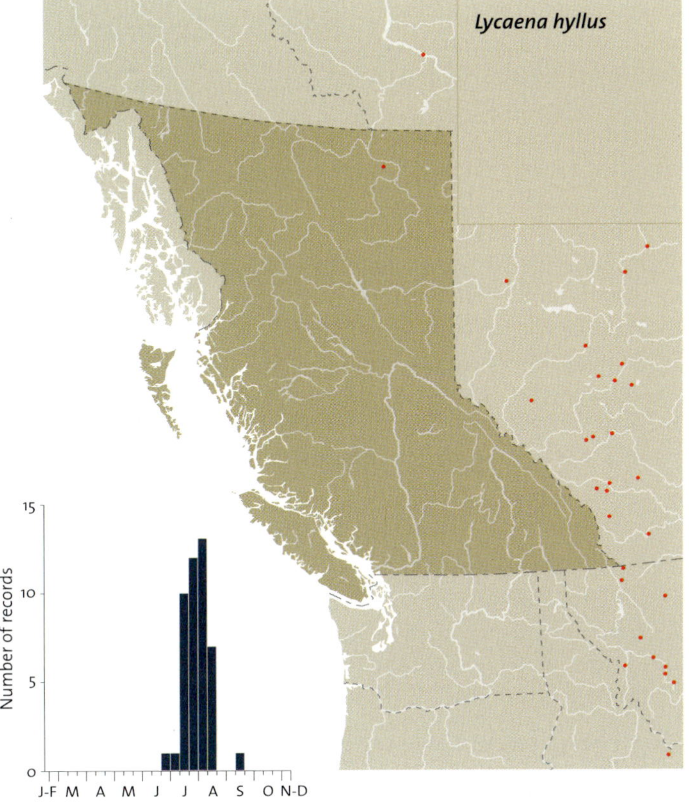

**Mature larva**

# DIONE COPPER

*Lycaena (Chalceria) dione* (Scudder, [1870])

**ETYMOLOGY:** The species name *dione* refers to Dione, a Titan who was the mother of Aphrodite in Greek myth. One characteristic of the Titans was their gigantic size. The common name "Dione Copper" reflects the fact that this is the largest copper in North America. The name "Great Copper" has been applied to both the southwestern copper, *L. xanthoides* Boisduval, 1852, and the eastern copper, *L. dione,* when the Dione Copper was regarded as a subspecies of the Great Copper (Holland 1898; Klots 1951). Since the name "Great Copper" was first applied to *L. xanthoides,* recent books have restricted usage to that species. Recent books, however, have also ignored the historically correct common name "Dione Copper," which was first used by Holland (1931) and has been used by all other authors until Scott (1986b). We do not accept the usage of Scott and subsequent authors (Opler 1999).

**ADULT:** The Dione Copper is the largest species of copper in BC. Unlike its sister species in Europe, the Large Copper, it has no copper colour on the wings but instead is greyish brown. The dorsal forewings have a submarginal band of orange surrounding black spots. This is narrow in males and wider in females. On the underside the ground colour of the wings is chalk white. The submarginal orange band of the hindwing is as conspicuous as on the upperside and continues as a faint band into the submargin of the forewing.

**IMMATURE STAGES:** The mature larva is green with a darker green or red dorsal stripe. The pupa is light pinkish brown

♂ D (3.1 CM)

♀ D (3.6 CM)

♂ V (3.1 CM)

with many dark spots (Opler and Krizek 1984).

**BIOLOGY:** The few observations of the Dione Copper in BC indicate that it flies in a prolonged emergence from mid-July to mid-August. Eggs are laid one at a time on the underside of foodplant leaves (Opler and Krizek 1984). Skinner (1893) first recorded a larval foodplant, *Rumex longifolius* ? = *obtusifolius* (Polygonaceae). Scott (1992) notes other *Rumex* sp. Klassen et al. (1989) record *Rumex aquaticus* var. *fenestratus* and *R. crispus* in the Canadian prairies. The BC population at Elizabeth Lake in Cranbrook has only *Polygonum amphibium* to utilize as a larval foodplant (JHS). BC adults nectar on introduced yellow sweet clover. The population is threatened by city park development, road development, and management of the lake water level by Ducks Unlimited. The Dione Copper was common when first observed in 1989, but numbers have dropped drastically. In 1999 only one adult was seen.

**SUBSPECIES:** None. The BC population is similar to Canadian Prairie populations, and represents, along with adjacent Montana and Idaho populations, relict populations of pluvial lakeshore habitat west of the Continental Divide. TL: Denison and New Jefferson, IA.

**RANGE AND HABITAT:** The only known BC population of the Dione Copper occurs on the north shore of Elizabeth Lake at Cranbrook. The adults fly in association with the presumed larval foodplant, which is found in the understorey of cattails on a broad shallow area near the shore.

**GENERAL DISTRIBUTION:** The Dione Copper is found from southeastern BC and the northern panhandle of ID east to MB, and east of the Rockies south to northern TX and OK in the prairies.

**CONSERVATION STATUS:** The Dione Copper is the most Endangered butterfly in BC (S1).

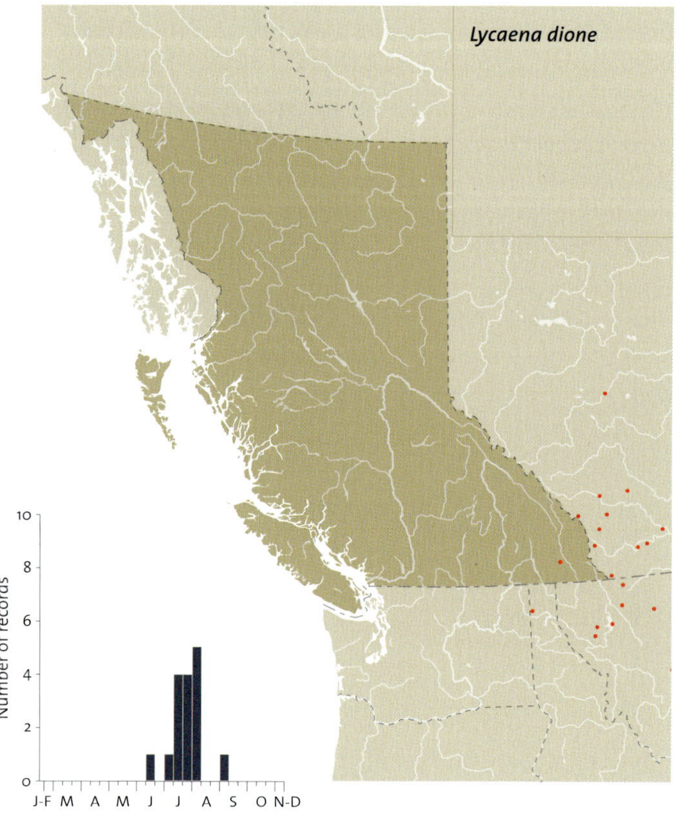

## RUDDY COPPER
*Lycaena (Chalceria) rubida* (Behr, 1866)

**ETYMOLOGY:** Both the species name, *rubida,* and the common name, Ruddy, refer to the bright red copper sheen on the upperside of the wings of the males. The subspecies *perkinsorum* is patronymic for the Perkins brothers. The common name was first used by Holland (1898).

**ADULT:** The wings of males have a bright, reddish copper colour on the upperside. In females the dorsal wing ground colour is brown. On the underside both sexes are a dirty grey white with small, black spots.

**IMMATURE STAGES:** Undescribed, except that Scott (1986a) describes the larva as brown with a dark red brown mid-dorsal line with yellow on each edge.

**BIOLOGY:** Adults of Washington populations have a long flight period, ranging from mid-May to early August in a

♂ D  (3.2 CM)

♀ D  (3.3 CM)

♂ V  (3.2 CM)

single generation per year. For any one population, adults appear to emerge at the same time that the local rabbitbrush is in bloom. They utilize rabbitbrush as a nectar source. The larval foodplants are various *Rumex* species (Brown et al. 1957; Funk 1975). Funk further observed the females laying eggs singly near the base of the plant, the eggs immediately dropping to the ground. Small black ants, *Formica altipetens* Wheeler, were seen taking the eggs away and even following the female as she landed on the top of the foodplant and walked to the base of the plant to lay an egg. It is not known whether the eggs were taken to the ant nests.

**SUBSPECIES:** WA populations are the subspecies *L. rubida perkinsorum* Johnson & Balogh, 1977 (TL: The Dalles, Wasco Co., OR).

**RANGE AND HABITAT:** The Ruddy Copper occurs south of BC in the Okanogan River valley of WA, and should eventually be found in the province.

**GENERAL DISTRIBUTION:** Eastern WA and southern AB south and east to the CA Sierras and northern NM, with disjunct populations centred in the Black Hills (Johnson and Balogh 1977).

**CONSERVATION STATUS:** Not yet known from BC.

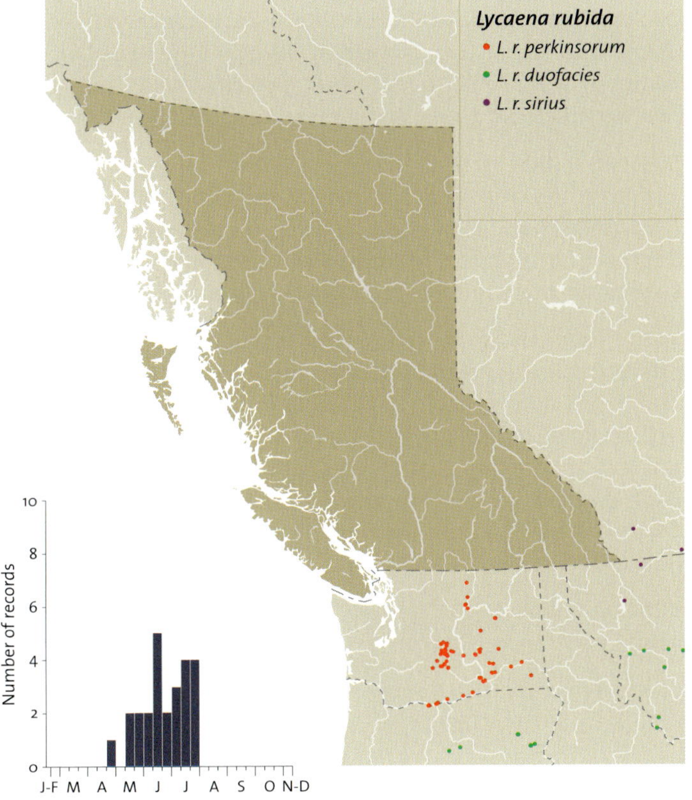


## BLUE COPPER
*Lycaena (Chalceria) heteronea* Boisduval, 1852

**ETYMOLOGY:** The species name *heteronea* comes from the Greek *heteros* (different). Boisduval noted that the females were like the copper species *sirius,* and the species name means that the females were like a copper and the males were like a blue. Boisduval described it as a blue (*Polyommatus* ), not a copper (*Lycaena*), and the etymology

of Bird et al. (1995), that it was the copper that was blue, cannot be correct. Blackmore (1927) apparently first switched to the common name Blue Copper from the older usage "Varied Copper." If so, this is the first common name of a North American butterfly coined by a resident of British Columbia.

**Blue Copper (*Lycaena heteronea heteronea*)**

♂ D (3.0 CM)

♀ D (3.2 CM)

♂ V (3.0 CM)

**ADULT:** The male Blue Copper is the only copper in which the upper surface of the wings is blue, like the blues of the subfamily Polyommatinae. It is larger than any blue that occurs in BC. Also, it is observed only in immediate proximity to the larval foodplant, where the only other blues are all about half as large. In females the dorsal wings are brown, with some infusion of blue at the base of the wings. Both sexes have the median spots on the underside of the wing reduced or absent and the ground colour is a dirty white.

**IMMATURE STAGES:** Both larvae and pupae of California populations are various shades of green and are camouflaged on the foodplant (Williams 1910).

**BIOLOGY:** The Blue Copper flies from late June at the lowest elevations to mid-August at the highest elevations (2,300 m). Williams (1910) recorded *Eriogonum* sp. as the larval foodplant, and Comstock (1927) confirmed this. Scott (1992) observed oviposition on *Eriogonum umbellatum* and *E. umbellatum* var. *majus*. Emmel and Emmel (1973) state that the egg is the over-wintering stage. In BC the species is associated with *Eriogonum* sp., but the BC populations have not been reared.

**SUBSPECIES:** BC populations of the Blue Copper are the nominate subspecies, *L. h. heteronea* Boisduval, 1852 (TL: Cavallo Point/Yellow Bluff, near Sausalito, Marin Co., CA [Emmel et al. 1998a]). This subspecies occurs in all of the species' distribution except southern California.

**RANGE AND HABITAT:** The Blue Copper is found from the Chilcotin grasslands south through the Southern Interior and the West Kootenay in sagebrush, ponderosa pine, and dry subalpine ridges where the larval foodplant is found. It is not recorded from the East Kootenay but should be found in the Flathead River drainage, where the foodplant, *Eriogonum* (Polygonaceae), occurs, along with other *Eriogonum*-feeding lycaenids.

**GENERAL DISTRIBUTION:** The Blue Copper is found from southern BC and AB to southern CA and northern NM.

**CONSERVATION STATUS:** Not of concern (S4).

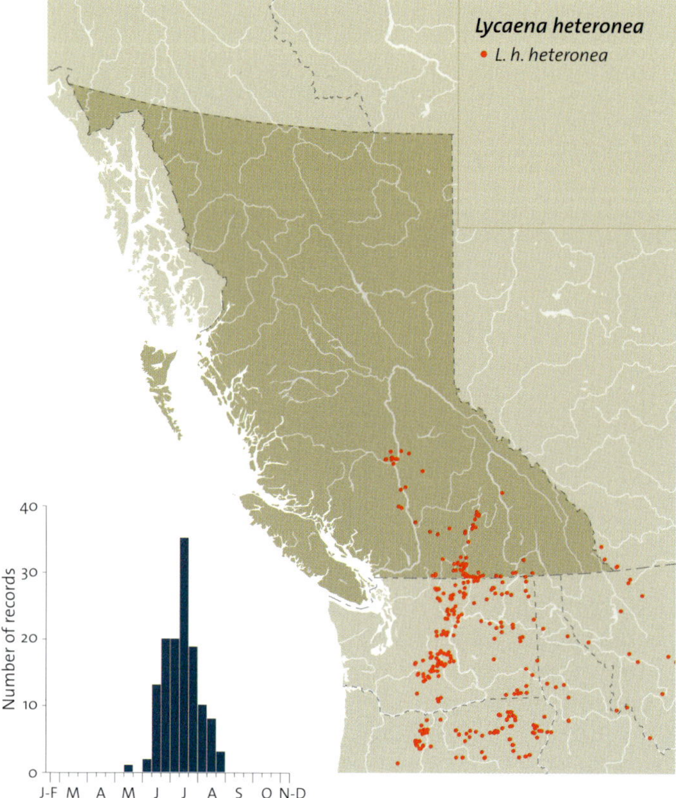

**Mature larva**

**Pupa**

# Dorcas Copper

*Lycaena (Epidemia) dorcas* W. Kirby, 1837

**Etymology:** The species name *dorcas* is derived from the Greek *derkomai* (to gleam or to flash like the eye), which refers to the quick, jerky flight of the butterfly (Reed 1871), and not *dorkas* (roe [deer]) (Bird et al. 1995). *Dorkas* is from the same root as *dorcas,* but refers to the bright eyes of the deer. Kirby (1837) first used the common name Dorcas Lycaena, and Holland (1931) reintroduced it in its modern form.

**Adult:** In the Dorcas Copper and the remaining species of coppers, the uppersides of the wings of males are purple instead of copper, blue, or brown. The females are brown on the uppersides of the wings, with a yellow/orange flush on

the forewings and rarely on the hindwings. The Dorcas Copper and the Purplish Copper are very similar in general appearance, with the undersides of the wings having a typical copper pattern. The Dorcas Copper is smaller than the Purplish Copper, with the submarginal orange area on its ventral hindwing less conspicuous (Fig. 64a).

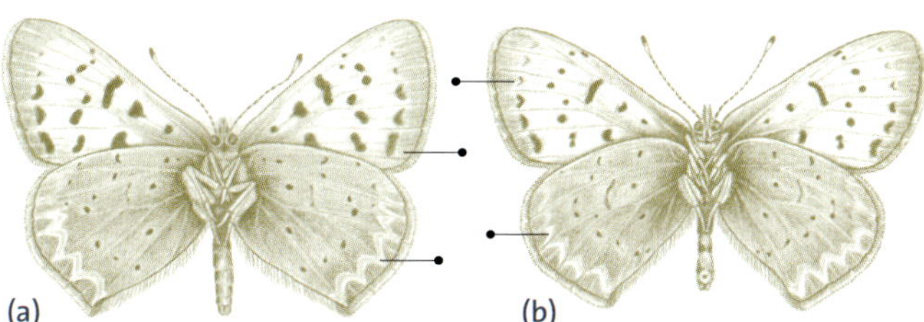

**64** Underside of the wings of *Lycaena* species: (a) *L. helloides*, (b) *L. dorcas*

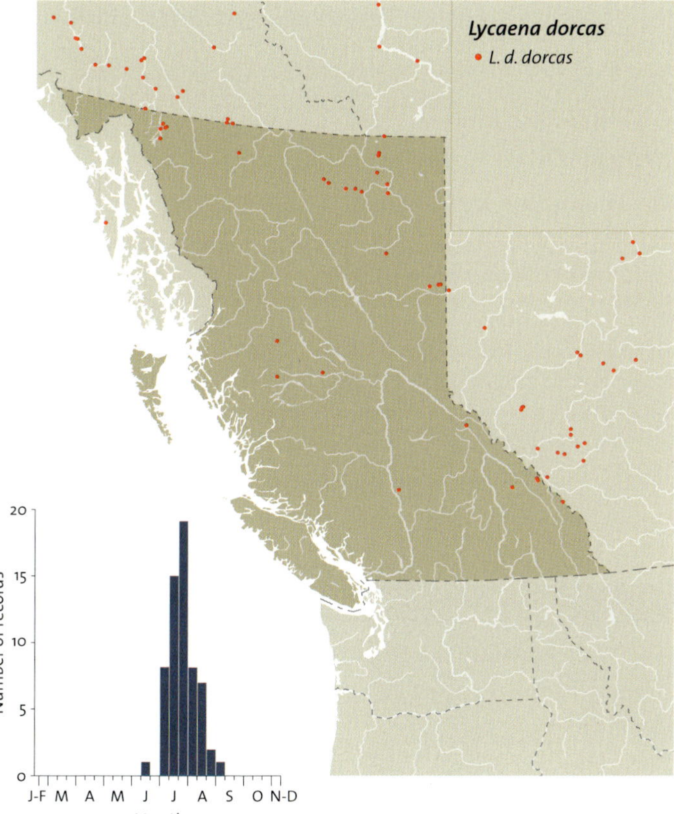



**Immature stages:** Newcomb (1910) described the immatures. The eggs are white when laid. The larva is pale green to blue green. The individual pupae were variously coloured from green to various dark patterns.

**Biology:** The Dorcas Copper is univoltine, flying from mid-July to early August. Newcomb (1910) first reared it on *Pentaphylloides floribunda* (Rosaceae), which the Dorcas Copper apparently uses throughout its range. This is an unusual larval foodplant family for North American coppers. Newcomb (1910) noted that the egg is attached to a leaflet, the species overwinters as an egg, the eggs hatch in April, and the Dorcas Copper completes its larval development and pupates just weeks before the adult emerges.

**Subspecies:** BC populations of the Dorcas Copper are the nominate subspecies, *L. d. dorcas* W. Kirby, 1837 (TL: The Pas, MB). It is found in boreal forest habitat from northern BC to PQ. In the Maritimes it is replaced by an endemic subspecies.

**Range and habitat:** The Dorcas Copper is found at lower elevations throughout the northern parts of BC in boreal forest habitat. In the central part of BC, it is often associated with bogs, such as the "Cranberry Bog" south of Valemont.

**General distribution:** The Dorcas Copper is found from BC east to NF and the Great Lakes states.

**Conservation status:** Not of concern (S5).

# Purplish Copper

*Lycaena (Epidemia) helloides* (Boisduval, 1852)

**ETYMOLOGY:** The species name *helloides* is from the Greek *hellus* (fawn). Either Boisduval misinterpreted the meaning of *dorcas* and in recognizing the similarity of *helloides* to *dorcas* implied the relationship, or he meant that it was like the European copper, *L. helle* (Opler and Krizek 1984). The first explanation seems more likely. The common name "Purplish Copper" refers to the purple colour of the dorsal wings of males (Holland 1898).

**ADULT:** The Purplish Copper is very similar in appearance to the Dorcas Copper but the two species do not occupy the same range. The Purplish Copper is larger and females have much more of the upperside ground colour yellow/orange rather than brown. There is also more orange in the submarginal band on the ventral hindwing of both sexes (Fig. 64a).

**IMMATURE STAGES:** The egg is a white, flattened sphere. The mature larva has a light brown head and apple green body with a yellowish spiracular line. The pupa is initially green. First-generation pupae turn dark fuscous (GAH).

**BIOLOGY:** In BC the Purplish Copper has two generations each year, with adults flying from late May to early July and again from mid-August to mid-September. Hardy (GAH) observed second-generation females in captivity laying eggs on

♂ D (2.8 cm)

♀ D (3.3 cm)

♂ V (2.8 cm)

*Polygonum persicaria* on 8–9 September. The eggs over-wintered and hatched on 5–12 April the following year. The larvae pupated between 6 and 22 June. The adults emerged between 27 June and 11 July. Jones (1940) reported *Polygonum amphibium* L. as a larval host on G.J. Spencer's authority, and oviposition on *P. amphibium* has been observed near Riske Creek by CSG. Hardy (GAH) reared it from *Rumex crispus*. Chambers (1963) and Scott (1992) record several *Rumex* species and *Polygonum* species as oviposition plants. In Colorado the Purplish Copper also overwinters in the egg stage.

**SUBSPECIES:** None. The type locality of the species is San Francisco, CA.

**RANGE AND HABITAT:** The Purplish Copper is found throughout southern BC in disturbed, roadside, and pasture habitat, and naturally in riparian situations. Throughout its range, the species has taken advantage of disturbed habitat.

**GENERAL DISTRIBUTION:** The Purplish Copper occurs from southern BC and AB south to Baja California and NM. It also ranges east to the Great Lakes of CAN and USA. There is evidence that the species has invaded the eastern part of its range in historic times (Opler and Krizek 1984).

**CONSERVATION STATUS:** Not of concern (S5).

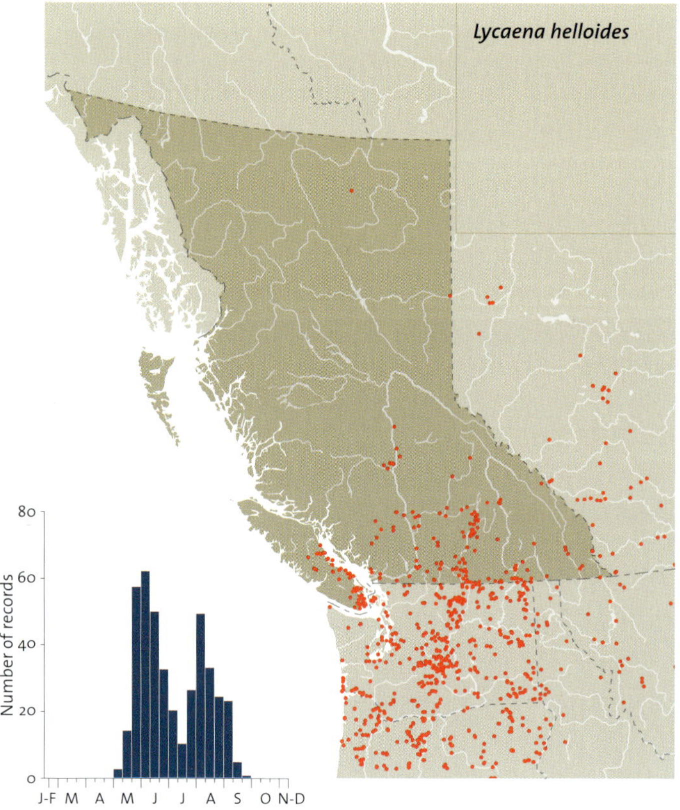

**Egg**

**Mature larva**

# Lilac-bordered Copper

*Lycaena (Epidemia) nivalis* (Boisduval, 1869)

**Etymology:** The species name *nivalis* is Latin for "of the snow." In his original description, Boisduval said that the species came from the region of the snows. The subspecies name *browni* honours F.M. Brown, the well-known Colorado butterfly authority. The common name refers to the lilac shading of the outer part of the ventral hindwing (Pyle 1981).

**Adult:** The Lilac-bordered Copper is easily distinguished from other butterflies by the appearance of the ventral hindwing, the outer half of which is lilac-coloured in the BC populations. The ventral hindwing also lacks the black spots found on most coppers and blues.

**Immature stages:** Newcomer (1964c) redescribed the immatures. The egg is bluish white. The mature larva is pale green with a rose/claret dorsal line outlined by a thin white line on each side. The pupa is straw yellow.

♂ D (2.8 cm)    ♀ D (3.0 cm)

♂ V (2.8 cm)

**Biology:** Adults of the Lilac-bordered Copper fly from late May to mid-August, depending on the elevation of the population but for about three weeks only at any one locality. The Lilac-bordered Copper over-winters as an egg and completes larval development the following spring. So far, only *Polygonum douglasii* has been recorded as a larval foodplant, in both California and Washington (Newcomer 1964c), but the species has been reared on *Rumex* sp. in the lab.

**Subspecies:** BC populations are *L. n. browni* dos Passos, 1938 (TL: Montpelier, ID).

**Range and habitat:** The Lilac-bordered Copper is found in the Okanagan Valley, at Rock Creek near Grand Forks, and as one historic record from Kaslo that has not been confirmed by recent observations. Adults are found on *Eriogonum* species, which are the nectar source.

**General distribution:** The Lilac-bordered Copper is found from extreme southern BC south to central CA and CO.

**Conservation status:** The Lilac-bordered Copper is of Special Concern in BC (S3).

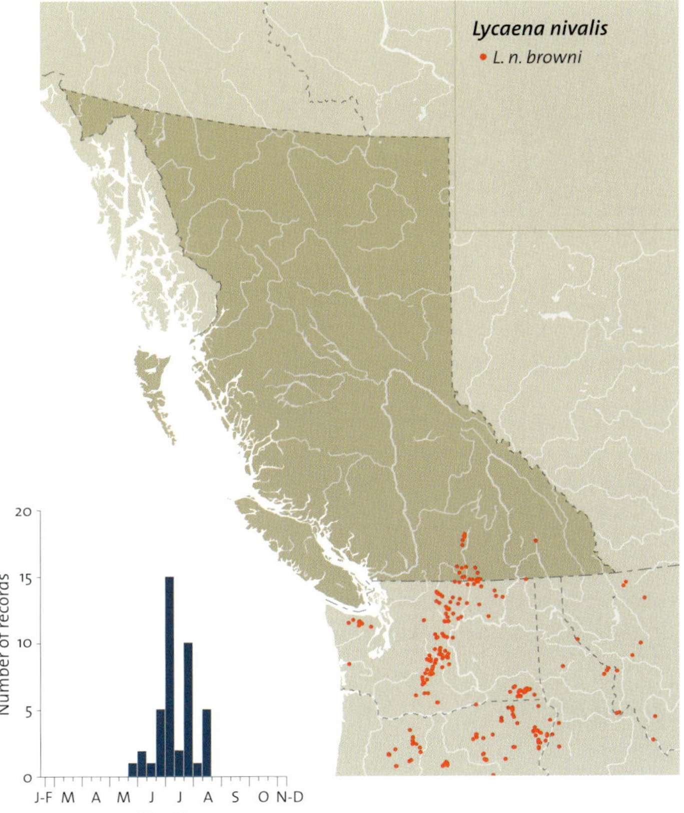

**Mature larva**

# Reakirt's Copper

*Lycaena (Epidemia) mariposa* (Reakirt, 1866)

**Etymology:** The species name *mariposa* is Spanish for butterfly. It probably refers to the fact that this species was described from the former Spanish territory of California. The subspecies *penroseae* was named after Mrs. Spencer Penrose of Colorado Springs in acknowledgment of her interest in Lepidoptera, and subspecies *charlottensis* was named for the Queen Charlotte Islands type locality. The common name acknowledges that Reakirt named the species (Holland 1898).

**ADULT:** Reakirt's Copper is easily distinguished from other small butterflies by the appearance of the ventral hindwing. It has a characteristic mottled grey pattern. On the upperside the wings of the male are a darker purple than those of other coppers. The females are brown on the upperside of the wings, with varying amounts of yellow/orange. The Charlotte Copper female has the most yellow and can be confused with females of the Small Copper in northwestern BC. The two species can be distinguished by the ventral hindwing pattern.

**IMMATURE STAGES:** Undescribed.

**BIOLOGY:** Reakirt's Copper flies from late June to early September, depending on elevation and the snow cover of the previous winter. There is only one generation each year. Pratt and Ballmer (1986) established *Vaccinium caespitosum* as the larval foodplant for Reakirt's Copper in northern California. In Burns Bog, BC, on 22–23 July 1991, four eggs were observed being laid on *Vaccinium uliginosum,* five on *V. oxycoccus,* and two on *Andromeda polifolia,* all of which are in the family Ericaceae (Richard Beard, pers. comm.).

**SUBSPECIES:** There are three subspecies in BC. The Charlotte Copper, *L. m. charlottensis* (Holland, 1930) (TL: Queen Charlotte Islands), is found on the Queen Charlotte Islands, in the northern Coast Ranges, and on the west coast of Vancouver Island. The nominate subspecies, *L. m. mariposa* (Reakirt, 1866) (TL: California), is found in the mountains of Vancouver Island, the Lower Mainland, and the Cascades. The Rocky Mountain subspecies, *L. m. penroseae* Field, 1938 (TL: Yellowstone Park, WY), occurs in the rest of the province.

Ssp. *mariposa*  ♀  D  (2.5 cm)

Ssp. *mariposa*  ♀  V  (2.5 cm)

Ssp. *charlottensis*  ♀  D  (2.7 cm)

Ssp. *charlottensis*  ♀  V  (2.7 cm)

Ssp. *penroseae*  ♀  D  (2.7 cm)

Ssp. *penroseae*  ♀  V  (2.7 cm)

**RANGE AND HABITAT:** Reakirt's Copper is found throughout BC in any forest opening or riparian situation where the larval foodplant, *Vaccinium* (Ericaceae), occurs.

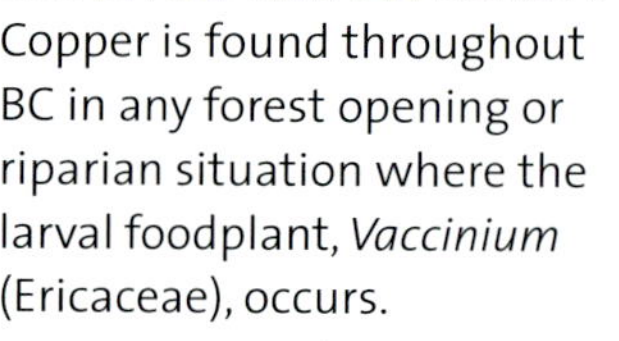

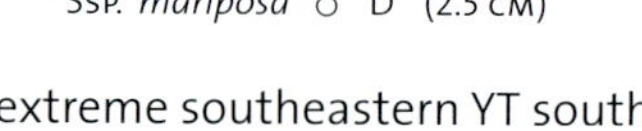

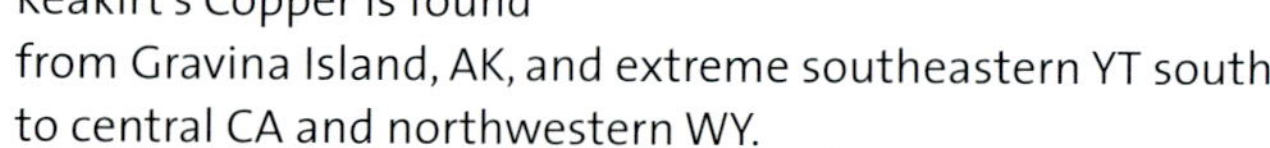

**GENERAL DISTRIBUTION:** Reakirt's Copper is found

Ssp. *mariposa*  ♂  D  (2.5 cm)

from Gravina Island, AK, and extreme southeastern YT south to central CA and northwestern WY.

**CONSERVATION STATUS:** Not of concern, with all three subspecies S5.

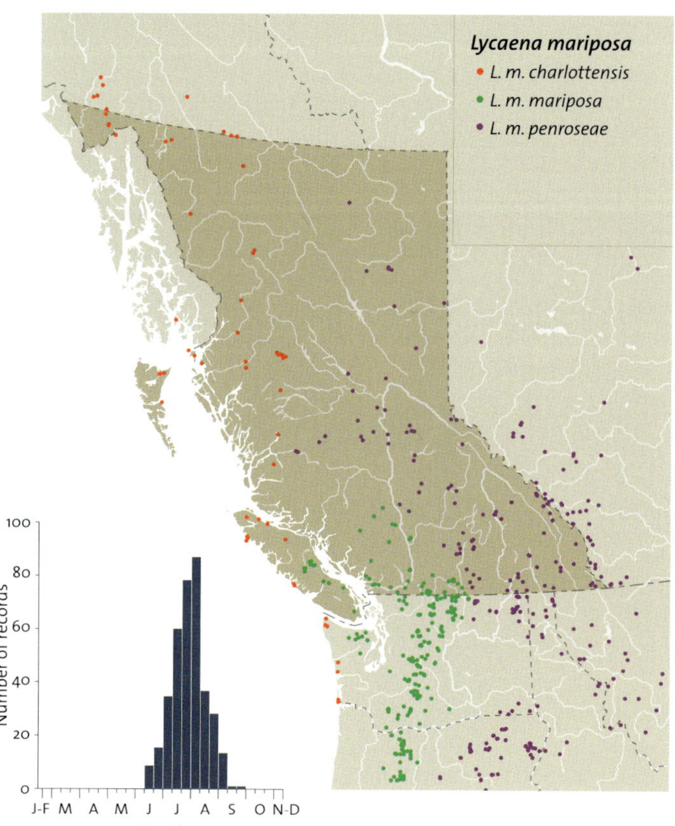

**Mature larva**

**Pupa**

The type genus of the subfamily is the European genus *Thecla*, named after a Roman virgin and martyr (Reed 1871). The common name "hairstreaks" is derived from the characteristic white "hairline" across the ventral hindwing. It is a very old English name that was in use before 1717 (Emmet 1990b).

British Columbia species of Theclinae all belong to one subgroup, the tribe Strymonini. This tribe is almost exclusively New World, with a huge radiation of species in the neotropics. The male genitalia are elongate with very small valves. The eyes are hairy. Males usually have scent patches on the dorsal forewings. Six genera are represented in the BC fauna. Only one, the genus *Strymon,* has Neotropical affinities. The larvae of the subfamily have honey glands but not eversible tubercles.

### GENUS *SATYRIUM* SCUDDER, 1876   HAIRSTREAKS

The name *Satyrium* is from the Latin *saturos* (Satyr), a goatlike woodland deity associated with Bacchus. The Satyrs were voluptuous dancers and this generic name draws attention to the sprightly flight of these hairstreaks (Emmet 1991). The common name is derived from the characteristic white "hairline" across the ventral hindwing (Fig. 65).

There are usually tails on the hindwings of species in this genus of hairstreaks. The aedeagus of the male is flared at the tip, with a serrated keel. The aedeagus has one or two cornuti, one of which is toothed. The pair of valves are close together at the base but very divergent at the ends. Clench (1961) provided the modern definition of the genus. He did not include in the genus the species *S. titus,* which has only one cornutus but is otherwise identical to the other species in the genus. Clench indicated that the genus was Holarctic, but authorities in the Palearctic recognize other genera for their fauna, such as *Strymonidia, Nordmannia,* etc. There are 15 species in this Nearctic genus, seven occuring in BC. The larvae feed on a wide variety of shrubs and perennials, including oaks (*Quercus*), willow (*Salix*), buckbrush (*Ceanothus*), chokecherry (*Prunus*), saskatoon (*Amelanchier alnifolia*), and, in one case, legumes.

## CORAL HAIRSTREAK
### *Satyrium titus* (Fabricius, 1793)

**ETYMOLOGY:** The name *titus* is a Latin proper name used in various contexts, with no clear relationship to the Coral Hairstreak. The subspecies name *immaculosus* refers to the small spots on the undersides of the wings. The common name was first used by Gosse (1859), and refers to the reddish pink (coral) spots on the ventral hindwing.

**ADULT:** The Coral Hairstreak is characterized by the submarginal row of coral spots on the ventral hindwing that can be almost contiguous. The spots extend into the ventral forewing as a faint row in southern BC populations and a strong row in Peace River populations. Otherwise the wings are a uniform brown.

**IMMATURE STAGES:** Saunders (1869d) first described the immatures, and Scudder (1889b) repeated the details with full credit. The egg is deep green with white projections on the chorion. The mature larva has a shiny black head. The body is dull green, with a dorsal dark green stripe on the thoracic segments. The dorsal surface of the abdominal segments is pinkish. The pupa is pale brown and covered with short, brown hairs, the typical appearance of a lycaenid pupa.

Ssp. *immaculosus* ♂ D (3.0 CM)

Ssp. *immaculosus* ♀ D (3.4 CM)

Ssp. *immaculosus* ♀ V (3.4 CM)

Ssp. *titus* ♀ V (2.9 CM)

**Biology:** Adults of the Coral Hairstreak from populations in the Southern Interior usually fly from late June to early August. There are, however, records of males from 20 May and 9 September, indicating that there may be a partial second generation. Elsewhere a second brood has been discounted (Opler and Krizek 1984). In the Peace River valley, the Coral Hairstreak flies in July. Throughout its North American range, it has been reared on a variety of wild cherry species, *Prunus* sp. The egg is laid in July or August and hibernates, hatching as the wild cherry is budding the following spring. The entire growth is completed in about two months. The butterfly is in the pupa for two or three weeks and then the adult emerges.

Mature larva

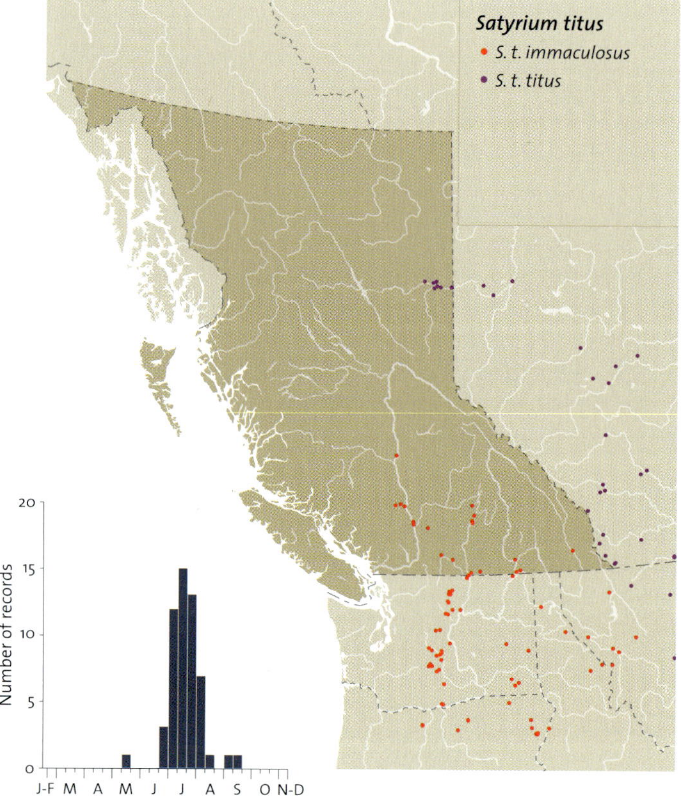


**Subspecies:** The subspecies name usually applied to the southern BC and Peace River populations is *S. t. immaculosus* (W.P. Comstock, 1913), the correct subspecies name for southern BC populations. The type locality of this taxon is on the west side of the Rockies in Utah, however, and should not be applied to Peace River and Canadian Prairie populations. The Peace River populations are the nominate subspecies, *S. t. titus* (Fabricius, 1793) (TL: "in Anglia" [NF]), with stronger coral markings.

**Range and habitat:** The Coral Hairstreak is found from the Chilcotin southeast through the Southern Interior and the extreme southwest of the West Kootenay. There are also disjunct populations in the Peace River Canyon from below Hudson's Hope to the AB border. The Coral Hairstreak is found in riparian situations and on the banks of large rivers where the larval foodplant, chokecherry (*Prunus virginiana*), is found.

**General distribution:** The Coral Hairstreak is found from the Peace River of BC east to NF, and from the Chilcotin south to northern CA and NM. In the east it extends south to GA, with disjunct populations in TX.

**Conservation status:** Subspecies *titus* is of Special Concern in BC (S3). Subspecies *immaculosus* is not of concern (S4).

## Behr's Hairstreak
*Satyrium behrii* (W.H. Edwards, 1870)

Behr's Hairstreak (*Satyrium behrii*)

**Etymology:** The species *behrii* was named in honour of the lepidopterist Dr. Herman Behr. He was the first significant resident California lepidopterist and curator of the Lepidoptera collection at the California Academy of Sciences, San Francisco, in the mid 1800s. The subspecies name *columbia* refers to the fact that it was described from BC. The common name was first used by Holland (1898).

**Adult:** Behr's Hairstreak is easily identified by the tawny orange of the upperside of the wings and the wide black area on the costal area of the dorsal forewing. The pattern of the ventral hindwing is also characteristic of the species.

**Immature stages:** Comstock (1928) described the immatures. The mature larva is green, matching the foodplant leaves. It has a narrow dorsal white line edged by a wider dark green area on each side.

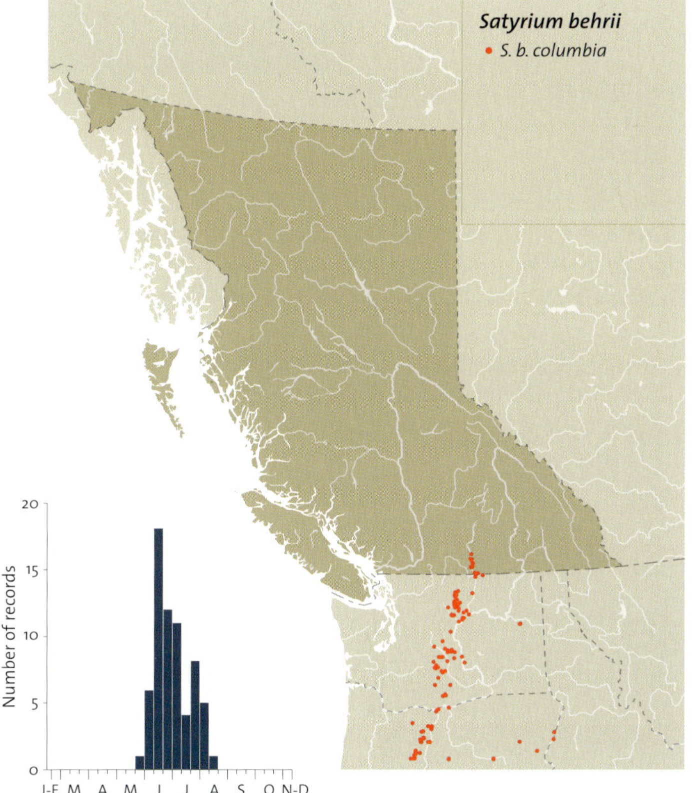

**BIOLOGY:** Behr's Hairstreak adults fly from June to early July, extending into late July if there has been a late spring. The species overwinters as an egg (Emmel and Emmel 1973) and completes development the following spring, spending a short time in the pupa (Comstock 1928). The BC foodplant is *Purshia tridentata* (JHS, field observations; Steven Ife, rearing). Scott (1992) records oviposition on *Cercocarpus montanus* in Colorado, a plant species that has not been recorded in BC.

**SUBSPECIES:** The BC subspecies, *S. b. columbia* (McDunnough, 1944), was described from specimens collected at Fairview [Oliver], BC. It ranges south in Washington along the east side of the Cascade Mountains.

**RANGE AND HABITAT:** Behr's Hairstreak is known only in the extreme southern Okanagan Valley in association with antelope brush, *Purshia tridentata,* the larval foodplant. Antelope brush is threatened by suburban and agricultural development, grazing management, and proposed casino development near Osoyoos. Behr's Hairstreak should be looked for in the East Kootenay, where the foodplant also occurs.

**GENERAL DISTRIBUTION:** Behr's Hairstreak ranges from the southern Okanagan Valley of BC through WA and then south to southern CA and NM.

**CONSERVATION STATUS:** Behr's Hairstreak is Threatened in BC (S2).

♂ D  (2.7 CM)

♀ D  (2.6 CM)

♂ V  (2.7 CM)

**Mature larva**

**Pupa**

# Sooty Hairstreak

*Satyrium fuliginosum* (W.H. Edwards, 1861)

**Etymology:** The name *fuliginosum* is derived from the Latin *fuliginosus* (sooty or painted black), in reference to the grey brown dorsal wing colour. The common name also reflects the dorsal wing colour. Holland (1898) first used the common name as "Sooty Gossamer-wing," which Pyle (1981) changed.

**Adult:** Both sexes of the Sooty Hairstreak are a uniform grey brown on the upperside of both wings. On the underside the wings have diffuse spots and a brown ground colour overlaid with white scales, giving it a sooty appearance. The Sooty Hairstreak can be confused with the females of some blues, especially Boisduval's Blue.

**Immature stages:** Undescribed.

**Biology:** The Sooty Hairstreak flies from late May to early July in what appears to be the most circumscribed flight period of any of the Southern Interior summer hairstreaks. Comstock (1933) and Scott (1992) both observed oviposition on *Lupinus* sp., but no one has described the life cycle. Shapiro et al. (1981) observed it ovipositing only on *L. croceus*

♂ D (2.8 cm)

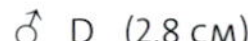

♀ D (2.9 cm)

♂ V (2.8 cm)

in northern CA, a plant not found in BC. Presumably the Sooty Hairstreak overwinters in the egg stage, like the other species in the genus.

**Subspecies:** BC populations are the subspecies *S. f. semiluna* Klots, 1930 (TL: Moose P.O., Jackson Hole, WY).

**Range and habitat:** The Sooty Hairstreak is almost as restricted in BC as Behr's Hairstreak. It is known from the lower Similkameen River valley as well as the South Okanagan. Whereas Behr's Hairstreak has historically occurred in large numbers, the Sooty Hairstreak is rarely seen in large numbers. Elsewhere the larvae feed on woody shrubs of the genus *Lupinus*. One recorded foodplant, *L. arbustus* Dougl. (Scott 1986a), is found in BC where the Sooty Hairstreak occurs.

**General distribution:** The Sooty Hairstreak is found from the southern Okanagan Valley and near Waterton Lake, AB, south to central CA and CO.

**Conservation status:** The Sooty Hairstreak is Endangered in BC (S1).

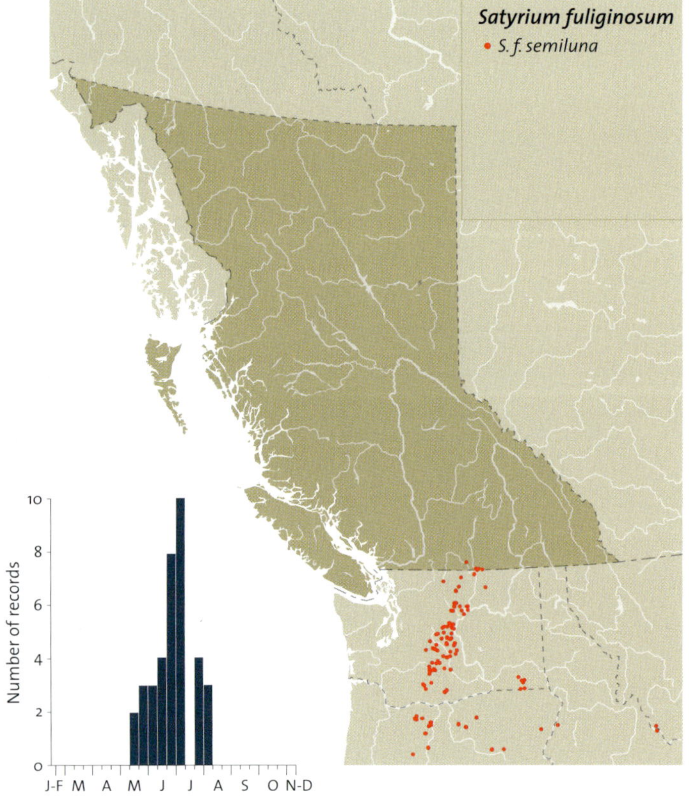

**Mature larva**

**Pupa**

# California Hairstreak
*Satyrium californicum* (W.H. Edwards, 1862)

**California Hairstreak (*Satyrium californicum*)**

**Etymology:** The species name *californicum* comes from the origin of the type specimens used to describe the species. The common name was first used by Comstock (1927).

**Adult:** The California and Sylvan Hairstreaks are very similar (see discussion under **Sylvan Hairstreak**). The submarginal area of the ventral forewing has orange spots along at least half of the submarginal line (Fig. 65a). The submarginal area of the ventral hindwing has orange spots on both sides of the blue spot and further orange spots along at least half of the submarginal line (Fig. 65a). The California Hairstreak has a very restricted range in BC, and it is only there that one needs to look for these almost microscopic characteristics.

♂ D  (2.7 cm)

♀ D  (3.1 cm)

♂ V  (2.7 cm)

**Immature stages:** Unknown except on oaks in California (Comstock 1933). The California larvae are likely very different in colour from the ones in BC because BC populations cannot feed on oak.

**Biology:** The California Hairstreak flies from early June to early August, with early flight dependent on an early spring. Comstock (1933) described the mature larva and noted the foodplant as oak, *Quercus* sp., which does not occur in the BC range of the species. Scott (1992) observed oviposition on *Prunus virginiana, Cercocarpus montanus,* and *Purshia tridentata.* In the southern Okanagan Valley, Shepard has observed the species in numbers on willow around a reservoir, and willow is a likely foodplant in BC. Guppy has seen it associated with mock-orange. In the same areas, it nectars on *Melilotus officinalis* (yellow sweet clover), an introduced plant (CSG; JHS).

**Subspecies:** None. The type locality of the species is Capell Creek, near Napa, CA.

**Range and habitat:** The California Hairstreak is more widely distributed than the Sooty or Behr's Hairstreaks, but is still not a common species. It is known from Lillooet to Grand Forks. At least in the southern Okanagan Valley, the California Hairstreak is found at willows surrounding water reservoirs and natural lakes and along meandering streams.

**General distribution:** The California Hairstreak is found from the Southern Interior of BC south to Baja California and CO.

**Conservation status:** The California Hairstreak is of Special Concern in BC (S3).

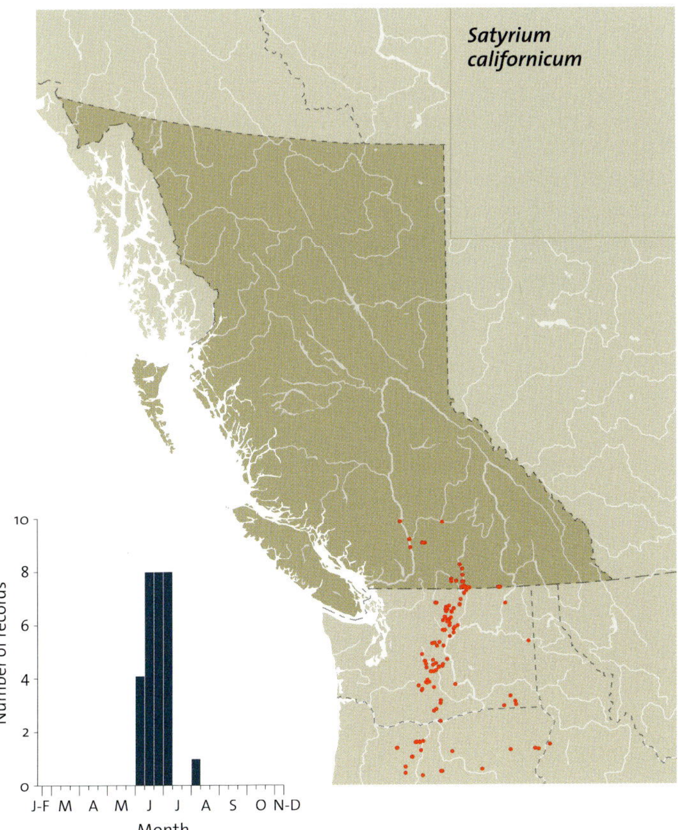

**Mature larva**

# Sylvan Hairstreak

*Satyrium sylvinum* (Boisduval, 1852)

**Etymology:** The species name *sylvinum* is derived from the Latin *silva* (woods). The common name "Sylvan Hairstreak" (Blackmore 1927; Comstock 1927) reflects the generally open woodland habitat. *S. s. nootka* was stated as having been named for the fact that this group of Aboriginals occupied some of the range of the butterfly. The Nootka Nation, the village of Nootka, and Nootka Sound are actually on the west side of Vancouver Island, outside the range of the butterfly.

**Adult:** The Sylvan and California hairstreaks are very similar in general appearance and cannot be told apart in the field unless closely observed while nectaring. In both, the ground colour of the uppersides of the wings is dark khaki and that of the undersides is khaki. The only consistent difference is the amount of orange spotting in the submarginal area on

♂ D  (2.9 cm)       ♀ D  (3.1 cm)

♂ V  (2.9 cm)

the undersides of both wings. The Sylvan Hairstreak lacks orange in the sub-marginal area of the ventral forewing. Orange spotting in the submargin of the ventral hindwing is limited to one conspicuous spot just lateral to the blue spot and a few orange scales elsewhere (Fig. 65b).

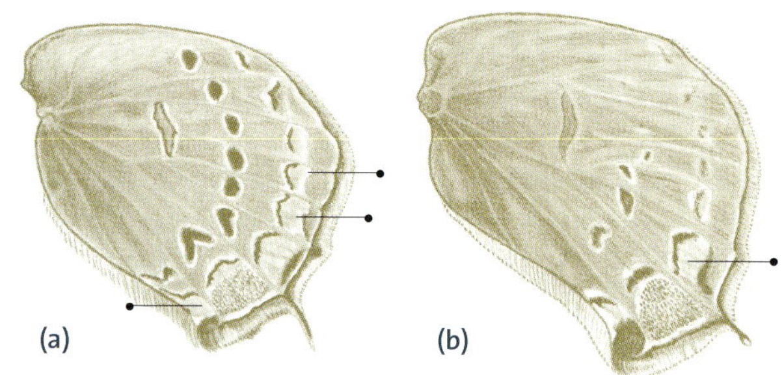

(a)                (b)

**65** Ventral hindwings of *Satyrium* species: (a) *S. californicum*, (b) *S. sylvinum*

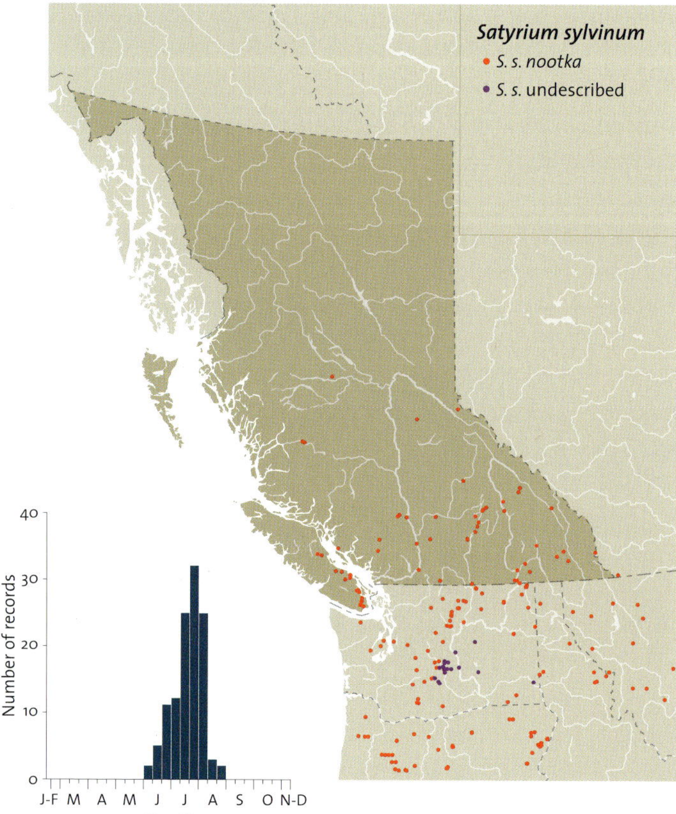



**Immature stages:** In California the mature larva is pale apple green with a thin dorsal white line and two diagonal white dashes laterally on each body segment (Comstock and Dammers 1934). Hardy (1962c), however, describes the mature larva as pale green with a dark green dorsal line flanked by yellow on the thorax and white on the abdomen, and with a second yellow line at the spiracles. The pupa is dark mahogany brown.

**Biology:** The Sylvan Hairstreak flies mainly from late June to mid-August. Comstock and Dammers (1934) first noted the foodplant as *Salix* sp. Hardy (1962c) reared the Vancouver Island populations on *Salix prolixa*. The butterfly overwinters as an egg and completes development the following spring Hardy (1962c).

**Subspecies:** *S. s. nootka* Fisher, 1998 (TL: Wellington [Nanaimo], BC) is the BC subspecies.

**Range and habitat:** The Sylvan Hairstreak is the most widely distributed *Satyrium* species in BC. It is known from near Bella Coola and Vanderhoof south throughout the province in riparian situations where the larval foodplant, *Salix* sp., occurs.

**General distribution:** The Sylvan Hairstreak ranges from central BC south to southern CA and NM.

**Conservation status:** Not of concern (S5).

**Mature larva**

# Striped Hairstreak
*Satyrium liparops* (Leconte, 1833)

**Etymology:** The species name *liparops* is derived from the Greek (shining eye). The subspecies name *fletcheri* honours James Fletcher, the first Dominion Entomologist in Canada. The common name "Striped Hairstreak" (Scudder 1889b) refers to the complex striping on the ventral wings.

**Adult:** The Striped Hairstreak has a dark brown ground colour on the upperside of the wings, with a large median tawny flush on the dorsal forewing of both sexes. On the underside the wings have the same ground colour, with slightly darker areas outlined by white lines on the lateral sides, hence the common name.

**Immature stages:** Saunders (1869d) first described the immatures. The mature larva is green with a darker green dorsal stripe.

**Biology:** The Striped Hairstreak flies from late June to late July. Mature larvae are found by early June. Pupation occurs

♂ D (2.6 cm)

♀ D (2.7 cm)

♂ V (2.6 cm)

in mid-June and the adult emerges within two weeks. The egg is the overwintering stage. The Striped Hairstreak was observed by Shepard perching on branches of saskatoon (*Amelanchier alnifolia*) where that was the only shrub, and saskatoon is thus the presumed larval foodplant. The Striped Hairstreak flies in the same habitat as the Coral Hairstreak in the Peace River region, but the Coral Hairstreak feeds on chokecherry.

**Subspecies:** BC populations are the northern Canadian Prairie subspecies *S. l. fletcheri* (Michener & dos Passos, 1942) (TL: "Manitoba").

**Range and habitat:** The Striped Hairstreak is found in BC only on the north banks of the Peace River Canyon and some of its tributaries. Males can be easily observed at the tops of the south-facing banks wherever chokecherry grows.

**General distribution:** The Striped Hairstreak is found east of the Rockies from the Peace River region of BC south to CO and east to NS. In the east it occurs south to TX and FL.

**Conservation status:** The Striped Hairstreak is of Special Concern in BC (S3).

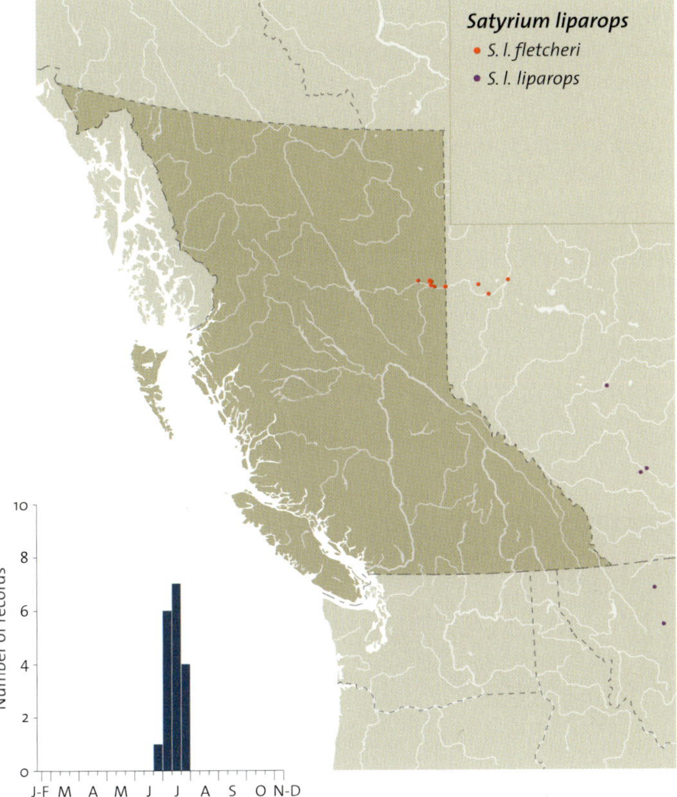

**Mature larva**

# Hedgerow Hairstreak

*Satyrium saepium* (Boisduval, 1852)

**Etymology:** The species name *saepium* is derived from the Latin *saepis* (a hedge or fence). The subspecies name *okanaganum* refers to the part of BC from which this subspecies was described. Both the common name (Holland 1898) and Latin name may reflect the use of *Ceanothus* species as the larval foodplant.

**Adult:** The Hedgerow Hairstreak has a uniform reddish brown ground colour on the upperside of the wings, with a very narrow black border. It is superficially similar to female *Mitoura* species, but they have a wider black border on both wings and a very different underside.

**Immature stages:** Comstock and Dammers (1933) characterize the egg as grey green. The mature larva is green with a line of light yellow chevron markings down both sides of the body. The pupa is brown and similar to other lycaenid pupae.

**Biology:** The Hedgerow Hairstreak flies in one generation from early June to late August, depending on how early warm weather begins and the elevation of any specific population. It presumably overwinters as an egg. After hatching the following spring, it quickly goes through the larval stages and pupates shortly before emerging as an adult. The larvae have been reared on various *Ceanothus* species elsewhere. The Forest Insect Survey has rearing records from *Ceanothus* in BC but did not record the species; Guppy, however, has reared them from larvae collected on *C. velutinus* near Lytton, BC.

♂ D (2.8 cm)

♀ D (3.1 cm)

♂ V (2.8 cm)

**Subspecies:** BC populations are the subspecies *S. s. okanaganum* (McDunnough, 1944) (TL: Peachland, BC).

**Range and habitat:** The Hedgerow Hairstreak is found from near Tweedsmuir Park south along the inner Coast Ranges to the international border, and east through the Southern Interior and West Kootenay. It usually occurs in ponderosa pine areas where the larval foodplant, *Ceanothus*, occurs, but also at higher elevations with open habitat.

**General distribution:** The Hedgerow Hairstreak is found from southern BC south to southern CA and NM.

**Conservation status:** Not of concern (S4).

**Mature larva**

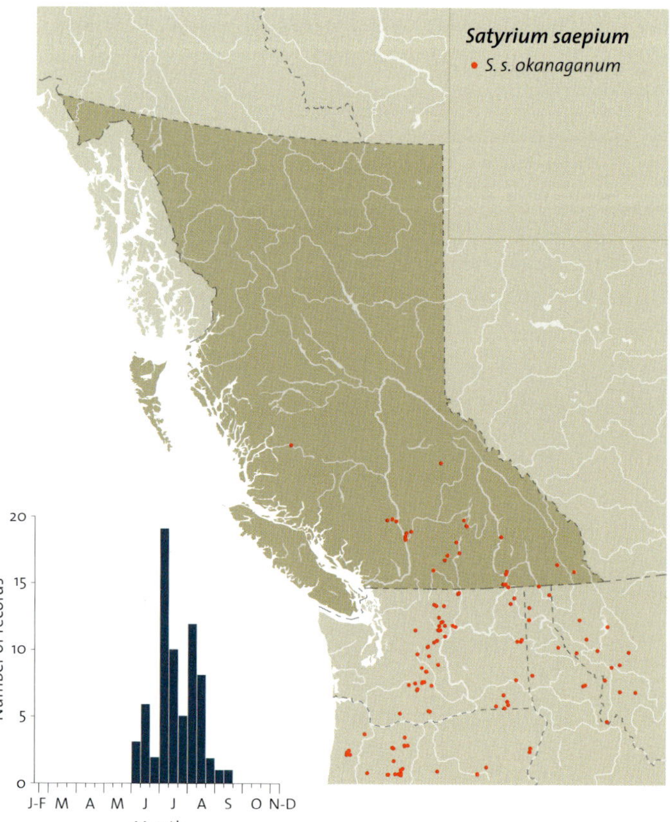

**Prepupal larva**

**Pupa**

The name *Callophrys* may be derived from the Greek *kallos* (beautiful) and *ophrys* (eyebrow). This likely refers to green scales on the eye (Emmet 1991). The common name refers to the green colour of the hindwings (Holland 1898).

This genus and the following three genera, *Loranthomitoura, Mitoura,* and *Incisalia,* are morphologically similar. Some authorities insist on combining them in one genus, *Callophrys.* Ballmer and Pratt (1992b) detail the anatomy of first instar larvae of these genera and show that this is a natural grouping but that the genera are separate. For all four, the male aedeagus is flared at the tip but lacks the ventral keel present in the genus *Satyrium.* There are two cornuti, both of which are dentate. The pair of valves are parallel along their entire length instead of divergent at the ends as in *Satyrium.* For all members of the genus *Callophrys,* the underside of the hindwing is some shade of green with a strong to weak white median line. The hindwings lack tails. The tips of the valves are not capped as in *Incisalia,* and the cornuti are slender. The genus is Holarctic, with at least seven species (Tilden 1963); two are found in BC (Clench 1963). The larvae of all BC members of this genus feed on buckwheat plants, *Eriogonum* spp.

## IMMACULATE GREEN HAIRSTREAK
*Callophrys affinis* (W.H. Edwards, 1862)

**ETYMOLOGY:** The species name *affinis* is Latin for neighbouring. Since Edwards described this on the same page and just after *C. viridis,* the name was probably meant to emphasize the close relationship of the two species. They were first included in the then broad genus *Thecla,* which included many species without green ground colour on the underside of the wings. The subspecies name *washingtonia* refers to the type material being from Washington state. The common name (Pyle 1981) refers to the lack of strong white markings on the underside of the hindwings.

♂ D  (2.9 CM)

♀ D  (2.6 CM)

♂ V  (2.9 CM)

**ADULT:** For the Immaculate Green Hairstreak, the undersides of the wings have an apple-green ground colour. The ventral hindwing has a median row of white spots arranged in a curved line. Often the white spots are weak or partially to almost completely lacking. The ground colour of the upperside of the female wing is tawny; that of the male is grey brown.

**IMMATURE STAGES:** Undescribed.

**BIOLOGY:** The Immaculate Green Hairstreak flies from early May to early June. The earlier flight dates are dependent on how early the warm spring weather begins. The Immaculate Green Hairstreak almost always occurs in the same low-elevation habitat as Sheridan's Hairstreak, but begins to fly two to three weeks later. No one has determined whether the Immaculate Green Hairstreak completes its larval stages and spends the late summer and winter as a pupa, like Sheridan's Hairstreak, or whether the larva only partially completes its development the first year and then completes it the following spring. The larvae feed on plants of the genus *Eriogonum,* in the buckwheat family.

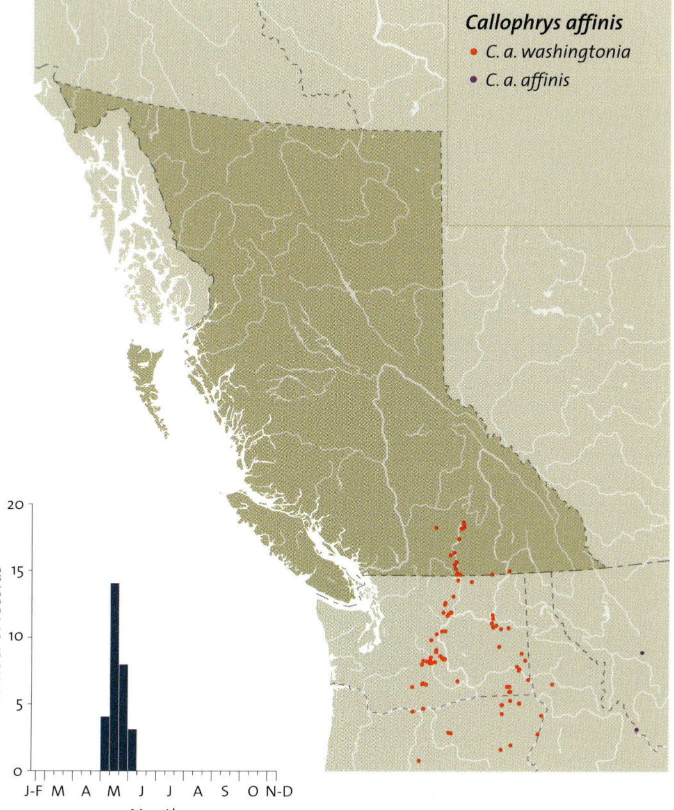

**Immaculate Green Hairstreak (*Callophys affinis*)**



SUBSPECIES: BC populations are *C. a. washingtonia* Clench, 1944 (TL: Alta Lake, Okanogan Co., WA). This subspecies ranges south to eastern WA and northeastern OR.

RANGE AND HABITAT: The Immaculate Green Hairstreak is found near Douglas Lake, east of Grand Forks, south of Trail, and along the entire Okanagan Valley floor in association with the larval foodplant, *Eriogonum* spp., usually in dry gullies.

GENERAL DISTRIBUTION: The Immaculate Green Hairstreak as defined by Scott (1986b) ranges from the Southern Interior and West Kootenay of BC south through CA and NM to northern MEX.

CONSERVATION STATUS: The Immaculate Green Hairstreak is of Special Concern in BC (S3).

# SHERIDAN'S HAIRSTREAK
*Callophrys sheridanii* (W.H. Edwards, 1877)

ETYMOLOGY: The species *sheridanii* was named for then Lieutenant General P.H. Sheridan at the request of the person who collected the type specimens, W.L. Carpenter. Sheridan was a famous American Civil War hero, especially remembered for Sheridan's Ride. Holland (1931) first used the common name. The subspecies name *newcomeri* honours E.J. Newcomer, an economic entomologist who, after retirement, made many contributions to butterfly knowledge in Washington and the rest of the American Pacific Northwest.

♂ D (2.3 CM)

♀ D (2.4 CM)

♂ V (2.3 CM)

ADULT: For Sheridan's Hairstreak, the undersides of the wings have a green ground colour. The ventral hindwing has a median row of white spots arranged in a straight line. The line of white spots is usually fused into one line, narrower in high-elevation populations and wider but often discontinuous in low-elevation populations, but always arranged in a straight line. The ground colour of the upperside of the wings is grey brown in both sexes.

IMMATURE STAGES: Unknown.

BIOLOGY: Sheridan's Hairstreak emerges two to three weeks before the Immaculate Green Hairstreak at the lower elevations where both occur. At these lower elevations it flies from mid-April to mid-May. For the higher-elevation (1,500–2,200 m) populations, the adults are active from late June to late July. Eggs hatch shortly after being placed on the foodplant and the larval development is very quick. The larvae usually begin feeding on the flowers and then shift to young leaves, presumably because that part of the plant is

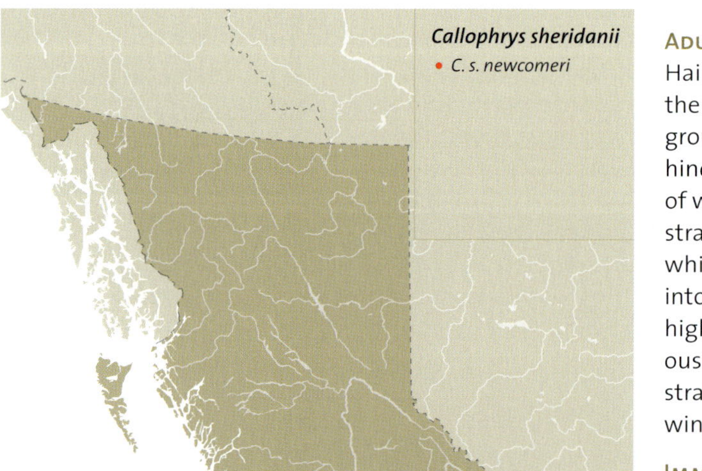

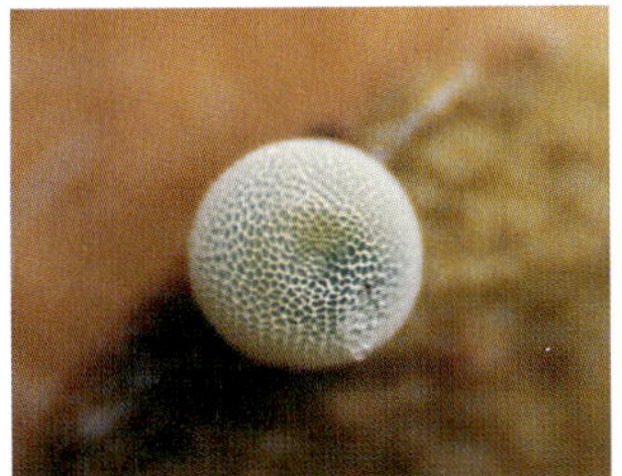

**Egg**

**First instar larva**

**Sheridan's Hairstreak (*Callophrys sheridanii*)**

tender, although toxins in older leaves have not been ruled out. The butterfly is in the pupal stage by June and over-winters as such. Hiruma et al. (1997) demonstrated that the pupa needs a period of 4°C exposure for adults to emerge. They also showed that 65% emerged within eight days when the temperature increased to early spring norms. The remaining individuals emerged over a prolonged period of 200 days. This staggered emergence is probably what accounts for the long adult flight of this species and other species in related genera.

**Subspecies:** BC populations are assigned to the subspecies *C. s. newcomeri* Clench, 1963 (TL: Mill Cr., Yakima Co., WA). We believe that the Washington butterfly atlas (Hinchliff 1996) and Clench (1963) misapplied the name *neoperplexa* Barnes & McDunnough, 1923 (TL: Eureka, UT) for Washington. Hopfinger, who was the source of Clench's original record, was notorious for labelling specimens "Brewster" when he collected all over the western half of Okanogan County in Washington. Clench's use of the name *neoperplexa* fits the high-elevation populations in northern Washington and southern BC, and those populations do not deserve sub-specific recognition.

**Range and habitat:** Sheridan's Hairstreak is found from Botanie Mountain near Lytton and Manning Provincial Park east to the AB border. The populations occur in two distinct habitats where the larval foodplant, *Eriogonum,* is found. At low elevations Sheridan's Hairstreak is found flying in the same habitat as the Immaculate Green Hairstreak, but it is more common and is also found at *Eriogonum* patches where the Immaculate Green Hairstreak is not found. Sheridan's Hairstreak is also found at 1,500–2,200 m on dry southern exposures where *Eriogonum* is found. Populations occurring in low valleys are reduced or lost, but high-elevation populations are stable.

**General distribution:** Sheridan's Hairstreak ranges from southern BC and the Rockies of southern AB south to central CA and NM.

**Conservation status:** Not of concern (S4).

## Genus *Loranthomitoura* Ballmer and Pratt, 1992   Mistletoe Hairstreaks

The genus *Loranthomitoura* is named after the larval foodplants, mistletoe (family Loranthaceae), and after the related genus *Mitoura*. The common name for the genus, "mistletoe hairstreaks," refers to the larval foodplants being mistletoe species; it is used here for the first time.

This genus and *Mitoura* are separated from *Callophrys* and *Incisalia* by the spatulate (flattened) cornuti and the presence of distinct tails on the hindwing. The ventral wing pattern of *Loranthomitoura* is typical of the Theclini, with strong postmedian white lines. Larvae feed on tree mistletoes, hence the name. For this reason adults are seen only when they are nectaring on flowers of perennial or annual plants at ground level or on low shrubs. They are never seen flying in an open meadow. The genus is Nearctic, with four species. Ballmer and Pratt (1992b) characterize the genus based on first instar larval morphology.

# Thicket Hairstreak
*Loranthomitoura spinetorum* (Hewitson, 1867)

**Thicket Hairstreak (*Loranthomitoura spinetorum*)**

♂ D (2.5 cm)

♀ D (2.7 cm)

♂ V (2.5 cm)

**ETYMOLOGY:** The species name *spinetorum* is derived from the Latin *spinetum* (thicket), in reference to the resemblance of the species to the European hairstreaks that inhabit thickets. The common name (Holland 1898) is a translation of the scientific name. Unfortunately *L. spinetorum* is not found in thickets and the common name has no utility.

**ADULT:** The Thicket Hairstreak and Johnson's Hairstreak are almost identical on the underside of the wings, and thus are difficult to discriminate without seeing the upperside of the wings. The ventral wing ground colour is a rich brown with a strong postmedian white line. The Thicket Hairstreak is grey blue on the upperside of the wings, whereas Johnson's Hairstreak is dark brown.

**IMMATURE STAGES:** Comstock and Dammers (1938) partially described the immatures. The egg is typical of lycaenids, without spicules on the surface. The mature larva has a yellow olive ground colour. The middle thoracic and several abdominal segments have an oblique white bar. Spiracles are pink with brown edges. The infra-stigmatal fold is bright orange, tinged above and below with magenta. The abdomen is covered with short, colourless hairs. The pupa is dark chestnut brown with white spiracles.

**BIOLOGY:** The Thicket Hairstreak flies from late April to mid-July, depending on elevation, snow melt, and how soon warm weather begins. Lab-reared specimens emerge throughout the winter if kept at warm temperatures (FIS). Eggs are laid and development is completed the same summer, with the butterfly overwintering as a pupa. Adults are seen only when they come down from the tops of the mature trees to obtain nectar, and are not often seen. The larvae are dependent on mistletoes growing on mature ponderosa pine, and are therefore threatened by logging pressure and suburban development, especially since a large portion of the habitat for ponderosa pine is on private land. Forest industry mistletoe eradication policies may eventually result in the Thicket Hairstreak becoming of conservation concern.

Larvae feed on all mistletoe species of the genus *Arceuthobium*, primarily when they parasitize pines (*Pinus*)

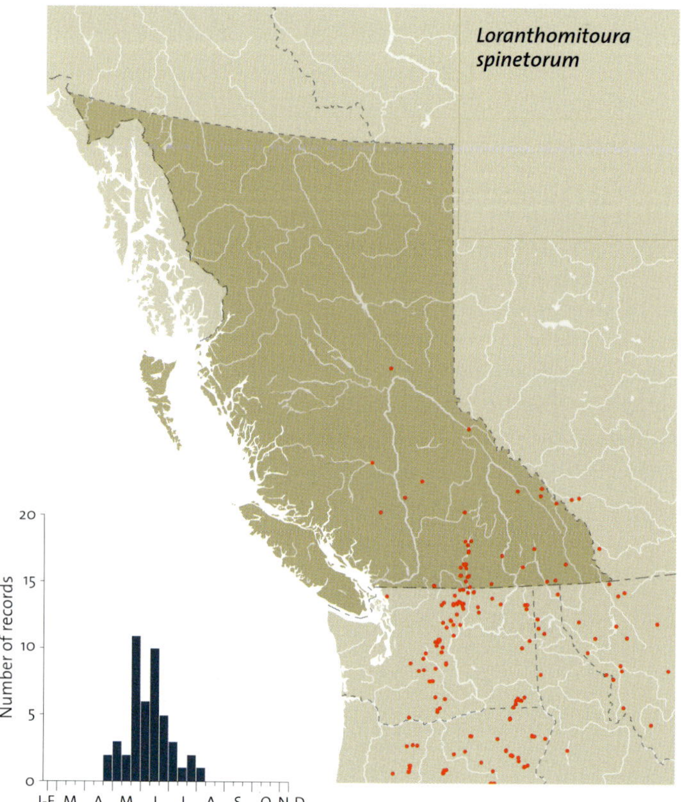

**Mature larva**

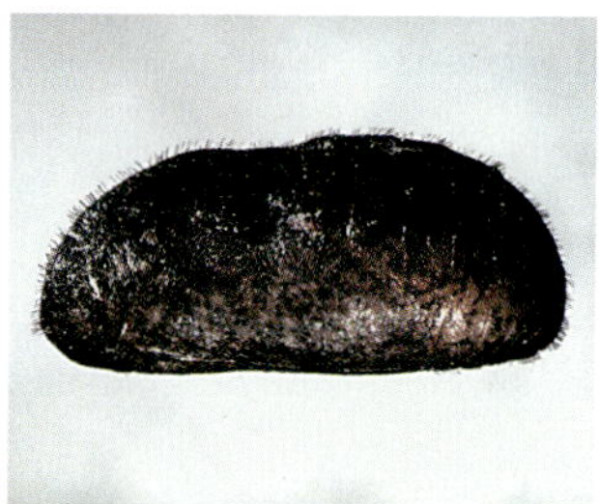

**Pupa**

but also on *Abies* spp. (Remington 1958; Shields 1966) and Douglas-fir (CSG). The only specific foodplant for BC is *Arceuthobium americanum* at 100 Mile House (FIS). Ponderosa pine is the normal host of the mistletoe larval foodplant, but occasionally populations of the Thicket Hairstreak exist on mistletoes growing on Douglas-fir, larch, or true fir (Pinaceae).

**SUBSPECIES:** None. The type locality of the species is vicinity Gold Lake Lodge, Sierra Co., CA (Emmel et al. 1998c).

**RANGE AND HABITAT:** The Thicket Hairstreak is found from central BC east and south through the Southern Interior and Kootenays in drier forest habitats at elevations up to 1,000 m.

**GENERAL DISTRIBUTION:** The Thicket Hairstreak is found from southern BC and adjacent AB south to Baja California and northern MEX.

**CONSERVATION STATUS:** Not of concern (S4).

## JOHNSON'S HAIRSTREAK
*Loranthomitoura johnsoni* (Skinner, 1904)

**ETYMOLOGY:** The species name *johnsoni* honours O.B. Johnson, a faculty member at the University of Washington in Seattle who trained the early Washington botanist and entomologist C.V. Piper of Washington State University. The common name was first used by Comstock (1927).

**ADULT:** Johnson's Hairstreak is identical to the Thicket Hairstreak except that the ground colour of the upperside of the wings is dark brown, not grey blue.

**IMMATURE STAGES:** Undescribed, but see photographs below.

**BIOLOGY:** Johnson's Hairstreak flies from late May to early July. Eggs are laid and the life cycle reaches pupation in a short time. The butterfly overwinters in the pupal stage. The species was first observed as larvae on mistletoe by C.V. Piper in Seattle (Skinner 1904). Since then various rearings by the

♂ D (3.2 CM)

♂ V (3.2 CM)

FIS and D.V. McCorkle (pers. comm.) have established that the larvae are usually found feeding on *Arceuthobium* spp. parasitizing western hemlock. Forest industry mistletoe eradication, and *Bt* spraying to eradicate Gypsy Moth introductions, will result in Johnson's Hairstreak becoming increasingly rare.

**SUBSPECIES:** None. The type locality of the species is Seattle, WA.

**RANGE AND HABITAT:** Johnson's Hairstreak is known only from southeastern Vancouver Island and the Lower Fraser Valley east to Hope. The larvae feed on mistletoes growing on western hemlock at elevations below 625 m.

**GENERAL DISTRIBUTION:** Johnson's Hairstreak is found from southwestern BC south to central CA west of the Cascade and Sierra Nevada mountains, and in a group of disjunct populations in Baker Co., OR, and adjacent ID.

**CONSERVATION STATUS:** Johnson's Hairstreak is Endangered in BC (S1S2).

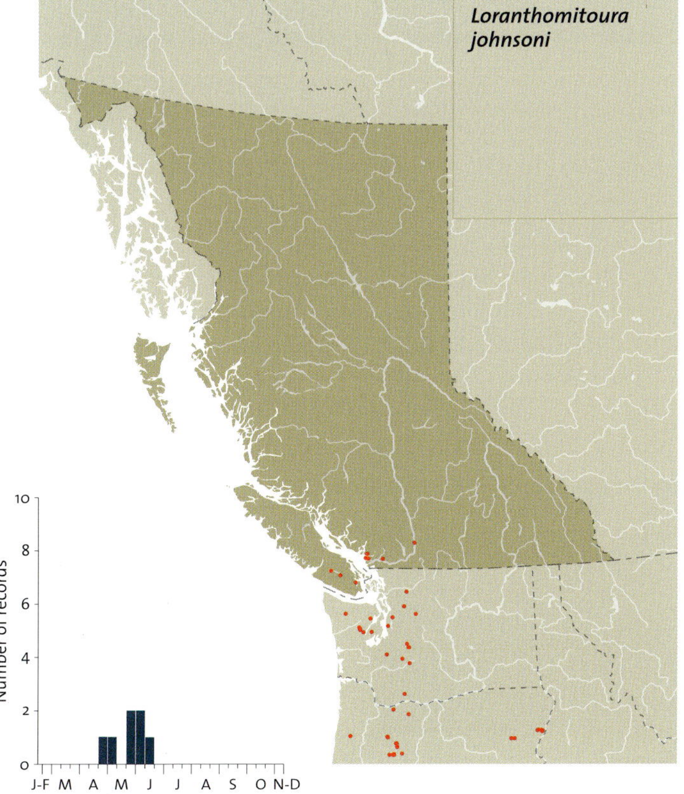

**Mature larva**

**Pupa**

The name *Mitoura* is derived from the Latin *mitos* (thread) and *oura* (tail); hence it refers to the threadlike tails. The common name for the genus is used here for the first time.

Adults of BC species in this genus are similar to *Loranthomitoura* but lack the strong, white median line on the ventral hindwing. Larvae of this genus feed on trees or shrubs related to western red cedar and juniper (Cupressaceae). For this reason adults are seen only when they are nectaring on flowers of perennial or annual plants at ground level or on low shrubs. They are not seen flying in open meadows. The genus is Nearctic, with two to nine species, depending on which authority one follows.

## CEDAR HAIRSTREAK

*Mitoura rosneri* Johnson, 1976

**ETYMOLOGY:** The species *rosneri* was named after Renate Rosner (Johnson 1976). The subspecies name *plicataria* is derived from the species name of western red cedar, *Thuja plicata*, with the addition of the Latin ending *aria* (around the) (Johnson 1976), thus "around western red cedar." The common name "Cedar Hairstreak," used here for the first time to contrast with the "Juniper Hairstreak," refers to the larval foodplant.

**ADULT:** The Cedar Hairstreak and the Juniper Hairstreak are two very closely related species that some recent authorities have combined into one species. Because these were combined with no discussion of the distinguishing characters pointed out by Johnson (1976), and because in the interior the two choose different native hosts and have consistently different wing colour patterns, we separate

Ssp. *rosneri* ♂ D  (2.5 CM)

Ssp. *rosneri* ♀ D  (2.6 CM)

Ssp. *rosneri* ♂ V  (2.5 CM)

Ssp. *plicataria* ♂ V  (2.6 CM)

them. Only protein or DNA analysis is likely to give a better definition of species in this difficult genus.

These two species are similar in appearance to the two *Loranthomitoura* species discussed previously, but are smaller. The ventral wing pattern is similar to that of *Loranthomitoura* but appears washed out. The upperside of the male wings has a dark brown ground colour with submarginal tawny areas. The upperside of the female wings

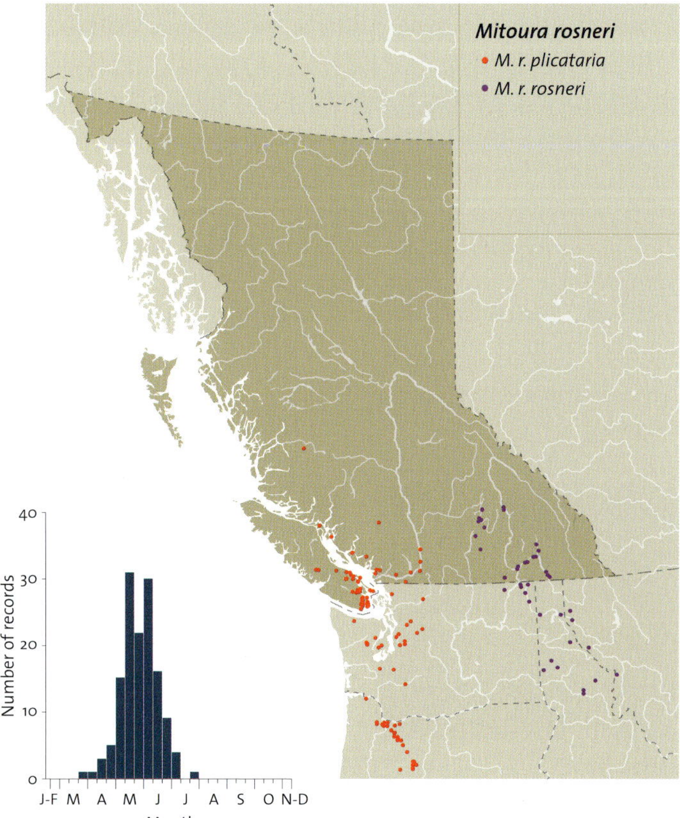

**Mature larva**

**Pupa**

Cedar Hairstreak (*Mitoura rosneri*)

dependent. On Vancouver Island, the Cedar Hairstreak can be seen nectaring on English daisies, oxeye daisies, and native composites. Eggs hatch shortly after being laid and quickly complete development. By midsummer the mature larva has pupated. The butterfly overwinters as a pupa. Western red cedar is the main native foodplant (Guppy 1954; Johnson 1976; FIS), and CSG has reared both mainland and Vancouver Island populations on both western red cedar and Rocky Mountain juniper. Shepard has noted oviposition on horticultural juniper shrubs at Nelson.

**SUBSPECIES:** Populations west of the Cascade and Coast ranges are *M. r. plicataria* Johnson, 1976 (TL: Cameron Lake, BC). The nominate subspecies, *M. r. rosneri* (TL: 2 miles south of Kaslo, BC) (= *byrnei* Johnson, 1976; TL: 5.6 mi. S. Emida, Benewah Co., ID), is found in the interior. The two subspecies range south to central OR and northern ID, respectively.

**RANGE AND HABITAT:** The Cedar Hairstreak is found across southern BC from Vancouver Island to the West Kootenay wherever western red cedar is found. It should also occur on the north coast and the southern Alaska panhandle, but has not been recorded there.

**GENERAL DISTRIBUTION:** The Cedar Hairstreak is found from southern BC south to OR, the ID panhandle, and northwest MT, wherever western red cedar is found.

**CONSERVATION STATUS:** Not of concern, with both subspecies S4.

has a tawny ground colour with wide black borders. The Cedar Hairstreak ventral hindwing has less contrast between basal and apical areas than that of the Juniper Hairstreak.

**IMMATURE STAGES:** Mature larvae are green with a broad diagonal white stripe on the side of each segment.

**BIOLOGY:** Adults of the Cedar Hairstreak fly from late April to early June on the coast and mid-May to late June in the interior. Lab-reared individuals emerge as early as February (FIS), which shows that emergence is temperature-

## JUNIPER HAIRSTREAK
*Mitoura siva* (W.H. Edwards, 1874)

Juniper Hairstreak (*Mitoura siva barryi*)

♂ D  (2.3 CM)

♀ D  (2.6 CM)

♂ V  (2.3 CM)

*barryi* was named after Aidan Barry (Johnson 1976). The common name "Juniper Hairstreak" (Comstock 1927) refers to the larval foodplant, junipers.

**ADULT:** The Juniper Hairstreak is identical to the Cedar Hairstreak except for the ventral hindwing. The whole pattern of the underside is crisper and the ground colour has a pink tone not found on the Cedar Hairstreak.

**ETYMOLOGY:** The species name *siva* likely refers to the Sivas area of Turkey. In the original description, *siva* is extensively compared to *castalis,* named after a fountain at the base of Mount Parnassius. Thus *siva* and *castalis* are similar but separate, like the Greek peninsula and Persia. Subspecies

 Mature larvae are green with a broad diagonal white stripe on the side of each segment.

**BIOLOGY:** Juniper Hairstreak adults fly from early May to late June. The eggs hatch soon after being laid and the mature larva pupates by midsummer. The butterfly overwinters as a pupa. The larvae feed on Rocky Mountain juniper (*Juniperus*

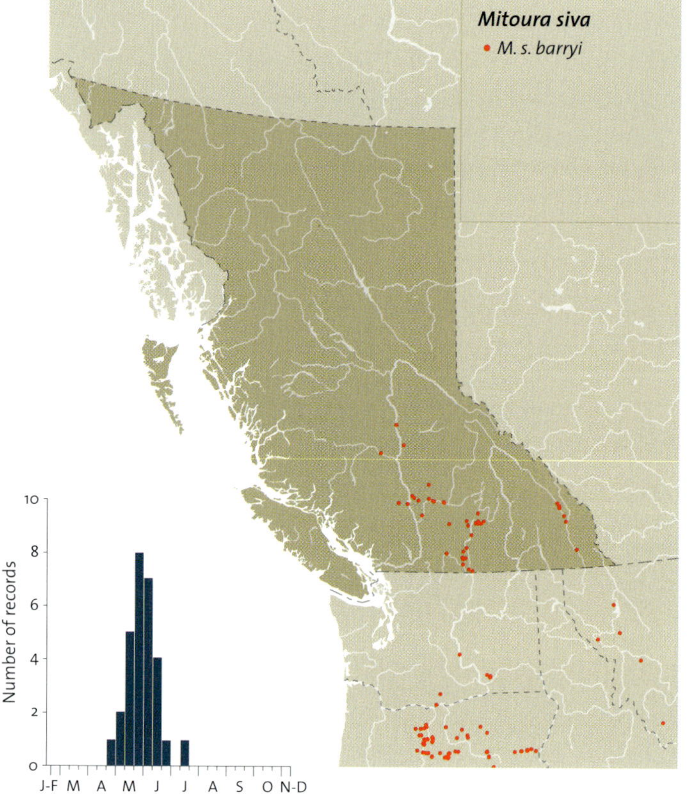

Mature larva

Pupa

*scopulorum*) in the wild (CSG; FIS), but CSG has also reared them in captivity on western red cedar.

**SUBSPECIES:** BC populations are the subspecies *M. s. barryi* Johnson, 1976 (TL: Union Co., OR), which ranges south through eastern WA to eastern OR.

**RANGE AND HABITAT:** The Juniper Hairstreak occurs in two disjunct sets of populations, one from the southeastern Chilcotin southeast to the Okanagan Valley and the other near Windermere, in the bottom of the Rocky Mountain Trench. It has always been found in association with Rocky Mountain juniper. Since this tree is more widespread than the known records for the Juniper Hairstreak, the butterfly may eventually be found between the two known groups of populations. If the Rocky Mountain junipers that the larvae depend on for food are destroyed to improve grazing, the species will be threatened.

**GENERAL DISTRIBUTION:** The Juniper Hairstreak is found from southwestern BC south to southern CA and northern MEX.

**CONSERVATION STATUS:** Not of concern (S4).

---

## GENUS *INCISALIA* SCUDDER, 1872  ELFINS

The name *Incisalia* is of unknown derivation. The common name "elfins" refers to their small size and flight habits that make them seem to magically appear and disappear. It was first used for the genus by Scudder (1875).

Species of the genus *Incisalia* lack tails on the hindwing and green colouring on the ventral hindwings. The tips of the valves are "capped," meaning that they have a terminal thickening not found in the genera *Callophrys*, *Mitoura*, or *Loranthomitoura*. The cornuti are neither slender nor spatulate. This is a Nearctic genus, with nine species. Six species occur in BC.

## BROWN ELFIN
*Incisalia augustinus* (Westwood, [1852])

**ETYMOLOGY:** The species name *augustinus* is a replacement name for *augustus* Kirby, and is derived from it. Kirby named *augustus* after Augustus, one of the Inuit with Sir John Franklin's expedition (Kirby 1837). Kirby (1837) used the common name "Augustus Thecla," which was not adopted by others. The common name "Brown Elfin" (Scudder 1875) refers to the plain brown dorsal and ventral wing colours.

**ADULT:** The Brown Elfin and all other elfins lack tails. On the upperside of the wings, BC elfins are very similar. The underside of the wings provides the only good species distinguishing characteristics. The underside of the Brown Elfin is a fairly uniform chocolate brown with typical elfin median and postmedian pattern. Since the publication of *The Butterflies of Canada*, we have located more populations of this species in central BC, south of Pine Pass to Clinton. There do not seem to be any intermediate specimens between the populations of the Brown Elfin and the Western Elfin. The two are allopatric, with closely adjacent populations, and appear to be another western/boreal species pair. Thus we re-elevate the Western Elfin, *I. iroides*, to species status. The Brown Elfin underside ground colour is chocolate brown, while that of the Western Elfin is reddish brown.

**IMMATURE STAGES:** Cook (1906) described the immature stages from New York. The egg is laid at the base of or in the flower buds. It is green when first laid but quickly changes

♂ D (2.3 cm)      ♀ D (2.6 cm)

♂ V (2.3 cm)

colour as the larva develops. The egg hatches within five days. The larva feeds on the flower and developing fruit. The entire development from egg to pupa takes one month, from mid-May to mid-June. Pupation occurs on the ground among dead plant material. The pupa remains dormant until the following spring.

**BIOLOGY:** The Brown Elfin flies from late April to mid-June. Cook (1906) demonstrated that Brown Elfin larvae feed on *Vaccinium* sp. Ziegler (1953) reared the Brown Elfin on *Arctostaphylos uva-ursi. Arctostaphylos* is the most likely BC foodplant.

**SUBSPECIES:** BC populations are the nominate subspecies (TL: 54° lat. [near Cumberland House, MB]).

**RANGE AND HABITAT:** The Brown Elfin is known from the Central Interior of BC. Since little collecting has been done in northern BC during the species flight period in early spring, the species will likely be recorded across central and northern BC, east of the coastal mountains.

**GENERAL DISTRIBUTION:** The Brown Elfin ranges from YT and central BC east to NF. In the east it is found south to the southern shores of the Great Lakes and in the Appalachians south to GA.

**CONSERVATION STATUS:** Not of concern (S4).

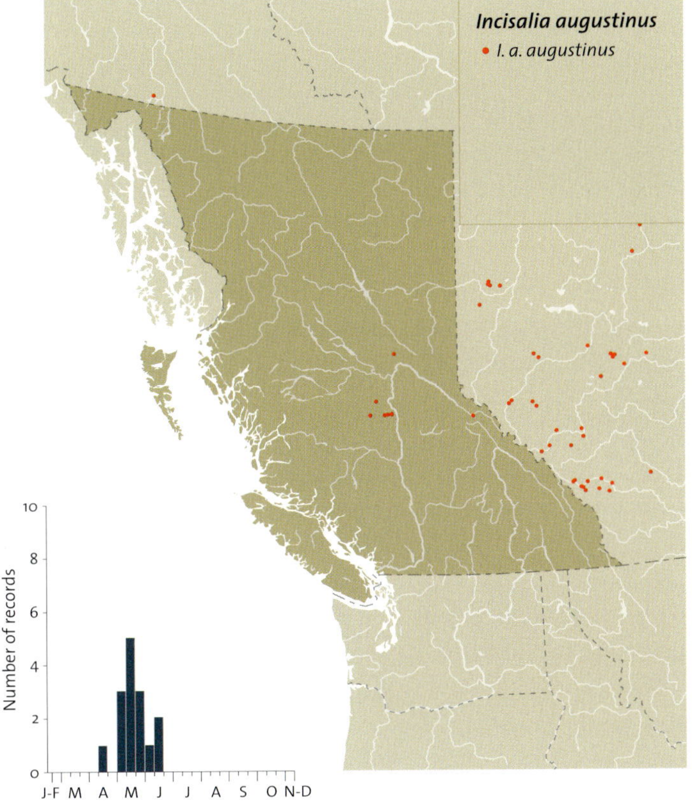



# WESTERN ELFIN
*Incisalia iroides* (Boisduval, 1852)

**ETYMOLOGY:** In his original description of *iroides*, Boisduval compared it to *irus,* thus the species name (like *irus*). Irus, the blind beggar of Homeric legends, was celebrated for his voracity. The common name "Western Elfin" (Comstock 1927) refers to the extreme western distribution of the species.

**ADULT:** The ground colour of the ventral hindwing of Western Elfins is a uniform reddish brown. By contrast, the Brown Elfin is chocolate brown.

**IMMATURE STAGES:** Comstock and Dammers (1933) described the egg as having a rich jade green colour. Hardy (GAH)

**Western Elfin (*Incisalia iroides*)**

♂ D (2.5 cm)

♀ D (2.7 cm)

♂ V (2.5 cm)

observed the eggs to be pale green, and described the mature larva as follows: head dark brown; body greenish yellow; some individuals with stripes edged dorsally with white; some with spiracular fold white. He found that larvae fed on flowers had a different ground colour from those fed on leaves.

**BIOLOGY:** The Western Elfin flies from mid-April to early June. The eggs are laid in the second half of April and hatch quickly; mature larvae pupate by early June on the coast (GAH). Hardy obtained oviposition in the lab on *Arctostaphylos uva-ursi,* and the hatched larvae fed on the surface of the leaves. Hardy also found larvae on salal, *Gaultheria shallon,*

and *Arbutus menziesii.* On Vancouver Island, oviposition was observed on a salal flower and the larva reared to maturity by A.G. Guppy (pers. comm.). The pupa stays dormant and is the overwintering stage. On the coast, the Western Elfin is usually associated with salal (Guppy 1956; JHS) and arbutus (*Arbutus menziesii*) (FIS). Jones (1938) records it on ocean-spray (*Holodiscus discolor*). In the interior the Forest Insect Survey has reared it on *Ceanothus sanguineus* at Adams River. It has also been reared on apple at Victoria (Jones 1939) and at Salmon Arm by Dennys (CNC). The interior populations appear closest to California populations, which also feed on *Ceanothus* sp. (Powell 1968).

**SUBSPECIES:** BC populations are the nominate subspecies, *I. i. iroides* (Boisduval, 1852); TL: Hwy. 70 at Soda Creek, Plumas Co., CA (Emmel et al. 1998a).

**RANGE AND HABITAT:** The Western Elfin is found across southern BC in open forests where *Arbutus menziesii, Gaultheria shallon,* or *Ceanothus* grow.

**GENERAL DISTRIBUTION:** The Western Elfin ranges from southern BC south to Baja California and northern MEX.

**CONSERVATION STATUS:** Not of concern (S5).

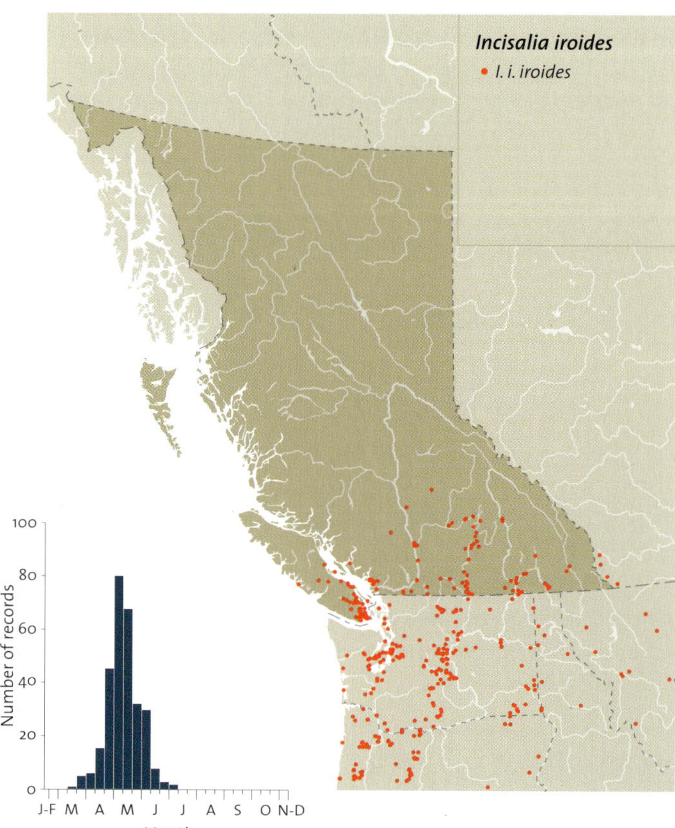

Egg

Mature larva

Mature larva



FAMILY LYCAENIDAE (GOSSAMER WINGS)

# Moss' Elfin
*Incisalia mossii* (Hy. Edwards, 1881)

**Moss' Elfin (*Incisalia mossii mossii*)**

Ssp. *mossii* ♂ D  (2.5 cm)

Ssp. *mossii* ♀ D  (2.6 cm)

Ssp. *mossii* ♂ V  (2.5 cm)

Ssp. *schryveri* ♂ V  (2.5 cm)

**Etymology:** The species *mossii* is named for a Dr. Moss of Esquimalt, BC. The subspecies *schryveri* was named after C.D. Schryver, a Denver lepidopterist. The common name was first used by Holland (1931).

**Adult:** For Moss' Elfin the ventral hindwing has a dark brown basal half and a grey postmedian area. The grey area is more contrasting for interior populations and obscure for Vancouver Island populations.

**Immature stages:** Hardy (1957) described the entire life history. The egg is a pale pastel green, and is 0.75 mm × 0.33 mm in size. The raised edges of the sculptured surface are covered with hyaline margins, giving the egg a hoary appearance. Mature fourth instar larvae are 16–17 mm long. The body is greenish yellow. Some individuals have white-edged dorsal stripes and/or a white spiracular fold. The pupa is 11 mm × 17 mm in size. It is pale yellow at first but soon changes to dark chocolate brown with a pale dorsal line flanked by a double row of small fuscous dots on each side. Spiracles are white.

**Biology:** Moss' Elfin flies from mid-April to late May. Early emergence depends on a warm, early spring. Eggs are usually laid at the base of the flower of the larval foodplant, *Sedum spathulifolium* (Hardy 1957). *S. lanceolatum* is probably used in the interior. Eggs hatch within days and the first instar larvae begin feeding on the flower bud. At the end of the second instar, the larva has consumed or left the flower head. The entire developmental period from egg to pupa takes no more than six weeks. The larvae crawl off the foodplant and pupate on the ground among plant debris. The pupae stay dormant until the following spring, when the next generation of adults emerge.

The number of Vancouver Island populations has been greatly reduced by the development of the Victoria region. In addition, the greatly increased deer population grazes the larval foodplant heavily (CSG). There are still viable populations on some bare

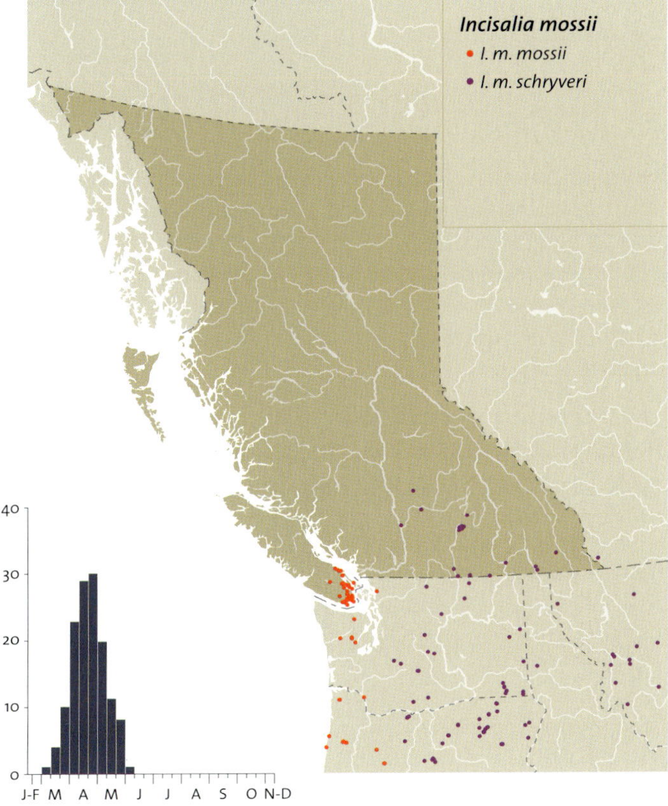

**Mature larva, ssp. *schryveri***

rock faces where the larval foodplant, *S. spathulifolium*, grows abundantly, but this habitat could be seriously damaged by climbers and hikers, such as along the trails on Mt. Finlayson.

**SUBSPECIES:** Vancouver Island populations are the nominate subspecies, *I. m. mossii* (Hy. Edwards, 1881) (TL: Esquimalt, BC). Interior populations are the Rocky Mountain subspecies, *I. m. schryveri* Cross, 1937 (TL: Chimney Gulch, CO).

## HOARY ELFIN
*Incisalia polia* (Cook & Watson, 1907)

**ETYMOLOGY:** The name *polia* is derived from the Greek *polios* (grey), and refers to the grey ventral wing colour. This grey or hoary colour is also reflected in the common name (Scudder 1875).

**ADULT:** The Hoary Elfin's ventral hindwings have a post-median and submarginal area of continuous grey colour. In addition, this is the only BC elfin where the uppersides of the wings is similar for the two sexes (dark brown in both).

**IMMATURE STAGES:** Cook (1908) partially described the immatures from New York. The egg is flat, with the micropyle in a deep pit. The surface of the egg is also deeply pitted.

**BIOLOGY:** The Hoary Elfin flies from mid-April to early June. Eggs are laid at the base of the flower buds of the larval foodplant, *Arctostaphylos uva-ursi* (Cook 1908). In BC the Hoary Elfin is always found in association with this plant.

**RANGE AND HABITAT:** Moss' Elfin is found on southeastern Vancouver Island and in the Southern Interior and West Kootenay. It is always found on dry, rocky or scree slopes where the larval foodplant, *Sedum* sp., grows.

**GENERAL DISTRIBUTION:** Moss' Elfin is found from southern BC south to southern CA and CO.

**CONSERVATION STATUS:** Subspecies *mossii* is of Special Concern in BC (S3). Subspecies *schryveri* is not of concern (S4).

♂ D (2.4 CM)

♀ V (2.7 CM)

♂ V (2.4 CM)

The eggs hatch quickly and the mature larvae pupate within six weeks. The pupae remain dormant until the following spring.

**SUBSPECIES:** The Rocky Mountain populations were differentiated as the subspecies *I. p. obscura* Ferris and Fisher, 1973, but we cannot separate BC material from the nominate subspecies (TL: Lakehurst, NJ).

**RANGE AND HABITAT:** The Hoary Elfin has been recorded at Atlin and in the Peace River region and in the Southern Interior and the Kootenays. It undoubtedly occurs throughout most of the province east of the Coast Ranges, but it emerges very early in the spring and has been overlooked. It occurs in mature pine stands where the larval foodplant, *Arctostaphylos uva-ursi*, grows.

**GENERAL DISTRIBUTION:** The Hoary Elfin ranges from central AK to NS. In the Rocky Mountains it is found south to CO and in the east south to the Great Lakes and VA in the Appalachians. Populations on the coast of northern CA, OR, and WA are disjunct.

**CONSERVATION STATUS:** Not of concern (S5).

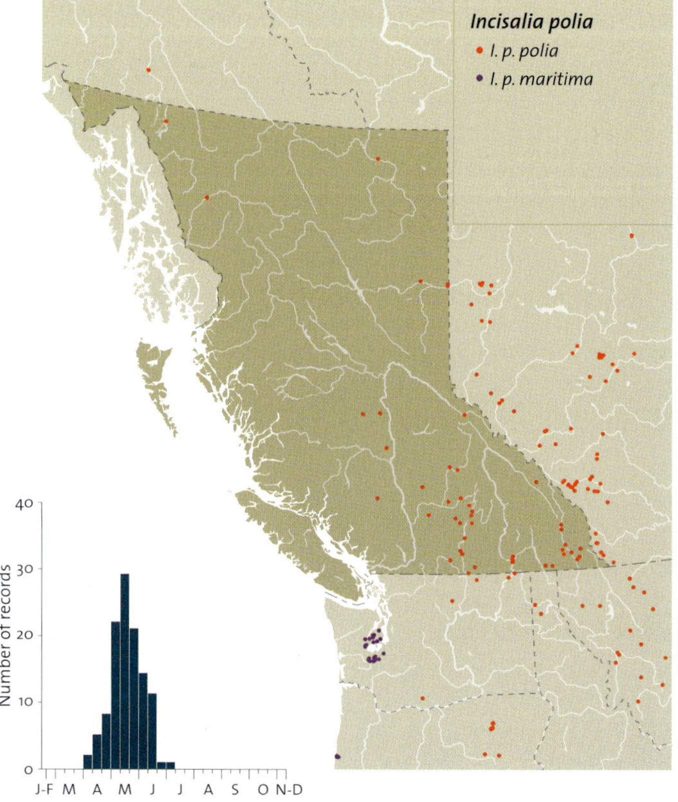



**Mature larva, ssp. *maritima***

# EASTERN PINE ELFIN
*Incisalia niphon* (Hübner, [1823])

**ETYMOLOGY:** The species name *niphon* is derived from the Greek *niphas* (snow or white), presumably in reference to the pattern of white lines on the ventral hindwings that was distinctive for the species when it was described. The subspecies name *clarki* honours Austin H. Clark, Curator of Marine Invertebrates, Smithsonian Institution, who also worked on butterflies. The common name (Pyle 1981) denotes the primarily eastern distribution of this pine-feeding elfin. This and the Western Pine Elfin are rare examples where an established common name of a well-known butterfly has changed. From Scudder (1875) to Holland (1931) these two elfins were called "banded elfins." Klots (1951) began the transition by calling the eastern species the Pine Elfin.

**ADULT:** The Eastern Pine Elfin has only recently been recorded for BC. It is similar to the Western Pine Elfin but differs by a character of the cell on the ventral forewing. Eastern Pine Elfins have two black bars in the cell, in contrast to the one black bar in Western Pine Elfins.

**IMMATURE STAGES:** Scudder (1889b) described the immatures. The pale green egg is flattened in profile and covered with

♂  D  (2.6 CM)

♀  D  (2.7 CM)

♀  V  (2.7 CM)

blunt conical tubercles. The mature larva has a yellowish head and pale green body with four longitudinal white stripes.

**BIOLOGY:** The only BC individuals were found in mid-June of 1999. Since the spring was very late, however, its normal flight period is likely early to late May. Elsewhere it feeds primarily on jack pine but has been reared on other pines (Layberry et al. 1998). In BC it is found in a stand of jack pine (CSG). Eggs are laid in May and hatch within 10 days. The larvae mature quickly, and pupation takes place by early July. The pupae remain dormant until the following spring, when the adults emerge (Scudder 1889b).

**SUBSPECIES:** BC populations are the Canadian subspecies *I. n. clarki* T.N. Freeman, 1938 (TL: Constance Bay, ON).

**RANGE AND HABITAT:** The species was just found in BC in 1999. Guppy collected it along the Liard Highway in a jack pine forest.

**GENERAL DISTRIBUTION:** The Eastern Pine Elfin is found from northeastern BC and southeastern NT east to NS. In the east it occurs south to FL.

**CONSERVATION STATUS:** The Eastern Pine Elfin is of Special Concern in BC.

Mature larva

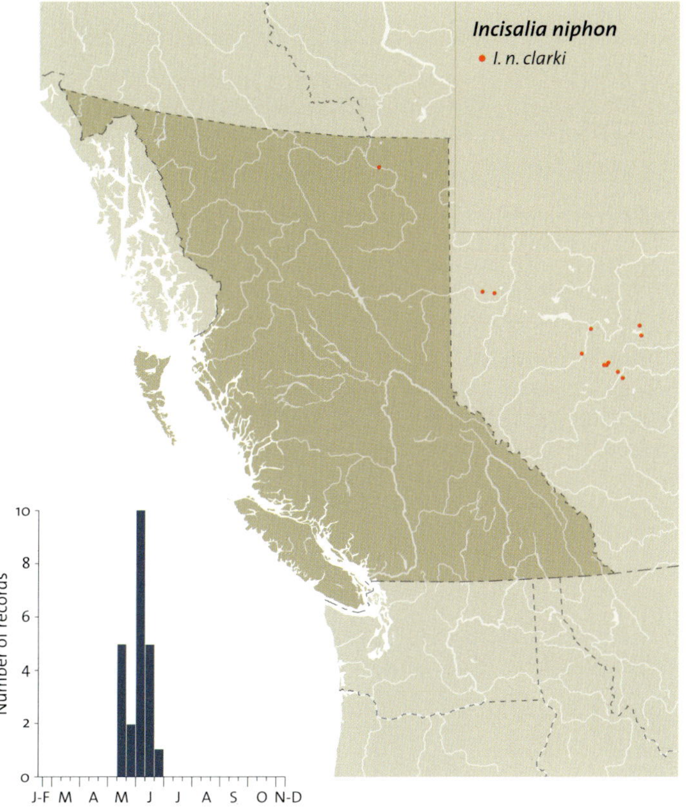

# WESTERN PINE ELFIN

*Incisalia eryphon* (Boisduval, 1852)

**ETYMOLOGY:** The derivation of the name *eryphon* is unknown. Perhaps it was coined because it rhymes with the species name of its closest relative, *niphon*. The common name (Pyle 1974) denotes the primarily western distribution of this pine-feeding elfin.

**ADULT:** The Western Pine Elfin is slightly larger than other elfin species. The pattern of the ventral hindwing is very distinctive. The submarginal area is a row of connected black chevrons accented by a contiguous lateral row of tawny chevrons.

**IMMATURE STAGES:** Hardy (1959a) described the BC immatures. The egg is 0.75 mm × 0.50 mm, and thus not as flat as the egg of Moss' Elfin. The mature larva has a brown head and rich velvet green body with two cream white stripes on each side. The body is overlaid with fine brown hairs.

**BIOLOGY:** Most individuals are seen flying from early May to early June. In any one season, the Western Pine Elfin emerges later than other *Incisalia* species. Adults take nectar from *Salix prolixa*. Eggs laid on 24 May at the base of needles of the larval foodplant, *Pinus contorta,* hatched on 1 June (Hardy 1959a). Young larvae fed on the base of the pine needles, eating through the needle, which dropped off. Later instars fed on the surface of needles. By 20 July the larvae were mature and pupated in the lab among debris at the bottom of the breeding cage. The adults emerged on 26–28 March

Ssp. *eryphon* ♂ D (2.7 CM)

Ssp. *eryphon* ♀ D (2.6 CM)

Ssp. *eryphon* ♂ V (2.7 CM)

Ssp. *sheltonensis* ♂ V (2.7 CM)

the following year. The Forest Insect Survey has also reared the Western Pine Elfin from *P. ponderosa* and *P. monticola*.

**SUBSPECIES:** Most BC populations are the nominate subspecies, *I. e. eryphon* (Boisduval, 1852); TL: Hwy. 70 at Soda Creek, Plumas Co., CA (Emmel et al. 1998a). Vancouver Island and Lower Fraser Valley populations are *I. e. sheltonensis* Chermock & Frechin, 1948 (TL: Shelton [near Olympia], WA).

**RANGE AND HABITAT:** The Western Pine Elfin is found from the Nass River and Fort Nelson, south throughout BC in mature pine stands.

**GENERAL DISTRIBUTION:** The Western Pine Elfin occurs from central BC south to southern CA and NM. Since Klots (1951) did not recognize this species in eastern North America, it may have invaded east to NB and New England in historic times.

**CONSERVATION STATUS:** Not of concern; subspecies *eryphon* is S5 and subspecies *sheltonensis* is S4.

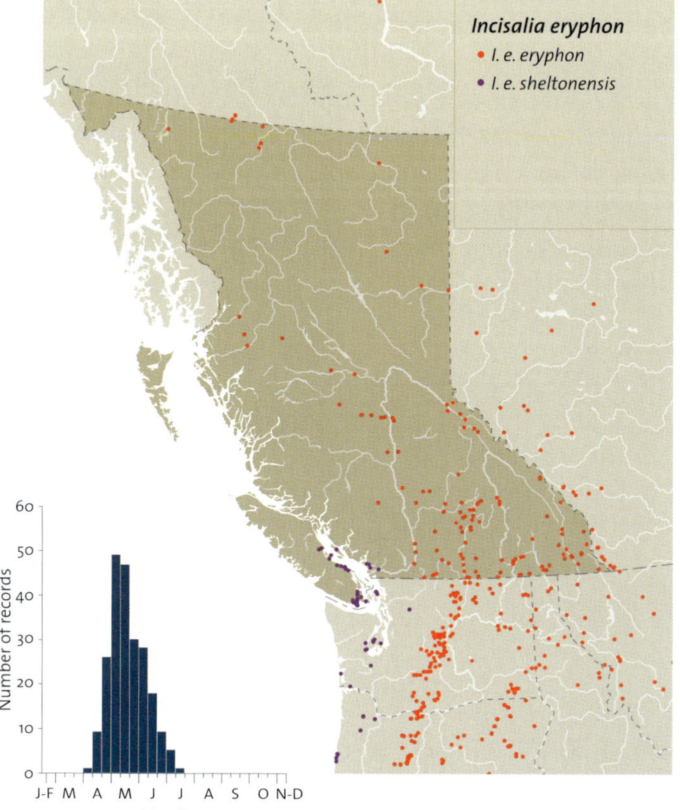

**Western Pine Elfin**
(*Incisalia eryphon eryphon*)

**Mature larva, ssp. *sheltonensis***

The name *Strymon* is derived from the Strymon River at the boundary of Macedonia and Thrace (Emmet 1991). The common name for the genus was recently coined by Opler (1998).

The genus *Strymon* differs from all other BC Theclinae in the structure of the male genitalia. The tip of the aedeagus is not flared or bent as in other genera in BC.

Our single species has tails and obscure scent patches on the male dorsal forewing, but other species in the genus may lack either tails or scent patches. This is a large genus, found primarily in the neotropics. There are 12 species in the southern USA with one ranging north to Canada. The genus uses a large variety of larval foodplants.

## GREY HAIRSTREAK
*Strymon melinus* Hübner, [1818]

**ETYMOLOGY:** The species name *melinus* is derived from the Greek *melinos* (grey), in reference to the grey wing colour. The subspecies *setonia* was named after the type locality, Seton Lake, BC. The subspecies name *atrofasciatus* (dark face) refers to the dark ground colour and heavy black spotting on the underside of the wings. The common name was first used by Scudder (1875).

**ADULT:** The Grey Hairstreak is the only tailed hairstreak found in BC that has a grey ground colour on the upperside of the wings. The underside of the wings is also distinctive. The spring generation has a grey ground colour and the summer generation has a chalky white ground colour. Both broods have a line of contrasting black postmedian spots on the underside of both wings.

SSP. *atrofasciatus* ♂ D  (2.4 CM)

SSP. *atrofasciatus* ♂ V  (2.4 CM)

SSP. *setonia* ♂ V  (2.7 CM)

**IMMATURE STAGES:** Because the Grey Hairstreak is a pest of beans, it has often been reared. Scudder (1889b) gives a good description of the immatures. The egg is 0.60 mm wide by 0.36 mm high, and is pea green. The mature larva is a dingy, velvety brown with no apparent markings. It is covered by long, bristly hairs up to 5 mm long. The size of the mature larva is 6.8 mm long by 2.4 mm wide. The typical lycaenid pupa is 9.0 mm long.

**BIOLOGY:** The Grey Hairstreak flies in two broods, from late April to early June and from mid-July to late August. In the Okanagan Valley there is at least a partial third generation. In BC the overwintering stage is the pupa (JHS). Larvae feed on a variety of native herbaceous growth, legume genera, and garden beans. Jones (1939) recorded it on fruits of

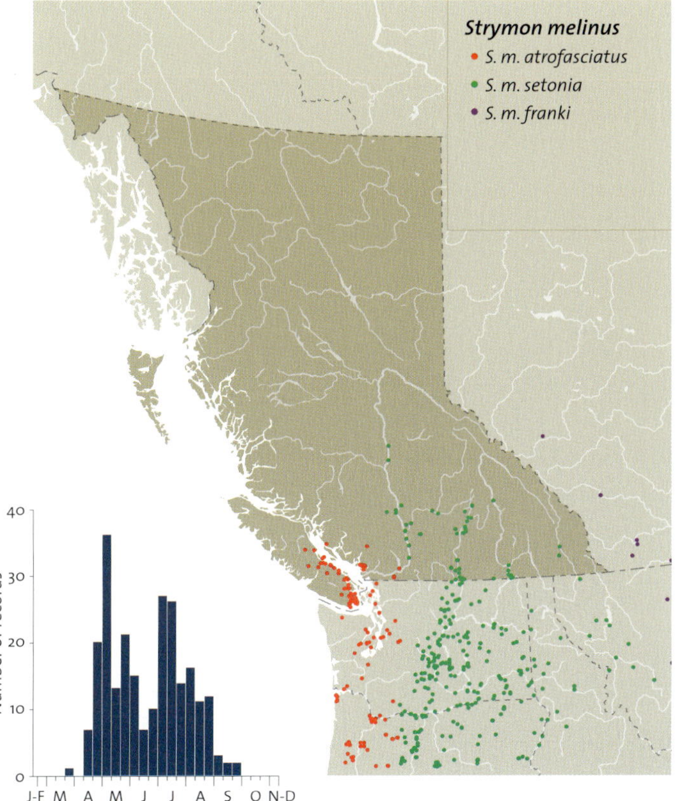

**Mature larva**

**Pupa**

**Grey Hairstreak (*Strymon melinus atrofasciatus*)**

raspberry. Guppy (1954) reared spring brood adults from Vancouver Island larvae on the native clovers *Trifolium oliganthum* and *T. willdenowii*. Guppy (1959) again reared spring brood adults but on a different native plant, the wild strawberry, *Fragaria vesca* ssp. *bracteata*. Guppy (1959) also reared second brood adults from *Anaphalis margaritacea*. In all of Richard Guppy's rearings, the larvae fed on flowers. Oviposition has been observed on *Trifolium oliganthum* at Bamberton (CSG) and second-generation larvae have been found feeding on *Gaultheria shallon* berries at Metchosin (AGG). Shepard (JHS) has reared larvae from second-brood adults on green bean pods in the Nelson area. The young larvae fed on the developing beans inside the pod, and the mature larvae fed on the surface of the maturing bean pod (JHS).

SUBSPECIES: Coastal populations are the subspecies *S. m. atrofasciatus* McDunnough, 1921 (TL: Wellington [Nanaimo], BC). In the interior the subspecies is *S. m. setonia* McDunnough, 1927 (TL: Seton Lake, [Lillooet], BC).

RANGE AND HABITAT: The Grey Hairstreak is found on Vancouver Island and east across the southern part of BC. It is found in the understorey of mature pine or Douglas-fir stands and in riparian situations in association with legumes, which are the native larval foodplants. It can also be a pest of cultivated beans.

GENERAL DISTRIBUTION: The Grey Hairstreak is found from southern BC across the southern fringe of CAN to NS, and south throughout the entire USA and MEX.

CONSERVATION STATUS: Not of concern, with subspecies *atrofasciatus* S4 and *setonia* S5.

The subfamily name is derived from the genus *Polyommatus* (many-eyed), an epithet of Argus (Emmet 1991), the name of a species in the genus. The blues derive their common name from the males of most species, including all those found in BC, having the uppersides of the wings covered with blue reflecting scales. Females of many BC species also have some blue scales at the base of the uppersides of the wings, but their prevailing ground colour is brown.

Blues have androconial scales on the dorsal forewings of males. The eyes can be hairy or hairless (smooth). The male aedeagus is short and blunt. The valves are usually well developed. All species except those in the genus *Everes* lack tails on the hindwings. The larvae of our blues, except *Agriades,* have honey glands and eversible tubercles.

For the Holarctic region as a whole, the blues are the best developed and most diverse of the various subfamilies. In BC, however, this subfamily does not have as many species as the Theclinae.

## TRIBE EVERINI TUTT, 1908

### GENUS *EVERES* HÜBNER, [1819]  TAILED BLUES

The origin of the name *Everes* is unknown (Emmet 1991). The common name "tailed blues" refers to the adults having small tails on the hindwings, which is unusual in temperate species of blues. Scudder (1875) first used the name "tailed blue" for the Eastern Tailed Blue.

This is the only genus of blues in BC with tails on the hindwings. It is also the only genus of blues with a well-developed uncus in the male genitalia. The eyes are smooth and the palpi are porrect and twice as long as the head. The genus is Holarctic, with a total of five species, two of which are found in the Nearctic and BC. The larvae of all species in the genus feed on perennial Fabaceae.

There are two very similar species of tailed blues in BC. Where the two species distributions overlaps, the species can be differentiated only by examination of the male genitalia (Fig. 66).

## EASTERN TAILED BLUE
*Everes comyntas* (Godart, [1824])

**ETYMOLOGY:** The origin of the name *comyntas* is unknown. The name "Eastern Tailed Blue" (Holland 1898) reflects the species' predominantly eastern North American distribution.

**ADULT:** See the discussion under **Western Tailed Blue**. The Eastern and Western tailed blues can be reliably distinguished only by examination of male genitalia (Fig. 66).

**IMMATURE STAGES:** The immatures are very similar to those of the Western Tailed Blue (Scudder 1889b). Lawrence and Downey (1967) described the immature stages in extreme detail.

**BIOLOGY:** The few records of the Eastern Tailed Blue in BC indicate that there is one generation from mid-June to mid-July. Nothing is known of the life cycle of any of the western disjunct populations of this species, but the life cycle of eastern populations has been well described by various workers (Scudder 1889b). This species has the same habits as the Western Tailed Blue. The eastern larval foodplant,

♂ D (2.5 CM)

♀ D (2.2 CM)

♂ V (2.5 CM)

♀ V (2.2 CM)

*Lespedeza capitata,* is not found in BC, but Lawrence and Downey (1967) record white clover (*Trifolium repens*) and red clover (*T. pratense*), which do occur in BC. The ability to utilize these weedy clovers that have spread throughout the west may account for the recently discovered British Columbia populations.

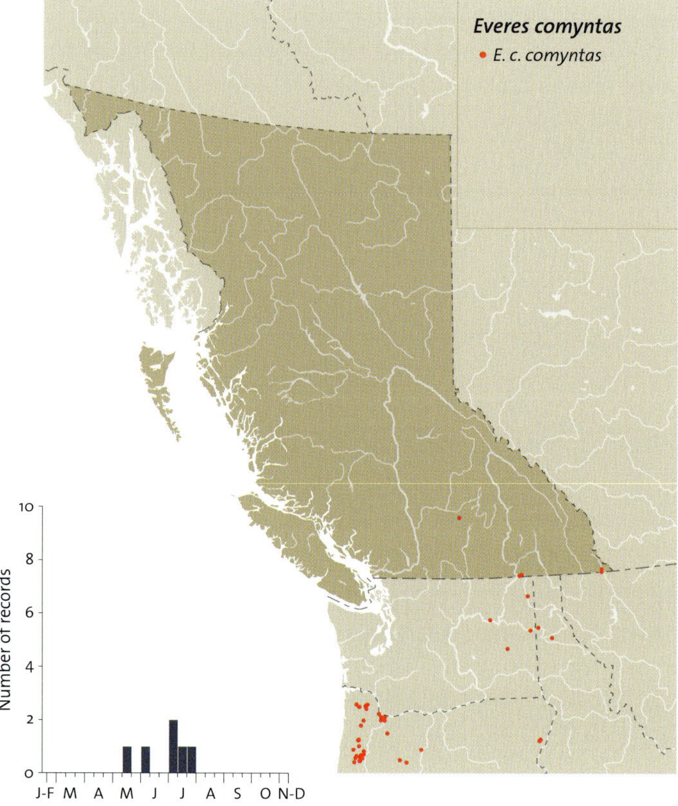

**Mature larva**

**SUBSPECIES:** BC populations are the nominate subspecies, *E. c. comyntas,* which occurs everywhere except the southern extreme of the species range (TL: "North America").

**RANGE AND HABITAT:** The Eastern Tailed Blue is known from only three populations in BC, one in the Flathead drainage, one near Vernon, and another near the mouth of the Pend-d'Oreille River, south of Trail. There are also scattered records in adjacent eastern Washington and the Idaho panhandle. Washington and Oregon authorities (pers. comm.) have speculated that the species was brought in by nursery stock. BC populations, however, are found in natural riparian situations with, in the case of the Flathead population, little or no human disturbance. This would indicate that the species is native to the Pacific Northwest. The isolated populations most likely represent remnants of a much larger preglacial or immediate postglacial distribution, similar to the distribution of the Dione Copper in the upper Columbia River drainage.

**GENERAL DISTRIBUTION:** The Eastern Tailed Blue is found as disjunct populations from the Willamette Valley, OR, south to central CA, the upper Columbia River drainage of WA, BC, and ID, and then from MB and CO east and south to TX and FL, with other populations in AZ/NM south into northern MEX.

**CONSERVATION STATUS:** The Eastern Tailed Blue is of Special Concern in BC (S3).

## WESTERN TAILED BLUE
*Everes amyntula* (Boisduval, 1852)

**ETYMOLOGY:** The origin of the name *amyntula* is obscure. It could be derived either from Amyntula, a dependent king of the Romans in Asia Minor, or from Amyntes, a dynastic name of the Macedonians. The common name "Western Tailed Blue" (Holland 1898) reflects the western North American distribution of the species.

**ADULT:** The Western and Eastern tailed blues are the only BC species of blues with tails. The Western Tailed Blue is slightly larger and has a somewhat whiter ground colour on the underside of the wings, but the two species can be reliably told apart only by dissection of the male genitalia (Fig. 66). Females can be identified only by association with males.

**IMMATURE STAGES:** Comstock and Dammers (1936) described the immature stages. The egg is green and typical of lycaenid eggs. The mature larva is grey green with white bristles along either side of the dorsal ridge. The body is covered with

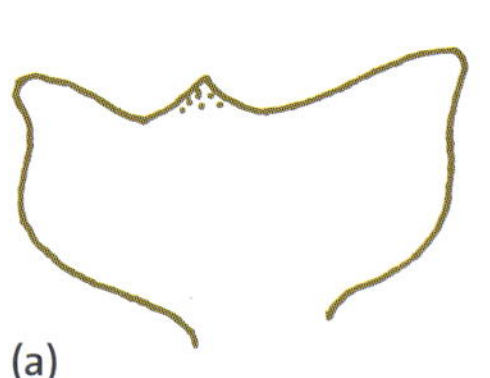
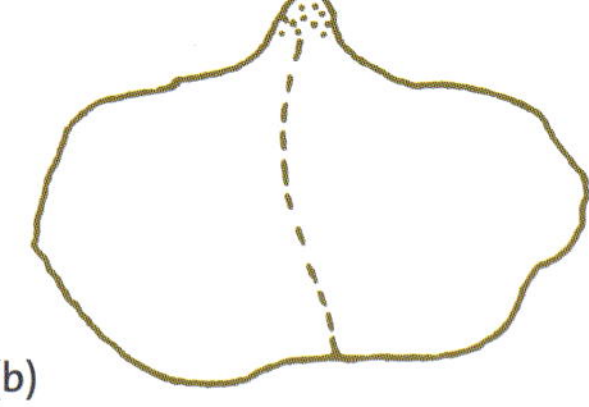



**66** Dorsal view of the uncus of *Everes* species: (a) *E. comyntas,* (b) *E. amyntula*

fine white hairs. Basally on each side there is a cream-coloured line. The pupa is yellow green with a white abdomen. By contrast, Hardy (GAH) found that Vancouver Island eggs were white. The larval body was pale green, with a dark green dorsal line and a pale cream spiracular line. The underside was a darker green. The hibernating larvae changed to a light brown.

Western Tailed Blue (*Everes amyntula*)

♂ D (2.6 cm)

♀ D (2.4 cm)

♂ V (2.6 cm)

♀ V (2.4 cm)

**BIOLOGY:** The Western Tailed Blue flies from late April to mid-June in the south and into July in the north; there is one generation per year. Hardy (GAH) observed oviposition on *Vicia americana* in late April. Hardy also found eggs on a stem of *Lathyrus nevadensis* var. *nuttallii,* near the flowers. The eggs hatched on 10 May and mature larvae quit feeding by 8 June. The larvae remained dormant through 12 October. Apparently Hardy did not get the larvae through the winter.

Outside BC, others have observed eggs laid at the base of developing pods of the larval foodplants, including *Astragalus* and *Lathyrus*. The newly hatched larva bores into a pod, seals the opening, and completes larval development undisturbed. In California, where there are two generations per year, Wright (1884) found that the second-generation larvae go into hibernation. He assumed that they did not pupate until the following spring.

**SUBSPECIES:** BC populations are the nominate subspecies; TL: 2 mi. W. Quincy, Plumas Co., CA (Emmel et al. 1998a).

**RANGE AND HABITAT:** The Western Tailed Blue is found throughout BC in open meadows, in the understorey of mature, open forests, and in riparian situations wherever the larval foodplants occur.

**GENERAL DISTRIBUTION:** The Western Tailed Blue occurs from central AK south through BC to Baja California and NM, and east to the Great Lakes.

**CONSERVATION STATUS:** Not of concern (S5).

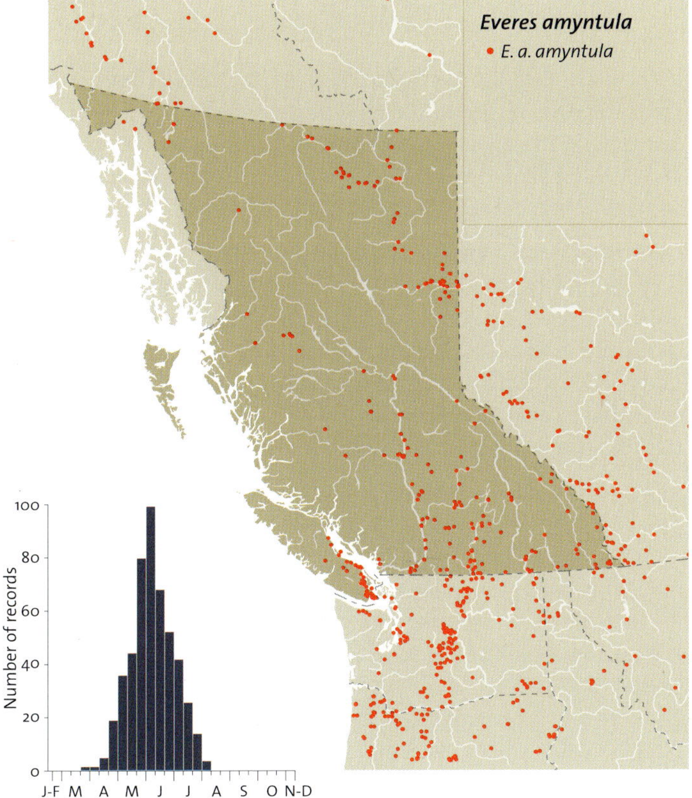

Mature larva

Pupa

## GENUS *CELASTRINA* TUTT, 1906  SPRING AZURES

At the time the genus *Celastrina* was named, the name *celastrina* was used for holly, the larval foodplant of the type species for this butterfly genus (Emmet 1991). The common name "spring azures" reflects the early spring flight period and the azure (blue) wing colour of males of this genus. The name was first used by Gosse (1840) for what is now subspecies *lucia* of the Boreal Spring Azure.

The genus *Celastrina* differs from all other blues in the structure of the male genitalia. The labides and vinculum are fused and developed into a broad plate extending to articulate with the valves. The eyes are hairy. Spring azures are the first blues to emerge from the pupa in the spring and are a sure sign that spring has arrived in the northern hemisphere. The genus is Holarctic, with one species across most of Europe and Russia and several in northeastern China and adjacent areas (Eliot and Kawazoe 1983). Until very recently all North American populations were regarded as one species, but Opler and Krizek (1984) recognize three, two of which are found in eastern deciduous forest habitat. Pratt et al. (1994) further defined overlapping subspecies and overlapping races of the three species but did not recognize any additional species. Several North American authors have split off the widely distributed Nearctic species *C. ladon* from the Palearctic species *C. argiolus,* but the only study to look at the genus as a whole treats *argiolus* and *ladon* as one Holarctic species (Eliot and Kawazoe 1983). Observations in BC show that the two "subspecies" of *ladon* occur together and must be regarded as separate species. The southern *C. echo* is closer in appearance to *C. argiolus* from the Palearctic. *C. ladon* is therefore kept as a separate Nearctic species. Recent unpublished DNA analysis of Pratt and Wright (in prep.) support this split into two species. Thus there are four species in North America and many more in the Palearctic.

## WESTERN SPRING AZURE
*Celastrina echo* (W.H. Edwards, 1864)

**Western Spring Azure (*Celastrina echo*)**

**ETYMOLOGY:** Edwards compared *echo* to the eastern species *pseudargiolus.* Thus the Western Spring Azure echoes its eastern counterpart. The common name, used here for the first time, refers to the western North American distribution and early spring flight period of the species.

**ADULT:** The upperside of the wings of the male Western Spring Azure are a uniform violet blue with a narrow white margin and no orange or dark areas on either wing. The underside of the wings is a dirty chalk white, with the pattern of dark spots varying from mostly inconspicuous to very contrasting. Every population has a variable pattern on the underside of the wings. The female is similar to the male except that the dorsal forewings have wide black margins and the dorsal hindwings have a row of black marginal spots. The blue is much reduced, but is violet blue.

**IMMATURE STAGES:** The literature on this species complex is very confusing, but apparently the immature stages of the Western Spring Azure have not been described except for last instar larvae (Ballmer and Pratt 1989). The ground colour of the mature larva can be white, pale pink, or pale green. Lateral lines or chevrons are absent, but there is a conspicuous transverse bar on the first abdominal segment.

**BIOLOGY:** The Western Spring Azure has two generations per year. The first flies from early April to early June and is abundant. The males congregate at mud puddles. There is a partial second generation in the warmer parts of southern BC, including southeastern Vancouver Island, near Hope, and the southern part of the Southern Interior. In California Comstock (1927) listed *Ceanothus* sp. as a larval foodplant. *Ceanothus* has been confirmed in the BC Southern Interior (FIS). On southeastern Vancouver Island, both Guppy (1954) and Hardy (GAH) have reared the Spring Azure on *Holodiscus discolor.* Hardy confined a female that had been walking on an inflorescence of *H. discolor.* It laid eggs on the inflorescence on 22–24 May; the eggs were a white, flattened sphere. The mature larvae had a black head with either an all green or purplish dorsal and green ventral body. Pupation occurred on 20–23 June. Regrettably Hardy did not record when the adults emerged.

Shepard observed adults of the Western Spring Azure in numbers along a long natural hedge of *Spiraea douglasii* at the back of sand dunes on Vancouver Island, but no oviposition was observed. This plant is found throughout the range of the species in BC and is likely a foodplant. In Colorado Scott (1992) found that adults with a wing pattern similar to that of BC populations oviposited primarily on *Jamesia americana,* a plant not found in our area. He also noted occasional oviposition on *Prunus* sp., and mentioned that that plant should be investigated for BC. Guppy (1954) found that the larvae fed exclusively on flowers. The larvae must pupate before winter since newly emerged adults are present in the earliest spring.

**SUBSPECIES:** Populations in BC are the nominate subspecies, *C. e. echo* (W.H. Edwards, 1864) (TL: San Francisco, CA) (= *nigrescens* Fletcher, 1903; TL: Kaslo, BC).

♂ D  (2.9 cm)

♂ V  (2.9 cm)

♀ D  (2.6 cm)

♀ V  (2.6 cm)

♀ summer form  D  (2.6 cm)

♀ summer form  V  (2.6 cm)

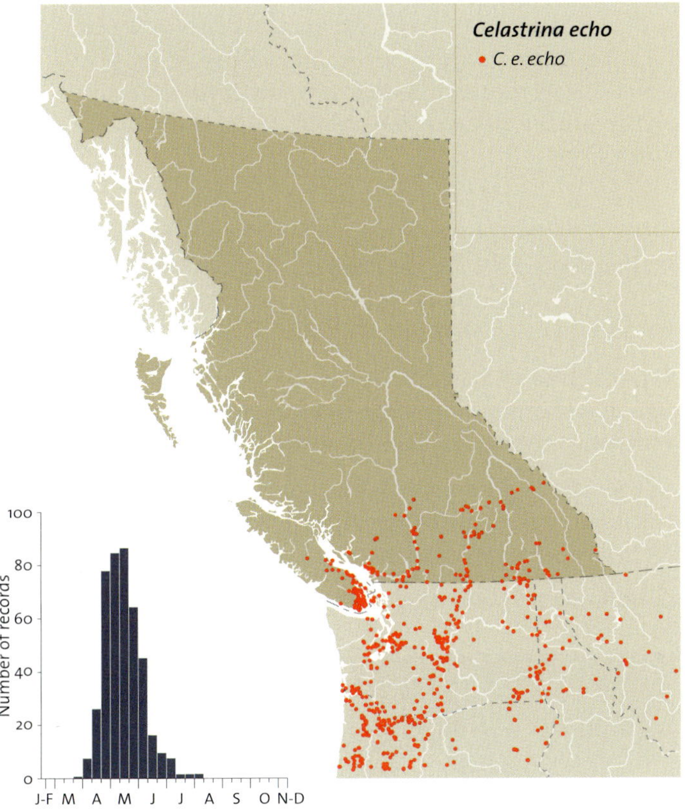



**RANGE AND HABITAT:** The Western Spring Azure is found throughout the southern fifth of BC, in riparian situations and along the edges of open meadows in most of southern BC. On southeastern Vancouver Island, it is also found associated with shrubs just at the back of stabilized sand dunes.

**GENERAL DISTRIBUTION:**
The Western Spring Azure is found from southern BC south to the tip of Baja California and south through the Rockies to northern MEX.

Mature larva

**CONSERVATION STATUS:**
Not of concern (S5).

# BOREAL SPRING AZURE
## *Celastrina ladon* (Cramer, [1780])

**ETYMOLOGY:** The name *ladon* is mostly likely derived from Ladon, a dragon who guarded the apple tree in classical Greek mythology. The subspecies name *lucia* is derived from the Latin *lucis* and more likely refers to the light blue colour of the upperside of the wings, not the underside as Bird et al. (1995) suggested. The common name refers to the northern boreal forest distribution and the early spring flight period of the species. It is used here for the first time, instead of the older name "Spring Azure," which has been applied to all four species that are now recognized for the genus in North America.

**ADULT:** The upperside of the wings of the male Boreal Spring Azure are a uniform sky blue. The blue on the upperside of the female wings is also sky blue. The overall pattern is very similar to that of the Western Spring Blue. The two species are separated because they occur together in two areas in the province, and we consider this sufficient criterion to separate them into two species. David Threatful found both species flying simultaneously near Chase, BC. Shepard has taken both species near Lytton, the Western Spring Azure at a lower elevation than the Boreal Spring Azure. Older literature says that the two taxa overlap broadly, but we have not found this to be true. The distribution of the two species follows that of other boreal/western species pairs that meet in BC. Some of this confusion in the older litera-ture results from synonymizing the name *quesnellii* (Cockle, 1910) with *nigrescens* (Fletcher, 1903). The population named *quesnellii* is the Boreal Spring Azure and the one named *nigrescens* is the Western Spring Azure.

**IMMATURE STAGES:** The immatures of the second brood in the southeastern USA were described by Edwards (1878) in some detail. The mature larva is typical of lycaenids and has the colour of the part of the plant that it is feeding on at the time, for example, green if feeding on leaves and white if feeding on flowers. When Edwards described the life cycle, he was the first in North America to notice that ants attended the larvae of the Lycaenidae.

♂ D  (2.7 CM)          ♀ D  (2.5 CM)

♂ V  (2.7 CM)          ♀ V  (2.5 CM)

**BIOLOGY:** The Boreal Spring Azure has one generation per year in BC. Adults fly from late April to early June. Nothing is known of the life history in BC. In boreal eastern North America, Pratt et al. (1994) report that *Vaccinium* and *Viburnum* are the usual larval foodplants. *Ledum, Prunus,* and *Cornus* are other eastern larval foodplants that may be utilized in BC.

**SUBSPECIES:** Our populations are the boreal forest subspecies *C. l. lucia* (W. Kirby, 1837) (TL: 54° lat. [near Cumberland House, MB]), and *quesnellii* (Cockle, 1910) (TL: Quesnel, BC) is a synonym. It ranges south to Quesnel, with isolated popula-tions south to near Lytton and Chase.

**RANGE AND HABITAT:** The Boreal Spring Azure is found throughout central and northern BC in riparian situations and along the edges of open meadows.

**GENERAL DISTRIBUTION:** The Boreal Spring Azure is Nearctic and is found across all of boreal North America from central AK to NF and south to northeastern TX and northern FL.

**CONSERVATION STATUS:** Not of concern (S5).

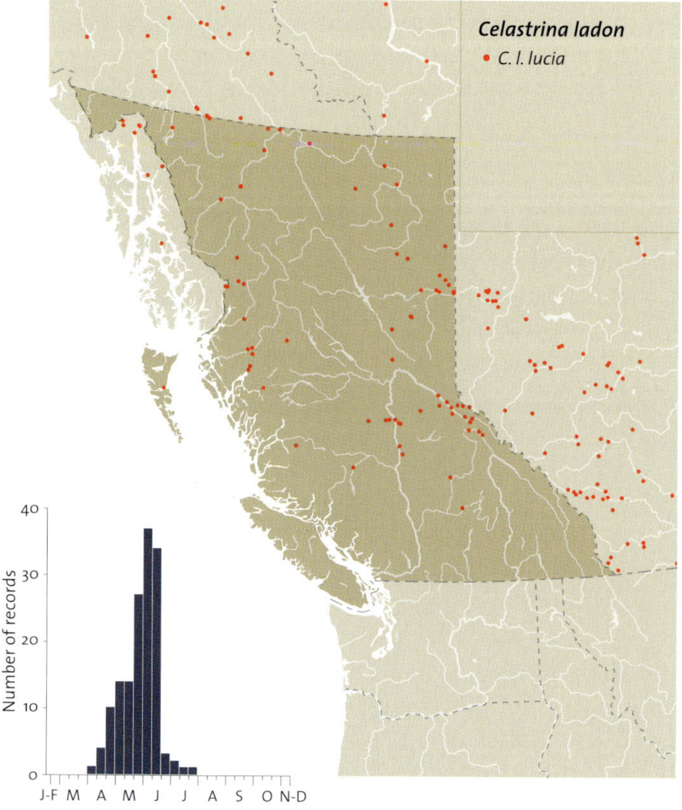

**Boreal Spring Azures (*Celastrina ladon lucia*) mating**

## Genus *Euphilotes* Mattoni, [1978]

The name *Euphilotes* is derived from the Greek prefix *eu* (true) and *Philotes,* the genus in which the butterflies were previously placed.

In this genus and the next, the eyes are smooth or with only a few hairs. Veins 7 and 8 of the forewing are joined at base. There are also genitalic differences. *Euphilotes* species are much smaller than *Glaucopsyche,* and feed on the plant genus *Eriogonum.* There are four Nearctic species in the genus *Euphilotes.*

Currently only one species is recorded for BC, *E. battoides.* A second species, *E. enoptes* (Boisduval, 1852), occurs very close to the BC border, in Okanogan County, WA. It is not possible to tell the two species apart by wing characters in our area. The critical part of the male genitalia is therefore shown in Fig. 67 to help identify the two species. There is also the remote possibility that a third species, *E. ancilla* (Barnes & McDunnough, 1918) with male genitalia identical to those of *E. enoptes* (Fig. 67a) will be found in extreme southeastern BC.

## Square-spotted Blue
*Euphilotes battoides* (Behr, 1867)

**Etymology:** The derivation of the name *battoides* is unknown. The subspecies name *glaucon* is derived from the Latin *glaucus* (blue grey), referring to the blue grey colour. The common name "Square-spotted Blue" refers to the square shape of the black spots on the ventral forewing. Holland (1931) accepted Comstock's (1927) use of Square-spotted Blue and dropped his earlier usage of "Behr's Blue" (Holland 1898), now used for a subspecies of the Silvery Blue.

♂ D  (2.6 cm)

♀ D  (2.5 cm)

♂ V  (2.6 cm)

**Adult:** The Square-spotted Blue has a wide postmedian orange line on the ventral hindwing. There is no adjacent area of distal submarginal metallic scales as in the Acmon Blue. The uppersides of the male wings are blue with black borders. The uppersides of the female wings have a rich brown ground colour, with the ventral postmedian orange area repeated. The female dorsal forewing has no orange. The valve of the male genitalia is shown in Fig. 67b.

**Immature stages:** Comstock and Dammers (1934) first described the immatures. The egg is bluish white. The

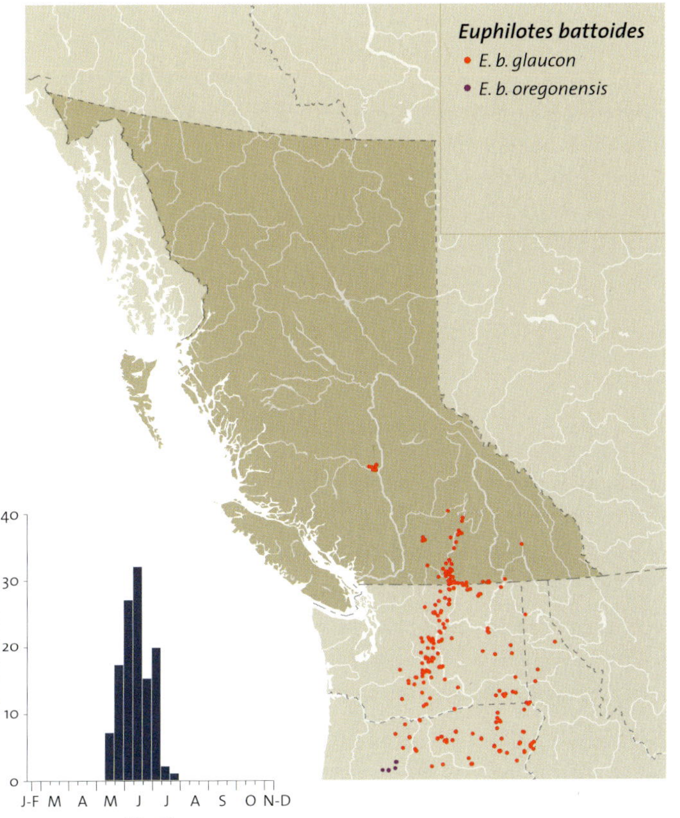


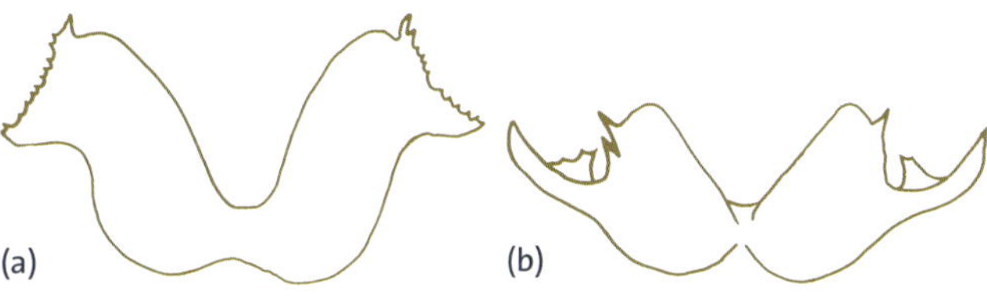

**67** Valves of *Euphilotes* species: (a) *E. ancilla,* (b) *E. battoides.*

mature larva is extremely variable in colour, from pale blue green through various yellow/green shades to a uniform pink. In the more common green form, there is a dorsal chocolate brown line and, laterally on each side, an elongated chocolate brown triangle with the longest tip pointed caudally. The mature larva is 11 mm long. The larval head is black and the spiracles dirty white. The pupa is 5–7 mm long and pale chestnut in colour, with dark chocolate stigmata. There are no hairs on the surface and there is no silken attachment.

**BIOLOGY:** Adults of the Square-spotted Blue fly from mid-May to early July, and into late July at the highest elevations. Eggs are laid singly on flowers of *Eriogonum* spp., the larval

Mature larva, ssp. *oregonensis*

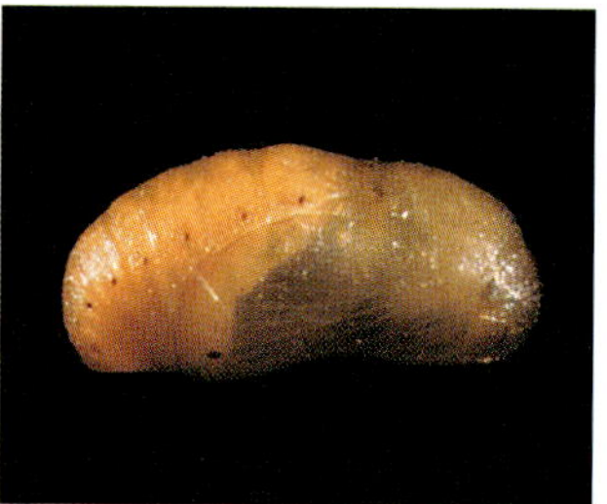
Pupa, from California

foodplant. Larvae feed only on the flowers, and pupation occurs in the drying flower heads, which remain on upright stems until the following summer. The Square-spotted Blue overwinters in the pupal stage and emerges as an adult the following spring. Emergence appears to be tied to the flower development of the foodplant, and not the first warm weather in spring.

**SUBSPECIES:** BC populations are the Great Basin subspecies, *E. b. glaucon* (W.H. Edwards, 1871) (TL: Storey Co., NV).

**RANGE AND HABITAT:** The Square-spotted Blue is found in the Southern Interior and the southwestern West Kootenay at most localities where the larval foodplant occurs, including isolated populations on high (1,500–2,200 m), exposed ridges.

**GENERAL DISTRIBUTION:** The Square-spotted Blue ranges from the Southern Interior of BC south to Baja California and NM.

**CONSERVATION STATUS:** Not of concern (S4).

Square-spotted Blues (*Euphilotes battoides glaucon*) mating

---

## GENUS *GLAUCOPSYCHE* SCUDDER, 1872

The name *Glaucopsyche* is derived from the Greek *glaucus* (grey blue) and *psyche* (the soul personified as a butterfly [Emmet 1991]). There does not appear to be a common name for the genus.

The genus *Glaucopsyche* is structurally similar to the preceding genus, with some genitalic differences. The species are significantly larger in size and lack the red markings of *Euphilotes*. The larvae feed on plants of the family Fabaceae, especially *Lupinus*. There are three Nearctic species, including the extinct *G. xerxes* (Boisduval, 1852) from San Francisco, CA, and three Palearctic species.

## ARROWHEAD BLUE
*Glaucopsyche piasus* (Boisduval, 1852)

**ETYMOLOGY:** The derivation of the species name *piasus* is unknown. The subspecies name *toxeuma* is Greek for "that which is shot" – in other words, an arrow (Brown 1971). The common name "Arrowhead Blue" (Holland 1898) refers to the light arrowhead-shaped median markings on the ventral hindwing.

**ADULT:** The Arrowhead Blue is the largest species of blue in BC. The upperside of the male wings is a uniform blue with black borders and a conspicuous white fringe. The females are similar except that the outer third of the wing is black. There are no orange markings in either sex. The median row of large, elongate white chevrons of the ventral hindwing is very distinctive for the species.

**IMMATURE STAGES:** Comstock (1927) illustrated the egg. It is twice as wide as it is tall, and highly sculptured. Otherwise, no one has published a description of the immatures.

**BIOLOGY:** Arrowhead Blue adults fly from mid-May to mid-June, and into July at the highest elevations (1,250 m). Emmel

**Arrowhead Blue (*Glaucopsyche piasus*)**

♂ D (3.3 CM)    ♀ D (3.4 CM)

♂ V (3.3 CM)

and Emmel (1973) list lupines (*Lupinus* sp.) and *Astragalus* sp. as larval foodplants. The butterfly occurs in association with lupines in BC.

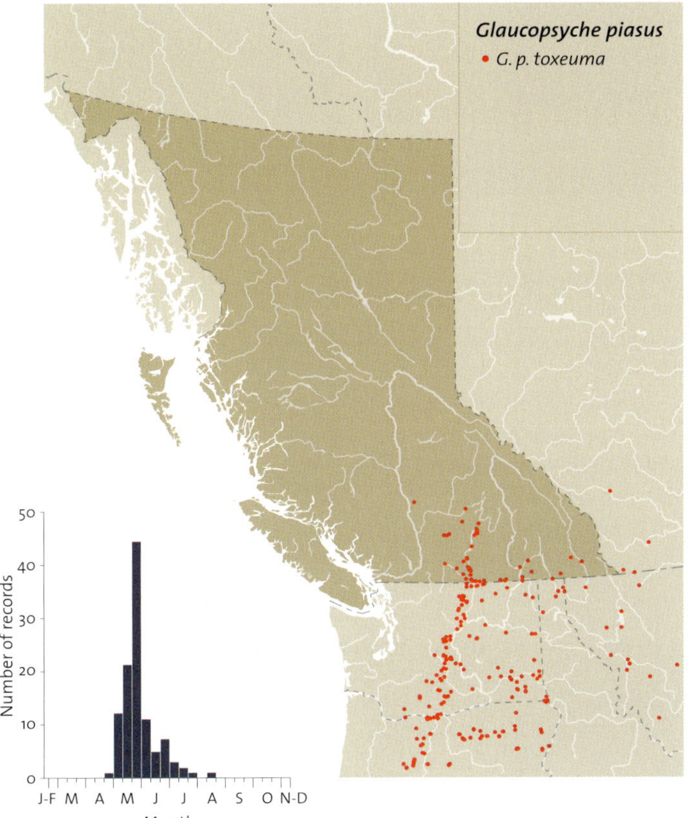



**SUBSPECIES:** BC populations are the subspecies *G. p. toxeuma* F.M. Brown, 1971 (TL: Summerland, BC).

**RANGE AND HABITAT:** The Arrowhead Blue occurs in the Southern Interior and the West Kootenay in sagebrush and ponderosa pine habitat.

**GENERAL DISTRIBUTION:** The Arrowhead Blue ranges from the Southern Interior of BC south to Baja California and NM.

**CONSERVATION STATUS:** Not of concern (S4).

**Pupa, from California**

**Mature larva, ssp. *piasus***

## SILVERY BLUE
*Glaucopsyche lygdamus* (Doubleday, 1841)

**ETYMOLOGY:** The name *lygdamus* refers to the Latin poet Lygdamus. This shows the close relationship of the Silvery Blue to its European counterpart, *G. alexis,* named for Alexis, a Greek poet. Subspecies *columbia* is named for the type locality, Fort Columbia, WA, on the Columbia River. Subspe-

cies *couperi* is named for William Couper of Anticosti, PQ, who supplied the type material. The common name "Silvery Blue" was first used by Gosse (1883), and refers to the silvery blue dorsal wing colour of males.

ADULT: For the Silvery Blue the upperside of the male wing is a "silvery" blue with a narrower black border than the Arrowhead Blue. The ground colour of the upperside of the female wings is brown, with a flush of blue scales at the base. Neither sex has any orange markings. The underside of the Silvery Blue is very diagnostic. The ground colour is a uniform light brown with strongly contrasting black spots bordered by white. There is no mottling or other variation to the uniform ground colour.

IMMATURE STAGES: Both Bower (1911) in the east and Williams (1908) described the immatures. Bower gave the best account. The egg is turban-shaped. It is green and the white chorion has a reticulated surface. The egg turns white before hatching. The mature larva is light green and covered with a granulated surface of white dots. The dorsal line is dark green, edged with light yellow green. The subspiracular line is conspicuously cream, edged with dark green.

BIOLOGY: The Silvery Blue flies from late April at low elevations to mid-July in subalpine meadows, but not for more than a month at any site, especially at dry, low elevations. Eggs are laid on the foliage of the larval foodplant and the larvae feed externally, attended by ants (Bower 1911). The larvae pupate in July and overwinter in ant nests (Tilden 1947). Bower (1911) states that the entire life cycle from egg laying to pupation takes only 26 days. Dolinger et al. (1973) found that populations of lupines that were intensely fed on by larvae of the Silvery Blue developed very high concentrations of alkaloids, thus discouraging further larval feeding.

Ssp. *columbia* ♂ D  (3.1 cm)

Ssp. *columbia* ♀ D  (2.9 cm)

Ssp. *columbia* ♂ V  (3.1 cm)

Ssp. *couperi* ♂ V  (2.6 cm)

Ssp. *couperi* ♂ D  (2.6 cm)

Ssp. *couperi* ♀ D  (2.6 cm)

SUBSPECIES: Populations in southern BC are the subspecies *G. l. columbia* (Skinner, 1917) (TL: Fort Columbia, [Okanogan Co.], WA). Northern populations are the boreal subspecies, *G. l. couperi* Grote, 1873 (TL: Anticosti Island, PQ). The northern subspecies is smaller.

Pupa, ssp. *columbia*

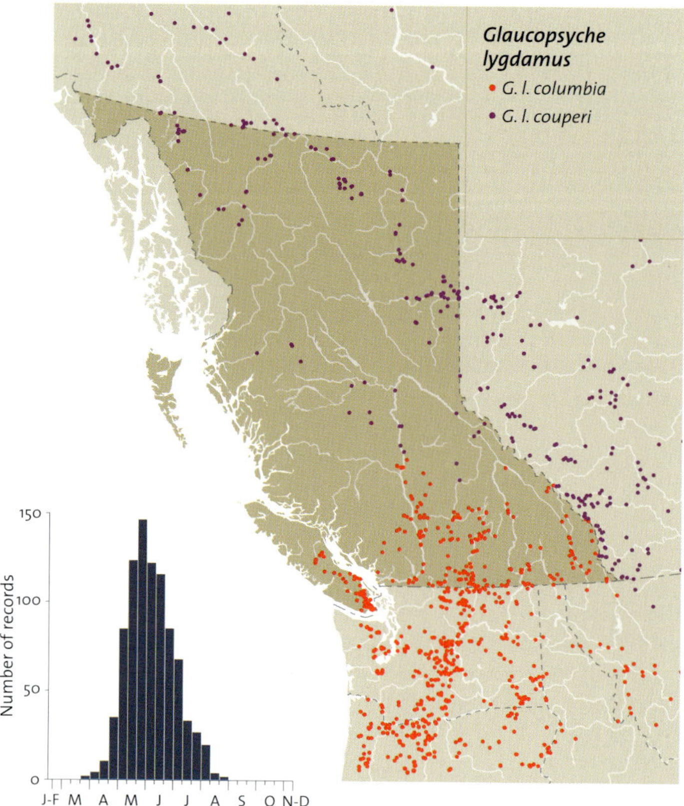

Mature larva, ssp. *columbia*



FAMILY LYCAENIDAE (GOSSAMER WINGS)

**Range and habitat:** After the flight period of the Spring Azure, this species becomes the common late spring to early summer blue throughout BC, except for the west coast of Vancouver Island, the Queen Charlotte Islands, and the North Coast. It is found in virtually any situation where wild legumes are found, from the lowest elevations to near timberline.

**General distribution:** The Silvery Blue ranges from western AK south through BC to Baja California and NM, and east to Labrador and NF. In the east it occurs south to the Great Lakes, with disjunct populations in the Ozarks and the Appalachians.

**Conservation status:** Not of concern, with both subspecies S5.

**Silvery Blue (*Glaucopsyche lygdamus columbia*)**

## Tribe Polyommatini Swainson, 1827

The following genus and the remaining blues belong to the tribe Polyommatini. They differ from the genera *Euphilotes* and *Glaucopsyche* in that veins 8 and 9 of the forewing are joined at the base. The eyes of BC species are smooth. The male genitalia are weakly sclerotized. The valves are elongate and usually with a modified costal process. Some North American authorities lump all North American genera of this tribe into one genus, but there are structural differences defining the five genera. Also, the genera appear to represent natural groupings, and we prefer to use the separate generic names, otherwise all are subsumed into an extremely large genus, *Plebeius*.

### Genus *Lycaeides* Hübner, [1819]

The name *Lycaeides* comes from *Lycaena,* the older generic name for blues, and *ides* (like), thus "like other blues."

This is the only genus of blues in BC with a strong row of submarginal reddish spots on the undersides of both wings. The falces is elongate and equal in length to the labides. The dorsal tip of the valve is serrate and the ventral tip is strongly hooked back to project dorsally.

There are six species in the genus *Lycaeides,* three Palearctic, two Nearctic, and one Holarctic. All North American species occur in BC.

The BC species of *Lycaeides* are very difficult to tell apart by wing characters where the species occur together (Fig. 69). The pertinent part of the male genitalia is illustrated in Fig. 68.

# NORTHERN BLUE

*Lycaeides idas* (Linnaeus, 1761)

**Northern Blue (*Lycaeides idas scudderi*)**

**ETYMOLOGY:** The name *idas* is likely derived from Ida, the name of two mountains in Greek mythology, thus referring to the mountain habitat of the species. The various subspecies names are derived as follows: *alaskensis* is from Alaska; *scudderi* honours Samuel Scudder of Harvard University; and *atrapraetextus* literally means "woven with black," referring to the fuscous or black border on the wings. The common name, dating from Klots (1951), refers to the predominantly northern distribution of the species.

**ADULT:** The Northern Blue, Anna's Blue, and Melissa's Blue are the only BC blues where the submarginal orange spots of the ventral hindwing extend into the ventral forewing. Northern Blue males are a uniform violet blue with obvious black submarginal spots on the dorsal hindwings. Females are a uniform brown, usually with submarginal orange spots on the dorsal hindwing that can extend to the forewing. Where the Northern Blue occurs near the very similar Melissa's Blue, the two species are difficult to tell apart in the field. The only clue is that the Northern Blue will usually be found above 1,000 m and Melissa's Blue below 1,000 m. Otherwise, the male genitalia are the only reliable character for separating males (Fig. 68). *Lycaeides anna* and *L. idas* have nearly identical genitalia. Females are identified by their association with males. In long series of females, the orange submarginal spots of the dorsal hindwing are much narrower for *L. idas* (Fig. 69b) than for *L. melissa* (Fig. 69a). Since the Northern Blue in North America has a very different larval foodplant (the heather family) and a very different ventral wing from *L. anna* and is allopatric to *L. anna,* we separate the two species. It is also likely that this species is not the same as the Palearctic *L. idas.*

**IMMATURE STAGES:** Undescribed. Apparently all descriptions of immatures of this genus under the name *scudderi* in the northeastern USA and Ontario refer to *L. melissa samuelis* Nabokov, 1944.

**BIOLOGY:** The Northern Blue flies in July and early August in one generation per year. Nielsen and Ferge (1982) observed the Northern Blue ovipositing on stems of dwarf bilberry,

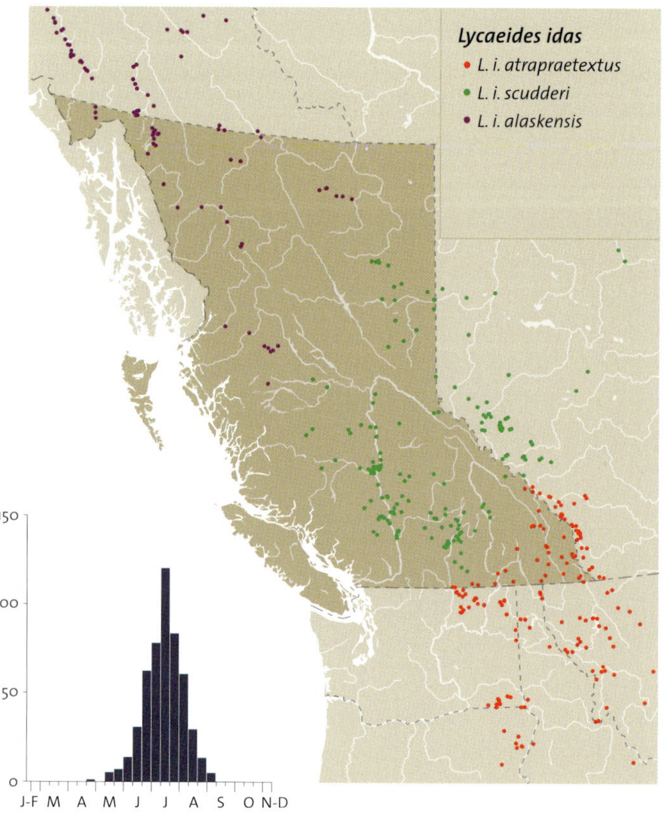



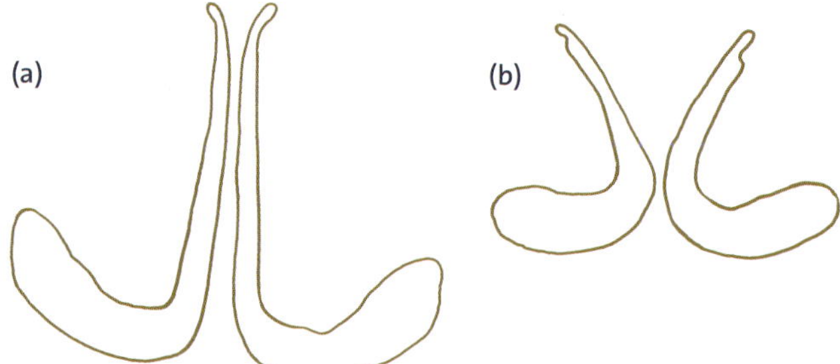



**68** Gnathos of *Lycaeides* species: (a) *L. melissa*, (b) *L. anna*

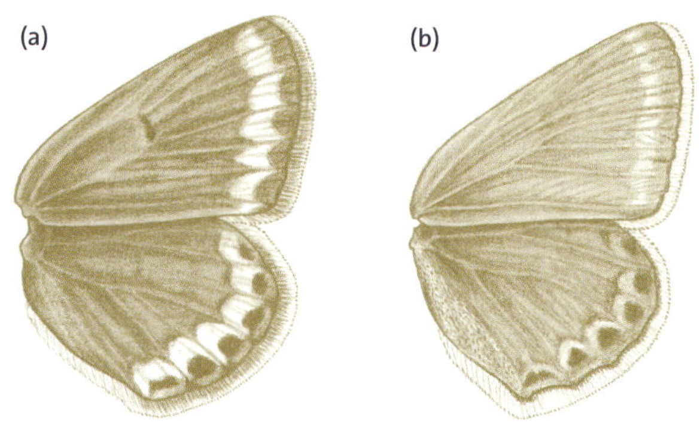



**69** Upperside of the wings of female *Lycaeides* species: (a) *L. melissa*, (b) *L. anna*

Ssp. *atrapraetextus* ♂ D (2.8 cm)

Ssp. *atrapraetextus* ♀ D (3.0 cm)

Ssp. *atrapraetextus* ♂ V (2.8 cm)

Ssp. *alaskensis* ♂ D (2.5 cm)

Ssp. *scudderi* ♂ D (2.6 cm)

Ssp. *scudderi* ♀ D (2.7 cm)

Ssp. *scudderi* ♂ V (2.6 cm)

Ssp. *alaskensis* ♀ D (2.7 cm)

Ssp. *alaskensis* ♂ V (2.5 cm)

*Vaccinium caespitosum*, in Wisconsin and noted earlier observations in Ontario on the same plant. This *Vaccinium* species occurs in BC and is the likely foodplant of *L. i. scudderi*. The overwintering stage is not known.

**Subspecies:** In northwestern BC the Beringian subspecies, *L. i. alaskensis* (Chermock, 1945) (TL: Fort Yukon, AK), with very blue females is found. In the boreal forest habitat of northeastern and central BC, the boreal subspecies, *L. i. scudderi* (W.H. Edwards, 1861) (TL: mouth of the Saskatchewan R., MB), occurs. It has females with some blue but the ventral pattern is much more distinct than that of the Beringian subspecies. It ranges south to near Lillooet, Salmon Arm, and Revelstoke. East of the Okanagan Valley bottom in southeastern BC, the Rocky Mountain subspecies, *L. i. atrapraetextus* (Field, 1939) (TL: Priest River, ID) [= *ferniensis* (Chermock, 1945); TL: Fernie, BC] is found. It is closest to subspecies *scudderi* but is smaller and darker.

**Range and habitat:** The Northern Blue is found throughout BC except the Queen Charlotte Islands, Vancouver Island, the North Coast, and the Cascades. It is found in open meadow situations to timberline and above. In the Southern Interior, where it comes in close proximity to Melissa's Blue, the Northern Blue is found above 1,000 m and Melissa's Blue below 1,000 m. The two species have never been found in the same locality.

**General distribution:** The Northern Blue is Holarctic. In North America it occurs from western AK east to PQ (Ungava) and NF in boreal habitat. It ranges south to the CA Sierras and northwest WY, with a disjunct set of populations in CO. In the east it is found south of the Canadian border only on the south shore of Lake Superior.

**Conservation status:** Not of concern, with all three subspecies S5.

**Mature larva**

## Anna's Blue
*Lycaeides anna* (W.H. Edwards, 1861)

**Etymology:** The species name *anna* is one of several Edwards species names based on female Christian names the source of which we cannot trace. (Another is *melissa,* the following species.) The subspecies *ricei* is named for Harold Rice, the Oregon butterfly collector who discovered the subspecies. We have resurrected the common name of the species from Comstock (1927) but use the possessive form because we believe that the butterfly is named for an unknown modern woman, not a classical figure.

**Adult:** This blue is characterized by the very white underside ground colour, which has a blue flush at the base of the ventral hindwing. The black spot pattern is very obscure compared with that in other species in the genus. The genitalia are nearly identical to those of *L. idas.* Besides the wing pattern differences, this species has a very different larval foodplant from *L. idas.*

**Immatures:** Undescribed.

**Anna's Blue (*Lycaeides anna vancouverensis*) female**

**BIOLOGY:** Anna's Blue flies in July and August at timberline. Adult flight time is controlled by how soon the high-elevation snow disappears in the spring. There were historic populations on Vancouver Island at low elevations that emerged earlier, but these populations have been extirpated. This species is associated with lupines on Vancouver Island, and G.R. Pratt (pers. comm.) confirmed that the only larval foodplants in California are *Lupinus* and *Lotus,* both legumes. By contrast the closely related *L. idas* utilizes plants in the heather family (Ericaceae) as larval foodplants.

**SUBSPECIES:** Mainland populations are the Cascade Mountains subspecies, *L. a. ricei* (Cross, 1937) (TL: Big Cultus Lake, Deschutes Co., OR). The Vancouver Island subspecies is a new

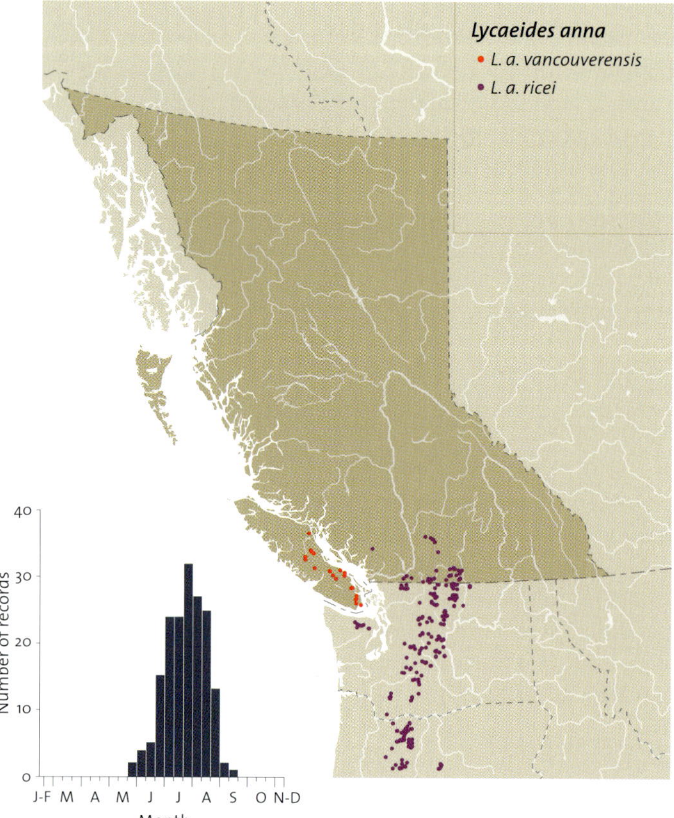

Ssp. *ricei* ♂ D  (3.0 CM)

Ssp. *ricei* ♂ V  (3.0 CM)

Ssp. *ricei* ♀ D  (2.8 CM)

Ssp. *ricei* ♀ V  (2.8 CM)

Ssp. *vancouverensis* ♂ D  HOLOTYPE (2.7 CM)

Ssp. *vancouverensis* ♂ V  HOLOTYPE (2.7 CM)

Ssp. *vancouverensis* ♀ D  (2.7 CM)

Ssp. *vancouverensis* ♀ V  (2.7 CM)

subspecies, which, in the older British Columbia literature, was identified as *L. a. anna* or *L. melissa* (Shepard 1964).

***Lycaeides anna vancouverensis* Guppy & Shepard, new subspecies**. The upperside of the male wings is a uniform blue. There is almost no evidence of black marks in the submarginal area. The ventral hindwing has a blue flush on the basal half, a characteristic of the species. The postmedian black spots on the underside of both wings are much larger than in *L. a. ricei*. The inner submarginal orange spots on the underside of both wings is much stronger than in *L. a. ricei*. In this way *L. a. vancouverensis* is closer in appearance to California subspecies of *L. anna* and *L. melissa*. Females have the same strong postmedian black spots and strong inner submarginal orange spots as the males. On the upperside of the female wings, the submarginal orange spots are always large on the hindwing and usually so on the forewing. This is in contrast to the subspecies *L. a. ricei,* where these submarginal orange spots are weak to absent on both surfaces of the wings. **Types**. Holotype: male, BC, Strathcona Park, Cream Lake, 1,400 m, 22 August 1988, C.S. Guppy. A label "Holotype / *Lycaeides anna* / *vancouverensis* Guppy & Shepard" is attached. The holotype is deposited in the Royal British Columbia Museum, Victoria, BC, CAN. Paratypes: 27 males, 23 females, same data as holotype (CSG); 2 males, 2 females, same data as holotype (JHS); 1 female, BC, Strathcona Park, Flower Ridge, 4,600 ft., 11 August 1974, C.S. Guppy (CSG); 2 males, 6 females, BC, Mt. Cokely Ski Hill Rd., 3 August 1995, J. and S. Shepard (JHS); 1 female, same locality, 26 July 1995, J. and S. Shepard (JHS); 6 males, 6 females, BC, Mt. Cokely summit, 1,600 m, 12 August 1977, C.S. Guppy (CSG); 1 female, BC, Sproat lake (north side), 10 June 1975, C.S. Guppy (CSG).

**RANGE AND HABITAT:** Anna's Blue is restricted to Vancouver Island, the Coast Range immediately around the Fraser Valley, and the Cascade Mountains.

**GENERAL DISTRIBUTION:** Anna's Blue is found from Vancouver Island, the Olympic Mountains, and the BC Cascades south

Anna's Blues (*Lycaeides anna ricei*) mating

through the Cascade and Sierra Nevada Mountains and the northern CA coast ranges.

**CONSERVATION STATUS:** Not of concern, with both subspecies S4.

## MELISSA'S BLUE

*Lycaeides melissa* (W.H. Edwards, 1873)

**ETYMOLOGY:** The name *melissa* has so many possible derivations that none is reliable. The most likely is that it is the Christian name of an unknown female, as Edwards used few names based on classical figures and several other of his

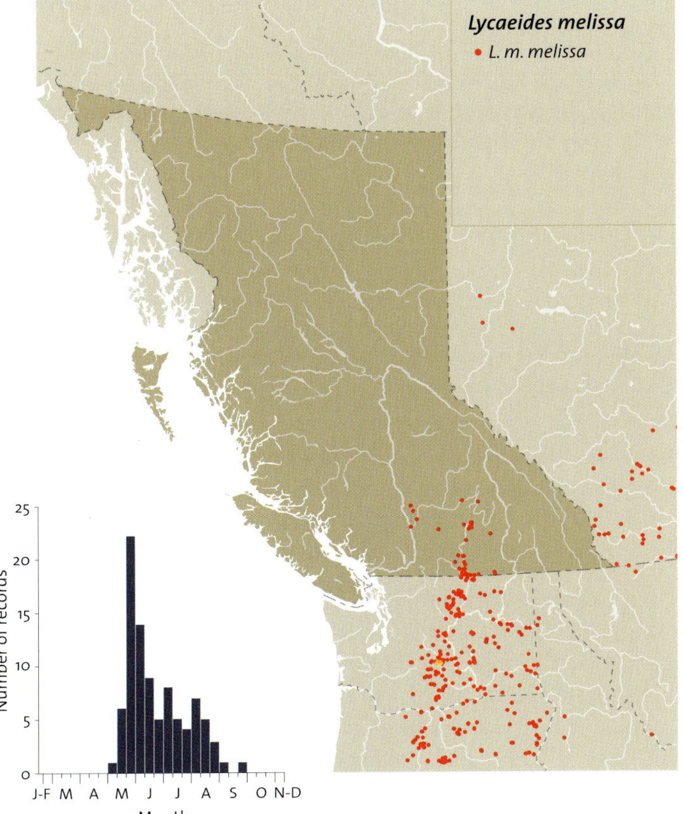

♂ D  (2.7 CM)

♀ D  (2.7 CM)

♂ V  (2.7 CM)

species names appear to be based on the given names of women, not on classical characters. The common name dates from Klots (1951) but it is emended to reflect the probable etymology of the name.

**ADULT:** See the discussion of the Northern Blue to distinguish Melissa's Blue. The male genitalia are illustrated (Fig. 68a). The underside wing ground pattern is always contrasting in Melissa's Blue. If one looks at large numbers of individuals of an adjacent Northern Blue population, there are always some individuals with a washed out wing pattern among the Northern Blues.

**IMMATURE STAGES:** Scudder (1889b) described the immatures as a subspecies of the Northern Blue. Comstock (1928) illustrated the egg of California populations. It is very flat

Melissa's blue (*Lycaeides melissa melissa*)

to the base of, *Lupinus* sp. Comstock (1929) recorded wild licorice (*Glycyrrhiza lepidota*) as a larval foodplant. Eggs laid by the first-generation adults hatch within one week and become second-generation adults by six weeks. The overwintering stage of immatures from the second generation is not known.

Mature larva, tended by ants

**SUBSPECIES:** BC populations are the nominate subspecies, *L. m. melissa* (TL: La Plata Peak, Lake Co., CO).

**RANGE AND HABITAT:** Melissa's Blue is found in the Southern Interior below 1,000 m in ponderosa pine and sagebrush habitat.

**GENERAL DISTRIBUTION:** Melissa's Blue is found from the Southern Interior of BC south to Baja California and northern MEX, and east of the Rockies from central AB east to MB. In the east, it is found south of the Great Lakes and in southern ON and NY, where an endangered subspecies, the Karner Blue, occurs (*L. m. samuelis* Nabokov, 1944) (TL: Karner, NY).

**CONSERVATION STATUS:** Not of concern (S4).

and light green. Comstock (1929) illustrated the larva and pupa.

**BIOLOGY:** Melissa's Blue flies in two generations each summer. The first generation is present from mid-May to early July, and the second generation from late July to early September. Comstock (1928) observed southern California females ovipositing on the base of, or on small pebbles next

---

## GENUS *PLEBEIUS* KLUK, 1780

In Latin *Plebeius* means "plebeian" or "common." The name *Plebeius* is derived from the Plebeji or the fifth of six phalanges in which Linnaeus placed all the small species, such as blues and skippers – thus, the common or smaller and less colourful species.

The labides of the male genitalia is elongated and usually blunt at the tip. The falces is half the length of the labides. The dorsal tip of the valve is toothed. The ventral tip is rounded and not developed to any particular shape, but projects caudad.

This genus has at least five Palearctic and one Nearctic species. The generic name *Plebejus* Kluk, 1802, is a synonym of *Plebeius* Kluk, 1780 (Paclt 1955; Karsholt and Razowshi 1996).

## GREENISH BLUE
*Plebeius saepiolus* (Boisduval, 1852)

Greenish Blue (*Plebeius saepiolus*)

**ETYMOLOGY:** The name *saepiolus* may be derived from the Latin *saepio* (to fence in) and *olus,* a diminutive ending – thus, a small fencing. The subspecies name *insulanus* refers to Vancouver Island. The subspecies name *amica* is Latin for friend, thus, a species related to *saepiolus*. The name "Greenish Blue" (Holland 1898) refers to the greenish shade of the ventral wing base in some subspecies.

**ADULT:** Despite the common name, there is no green on the wings of either sex of the Greenish Blue in BC. On the upperside, the male wing is a uniform blue with wide black margins and white fringe. The upperside wing ground colour of the female is brown, with sometimes a flush of blue scales at the base. Sometimes the female hindwing has a weak submarginal row of tawny spots. On the underside, the male is chalk white and the female brown. Both have matching rows of postmedian and submarginal black spots.

**IMMATURE STAGES:** Undescribed.

**BIOLOGY:** Adults of the Greenish Blue are observed from late May to July, with a few records in early August. The species is assumed to be univoltine. Emmel and Emmel (1973) record several species of clovers, *Trifolium* spp., as larval foodplants and state that the butterfly overwinters as an immature larva. In southern BC and the adjacent USA, the Greenish Blue is often found along abandoned or infrequently used dirt roads in association with introduced clover (*Trifolium pratense*). In Colorado, Sharp and Parks (1973) did a mark-release-recapture study showing that the Greenish Blue did not disperse well and stayed close to the larval foodplant, *Trifolium hybridum,* which was also the adult nectar source. This lack of dispersion may account for the loss of Vancouver Island populations, if the native foodplant was eliminated and there were no introduced clovers nearby for the species to utilize. One reference from the early literature indicated that clovers were not the only larval foodplant for the Greenish Blue. Fletcher (1908) stated that he saw females ovipositing on flower buds of *Hedysarum boreale* at Kenis-tino, AB, on 25 July 1907. Since this plant genus is not found on Vancouver Island, either the now uncommon native *Trifolium* or other legumes were the larval foodplant of *P. s. insulanus*.

Ssp. *insulanus* ♂ D (2.7 CM)

Ssp. *insulanus* ♂ V (2.7 CM)

Ssp. *insulanus* ♀ D (2.8 CM)

Ssp. *insulanus* ♀ V (2.8 CM)

Ssp. *amica* ♀ D (2.8 CM)

Ssp. *amica* ♂ V (2.4 CM)

**SUBSPECIES:** Vancouver Island populations are the endemic subspecies, *P. s. insulanus* Blackmore, 1919 (TL: Victoria, BC). Mainland populations are the boreal subspecies, *P. s. amica* (W.H. Edwards, 1863) (TL: Fort Simpson, NWT).

**RANGE AND HABITAT:** The Greenish Blue is found throughout BC except in the immediate coastal regions. It is associated with native clovers but is also found in old fields and the sides of old dirt roads, where the species utilizes introduced short, white clovers.

**GENERAL DISTRIBUTION:** The Greenish Blue occurs from central BC south through BC to southern CA and NM and east to NS. In the eastern USA, it is known only from the Great Lakes states.

**CONSERVATION STATUS:** The endemic Vancouver Island subspecies *insulanus* is Endangered, known only from historical records (Shepard 2000) (SH) and may be globally Extinct. Subspecies *amica* is not of concern (S5).

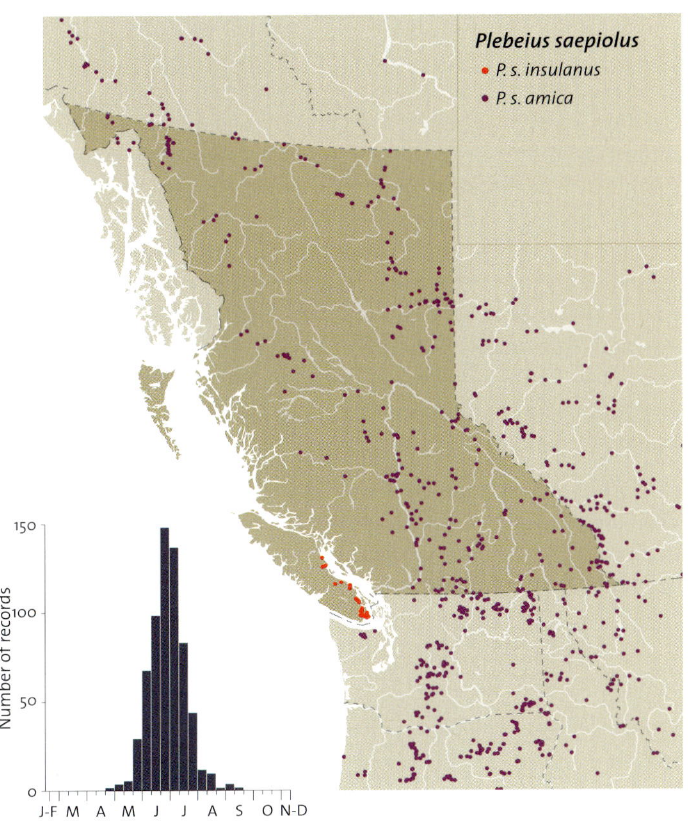

**Mature larva, ssp. *saepiolus***

## Genus *Icaricia* Nabokov, 1944  Blues

When Nabokov (1944) coined this generic name, he said it was allied to the European genus *Aricia* Rechenbach, 1817. Considering Nabokov's love of word games, the generic name *Icaricia* is a play on that relationship and the name of the type species, *icarioides*.

The labides of the male genitalia is stout and often toothed. The falces is short, poorly sclerotized, and often inconspicuous. The dorsal tip of the valve is elongate and enlarged except in the species *Icaricia shasta* (W.H. Edwards, 1862), which does not occur in our fauna.

This Nearctic genus contains five species. Two occur in BC. Larvae feed on either Fabaceae or Polygonaceae.

## Boisduval's Blue
### *Icaricia icarioides* (Boisduval, 1852)

**Etymology:** The name *icarioides* is derived from the name of the European species of blue, *icarus,* meaning that the American species was similar to the European one. Subspecies *blackmorei* is named for E.H. Blackmore, the well-known BC lepidopterist who supplied the type material to Barnes and McDunnough. The subspecies name *montis,* Latin for mountain, refers to the mountain type locality. The subspecies name *pembina* was used in reference to what was thought to be the type locality, Lake Winnipeg; Pembina is a local geographic name applied to the Red River valley. The common name, "Boisduval's Blue" (Holland 1898), acknowledges Boisduval, who first named this species along with many other species from California specimens.

**Adult:** Boisduval's Blue is difficult to characterize. The blue on the upperside of the male wing is not clear but some-

**Boisduval's Blue (*Icaricia icarioides pembina*)**

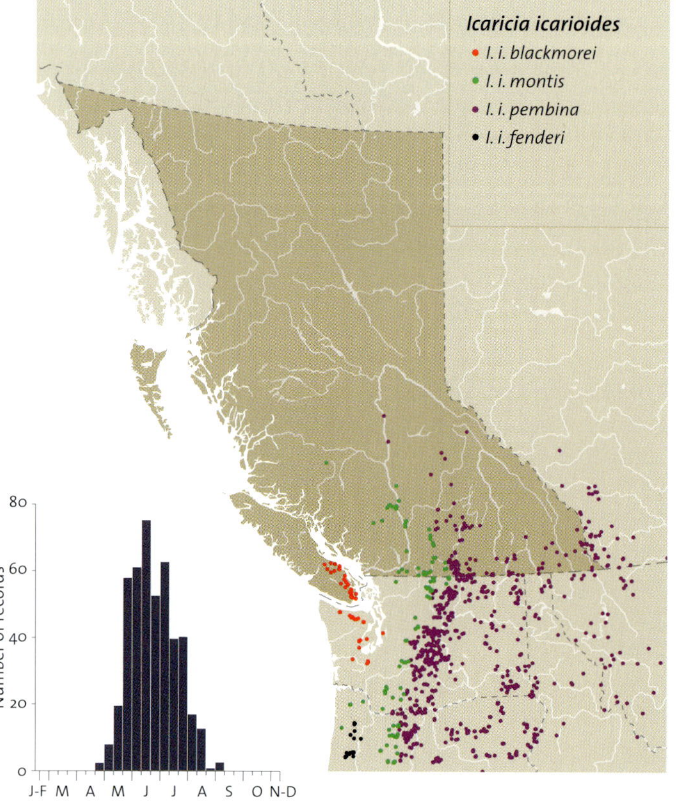



what washed out. The male underside is brown, heavily overlaid with white scales that give it the appearance of the underside of the Sooty Hairstreak, which is not blue on the upperside. The female upperside is brown, with usually a flush of blue scales at the base of the forewing. The ground colour of the female underside is brown. For Vancouver Island individuals, the female ventral hindwing has a median row of white spots. For mainland populations, these spots are large with small black centres. The ventral forewing of Vancouver Island females has a median row of black spots surrounded by white scales. These spots are much larger in mainland populations.

**Immature stages:** Comstock and Dammers (1935) described the egg as green overlaid by the white papillae of the chorion. The mature larva is green with three inconspicuous diagonal white lines laterally on each body segment. The body is covered with short white hairs. The pupa is green on the head, thorax, and wings, and chestnut red with green spots on the abdomen.

**Biology:** Boisduval's Blue is univoltine and flies from mid-May at lower elevations to mid-August at higher elevations, but only for a month at any one locality in any one season. Newcomer (1911) partially described the life history of the

Ssp. *blackmorei* ♂ D (2.9 cm)    Ssp. *blackmorei* ♀ D (3.2 cm)

Ssp. *blackmorei* ♂ V (2.9 cm)    Ssp. *pembina* ♂ V (2.9 cm)

Ssp. *pembina* ♂ D (2.9 cm)    Ssp. *pembina* ♀ D (2.9 cm)

Ssp. *montis* ♂ D (3.0 cm)    Ssp. *montis* ♀ D (2.8 cm)

California Sierran sub-
species. Eggs
are laid in equal numbers
on leaves and seed pods of
lupines. The eggs hatch
within 10 days and develop
into mature second instar
larvae before hibernating.
Newcomer (1911) also
mentioned collecting a

Ssp. *montis* ♂ V (3.0 cm)

mature larva in August, apparently from lower elevation; it
pupated in late August and emerged the following spring.

Comstock and Dammers (1935) reared the southern California mountain subspecies and found that it also overwintered
as mature second instar larvae. They observed pupation at
the base of the larval foodplant, *Lupinus* sp. The pupal stage
lasted three weeks before the adults emerged. Downey
(1962a) documented extensive myrmecophily, or ant attendance, upon Boisduval's Blue larvae. He recorded 11 species of
ants from six genera, and often two ant species for one
population of Boisduval's Blue. Downey (1962b) recorded the
genus *Trichogramma* (Hymenoptera: Trichogrammatidae) as
an egg parasitoid from 50% of the 53 populations studied.
He also recorded one instance of larval parasitism by
*Apanteles theclae* Riley (Hymenoptera: Braconidae). Newcomer (1911) noted tachinid fly eggs on the larvae but did not
rear them to maturity. Boisduval's Blue has been recorded
feeding as larvae on 38 species or varieties of lupines across
western North America, but for any one population of
Boisduval's Blue, only one type of lupine is used for oviposition (Downey and Fuller 1961; Downey and Dunn 1964).
Barnes and McDunnough (1919) reported that Blackmore
associated Boisduval's Blue with *Lupinus arcticus.*

**SUBSPECIES:** Vancouver Island populations are the subspecies *I. i. blackmorei* (Barnes & McDunnough, 1919)
(TL: Goldstream, BC). Populations from the southern Coast
Range and the adjacent Cascades are *I. i. montis* (Blackmore,
1923) (TL: Mt. McLean, near Lillooet, BC). Populations from the
Okanagan Valley eastward are the Rocky Mountain subspecies, *I. i. pembina* (W.H. Edwards, 1862) (TL: Ravilli Co., MT).

**RANGE AND HABITAT:** Boisduval's Blue occurs across southern
BC wherever *Lupinus* spp., especially hair-leafed species,
occur. Low-elevation Vancouver Island populations have not
been seen since the 1960s. Most likely the spread of Scotch
broom and fire suppression choked out the native lupine,
*Lupinus latifolius,* on which the larvae feed. Subalpine
clearcut logging has opened up more habitat for new lupine
growth, however, and the butterfly is currently flourishing in
some subalpine areas.

**GENERAL DISTRIBUTION:**
Boisduval's Blue is found
from southern BC to
southern SK and south to
Baja California
and northern MEX.

**CONSERVATION STATUS:**
Subspecies *blackmorei* is of
Special Concern in BC (S3).
Subspecies *montis* (S4) and
*pembina* (S5) are not of concern.

**Mature larva, ssp. *montis***

# ACMON BLUE
*Icaricia acmon* (Westwood & Hewitson, [1852])

**Acmon Blue (*Icaricia acmon*)**

♂ D  (2.6 cm)

♀ D  (2.6 cm)

♂ V  (2.6 cm)

**ETYMOLOGY:** Acmon was one of the two Cercopes, a pair of mischievous dwarfs who plagued Hercules; hence *acmon* is a dwarf blue. The subspecies name *lutzi* honours Frank Eugene Lutz, a co-worker of dos Passos at the American Museum of Natural History. The common name was first used by Holland (1898).

**ADULT:** The Acmon Blue is the easiest blue to identify. On the ventral hindwing is a submarginal row of orange spots. Outside the orange spots there is a row of metallic spots not seen in any other BC species of blue. On the upperside the male wings are blue with a row of submarginal orange spots. The upperside of the female wings is brown with blue scales on the basal portion of both wings. Again there is a strong row of submarginal orange spots.

**IMMATURE STAGES:** Emmel and Emmel (1973) described the immatures from Dammers' paintings. The egg is pale green. The mature larva is a dirty yellow with variable lateral markings. The body is covered with short white hairs. There is a dorsal, darker green line. The pupa is initially brown with a green abdomen, whereas a pupa reared from larvae by Shepard changed to dark brown.

**BIOLOGY:** The Acmon Blue is univoltine, flying from mid-May at the lowest elevations to mid-August where *Eriogonum* occurs on open ridges at 1,900 or more metres. Sometimes there is a second generation in the South Okanagan. In BC, the Acmon Blue is always found in association with *Eriogonum* sp., and occurs wherever the foodplant is found in the interior. Goodpasture (1973) described the life cycle. Eggs are laid individually on leaves and flowers of the foodplant. The species overwinters as a second- or third instar larva in California, where there are at least two broods, but may overwinter at some other stage in BC. Diapause is dependent on day length, and larvae that would normally hibernate will not do so if raised in the lab with 16 hours of daylight. Shepard found that a large number of field-collected larvae yielded tachinid flies rather than butterflies after pupation (JHS).

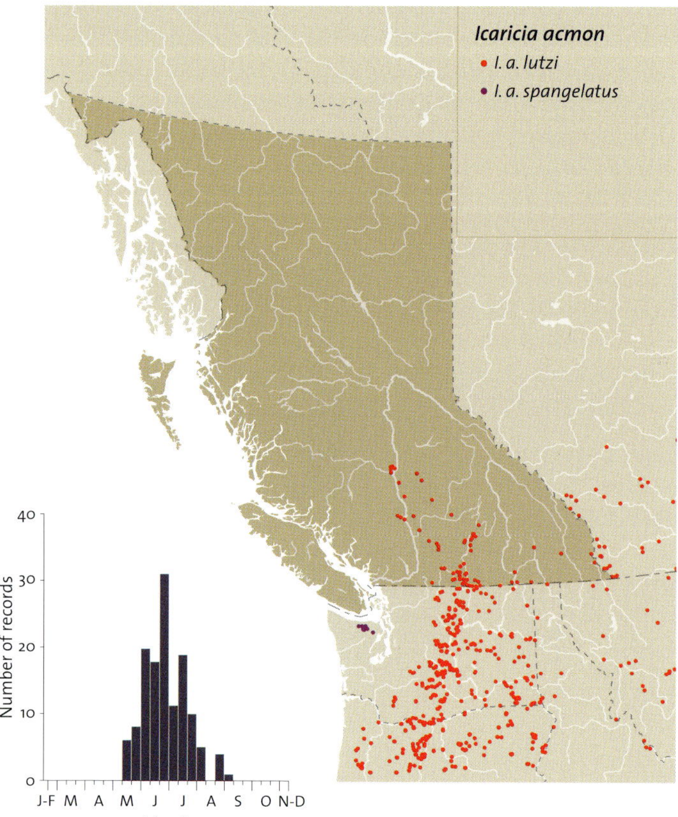

**Mature larva, ssp. *acmon***

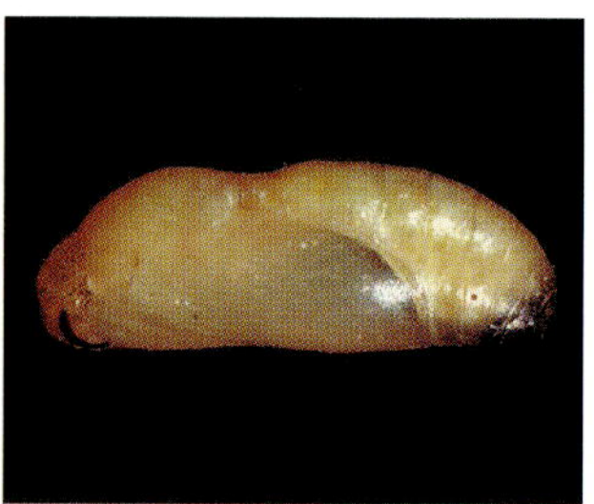

**Pupa, ssp. *acmon***

**Subspecies:** BC populations are the northern Great Basin subspecies, *I. a. lutzi* (dos Passos, 1938) (TL: Montpelier, ID).

**Range and habitat:** The Acmon Blue is found from the southeastern Chilcotin southeast through the Southern Interior and southern Kootenays. It is found both in valley bottoms in sage and ponderosa pine habitat and at or near timberline wherever the larval foodplant, *Eriogonum* sp., grows.

**General distribution:** The Acmon Blue occurs in the Southern Interior of BC and east to SK. From BC it occurs south to Baja California and mainland MEX.

**Conservation status:** Not of concern (S4).

## Genus *Vacciniina* Tutt, 1909

The name *Vacciniina* is derived from the larval foodplant, *Vaccinium,* which includes bog cranberry and blueberry. Since this is a monotypic genus, it does not require a common name other than that of the one species.

The labides and falces of the male genitalia are subequal in length. The labides is not cone-shaped. The falces is blunt at the tip.

The genus *Vacciniina* is Holarctic, with one species.

## Cranberry Blue
*Vacciniina optilete* (Knoch, 1781)

**Etymology:** The name *optilete* is of unknown derivation. Subspecies *yukona* is named after the upper Yukon River region in Alaska. The subspecies was called the Yukon Blue by Holland (1931) because he considered it to be a separate species.

**Adult:** The dorsal wings of the Cranberry Blue male are a dark violet blue. The female upperside ground colour is brown, with a strong overlay of blue scales that can mask

♂ D  (2.3 cm)        ♀ D  (2.6 cm)

♂ V  (2.3 cm)

the brown. In this case, the submarginal area is a washed-out blue. The underside of the wings of both sexes is grey with the typical black spots. There is one orange spot in the submargin of the hindwing.

**Immature stages:** Undescribed for North America. Henriksen and Kreutzer (1982) described the species in Scandinavia. The egg is white. The mature larva is green with a darker green dorsal stripe. There are pale yellow, violet-bordered lateral lines, and the body is covered with short, reddish hairs. The pupa is pale green with reddish hairs.

**Biology:** The Cranberry Blue is seen flying from early July to mid-August. Nothing is known of its biology in North America, but the Scandinavian subspecies is described by Henriksen and Kreutzer (1982). The eggs are laid on the larval foodplant, *Vaccinium uliginosum,* and other Ericaceae. Eggs hatch within the week and larvae partially develop before hibernating. The larvae complete development the following spring. The adult emerges after a short pupal period.

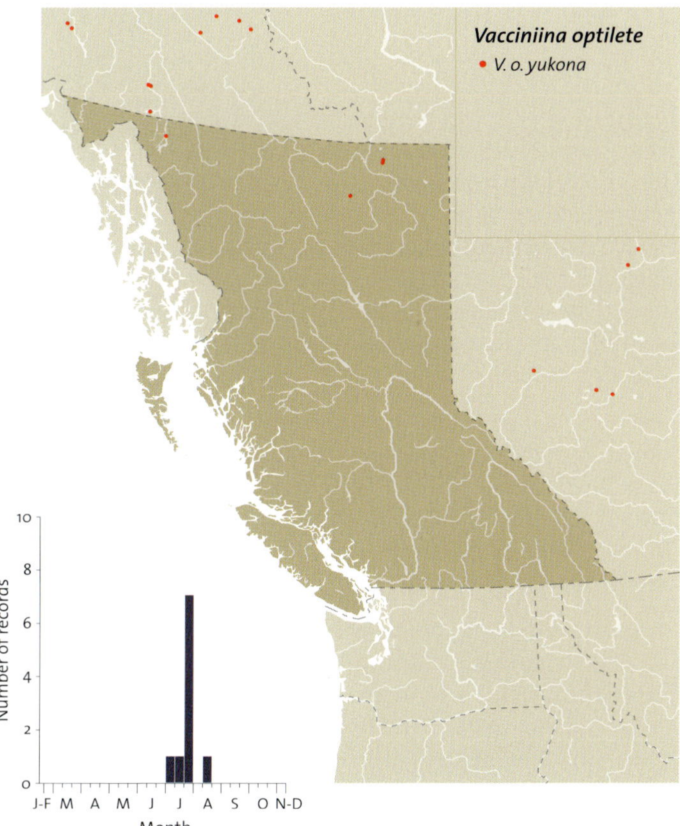

**Subspecies:** BC populations are the eastern Beringian subspecies, *V. o. yukona* (Holland, 1900) (TL: between Mission and Forty-mile Creeks, AK).

**Range and habitat:** The Cranberry Blue is known only from extreme northern BC, but should be found in areas where the foodplant, *Vaccinium,* occurs at or near timberline and in bogs at lower elevations.

**General distribution:** The Cranberry Blue ranges from northern Scandinavia east through polar Russia to AK, YT, southwestern NT, and northern BC to MB.

**Conservation status:** The Cranberry Blue is of Special Concern in BC (S3).

## Genus *Agriades* Hübner, [1819]

The name *Agriades* most likely means "like *argus*," another European blue. The original Argus, from the myth of Zeus and Io, had 100 eyes, a reference to the many spots on the underside of the wings (Emmet 1991).

The major structural differences that have defined this genus are the falces and labides of the male genitalia. The falces is elongate and extends to just beyond the labides. In this characteristic, the genus is closest to *Lycaeides* and *Vacciniina,* and all three are far removed from *Icaricia* and *Plebeius.*

The genus *Agriades* has four species: one Palearctic, one Holarctic, and two Nearctic species restricted to California and southwestern Oregon. Emmel and Emmel (1998) clearly define two Californian species, but they do not clearly show that other taxa are not all *A. glandon.* Both European and North American authorities have tried to recognize more than one species within the species *A. glandon,* but so far no conclusive evidence has been published.

## Arctic Blue
*Agriades glandon* (de Prunner, 1798)

**Etymology:** The derivation of the species name *glandon* is unknown. The subspecies name *megalo* is derived from the Greek *mega* (great). At the time this subspecies was

described, it was considered much larger than its relatives. The subspecies name *lacustris* may refer to lacustrine (water-deposited) soils near the type locality. The name "Arctic Blue" (Klots 1951) refers to the butterfly's habitat being frequently arctic or alpine.

**Adult:** The Arctic Blue is easily distinguished by the ventral hindwing, which has a postmedian area of almost fused white spots. Both sexes have this characteristic. The upperside of the male wings is steel blue. In northern populations, the blue is interrupted by areas of dark scales. Females in southern populations have a brown ground colour. On the upperside of the female forewings there is a median row of white spots with small black centres. Females of northern population have this same pattern overlaid by steel-blue scales.

**Immature stages:** Hardy (1963) partially described the immature stages of a Vancouver Island population of the Arctic Blue. The egg is white, flattened, and turban-shaped. The second instar larva has a rich, vinaceous purple head. The body has a pair of faint subdorsal lines on a honey-coloured ground colour.

**Biology:** Peace River populations of the Arctic Blue fly from late May to late June. Other northern BC populations fly from early June to early August, depending on the timing of the spring snow melt. Southern alpine populations fly from late June to late August, again depending on the timing of

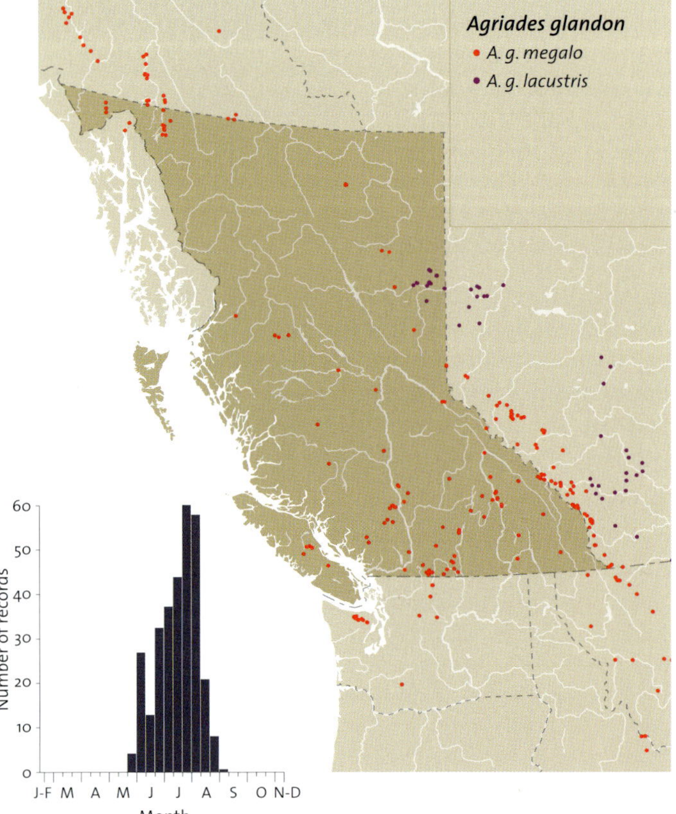

the snow melt. All populations are univoltine. Hardy (1963) documented oviposition and larval feeding on *Saxifraga bronchialis* at Mt. Becher, Vancouver Island. Eggs were laid on 1 August and hatched on 13 August. Second instar larvae entered hibernation on 11 September. J. Pelham (pers. comm.) has observed oviposition on *Saxifraga tricuspidata* in the Yukon. Adults of the Arctic Blue have been associated with *Saxifraga* throughout most of BC. It is not clear whether this affiliation is related to oviposition or only to nectaring. In the Peace River area, J. Pelham (pers. comm.) has observed the Arctic Blue ovipositing on *Oxytropis* sp., but Shepard has seen it always in association with *Saxifraga* sp. in the same area. Scott (1992) has observed Colorado Arctic Blues ovipositing on species of *Androsace* that also occur in BC alpine habitat. Thus *Saxifraga* may be only a nectar source in parts of BC.

**SUBSPECIES:** Populations in most of BC are the subspecies *A. g. megalo* (McDunnough, 1927) (TL: Mt. McLean, Lillooet, BC). In these populations, the upperside of the wings of females have obvious postmedian white spots on forewings and hindwings. Populations from the Peace come close in appearance to the Canadian Prairie subspecies, *A. g. lacustris* (T. Freeman, 1939) (TL: Norway House, MB). These populations have females with faint postmedian white spots on both upper wings. The ground colour is the same grey. Bird et al. (1995), Klassen et al. (1989), and Layberry et al. (1998) erred in applying the name *A. g. rustica* (W.H. Edwards, 1865) (TL: Empire, Clear Creek Co., CO) to the Canadian Prairie populations. Hooper (1986), correctly, did not use the taxon *rustica* for Saskatchewan. In the southern Rocky Mountains subspecies, *A. g. rustica,* the upper surface of the female wings is a uniform chocolate brown. The Prairie subspecies is grey with a visible pattern of spots, similar to *megalo*.

**RANGE AND HABITAT:** The Arctic Blue is found throughout BC except for the Northern Mainland Coast and the Queen

Ssp. *megalo* ♂ D (2.5 CM)

Ssp. *megalo* ♀ D (2.5 CM)

Ssp. *megalo* ♂ V (2.5 CM)

Ssp. *lacustris* ♂ V (2.5 CM)

Charlotte Islands. In southern BC, the Arctic Blue is found above timberline. In northern BC, it is found at elevations from 500 to 2,000 m.

**GENERAL DISTRIBUTION:**
The Arctic Blue is found in central and northern AK and

Ssp. *lacustris* ♀ D (2.3 CM)

east across most of the Canadian Arctic to NF. In the west it ranges south to northern WA and southeastern MB, and south in the Rockies to NM.

**CONSERVATION STATUS:** Subspecies *lacustris* is of Special Concern in BC (S3). Subspecies *megalo* is not of concern (S5).

# Family Riodinidae Grote, 1895   Metalmarks

The common name is derived from the shiny metallic spots or streaks that are present on many species, although not the BC species.

Metalmarks are a group with only a few species worldwide except in the neotropics, where they represent a significant proportion of the fauna.

The metalmarks are closely related to lycaenids but differ in several respects. The forelegs of the male have lost tarsal segments and are not functional for walking; by contrast, the forelegs of male lycaenids have fused tarsal segments. The hindwing of metalmarks has a precostal vein that is not present in lycaenids. The male genitalia of metalmarks have a well-developed uncus. The larvae lack both honey glands and eversible tubercles, and this distinguishes them from all lycaenids except the Lycaeninae (coppers). Unlike lycaenids, metalmarks cannot retract the head. Metalmark larvae have five instars, whereas those of lycaenids have four. Another important characteristic of metalmark larvae is the spiracle of the first abdominal segment, which is displaced anteroventrally (Ballmer and Pratt 1989).

Mormon Metalmark (*Apodemia mormo*)

The derivation of the name *Apodemia* is unknown. The common name is probably derived from the name of the family, since metallic markings are notably absent in this genus.

*Apodemia* is a Nearctic genus of 10 species. Most are found in the American southwest and/or northern Mexico. One species is widespread in western North America and just reaches BC.

## MORMON METALMARK
*Apodemia mormo* (C. & R. Felder, 1859)

**ETYMOLOGY:** The species name *mormo* referred to the Mormons of Utah because C. and R. Felder thought that the type locality was Salt Lake, Utah. The common name was first used by Holland (1898) in the form "The Mormon."

**ADULT:** The Mormon Metalmark is like no other butterfly in BC and can be easily identified by referring to the photographs of the adults.

**IMMATURE STAGES:** The larvae of the only species of metalmark in BC can be distinguished from those of BC lycaenids using the family characteristics given earlier.

**BIOLOGY:** The Mormon Metalmark has one generation in BC. Adults are seen from mid-August to late September. Further south in the species' range, it overwinters as a larva, feeding whenever the temperature is above 13°C (Ballmer and Pratt 1989). Since such temperatures are rarely reached in the Keremeos area in winter, it must hibernate, either as an egg or as an early instar larva. The populations in the Keremeos

♂ D  (3.0 cm)                    ♂ V  (3.0 cm)

area are found on *Eriogonum niveum* (CSG; JHS). In 1989, adults at the population near Keremeos were at peak abundance from 11 to 20 August, and adults were observed until 1 October 1989 (Huber Moore, pers. comm.).

**SUBSPECIES:** BC populations are the nominate subspecies, *A. m. mormo* (C. & R. Felder, 1859) (TL: Davis Creek Park, Washoe Co., NV [Emmel et al. 1998d]), which occurs south through the Columbia Basin and northern Great Basin.

**RANGE AND HABITAT:** The Mormon Metalmark is historically known only from the southern Okanagan Valley and near Keremeos, where *Eriogonum* and rabbitbrush occur together. It has just recently (1999) been rediscovered in the Okanagan Valley. A group of closely adjacent populations near Keremeos are still strong within a very limited area, but the habitat needs to be protected from detrimental changes.

**GENERAL DISTRIBUTION:** The Mormon Metalmark is found in the southernmost part of the Southern Interior of BC and southern SK, south to Baja California and northern MEX.

**CONSERVATION STATUS:** The Mormon Metalmark is Endangered in BC (S1).

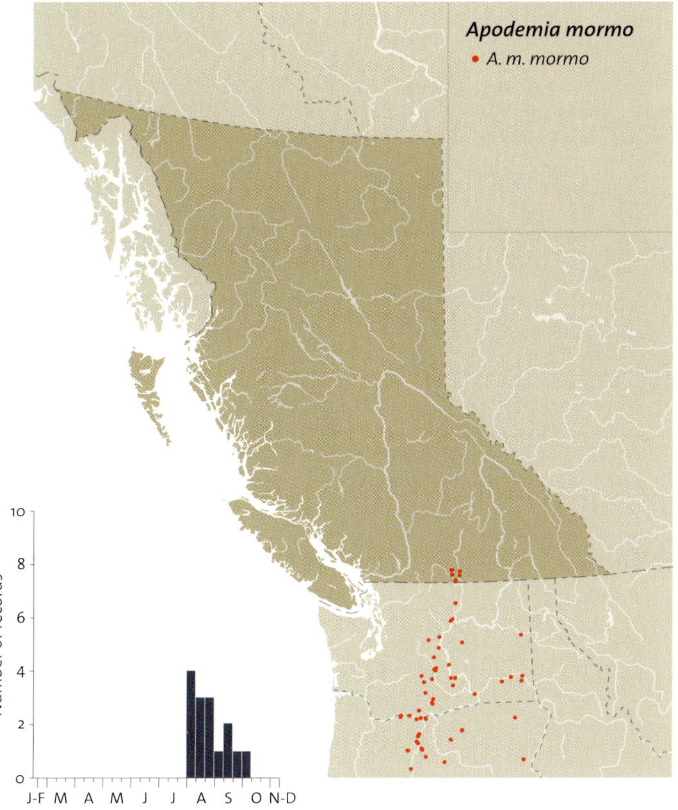

**Mature larva**

**Pupa**

# FAMILY NYMPHALIDAE SWAINSON, 1827  BRUSHFOOTS

The family name is derived from the name of the type genus, *Nymphalis*. The common name "brushfoots" refers to the reduced, hairy, "brushlike" forelegs.

The brushfoots are variable in colour, but the majority are medium-sized to large butterflies with wings that are orange brown with black markings. Some species are very small, and some have other wing colours. There are about 3,900 species worldwide, mostly in the tropics.

The forelegs of brushfoots are strongly reduced to hairy brushes that are not functional for walking. In addition, the eyes are not indented next to the antennae, and the face is as wide as it is tall between the eyes. Lycaenids and riodinids also have strongly reduced forelegs, but their eyes are indented next to the antennae and the face is much taller than it is wide (Weller et al. 1996).

**Anicia Checkerspot (*Euphydryas anicia*)**

The subfamily name is derived from the type genus, *Nymphalis*. The common name "anglewings" is derived from the angular, irregular outer edge of the wings of many of the butterflies in this subfamily.

Anglewings are medium to large butterflies with broad wings that have "ragged" edges. In BC most species are either orange brown with black markings on the upperside, and leaf- or bark-coloured on the underside, or black with white markings. They typically have a moderately fast and direct flight pattern, and frequently land on the ground or on vegetation after a short flight.

Antennae are long and rigid, with a well-defined but narrow club. The discal cells of both the forewing and hindwing are closed in most species. The legs of many species are extremely hairy, especially the male foreleg tarsus.

Eggs are barrel-shaped, reticulate, and strongly ribbed in BC species. The larvae have dorsal, lateral, and subspiracular rows of branching spines, and have no tails. The pupae have twin prominences on the head, and either dorsal tubercles or a dorsal lobe on the second abdominal segment. In BC all species either hibernate as adults and/or migrate into the province in spring from the southern USA and northern Mexico.

## GENUS *POLYGONIA* HÜBNER, [1819]  ANGLEWINGS

The name *Polygonia* comes from the Greek *polygonos* (many-angled), and it is very appropriate as the wings of these insects are adorned with many indentations that produce an almost ragged appearance. The common name was first used by Holland (1898) in reference to the Latin name and the many angles along the edge of the wings.

Anglewings are generally medium-sized butterflies. The wings are orange brown with black markings on the upperside, and bark- or leaf-patterned on the underside. The edges of the wings are ragged in appearance. Males and females usually have quite different patterns on the underside of the wings, with the female pattern being plainer. There are about 15 species worldwide.

Eggs are laid singly on the underside of leaves of the foodplants. The eggs are cream in colour, later turning dark as the larva matures inside. Mature larvae are variably coloured but usually resemble bird droppings; they have numerous branching spines. Anglewing adults hibernate in sheltered areas such as hollow trees or stumps, debris piles, house crawl spaces, or barns.

## SATYR ANGLEWING
*Polygonia satyrus* (W.H. Edwards, 1869)

**ETYMOLOGY:** The name *satyrus* refers to Satyrs, Greek forest deities who were half man, half goat and lived in forests. Thus the Latin and common names refer to the open forests inhabited by this butterfly. The common name was first used by Holland (1898).

**ADULT:** Satyr Anglewings are the most common anglewing in most areas of BC. The upperside of the wings is brightly golden brown; only the Grey Comma is more golden. Males are darker on the upperside of the wings than females, and the underside of the wings is a variegated grey brown. Females are tan-coloured on the underside of the wings; the basal half of the wings is darker than the outer half.

**IMMATURE STAGES:** Eggs have 10–12 vertical ribs (Scott 1988) and are whitish, turning lead-coloured with the head showing as a jet-black spot on top at maturity. Mature larvae have black angular heads that are bilobed, with a spiny tubercle at the top of each lobe. The body is black with a dorsal row of spines and three rows of spines on the sides. There is a broad white or green white band on the back that includes the dorsal and subdorsal rows of spines. The white

**Satyr Anglewing (*Polygonia satyrus*)**

band has a fine V-shaped black mark around each dorsal spine. The bottom row of spines is also green white along the middle portion of the body. In mature larvae in Colorado, the white areas are yellow in colour (Scott 1988). Pupae are tan or straw-coloured and sometimes yellowish on the back, rarely brown all over. The projection in the middle of the back of the thorax, and the projections in a row down the back of the abdomen, are much taller than in our other anglewings (Edwards 1868–72; Scott 1988; GAH).

**BIOLOGY:** Satyr Anglewings are bivoltine on the south coast, probably bivoltine in the Southern Interior, and likely univoltine elsewhere in BC. South coast adults that hibernate lay eggs on young stinging nettle in April and May. These eggs hatch within 7–10 days, and the next generation of adults emerges about two months later, in June. More eggs are laid, and adults emerge in July and August. These adults do not reproduce immediately but instead hibernate. Larvae are easily found on stinging nettle in leaves folded into "tents." Both males and females feed on tree sap in the fall and sap from broken willow branches in early spring; they also mud-puddle (CSG; JHS).

Stinging nettle (*Urtica dioica*) and hops (*Humulus lupulus*) are used as larval foodplants in BC, with eggs being laid on the stems and the underside of the leaves (Dyar 1904b; Harvey 1908; Jones 1933; CSG; GAH). Hops are grown in the Lower Fraser Valley of BC.

♂ D (5.2 cm)

♂ V (5.2 cm)

♀ V (5.3 cm)

**SUBSPECIES:** None. The type locality of the species is Empire, CO. The subspecies name *neomarsayas* dos Passos, 1969 (TL: Salmon Meadows, Okanogan Co., WA) is a synonym of the nominate subspecies.

**RANGE AND HABITAT:** Satyr Anglewings occur throughout BC in riparian areas and other moist habitats that support stinging nettle, especially open deciduous forests. They could become a common suburban butterfly if patches of stinging nettle are encouraged on vacant land and in out-of-the-way areas of parks and gardens.

**GENERAL DISTRIBUTION:** Satyr Anglewings are found across southern CAN and northern USA. They are also found south to CA and NM, almost to the Mexican border.

**CONSERVATION STATUS:** Not of concern (S5).

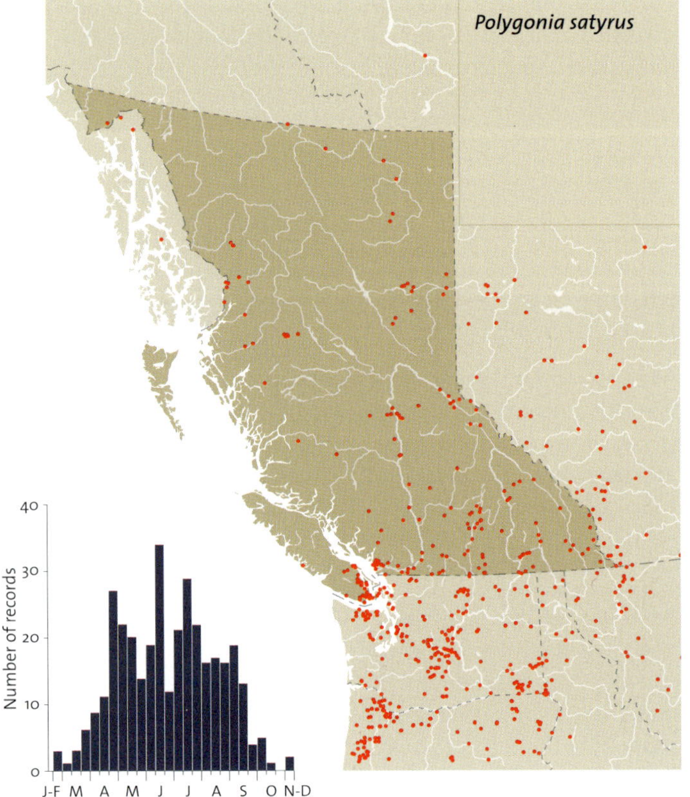

**Eggs**

**Mature larva**

**Pupa**

# Green Comma
*Polygonia faunus* (W.H. Edwards, 1862)

Green Comma (*Polygonia faunus*) female

♂ D (4.6 cm)

♂ V (4.6 cm)

♀ V (4.8 cm)

**Etymology:** The name *faunus* refers to a faun (a woodland deity), either Pan himself or, more often, one of his attendants (Emmet 1991), in reference to the butterfly's woodland habitat. The subspecies name *rusticus* is Latin, meaning "in the country." The common name was first used by Gosse (1840) in reference to the characteristic greenish markings on the underside of the wings, as well as the white "comma" on the ventral hindwing.

**Adult:** Green Commas are smaller and darker than any other anglewing except Hoary Anglewings, and are the second most common anglewing in BC. They are characterized by the green spotting on the underside of the wings of the males, which is sometimes barely discernible in females. Males have a variegated pattern on the underside of the wings, whereas females are more evenly brown.

**Immature stages:** Eggs are green, with 10–12 vertical ribs. Mature larvae have black bilobed heads, with short black spines at the top of the lobes. The body has a dorsal row of branched spines, and three lateral rows along each side. The front half of the body is bright buff orange, and the rear half is pure white. Along the sides are two wavy orange lines that unite irregularly, above which are faint black lines. There are black dots in the gaps between the segments; the gaps are buff (front half of body) or white (rear half of body). The spiracles are black and ringed with white; the underside is pinkish tan. Pupae are light brown, often with a reddish flush on top of the front of the abdomen, or dark grey. The pupae are longer and slimmer, with front projections twice as long, compared with the pupae of our other anglewings. The upper tubercles are silvered (Edwards 1868–72, 1874–84; Scott 1988; CSG). The larval description by Sugden (1970) is apparently in error.

**Biology:** Green Commas are univoltine. In southern BC adults that hibernate lay eggs on birch or willow in April and May. Oviposition on birch sometimes occurs in small sunlit

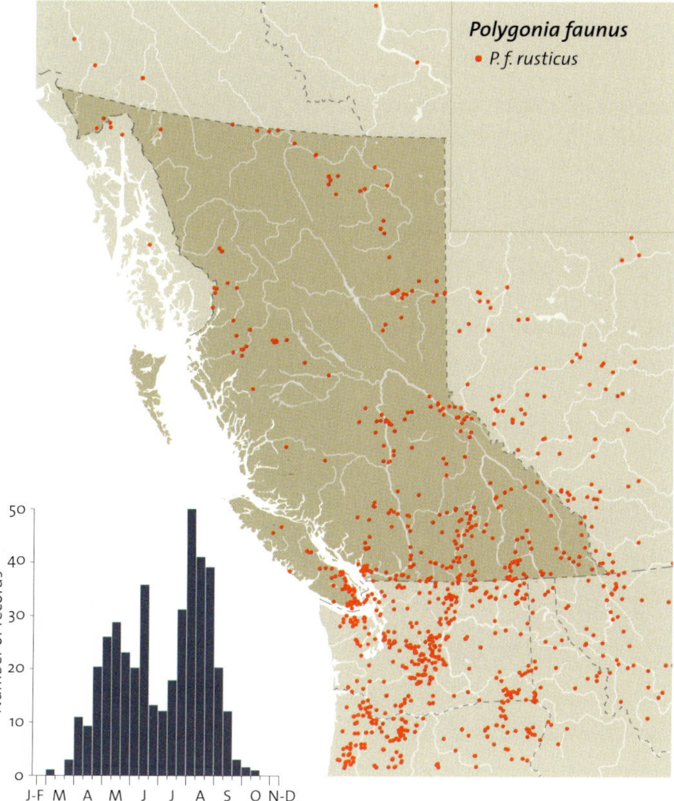

Mature larva

Pupa

patches within moderately dense forest (CSG). These eggs hatch within in a week or so, and the next generation of adults emerges in July. These adults do not reproduce immediately, but hibernate instead. Adults feed on poplar sap in the fall, and sap from broken willow branches in early spring (CSG; JHS). Adults hibernate regularly in woodpiles near Prince George (Jack McGhee, pers. comm.), debris piles, crawl spaces, barns, hollow trees and stumps, and other similar sheltered areas.

Larval foodplants in BC include paper birch, mountain alder, green alder, and willow (Harvey 1908; Sugden, 1970; CSG; FIS). Outside BC additional foodplants include *Betula lenta, Ribes inerme, Salix humilus, S. bebbiana,* and *Rhododendron occidentale* (Edwards 1868–72, 1884; Scudder 1889a; Caulfield 1875; Ferris and Brown 1981; Scott 1992).

**SUBSPECIES:** Green Commas in BC are subspecies *rusticus* (W.H. Edwards, 1874) (TL: Big Trees, Calaveras Co., CA). In the mountains of northern BC, they tend to be smaller, darker, and greyer, tending towards subspecies *arcticus* Leussler, 1935 (TL: base of Black Mt., 30 miles SW of Aklavik, NT). Subspecies *arcticus* is more similar to subspecies *hylas* Edwards, 1872 (TL: Colorado) than to subspecies *rusticus* (dos Passos 1977).

**RANGE AND HABITAT:** Green Commas are found in riparian and open deciduous and coniferous forests throughout BC.

**GENERAL DISTRIBUTION:** Green Commas occur from central AK south and east across the Canadian boreal forest to NS, south to CA and CO in the west, and south along the Appalachians in the eastern USA.

**CONSERVATION STATUS:** Not of concern (S5).

## ZEPHYR ANGLEWING
*Polygonia zephyrus* (W.H. Edwards, 1870)

**ETYMOLOGY:** Zephyrus was the personification of the west wind; thus the name refers to this butterfly's western distribution. The common name was first used by Holland (1898) as a direct transliteration of the species name.

**ADULT:** Zephyr Anglewings are characterized by two-toned grey undersides to the wings, with greenish yellow sub-marginal spots that have dark centres. Males and females are very similar. Compared with males, females are slightly larger and slightly brighter orange brown on the upperside

♂ D  (4.9 CM)

♂ V  (4.9 CM)

♀ V  (4.8 CM)

of the wings, and appear slightly more faded on the underside of the wings.

**IMMATURE STAGES:** Mature larvae have bilobed shiny black heads, with black spines on each lobe and with white markings and white hairs. The body has a dorsal row of spines and three lateral rows of spines on each side. They are black, with a pale yellow line at the front and rear of segments 3–5 on

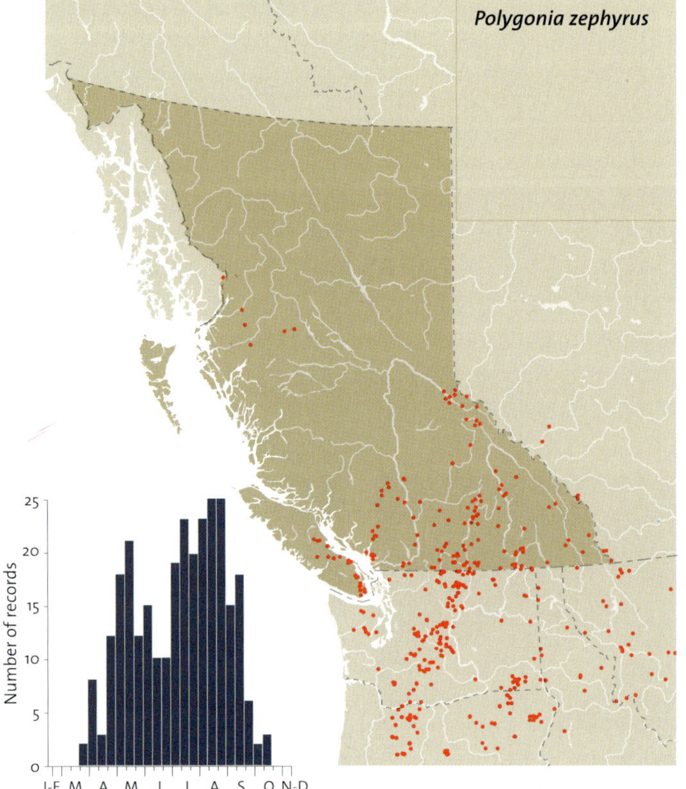

Mature larva

Pupa



FAMILY NYMPHALIDAE (BRUSHFOOTS)

the back, and with a faint pale yellow line down the back of the thorax. The front half of the back is reddish, including the dorsal and subdorsal rows of spines. The rear half of the back is white, including the three rows of spines, except for a black arrow on the top of each segment. All the other spines are black, and on the sides are many pale tubercles. The underside is dark brown. Pupae can be either dark- or light-coloured; dark pupae appear to be females and light pupae males (Edwards 1874–84; GAH).

**BIOLOGY:** Zephyr Anglewings are univoltine, and mate and lay eggs early in spring, after hibernation. These eggs hatch within a week or so, and the next generation of adults emerges in July or August. These adults do not reproduce immediately but hibernate instead in sheltered areas such as hollow trees, stumps, and debris piles.

## HOARY ANGLEWING
*Polygonia gracilis* (Grote & Robinson, 1867)

**ETYMOLOGY:** *Gracilis* is Latin for slender, slight, perhaps referring to the small size of the species. The common name was first used by Scudder (1874b) in reference to the flecked grey (hoary) colour of the outer part of the underside of the wings.

**ADULT:** Hoary Anglewings are very similar to Zephyr Anglewings, but are smaller. The upperside of the wings has smaller black markings. The underside of the wings is a darker grey. As with Zephyr Anglewings, males and females are very similar in appearance.

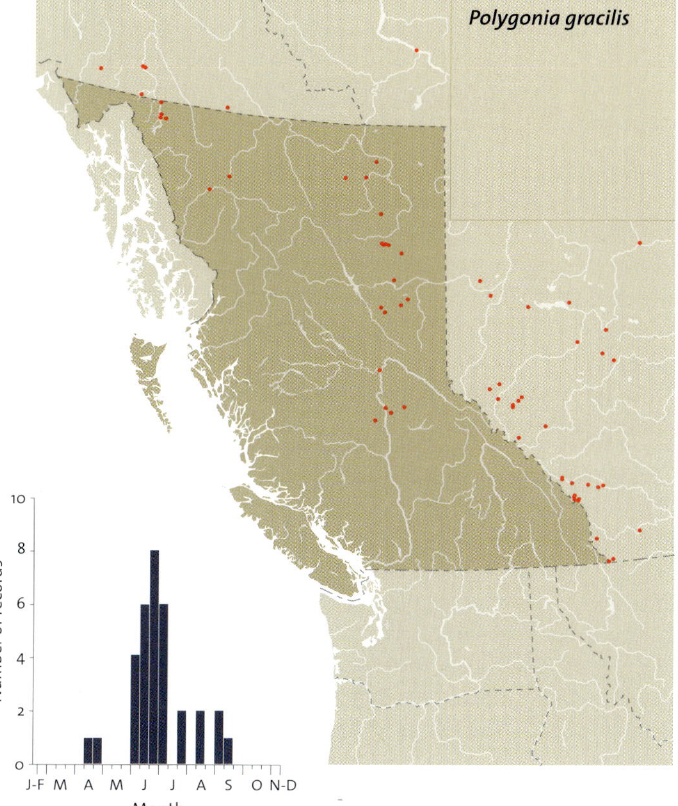


Larval foodplants in BC are *Ribes,* including red-flowering currant and white rhododendron (Harvey 1908; Sugden 1970; ACJ; CSG; GAH). Outside BC larval foodplants include *Ribes cereum* and *R. inerme* (Emmel and Emmel 1974; Ferris and Brown 1981; Scott 1992).

**SUBSPECIES:** None. The type locality of the species is Virginia City, Storey Co., NV.

**RANGE AND HABITAT:** Zephyr Anglewings occur across southern BC at all elevations, but most commonly in sub-alpine coniferous forest with white rhododendron.

**GENERAL DISTRIBUTION:** Zephyr Anglewings occur from southern BC and AB south to CA and NM in the west.

**CONSERVATION STATUS:** Not of concern (S5).

♂ D  (4.5 cm)          ♂ V  (4.5 cm)

**IMMATURE STAGES:** Mature larvae from Clemina, BC, are similar to those of *P. zephyrus* except that in a dark variant, pale portions are washed with pale buff and brown (Sugden 1970).

**BIOLOGY:** Hoary Anglewings are univoltine, with adults mating and laying eggs after hibernation, in May and June. These eggs hatch within in a week or so, and the next generation of adults emerges in July and August. These adults do not reproduce immediately, but instead hibernate during winter.

Larval foodplants are willows at Clemina, BC (Sugden 1970). CSG reared two Hoary Anglewings from larvae on Sitka alder north of Wonowon, at Mile 110 of the Alaska Highway. Wild red currant is used as a foodplant in Alaska (Bird et al. 1995).

**SUBSPECIES:** None. The type locality of the species is Mt. Washington, NH. Some authors consider that the Zephyr Anglewing is a subspecies of the Hoary Anglewing.

**RANGE AND HABITAT:** Hoary Anglewings occur in openings in coniferous forests and along river banks across northern BC.

**GENERAL DISTRIBUTION:** Hoary Anglewings occur from central AK south to central BC, and east across boreal Canada to the Maritimes and the northern USA.

**CONSERVATION STATUS:** Not of concern (S5).

# OREAS ANGLEWING
*Polygonia oreas* (W.H. Edwards, 1869)

**ETYMOLOGY:** *Oreas* is Latin for mountain nymph, and the subspecies name *silenus* refers to Silenus, the leader of Satyrs in Greek mythology and foster father of Bacchus. Both names continue the theme of Greek nymphs in this genus. Subspecies *threatfuli* is named for David L. Threatful, who collected the holotype and who has contributed quietly but very significantly to our knowledge of the butterflies of BC. The common name was first used by Holland (1931).

**ADULT:** Oreas Anglewings are similar to Zephyr Anglewings in the upperside of the wings. The hindwing margin has obvious patches of white fringe hairs, which are also present along the outer margin of the forewing. The basal half of the ventral wing surface is dark blackish grey, and the outer half is lighter in colour, similar to Grey Anglewings. Males and females are similar in coloration, although females tend to be larger and more brightly orange brown on the upperside.

**IMMATURE STAGES:** Mature larvae have black bodies, and the front edge of each segment has a dorsal line of pale yellow on the thorax and white on the abdomen. A more or less distinct pale yellow line runs down the middle of the back of the thorax (Edwards 1874–84). In contrast, mature larvae of subspecies *oreas* from California are dark brown. The front half of the back is yellowish orange, with the middle area orangish yellow and the rear third pale yellowish. There are blackish "V" marks around the mid-dorsal spines of the rear half, with a larger brown mid-dorsal triangle within each "V"

Ssp. *silenus* ♂ D  (4.7 CM)

Ssp. *threatfuli* ♂ D  HOLOTYPE (5.0 CM)

Ssp. *silenus* ♂ V  (4.7 CM)

Ssp. *threatfuli* ♂ V  HOLOTYPE (5.0 CM)

mark. There are black and cream rings around the body between segments, and a wavy orange line above the spiracles containing orange spines. The head is black with orange markings and two spiny black horns (Scott 1986b).

**BIOLOGY:** Oreas Anglewings are univoltine, and hibernate as adults. Populations are usually small and the species is seldom seen. The larval foodplants of subspecies *silenus* on Vancouver Island include garden *Ribes* at Tofino, wild *Ribes* near Victoria, and *Ribes divaricatum* near Cobble Hill (Jones 1938; AGG; GAH). *Ribes divaricatum* is used in California (Emmel et al. 1971). Subspecies *threatfuli* occurs in association with *Ribes* at Vernon, BC.

**SUBSPECIES:** Subspecies *silenus* (W.H. Edwards, 1870) (TL: Portland, OR) is characterized by the extremely dark brown colour of the underside of the wings, which is unlike any other anglewing in BC. It occurs on Vancouver Island and the coastal mainland, north to Bella Coola on the coast and to Quesnel in the interior. From the Okanagan Valley east through the Kootenays to southeastern Alberta, and south through Washington, is an undescribed subspecies, for which we provide a name here.

***Polygonia oreas threatfuli*** **Guppy & Shepard, new subspecies.** *Polygonia oreas threatfuli* is similar to subspecies *silenus* on the upperside. The upperside is much more darkly marked, and has smaller yellow submarginal spots on the dorsal hindwing, than *nigrozephyrus* Scott, 1984. On the underside of the wings, the basal half is dark blackish grey, and the outer half lighter grey. The overall ventral coloration is much lighter and greyer than that of subspecies *silenus,* and the reddish edge to the wings of *silenus* is absent from *threatfuli.* The ventral wings are darker, and grey rather than brown,

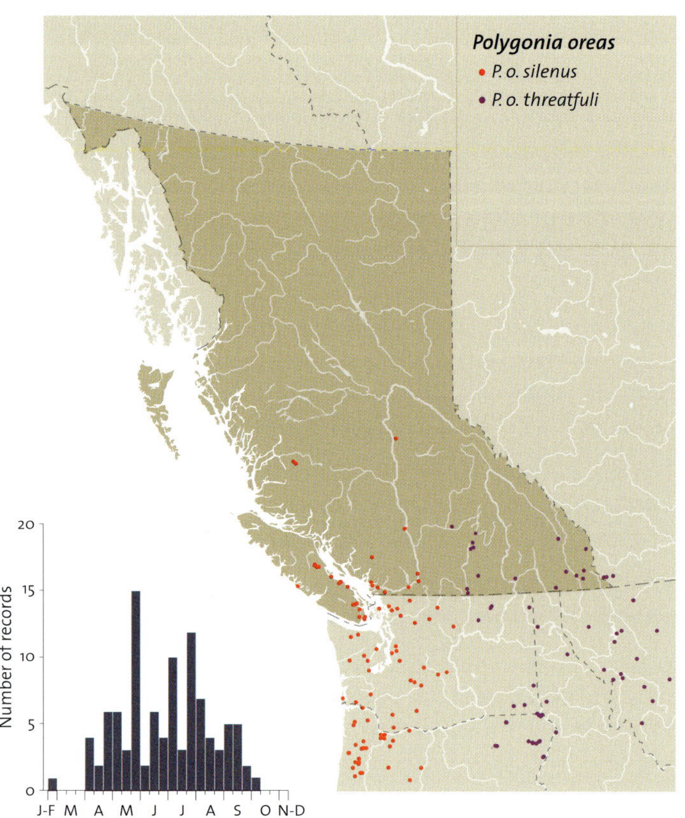

compared with those of the nominate subspecies. The wing margin is less strongly scalloped than in *silenus,* similar to that of *nigrozephyrus,* and slightly less scalloped than that of the nominate subspecies. **Types**. Holotype: female, BC, Vernon, Kalamalka L. Prov. Park, 21 September 1982, D.L. Threatful; a label "HOLOTYPE / *Polygonia oreas / threatfuli* Guppy & Shepard" is attached. The holotype is deposited in the Royal British Columbia Museum, Victoria, BC, CAN. Paratypes: 1 female, BC, Enderby, 7 Sep 1921, J. Wynne (UBC); 1 female, BC, Enderby, 20 Aug 1921, J. Wynne (UBC); 3 males, BC, Enderby, 4 Sep 1921, J. Wynne (UBC); 1 male, 1 female, BC, W. Crescent Valley, 23 May 1967, J.H. Shepard (JHS); 1 female, BC, W. Crescent Valley, 5 June 1967, J.H. Shepard (JHS); 1 male, BC, Natal, 18 mi. W, 17 June 1968, J.H. Shepard (JHS); 1 male, BC, near Yahk, 3 mi. W. junction Hwy 3/95, 16 June 1967, J.H. Shepard (JHS).

**RANGE AND HABITAT:** Oreas Anglewings occur on the coast south of Bella Coola, west across central and southern BC to the Rockies. Subspecies *silenus* occurs in moist forest habitats from sea level to subalpine habitats, from the south coast north to Quesnel in the Central Interior. Subspecies *threatfuli* occurs in moist sites within dry grassland environments, and also in open forest. It has apparently always been

Mature larva (ssp. *silenus*)          Pupae

rare in BC, and its habitat is being degraded by cattle grazing in grassland riparian areas.

**GENERAL DISTRIBUTION:** Oreas Anglewings occur from the south coast of BC east to southwestern AB, south to northern CA in the west and UT, WY, and CO in the east. The locality label of a subspecies *silenus* specimen from "Klotassin River" in the northern YT (Ferris et al. 1983; Layberry et al. 1998) is unlikely to be correct, although the recent discovery of subspecies *silenus* near Quesnel, BC, provides some support for the YT record.

**CONSERVATION STATUS:** Subspecies *threatfuli* is of Special Concern in BC (S3). Subspecies *silenus* is not of concern (S4).

# GREY COMMA
*Polygonia progne* (Cramer, [1776])

**ETYMOLOGY:** Progne was a daughter of Pandion, King of Athens (Reed 1870). The common name was first used by Gosse (1840) in reference to the grey brown colour of the underside of the wings. American publications have since consistently misspelled the name as "Gray" Comma.

**ADULT:** The Grey Comma is more brightly orange brown on the upperside than any of the other anglewings, and as a result can be recognized in flight. The underside of the wings is grey brown and tan. The ventral hindwing "comma" is very faint, and the base of the ventral wings is very dark. The ventral forewing has a pale tan central band.

**IMMATURE STAGES:** Eggs are green and conical. Mature larvae are buff-coloured, with black and pale buff stripes across the back of each segment. There is a V-shaped reddish mark around the base of each of the spines down the middle of the back, a reddish bar at an angle in front of each of the dorsolateral spines, and a reddish patch at the base of each lateral spine. The spines down the back are white with reddish yellow bases, the dorsolateral spines are mostly white (black near the front and at the rear), the lateral spines are black with reddish yellow bases, and the ventrolateral spines are white with yellow bases. The black spiracles are surrounded with reddish yellow ovals. The head is black with reddish yellow markings and two spiny horns. Pupae have dull green heads, the wings are pinkish white, with a broad pale band across them, and a darker green stripe occurs

♂ D  (4.9 CM)          ♀ D  (4.6 CM)

♂ V  (4.9 CM)          ♀ V  (4.6 CM)

down each side of the abdomen; the remaining areas are mottled pinkish brown, other than a pink stripe down the back of the abdomen and some oblique pink lines on the side of the abdomen. Both larval and pupal colours are highly variable (Edwards 1880).

**BIOLOGY:** Grey Commas are univoltine in BC, although they are reportedly bivoltine elsewhere in North America

(Layberry et al. 1998); the adults hibernate. Larvae feed exposed on the leaf bottom or side, or on the leaf petiole. Mature larvae rest with the front part of the body arched and turned at a right angle to the rest of the body, and with the last three segments of the abdomen angled upward into the air. Only the middle prolegs are used to grasp the leaf. When disturbed, the larvae flail their heads and tails. Eggs and larvae have been found on wild gooseberry (*Ribes rotundifolium*) and elm (*Ulmus*) in eastern North America,

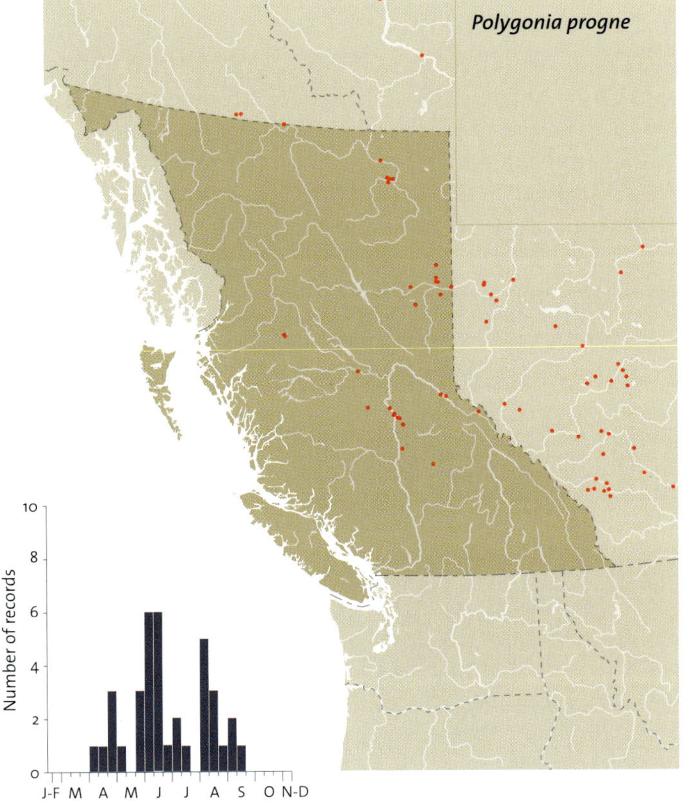

Grey Comma (*Polygonia progne*)

and have been a pest on cultivated currants in Illinois and Ontario (Edwards 1880; Scudder 1889a). Adults feed on poplar tree sap in the fall (CSG).

**SUBSPECIES:** None. The type locality of the species is New York. Some authors consider the Oreas Anglewing to be a subspecies of the Grey Comma because of the similarity of the underside of the hindwings, because the distributions are complementary, and because they use the same larval foodplants. The dorsal wing pattern is very distinctly different, however, and both species occur together near Quesnel, BC, in Guppy's front yard.

**RANGE AND HABITAT:** Grey Commas occur in forest openings and riparian areas in central and northeastern BC.

**GENERAL DISTRIBUTION:** Grey Commas occur from central and northeastern BC east across boreal Canada to NF, and south to the central USA east of the Rockies.

**CONSERVATION STATUS:** Not of concern (S5).

## GENUS *RODDIA* KORSHUNOV, 1995  TORTOISESHELLS

The genus *Roddia* was named in honour of the Russian entomologist Evgenii Georgevich Rodd (Korshunov and Gorbunov 1995). The common name "tortoiseshells" is shared with the genera *Nymphalis* and *Aglais* because the species in all three genera were originally placed in the genus *Aglais*.

There is only a single species in this genus, the Compton Tortoiseshell, *Roddia l-album,* although it may eventually be demonstrated that the North American subspecies *j-album* is actually a separate species from the Asian *l-album*. *Roddia l-album* has traditionally been called *Nymphalis vau-album* (Denis & Schiffermüller, [1775]). Compton Tortoiseshells are large, golden brown butter-

flies that are frequently very abundant in the BC interior, especially in the fall.

The genus *Roddia* Korshunov was described in 1995, with *Papilio l-album* Esper, 1780 as the type species (Korshunov and Gorbunov 1995). *Roddia* is more closely related to *Polygonia* than to *Nymphalis,* according to comparisons of mitochondrial and nuclear DNA (Soren Nylin, pers. comm.). This is supported by several characteristics it has in common with *Polygonia:* the white "comma" marking on the ventral hindwing, the sexual dimorphism in ventral wing colour, the pair of branched spines on the top of the head of the larva, and the branched mid-dorsal spines on all abdominal segments of the larva. As a result *Roddia* could be



FAMILY NYMPHALIDAE (BRUSHFOOTS)

treated either as a subgenus of *Polygonia* or as a full genus. We have chosen to treat *Roddia* as a full genus because of differences from *Polygonia,* such as wing shape and genitalia that more closely resemble those of *Nymphalis.*

The species name *l-album* was generally used until dos Passos (1964) and Higgins and Riley (1970) treated the species as *vau-album* (Denis & Schiffermüller, [1775]). The North American *j-album* was treated as a separate species until dos Passos (1964) placed it as a subspecies of *vau-album.* Koçak (1981) concluded that the species name *vau-album* is a *nomen nudum,* or "naked name," a name not attached to a specific taxon, by Article 12.1 of the International Code of Zoological Nomenclature. The first available valid name for the species is the traditionally used *l-album* Esper, 1780, which Koçak recommended be returned to use. Kudrna (1986) agreed with this approach, although Sattler and Tremewan (1984) and

Sattler (1989) disagreed on the grounds that *vau-album* was the accepted (modern) usage. Karsholt and Razowski (1996) followed Sattler's recommendation that all Denis and Schiffermüller names be retained "as the opposite view will have incalculable consequences for the stability of the nomenclature of European Lepidoptera." We disagree with this uncritical approach, preferring that of Kudrna (1986), in which each name is looked at individually, retaining only those Denis and Schiffermüller names that actually need to be retained for nomenclatural stability. The name *vau-album* is not one of those names.

The name *l-album* was in use for the 184 years between its description by Esper in 1780 and its rejection by dos Passos (1964), whereas the use of *vau-album* has still been disputed (e.g., Koçak 1981; Kudrna 1986) in the few decades since dos Passos (1964) and Higgins and Riley (1970). As a result of these changes in genus and species, *Nymphalis vau-album* is correctly called *Roddia l-album.*

## COMPTON TORTOISESHELL
*Roddia l-album* (Esper, 1780)

**ETYMOLOGY:** The name *l-album* is derived from the L-shaped white marking on the ventral hindwing and the Latin *album* (white). Subspecies *watsoni* was named for Frank Edward Watson of the American Museum of Natural History. The common name was first used by Gosse (1840) in reference to the town of Compton, Quebec, where he studied the species.

♂ D (6.0 CM)

♂ V (6.0 CM)

♀ V (6.4 CM)

♀ grey form V (6.2 CM)

The European subspecies is called the "False Comma" (Higgins and Riley 1970), a name that could reasonably be applied to the species as a whole.

**ADULT:** Compton Tortoiseshells are large orange brown butterflies with black markings. The underside of the wings has the pattern of lichens or dead leaves. The only somewhat similar butterflies are anglewings, which are much smaller. Males have a strongly variegated grey pattern on the underside of the wings; females have a plain brown wing underside, with the pattern of the males only weakly present. Male forelegs are very hairy, whereas the female foreleg is much less hairy, especially on the inner side.

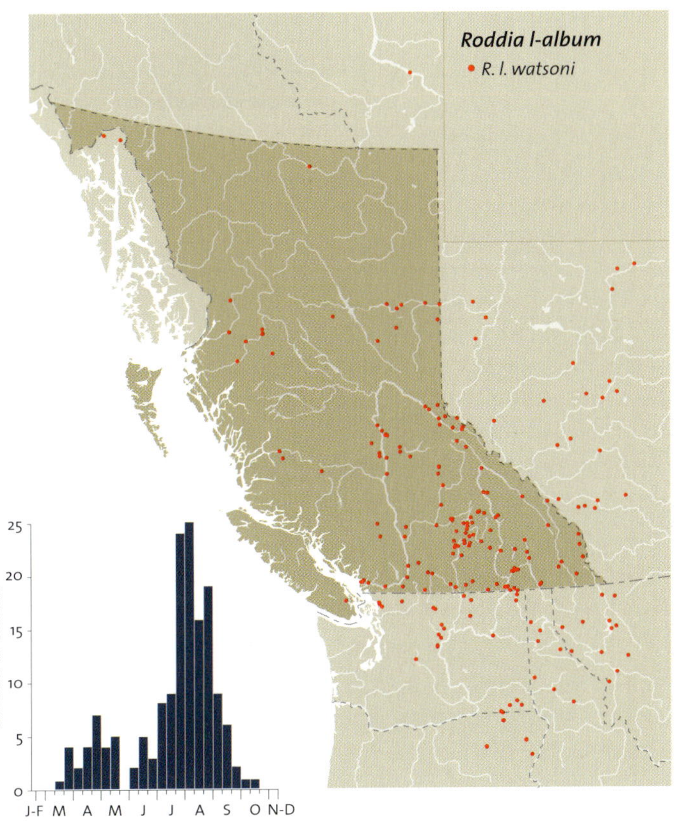

**IMMATURE STAGES:** Eggs are pale green and cylindrical, and have 11 heavy vertical ribs with many fine cross-ribs between them. First instar larvae are tan with a shiny black bilobed head, many black tubercles from which black hairs project, and dark brown legs. Mature larvae from Oliver, BC, have a dull black bilobed head covered with raised white tubercles tipped with hairs, and two spiny black tubercles rise from the top of the head. The body above the spiracles is initially black except for the last two segments, which are red brown, and gradually turns to light yellow brown blotched with darker brown as the larva matures. Down the back is a double row of white dashes, and there are numerous pale brown to white speckles and streaks on the sides. The white dashes down the back and the pale brown to white speckles and streaks on the sides all turn pale green as the larva matures. There is a thin lateral line of white dashes. The body below the lateral line is initially pale grey and brown, but turns pale green with pale brown blotches as the larva matures. There are spiny black scoli with black hairs on the back, which turn yellow brown as the larva matures. Pale brown green scoli on the sides turn pale green with white hairs. Pupae are pale green with pinkish highlights, and three pairs of silver-gold spots on each side of the back of the thorax (CSG). This is in reasonable agreement with the descriptions of Denton (1889) for larvae from Massachusetts, and Dawson (1889) for larvae from Montreal. The descriptions of mature larvae in the literature describe them as being pale green (e.g., Layberry et al. 1998). This is apparently derived from Lintner's description of a single prepupal larva (Scudder 1889a), which ignored the original descriptions of Denton and Dawson, and that single description was then repeatedly reproduced for more than 110 years.

**BIOLOGY:** Compton Tortoiseshells are univoltine, with adults hibernating. They are in flight from late July until they enter hibernation in October, and again from the time they leave hibernation in March until May. In the fall they commonly occur in very large numbers sitting in the middle of

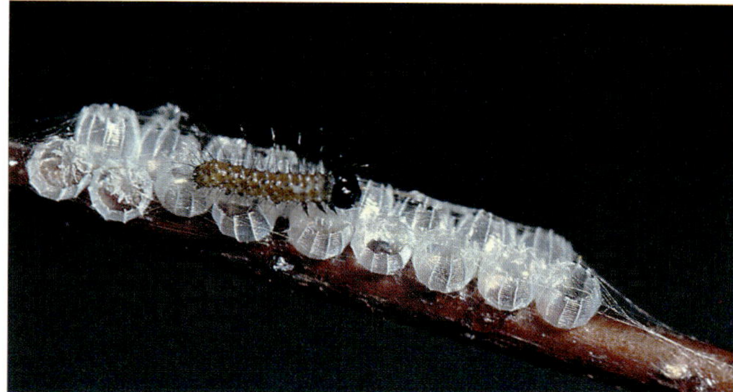

**Hatched eggs and first instar larva**

**Mature larva**

**Prepupal larva**

backroads and highways, with vehicles causing high mortality. Compton Tortoiseshells coming out of hibernation in a Quesnel barn in March 1999 consisted of 17 males and 28 females, suggesting higher male than female mortality (CSG). Oviposition occurs in April. Eggs are laid in small groups or short lines near the tip of twigs of birch

**Pupa**

whose leaves are just beginning to expand. Females bask head up in the sun while ovipositing and between oviposition episodes. When the eggs hatch, the larvae remove a neat disc from the top of the egg, and do not eat the egg chorion. The first and second instars are gregarious, whereas the third to fifth instars are solitary. All larval instars produce silk. In the fall near Attachie in the Peace River area of BC, JHS observed a Compton Tortoiseshell flying directly to a source of tree sap as though returning to a known site, and feeding there. Nectar sources were abundant, hence the butterfly was apparently choosing sap over nectar.

The only larval foodplants known in BC are water birch and paper birch (CSG; FIS). Outside BC, willow and trembling aspen have also been reported (Denton 1889; Holland 1931; Ferris and Brown 1981).

**SUBSPECIES:** Subspecies *watsoni* (G.C. Hall, 1924) (TL: Sicamous, BC) occurs in BC. It is more brown (less grey) on the underside of the wings than the eastern subspecies,

**Compton Tortoiseshell (*Roddia l-album*)**



FAMILY NYMPHALIDAE (BRUSHFOOTS)

*j-album* (Boisduval & Leconte, [1833]) (TL: vicinity New York, Philadelphia, and New Harmony, IN).

**RANGE AND HABITAT:** Compton Tortoiseshells occur throughout the BC mainland east of the coastal mountains, with occasional strays on the coast. They are most common in forest openings, and can be extremely abundant along logging roads in the fall.

**GENERAL DISTRIBUTION:** Compton Tortoiseshells occur in Eurasia, and in North America from western BC across boreal CAN and the northern USA to NF. They frequently migrate far north into the arctic, and far south into the southern USA.

**CONSERVATION STATUS:** Not of concern (S5).

## GENUS *NYMPHALIS* KLUK, 1780  TORTOISESHELLS

The name *Nymphalis* is linked to the Nymphales (nymphs), the name Linnaeus gave to a group of butterflies primarily consisting of present-day Nymphalinae. Both names were derived from the Greek *numphe* (a bride; a nymph all too often made a bride against her will by a lascivious god). The nymphs were lesser deities associated with springs, groves, mountains, and so on (Emmet 1991), a theme that was carried forward into genera such as *Polygonia* and *Cercyonis*. Linnaeus may have been referring to the forested habitats of many of the species in his Nymphales. The common name "tortoiseshells" refers to the resemblance of the upperside of the wings of the butterflies to the mottled yellowish brown "tortoiseshell" of some sea turtles (Comstock and Comstock 1915) that is sometimes used for making combs. The name is shared with the genera *Aglais* and *Roddia* because the species in all three genera were originally placed in genus *Aglais*.

The butterflies in this genus are highly variable in wing pattern and colour, but they all share the same basic irregular wing margins, which includes a lobe on the hindwing resembling a small tail. Comparison of the dorsal and ventral wing patterns of each species shows that they share the same original pattern. Compton Tortoiseshells are now placed in the genus *Roddia,* as discussed earlier. Some authors also incorrectly place Milbert's Tortoiseshells in the genus *Nymphalis* rather than *Aglais*.

Tortoiseshells hibernate in hollow stumps and logs, debris piles, unheated buildings such as cabins and barns, and rock piles. There are seven species of *Nymphalis* worldwide.

## CALIFORNIA TORTOISESHELL
*Nymphalis californica* (Boisduval, 1852)

**ETYMOLOGY:** The name *californica* refers to the species having been named from California. The common name was first used by Holland (1898).

**ADULT:** California Tortoiseshells cannot be confused with any other butterfly in BC. Females are slightly larger and lighter in colour than males, and the forelegs of males are hairier than those of females.

**IMMATURE STAGES:** Mature larvae are deep velvet black. On each segment are five bright steel blue tubercles, each with branching spines. Between the spines are small white dots, and at the base of the dorsal spines are bright yellow patches that form a yellow dorsal line. Pupae are ash grey with a bluish tint, and the abdomen is grey brown. The head has two blackish projections, and the thorax is mottled with brown. There are black dots on the abdomen (Comstock 1927).

**BIOLOGY:** California Tortoiseshells are migratory, and migrate into BC in spring (Coles 1948). In years when large numbers migrate into BC, the larval foodplant, *Ceanothus* sp., can be

D  (5.9 CM)    V  (5.9 CM)

completely defoliated over large areas. This occurred in the Kootenays in 1912, and in Lillooet and Kaslo in 1918 (Middleton 1913; Ross 1913; Cockle 1920; Anderson 1923). California Tortoiseshells hibernate in BC in piles of lumber, woodpiles, and cool buildings (Ross 1913; Cockle 1920; Hardy 1947; Downes 1948), with adults emerging in the spring (Hardy 1961, 1962b). The species is very rare or absent from BC in many years, suggesting that winter survival is very low in some years and that migrants from the south may be required to maintain BC populations. In California they migrate north and east in May and June, and south and west in September and October (Shapiro 1976c). The adults have been seen migrating up mountains in California in midsummer, possibly in search of larval foodplants (Tilden 1962).

Larvae feed on *Ceanothus* species, usually *C. sanguineus,* in BC (Harvey 1908; Cockle 1920; Dyar 1904b; FIS; CSG). California Tortoiseshells found in coastal BC tend to be larger and paler than those in the interior. The coastal BC California Tortoiseshells are all migrants, and they cannot successfully breed there because their larval foodplant occurs only in very small, isolated populations on the coast. California Tortoiseshells successfully produce one new generation in the summer in the interior, and those butterflies are smaller and darker than the coastal ones. This is the phenotype that was named subspecies *herri* in Colorado.

Exudations from freshly opening young needles of true firs, *Abies,* are a major food source for adults in California (Wright 1891) and the butterflies feed on fir sap on Vancouver Island in October (Danby 1891). They also mud-puddle and feed on overripe fruit such as blackberries and apples (Hardy 1961).

Eggs

Third instar larva

Mature larva

Pupa

Spring migrants use early spring flowering bulbs such as crocuses as nectar sources (CSG).

**Subspecies:** None. The type locality of the species has been restricted to near Belden, Plumas Co., CA (Emmel et al. 1998a). It is unlikely that any genetic differentiation is possible, given the strong migratory movements that would constantly mix populations. The subspecies name *herri* Field, 1936 is sometimes used for Rocky Mountain populations, but the *herri* wing pattern is actually a variation resulting from maturation in the mountain environment rather than being due to genetic differences.

**Range and habitat:** California Tortoiseshells occur across southern and central BC as migrants. They can occur in any habitat and at any elevation.

**General distribution:** California Tortoiseshells occur from southern BC south to the northern MEX border, west of the Great Plains. Migratory individuals or temporary populations occur east across southern CAN and northern USA to NY.

**Conservation status:** The California Tortoiseshell is not of conservation concern. The species' rank is S4BSZN, indicating that it breeds in BC, and there are also non-breeding immigrants in areas of the province without *Ceanothus.*

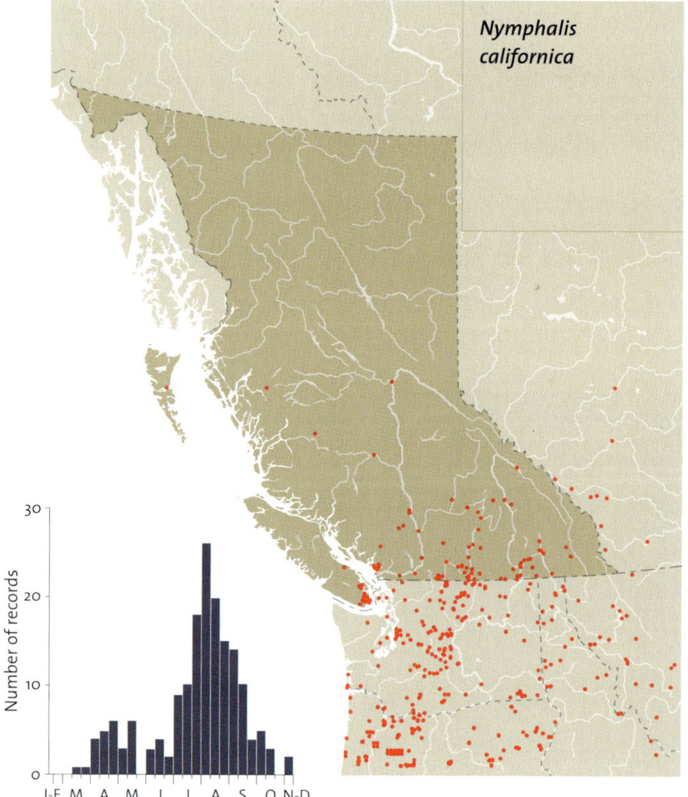

# Mourning Cloak

*Nymphalis antiopa* (Linnaeus, 1758)

**Mourning Cloak (*Nymphalis antiopa*) feeding on sap from birch stump**

D (6.1 cm)

V (6.1 cm)

**Etymology:** The species name *antiopa* refers to Antiope, who gave birth to twin sons, Amphion (a celebrated Greek musician [Reed 1870]) and Zethus, by Zeus. She later married Lycus, king of Thebes (Emmet 1991). The common name "Mourning Cloak" was first used by Harris (1862) and is translated from the German *Trauermantel* (Behrens 1874). The name refers to the traditional colour of a cloak worn during mourning. Scudder (1874b) disapproved of this name because it was not of British derivation, and preferred the British name "Camberwell Beauty" that Gosse (1840) first used in North America.

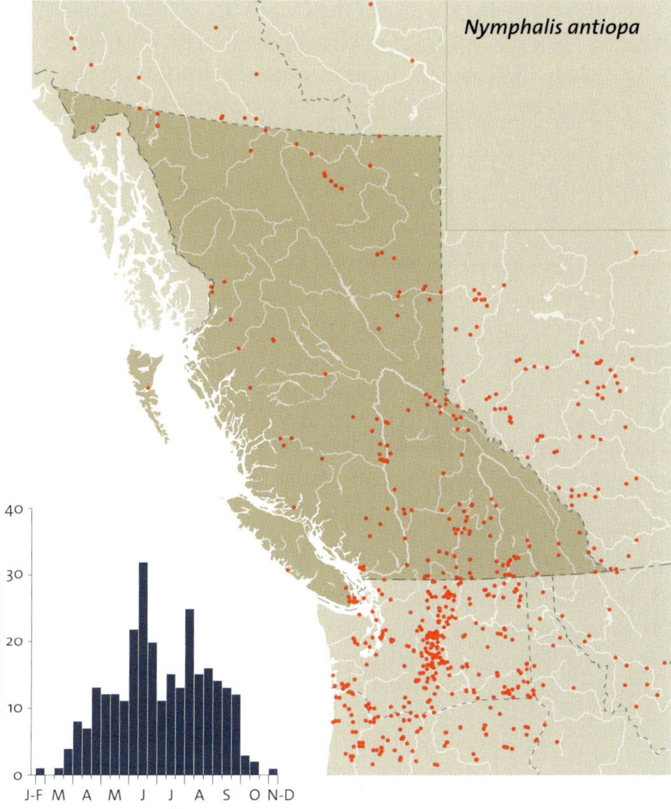



**Adult:** Mourning Cloaks cannot be confused with any other butterfly in BC, with their deep maroon to almost black colour, with a yellow border that fades to white with age. Males and females are almost identical, but the forelegs of males are hairier than those of females.

**Immature stages:** Mature larvae are black with rows of tiny white dots. Down the back is a row of bright red spots. There are rows of tubercles with black spines down the back and sides. The prolegs are dull red. Pupae are dark yellow brown, with blackish markings and reddish patches (Saunders 1869c; Scudder 1893; Sugden 1970).

**Biology:** Mourning Cloaks are univoltine in BC but are bivoltine in eastern Canada and Colorado (Brown et al. 1957; Layberry et al. 1998). Eggs are laid in cylindrical clusters surrounding a willow twig in the spring. The larvae live in an aggregation until the last instar, when they gradually disperse. There are typically about 100 eggs in a cluster, so the aggregations of larvae cause considerable defoliation of the willow plant on which they are feeding. Mature larvae drop or crawl off their larval foodplant, and wander considerable distances before pupating. They are frequently seen crossing streets and sidewalks at this time. Adults emerge from the pupae a couple of weeks after pupation, in July at low elevations.

At Tofino the first Mourning Cloak was collected hibernating in a woodpile in the mid-1930s; they were unknown before this. A few years later a massive population outbreak occurred in which all the willows on a small island in Tofino Inlet were completely defoliated and thousands of larvae crawled over the island searching for food. The outbreak has not been repeated, and Mourning Cloaks have remained present but uncommon around Tofino (AGG). The population outbreak may have resulted from colonization of the area by the species in response to land clearing, which promoted willow growth in an area originally dominated by conifers. Parasites may initially have been absent or rare, permitting a population outbreak of the butterfly.

Mourning Cloaks feed on poplar tree sap prior to hibernation, and hibernate in various sheltered locations, such as lumber piles and debris piles (CSG; JHS). When they emerge from hibernation on warm days in early spring, they feed on sap from stumps of recently cut trees (Saunders 1869c; CSG). Adults also feed on tree sap or in orchards on fruit juices, and are killed by orchard sprays (Atkinson 1978). Adults sometimes live a full year, with ragged hibernated adults still in flight when their freshly emerged offspring appear in flight (Cockle 1915).

Mourning Cloaks migrate in Europe, where it has long been known that they migrate to Great Britain but are not permanent residents (Williams 1949a, 1949b). They also migrate south in the fall in eastern North America, using gliding and soaring flight between periods of flapping flight, in a manner similar to that of Monarchs (Gibo 1981). There is no indication that Mourning Cloaks migrate in western North America.

Larval foodplants in BC are willows (*Salix scouleriana, S. babylonica, S. alba,* and weeping willow), black cottonwood, trembling aspen, and ornamental elm (Harvey 1908; Jones 1942; Sugden, 1970; CSG; FIS). Outside BC other foodplants include *Celtis occidentalis, Onobrychis vicifolia, Salix exigua,*

Eggs

Mature larva

Pupa

*Salix jeppsoni,* and *S. bebbiana* (Shapiro et al. 1981; Byers and Richards 1986; Scott 1992).

Mourning Cloak (*Nymphalis antiopa*)

**SUBSPECIES:** The nominate subspecies occurs throughout BC. Subspecies *hyperborea* (Seitz, 1914) (TL: "extreme northern Alaska") does not occur in BC, and is a synonym of the nominate subspecies (Layberry et al. 1998). Specimens from Fairbanks, AK, considered to be *hyperborea* by Shapiro (1981c, 1981d), are identical to those throughout BC.

**RANGE AND HABITAT:** Mourning Cloaks are found throughout BC at all elevations. They are most often found in forest openings, especially along wet riparian areas in which willow grows.

**GENERAL DISTRIBUTION:** Mourning Cloaks occur throughout much of Eurasia, and throughout subarctic, temperate, and subtropical North America south to northern South America.

**CONSERVATION STATUS:** Not of concern (S5).

### GENUS *AGLAIS* DALMAN, 1816  TORTOISESHELLS

The name *Aglais* is from the Greek *aglaos* (beautiful) or *aglaia* (beauty) (Emmet 1991). The common name "tortoiseshells" refers to the upperside of the wings of the butterflies resembling the mottled yellowish brown "tortoiseshell" of some sea turtles (Comstock and Comstock 1915), which is sometimes used for making combs. The name is shared with the genera *Nymphalis* and *Roddia* because the species in all three genera were originally placed in the genus *Aglais*.

The genus *Aglais* is defined on the basis of differences in genitalia compared with *Nymphalis,* such as an elongate penis and a long saccus (Layberry et al. 1998). In addition, the mitochondrial DNA comparisons of Soren Nylin (pers. comm.) indicate that *Aglais* is sufficiently distant from *Nymphalis* to be treated as a separate genus. There are four species worldwide.

There is only one species in the genus native to North America, the Milbert's Tortoiseshell. A related European species, the Small Tortoiseshell (*Aglais urticae*), has recently been accidentally introduced to North America near New York and may eventually become more widely established. A previous introduction in 1875 near Waterton, MA (Anonymous 1875) apparently did not become established.

# Milbert's Tortoiseshell
*Aglais milberti* (Godart, [1819])

**Milbert's Tortoiseshell (*Aglais milberti*)**

D  (4.8 cm)

V  (4.8 cm)

**Etymology:** The species *milberti* was named for Mr. Milbert, a friend of Godart who collected North American butterflies in 1826. His collection is now in the Museum of Natural History in Paris (Opler and Krizek 1984). Harris (1862) first referred to it as "Milbert's Butterfly."

**Adult:** The wing pattern of the Milbert's Tortoiseshell cannot be confused with that of any other butterfly. Males and females are almost identical, but male forelegs are hairier than female forelegs.

**Immature stages:** Eggs are green and conical, with a flattened base and a rounded top, and 8–9 vertical ribs. First instar larvae are yellow green, with rows of black tubercles. Each tubercle has a long black barbed hair. Mature larvae are black on the top half, and yellow green on the bottom half. They are thickly sprinkled with small yellow white or white dots and fine whitish hairs, producing an overall greyish appearance. The body is covered with branching black spines. There is a greenish yellow line down each side, with another line of brighter orange yellow dashes above it. The spines coming from the greenish yellow line are also greenish yellow. The underside of the body is greyish green. Pupae are generally brown with numerous brilliant metallic gold points, but are highly variable. Pupae that are parasitized by ichneumonid wasps have dull red abdomens, and the thorax and head are polished gold tinged with green (Saunders 1869c; Gosse 1883; Edwards 1885c).

**Biology:** Eggs are laid in masses on stinging nettle leaves, and the larvae form an aggregation for feeding. They feed in aggregations and skeletonize the nettle leaves through the first three instars. Some webbing is spun, apparently to provide a secure foothold. Fourth instar larvae begin dispersing and cut, fold, and roll leaves around themselves in the same way as the other nettle-feeders: Satyr Anglewings, West Coast Ladies, and Red Admirals. More than one larva may inhabit a single folded leaf, and larvae of more than one nettle-feeding butterfly species may share a single leaf. Larvae in the fifth and final instar disperse and feed in the open, without rolling leaves (Edwards 1885c; CSG).

Milbert's Tortoiseshells are univoltine in BC but are bivoltine or trivoltine in eastern Canada and Colorado (Layberry et al. 1998; Brown et al. 1957). Adults emerging in July may aestivate until fall and then hibernate for the winter (Guppy 1955). They overwinter successfully near Vernon (Venables 1913), and it is likely that they do so throughout BC. They may also hibernate as both larvae and pupae in Colorado (Ferris and Brown 1981). Shapiro (1974a, 1979) suggests that in California *milberti* adults hibernate at low elevations on the coast and breed there in the spring, with the resulting offspring migrating to higher elevations. The upward movement appears to occur in the Southern Interior of BC as well, but with no indication of a fall movement back to low elevations. Population densities are usually too low to permit monitoring of elevational movements.

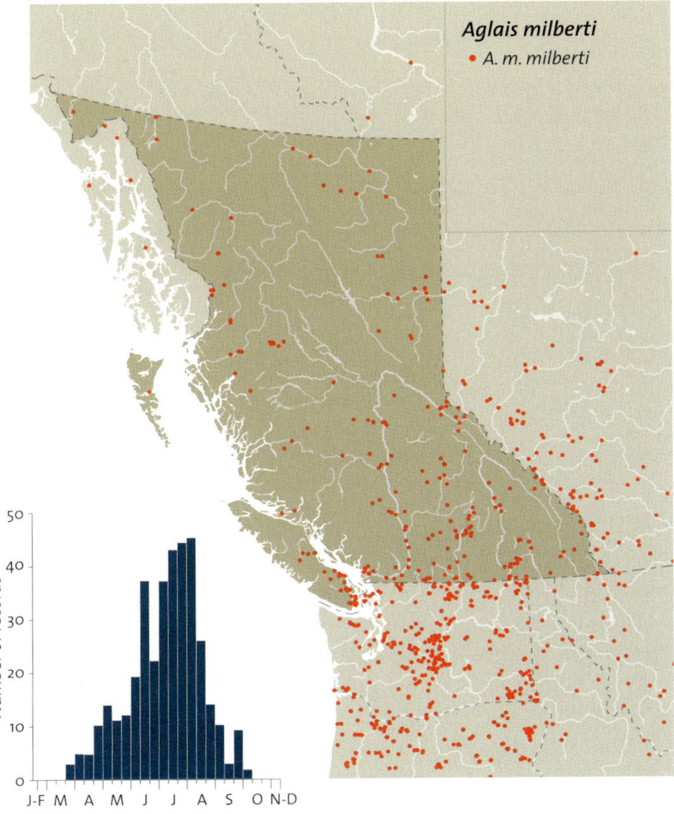

Larvae suffer an extremely high rate of parasitism (Edwards 1885c). All of the about 50 Milbert's Tortoiseshell larvae collected on southern Vancouver Island (Guppy 1951), and all but one of about 100 larvae collected at Botanie Valley near Lytton (CSG), were parasitized by tachinid flies.

Stinging nettle (*Urtica dioica*) is the larval foodplant in BC (Dyar 1904b; Harvey 1908; Jones 1933; Guppy 1951; Atkinson 1978; CSG; FIS) and in most other places. In Colorado, willows and wild sunflowers (*Helianthella*) are reportedly used in addition to stinging nettle (Brown et al. 1957), which seems improbable.

**SUBSPECIES:** The nominate subspecies, *A. m. milberti* (Godart, [1819]) (TL: Philadelphia, PA) occurs throughout BC.

**RANGE AND HABITAT:** Milbert's Tortoiseshells are found throughout BC at all elevations. In August they are frequently numerous in mountain meadows, to which they have flown for nectar from the low-elevation areas where breeding occurs. Breeding occurs through northern BC, with numerous larvae found near the Tatshenshini River in the St. Elias Mountains of extreme northwestern BC (CSG).

**GENERAL DISTRIBUTION:** Milbert's Tortoiseshells occur from southeastern AK east across the southern half of CAN to NF. In the west they occur south to CA and NM almost to the Mexican border, and east of the Rockies south to the central USA.

**CONSERVATION STATUS:** Not of concern (S5).

Eggs

First instar larva

Pupa

Mature larva

## GENUS *VANESSA* FABRICIUS, 1807  LADIES

The name *Vanessa* is probably from Swift's poem of Cadenus and Vanessa (Reed 1870; Emmet 1991). The common name "ladies" is derived from that of the most common species in the genus, the Painted Lady.

Ladies in BC are predominantly medium-sized to large orange brown butterflies with black markings, except for the Red Admiral, which is black with red and white markings. Adults are strong, fast erratic flyers that are most easily approached when they are nectaring. Thistles, alfalfa, and asters are favourite nectar sources (CSG), probably because ladies are most abundant in mid to late summer, when these are the predominant nectar sources.

All four of the ladies in BC are migratory and without permanent populations in BC. There are nine species worldwide.

## AMERICAN LADY
*Vanessa virginiensis* (Drury, 1773)

**ETYMOLOGY:** Drury named the species *virginiensis* because the earliest description was by Petiver (1702–05), who received a specimen from Virginia and called it *Papilio Virginiana* (Emmet 1991). The common name "American Painted Lady" was first used by Abbott and Smith (1797); it has since been shortened to "American Lady" and refers to the North American distribution of the species.

**ADULT:** American Ladies are characterized by two large submarginal eyespots on each ventral hindwing. The dorsal hindwings have four submarginal eyespots, with the first and last spots larger than the inner two spots. The dorsal hindwings and the basal half of the forewings are almost entirely orange brown, without most of the black markings of the Painted Lady and the West Coast Lady. Males and females are very similar in appearance.

**IMMATURE STAGES:** Eggs are barrel-shaped with 13–16 high vertical ribs; they are yellow green. Mature larvae have black heads. The body is velvet black and covered with black spines, with delicate transverse yellowish lines across each segment. At the front of the base of each dorso-lateral spine is a large round silver white spot. Pupae are usually a dull grey white marked with brown or olive brown, but are sometimes golden green marked with purple. The darker markings form an irregular broad band along the length of the sides, and there are orange-tipped tubercles (Scudder 1893).

**BIOLOGY:** American Ladies are strictly migratory into BC, where they rarely reproduce. Apparently they do not success-

D  (4.9 cm)

V  (4.9 cm)

fully hibernate as adults in BC, and are rarely encountered. The larvae live in silk nests among the leaves of the foodplants, and pupation often occurs in the nest. Both pupae and adults hibernate (Scudder 1893).

Outside BC larval foodplants include *Anaphalis margaritacea, Antennaria parvifolia, A. plantaginifolia, Artemisia ludoviciana, Gnaphalium palustre, G. bicolor, G. obtusifolium, G. polycephalum, G. purpureum, Helianthus, Myosotis,* and *Senecio bicolor* var. *cineraria* (Scudder 1889a; Emmel and Emmel 1962; Shapiro 1975b; Ferris and Brown 1981; Scott 1992). J.W. Cockle's record of nettle being used in BC (Jones 1936) is undoubtedly an error.

**SUBSPECIES:** None. The type locality of the species is "Virginia."

**RANGE AND HABITAT:** American Ladies occur sporadically across southern BC as rare migrants. Although they have been found only near Vancouver and Victoria and in the Kootenays, they could potentially occur in any open habitat in southern BC.

**GENERAL DISTRIBUTION:** American Ladies occur from southern CAN south to northern South America. In the west, breeding populations occur north to about Seattle, with migratory individuals occasionally straying into southern BC. Strays also turn up to the west in Hawaii, and to the east in western Europe.

**CONSERVATION STATUS:** Rare accidental migrants (SAN).

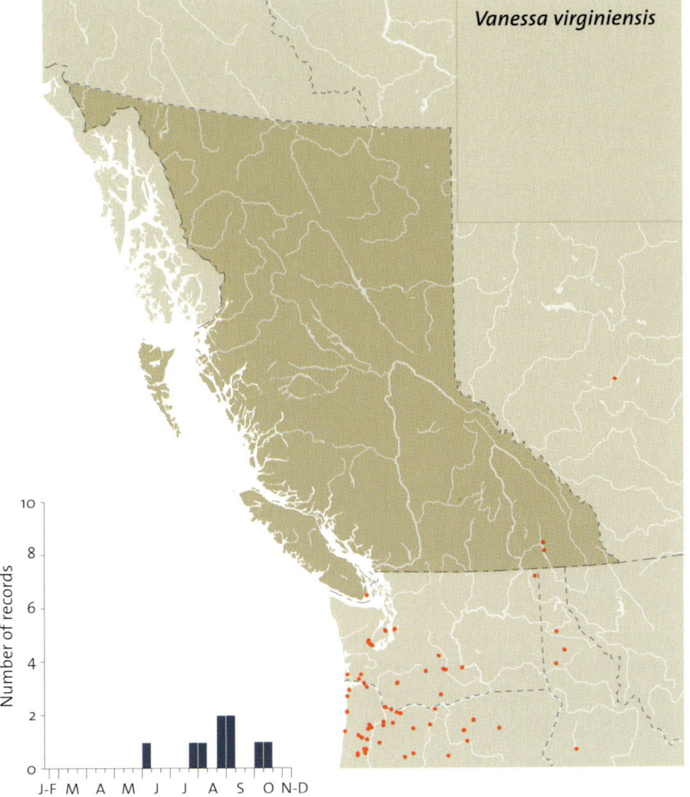



## PAINTED LADY
*Vanessa cardui* (Linnaeus, 1758)

**ETYMOLOGY:** The name *cardui* is derived from *Carduus,* the thistle genus, because Linnaeus knew that they were the larval foodplants (Emmet 1991). The common name is a very old English name (Emmet 1991). It was first used in North America by Ross (1873), and may refer to the pink colours of live butterflies, which resemble makeup.

**ADULT:** Painted Ladies have a white subapical dorsal forewing spot, unlike the West Coast Lady, in which the spot is orange brown. The subapical spot is the large pale spot second back from the wing apex, along the front margin of the forewing. Males and females are very similar in appearance.

D  (5.9 cm)

V  (5.9 cm)

**Painted Lady (*Vanessa cardui*)**

**Eggs**

**Mature larva**

**Pupa**

**IMMATURE STAGES:** Eggs are light green and barrel-shaped with vertical ribs. First instar larvae are dark brown, flecked with lighter brown and grey, and with an irregular white line along each side of a dark dorsal line, and an irregular grey lateral line. Mature larvae are greyish brown, variegated with yellow and black. The second to fifth segments, and the last segment, are black with many whitish dots, from which fine white hairs arise. There is a line of dashes down the black, which is white on the thorax and yellow on the abdomen. The body is covered in spines, varying in colour from yellowish to brownish white and tipped with black. There is a pale yellowish dashed line down each side. The spiracles are black, with a dull yellow ring around each one. Individual larvae are highly variable in colour. The pupa is a beautiful dusky copper colour, with a metallic shine when turned under a light (Saunders 1869d; CSG).

**BIOLOGY:** Painted Ladies are strongly migratory, with permanent populations breeding year-round in the deserts of the southern USA and northern Mexico. In spring they migrate north, with the migrants estimated as several billion individuals in some years. Only a tiny proportion of the migrants reach BC, hence none are seen in some years and thousands are seen in other years. Apparently only one generation of offspring is produced in BC, with the first adults emerging in July. They do not appear to hibernate as adults or larvae, so BC must be recolonized by migrants each summer. Hardy (1953, 1959b) summarizes some of the "migration years" of that time on Vancouver Island and other areas of BC. Painted Ladies appear to hibernate successfully in Colorado (Scott 1992), and one early March record in BC indicates successful hibernation. There is currently no evidence that Painted Ladies migrate south from BC in the fall. There are sporadic reports of southward migrations elsewhere, such as in northern California (Shapiro 1980b). Populations in France and North America are genetically very similar (Shapiro and Geiger 1989), suggesting that some migration across the Atlantic may occur.

Larval foodplants recorded outside BC are in many different plant families, but members of the Asteraceae, especially thistles, are the most frequently used. In BC bull thistle and Canada thistle are the most commonly used larval foodplants, and *Gnaphalium*, *Arctium lappa*, and *Eriophyllum lanatum* are also used (Harvey 1908; Jones 1933, 1938; Hardy 1959b; ACJ; AGG; CSG). Outside BC other larval foodplants include *Alcea rosea*, *Amsinckia douglasiana*, *Anaphalis margaritacea*, *Arctium lappa*, *Borago officinalis*, *Carduus nutans*, *Centaurea solstitialis*, *Cirsium arvense*, *C. callilepis* var. *oregonense*, *C. canescens*, *C. hydrophilum*, *C. ochrocentrum*, *C. parryi*, *C. scopulorum*, *C. undulatum*, *C. vulgare*, *Cnicus benedictus*, *Helianthus*, *Lupinus argenteus*, *Malva neglecta*, *Onopordum acanthium*, *Senecio bicolor* var. *cineraria*, *Silybum marianum*, and many other plants (Scudder 1889a; Keji 1951; Shapiro 1975b; Byers et al. 1984; Scott 1992). When the preferred larval foodplants have been defoliated by a large

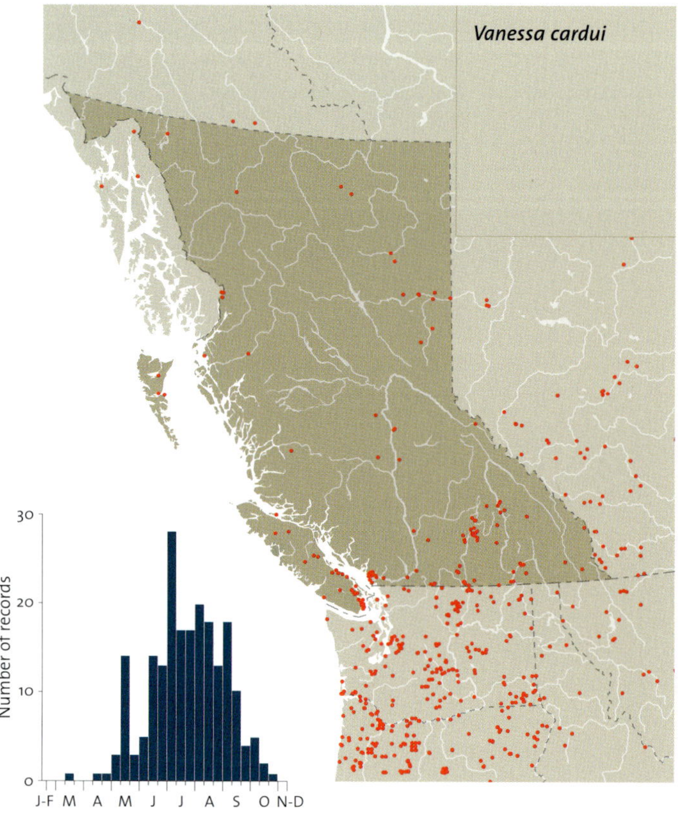

population of larvae, the larvae will also feed on many other plants, including *Brassica* crops (Byers et al. 1984). Larvae feed within characteristic silk webs among the leaves of the foodplants. Once familiar with their appearance, the observer can identify the webs with certainty even after the larvae have left.

**Subspecies:** None. The type locality of the species is Sweden.

**Range and habitat:** Painted Ladies occur across southern BC in open areas at all elevations, and less commonly in central and northern BC. They are one of the few butterflies seen on the Queen Charlotte Islands, and were first sighted there in 1952 (Hardy 1953).

**General distribution:** Painted Ladies are probably the most widely distributed butterfly in the world, occurring throughout most of the northern hemisphere and parts of the southern hemisphere. In the western hemisphere they are permanent residents from the extreme southern USA south to Venezuela. From these subtropical and tropical areas they migrate north each year as far as the arctic.

**Conservation status:** Common breeding immigrants (SZB).

## West Coast Lady
*Vanessa annabella* (Field, 1971)

**Etymology:** The derivation of the name *annabella* was not specified by Field (1971), and is not obvious. The common name was first used by Holland (1898) in reference to the western North American distribution of the species.

**Adult:** West Coast Ladies are smaller than the other ladies. They are most similar to the Painted Lady, but have an orange subapical forewing spot, instead of the white subapical spot found in Painted Ladies and American Ladies. The subapical spot is the large pale spot second back from the wing apex, along the front margin of the forewing. Males and females are very similar in appearance.

**Immature stages:** Eggs are light green and barrel-shaped, with 10–14 vertical ribs. First instar larvae have shiny black

D  (4.5 cm)    V  (4.5 cm)

heads and greyish brown bodies. There are three pairs of light yellow spots on the back. Mature larvae have brownish black heads with bronze highlights and sometimes vertical whitish tan stripes. The bodies vary from black to greenish white or greyish white, including brown and tan shades. There are various coloured markings, ranging from dark rusty reds and oranges to yellow or various browns and tans. There are black to whitish spines over the body. Pupae are roughly cylindrical, with a row of raised points down the midline of the abdomen, and two more rows on each side of the midline. There is a raised point in the middle of the thorax. The colour ranges from tan to mottled dark brown, sometimes with a greenish golden cast. There is a pair of large white spots on the top of the back of the thorax, and a pair of small white spots on the back of the front of the abdomen (Skinner 1889).

**Biology:** West Coast Ladies are in flight from July to October, and again in the spring from March to June. It is unlikely that they successfully overwinter in BC as suggested by Treherne (1915), because of the lateness of the first recorded dates of appearance (late April). There is only a single brood, and the numbers of adults seen vary greatly from year to year. Eggs are laid from April to June, and develop into adults by late July. After the adults emerge in July and August, many, especially in the Southern Interior, fly up to subalpine and mountain meadows and join the throngs of Painted Ladies nectaring there. Males hilltop throughout the year, especially in late winter and spring in California. Males chase each

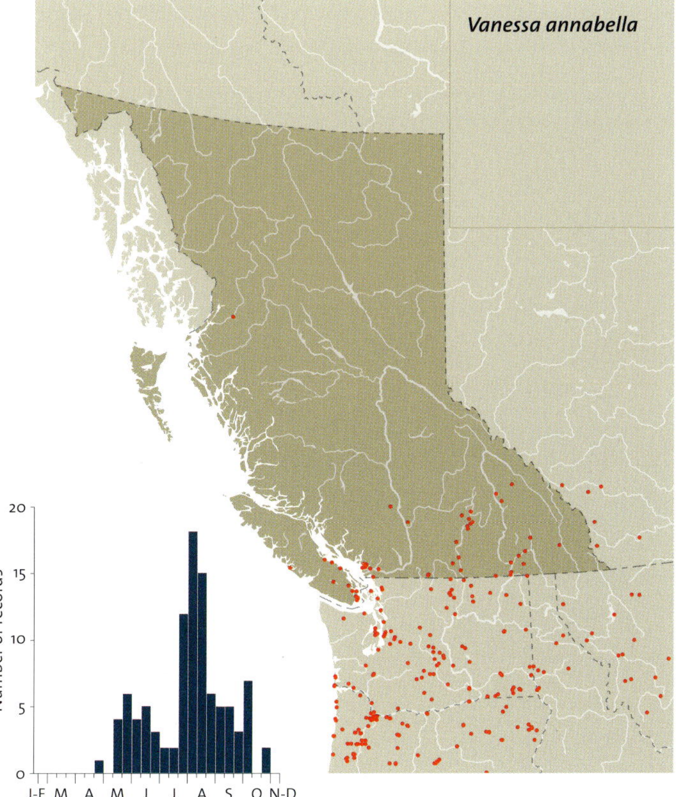

other, other butterflies but especially other ladies, other large insects, and birds (Dimock 1973, 1978). The swarms of Painted Ladies, West Coast Ladies, and sometimes Red Admirals in the subalpine meadows become quite annoying when one is trying to spot other late summer butterflies.

Eggs are laid singly on the upperside of the leaves. The first instar larva makes a web shelter on the top of the leaf, and places its frass next to the webbing as camouflage. By the fourth instar, the leaf is rolled upward (sometimes downward) and tied together with silk. The fifth instar larva may wrap several leaves together in a nest. Pupation occurs within a folded leaf or leaves of the larval foodplant or other plants, or the pupa may be exposed on twigs or branches. Parasitism by tachinid flies is frequent, with maggots emerging from mature larvae or pupae (Dimock 1978).

The known larval foodplants in BC are stinging nettle and garden hollyhock (Dyar 1904b; Harvey 1908; Gibson 1915; Jones 1936; CSG). Outside BC larval foodplants include

**Mature larva**      **Pupa**

various Malvaceae, including *Malva nicaeensis, M. parviflora, M. rotundifolia, Malvella leprosa, Sidalcea glaucescens, S. oregana,* and *Sphaeralcea ambigua* (Dimock 1973, 1978; Shapiro 1975b; Shapiro et al. 1981).

The separation of the North American *Vanessa annabella* as a distinct species from the South American *V. carye* is supported by protein differences shown through electrophoresis. The protein differences suggest that about 3 million years ago, when the Isthmus of Panama was formed, linking North and South America, *V. carye* colonized North America and then differentiated to form a new species, *V. annabella*. It is equally possible that the reverse pattern of colonization and differentiation occurred (Shapiro and Geiger 1989).

**SUBSPECIES:** None. The type locality of the species is the first valley west of Arroyo Verde Park, Ventura, CA.

**RANGE AND HABITAT:** West Coast Ladies occur across southern BC and on the coast to the Nass River in open areas at all elevations.

**GENERAL DISTRIBUTION:** West Coast Ladies occur from BC and AB, west of the prairies, south to Guatemala with occasional strays into the prairies.

**CONSERVATION STATUS:** Common breeding immigrants (SZB).

**West Coast Lady (*Vanessa annabella*)**

## RED ADMIRAL
*Vanessa atalanta* (Linnaeus, 1758)

**ETYMOLOGY:** Atalanta was a celebrated Greek beauty who made all her lovers run races with her on penalty of death if they could not catch her (Reed 1870; Emmet 1991). The subspecies name *rubria* is from the Latin *ruber* (red), for the red markings on the wings or perhaps in reference to the common name. The common name "Admiral" was already in use in England in 1699, when Petiver used it for a small group of butterflies, including what are now known as Red Admirals and White Admirals. From 1747 to 1819, and in 1945 by E.B. Ford, the name was corrupted to "Admirable," otherwise "Admiral" has been used. "Red" and "White" were added to the name "Admiral" to distinguish the two British species (Bretherton and Emmet 1990). Ross (1873) was the first to use the common name "Red Admiral" in North America.

D  (5.5 CM)      V  (5.5 CM)

**ADULT:** Red Admirals are unmistakable, with their black upper wings marked with bold red stripes and white spots. Males and females are very similar in appearance. Male forelegs are much hairier than those of females.



FAMILY NYMPHALIDAE (BRUSHFOOTS)

 Eggs are light green and barrel-shaped, with about 9 vertical ribs. First instar larvae are greenish brown, with 10 rows of black hairs and with a black head (Edwards 1882b). Mature larvae are highly variable, but their basic coloration is grey, greenish, cream, cocoa, or black. They have minute light spots on the darker background. All the colour forms have a lateral band of greenish yellow or creamish patches (R. Ashton, pers. comm.). There are seven rows of long branching spines, usually black but sometimes pale. Red Admiral larvae have numerous stiff white hairs on the head, which are black in Painted Lady larvae (Dimock 1978). Apparently in Britain the nominate subspecies has mature larvae that are usually grey green to whitish (rarely black), with an indistinct lateral stripe. Pupae are similar to those of other *Vanessa,* the colour varying from reddish grey to greenish grey with a bronze sheen (Edwards 1882b).

**Biology:** Red Admirals in Colorado hibernate as both pupae and adults (Brown et al. 1957). There is one record for 6 March 1984 near Sooke, BC (Gwyn Owen, pers. comm.), which must be of successful hibernation because that date is too early for a migrant from the south. From mid-April to June adults migrate into BC and lay eggs, with the next generation of adults emerging in late July and being in flight into October. Whether a southward migration occurs is unknown. Red Admirals will sometimes feed on the juice of split fruit such as plums (CSG). First instar larvae fold a small, just expanding leaf upward around themselves, and feed within. Second to fifth instar larvae fold nettle leaves downward into a

Egg

Mature larva

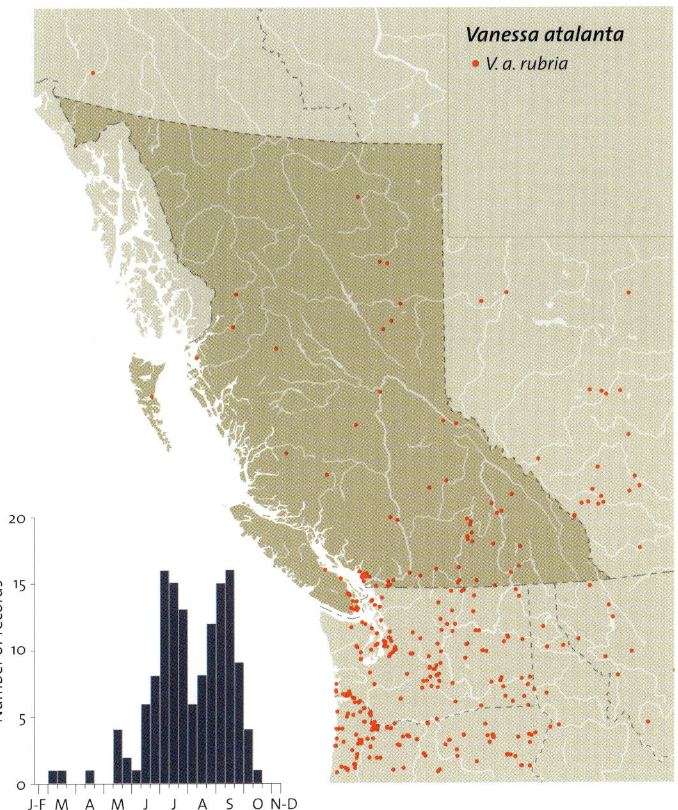
Pupa

"tent" and feed from within. In forming the tent they usually cut the midrib of the leaf near the base and cut the leaf membrane on both sides of the midrib, so that the leaf droops downward, and then fold the leaf around the lower surface. They sometimes fold a large nettle leaf and pupate within that (Edwards 1883, 1885b).

The primary larval foodplant in BC is stinging nettle (Harvey 1908; CSG). The only other member of the Urticaceae that occurs in BC, *Parietaria pennsylvanica,* is used in New York (Shapiro and Shapiro 1974). Hop (*Humulus lupulus*) is grown in the Fraser Valley and is used as a larval foodplant outside BC, where *Boehmeria cylindrica* and *Parietaria debilis* are also used (Scudder 1889a).

**Subspecies:** *V. a. rubria* (Fruhstorfer, 1909) (TL: Mexico) is the North American subspecies, which has a narrower submarginal white bar on the forewing than the European nominate subspecies. The European subspecies is frequently found in "butterfly houses," and may occasionally escape into the wild in BC.

**Range and habitat:** Red Admirals occur occasionally as migrants in central and northern BC except in the extreme northwest, and regularly as migrants in southern BC. They breed across southern and central BC in moist open or partially wooded areas, and in scattered localities further north. Preservation of patches of stinging nettle in out-of-the-way areas of parks and gardens would help maintain Red Admirals as urban butterflies.

**General distribution:** Red Admirals occur throughout non-arctic North America south to Guatemala. They also occur throughout most of Europe and much of Asia.

**Conservation status:** Common breeding immigrants (SZB).

The type genus of this subfamily is *Argynnis,* a European genus of fritillaries. Argynnis is a surname of Venus, from the temple erected in her honour by Agamemnon on the death of his favourite, Argynnus (Reed 1870). Emmet (1991) suggests that there is a second meaning or pun intended by Fabricius, as the Greek *arquros* means "silver," thus the silver spots on the underside of the wings. The origin of the common name "fritillaries" is obscure, but it was in use as early as 1717 in British books (Emmet 1990). The most likely explanation is that the butterflies resemble the lily genus *Fritillaria.* Several species in this genus have an orange and black pattern like the butterflies. In Europe the name "fritillary" is also applied to species in the subfamily Melitaeinae, which in North America are called checkerspots.

Fritillaries have a checkered pattern of black and tawny orange rectangular to square spots on the upperside of both wings. This led many early taxonomists to combine fritillaries with the following subfamily, Melitaeinae, but there are important structural differences. The hindwing cell is closed in the subfamily Argynninae (Warren 1944). The second segment of the labial palpi is elongated, thickened, and covered with both hairs and bristles. The uncus of the male genitalia is well developed. There is no overall treatment of the subfamily, but Warren (1944), dos Passos and Grey (1945), and Warren et al. (1946) cover the world genera and North American *Speyeria.*

**Northwestern Fritillary (*Speyeria hesperis beani*)**

### GENUS *EUPTOIETA* DOUBLEDAY, [1848]   FRITILLARIES

The name *Euptoieta* is derived from the Greek *eu* (true) and *ptoietos* (terror or truly terrifying). This is in reference to the type species of the genus, *claudia,* for the Roman emperor Claudius I. Claudius, Tiberius, Nero, and Caligula were infamous for their terrifying reigns in what is known as the Julia-Claudian dynasty. The common name "fritillaries" is shared with the rest of the subfamily.

Adults of the genus *Euptoieta* lack silver spots on the underside of the wings. The underside ground colour is a diffused tan to grey, and more closely resembles some European woodnymphs (Satyrinae). The upperside is that of a typical fritillary. There are only two species in this Neotropical genus, one of which is an occasional migrant to southeastern BC. Biologically and morphologically, this genus is a connecting link between the genus *Speyeria* in the Nearctic and the well-known Neotropical genus *Heliconius* Kluk, 1780.

# Variegated Fritillary
*Euptoieta claudia* (Cramer, [1775])

**Etymology:** The name *claudia* refers to Claudius I, a Roman emperor on whom the Public Broadcasting Service series *I, Claudius* was based. The common name was first used by Gosse (1859), and refers to the variegated colour pattern on the ventral wings.

**Adult:** The Variegated Fritillary is the only fritillary in which the outer margin of the forewing is concave rather than convex. There are no silver spots on the undersides of the wings.

♂ D (5.4 cm)   ♂ V (5.4 cm)

**Immature stages:** The mature larva is similar to *Speyeria* larvae, but the spines are longer and the body colour is brick red, not black as in the *Speyeria* (P.A. Hammond, pers. comm.). The ground colour of the pupa is white, and two lines of protuberances are brown.

**Biology:** Adults of the Variegated Fritillary have been seen, as rare individuals, from 17 July to 4 September. Outside BC, in late summer and early fall the larvae of the Variegated Fritillary can sometimes be seen feeding on garden pansies. They stay on the foodplant during the day. This species never overwinters, as it is intolerant of freezing or even near-freezing temperatures. It is unlikely that it even attempts to breed in BC.

**Subspecies:** None. The type locality of the species is "Jamaica."

**Range and habitat:** The Variegated Fritillary is an uncommon migrant found at all elevations in southeastern BC. It has no habitat preferences in the province because it does not breed here.

**General distribution:** The Variegated Fritillary is found as a migrant from southeastern BC across CAN to PQ. It is a permanent resident from southern AZ east to NC and south.

**Conservation status:** The Variegated Fritillary is a rare non-breeding migrant (SZN).

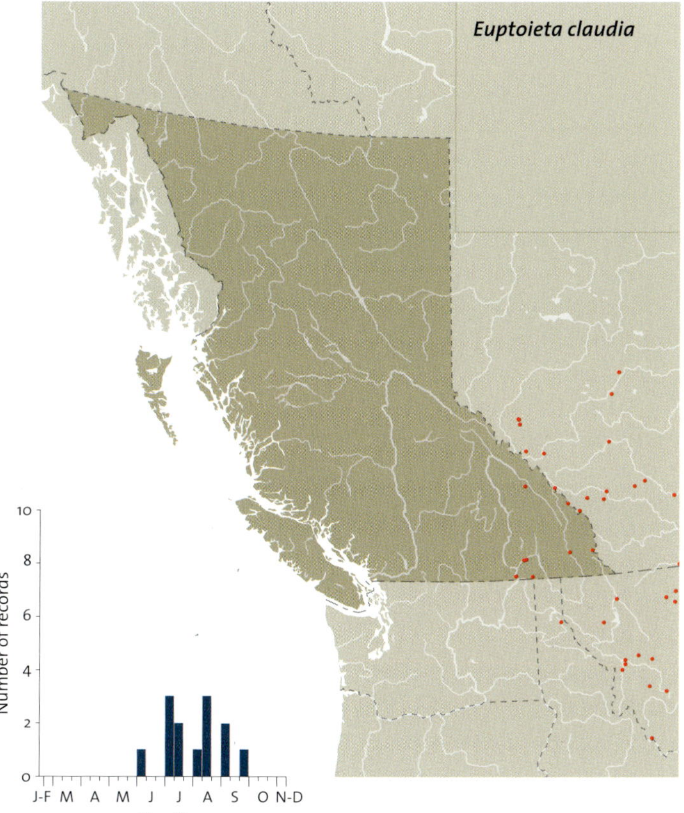



## Genus *Speyeria* Scudder, 1872 — Greater Fritillaries

The genus *Speyeria* is named for the German lepidopterist Adolph Speyer (1812–92). The name "greater fritillaries" refers to the large size of the species in this genus, in contrast to the lesser fritillaries in the genera *Boloria* and *Clossiana*.

At least some populations of all species of *Speyeria* in BC have individuals with silver spots on the ventral hindwing (Fig. 70). By contrast, only one species of *Clossiana* has these silver spots. The genus is entirely Nearctic, with 14 recognized species, 8 of which are found in BC. Two other species, *S. coronis* (Behr, 1864) and *S. egleis* (Behr, 1862) occur immediately south of the BC border in Washington or Montana, and might eventually

be recorded in the province. Dos Passos and Grey (1947) produced the definitive treatment of the genus. In this genus, and all genera in the subfamily except *Boloria* and *Clossiana,* the aedeagus is open at the proximal end. Dos Passos and Grey (1947) reduced the number of recognized species from more than 100 species to 13 species, and reduced the other species names to either subspecies or synonyms. The dos Passos and Grey paper, Gunder (1929b), Davenport (1941), and Nabokov (1949) set the standard for our modern species concepts for North American butterflies. P.A. Hammond (pers. comm.) has provided the information on the biology and appearance of the larvae.

# Great Spangled Fritillary

*Speyeria cybele* (Fabricius, 1775)

**Etymology:** The species name *cybele* refers to Cybele, the wife of Coronus and mother of the Olympian gods. The subspecies name *leto* refers to another Greek mythological figure who was either the first wife or mistress of Zeus, depending on which story one credits, and the mother of Apollo and Artemis. The subspecies name *pseudocarpenteri* is derived from the Latin prefix *pseudo* (false) and *carpenteri*, another subspecies of Great Spangled Fritillary. The subspecies was confused with *carpenteri*. The common name was first used by Gosse (1840), and refers to the butterfly's large size and bright ventral spotting.

**Adult:** Adults of the Great Spangled Fritillary are larger than those of any other *Speyeria* species. On the ventral hindwings there is a wide creamy coloured band between the postmedian silver spots and the submarginal silver spots. The southern subspecies, *leto,* is our largest fritillary; in addition, the females are the only dimorphic females of *Speyeria* in BC. The northern subspecies, *pseudocarpenteri,* is smaller than some southern *Speyeria* but larger than the other *Speyeria* species in northeastern BC. The sexes are not dimorphic.

**Immature stages:** The mature larva is black. The body is covered with protuberances that bear spines. There are no mid-dorsal narrow yellow stripes.

**Biology:** The Leto Fritillary in southern BC flies from late June to late August; there is one generation each year. The Prairie

Ssp. *leto* ♂ D (6.6 cm)

Ssp. *leto* ♀ D (6.9 cm)

Ssp. *leto* ♂ V (6.6 cm)

Ssp. *pseudocarpenteri* ♀ V (6.6 cm)

Ssp. *pseudocarpenteri* ♀ D (6.6 cm)

Fritillary flies in July in the Peace River region. Eggs are laid at the base of the foodplant, *Viola* sp. They hatch and the first instar larvae overwinter. Larvae begin feeding the following spring, as soon as the foodplant has leafed out.

**Subspecies:** There are two subspecies of the Great Spangled Fritillary in BC. In the Southern Interior and the Kootenays, one finds the Leto Fritillary, *S. c. leto* (Behr, 1862) (TL: Carson City, NV). The Leto ranges south to central California and northwestern Wyoming. In the Peace River area one finds the Prairie Fritillary, *S. c. pseudocarpenteri* (F. & R. Chermock, 1940) (TL: Sand Ridge, MB). The Prairie Fritillary occurs east to Manitoba and across northern Montana and North Dakota.

**Range and habitat:** The Great Spangled Fritillary occurs in the Southern Interior, the Kootenays, and the Peace. In the south it is found primarily in ponderosa pine and Douglas-fir areas up to 1,000 m. The southern populations, the Leto Fritillary, occur in xeric meadows and adults nectar primarily on Canadian thistle and other *Cirsium* species. In the Peace, it

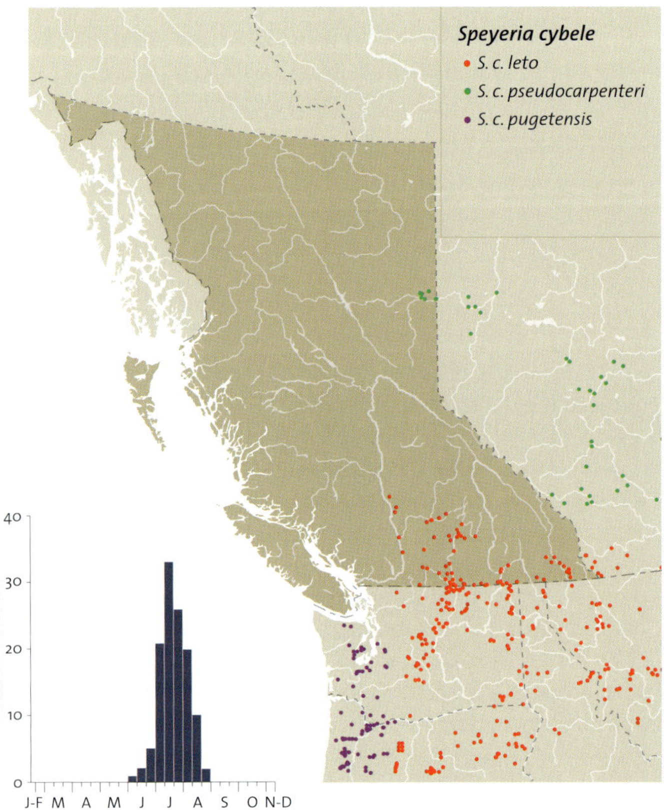

**Mature larva, ssp. *pugetensis***

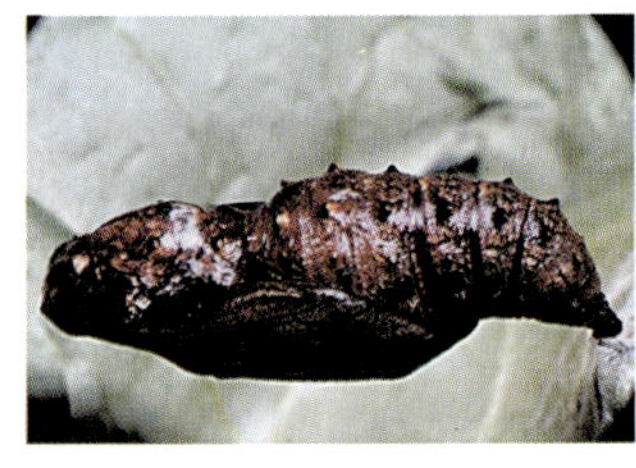

**Pupa, ssp. *pugetensis***

is found at the edges of and in open, mature aspen wood-land. The Prairie Fritillary occurs in more mesic habitat and has not been observed nectaring like fritillaries in the south.

**GENERAL DISTRIBUTION:** The Great Spangled Fritillary ranges across southern CAN from BC to NS. In the west, it occurs south to central CA and northern NM. In the east it occurs south to northwestern GA, with isolated populations in TX and LA.

**CONSERVATION STATUS:** Subspecies *pseudocarpenteri* is of Special Concern in BC (S3). Subspecies *leto* is not of concern (S5).

## APHRODITE FRITILLARY
*Speyeria aphrodite* (Fabricius, 1787)

**ETYMOLOGY:** The name *aphrodite* refers to the Greek goddess of love and is derived from the fable of her having sprung from the sea foam, Aphros (Reed 1870). Subspecies *manitoba* is named for the province of Manitoba, and subspecies *columbia* is named for British Columbia. Subspecies *whitehousei* is named after Frank C. Whitehouse, of dragon-fly fame, who was living in Cranbrook at the time he pro-vided Gunder with the type material for the subspecies. Kirby (1837) first used the common name "Aphrodite Argynnis," which Harris (1862) modified to "Aphrodite butterfly" (Scudder 1889b), and Holland (1898) changed it to the modern form.

**ADULT:** Adults of the Aphrodite Fritillary have a unique characteristic on the ventral hindwing that separates them from all other *Speyeria,* including the one it is confused with most easily, *S. hesperis*. Between vein $M_3$ and $CuA_1$, the postmedian black spot is always surrounded by at least a faint black circle or halo (Fig. 70a).

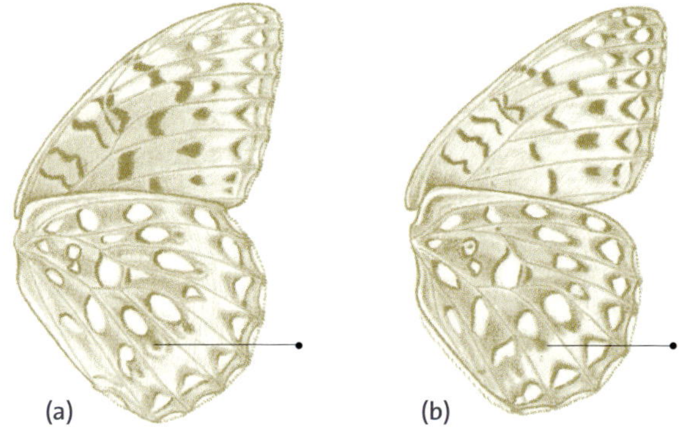


**70** Underside of the wing of *Speyeria* species: (a) *S. aphrodite*, (b) *S. hesperis*

**IMMATURE STAGES:** The mature larva is black. The body is covered with protuberances that bear spines. There are no mid-dorsal narrow yellow stripes.

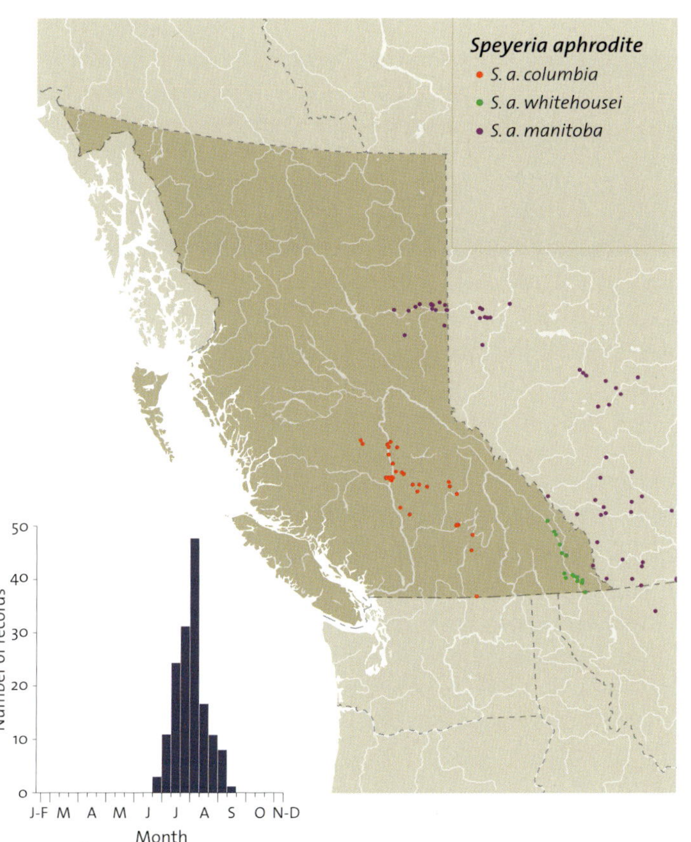

Ssp. *columbia* ♂ D (5.4 CM)

Ssp. *columbia* ♂ V (5.4 CM)

Ssp. *whitehousei* ♂ D (6.3 CM)

Ssp. *whitehousei* ♂ V (6.3 CM)

Ssp. *manitoba* ♂ D (5.2 CM)

Ssp. *manitoba* ♂ V (5.2 CM)

**BIOLOGY:** Adults of the Manitoba Fritillary fly from mid-July to mid-August. The Columbia Fritillary flies from late June to early September. Whitehouse's Fritillary flies from mid-July to mid-September. Eggs are laid at the base of the foodplant, *Viola* sp. They hatch and the first instar larvae overwinter. Larvae begin feeding the following spring, as soon as the foodplant has leafed out.

**SUBSPECIES:** The Cariboo populations are the Columbia Fritillary, *S. a. columbia* (Hy. Edwards, 1877) (TL: Lac la Hache, BC). The East Kootenay populations are Whitehouse's Fritillary, *S. a. whitehousei* (Gunder, 1932) (TL: Jaffray, BC). The Peace populations are the Manitoba Fritillary, *S. a. manitoba* (F. & R. Chermock, 1940) (TL: Sand Ridge, MB).

**RANGE AND HABITAT:** The Aphrodite Fritillary occurs in BC as three disjunct sets of populations in the Cariboo, the East Kootenay, and the Peace. The Cariboo and Peace populations are associated with mesic meadows in aspen woodland habitat. The East Kootenay populations occur at the bottom of the Rocky Mountain Trench in very xeric habitat, and are impacted by both grazing and suburban development. Adults of the Cariboo and East Kootenay populations are commonly found nectaring at thistles (*Cirsium*) but the Peace populations have not been seen nectaring.

**Mature larva, ssp. *whitehousei***

**GENERAL DISTRIBUTION:** The Aphrodite Fritillary is found from BC east to NS. In the west it occurs south to AZ and NM. In the east, it occurs south to the northeastern USA and south in the Appalachians to northern GA.

**CONSERVATION STATUS:** Subspecies *whitehousei* and *manitoba* are of Special Concern in BC (both S3), and subspecies *columbia* is not of concern (S5).

## ZERENE FRITILLARY
*Speyeria zerene* (Boisduval, 1852)

**ETYMOLOGY:** The species name *zerene* is derived from the Latin *zerena* (parched or dried up), in reference to the dry southern California type locality. Subspecies *garretti* was named after C.B. Garrett, an ardent collector in the Cranbrook area who supplied Gunder with thousands of specimens from the East Kootenay. The subspecies name

Ssp. *sitka* ♂ D HOLOTYPE

Ssp. *sitka* ♀ D

Ssp. *sitka* ♂ V HOLOTYPE

Ssp. *sitka* ♀ V

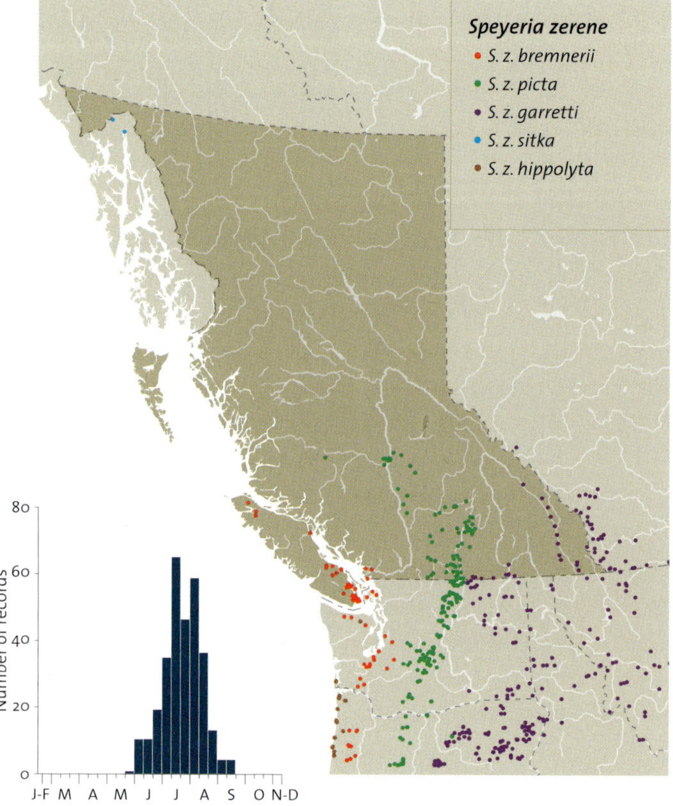


*picta* is from the Latin *pictus* (coloured), in reference to the brown cast to the ventral wing colour. Subspecies *bremnerii* was named after Dr. Bremner of *HMS Zealous*. The name of the newly described subspecies *sitka* was chosen to indicate the restricted Alaskan range of the subspecies. The common name was first used by Holland (1898) as "Zerene."

**ADULT:** The Zerene Fritillary is the most variable of BC fritillary species, and thus the hardest to characterize. Wherever it occurs in BC, it is larger than any of the species still to be discussed. On the ventral hindwing, the basal and median area is a chocolate brown colour that varies in intensity between subspecies. Comparing specimens with the photographs provided is the best way to determine the species.

Ssp. *bremnerii* ♂ D (5.8 cm)

Ssp. *bremnerii* ♂ V (5.8 cm)

Ssp. *garretti* ♀ D (6.1 cm)

Ssp. *garretti* ♀ V (6.1 cm)

Ssp. *picta* ♂ D (5.6 cm)

Ssp. *picta* ♂ V (5.6 cm)

*S. coronis* ♂ D (5.4 cm)

*S. coronis* ♀ V (6.6 cm)

**IMMATURE STAGES:** The mature larva is black. The body is covered with protuberances that bear spines. The bases of the spines are black on the first two rows dorsally and yellow on the third and fourth rows (Hardy 1958a). There are two mid-dorsal narrow yellow stripes. *Speyeria zerene* and *S. callippe* are identical in the larval stages. The egg is 1.0 mm × 0.60 mm and dull cream (Hardy 1958a).

**BIOLOGY:** Southern Interior and Kootenay populations of the Zerene Fritillary have a very protracted flight, from late May to September. This is because the females retreat to high elevations during the dry summer and fly back to the larval foodplant habitat to lay eggs when cooler weather returns in late August or early September. Extant populations of Bremner's Fritillary fly from early July to late August, depending on the elevation of the populations. Mating takes place immediately after adult females emerge from the pupa (JHS). Eggs are laid at the base of the foodplant, *Viola* sp.

(Hardy 1958a). Guppy (1956) observed that eggs kept in captivity did not hatch until the following February. Hardy (1958a) found, however, that the eggs hatched in August and the first instar larvae hibernated until the following spring, when the larvae begin feeding as soon as the foodplant has leafed out. The larvae are very gregarious in the early instars and mostly solitary when mature. Hardy reared them on *Viola palustris* on Vancouver Island.

**SUBSPECIES:** The Vancouver Island and Gulf Islands populations are Bremner's Fritillary, *S. z. bremnerii* (W.H. Edwards, 1872) (TL: San Juan Is., WA), which occurs south in the Puget Trough and the Willamette Valley to near Corvallis, OR. This is the darkest of the described subspecies. The Alaska panhandle subspecies (a new subspecies) is smaller and somewhat paler; it is expected that *S. z. sitka* will eventually be found in BC. The Painted Fritillary, *S. z. picta* (McDunnough, 1924) (TL: Aspen Grove, BC), is restricted to the eastern slope of the Cascades and the southern Cariboo. It is smaller than either *bremnerii* or *garretti*. The Washington butterfly atlas (Hinchliff 1996) does not distinguish the Painted Fritillary from the much more widespread Garrett's Fritillary, *S. z. garretti* (Gunder, 1932) (TL: Cranbrook, BC), which occurs from the Okanagan Valley east to the Rocky Mountain Trench. Garrett's Fritillary occurs south to northeastern Oregon and northwestern Wyoming. It is the largest subspecies and has the lightest ground colour in the basal and median areas of the ventral hindwings. In Washington this species is often confused with the Coronis Fritillary, *S. coronis,* which has not yet been recorded in BC.

***Speyeria zerene sitka* P.C. Hammond, J.L. Harry & D.V. McCorkle, new subspecies. Male:** Forewing length (*n* = 9) 23–26 mm (average 25 mm). Dorsal wing surface medium orange to yellow orange with black spots and bars, heavy dark basal suffusion present, and veins of forewing thickly covered with dark scales. Ventral forewing with yellow orange ground colour, brown patches around the silver postmedian spots of the subapical area. Ventral hindwing with a red-brown disc covered with some yellow suffusion,

**Zerene Fritillary (*Speyeria zerene picta*)**

and with a wide yellow submarginal band. Spots brightly silvered and narrowly outlined basally with black scales. Median spots small and round, submarginal spots also small and rounded to flattened. Marginal border red brown like disc. **Female**: Forewing length (*n* = 4) 25–28 mm. Similar to the male, but veins of forewing not thickly covered with dark scales. Ventral forewing with red orange ground colour.

Grey and Moeck (1962) associated this disjunct Alaskan subspecies with *S. z. hippolyta* (W. H. Edwards, 1879) (TL: Oceanside, Tillamock Co., OR) of the OR and WA coast because of the small wing size of the two subspecies. However, *S. z. hippolyta* usually has a much darker, reddish brown disc and a narrow yellow submarginal band on the ventral hindwing. Between these two subspecies, a third subspecies, *S. z. bremnerii,* is found from Oregon east of the coast and north to the north end of Vancouver Island. *S. z. bremnerii* is much larger, with male forewing length 27–33 mm and female forewing length 30–36 mm. A few specimens of *S. z. bremnerii* from western Washington have the bright red-brown disc and wide yellow submarginal band of *S. z. sitka.* Thus, it is plausible that *S. z. sitka* is derived from *S. z. bremnerii.*

**Types.** Holotype: male, AK, Haines, 11 August 1949, G.E. Pollard. The holotype is deposited in the California Academy of Sciences, San Francisco, CA, USA. Allotype: female, same data. Paratypes: 1 male and 3 females, same data (CAS); 1 male, AK, Mile 18 Haines Hwy., 7 July 1972, J.L. Harry (JLH); 4 males, AK, Haines, 2.1 miles SW at Mile 26 Haines Hwy., 23 July 1999, J.L. Harry (JLH); 2 males, same locality, 31 July 1999 (JLH).

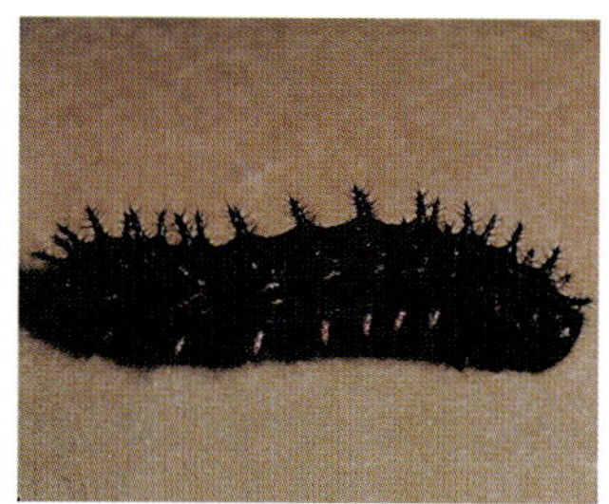

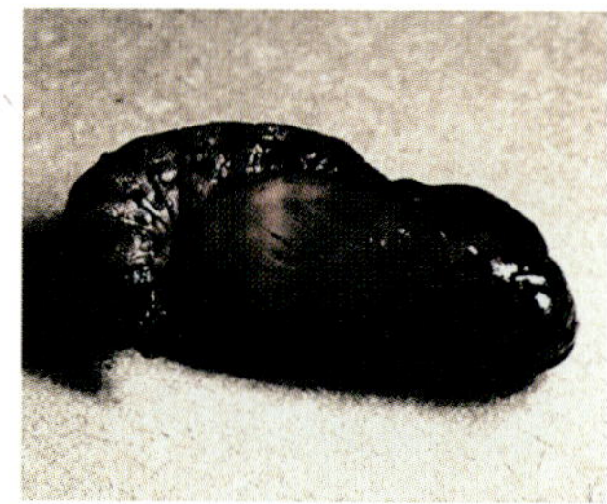

Mature larva, ssp. *sitka*

Pupa, ssp. *sitka*

**RANGE AND HABITAT:** The Zerene Fritillary occurs across southern BC. There are, however, only three records from the Lower Mainland, which may represent strays from known Saltspring Island or Orcas Island (Mt. Constitution) populations. On southern Vancouver Island and the Gulf Islands, the Zerene Fritillary is found in mesic meadows and xeric meadows with permanent springs that have not been invaded by Scotch broom. In the Cariboo and the eastern slope of the Cascade Mountains, the Zerene Fritillary is associated with mesic meadows in Douglas-fir habitat. Further east it is found in xeric meadows in sagebrush, ponderosa pine, and dry, low-elevation Douglas-fir habitat.

**GENERAL DISTRIBUTION:** The Zerene Fritillary is found from the top of the AK panhandle (Haines) and from southern BC east to southwestern SK and south to central CA, NV, UT, and NM.

**CONSERVATION STATUS:** Subspecies *bremnerii* is of Special Concern in BC (S3), and subspecies *picta* and *garretti* are not of concern (S4).

## CALLIPPE FRITILLARY
*Speyeria callippe* (Boisduval, 1852)

**ETYMOLOGY:** The species name *callippe* is derived from a famous statue of Aphrodite and refers to either the resemblance to the Aphrodite Fritillary or the beauty of the Callippe Fritillary. The subspecies name *semivirida* is derived from the Latin *semi* (half) and *viridis* (green), in reference to the partly green ventral hindwing surface. The common name was first used by Holland (1898).

**ADULT:** The common and widely distributed subspecies *S. c. semivirida* is distinguished by the fact that the ground colour of the ventral hindwing is overcast with green. The new subspecies described below cannot be confused with any other BC *Speyeria* species. It could, however, be confused with *S. egleis,* a species not yet known from BC.

**IMMATURE STAGES:** The mature larva is black. The body is covered with protuberances that bear spines. There are two mid-dorsal narrow yellow stripes. *S. callippe* and *S. zerene* are identical in the larval stages.

**BIOLOGY:** The Callippe Fritillary has a very protracted flight period, from mid-May to late August. The females retreat to high elevations during the dry summer and fly back to the larval foodplant to lay eggs when cooler weather returns in late August. Eggs are laid at the base of the foodplant, *Viola* sp., or at the base of sagebrush (*Artemisia*) (Durden 1965). They hatch and the first instar larvae overwinter. Larvae begin feeding the following spring, as soon as the foodplant has leafed out.

**SUBSPECIES:** The Green Fritillary, *S. c. semivirida* (McDunnough, 1924) (TL: Aspen Grove, BC) is found in southeastern BC. The ventral hindwings have a green ground colour. In the Chilcotin Fritillary, described below, the ventral hindwings have a brown ground colour.

***Speyeria callippe chilcotinensis* Guppy & Shepard, new subspecies.** In both males and females of *S. c. chilcotinensis* the ground colour of the ventral hindwings is dark chocolate brown, rather than the green of *S. c. semivirida.* Also, the ground colour of the ventral forewings of both sexes is primarily a reddish brown. *S. c. semivirida,* by contrast, has fewer reddish brown scales and the ground colour is much lighter. **Types**. Holotype: female, BC, Riske Creek, 23 July 1981, J. and S. Shepard. A label "Holotype / *Speyeria callippe / chilcotinensis* Guppy & Shepard" is attached. The holotype is deposited in the Royal British Columbia Museum, Victoria, BC, CAN. Paratypes: 2 males, BC, Riske Creek, 27 June 1970, J. and S. Shepard (JHS); 3 females, same locality, 23 July 1981, J. and S. Shepard (JHS); 1 male, 1 female, BC, Pavilion, 10 mi. N., 18 July 1981, J. and S. Shepard (JHS); 2 males, BC, 70 Mile House, 1 July 1972, S. Shepard (JHS); 2 males, BC, Jesmond Lookout, el. 6,000', 29 July 1972, S. Shepard (JHS); 1 male, BC, Riske Creek, Deer Park Ranch on Moon Road, 21 June 1997, A.I. Fischer (CSG); 1 male, BC, Riske Creek, Becher's Prairie, 24 June 1997, A.I. Fischer (CSG); 1 male, BC, Riske Creek, Moon Road, 24 June 1997, A.I. Fischer (CSG); 1 male, BC, Riske Creek, Hwy 20, 25 June 1997, A.I. Fischer (CSG); 3 males, BC, Riske Creek, road to Farwell Canyon, 12 July 1997, A.I. Fischer (CSG); 1 female, BC, Williams Lake, south of town, 12 July 1997, A.I. Fischer (CSG); 1 male, BC, Riske Creek, west of road to Farwell Canyon, A.I. Fischer (CSG); 3 males, 1 female, BC, Riske Creek, Becher's Prairie, East Lake, 12 July 1997, A.I. Fischer (CSG); 3 males, BC, Riske Creek, Cotton Road, 12 July 1997, A.I. Fischer (CSG); 2 males, 1 female, BC, Riske Creek, Doc English Gulch, 20 July 1997, A.I. Fischer (CSG); 1 female, BC, Riske Creek, Stack

Ssp. *chilcotinensis* ♂ D HOLOTYPE (5.9 CM)

Ssp. *chilcotinensis* ♂ V HOLOTYPE (5.9 CM)

Ssp. *chilcotinensis* ♀ D (5.9 CM)

Ssp. *chilcotinensis* ♀ V (5.9 CM)

Ssp. *semivirida* ♂ D (5.5 CM)

Ssp. *semivirida* ♂ V (5.5 CM)

Valley Road, 10 August 1997, A.I. Fischer (CSG); 1 male, BC, Hanceville, 25 km east at Harper Lake, 12 July 1997, A.I. Fischer (CSG).

**RANGE AND HABITAT:** The Callippe Fritillary is found from the eastern Chilcotin southeast to Osoyoos and east to near the AB border. It is usually found in slightly less xeric habitat than the Zerene Fritillary, and not in sagebrush habitat.

**GENERAL DISTRIBUTION:** From BC east to MB, and south to Baja California, MEX, and CO.

**CONSERVATION STATUS:** Not of concern for both subspecies (S4).

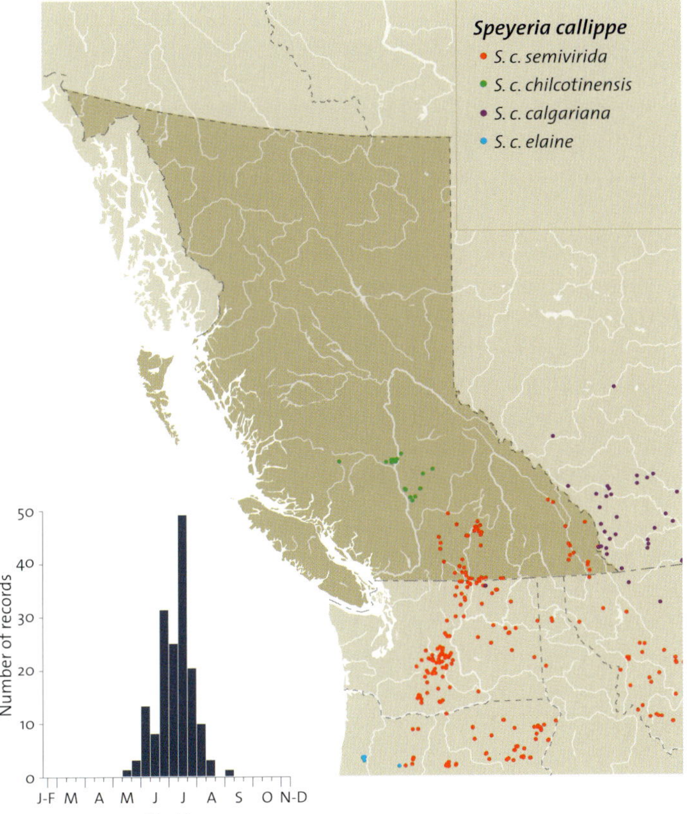

**Mature larva, ssp. *semivirida***

**Pupa, ssp. *semivirida***

# ATLANTIS FRITILLARY

*Speyeria atlantis* (W.H. Edwards, 1862)

**ETYMOLOGY:** The name *atlantis* is derived from Atlantis, a mythical island in the Atlantic Ocean. Subspecies *hollandi* is named in honour of W.J. Holland, whom the Chermock brothers characterized as "too well known to require further explanation." Holland wrote the first book covering the entire Canadian and American butterfly fauna in two editions, 1898 and 1931. The common name was first used by Klots (1951).

**ADULT:** The Atlantis Fritillary and the Northwestern Fritillary are sometimes hard to distinguish. Dos Passos and Grey (1947) considered them one species, but recent Canadian books distinguish two species. The Atlantis Fritillary has a more northern distribution; the basal and median portions of the ventral hindwings are a dark, almost black, chocolate colour. The fact that the two species occur together throughout the range of *S. hesperis* is the reason for regarding the Atlantis and Northwestern fritillaries as two species.

♂ D  (5.1 CM)

♂ V  (5.1 CM)

**IMMATURE STAGES:** The mature larva is black. The body is covered with protuberances that bear spines. There are two mid-dorsal narrow yellow stripes.

♀ V  (5.5 CM)

**BIOLOGY:** The Atlantis Fritillary flies from early July to mid-August. Eggs are laid at the base of the foodplant, *Viola* sp. They hatch and the first instar larvae overwinter. Larvae begin feeding the following spring, as soon as the foodplant has leafed out.

**SUBSPECIES:** BC populations from the Peace River region are Holland's Fritillary, *S. a. hollandi* (F. & R. Chermock, 1940) (TL: Riding Mountain, MB). Populations from southeastern BC are tentatively also assigned to this subspecies, but may prove to be just a very dark variety of *S. hesperis* (P. Hammond, pers. comm.).

**RANGE AND HABITAT:** The Atlantis Fritillary is found in a few isolated populations in eastern BC, from the Peace River region south to the West Kootenay. In the Peace it and the closely related Northwestern Fritillary occur together in aspen woodland habitat. In the Peace and east, the Atlantis Fritillary occurs in boreal habitat. In southeastern BC it is found in very moist habitat is association with *Clossiana selene* and several skipper species.

**GENERAL DISTRIBUTION:** The Atlantis Fritillary is found from eastern BC east to NF and in the northeastern USA in boreal habitat. In the west it occurs south of the Canadian border in northeastern WA, northern ID, northwest MT, CO, and SD.

**CONSERVATION STATUS:** Not of concern (S5).

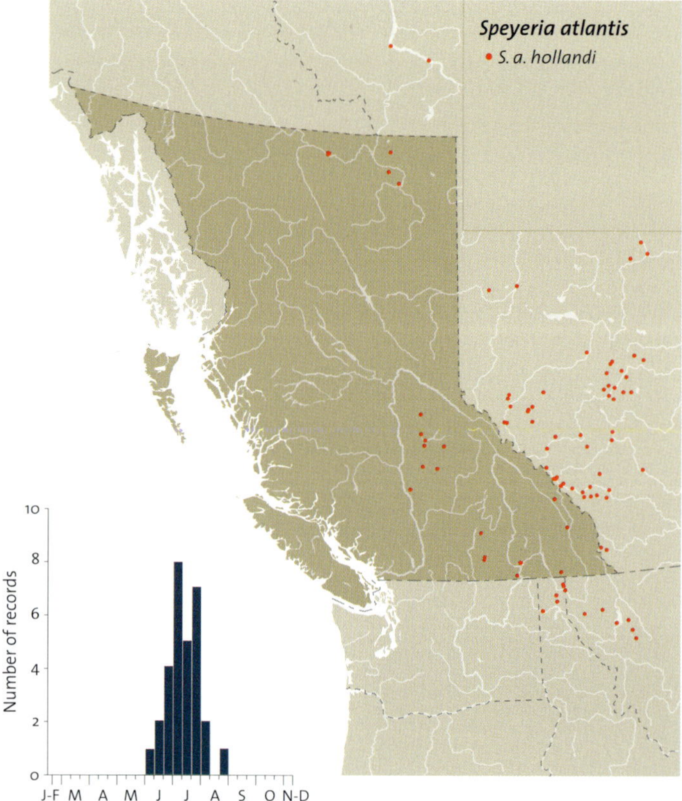

# Northwestern Fritillary
*Speyeria hesperis* (W.H. Edwards, 1864)

**Northwestern Fritillary (*Speyeria hesperis beani*)**

Ssp. *beani* ♂ D  (5.4 cm)

Ssp. *beani* ♂ V  (5.4 cm)

Ssp. *helena* ♀ D  (5.1 cm)

Ssp. *helena* ♀ V  (5.1 cm)

*S. egleis* ♂ D  (5.5 cm)

*S. egleis* ♂ V  (5.5 cm)

**ETYMOLOGY:** The species name *hesperis* is Latin for western, in reference to the western distribution of the species. Subspecies *beani* is named after T.E. Bean, an early lepidopterist who extensively collected the Lake Louise area near Banff, the type locality. The etymology of the subspecies name *helena* was not given in the original description. The common name refers to the species first being thought to occupy only the former Canadian Northwest area policed by the North-West Mounted Police, the precursor of the RCMP. It was coined by Klassen et al. (1989).

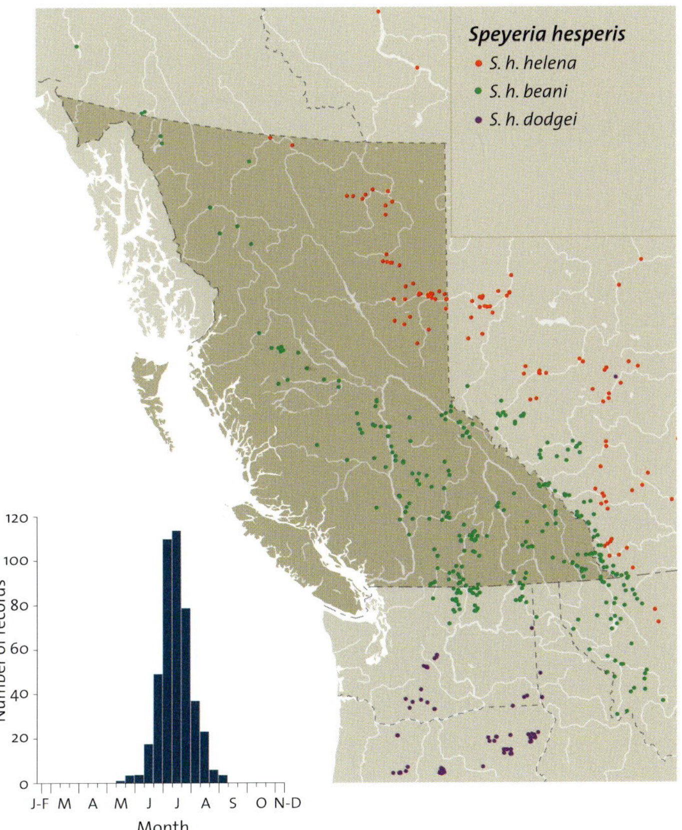



**ADULT:** In northeastern BC, this species can be confused with the Aphrodite Fritillary (see the discussion of the latter species for distinguishing characters). In the Southern Interior and Kootenays the ground colour of the hindwing is very variable, from light to dark with a hint of red. In the south it can be confused with the Atlantis Fritillary, which is found in much moister habitat along with *Clossiana selene* and other bog or wetland species. The ground colour of the ventral hindwing of the Atlantis Fritillary is almost black, with no red. An aberrant specimen of the Northwestern Fritillary was erroneously identified as the Great Basin Fritillary, *S. egleis* (Behr 1862), which, however, may eventually be recorded in British Columbia.

**IMMATURE STAGES:** The mature larva is black. The body is covered with protuberances that bear spines. There are no mid-dorsal narrow yellow stripes.

**BIOLOGY:** The Northwestern Fritillary flies from late June to early August in the Peace River region. Bean's Fritillary has a protracted flight period, from mid-June to early September. Adults do not generally leave the larval foodplant habitat, unlike the Zerene and Callippe Fritillaries. Shepard has observed females at the top of Plateau Mountain in Alberta, at an elevation of 2,500 m; this was during a year of extreme drought on the Alberta prairie. Eggs are laid at the base of the larval foodplant, *Viola* sp. They hatch and the first instar larvae overwinter. Larvae begin feeding the following spring, as soon as the foodplant has leafed out.

**SUBSPECIES:** The Peace populations are the larger subspecies and are assigned to the prairie subspecies, the Northwestern Fritillary, *S. h. helena* dos Passos & Grey, 1957 (TL: Edmonton, AB). The rest of the BC populations have traditionally been assigned to Bean's Fritillary, *S. h. beani* (Barnes & Benjamin, 1926) (TL: Banff, AB) (= *brico* Kondla, Scott, and Spomer, 1998; TL: Castle Cr. Rd., near McBride, BC), but this subspecies is weakly differentiated from *S. h. helena*. The current International Code of Zoological Nomenclature rules uphold the replacement name *helena* dos Passos & Grey, 1957 for the homonym *Argynnis lais* W.H. Edwards, 1884.

**RANGE AND HABITAT:** The Northwestern Fritillary is found throughout BC east of the Cascade and Coast mountains. In the extreme north it is found only in very moist meadows associated with hot springs such as Atlin Hot Springs. Further south it is found in mesic meadows; the Atlantis Fritillary is found in moister meadows in association with *Clossiana selene*.

**GENERAL DISTRIBUTION:** The Northwestern Fritillary occurs from YT southeast in isolated populations to northwestern BC and from there east to MB. It ranges south of BC and AB to CA and CO.

**CONSERVATION STATUS:** Not of concern, with subspecies *helena* S4 and *beani* S5.

## HYDASPE FRITILLARY
*Speyeria hydaspe* (Boisduval, 1869)

**ETYMOLOGY:** The species name *hydaspe* is derived from the Hydaspes River, the ancient name for the Jhelum River, which arises in the mountains of Kashmir where Alexander the Great defeated Porus in 326 BC – thus, a fritillary of the mountains. The subspecies name *rhodope* comes from Rhodopis, or "the rosy-cheeked," an epithet for a famous Greek courtesan, Doricha. It refers to the reddish cast to the ventral hindwings. The subspecies name *minor* refers to the small size of subspecies. The subspecies name *sakuntala* is of unknown origin. The common name was first used by Comstock (1927).

Ssp. *rhodope* ♂ D (5.2 CM)

Ssp. *rhodope* ♂ V (5.2 CM)

Ssp. *minor* ♂ V (4.7 CM)

Ssp. *sakuntala* ♂ V (4.8 CM)

**ADULT:** The ground colour of the ventral hindwing of the Hydaspe Fritillary is dark brown mixed with red, giving rise to the common name "Purple Fritillary."

**IMMATURE STAGES:** The mature larva is black. The body is covered with protuberances that bear spines. There are no mid-dorsal narrow yellow stripes.

**BIOLOGY:** The Hydaspe Fritillary flies from mid-June to mid-August, depending on the elevation of the breeding population. At high elevations in the mountains of Vancouver Island, the Cascades, and the south Coast Ranges, the species will fly as late as the first week of September if the high snowpack of the previous winter delays larval development. Eggs are laid at the base of the foodplant, *Viola* sp. They hatch and the first instar larvae overwinter. Larvae begin feeding the following spring, as soon as the foodplant has leafed out.

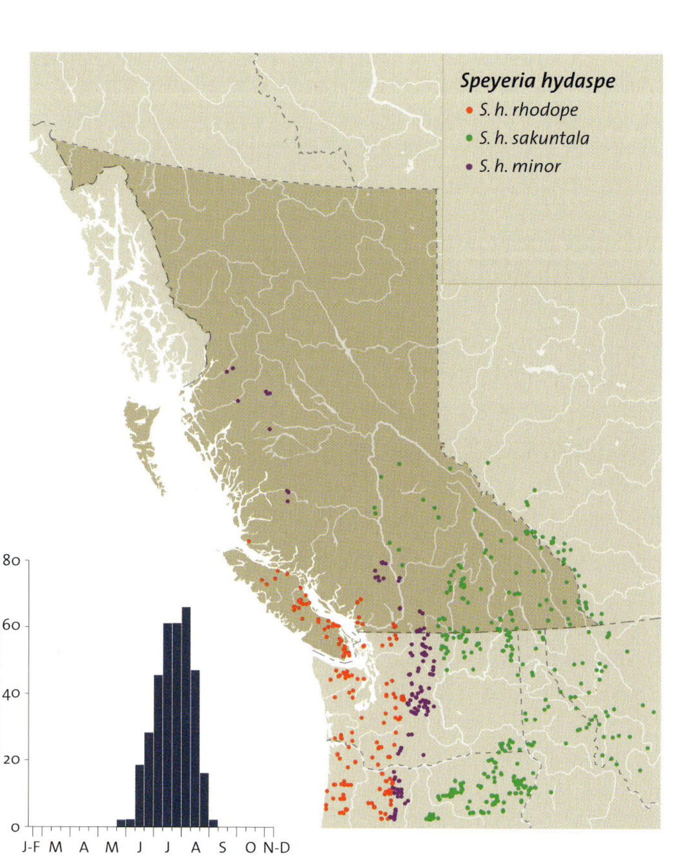

**Hydaspe Fritillaries (*Speyeria hydaspe rhodope*) mating**

**Mature larva, ssp. *rhodope***

**SUBSPECIES:** The Rhodope Fritillary, *S. h. rhodope* (W.H. Edwards, 1874) (TL: Fraser River Lowlands, BC), is found on Vancouver Island and in the Lower Fraser Valley. The Minor Fritillary, *S. h. minor* dos Passos & Grey, 1947 (TL: Mt. McLean, BC), is a small form of the Hydaspe Fritillary found at higher elevations along the east side of the Cascade and Coast mountains from the Nass River south to Manning Provincial Park. The Sakuntala Fritillary, *S. h. sakuntala* (Skinner, 1911) (TL: Kaslo, BC), occurs in the southern Cariboo, the Southern Interior, and the Kootenays. It has the widest altitudinal range.

**RANGE AND HABITAT:** The Hydaspe Fritillary is found on Vancouver Island, in the Lower Mainland, on the east side of the Cascade and Coast mountains from the Nass River south to Manning Provincial Park, and in the south Cariboo, the Southern Interior, and the Kootenays. It occurs in mountain meadows from 800 to 1,900 m, with males commonly hilltopping at peaks to 2,300 m.

**GENERAL DISTRIBUTION:** The Hydaspe Fritillary is found from southern BC and the Rocky Mountains of AB south to CA in the Sierras and northern NM in the Rockies. It has the most limited distribution of any *Speyeria* that occurs in BC.

**CONSERVATION STATUS:** Not of concern, with subspecies *rhodope* S4 and subspecies *sakuntala* and *minor* both S5.

## MORMON FRITILLARY
*Speyeria mormonia* (Boisduval, 1869)

**ETYMOLOGY:** The species name *mormonia* refers to the Mormons of Utah, because Boisduval believed that the type specimens came from Salt Lake, Utah. Subspecies *bischoffii* is named for the collector of the type series, M. Bischoff. Bischoff sold Spencer F. Baird of the Smithsonian Institution vertebrates and other natural history specimens collected in Alaska and on the Russian side of the Bering Strait at what is now Telegraph Camp, Russia.

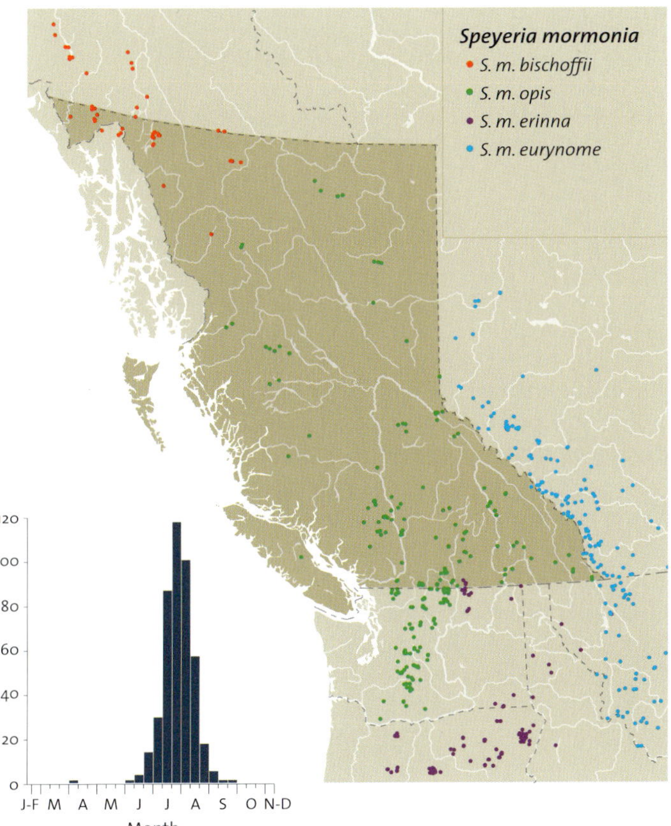

Ssp. *opis* ♂ D (4.6 cm)

Ssp. *opis* ♂ unsilvered V (4.9 cm)

Ssp. *opis* ♂ silvered V (4.6 cm)

Ssp. *erinna* ♂ V (4.8 cm)

Ssp. *bischoffii* ♂ V (4.6 cm)

Mormon Fritillaries (*Speyeria mormonia opis*) mating

**Fourth instar larva, ssp. *opis***

adult flight is the latest of any BC fritillary. The Mormon Fritillary flies from mid-July to late August. If snowmelt is delayed, it can fly until early September. Eggs are laid at the base of the larval foodplant, *Viola* sp. They hatch and the first instar larvae overwinter. Larvae begin feeding the following spring, as soon as the foodplant has leafed out.

He collected with the permission of the Russians just before the American purchase of Alaska. The subspecies name *opis* is derived from Opis or Ops, the sister and wife of Saturn and goddess of abundance on the earth. The subspecies name *erinna* comes from Erinna, a celebrated Greek poet and friend of Sappho and Mytilene. The common name was first used by Comstock (1927).

**ADULT:** The Mormon Fritillary is our smallest fritillary. When silvered, the ventral hindwings appear green, but it is a green not confused with that of the Callippe Fritillary. When unsilvered, the ventral hindwings appear brown but have a distinctive arrangement of the pattern that is not easily confused with that of other fritillaries.

**IMMATURE STAGES:** The mature larva is black. The body is covered with protuberances that bear spines. There are two mid-dorsal narrow yellow stripes. The spines are shorter than those of other *Speyeria* larvae that have the narrow yellow stripes.

**BIOLOGY:** The larval habitat of the Mormon Fritillary occurs at higher elevations than that of other *Speyeria* species, so the

**SUBSPECIES:** The unsilvered subspecies, Bischoff's Fritillary, *S. m. bischoffii* (W.H. Edwards, 1870) (TL: Sitka, AK), is found in extreme northwestern BC and the adjacent Alaska panhandle. In most of the rest of the species' range in BC is found a predominantly silvered subspecies, the Opis Fritillary, *S. m. opis* (W.H. Edwards, 1874) (TL: Bald Mt. [between Likely and Barkerville], BC) [= *jesmondensis* dos Passos & Grey, 1947, TL: Jesmond, BC; and *washingtonia* (Barnes & McDunnough, 1913), TL: Paradise Valley, Mt. Rainier, WA]. In the extreme south of the West Kootenay, another unsilvered subspecies, the Erinna Fritillary, *S. m. erinna* (W.H. Edwards, 1883) (TL: Spokane Falls, WA), is found.

**RANGE AND HABITAT:** The Mormon Fritillary is found commonly throughout BC east of the crest of the Cascade and Coast mountains, but has not been recorded in the northeastern lowlands. In the south it is usually found above 1,250 m, with the males hilltopping to 2,300 m.

**GENERAL DISTRIBUTION:** The Mormon Fritillary ranges from southeastern AK east through BC to MB and western MN. It occurs south to central CA and northern NM, with disjunct populations in the White Mountains of AZ.

**CONSERVATION STATUS:** Subspecies *erinna* is Threatened in BC (S1S2). Subspecies *bischoffii* (S4) and *opis* (S5) are not of concern.

## GENUS *BOLORIA* MOORE, [1900] ANGLED LESSER FRITILLARIES

The name *Boloria* is derived from the Greek *bolos* (fishing net) and refers to the reticulate or checkerboard wing pattern (Emmet 1991). The common name "angled lesser fritillaries" refers to their small size and to the angled margin of the hindwing. The common name is used here for the first time.

Palearctic workers and their popular books restrict the use of the genus *Boloria* to that of Warren (1944) and use a second genus, *Clossiana,* for most species in this group. We accept that approach. As used by Palearctic authors,

the genus *Boloria* is defined by the shape of the valves. The dorsal or superior distal process has two distinct and heavily chitinized projections (Shirôzu and Yamamoto 1953). The genus *Clossiana* has only one (Warren 1944). Grey (1989) further points out that the uncus of *Boloria* is heavily spiculate or covered with short, thick hairs. In addition, the hindwing margin has a characteristic sharp angle at the $M_3$ vein. Both *Boloria* and *Clossiana* have a bifid uncus, a tribal character. As defined by Palearctic workers, there are two to four species, only one of which occurs in North America.

# Mountain Fritillary

*Boloria napaea* (Hoffmansegg, 1804)

**ETYMOLOGY:** The species name *napaea* means "a wooded vale or valley," and likely refers to the habitat that Hoffmansegg thought the species inhabited. Subspecies *alaskensis* is named for Alaska, where the type specimens originated. The common name (Higgins and Riley 1970) refers to the species' alpine and subalpine habitat.

**ADULT:** The Mountain Fritillary is easily recognized by the small size of the butterfly and the angled margin of the hindwings.

**IMMATURE STAGES:** Undescribed for North America.

**BIOLOGY:** The Mountain Fritillary flies from late June to early August, with peak flight in mid-July. There is one generation each year. Eggs are laid and hatch within 10 days (JHS). The

♂  D  (3.3 cm)

♀  D  (4.1 cm)

♂  V  (3.3 cm)

♀  V  (4.1 cm)

overwintering stage is unknown. *Vaccinium* is the likely North American larval foodplant.

**SUBSPECIES:** The BC subspecies is the Alaskan Fritillary, *B. n. alaskensis* (Holland, 1900) (TL: Forty-Mile/Mission Creeks, AK). This subspecies ranges throughout all of Alaska except the northeast coast, central and southern YT, northern BC, and the Wilmore Wilderness Park area of AB.

**RANGE AND HABITAT:** The Mountain Fritillary occurs in mountains across the northern fourth of BC and south in the Rockies into AB. Its habitat is moist meadows, usually along streams at or near timberline.

**GENERAL DISTRIBUTION:** The Mountain Fritillary is found throughout AK, along the north coast of YT and NT to the west coast of Hudson Bay. It ranges south through central and southern YT to the AB Rockies near Adams Lookout, Wilmore Wilderness Park. There are also disjunct populations in the Wind River Range of WY. The species is Holarctic, occurring from northern Europe to northeastern Siberia.

**CONSERVATION STATUS:** Not of concern (S4).

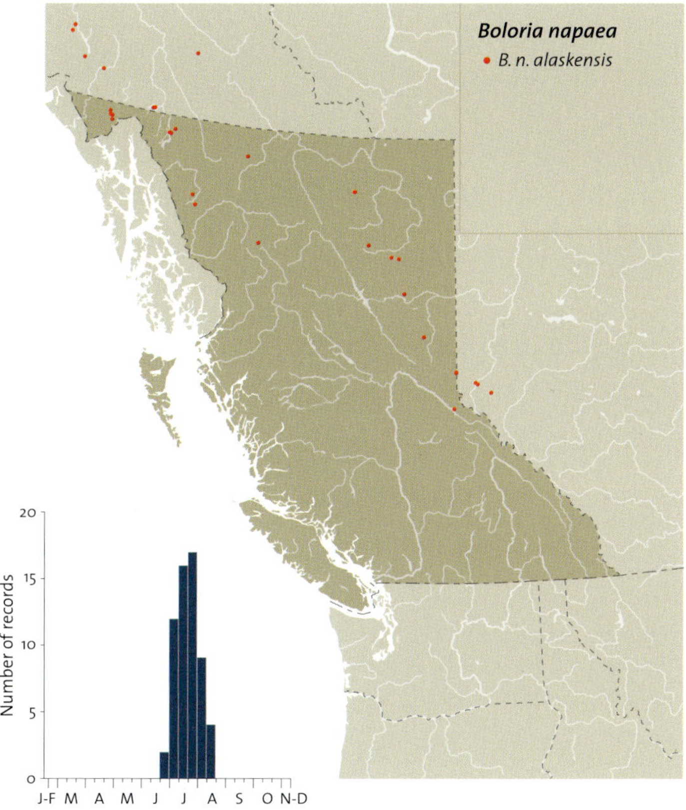

Reuss (1922) stated that he named the genus *Clossiana* for the recognized entomologist Herr Adolf G. Closs, but it appears that Closs was only a minor worker on Lepidoptera. The common name "lesser fritillaries" refers to the small size compared with *Speyeria*.

Under the restricted generic usage of *Boloria,* we state why we recognize the genus *Clossiana* and define the genus. On the upperside, the wings are very similar to those of *Speyeria*. Some males of one species of *Speyeria,* *S. mormonia,* are as small as the largest females of our largest *Clossiana, C. tritonia*. Only one species, *Clossiana selene,* has silver spots on the ventral hindwing. This genus is Holarctic, with at least 21 species; 13 are found in North America and 12 of these occur in BC. Nine BC species are Holarctic. The 4 temperate species, 3 in BC, feed on violets (*Viola*) but the northern species do not. There has been much confusion in the literature regarding larval foodplants, and we discuss only those verified by Shepard (1975) and later.

## BOG FRITILLARY
*Clossiana eunomia* (Esper, 1799)

**ETYMOLOGY:** The name *eunomia* is derived from Eunomia of the Horae, who in Homer were handmaidens of Zeus who presided over the seasons. Another interpretation is that Eunomia was a daughter of Zeus and Themis. In this version Eunomia stands for good order. Subspecies *dawsoni* was named after "Mr. Horace Dawson, who by his conscientious collecting has greatly added to our knowledge of the lepidopterous fauna of Northwestern Ontario" (Barnes & McDunnough 1916). The subspecies name *nichollae* honours Mrs. Nicholl, an English collector, who travelled through the high Rocky Mountains of BC and Alberta in 1904 and 1906. The common name "Bog Fritillary" (Klots 1951) was coined because where it was first known in eastern North America, it inhabited the fringes of bogs.

SSP. *dawsoni* ♂ D (3.8 CM)

SSP. *dawsoni* ♂ V (3.8 CM)

SSP. *nichollae* ♂ D (3.3 CM)

SSP. *nichollae* ♂ V (3.3 CM)

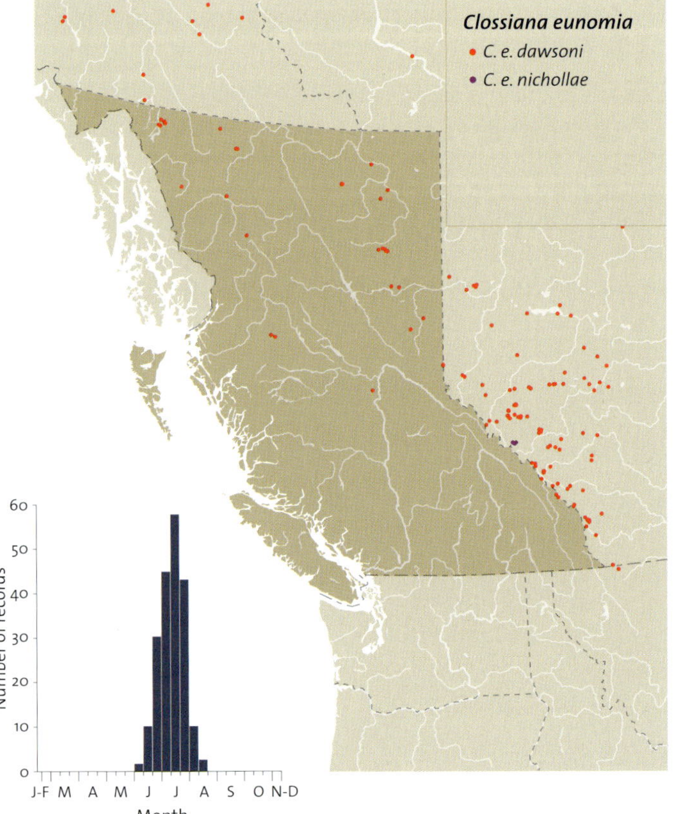


**ADULT:** The Bog Fritillary is easily distinguished by the ventral hindwing. The postmedian row of spots are an even row. The spots are almost perfect circles, cream-coloured and edged with black. Subspecies *C. e. nichollae* is a dark form from high elevations in the Rockies.

**IMMATURE STAGES:** Undescribed.

**BIOLOGY:** The Bog Fritillary is one of four species found together in high mountain habitat in the Rocky Mountains and boreal habitat across Canada. It is the third of the four species to emerge, after *C. freija* and *C. frigga,* and before *C. chariclea*. It is presumed that the peak flight of the four species is staggered to avoid interspecific matings (JHS). There is much comment in the literature that this species has a very short adult flight period, but such is not the case in western Canada. Mark-release-recapture studies by Shepard (JHS) showed that the Bog Fritillary had a flight period of normal length and was equal to *C. chariclea*. The flight is from mid-June to early August with a peak in early July, depending on elevation and high-elevation snow cover.

**SUBSPECIES:** The BC populations are the widespread boreal subspecies *C. e. dawsoni* (Barnes & McDunnough, 1916) (TL: Hymers, ON). The subspecies *C. e. nichollae* (Barnes & Benjamin, 1926) is currently known only from near the Columbia Ice Field in Alberta, but adjacent habitat in BC is totally unexplored and the subspecies may be found in BC. Nichollae's Fritillary seems to be genetically distinct, not just a high-elevation form.

**RANGE AND HABITAT:** The Bog Fritillary is found east of the Coast Ranges in the northern half of BC and in the Rockies along the AB border. The habitat is spruce forest meadow openings along streams, and around the fringes of bogs and small, mature glacial lakes at the southern edge of its distribution.

**GENERAL DISTRIBUTION:** The Bog Fritillary is found from AK and across most of CAN except the Canadian archipelago, southern BC, and the southern prairies. In the lower USA, it occurs only just south of Lake Superior and in disjunct populations in WY and CO. The species is Holarctic, occurring in northern Europe and across non-arctic Russia.

**CONSERVATION STATUS:** Not of concern (S5).

## SILVER-BORDERED FRITILLARY
*Clossiana selene* (Denis & Schiffermüller, [1775])

**ETYMOLOGY:** Selene was the Greek goddess of the moon. Both the Latin name and the common name refer to the silvery spots on the ventral hindwings, which shimmer like the moon. The subspecies name *atrocostalis* is derived from the Latin *atrox* (dark) and the costa or leading edge of the wing, for the darker margins of the wing compared with those of the subspecies from the southeastern United States, *myrina*. The common name was first used by Scudder (1875) in reference to the silvery spots along the ventral hindwing border.

**ADULT:** The Silver-bordered Fritillary is the only lesser fritillary with silver spots on the underside of the wings. The species often has melanic, aberrant individuals.

♂ D (3.9 CM)

♂ V (3.9 CM)

**IMMATURE STAGES:** Scudder (1889a) first described the immatures in detail. The egg is dull yellow and taller than it is broad; it has strong reticulation on the chorionic surface. The mature larva is dark brown/green with a pale, wide dorsal band. The body is covered with elaborately branched black tubercles. This is the general appearance of all *Clossiana* larvae, with some variation in colour pattern.

**BIOLOGY:** Adults of the Silver-bordered Fritillary fly from late May to early September in the Okanagan Valley. There are at least two complete broods in the south, with peak flights in mid-June and late July. A partial third brood flies in early September. At Atlin in northern BC, there is apparently only one generation per year. The larvae feed on whatever violets

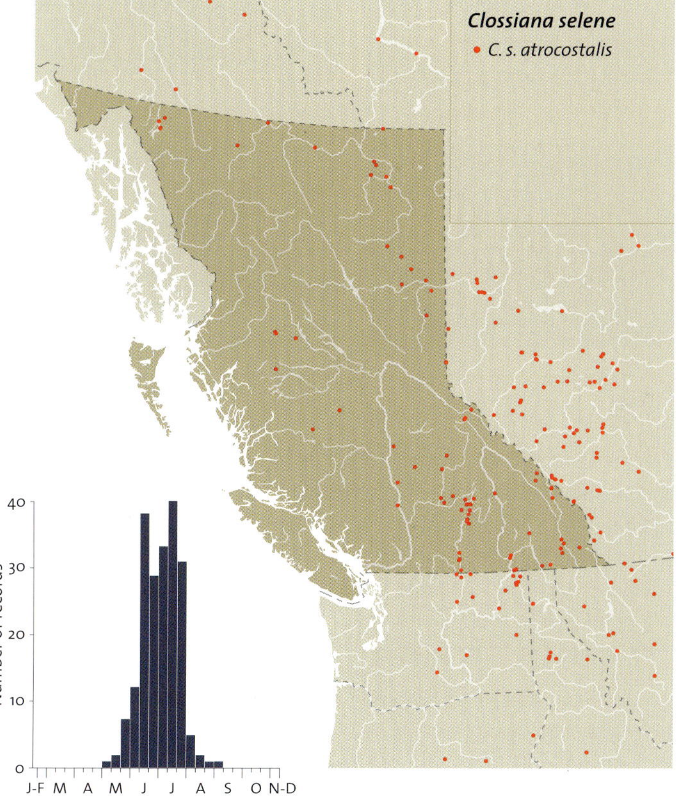

**Silver-bordered Fritillary (*Clossiana selene*)**

(*Viola* sp.) are available at any one site. For one population in the West Kootenay (JHS), three species of violets are available and the larvae feed on all three. Shepard has found in mass rearings that some individual larvae from first-brood eggs take much longer to develop. This staggered larval development may be the reason for an apparent partial third generation. In years of poor summer weather and for more northern populations, these larvae may be the overwintering stage. Scudder (1889a) observed this same delayed development in the northeastern United States.

**SUBSPECIES:** The BC subspecies is the Dark-bordered Fritillary, *C. s. atrocostalis* (Huard, 1927) (TL: Chicoutimi, PQ). This is a widespread boreal forest subspecies.

**RANGE AND HABITAT:** The Silver-bordered Fritillary is found throughout BC east of the Coast Range and Cascade Mountains. It is always found in sphagnum bog and fen situations in the south. In the far north it has been recorded only at bogs surrounding hot springs, such as Atlin or Liard Hot Springs. In the south various skippers are always found in the habitat along with the Silver-bordered Fritillary, for example, *Polites mystic, P. peckius, Carterocephalus palaemon,* and usually *Oarisma garita.* Also in the south, the Atlantis

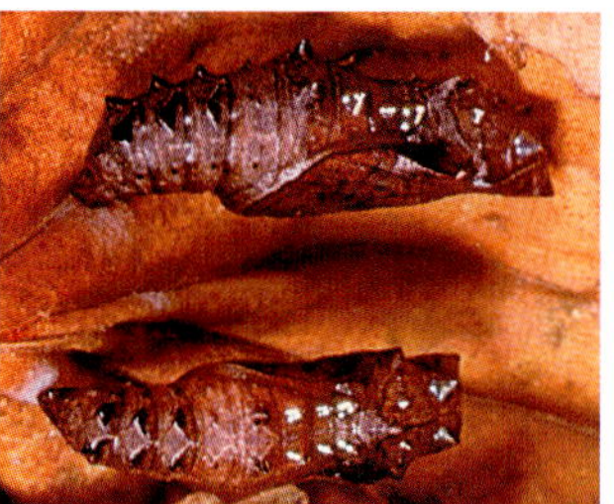

**Mature larva**

**Pupae**

Fritillary is often found in the Silver-bordered Fritillary habitat. The elevational range of this habitat is 660–1,000 m.

**GENERAL DISTRIBUTION:** The Silver-bordered Fritillary ranges from central AK across boreal CAN to NF. In the west it occurs south to central OR and northern NM. At the southern fringes of its western range, it is found at isolated springs or around spring-fed lakes with permanently moist meadows. In the eastern USA it is found south to latitude 40°N. The species is Holarctic, occurring in northern Europe and across southern Russia and southern Kamchatka.

**CONSERVATION STATUS:** Not of concern (S5).

## MEADOW FRITILLARY
*Clossiana bellona* (Fabricius, 1775)

**ETYMOLOGY:** In Roman mythology, Bellona was the goddess of war and either the wife or sister of Mars. Subspecies *jenistorum* was named for Mr. and Mrs. Harry E. Jenista, the

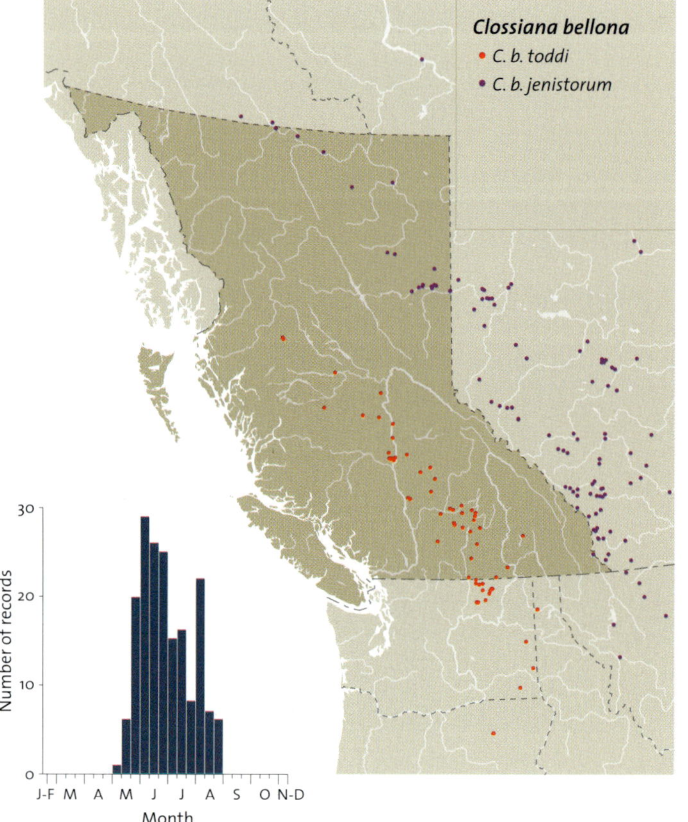

Ssp. *toddi* ♂ D (4.4 CM)

Ssp. *jenistorum* ♂ D (4.3 CM)

Ssp. *toddi* ♂ V (4.4 CM)

Ssp. *jenistorum* ♂ V (4.3 CM)

frequent field companions of Stallings and Turner. Subspecies *toddi* honours W.E. Clyde Todd, who collected the type series when he was in the first party of Europeans to cross Labrador from Three Rivers, PQ, to Ungava on Davis Strait. The name "Meadow Fritillary" was first used by Scudder (1875) and reflects the species' preference for old field meadows in the eastern USA.

**ADULT:** The Meadow Fritillary is very easily distinguished by the fact that the tip or apex of the forewing is squared off, not rounded as in all other lesser fritillaries.

**Immature stages:** Scudder (1889a) first described the immatures in detail. They are like those of the Silver-bordered Fritillary, but have some purple coloration not found on the latter.

**Biology:** The Meadow Fritillary flies from late May to late August in two broods, with peak flight in mid-June and early August in the south. In the Peace there appears to be only one generation per year. There is no evidence of a third generation. The species overwinters as early instar larvae. It has been recorded feeding on various violets (Scudder 1889a). In BC it is always associated with *Viola canadensis* (JHS).

**Subspecies:** Jenista's Meadow Fritillary, *C. b. jenistorum* [justified emendation] Stallings & Turner, 1947; TL: Rivercourse, AB (Kondla, 1996), is the prairie subspecies occurring in northeastern BC. The central BC populations are similar to the eastern boreal Canadian subspecies *C. b. toddi* (Holland, 1928) (TL: St. Margarets R., PQ).

**Range and habitat:** The Meadow Fritillary is essentially a species occurring in eastern temperate habitat. In BC it occurs as two disjunct sets of populations, one in northeastern BC and the other through the western Cariboo and the Okanagan Highlands. In both areas it is found in open meadows adjacent to aspen woodlands and in open, mature aspen woodlands.

**General distribution:** The Meadow Fritillary occurs from eastern BC east to Labrador and NS. In the west, the Cariboo/Okanagan Highlands form occurs in a few isolated populations in eastern WA, northeastern OR, and northern ID, and then in disjunct populations in WY and CO. In the east it is common south to latitude 38°N, often in old-growth meadows.

**Conservation status:** Not of concern, with both subspecies S4.


## Frigga Fritillary
*Clossiana frigga* (Thunberg, 1791)

**Etymology:** The Frigga Fritillary and most of the other species that were first described from Scandinavia were named after Norse mythological characters to emphasize their northern distribution in Europe. Frigg was the wife of Odin and queen of the gods. The subspecies name *saga* refers to the Scandinavian oral history, or Sagas. The common name was first used by Holland (1898).

♂ D  (3.9 cm)   ♂ V  (3.9 cm)

**Adult:** The Frigga Fritillary is close in general appearance to *C. bellona*, *C. epithore*, and *C. improba*. It is closest to *C. improba* but twice as large. The Frigga Fritillary has a large basal rectangular white area on the ventral hindwing that is not found on *C. epithore*.

**Immature stages:** Undescribed.

**Biology:** The Frigga Fritillary is the second of the four species to emerge in boreal habitat. The flight period is from late May to late July in one generation, with peak flight in mid-June. The hibernation stage is late instar larvae. Shepard (1975) reared the species on willows (*Salix*) in Alberta. In Europe the species feeds on *Rubus*.

**Subspecies:** BC populations are the Saga Fritillary, *C. f. saga* (Staudinger, 1861) (TL: Labrador), a widespread boreal subspecies.

**Range and habitat:** The Frigga Fritillary occurs rarely throughout most of BC east of the Cascade and Coast mountains. It occurs in the same habitat as the Bog Fritillary but not at as many sites. These two species, along with the Freija Fritillary and the Arctic Fritillary, are found together across boreal Canada.

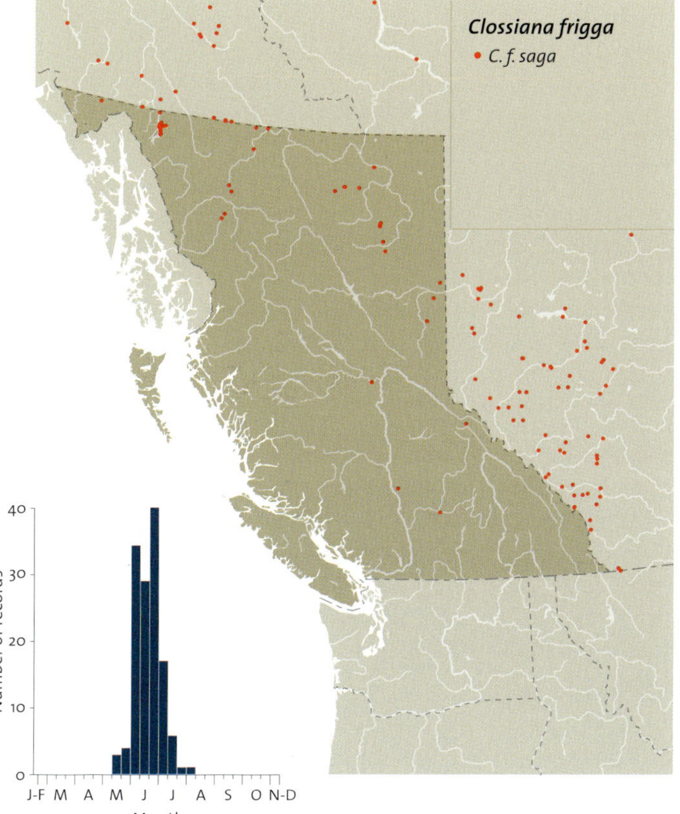

WY and CO. The species is Holarctic, occurring in Scandinavia
and across Russia except the southwest and Kamchatka.

**CONSERVATION STATUS:** Not of concern (S5).

## DINGY FRITILLARY
*Clossiana improba* (Butler, 1877)

**Dingy Fritillary (*Clossiana improba*)**

♂ D (3.4 CM)

♀ D (3.6 CM)

♂ V (3.4 CM)

♀ V (3.6 CM)

**ETYMOLOGY:** The species name *improba* is from the Latin
*improbus,* which has many meanings, including "inferior,"
for the fact that it is a small version of *C. frigga.* In the early
literature the two were often regarded as one species.
Subspecies *youngi* was named for the Reverend S. Hall Young,

a Moravian missionary in Alaska in 1899 who collected for
Holland because of their mutual church work. Holland was a
famous preacher in Pittsburgh, where he was curator of the
Carnegie Museum of Natural History. The common name
was first used by Klots (1951) in reference to the dark, dusky
wing colour of the nominate subspecies.

**ADULT:** The Dingy Fritillary is our smallest fritillary. It is very
similar to the Frigga Fritillary in wing pattern, but the two
never fly together.

**IMMATURE STAGES:** Undescribed except for the disjunct
subspecies *C. i. acrocnema* Gall and Sperling, 1980 of Colorado and Wyoming. Scott (1982) described the immatures in
detail. The egg is tan. The head of the mature larva is dark
brown. The body is brown with various lighter and darker
lines and a conspicuous cream dorsal line. The tubercles or
scoli are reddish brown.

**BIOLOGY:** Dingy Fritillaries fly from early June in the far north
to early August in the Rockies, with peak flight in mid-July for
all populations. There is one generation per year. Scott (1982)
reared the species on *Salix reticulata nivalis.* In BC it is always
found on ridges above timberline in association with similar
dwarf willows that barely grow above ground to avoid the
cold winds of the habitat.

**SUBSPECIES:** The BC subspecies, Young's Fritillary, *C. i. youngi*
(Holland, 1900) (TL: Forty-Mile/Mission Cr., AK), occurs in
central Alaska, central and southern Yukon, and across

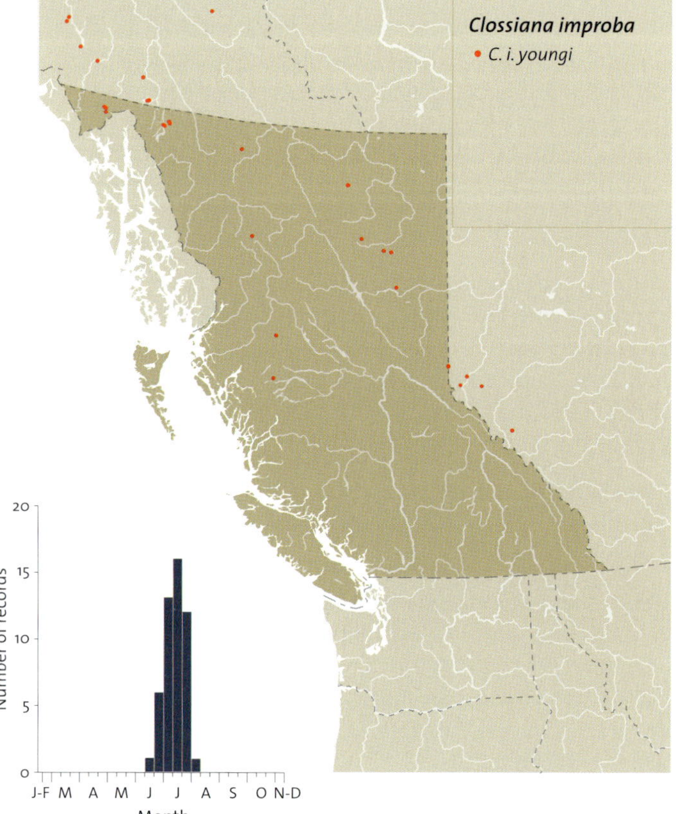

northern BC, just entering Alberta in the Rockies south to Hinton.

**RANGE AND HABITAT:** The Dingy Fritillary is found in the northern half of BC only above timberline in alpine tundra.

**GENERAL DISTRIBUTION:** The Dingy Fritillary ranges from AK east across arctic and alpine CAN, including the southern Canadian archipelago to Baffin Island but not Labrador. There are two disjunct populations, one in Wind River Range, WY, and the other in the San Juan Mountains, CO. The species is Holarctic, occurring in the far north of Norway and Sweden and the far north of Russia to eastern Siberia.

**CONSERVATION STATUS:** Not of concern (S4).

## WESTERN MEADOW FRITILLARY
*Clossiana epithore* (W.H. Edwards, [1864])

**ETYMOLOGY:** The species name *epithore* is derived from the Greek *epi* (similar to) and *thore,* the species name of the similar-looking Thor's Fritillary of Europe. Thor was the son of Odin, god of thunder and defender of other gods when they got into trouble. The subspecies *sigridae* is named for Sigrid Marie Shepard in recognition of the 39 years she has worked with her husband on butterflies. The subspecies *chermocki* is named for F.H. Chermock. The common name "Western Meadow Fritillary" (Comstock 1927) refers to the species' western North American distribution and the similarity in habitat choice and general wing pattern to the Meadow Fritillary, *C. bellona.*

**ADULT:** The Western Meadow Fritillary is closest in appearance to the Meadow Fritillary but lacks the blunt forewing apex of the latter.

**IMMATURE STAGES:** Edwards (1892) gave the only published account of the early stages. This paper has been overlooked

SSP. *sigridae* ♂ D HOLOTYPE (3.6 CM)

SSP. *sigridae* ♂ V HOLOTYPE (3.6 CM)

SSP. *sigridae* ♀ D (3.5 CM)

SSP. *sigridae* ♀ V (3.5 CM)

SSP. *chermocki* ♂ D (4.1 CM)

SSP. *chermocki* ♂ V (4.1 CM)

by all subsequent workers, including Shepard (1975). The fourth instar larva is grey with a dark dorsal line and lower lateral lines, and the area between cross-marked with black; the basal stripe is red. The spines are russet, except those on the second and third thoracic segments and last abdominal segment, which are black. Edwards noted that this species is quite different from *C. bellona.* Mr. Koeble supplied eggs from Spokane, WA.

**BIOLOGY:** The Western Meadow Fritillary flies from early May to early September, depending on elevation and latitude, but with only one generation per year and only for about three weeks for any one population. Where it is sympatric with either *C. selene* or *C. bellona,* the peak flight is between the two broods of the sympatric species. This occurs at about 1,000 m in the Southern Interior and the West Kootenay.

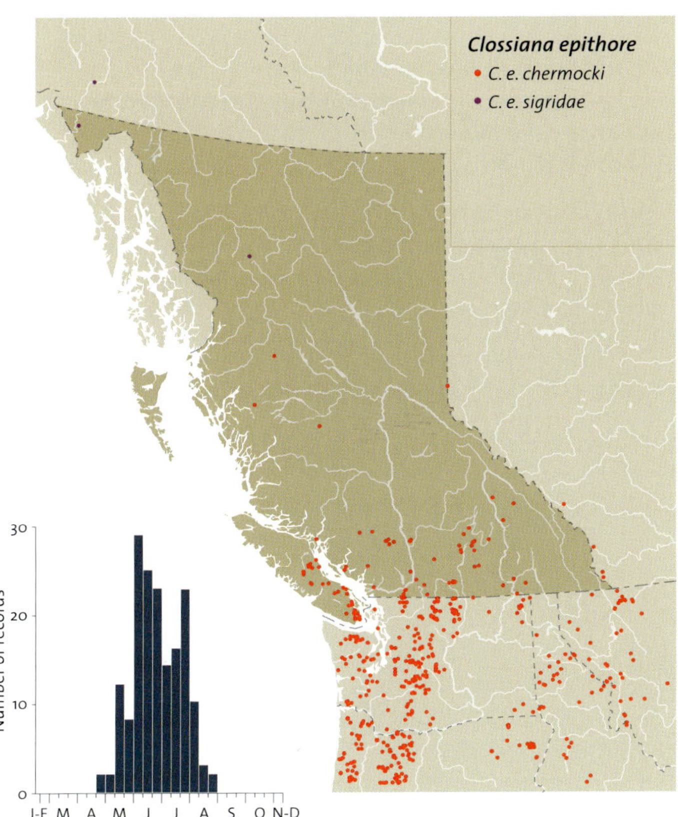

**Western Meadow Fritillary (*Clossiana epithore chermocki*)**

Eggs hatch in early July and grow to fourth instar larvae by mid-August before overwintering (Edwards 1892; JHS). Shepard (1975) has reared the Western Meadow Fritillary on various *Viola* species in BC and California, but the species is not associated with *Viola canadensis,* the violet usually associated with *C. bellona* (JHS) in BC and Alberta.

**SUBSPECIES:** The southern and central BC populations are the subspecies *C. e. chermocki* E. & S. Perkins, 1966 (TL: 2.9 mi. E Dolph, Yamhill Co., OR) (= *borealis* E. Perkins, 1973; = *uslui* Kocak, 1984), which ranges from northern coastal California north to New Aiyansh, BC (Shepard and Shepard 1974) and from southeastern BC to northwestern WY (Yellowstone). The populations in the Coast Ranges from Mt. Klapan north and the adjacent Alaska panhandle and Yukon are the new subspecies *C. e. sigridae* (described below). The type locality of the name *borealis* is in the Cascade Mountains of BC and was erroneously applied to all *C. epithore* populations from northern and eastern BC and southeast to central Idaho by E. Perkins. These interior populations in the southeast of BC are not separable from southwestern BC populations. Even if they were, the type locality of *borealis* is within the expected range of a western/Cascade subspecies, not a Rocky Mountain subspecies.

*Clossiana epithore sigridae* **Shepard, new subspecies**. In both males and females the spot between vein M$_1$ and vein M$_2$ in the postmedian row on the dorsal hindwings is elongate, not rounded as in other subspecies, and is pointed towards the base of the wing. The adjacent postmedian spot above vein M$_1$ is usually also elongated. These elongated spots are not found on specimens of the other *C. epithore* subspecies unless the specimen is clearly an aberrant one. The basal portion of the ventral hindwings, inside the median band of dull yellow connected spots, is a uniform dark reddish brown. By contrast, in subspecies *chermocki,* this basal area is heavily overlaid by yellow scales to the point that the median row of yellow spots does not stand out. Males of subspecies *sigridae* (*n* = 23) have an average wing length of 18.2 mm, females (*n* = 12) 20.0 mm. This is significantly smaller than in even the closest populations (New Aiyansh) of subspecies *chermocki,* where males (*n* = 5) average 20.0 mm and females (*n* = 2) average 21.5 mm. **Types**. Holotype: male, BC, St. Elias Mts., Tats Lake, 770 m, 22 July 1992, leg. C.S. Guppy. A label "Holotype /*Clossiana epithore* / *sigridae* Shepard" is attached. The holotype is deposited in the Royal British Columbia Museum, Victoria, BC, CAN. Paratypes: 20 males, 8 females, same data (RBCM); 1 male, 1 female, same data (JHS); 1 male, 3 females, BC, Mt. Klapan Rd., km. 109, el. 1,400 m, 11 July 1989, leg. J. and S. Shepard (JHS).

**RANGE AND HABITAT:** The Western Meadow Fritillary occurs in the Coast Ranges in northern BC and across southern BC except the Chilcotin. On the coast, the species occurs in coniferous forest meadows from almost sea level to sub-alpine meadows. In the south Cariboo, Southern Interior, and Kootenays, it occurs from 1,000 to 2,170 m in mountain meadows. This species and the Silver-bordered Fritillary often fly together at 1,000 m, the Western Meadow Fritillary along the stream banks and the Silver-bordered Fritillary in adjacent boggy areas. Above 1,000 m, the skipper *Carterocephalus palaemon* is usually associated with the Western Meadow Fritillary, but not the other skippers found in the same habitat as the Silver-bordered Fritillary. The Western Meadow Fritillary is never found in close association with the Meadow Fritillary, which is found in drier aspen woodlands.

**GENERAL DISTRIBUTION:** The Western Meadow Fritillary is found from extreme southwestern YT south through BC to central CA in the coast ranges and the Sierras, and south to northwestern WY.

**CONSERVATION STATUS:** Subspecies *sigridae* is of Special Concern in BC (S3). Subspecies *chermocki* is not of concern (S5).

# Polar Fritillary
*Clossiana polaris* (Boisduval, [1828])

**ETYMOLOGY:** The name *polaris* refers to the species' northern or polar distribution. This species is very accurately named as it occurs on the northern tip of Ellesmere Island, further north than any other Canadian butterfly except *C. chariclea*, which occurs as far north. The common name was first used by Holland (1898).

**ADULT:** The Polar Fritillary is easy to identify. The postmedian row of spots on the ventral hindwings is in a regular row. The spots are black in the centre and surrounded by a wide white area.

**IMMATURE STAGES:** Undescribed.

**BIOLOGY:** The Polar Fritillary flies from mid-June to late July, with one generation per year. In some parts of the species' range, including BC, it flies only in odd-numbered years, meaning that it takes two years to complete one generation. Larvae collected on *Dryas* species in the Canadian Arctic (CNC) suggest that this is the foodplant. Other *Clossiana* larvae collected at the same localities in the far north on *Salix* must have been *C. chariclea*, as *Salix* is its known foodplant. No other species of *Clossiana* were collected at these localities as adults.

♂ D  (4.1 cm)

♀ D  (4.0 cm)

♂ V  (4.1 cm)

♀ V  (4.0 cm)

**SUBSPECIES:** BC populations of the Polar Fritillary are the nominate subspecies, *C. p. polaris* (Boisduval, [1828]) (TL: Cape Nord, Norway).

**RANGE AND HABITAT:** The Polar Fritillary is found only in the highest and driest mountains of northern BC where the foodplant is common.

**GENERAL DISTRIBUTION:** The Polar Fritillary is found from AK east through YT, NT, and Labrador, with outlying southern populations at the northern edge of most provinces. The Polar Fritillary and the Arctic Fritillary are found further north than any other butterfly in the Canadian archipelago. The Polar Fritillary is Holarctic, occurring in northern Norway and across extreme northern Russia to eastern Siberia.

**CONSERVATION STATUS:** Not of concern (S4).

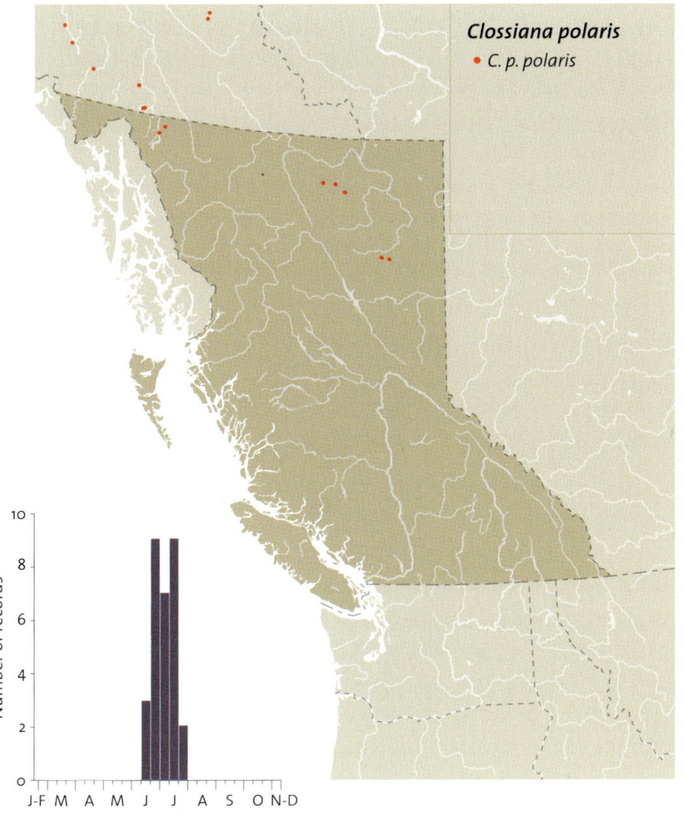

**Polar Fritillary (*Clossiana polaris polaris*)**

## Albert's Fritillary
*Clossiana alberta* (W.H. Edwards, 1890)

**Etymology:** It has been suggested that the species name *alberta* refers to the type locality near Laggan [vicinity Lake Louise], in what was then known as Alberta Territory (Bird et al. 1995). This is very unlikely as Edwards named less than 10 of his many species names for geographic localities. The year after the species was described, however, Edwards named a synonym of *C. tritonia, C. victoria* (W.H. Edwards, 1890) (TL: Laggan [vicinity Lake Louise]). Edwards most likely meant to give a permanent reminder of the enduring love of Queen Victoria and Prince Albert, as these two names were given just after the famous Golden Jubilee celebrations. He certainly meant this when he described *C. victoria*. What more fitting memorial than that two species of alpine butterflies are named Albert and Victoria and fly together forever? An alternative hypothesis is that *alberta* is meant for the second Christian name of Princess Louise, after whom Lake Louise was named. In either case, we amend the common name as first coined by Holland (1898).

**Adult:** Adults of Albert's Fritillary appear to be melanic and greasy, especially the females. This, however, is the normal appearance of the species and is an easy way to distinguish it from other *Clossiana* species that occur with it in BC. Most books discuss this species in conjunction with *C. tritonia* because they have the same habitat, but the two species are not closely related. Dubutolov and Shepard (in prep.) are

♂ D (3.9 cm)

♀ D (4.4 cm)

♂ V (3.9 cm)

revising the species in these two sections of *Clossiana*. *Clossiana alberta* is closely related to both *C. polaris* and *C. erda* (Christoph, 1893).

**Immature stages:** Undescribed.

**Biology:** Albert's Fritillary has a very restricted flight period, from early July to early August, with peak flight in late July. Some populations occur only in even-numbered years and others only in odd-numbered years. Thus it takes two years for the species to complete one life cycle from adult to adult. Bean obtained eggs from females confined with *Dryas octopetala* and other plants; most eggs were laid on the *Dryas* (Edwards 1887–97). The eggs were laid 20 July and hatched on 30 July. The larvae did not feed. Shepard has also observed the species in association with *Dryas*, the larval foodplant of the closely related and allopatric species *C. polaris*. It is assumed that *Dryas* is also the foodplant of *C. alberta*, but it remains to be positively proven. Shepard could not get females to lay eggs on *Dryas octopetala*.

**Subspecies:** None. The type locality of the species is Laggan [vicinity Lake Louise], AB.

**Range and habitat:** Albert's Fritillary is known only from extreme eastern BC along the AB border at and above timberline in the Rockies. Bird et al. (1995) and others who attribute this species to other areas in BC are in error.

**General distribution:** Albert's Fritillary has the most restricted total world distribution of any butterfly that occurs in BC. It is known only from the Rocky Mountains from northern AB and BC south to Glacier National Park, MT. Wyatt described a subspecies of *erda* from the Tschukotka Mountains of northeastern Siberia as belonging to this species. Shepard has examined colour photographs of the types of *C. alberta kurenzovi*, and they proved to be a subspecies of *C. erda*, not *C. alberta*.

**Conservation status:** Albert's Fritillary is of Special Concern in BC (S3).

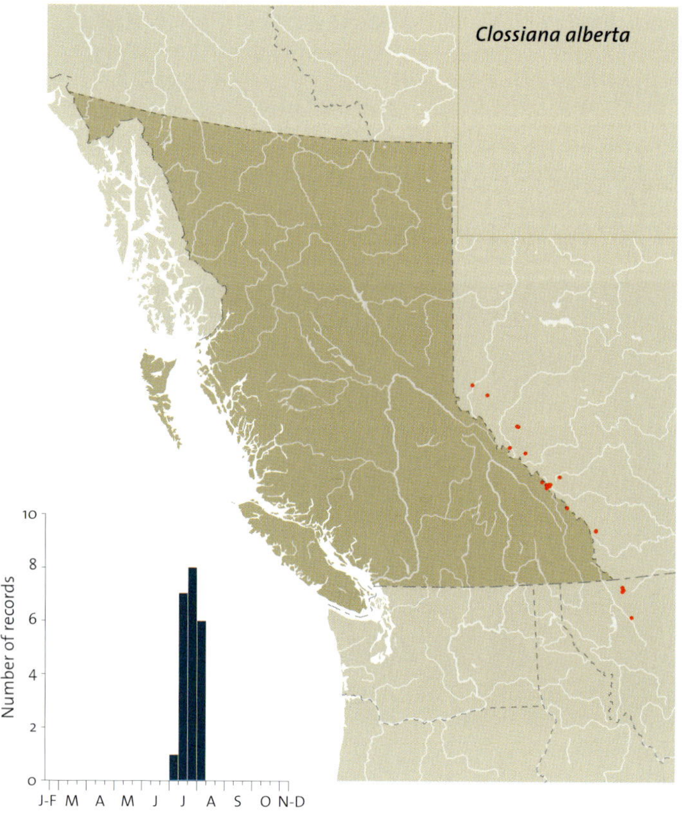

# Freija Fritillary
*Clossiana freija* (Thunberg, 1791)

Freija Fritillary (*Clossiana freija*)

♂ D (3.7 cm)

♀ D (3.8 cm)

♂ V (3.7 cm)

**Etymology:** The name *freija* is derived from Freya, the Norse goddess of love and fertility; this is another *Clossiana* species first described from Scandinavia. Kirby (1837) used the common name Freya Argynnis, which Klots (1951) was probably unaware of when he changed the common name from "Lapland Fritillary," which conformed to British usage, to Freiya Fritillary.

**Adult:** The Freija Fritillary is easily distinguished by the central white spot in the median row of spots on the ventral hindwings. This spot is large and triangular, with the distal point of the triangle almost touching the more distal irregular white line. The only other species so marked is *C. natazhati*.

**Immature stages:** Undescribed.

**Biology:** The Freija Fritillary is the earliest of the four boreal forest *Clossiana* species to fly, with stray males seen as early as late April. At high elevations in the north, stray females can be seen as late as early August. The peak flight period of most populations is late May to early June, and all but worn females are gone before *C. frigga* begins to fly. In the Palearctic the larvae are known to feed on various Ericaceae (the heath family). In North America the only confirmed foodplant is *Vaccinium caespitosum* (Shepard 1975).

**Subspecies:** BC populations are the nominate subspecies, *C. f. freija* (Thunberg, 1791) (TL: Vestrobothnia and Lapland).

**Range and habitat:** The Freija Fritillary is found throughout BC east of the Cascade and Coast mountains in mountainous and boreal habitat. It is found in the same habitat as the Frigga and Bog fritillaries, but also in more xeric habitat and pine forests.

**General distribution:** The Freija Fritillary is found from AK across all of CAN except the west coast rainforests, the southern prairie, and the far northern Canadian archipelago. It is also found in disjunct populations in the southern Rocky Mountains of the USA. The species is Holarctic, occurring in Scandinavia and across Russia to eastern Siberia, except in southwestern European Russia.

**Conservation status:** Not of concern (S5).

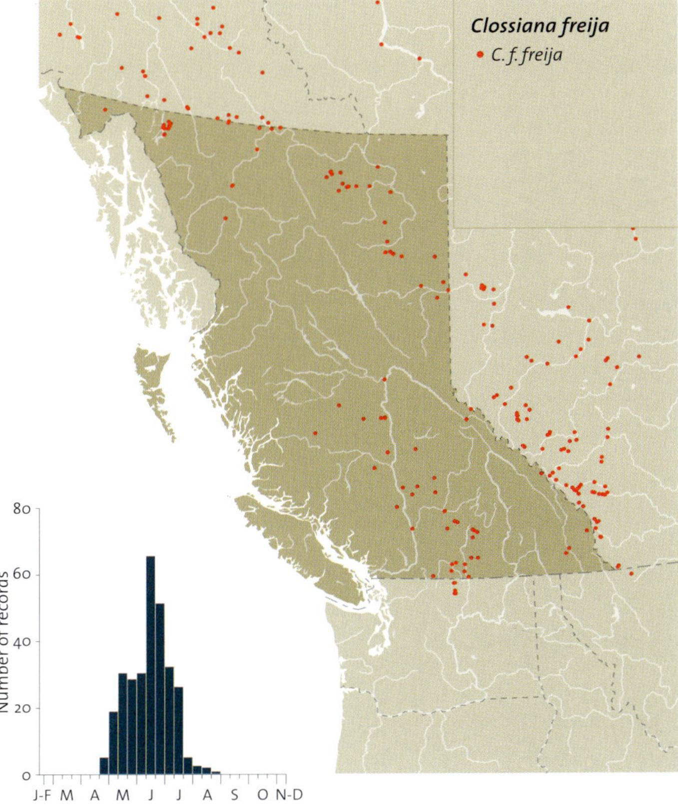

# Beringian Fritillary
### *Clossiana natazhati* (Gibson, 1920)

**ETYMOLOGY:** The species name *natazhati* is derived from the type locality, Mt. Natazhat, Yukon Territory. Subspecies *nabokovi* is named after Vladimir Nabokov (Stallings and Turner 1947), the eminent Russian writer and lepidopterist. The type specimens were from the Harvard Museum of Comparative Zoology, where Nabokov was then curator. The common name was first used by Layberry et al. (1998) and adopted by Opler (1999).

**ADULT:** The Beringian Fritillary has the same pattern on the underside as the Freija Fritillary, with the large, elongate median triangular spot. The Beringian Fritillary is almost twice as large, however, especially the subspecies *nabokovi*. The nominate subspecies has a somewhat dark appearance, similar to *Clossiana eunomia nichollae*. The subspecies *C. n. nabokovi* is extremely dark to the point of being almost black, with the wing pattern obscured. *C. natazhati* and *C. freija* have identical genitalia, and the former is thought to have recently evolved as a species from the latter (Shepard et al. 1998).

**IMMATURE STAGES:** Undescribed.

**BIOLOGY:** The few records for the Beringian Fritillary show that it can fly from early June to early August. In any one year, however, it flies for only about one month. The only instance when it has been taken twice in the same year at one locality was at Stone Mountain Provincial Park, where it was found flying at the same spot in both mid-June and

Ssp. *natazhati* ♂ D (4.1 cm)

Ssp. *natazhati* ♀ D (4.4 cm)

Ssp. *natazhati* ♂ V (4.1 cm)

Ssp. *nabokovi* ♂ V (4.3 cm)

Ssp. *nabokovi* ♂ D (4.3 cm)

Ssp. *nabokovi* ♀ D (4.6 cm)

mid-July 1989. By contrast, in 1999 it was not seen at Stone Mountain Provincial Park until early August, when both males and females were just emerging, because of the high snow cover of the previous winter and the resulting late snow melt at high elevations in the northern Rocky Mountains. In 1999 Shepard succeeded in getting the females to lay eggs in captivity on *Dryas integrifolia* but not *D. drummondii* when females were offered both at the same time. Eggs were also not laid on other plants, especially *Salix* species. Females were seen nectaring almost exclusively on the cushion plant, moss campion (*Silene acaulis* L.) (JHS).

**SUBSPECIES:** The one known BC population is the subspecies *C. n. nabokovi* (Stallings & Turner, 1947) (TL: Alaska Hwy. Mile 392, BC) (Shepard and Kondla 1993; Shepard et al. 1998). It is likely that the nominate subspecies (TL: Mt. Natazhat, YT) will eventually be found near Atlin, BC.

**RANGE AND HABITAT:** The Beringian Fritillary is currently known from BC only at Stone Mountain Provincial Park near Summit Lake, but the nominate subspecies should also be found in the high mountains northwest from Atlin because it is on adjacent Montana Mountain, near Carcross, YT (Shepard et al. 1998).

**GENERAL DISTRIBUTION:** The North American distribution of this species includes disjunct populations in YT and western NT, and one BC population.

**CONSERVATION STATUS:** Subspecies *nabokovi* is of Special Concern in BC (S3).

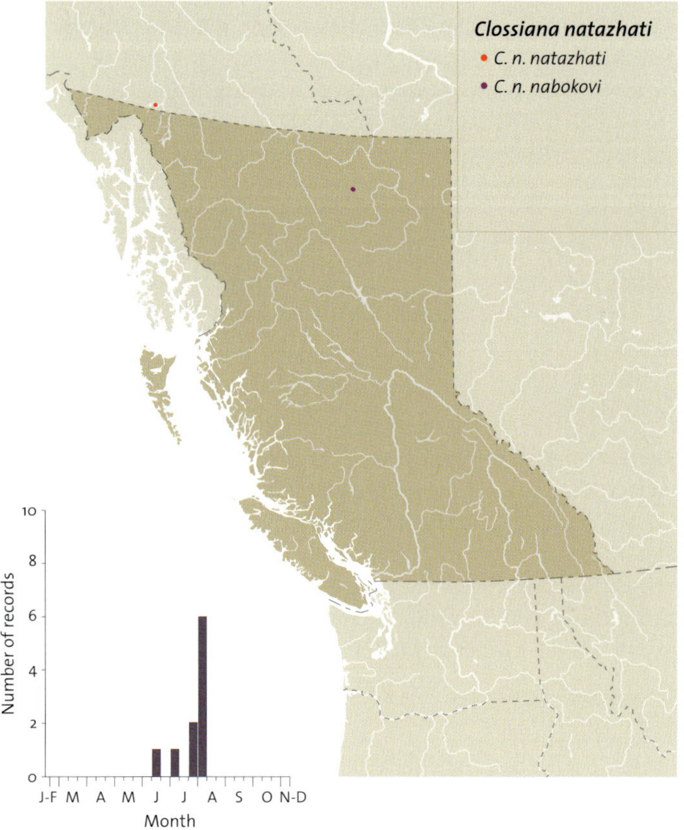

# Boeber's Fritillary
*Clossiana tritonia* (Boeber, 1812)

**ETYMOLOGY:** The species name *tritonia* is derived from Triton, the son the Poseidon and Amphitrite. The only attribute of the god shared by the butterfly is the large size. Boeber's Fritillary is the largest of the lesser fritillaries. The subspecies *astarte* is named for the Greek goddess Astarte, whose name represents Ishtar, the female counterpart of Baal in Assyro-Babylonian mythology. These two were the pre-eminent gods of the Phoenicians. Astarte was represented by the crescent moon, and she was worshipped in groves of trees. What possible relationship this has to the butterfly, other than that it was the use of yet another classical god's name for a butterfly, eludes our imagination. The subspecies *distincta* was so named because Arthur Gibson, the author of the name, was convinced that he was naming a distinct species. Time has shown that it is just a subspecies (Shepard 1975). The common name is coined to give Johann Boeber his long-overdue recognition for first naming this species.

**ADULT:** Boeber's Fritillary and the two BC subspecies are easily identified by referring to the photographs of the adults.

**IMMATURE STAGES:** Undescribed.

**BIOLOGY:** Boeber's Fritillaries fly from early July to early August. In the Coast Ranges of BC, the species flies only in even-numbered years (Phair 1919), like the Washington populations. In the Rockies it has been recorded in both even- and odd-numbered years, but there are not enough

SSP. *astarte* ♂ D  (4.3 CM)

SSP. *distincta* ♂ D  (4.1 CM)

SSP. *astarte* ♂ V  (4.3 CM)

SSP. *distincta* ♂ V  (4.1 CM)

observations of any one population to determine whether individual populations fly every year. Shepard (1975) has reared Boeber's Fritillary on *Saxifraga bronchialis* in the Canadian Rocky Mountains, and this plant species has also been recorded for the butterfly in Siberia. This *Saxifraga* occurs throughout the range of the nominate subspecies. *C. t. distincta* is associated with *Saxifraga tricuspidata* (JHS), a close relative of *S. bronchialis* in the large genus *Saxifraga*. Larvae reared in the lab have two obligate diapause states, as early and mature larvae (JHS). Thus, even if the species is found at some localities every year, those localities must contain two temporally isolated populations.

**SUBSPECIES:** Most populations in BC are the subspecies *C. tritonia astarte* (Doubleday, [1847]); TL: near Rock Lake, AB (Shepard 1984). Near Atlin, the subspecies *C. t. distincta* (Gibson, 1920) (TL: Harrington Cr., YT) is found. This second subspecies is found from east of Fairbanks through central and southern YT to Atlin, BC.

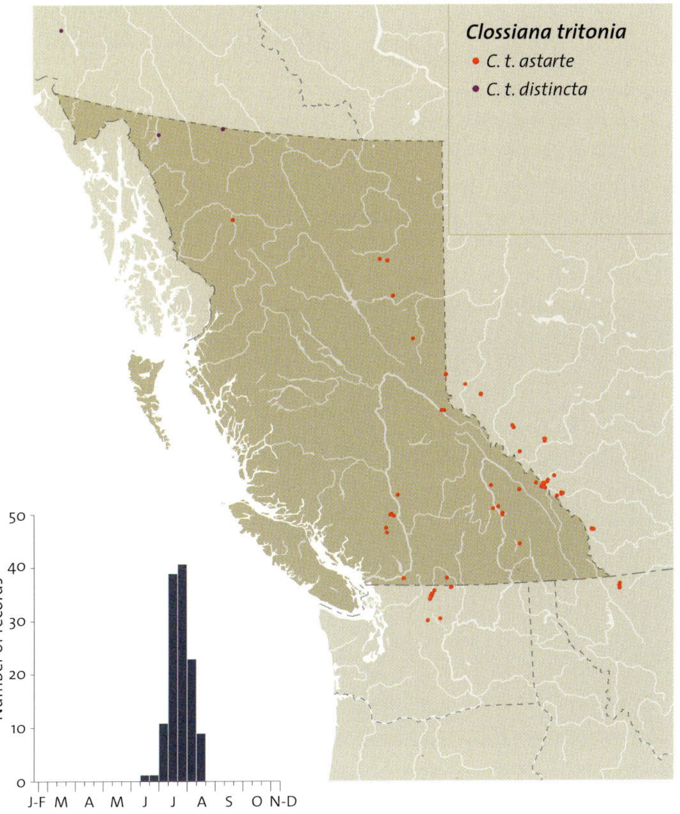

**Boeber's Fritillary (*Clossiana tritonia distincta*)**

**RANGE AND HABITAT:** Boeber's Fritillary occurs above timberline on scree and rock cliffs. It is found in northern BC except for the northeastern boreal spruce forest area, and in southern BC in all alpine areas except Vancouver Island.

**GENERAL DISTRIBUTION:** Boeber's Fritillary is found throughout eastern Siberia, northern and central AK, YT, BC, and the Rocky Mountains of AB. South of BC and AB it is known only from Okanogan Co., WA, and Glacier National Park, MT. Recent authors have tried to separate the subspecies into species, but have not been able to show any consistent morphological differences and no sympatric populations.

Shepard and Dubutolov (Novosibirsk) have exchanged drawings of the male genitalia of all subspecies in this complex. They have concluded that there are no significant differences between any of the Siberian and North American taxa in this complex, and that they should be regarded as one species. Since the name *tritonia* has been in constant use, at least in Russia, it is the correct species name according to International Code of Zoological Nomenclature rules.

**CONSERVATION STATUS:** Subspecies *distincta* is Vulnerable in BC (S3). Subspecies *astarte* is not of concern (S5).

## ARCTIC FRITILLARY
*Clossiana chariclea* (Schneider, 1794)

**ETYMOLOGY:** Chariclea was the heroine of the first Greek romance, *Aethiopica,* written by Heliodorus in the third century AD. The name *grandis* (large) refers to the large size of the subspecies (Barnes & McDunnough 1916). The subspecies *rainieri* is named for the type locality, Mt. Rainier, WA. The name of subspecies *butleri* honours Arthur Gardiner Butler, a well-known British lepidopterist of the period with whom Edwards corresponded. The common name "Arctic Fritillary" is well chosen, as this and the Polar Fritillary occur further north in North America than any other butterfly. As with the Bog Fritillary, Klots (1951) adopted the British usage when he showed that some North American forms previously regarded as species were really subspecies and part of one Holarctic species.

Ssp. *rainieri* ♂ D (4.0 cm)

Ssp. *rainieri* ♀ D (3.7 cm)

Ssp. *rainieri* ♂ V (4.0 cm)

**ADULT:** The adults of the Arctic Fritillary are the most variable of any *Clossiana*. After other species are eliminated, and with reference to the photos, this species can be determined with confidence. It is most often confused with the Freija Fritillary but lacks the large median triangular spot on the ventral hindwings. It is the fourth of the four boreal species to emerge, and flies from late July onward. Only a

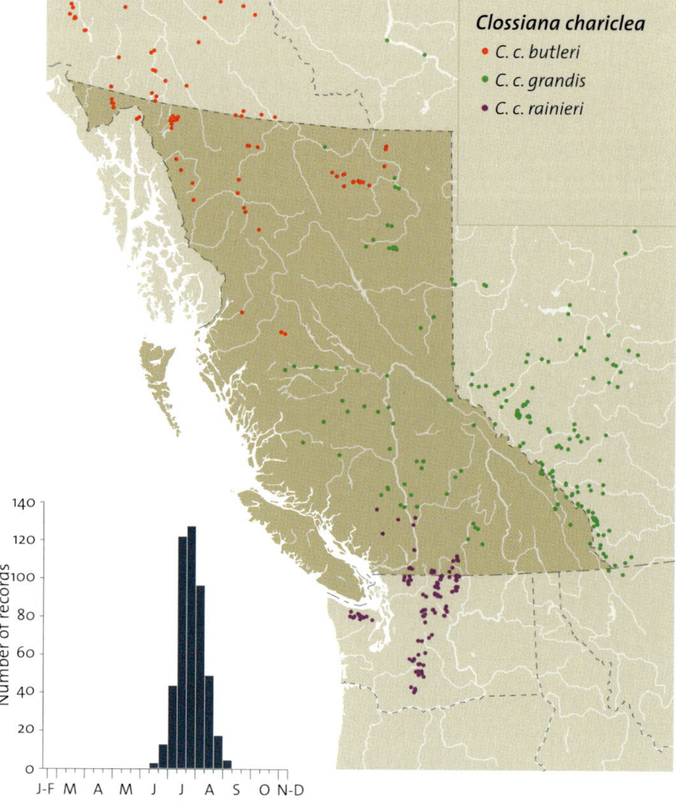

Arctic Fritillaries (*Clossiana chariclea rainieri*) mating

Ssp. *grandis* ♂ D (3.5 cm)

Ssp. *grandis* ♀ D (3.9 cm)

Ssp. *grandis* ♂ V (3.5 cm)

Ssp. *butleri* ♂ V (3.9 cm)

Ssp. *butleri* ♂ D (3.9 cm)

Ssp. *butleri* ♀ D (3.8 cm)

**Arctic Fritillary (*Clossiana chariclea grandis*)**

very occasional late-emerging female Freija Fritillary would be flying at the same time in the same habitat. In most of the historical North American literature, this species is treated as two species, *C. chariclea* and *C. titania*. Shepard (1998), however, showed that all North American forms are *C. chariclea* and that *C. titania* is a species restricted to the Palearctic.

**IMMATURE STAGES:** Undescribed.

**BIOLOGY:** The Arctic Fritillary flies in one brood from late July to early September. The eggs hatch shortly after being laid, but the larvae do not feed and instead go into hibernation (Shepard 1975). Pelham (Shepard 1975) found last instar larvae on *Polygonum bistortoides* in the Olympic Mountains of Washington. Larvae in the Canadian National Collection collected in the far north were found on *Salix* species, which is the presumed larval foodplant in most of the North American distribution of the species.

**SUBSPECIES:** The populations from southwestern BC are the Tacoma Fritillary, *C. c. rainieri* (Barnes & McDunnough, 1913) (TL: Mt. Rainier, WA). The populations from central BC and northeastern BC are the boreal subspecies, *C. c. grandis* (Barnes & McDunnough, 1916) (TL: Hymers, ON). The populations from northwestern BC are best ascribed to *C. c. butleri* (W.H. Edwards, 1883) (TL: Cape Thompson, AK). These northwestern BC populations are intermediate between *butleri* and *grandis*, however.

**RANGE AND HABITAT:** The Arctic Fritillary is found throughout BC east of the Coast Mountains. In southwestern BC the species occurs west of the Cascade and Coast mountains to near Whistler, from above 1,500 m to above timberline. The central and northeastern BC populations occur at moderate elevations in the same habitat as *C. freija, C. frigga,* and *C. eunomia*. These four species occur together across boreal North America and in the Palearctic.

**GENERAL DISTRIBUTION:** The Arctic Fritillary ranges from AK across all of CAN except some of the western Canadian archipelago and the southern prairies. It occurs south to the WA Cascades and northern NM in the west, and MN and NH in the east. The species is Holarctic, occurring from northern Scandinavia across northern Russia to eastern Siberia and Kamchatka.

**CONSERVATION STATUS:** Not of concern, with subspecies *butleri* S5, *grandis* S5, and *rainieri* S4.

The type genus of this subfamily is the European genus *Melitaea*. Emmet (1991) enumerates the various etymologies that have been suggested for this generic name, and suggests that Fabricius' love of word play is again the explanation. Thus it is most likely derived from the Greek *meli* (honey), because species in this genus are great lovers of nectar. The common name "checkerspots" is derived from the checkerboard pattern of the upperside of the wings.

The overall pattern is rectangular to square black spots with alternate tawny to brick red spots. The hindwing cell is open or partly closed in the checkerspots. The palpi are variable but the second segment lacks bristles. The uncus of the male genitalia is absent or vestigial. Higgins (1941, 1955, 1960, 1978, 1981) provided revisions of much of the subfamily. Worldwide there are at least 200 species, most of which are Neotropical.

Northern Checkerspot (*Charidryas palla*)

## Genus *Phyciodes* Hübner, [1819]   Crescents

The name *Phyciodes* may come from the Greek *phykos* (painted or "covered with cosmetics"), in reference to the complex ventral wing pattern. The common name "crescents" (Gosse 1840) refers to the crescent-shaped spot in the centre of the ventral hindwing margin.

*Phyciodes* in the limited sense used here is separated from the Neotropical genera *Anthanassa* Scudder and *Eresis* Boisduval by the presence of two to four in-curved hooks on the posterior tip of the tegumen of the male genitalia. The saccus is present in the male genitalia. The genus is Nearctic and contains nine species. The larvae of BC species feed on asters (*Aster*) or thistles (*Cirsium*).

# Pearl Crescent
### *Phyciodes tharos* (Drury, [1773])

**Pearl Crescent (*Phyciodes tharos pascoensis*)**

Ssp. *pascoensis* ♂ D (3.2 cm)

Ssp. *pascoensis* ♀ D (3.8 cm)

Ssp. *pascoensis* ♂ V (3.2 cm)

Ssp. *cocyta* ♂ V (3.1 cm)

Ssp. *cocyta* ♂ D (3.1 cm)

Ssp. *cocyta* ♀ D (3.5 cm)

**Etymology:** The etymology of *tharos* is unknown. The subspecies *pascoensis* is named after Pasco, WA. This was not stated as the type locality in the original description or when Tilden (1975) selected a lectotype. The actual type locality was Ainsworth (near Pasco), but that was across the railroad tracks and contained the local "houses of ill repute." The common name (Gosse 1840) is derived from the colour of the crescent spot.

**Adult:** Similar to the Tawny Crescent, as discussed under that species. There has been extreme confusion in the literature as to the correct species name for this species and whether it is one or two species. Most of the arguments for two species emphasize that the southern forms are multi-voltine and the northern ones univoltine. East of the Continental Divide there appear to be corresponding wing pattern differences. In BC the southern, multivoltine populations are more similar to the eastern univoltine populations. A recent paper by Porter and Mueller (1998) shows that the two behave as one species and suggests that the conservative approach of recognizing one species be adapted. For this book we take that approach but realize that the question is far from resolved. No matter what the outcome, no one has ever suggested that BC populations represent more than one species.

**Immature stages:** The eggs are pale green and are laid in a cluster on the larval foodplant (Bird et al. 1995). The larva is chocolate brown with cream markings. The pupae are cream with brown streaks.

**Biology:** The Pearl Crescent is univoltine in the Peace River area but at least bivoltine across the Southern Interior and the Kootenays. The larvae feed on asters.

**Subspecies:** The northern subspecies, the Northern Pearl Crescent, *P. t. cocyta* (Cramer, [1777]) (TL: Cape Breton, NS), occurs in northern BC. In the south the Pasco Crescent,

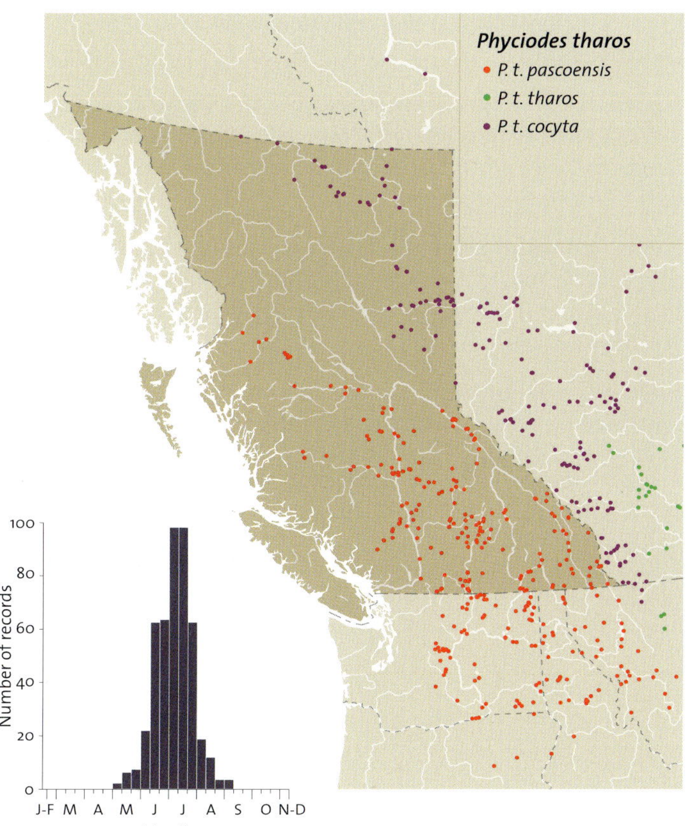

*P. t. pascoensis* Wright, 1905 (TL: Pasco, WA) is found. The subspecies are weakly differentiated.

**RANGE AND HABITAT:** The Pearl Crescent is found east of the Cascade and Coast ranges throughout BC, except the area north and west from the Nass River. The habitat varies from old fields to native meadows and mature aspen woodland, but always in mesic situations.

**GENERAL DISTRIBUTION:** The Pearl Crescent ranges from the YT east to NF and south to CO and the southeastern USA .

**CONSERVATION STATUS:** Not of concern; both subspecies are S5.

## TAWNY CRESCENT
*Phyciodes batesii* (Reakirt, 1865)

**ETYMOLOGY:** The species *batesii* was named after Henry W. Bates, Reakirt's friend (Reakirt 1865) and the lepidopterist for whom Batesian mimicry was named. Subspecies *lakota* is named after the Lakota Indians of Nebraska (Scott 1994). The common name was first used by Scudder (1875), and refers to the characteristic tawny spot on the border of the ventral hindwing.

**ADULT:** In the Tawny Crescent, the crescent spot of the ventral hindwing margin is the same colour as the overall ground colour, as opposed to the Pearl Crescent, which appears similar but which has a white crescent in females and in which the crescent is obscured by a continuous purple flush in males. The tips of the antennae are black in the

♂ D (3.0 см)

♀ D (3.7 см)

♂ V (3.0 см)

Tawny Crescent and black and orange in the Pearl Crescent in BC.

**IMMATURE STAGES:** Eggs are laid in batches and the larvae are gregarious in a web during the first two instars.

**BIOLOGY:** Adults fly in a single brood from late June to late July in the Peace River area. The species overwinters as a third instar larva (Bird et al. 1995). The Tawny Crescent feeds on native asters.

**SUBSPECIES:** Scott has recently described the Lakota Crescent, *P. b. lakota* Scott, 1994 (TL: Pine Ridge, NE). This subspecies ranges from NE north to western ON and across the prairie to the Peace River region of BC.

**RANGE AND HABITAT:** The Tawny Crescent is found on the Liard River Road and the Peace River from Attachie east to the AB border. It is found in mature, open aspen woodland and adjacent mesic meadows.

**GENERAL DISTRIBUTION:** The Tawny Crescent is found from northeast BC across Canada to NF and south to western NE and the Appalachians in isolated colonies to northern GA. In the west, it is known from the Bighorn Mountains of WY and from Carbon Co., WY, south to AZ in disjunct populations.

**CONSERVATION STATUS:** The Tawny Crescent is of Special Concern in BC (S3).

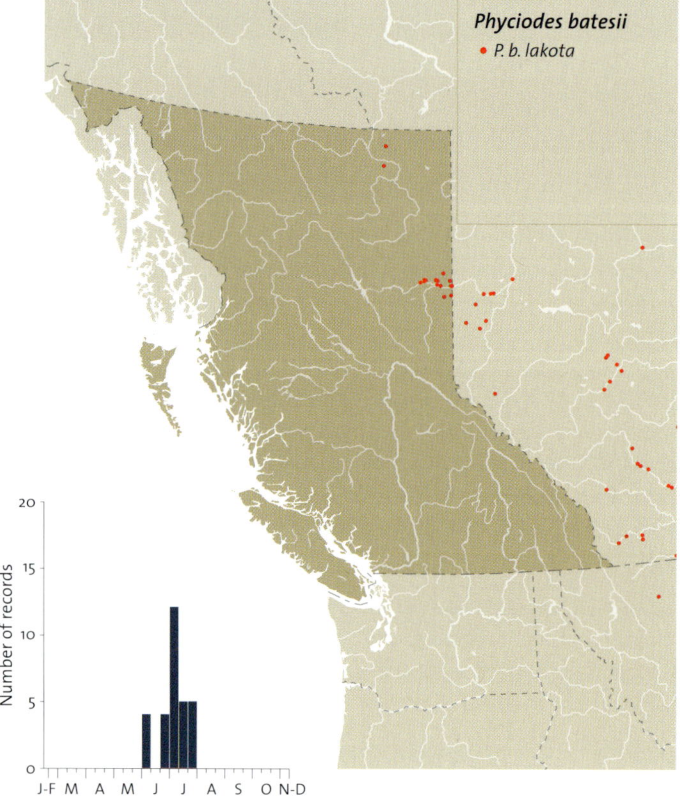

# Field Crescent

*Phyciodes pratensis* (Behr, 1863)

**Field Crescent (*Phyciodes pratensis pratensis*)**

♂ D  (3.4 cm)

♀ V  (3.4 cm)

♂ V  (3.4 cm)

**ETYMOLOGY:** The name *pratensis* is Latin, meaning "found in the field." The common name is a direct translation of the Latin. It was first used by Comstock (1927), replacing the older name "Meadow Crescent" (Holland 1898).

**ADULT:** Where the Field Crescent is sympatric with the Pearl Crescent, it is sometimes confused with the latter, especially females. The male Field Crescent lacks a purple flush on the margin of the ventral hindwing, and the crescent is the same colour as the ground colour of the Tawny Crescent. Both sexes have much more black in the median region of the dorsal hindwings than the Pearl Crescent.

**IMMATURE STAGES:** The larvae are a darker brown than the two preceding species of aster-feeders. Emmel and Emmel (1973) describe the mature larva as black with numerous branching spines. There is a dorso-lateral greyish white band, and a yellowish brown band below the spiracles.

**BIOLOGY:** Adults fly from late May to early August, depending on elevation and latitude, but there is only one generation per year. The seasonal bar graph clearly shows one peak in flight in early July. Cockle reared the Field Crescent on *Aster* sp. in the Kootenays (Harvey 1908). Elsewhere it has been reared from a variety of asters. The eggs are laid in a cluster on the foodplant.

**SUBSPECIES:** BC populations are the nominate subspecies, *P. p. pratensis* (Behr, 1863) (TL: San Francisco, CA). Scott (1994) attempted to separate the southwestern YT and adjacent AK populations into a separate subspecies, but was apparently unaware that the species occurs continuously south from southwestern Yukon with no known distinct break that would justify setting up a separate subspecies.

**RANGE AND HABITAT:** The Field Crescent is found throughout BC east of the Coast Ranges. West of the Coast Ranges and the Cascades, it is known only from southeastern Vancouver Island, with the only known extant population being in the Nanaimo River estuary, and from the adjacent mainland.

**GENERAL DISTRIBUTION:** The Field Crescent occurs from eastern AK south through BC and the Rockies of AB to CA, AZ, and NM.

**CONSERVATION STATUS:** Not of concern (S5).

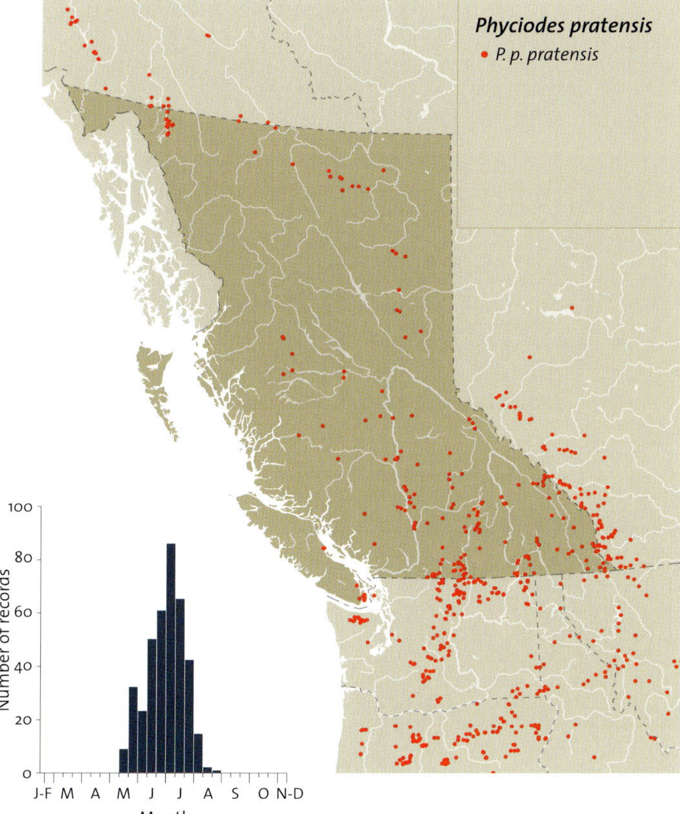

**Mature larva**

## Pale Crescent

*Phyciodes pallidus* (W.H. Edwards, 1864)

**Etymology:** The species name *pallidus* is from the Latin *pallidus* (pale) because it was paler than *P. mylitta*. Subspecies *barnesi* is named for William Barnes of Decatur, Illinois. Barnes built up the first truly continental collection of Lepidoptera, numbering more than 400,000 specimens at his death. It encompassed all families of Lepidoptera, not just butterflies. When this collection was purchased by the Smithsonian Institution in the early 1930s, the Smithsonian became the pre-eminent North American Lepidoptera collection. The common name was first proposed by Scott (1986), and is a direct translation of the Latin name into English.

**Adult:** Adults of this species and its sister species, *P. mylitta,* differ from others in the genus. They show the primitive checkerboard upperside wing pattern of the subfamily. The Pale Crescent looks essentially like the Mylitta Crescent, but is about twice as large, with a crisper underside pattern.

**Immature stages:** Undescribed.

**Biology:** The Pale Crescent flies in one generation from mid-May to early July. Where it flies sympatrically with the double-brooded Mylitta Crescent, the peak flight of the Pale Crescent occurs between the two broods of the Mylitta Crescent. The overwintering stage is not known, but is

♂ D (4.1 cm)

♀ D (4.4 cm)

♂ V (4.1 cm)

suspected to be the fourth instar larva. Larvae have been reared from *Cirsium undulatum* in the Chilcotin, BC (Anna Roberts).

**Subspecies:** BC populations are currently differentiated as the Great Basin subspecies Barnes' Pale Crescent, *P. p. barnesi* Skinner, 1897 (TL: Glenwood Springs, CO), but are only weakly, if at all, different from the nominate subspecies (TL: Flagstaff Mt., Boulder Co., CO).

**Range and habitat:** The Pale Crescent is the most restricted member of the genus *Phyciodes* in BC. It is known only from the most xeric habitat in the Chilcotin, the Southern Interior, and the Rocky Mountain Trench, south of Windermere.

**General distribution:** The Pale Crescent is found from extreme southern interior BC south to AZ and CO.

**Conservation status:** Not of concern (S4).

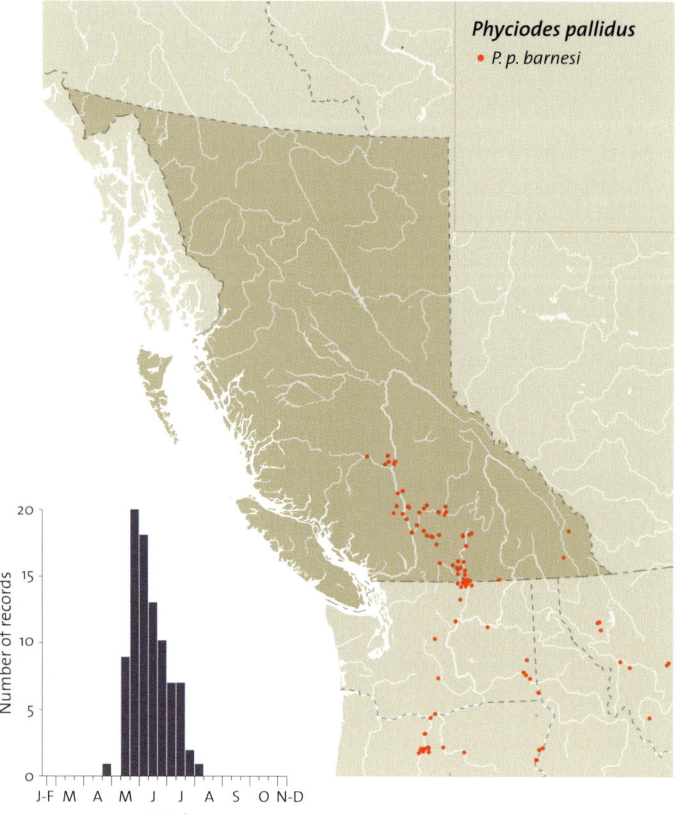


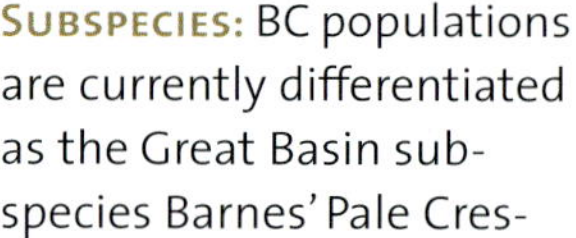

Eggs

Mature larva

Pupa

Pale Crescent
(*Phyciodes pallidus*)

# Mylitta Crescent

*Phyciodes mylitta* (W.H. Edwards, 1861)

**Etymology:** Mylitta is another name for Belit, the Assyro-Babylonian wife of Bel and mother of the gods. Perhaps Edwards had the etymology of *Clossiana tritonia astarte* in mind when he chose this species name. Holland (1898) coined the common name.

**Adult:** The Mylitta Crescent is very similar to the Pale Crescent but much smaller. The pattern on the underside is less crisp.

**Immature stages:** The egg is pale green when laid, but changes to dark grey before the larva emerges. It is 0.50 mm across. The mature larva has a shiny black head. The body is dull black with a dark dorsal line and white spots on the anterior base of prominent hairy spines. The spiracular band consists of two thin yellow lines (Hardy 1964).

**Biology:** There are two generations of the Mylitta Crescent, one flying in late April and May and the other in July and August. Hardy (1964) reared the newly established Vancouver Island populations of the Mylitta Crescent on *Cirsium arvense*. The larvae produced by the second-brood adults hibernated as fourth instar larvae on Vancouver Island.

♂ D  (3.3 cm)

♀ D  (3.4 cm)

♂ V  (3.3 cm)

**Subspecies:** BC populations are the nominate subspecies, *P. m. mylitta* (W.H. Edwards, 1861) (TL: San Francisco, CA), which occurs everywhere except the extreme southern end of the species range, where several disjunct sets of populations have been given subspecies status.

**Range and habitat:** The Mylitta Crescent occurs across southern BC in the xeric habitat of the Pale Crescent, but also in more mesic habitat on Vancouver Island, the Lower Mainland, and the West Kootenay. At least on Vancouver Island (Hardy 1962a; Shepard 1977), the Mylitta Crescent was not present prior to European settlement, and was not recorded until the 1950s. There are also no Lower Mainland records before 1902 or before the introduction of Canadian thistle. This is the only known larval foodplant on Vancouver Island or the Lower Mainland, where the Mylitta Crescent is seen only in disturbed habitat. Thus the species may also not be native to the Lower Mainland.

**General distribution:** The Mylitta Crescent ranges from southern BC south to Baja California and Sonora, MEX.

**Conservation status:** Not of concern (S5).

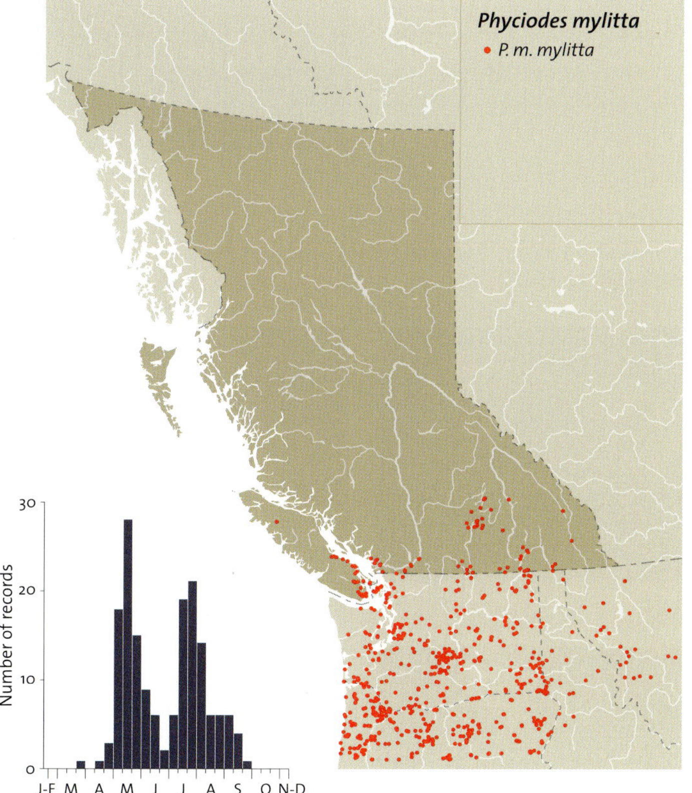

**Mature larva**

**Pupa**

The name *Charidryas* is derived from the Greek *chara* (joy or delight) and *dryas* or *Dryad* (wood nymph) (Emmet 1991). Hence these checkerspots are delightful Dryads. The name "checkerspots" is derived from the checkerboard pattern of the upperside of the wings.

Like the crescents, the *Charidryas* checkerspots have the saccus present in the male genitalia. The BC species are larger than BC species of crescents except the Pale Crescent, which is as large as *Charidryas* checkerspots. The coloured spots in the dorsal wing pattern are more orange than those of *Phyciodes*. The larvae feed on various composites.

## NORTHERN CHECKERSPOT
*Charidryas palla* (Boisduval, 1852)

**ETYMOLOGY:** The name *palla* is from the Latin *pallens* (pale). It refers to the pale female underside (Boisduval 1852). The subspecies is named after Calydon, a town in Aetolia. The common name "Northern Checkerspot" (Holland 1898) reflects the northern distribution of the species.

**ADULT:** The Northern Checkerspot is usually not found in the same habitat as other *Charidryas* species in BC, thus the species is easily identified by the generic characteristics. It is usually found below 1,250 m, whereas the other two species are found from 1,250–2,200 m.

**IMMATURE STAGES:** The mature larva is dull black, with five rows of branching tubercles (Emmel and Emmel 1973).

**BIOLOGY:** The Northern Checkerspot flies from late June to early August. Guppy found eggs on native *Aster* at Botanie

♂ D (3.9 cm)

♀ D (3.6 cm)

♂ V (3.9 cm)

Valley and reared them on cultivated asters. Emmel and Emmel (1973) report another composite plant genus, *Chrysothamnus,* as a larval foodplant.

**SUBSPECIES:** BC populations are the Calydon Checkerspot, *C. p. calydon* (Holland, 1931) (TL: Turkey Cr. Jct., CO), a Rocky Mountain subspecies.

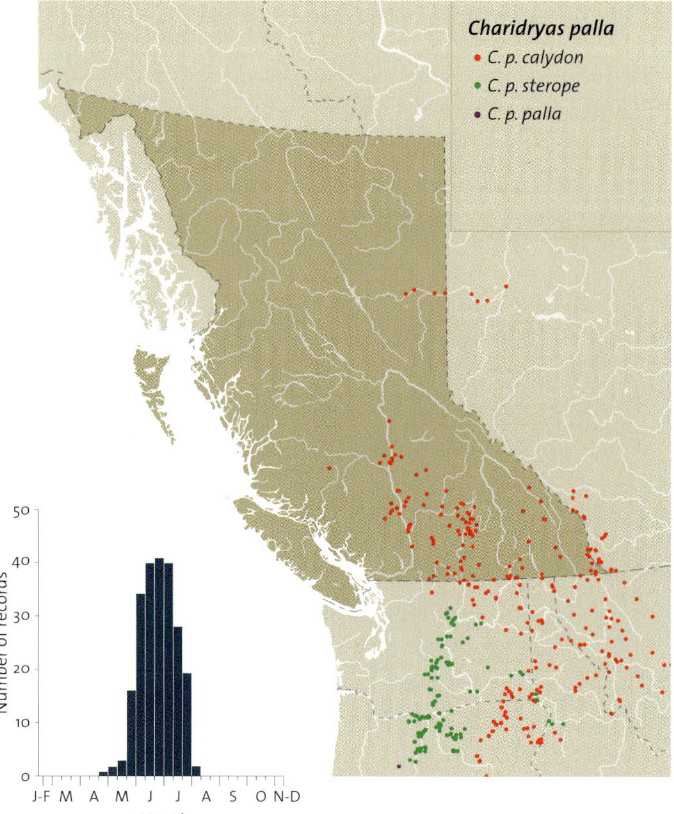

**Eggs**

**First instar larvae**

**Prepupal larva**

**Pupa**

**Northern Checkerspot (*Charidryas palla*)**



RANGE AND HABITAT: The Northern Checkerspot is found in the Chilcotin, Southern Interior, and Kootenays in mesic meadows and riparian habitat in low valleys up to 1,500 m in open, drier slopes. It is also known from the Peace River area. At higher elevations it is replaced by either Whitney's Checkerspot or Hoffman's Checkerspot.

GENERAL DISTRIBUTION: The Northern Checkerspot occurs from southern BC and the BC Peace region south to southern CA and CO.

CONSERVATION STATUS: Not of concern (S5).

## WHITNEY'S CHECKERSPOT
*Charidryas whitneyi* (Behr, 1863)

ETYMOLOGY: The species and common names honour Josiah Dwight Whitney, who was state geologist of California when Behr described the butterfly. He was simultaneously a Harvard University faculty member and wrote a book on the plants of California. The subspecies name *damoetas* is the name of a shepherd in Theocritus and Virgil, hence, in pastoral poetry, a rustic. The common name was first used by Holland (1898).

ADULT: Whitney's Checkerspot is found only above timberline and, for that reason alone, cannot be confused with other *Charidryas* species. The median and postmedian areas on the dorsal forewing have fewer black markings than the other two BC *Charidryas* species. On the underside of the

♂ D (3.6 CM)

♀ D (3.8 CM)

♂ V (3.6 CM)

hindwings the light-coloured spots are much whiter than for other *Charidryas* species. We agree with Emmel et al. (1998d) that the taxon *whitneyi* is the correct species name for what all earlier books refer to as *damoetas*.

IMMATURE STAGES: Undescribed.

BIOLOGY: Whitney's Checkerspot flies from mid-July to early August in normal years, but until early September at Mt. McLean near Lillooet, when snow melt is delayed. Nothing is known of the larval foodplant or overwintering stage.

SUBSPECIES: The BC populations are the Rocky Mountain subspecies, *C. w. damoetas* (Skinner, 1902) (TL: Williams River Range, CO) (= *altalus* Scott, 1998).

RANGE AND HABITAT: Whitney's Checkerspot is definitely known only from near Lillooet, in isolated peaks east of Anahim Lake, and on Watch Mountain west of Windermere, but it should also be present in the high Rocky Mountains along the AB border.

GENERAL DISTRIBUTION: Whitney's Checkerspot is known from the three BC populations, from Jasper south to central CO in the alpine areas of the Rockies, and from a disjunct set of populations in the alpine Sierras of CA.

CONSERVATION STATUS: Not of concern (S4).

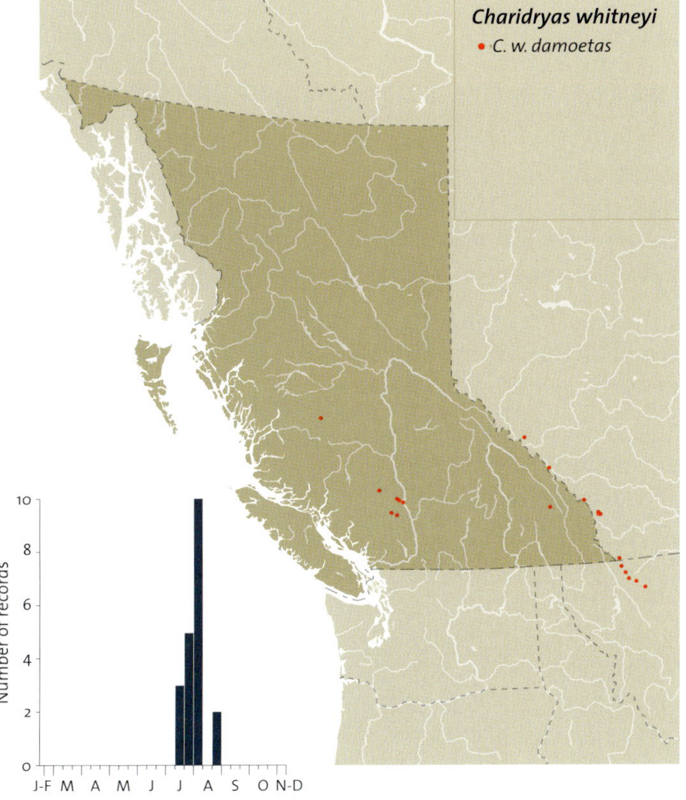

# Hoffman's Checkerspot
## *Charidryas hoffmanni* (Behr, 1863)

**Etymology:** The species *hoffmanni* was named for Charles Frederick Hoffman, a member of the California State Geological Survey from 1858 to 1874 (Emmel et al. 1998d). Bauer (1959) gives no clue to the etymology of the subspecies *manchada*. The common name was first used by Holland (1898).

**Adult:** Hoffman's Checkerspot is difficult to distinguish from the Northern Checkerspot. On the dorsal hindwing, the median row of light spots is almost white. In Northern Checkerspots this light area is cream to reddish.

**Immature stages:** Newcomer (1967a) described the immatures in detail. The egg is light green and 0.6 mm wide. The mature larvae are black with white spots on the uppersides and top. The line above the spiracles is cream, and below that the body is brown. Above the spiracular line the tubercles with numerous branches are black ringed with

♂ D (3.5 cm)

♀ D (3.9 cm)

♂ V (3.5 cm)

white at the base. Below the spiracular line, the tubercles are brown. The ground colour of the pupa is pearly white, with an intricate pattern of brown/black markings.

**Biology:** Hoffman's Checkerspot flies from late June to late July in BC. The earliest records on the flight season bar graph are for Washington. Newcomer (1967a) reared Washington populations. Eggs were laid in a mass and early instar larvae were found clustered on the leaves of *Aster conspicuus* Lindley. They remained clustered and fed until the third instar before hibernating. In mid-May the following year, fourth instar larvae were found feeding individually on the same aster. They began to pupate in late May. The pupal stage lasted only 10 days before adults began to emerge.

**Subspecies:** The BC subspecies is the Manchada Checkerspot, *C. h. manchada* (Bauer, 1960) (TL: Tumwater Cyn., WA). It ranges south from Manning Provincial Park through the Washington Cascades to, but not across, the Columbia River.

**Range and habitat:** Hoffman's Checkerspot is known only from the Cascade Mountains in Manning Provincial Park, but should occur east to near Keremeos. It is found from 1,250 to 1,900 m in meadows associated with dry Douglas-fir habitat.

**General distribution:** The species ranges from Manning Provincial Park, BC, south through the Cascades and Sierras in similar habitat below timberline.

**Conservation status:** Hoffman's Checkerspot is of Special Concern in BC (S3).

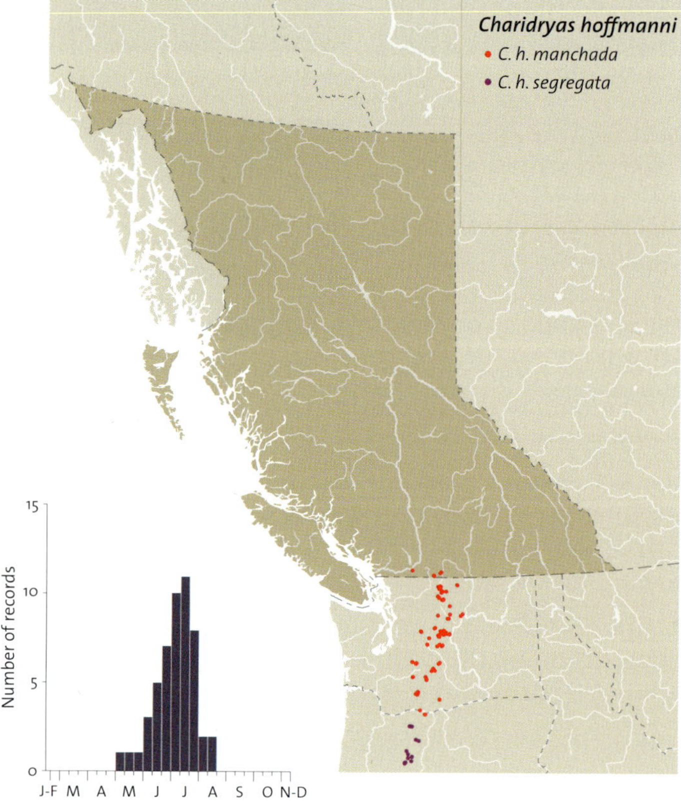

The name *Euphydryas* is derived from the Greek *euphys* (a goodly shape) and *dryas* (a dryad or wood nymph) (Emmet 1991), hence these checkerspots are goodly shaped wood nymphs. The common name "checkerspots" is derived from the checkerboard pattern of the upperside of the wings.

The *Euphydryas* checkerspots differ from *Charidryas* checkerspots by having brick red, not tawny, spots alternating with the black spots. The male genitalia lack the saccus. This is a Holarctic genus, with five Nearctic species and four Palearctic species. None of the individual species are Holarctic. The larvae feed on a variety of plants. Adults of three western North American species, all of which are in BC, are very hard to distinguish without reference to the male genitalia. These three species are in the subgenus *Euphydryas* (= *Eurodyras* Higgins = *Occidryas* Higgins). The fourth BC species, *E. gillettii*, is in the subgenus *Hypodryas*. All species worldwide are in one or the other of the two North American subgenera.

Fig. 71 illustrates the one part of the male genitalia, the harpe, that is diagnostic for species. The short arms of the harpe, one rounded and toothed and the other flattened out, are diagnostic for the subgenus *Hypodryas*. In the subgenus *Euphydryas,* at least one arm of the harpe is longer than the body of the harpe. This subgeneric classification is supported by recent DNA analysis of worldwide Melitaeinae by Zimmermann et al. (2000), who prefer subgenera to genera. Once this paper has been examined closely, however, others may choose to recognize two genera instead of one for the North American fauna.

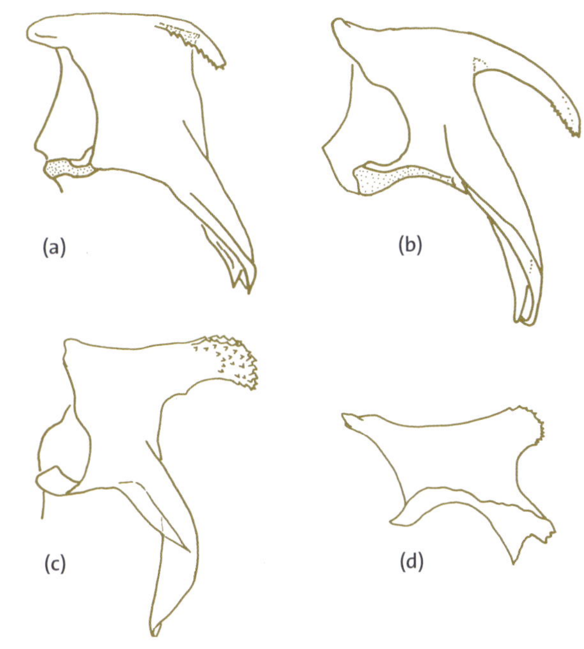

**71** Harpe, located on the inner surface of valves, of *Euphydryas* species: (a) *E. chalcedona*, (b) *E. anicia*, (c) *E. editha*, (d) *E. gillettii*

## CHALCEDON CHECKERSPOT
*Euphydryas (Euphydryas) chalcedona* (Doubleday, [1847])

**ETYMOLOGY:** The species name *chalcedona* is derived from Chalcedon, a town on the Bosporus opposite the Byzantium side, known as the city of the blind because the founders did not settle on the Byzantium side. The subspecies name *paradoxa,* Latin for paradox, was coined by McDunnough (1927) because he found the taxon very confusing. Holland (1898) coined the common name.

**ADULT:** The harpe of the male genitalia is the only reliable character for identifying the species (Fig. 71a). The dorsal arm is short and at a right angle to the longer ventral arm. This character can be seen only in genitalia kept in vials. If the preparation is mounted on a slide, the character is distorted. This may account for the confusion in the literature, with several authors treating the Chalcedon and Anicia checkerspots as one species (Layberry et al. 1998; Scott 1980). For a further description, see the "Adult" discussion under **Anicia Checkerspot**.

**IMMATURE STAGES:** Emmel and Emmel (1973) illustrate and describe the immatures of the nominate subspecies. The egg

♂ D  (3.3 CM)

♀ D  (4.3 CM)

is lemon yellow. The mature larva is black with whitish dorsal and lateral bands along the body. The entire body is covered with small tubercles that are white with white hairs.

**BIOLOGY:** The adults formerly flew in late May and June at lower elevations on

♂ V  (3.3 CM)

Vancouver Island. On the mainland, populations occur from 1,250 to 1,900 m in the south, but at lower elevations in the north. All fly from early July through early August in normal years, and later if snow melt is delayed. Nothing is known of the life cycle in BC, but the adjacent Olympic Mountains populations feed on *Penstemon* (J. Pelham, pers. comm.), which is a known larval foodplant further south. The species overwinters as larvae in California (Emmel and Emmel 1973), and presumably do the same in BC. As in other checkerspots,

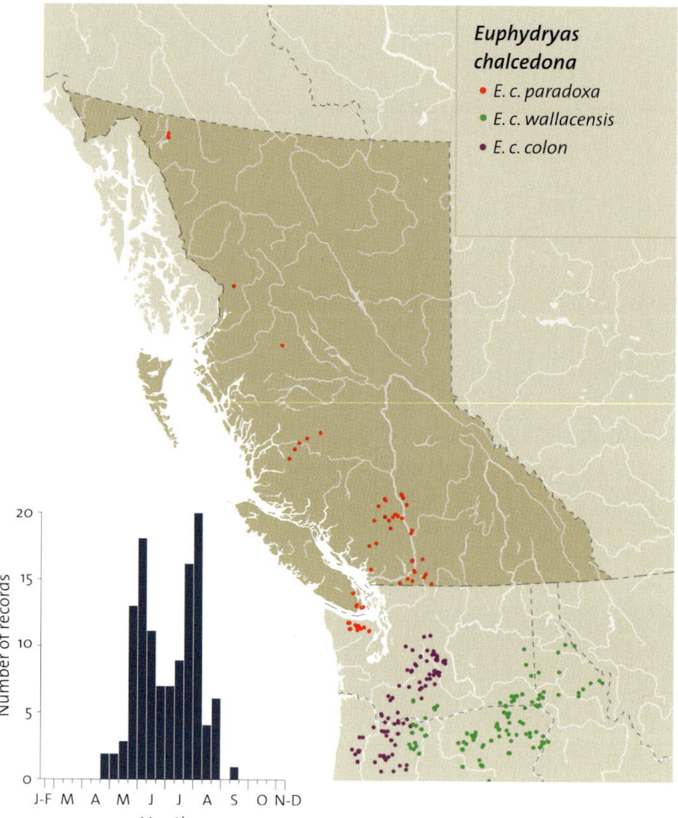


the eggs are laid in clusters and the early instar larvae are gregarious.

**SUBSPECIES:** The Western Checkerspot, *E. c. paradoxa* McDunnough, 1927 (TL: Seton Lake, BC) is the only BC subspecies. All previous authors have ascribed Vancouver Island populations to *E. c. perdiccas* (W.H. Edwards, 1881) (TL: Tenino, Thurston Co., WA), which is a synonym of *E. c. colon* (W.H. Edwards, 1881) (TL: Kalama, Cowlitz Co., WA). The type locality of *E. c. perdiccas* was erroneously restricted to Tenino, and it is herein re-restricted by Shepard to Kalama, Cowlitz Co., WA. *E. chalcedona* has never been collected at or near Tenino, despite Brown (1966). The BC subspecies also occurs in the Olympic Mountains of Washington.

**RANGE AND HABITAT:** The Chalcedon Checkerspot is found in BC from Atlin south in a narrow band along the Coast Ranges to the WA border at Mt. Cheam, and formerly in the Malahat, Vancouver Island. From what can be discerned from museum specimens, the Vancouver Island populations occurred on bare hillsides from near sea level to 1,250 m. The Cascade and Coast populations occur from 1,250 to 1,900 m in the south and along the eastern shore of Atlin Lake in the north.

**GENERAL DISTRIBUTION:** The Chalcedon Checkerspot is found from Atlin, BC, south in a narrow band along the coastal mountains and on Vancouver Island, then south to southern CA and adjacent AZ and south from Wallace, ID, to at least northeastern OR and central ID as the subspecies *E. c. wallacensis*. Other records in the American Rockies (Stanford and Opler 1993) are *E. anicia*. The amended USA distribution, as shown on the distribution map, is based on several hundred genitalia dissections (JHS).

**CONSERVATION STATUS:** Not of concern (S4).

## ANICIA CHECKERSPOT
*Euphydryas (Euphydryas) anicia* (Doubleday, [1847])

**ETYMOLOGY:** The name *anicia* is derived from Anicia Iuliana, an aristocrat of Constantinople (461–527 AD). Anicia was a Chalcedonic Christian, and the city of Chalcedon is across the Bosporus from Constantinople. Thus the names for the Anicia and Chalcedon checkerspots indicate the close resemblance of the two species. The subspecies *hopfingeri* was named in honour of T.C. Hopfinger, the best known of pre-Second World War lepidopterists in Washington state, a charter member of the Lepidopterists' Society, and the first Pacific Northwest coordinator for the annual Season's Summary of the society. Pyle (1981) coined the common name after it was established that Doubleday, not Hewitson, was the author of the species name.

**ADULT:** In most of BC only the Anicia Checkerspot is found. Where it overlaps with the Chalcedon Checkerspot along the

**An uncommon dark colour form of the Anicia Checkerspot (*Euphydryas anicia*)**

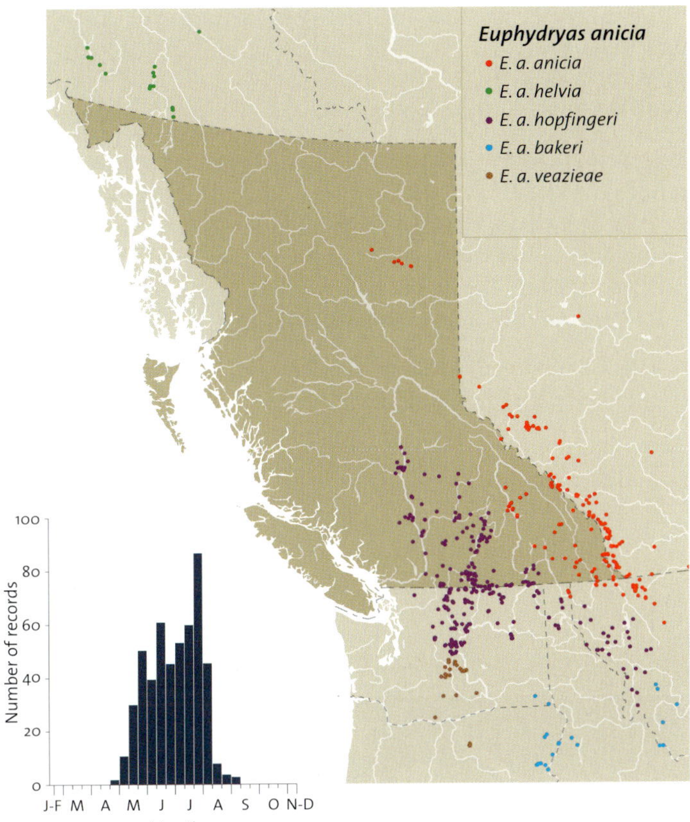

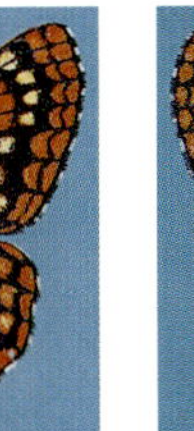

Ssp. *anicia* ♂ D  (4.0 cm)          Ssp. *anicia* ♀ D  (4.1 cm)

Ssp. *anicia* ♂ V  (4.0 cm)          Ssp. *hopfingeri* ♂ V  (3.7 cm)

Ssp. *hopfingeri* ♂ D  (3.7 cm)      Ssp. *hopfingeri* ♀ D  (4.4 cm)

crest of the southern Coast Ranges it can be reliably distinguished only by reference to the genitalia (Fig. 71b). The dorsal arm of the harpe is longer than that of any other *Euphydryas* species. The two arms of the harpe are almost parallel, whereas in other *Euphydryas* species the two arms are at right angles to each other. When the Anicia and Edith's checkerspots are found together in the Southern Interior and the Kootenays, the Anicia Checkerspot differs from Edith's Checkerspot in that the postmedian row of red spots on the ventral hindwing is not divided by a black line. The Anicia Checkerspot differs from the Chalcedon Checkerspot in the ventral forewing: the postmedian row of apical white spots is not accented on the inner side by a wide, black area (Fig. 72). These wing characters are not completely reliable,

however, especially for distinguishing Anicia from Chalcedon Checkerspots.

**Immature stages:** Venables (1912) partially described the immatures in BC. Eggs are lemon yellow. Early instar larvae are black with a dull white dorsal band.

**Biology:** Adults fly from late May to early July at low elevations. Alpine populations fly from early July to mid-August, depending on when the snow is gone for the summer. In June there is usually a cool, damp period that delays the emergence of a portion of low-elevation populations. When these individuals emerge, they appear darker than those that emerged earlier. In the West Kootenay and adjacent Pend Oreille Co., WA, these dark individuals look similar to *E. c. wallacensis* Gunder, 1928, but that taxon does not occur in BC, Idaho north of Wallace, or Pend Oreille Co., WA.

Venables (1912) found that eggs were laid in a clump on snowberry (*Symphoricarpos*), but this cannot be the only larval foodplant in the province as snowberry does not occur throughout the BC range of the Anicia Checkerspot. Larvae refused to eat after moulting to the third instar, which is the apparent overwintering stage. Hibernating checkerspot larvae found in webs on *Penstemon* on Crater Mountain, near Keremeos, BC, were probably Anicia Checkerspots (CSG).

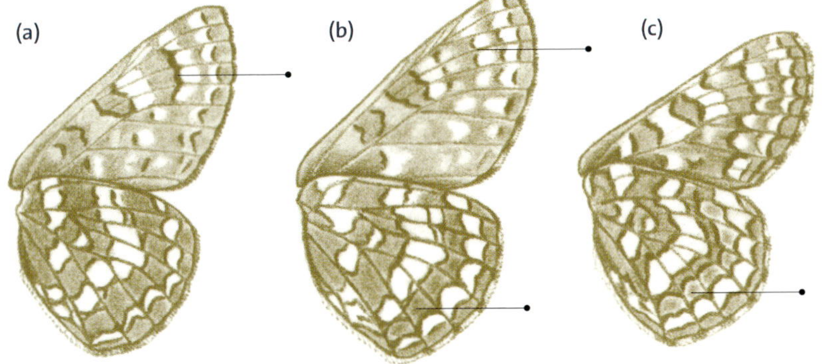


**72**  Underside of the wings of *Euphydryas* species: (a) *E. chalcedona*, (b) *E. anicia*, (c) *E. editha*

**Mature larva**

**Pupa**

**SUBSPECIES:** Populations in the Chilcotin, Southern Interior, and West Kootenay are the darker subspecies, Hopfinger's Checkerspot, *E. a. hopfingeri* Gunder, 1934 (TL: Black Canyon, Okanogan Co., WA). Populations from the East Kootenay and northeastern BC are the nominate subspecies, *E. a. anicia*

(Doubleday, [1847]); TL: nr. Rock Lake, AB (Shepard 1984), which has a very red ground colour. *E. a. helvia* (Scudder, 1869) (TL: Ramparts, Yukon R., AK) should eventually be recorded from the Atlin region.

**RANGE AND HABITAT:** The Anicia Checkerspot is the widest-ranging BC species of *Euphydryas*. It is found throughout BC east of the Coast and Cascade mountains except for the northeast quarter. It can occur at any elevation from the bottoms of valley floors to above timberline in the south, and in dry pine or aspen woodlands in the north.

**GENERAL DISTRIBUTION:** The Anicia Checkerspot occurs from central AK south to central CA, AZ, NM, and northern MEX.

**CONSERVATION STATUS:** Not of concern, with subspecies *anicia* S4 and *hopfingeri* S5.

## EDITH'S CHECKERSPOT
*Euphydryas (Euphydryas) editha* (Boisduval, 1852)

**ETYMOLOGY:** The species *editha* is presumably named after an unknown woman named Edith. Subspecies *taylori* was named after the Reverend George W. Taylor, the first well-known BC lepidopterist and a noted authority on geometrid moths (inchworms). Subspecies *beani* was named after the early lepidopterist Thomas E. Bean, who collected for William H. Edwards near Lake Louise, Alberta. The common name was first used by Holland (1898).

**Edith's Checkerspot (*Euphydryas editha taylori*)**

**ADULT:** The dorsal arm of the harpe of the male genitalia is covered with small spines (Fig. 71c), a unique character for BC *Euphydryas* species. For a further description, see the "Adult" discussion under **Anicia Checkerspot**.

**IMMATURE STAGES:** Emmel and Emmel (1973) describe the immatures in California. The egg is yellow. The mature larva is black; the dorsal and lateral lines are orange, not white as for the Chalcedon Checkerspot.

**BIOLOGY:** Low-elevation populations on Vancouver Island fly from mid-April to mid-May. Interior alpine populations fly from late June to early August. Danby

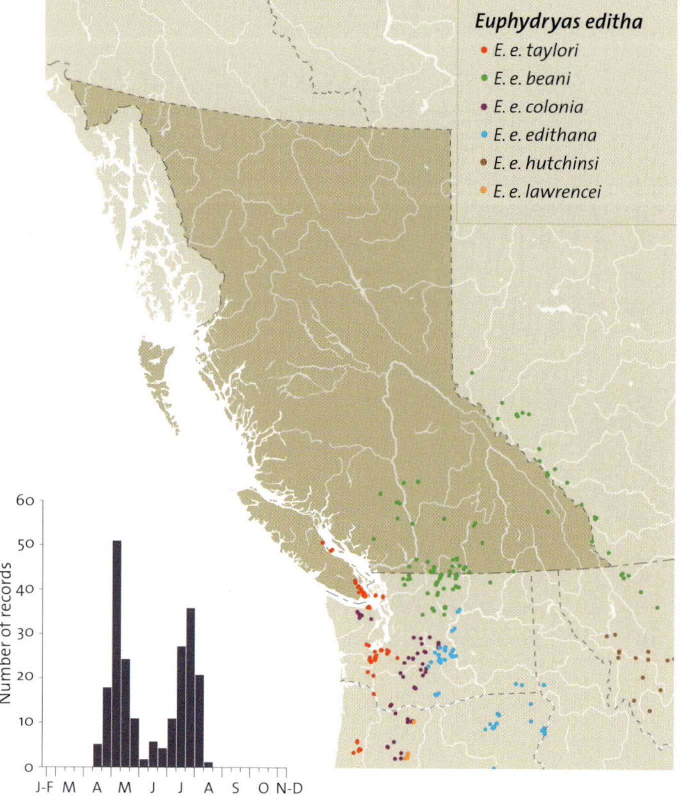

**Eggs, ssp. *taylori***

(1890) first observed the larvae at Beacon Hill Park on the flats above the sea cliffs. They were feeding on the introduced rib-wort plantain, *Plantago lanceolata.* The larvae feed until the fourth or fifth instar, and then hibernate until the following spring, when they mature, pupate, and emerge as adults by mid-April (Shepard 2000a). In 1972 Shepard (2000a) found larvae from the Hornby Island population feeding on the native plantain, *P. maritima,* as well as *P. lanceolata.* The adults on Hornby Island utilize spring gold, *Lomatium utriculatum,* as the primary nectar source (Shepard 2000a). At the site of the Beacon Hill and Uplands Parks extinct populations, there are virtually no spring gold plants left, but the introduced rib-wort plantain is still plentiful. Thus the loss of an adult nectar source is probably the most important reason for extinction.

Parmesan (1996) conjectured that BC populations have become extinct because of global warming. All but one of the extinct populations, however, became extinct well before global warming was a significant phenomenon. The last to become extinct, in the early 1990s, did so because its habitat was overrun by Scotch broom (CSG). This habitat was a powerline right-of-way that had been kept clear as a Christmas tree farm. After the tree farm was abandoned, BC Hydro adopted a policy of encouraging Scotch broom growth to reduce tree regeneration under powerlines. The one healthy population is on Hornby Island, where there is no significant Scotch broom. The local community is monitoring Scotch broom so that it does not invade the habitat of Taylor's Checkerspot.

**SUBSPECIES:** The Vancouver Island populations are Taylor's Checkerspot, *E. e. taylori* (W.H. Edwards, 1888) (TL: Victoria [Beacon Hill], BC). This subspecies ranges south to the southern Willamette Valley, OR, but is now nearly extirpated from Oregon and Washington. The interior subspecies is Bean's Checkerspot, *E. e. beani* (Skinner, 1897) (TL: Laggan [vicinity Lake Louise], AB). This subspecies ranges south to northern WA, northern ID, and northwestern MT.

**RANGE AND HABITAT:** Edith's Checkerspot occurs in two distinct areas of the province. One area is on southeastern

Ssp. *taylori* ♂ D (4.0 cm)

Ssp. *taylori* ♀ D (4.1 cm)

Ssp. *taylori* ♂ V (4.0 cm)

Ssp. *beani* ♂ V (3.6 cm)

Ssp. *beani* ♂ D (3.6 cm)

Ssp. *beani* ♀ D (3.4 cm)

Vancouver Island in lowland dry meadows near sea level. The Vancouver Island populations have been reduced to one known population on Hornby Island. The other area is at and above timberline in the Southern Interior and the Kootenays.

**GENERAL DISTRIBUTION:** Edith's Checkerspot occurs from southern BC south to Baja California, UT, and western CO.

**CONSERVATION STATUS:** Subspecies *taylori* is Endangered in BC (S1). Subspecies *beani* is not of concern (S5).

## GILLETT'S CHECKERSPOT
*Euphydryas (Hypodryas) gillettii* (Barnes, 1897)

**ETYMOLOGY:** The species *gillettii* was named for William Barnes' wife, Charlotte Gillett. In the same paper, Barnes described *Speyeria cybele charlottii,* also honouring his wife. The common name was first used by Holland (1931).

**ADULT:** Both the wings and the male genitalia are very distinct for this species, which cannot be confused with any other BC butterfly. The harpe of the male genitalia is distinct enough to at least warrant placement of this butterfly in a separate subgenus, if not genus. The upper arm of the harpe is wide and flattened, not rounded and pointed.

**IMMATURE STAGES:** Williams et al. (1984) describe the immatures in detail. The egg is yellow green and round, and has about 22 vertical ribs. The mature larva is black with a dorsal lemon yellow band and a lateral white band with black spiracles. The rest of the body is dark brown.

**BIOLOGY:** The few BC records of Gillett's Checkerspot indicate that it flies from mid-June to late July, but the flight period appears to vary. Williams et al. (1984) describe the biology in detail. Eggs are laid in clusters on the underside of leaves of the larval foodplant, *Lonicera involucrata,* in WY. Winter diapause can occur in the second, third, or fourth larval instars, which may account for the long flight period of adults in BC. The females will oviposit only on a foodplant that is in direct sunlight (Williams 1988).

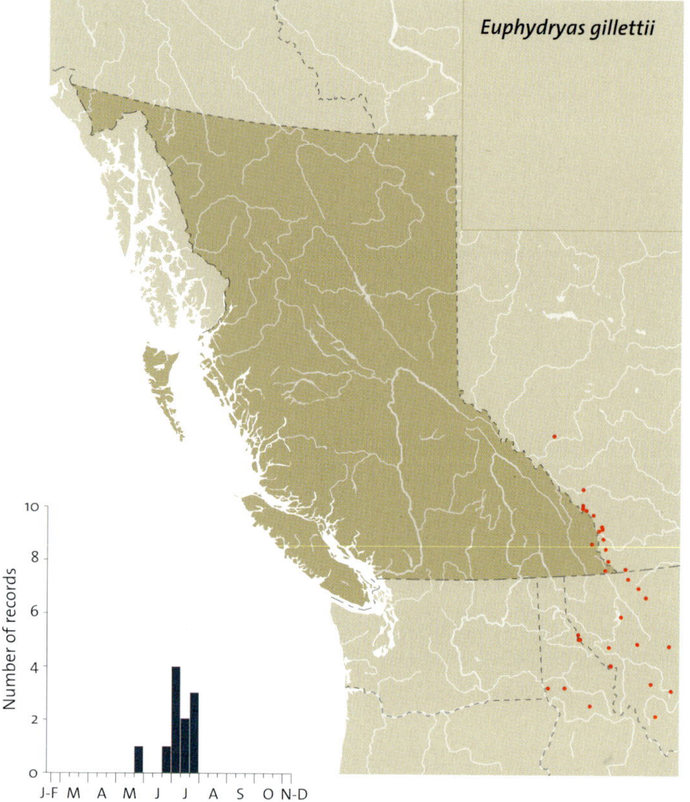

♂ D (3.8 cm)

♀ D (4.5 cm)

♂ V (3.8 cm)

**SUBSPECIES:** None. The type locality of the species is Yellowstone National Park, WY.

**RANGE AND HABITAT:** Gillett's Checkerspot is very restricted in its BC range. It has so far been recorded only from the Michel Creek and Flathead River drainages in extreme southeastern BC, where it is found in open, riparian situations in association with the foodplant.

**GENERAL DISTRIBUTION:** Gillett's Checkerspot has a very restricted distribution from the southern AB Rocky Mountain foothills and extreme southeastern BC at around 1,500 m south through ID and western MT to northwestern WY.

**CONSERVATION STATUS:** Gillett's Checkerspot is Threatened in BC (S2).

Both the subfamily name and the common name are derived from the type genus, *Limenitis* (below).

Admirals are medium to medium-large butterflies, with wing margins slightly scalloped and the upperside usually black marked with white spots and bands. The two sexes are quite similar. The antennae are gradually expanded and with scarcely developed clubs, and are more than half the length of the forewing. The forewing cell is open, or closed only by a vestigial vein.

Larvae have paired dorsal lobes or fleshy, club-shaped processes on several segments, and the vertex of the head is bifid. The pupa has twin cephalic pointed projections, and a characteristic dorsal lobe on the second abdominal segment.

Adults fly in a fairly straight, deliberate manner, and males characteristically perch and patrol small forest openings or edges of forests. Hibernation is usually as young larvae in a silk and rolled-leaf hibernaculum.

## GENUS *LIMENITIS* FABRICIUS, 1807 WHITE ADMIRALS

The name *Limenitis* is derived from the Greek *limenitis* (harbour keeping), an epithet applied to deities who protected harbours. Fabricius may have derived the name from the fact that the first specimen of an admiral came from the harbour town of Leghorn (Emmet 1991). More probably, however, it is derived from the male's defence of a territory based on a favourite perch site, analogous to an admiral protecting a harbour. The common name "admiral" may refer to one function of an admiral being to protect harbour towns, in much the same way as the male butterflies protect a favourite perch site. Since the common name pre-dated the Latin name, the Latin name may be derived from the common name. Holland (1898) first used the common name "white admirals" for the genus.

Admirals in BC are black with a broad white band across the upperside of the wings, except for the Viceroy, which is orange brown with black markings. Admirals are medium-sized to large butterflies. They are strong, fast fliers, but can usually be observed while perching and slowly patrolling their territories.

All the admiral species hybridize in nature where their distributions overlap, and hybrids between Lorquin's Admirals and White Admirals are common in southern

BC. During mating of all species in this genus, it is usual for the males to exhibit mid-valval flexion of the genitalia. Both valves can be seen to flex inward, perhaps stimulating the female (Platt 1979).

The generic name *Basilarchia* has been used by some authors for the admirals in BC. There are no significant structural differences in adults or larvae between *Limenitis* and *Basilarchia* (Layberry et al. 1998), hence we treat *Basilarchia* as a synonym of *Limenitis*. There are about 50 species of *Limenitis* worldwide.

♂ D (6.1 CM)    ♂ V (6.1 CM)

**Hybrid, *Limenitis lorquini* × *arthemis* male, matching the original description of *Limenitis lorquini burrisonii*, except for the small red tips to the forewings**

# White Admiral

*Limenitis arthemis* (Drury, [1773])

**Etymology:** The species name *arthemis* refers to Artemis, a Greek goddess who was a huntress and who was associated with Apollo and many aspects of women's lives through her association with the moon. The subspecies name *rubrofasciata* is derived from the Latin *ruber* (red) and *fasciae* (a swathe). This is in reference to the red submarginal spots on the underside of the hindwings that form a continuous red band, rather than the separate red spots of the nominate subspecies (Barnes and McDunnough 1916). The similar *Ladoga camilla* (Linnaeus) in England was first referred to as the White Leghorn Admiral by Petiver in 1702–05, which he changed to White Admiral in 1717 (Pollard and Emmet 1990). Klots (1951) was the first in North America to use the common name "White Admiral" for *L. arthemis*. Both species have a prominent white band across black wings, hence the "white" portion of the name. For the origin of "admiral," see the **Red Admiral**.

**Adult:** White Admirals are large black butterflies with a broad white band across the middle of the wings. The tips of their forewings are black, and there is a distinct lobe on the hind margin of the hindwing. The white bands crossing the wings are almost continuous, with only thin black veins breaking them into spots. The dorsal hindwings of White Admirals have a band of orange spots outside the white band, and usually a band of blue spots outside that. The ventral wing colour is deep red brown and lacks white in the basal area, as in the coastal subspecies of the Lorquin's

♂ D, 5 instars  (6.0 см)

♂ D, 4 instars  (4.9 см)

♂ V  (6.0 см)

♀ D  (6.6 см)

Admiral. Along the ventral margin of the wings is a double line of pale blue. Females are larger and have enlarged wing markings compared with males. Hybridization between White Admirals and Lorquin's Admirals is common where their ranges overlap, with the hybrids having a mixture of the characters of the two species.

**Immature stages:** Eggs are round and pale green. Mature larvae have pale brown heads, bilobed with one pair of modified scoli on the top of each lobe and short pale setae on prominent hairs. The body is dark yellowish brown or olive green, with the back mostly pale mauve or white, occasionally suffused with pale pink (Sugden 1970).

**Biology:** White Admirals are partially bivoltine in BC, and this may be generally true throughout their Canadian range (Ferris and Brown 1981). Eggs are laid precisely on the margin

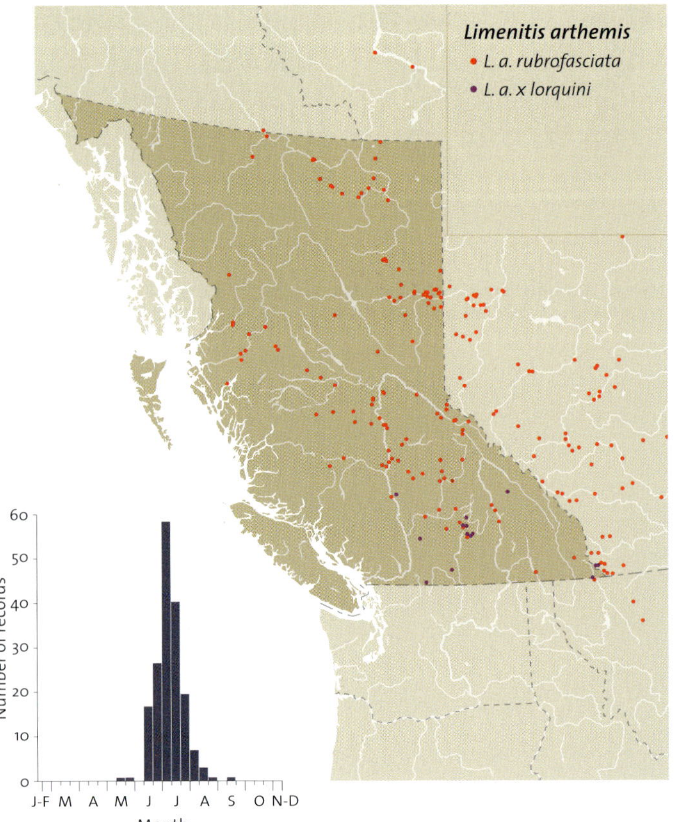

**White Admiral (*Limenitis arthemis rubrofasciata*)**

of leaves of the larval foodplants. The larvae build a dung and silk projection out from the tip of the midrib, or less frequently out from other major leaf veins, and they rest on these projections when not feeding. This occurs for the first several instars. Second instar larvae roll the sides of the base of a leaf upward, build a very solid silk cylinder the length of their body within the leaf roll, and cut away the remainder of the leaf. They then hibernate in this hibernaculum. In spring the larvae emerge from hibernation and mature through to adults (CSG). In captivity, about half of the larvae from near Quesnel matured directly through to adults without hibernation, which would have produced a partial second generation in the wild. In these second broods, most males pupate after the fourth instar, and all females pupate after five instars. The males from these fourth instar larvae are smaller and less clearly marked than those from fifth instar larvae. The mid-September record shown in the flight season graph is probably a second-brood individual.

Males perch in the sun on trees and tall shrubs, mostly 1–3 metres above the ground and along flyways likely to be used by females. They fly out on patrol and to investigate butterflies and other objects flying past. Males favour particular perches, but change perches periodically. When they fly out to investigate other butterflies, interactions with other male admirals last twice as long as interactions with other species. Males fly aggressively at other males until the intruder leaves the territory of the resident; the resident then returns to its perch (Lederhouse 1993). Adults feed on carrion such as old carcasses and fish offal (CSG).

Trembling aspen and willow species are larval foodplants in BC (Sugden 1970). Outside BC larval foodplants include hawthorn, birch, and other poplars (Scudder 1889a).

**Subspecies:** The subspecies in BC is *L. a. rubrofasciata* (Barnes & McDunnough, 1916) (TL: Alberta, Saskatchewan, and Manitoba).

**Range and habitat:** White Admirals occur throughout the northern and central interior of BC south to the northern Okanagan Valley, mostly in deciduous and mixed forests.

Eggs

First instar larva on frass and silk projection

Hibernaculum

Mature larva

Pupa

White Admiral (*Limenitis arthemis*) hanging from pupal case

**General distribution:** White Admirals range from central AK and YT across northern BC and east to NS. They are found throughout the eastern USA, as the subspecies known as the Red-Spotted Purple.

**Conservation status:** Not of concern (S5).

## Viceroy

*Limenitis archippus* (Cramer, [1776])

**Etymology:** The species name *archippus* refers to a king of ancient Italy (Reed 1870), or to Archippus, a Greek poet or comedian (Opler and Krizek 1984). Subspecies *idaho* is named for the Pliocene Lake Idaho, with which the distribution is associated (Austin 1998b). The common name was first used by Scudder (1874b) in reference to the species' mimicry of the Monarch. A viceroy rules a province or colony with the authority of his monarch, and hence mimics the monarch's authority.

**Adult:** Viceroys can be confused only with the butterfly they mimic, the Monarch. They can most readily be distinguished from Monarchs by the black line across the middle of their hindwing, which is absent in Monarchs. Females are larger and brighter than males.

**Immature stages:** Eggs are pale yellow when laid, and then turn grey. Mature larvae have strongly bilobed pale green heads, with two dull white lines down the front, and with a number of small green and greenish white tubercles. The body is a deep, rich green colour, with patches and streaks of dull white. There is a tubercle on each side of each segment along the back, with those on the middle thoracic segment being elongated into a long brownish horn. A white "saddle" covers the middle of the back of the abdomen. There are various other white, black, brown, and green markings. The pupa is a mix of brown, grey, pinkish, and white blotches, with a large keel on the back (Saunders 1869c; Bethune 1874).

**BIOLOGY:** Viceroys were bivoltine in BC. The few specimens available suggest that the adults of the first brood flew in May and early June, and those of the second brood from the last week of July through August. Eggs are laid singly or in groups of two or three near the tip of the leaf, usually on the underside but sometimes on the upperside of the leaf (Bethune 1874). The second or third instar larvae of the second brood hibernate in a hibernaculum formed of silk and rolled leaf (Saunders 1869c; Bethune 1874). Viceroys apparently switched from their native larval foodplants to cultivated apple, and were extirpated from BC and northern Washington when pesticide spraying to control coddling

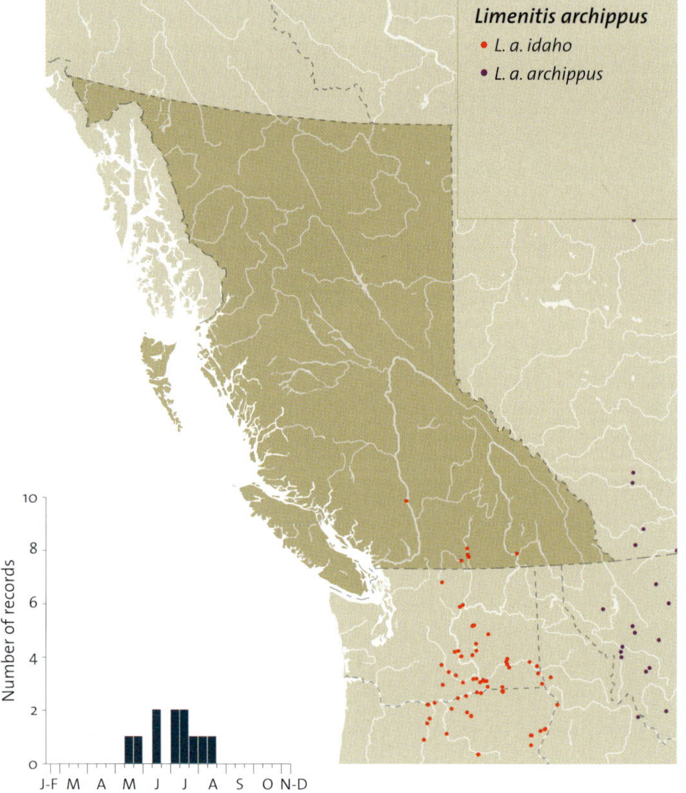

♂ D (6.3 CM)

♂ V (6.3 CM)

moth and other apple orchard pests started (Guppy et al. 1994). The last Viceroy in BC was recorded at Lillooet in 1930.

Viceroys were originally thought to be Batesian mimics of Monarchs, with the supposedly palatable Viceroys being protected from predation by their resemblance to the unpalatable Monarch. Recently, however, they have been shown to be Müllerian mimics of Monarchs, that is, Viceroys are as unpalatable as the Monarchs they resemble (Ritland and Brower 1991; Ritland 1995). Birds that attack either a Viceroy or a Monarch will learn to associate its colour pattern with unpalatability, and will then tend to avoid attacking both butterfly species in the future. Once a bird has had a bad experience when attempting to eat one individual of either species, both species are protected from bird predation by the close similarity of their wing patterns.

Willow, native crab apple, and cultivated apple were probably used as larval foodplants in BC. Larvae feed on willow in Yakima County, WA, and in Ontario and Connecticut (Saunders 1869c; Munroe 1948; Newcomer 1964a). Viceroys in eastern North America also use poplar, and are common on cultivated apple, plum, and cherry in orchards (Ferris and Brown 1981). Re-establishment of Viceroys in BC is possible, but will not be successful with continued insecticide spraying of apple orchards.

**SUBSPECIES:** Subspecies *idaho* Austin, 1998 (TL: Little Salmon River Valley, 3.1 mi. south of Jackpot, Elko Co., NV) formerly occurred in BC (Austin 1998b).

**RANGE AND HABITAT:** Viceroys formerly occurred across the Southern Interior of BC, with the northernmost record being from Lillooet.

**GENERAL DISTRIBUTION:** Viceroys occur from northern AB and adjacent NT, and formerly southern BC, east to NS and south to MEX.

**CONSERVATION STATUS:** Extirpated from BC (SX).

# Lorquin's Admiral
## *Limenitis lorquini* (Boisduval, 1852)

**Lorquin's Admiral (*Limenitis lorquini*)**

**ETYMOLOGY:** The species *lorquini* was named for Pierre Joseph Michel Lorquin, an early French butterfly collector in California who supplied Boisduval with many of the specimens for Boisduval's California butterfly names (Emmel et al. 1998a). Subspecies *itelkae* is named for Aud Itelke Fischer, one of the artists for this book. Subspecies *ilgae* is named for Aud's sister, Britta Ilge Fischer. Aud lives in the interior of BC, whereas Britta lives on the coast. The use of names of sisters symbolizes the relatedness of the interior and coastal subspecies. The common name "Lorquin's Admiral" was first used by Holland (1898).

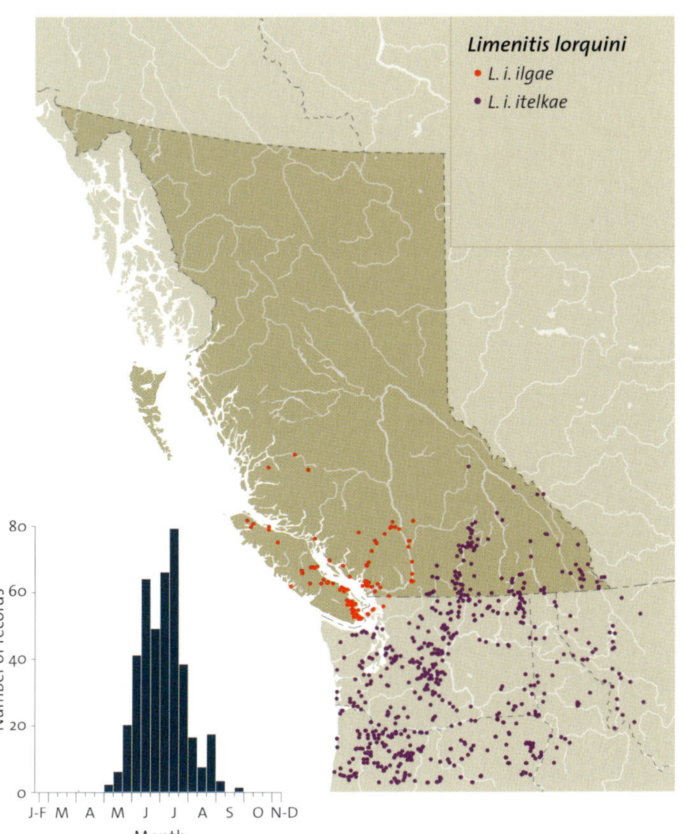

Ssp. *itelkae* ♂ D HOLOTYPE  (5.6 CM)

Ssp. *itelkae* ♂ V HOLOTYPE  (5.6 CM)

Ssp. *itelkae* ♀ D  (6.4 CM)

Ssp. *itelkae* ♀ V  (6.4 CM)

Ssp. *ilgae* ♂ D HOLOTYPE  (5.7 CM)

Ssp. *ilgae* ♂ V HOLOTYPE  (5.7 CM)

Ssp. *ilgae* ♀ D  (6.5 CM)

Ssp. *ilgae* ♀ V  (6.5 CM)

**ADULT:** Lorquin's Admirals are large black butterflies with a broad white band across the middle of the wings. The forewings have orange red tips on both the dorsal and ventral surfaces. The white bands crossing the wings are broken into distinct spots by wide black veins. Females are larger and have more well developed markings than males. Hybrids between Lorquin's Admirals and White Admirals have a mixture of the characters of the two species.

**IMMATURE STAGES:** Eggs are thimble-shaped, with deep hexagonal pitted cells; the edges or "corners" where the cells meet are drawn out into short, fine glassy hairs. They are pale green and the surfaces of the pits have a shiny, glassy effect (GAH). First instar larvae have a very large dark brown head and an olive green body with two pairs of dark tubercles on the thorax and two pairs on the abdomen (GAH). In mature larvae from Vernon, the head is pale lilac to pale mauve tan, and is bilobed with one pair of modified scoli on top of each lobe. The body is dark brown purple to grey mauve, with the back mostly white, washed with pale mauve. There is a mauve white line on each side of the abdomen (Sugden 1970). Pupae have a large keel projecting from the back of the thorax. The wings and the back of the abdomen are dark olive grey; the thorax is dull purplish and mottled with white. The rest of the abdomen is dirty white, shaded with grey and black on the sides and in a broad double ventral band (Dyar 1891).

BIOLOGY: Lorquin's Admirals are primarily univoltine, but have a partial second generation. The first generation commences flight in mid to late May, and the partial second generation extends the flight period to early August. In the Southern Interior there is a fragmentary third brood at least in some years, which is in flight in September and October. In California part of the offspring of the first brood, and all the offspring of the second brood, hibernate as second instar larvae (Dyar 1891), which also occurs in BC. Whether eggs laid by the fragmentary third brood successfully develop into hibernating larvae is unknown. The second instar larvae hibernate in a hibernaculum built of rolled leaf and silk. Larvae build a long, thin projection out from a leaf vein, as described for the White Admiral. Sap exuded by willow stems being bored into by beetle larvae are fed on by both male and female Lorquin's Admirals, and both sexes also mud-puddle (CSG).

Lorquin's Admiral males defend territories. They prefer a territory consisting of a bare or grassy patch on a south-facing slope, with dense shrubbery or trees at the upper end. The butterflies frequently settle on shrubs or trees that are thermally optimal due to warm updrafts and probably local solar heating. Lorquin's Admiral males are frequently seen elsewhere, but the optimal territory types are always

Lorquin's Admiral (*Limenitis lorquini ilgae*)

Egg, ssp. *ilgae*

Hibernaculum, ssp. *ilgae*

Pupa, ssp. *ilgae*　　　　　Mature larva, ssp. *ilgae*

reoccupied rapidly after the resident is removed (Guppy 1970).

Larval foodplants in BC are black cottonwood, trembling aspen, willow, cotoneaster, garden apple, Oregon crab apple, ornamental Siberian crab apple, hardhack, saskatoon, chokecherry, bitter cherry, and hawthorn (Dyar 1891; Harvey 1908; Jones 1935, 1942, 1943; Sugden 1970; Emmel et al. 1971; ACJ; CSG; FIS). Larvae formerly defoliated young apple trees in Yakima Co., WA (Newcomer 1964a), which is now prevented by orchard spraying.

SUBSPECIES: The subspecies name *burrisonii* Maynard, 1891 (TL: here restricted to the vicinity of Landsdowne, BC) has previously been applied to *Limenitis lorquini* in BC, Washington, and Oregon. The settlement of Landsdowne existed from 1889 to at least 1892 where Armstrong is now located, with the Armstrong family being among the first settlers (Williams 1889, 1892). *L. lorquini* and *L. arthemis* are found in the area, and hybrids between the species frequently occur. *L. l. burrisonii* was named by Maynard (1891), with a reasonably detailed description. He then stated that "there is considerable variation from the type towards typical *lorquini* ..." This demonstrates that the description was of a single specimen, which is the holotype. *L. lorquini* adults in the vicinity of the type locality never have the combination of

characters given in the original description, but *lorquini* ×
*arthemis* hybrids are always very similar to that description.
The holotype, as originally described, is phenotypically
intermediate between interior BC *L. lorquini* and *L. arthemis,*
and was a hybrid (see p. 313). This is consistent with the
conclusion of Layberry et al. (1998). The name *burrisonii* is
therefore unavailable but enters into homonymy (ICZN
Article 23.8) and is not available for reuse.

The name *maynardi* Field, 1936 (TL: Vancouver, BC), which
might have applied to the southwestern BC populations, was
proposed as a "transitional form" of *L. l. burrisonii* (Field 1936).
As a form name, *maynardi* is unavailable for use as a subspecies name, contrary to Perkins and Perkins (1966). As a
result there are no valid subspecies names to apply to the
*L. lorquini* populations in BC; names and descriptions for two
subspecies are provided here.

***Limenitis lorquini itelkae* Guppy, new subspecies**. *Limenitis
lorquini itelkae* has smaller orange red forewing tips, the
wing tips are a darker red, and the white band on the
upperside of the wings is narrower than in the nominate
subspecies. It has a paler ventral wing surface and larger
white markings, and more of them, than subspecies *ilgae*
(below). The red brown ground colour of the hindwing
underside is lighter than in subspecies *ilgae,* and there is a
cluster of white markings in the basal area of the hindwing
in subspecies *itelkae* that is missing from subspecies *ilgae*.
**Types**. Holotype: male, BC, Keremeos, 1 mile N of Hwy 3A on
Mt. Apex Rd., 8 June 1982, C.S. Guppy; a label "Holotype /
*Limenitis lorquini / itelkae* Guppy" is attached. The holotype
is deposited in the Royal British Columbia Museum, Victoria,
BC, CAN. Paratypes: 8 males, 1 female, BC, 8 km. E Cawston,
1 July 1978, J. and S. Shepard (JHS); 1 female, BC, Kilpoola L.,
1 July 1996, J. and S. Shepard (JHS); Osoyoos, Anarchist
Mountain, 1 June 1975, C.S. Guppy (CSG).

***Limenitis lorquini ilgae* Guppy, new subspecies**. In *Limenitis
lorquini ilgae,* the upperside of the wings is similar to that of
subspecies *itelkae,* except for somewhat wider black veining

separating the white spots composing the white band across
the wings. The extent and variability of the orange red forewing tips is similar for the two subspecies. In subspecies
*ilgae,* the ventral hindwing surface is a darker, richer red
brown ground colour compared with subspecies *itelkae,* and
there are no white markings at the base of the ventral hindwing. **Types**. Holotype: male, BC, Bamberton, Jones Creek
Road to Oliphant Lake, 9 July 1988, C.S. Guppy; a label reading
"Holotype / *Limenitis lorquini / ilgae* Guppy" is attached. The
holotype is deposited in the Royal British Columbia Museum,
Victoria, BC, CAN. Paratypes: 6 males, 1 female, same data as
holotype (CSG); 1 male, BC, Bamberton, 0.5 km south of
cement plant, 23 June 1989, C.S. Guppy (CSG); 1 male, BC,
Duncan, Mt. Prevost summit, 12 July 1987, C.S. Guppy (CSG);
2 males, 1 female, BC, Malahat, 17-Mile Road, 1 July 1987, C.S.
Guppy (CSG); 1 female, BC, Mt. Benson, headwaters of Wolf
Creek, 24 June 1989, C.S. Guppy (CSG); 2 males, BC, Mt.
Brenton Road, 16 July 1995, J. and S. Shepard (JHS); 2 males,
BC, Uplands Park, Vanc. Isl., 2 June 1995, Jon H. Shepard (JHS);
1 female, BC, powerline at Shawnigan L. – Mill Bay Road, 24
May 1995, Jon H. Shepard (JHS); 1 male, BC, Fitzgerald, 26 June
1995, C.S. Guppy (JHS).

**Range and habitat:** Lorquin's Admirals occur across southern BC in deciduous forests and riparian habitats. Subspecies
*itelkae* occurs east of the Cascade Mountains, from Manning
Provincial Park east across the Okanagan Valley and Kootenays
to AB. Subspecies *ilgae* occurs west of the Cascade Mountains, including the Fraser River north to Lillooet, the Coast
Range north to Bella Coola and Tweedsmuir Provincial Park,
and west to Vancouver Island.

**General distribution:** Lorquin's Admirals occur from
southern BC and extreme southwestern AB south to Baja
California and central ID. Subspecies *itelkae* extends from BC
and AB southward east of the Cascades through WA, OR,
northern ID, and northwestern MT. Subspecies *ilgae* occurs in
southwestern BC west of the Cascade Mountains.

**Conservation status:** Not of concern, with both subspecies S5.

The subfamily name is derived from the name of the European type genus, *Satyrus,* which in turn is derived from the Greek god Satyr, half man and half goat (Reed 1871a). This may be in reference to the butterfly's bouncy flight pattern, resembling Satyr bounding through the forests he inhabited.

Satyrs are small to medium-sized grey, brown, or black butterflies that frequently have small eyespots on the wings. They are sexually dimorphic to a greater or lesser extent. The antennae are variable, sometimes with clearly developed clubs but usually with gradually expanded and scarcely developed clubs. The forelegs are extremely reduced. The forewing has 12 veins, and one vein at the base of the forewing is swollen.

Eggs are squatly cylindrical, with ridged and pitted sides. Larvae are slender, thinly haired, slightly tapered towards both ends, and with a short forked tail. The colours of the immature stages of most species are geographically variable, and details are unknown for BC. Pupae are smooth, have rigid abdominal segments, and are either suspended or lie free among grass and debris.

Adults fly very erratically and, in the slower-flying species, have a bouncy flight pattern. The butterflies usually hibernate as young larvae.

### GENUS *COENONYMPHA* HÜBNER, [1819]   RINGLETS

The name *Coenonympha* is derived from the Greek *koinos* (shared in common) and *numphe* (a nymph), possibly meaning a genus containing nymphalid butterflies with widespread distribution (Emmet 1991). The generic common name "ringlets" was first used by Holland (1898) in reference to the small "ringlets" or eyespots on the wings of most species.

Ringlets are small white, orange brown, or grey butterflies with the peculiar bouncing flight pattern characteristic of many other Satyrinae.

In North America, the genus *Coenonympha* has been treated in a variety of ways by various authors, with the Hayden's Ringlet, *C. haydeni* (W.H. Edwards, 1872), being the only consistently recognized species. Some authors have considered all North American *Coenonympha* except *haydeni* to be subspecies of the European *C. tullia* Müller, 1764; others have recognized five species in addition to *haydeni.* Most recently, Layberry et al. (1998) considered all *Coenonympha* in Canada to be *C. tullia,* except for *C. nipisiquit* McDunnough, 1939 in coastal Quebec and New Brunswick.

We have determined that the northwestern BC *Coenonympha* in the *C. tullia* group have male genitalia that are significantly different from the *C. tullia* group in the rest of North America (Fig. 73b). The male genitalia of northwestern BC *Coenonympha* are similar to those of European *C. tullia,* hence we apply the name *C. tullia* to

the northwestern BC populations. The remaining North American populations in the *C. tullia* group are *C. california* Westwood, [1851], with *C. nipisiquit* as a possible additional species.

North American *Coenonympha* species are therefore *C. haydeni, C. tullia, C. california,* and possibly *C. nipisiquit. C. haydeni* is a clearly defined species that needs no further discussion. We do not have enough data to adequately assess the status of *C. nipisiquit. C. tullia* occurs in Europe, Asia, and northwestern North America. *C. california* occurs in North America from the Mackenzie River east to Newfoundland, and, in the west, south to California. There are about 22 species in the genus worldwide.

All ringlets hibernate as larvae; the hibernating instar varies within as well as between populations. Larval foodplants are grasses, but whether specific grasses are chosen is unknown.

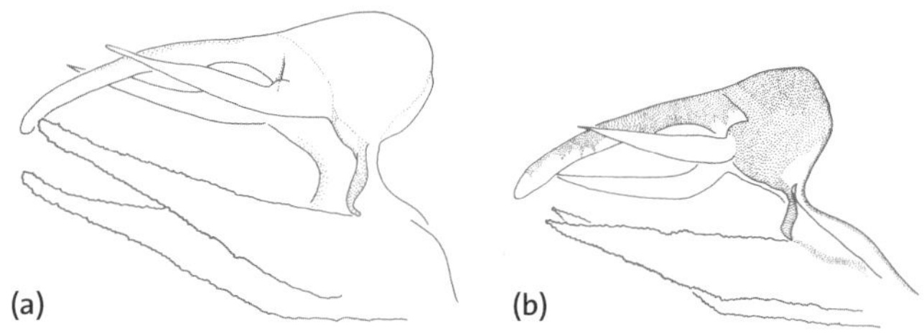


**73** Male genitalia of North American *Coenonympha* species: (a) *C. california,* (b) *C. tullia*

# Northern Ringlet

*Coenonympha tullia* Müller, 1764

**Etymology:** The species name *tullia* is derived from Tullia, the feminine form of Tullius, a common Roman name, as in Marcus Tullius Cicero. Roman names are commonly used for species in this genus (Emmet 1991). Subspecies *yukonensis* was named for its Yukon type locality. The common name "Northern Ringlet" is used here for the first time, in reference to its northern distribution.

**Adult:** Northern Ringlets are small, grey brown butterflies with eyespots reduced or absent. Females have orange brown areas on the upperside, and males have tints of orange brown. Male genitalia are short and stocky, with the uncus up to 1.5 times longer than the tegumen. The valves and brachia are broad and slightly sinuous. Male genitalia are grey, shading to black when heavily sclerotized (Fig. 73b). The genitalia of European populations are similar, but the uncus may be almost twice the length of the tegumen.

♂ D  (3.1 cm)

♀ D  (3.6 cm)

♂ V  (3.1 cm)

The only similar species in BC is the Common Ringlet, which is tan or orange brown in colour in both sexes.

**Immature stages:** Undescribed in North America.

**Biology:** Northern Ringlets are univoltine in North America, and fly in June and July. The larvae hibernate (Tuzov 1997). Larval foodplants are unknown in North America. In Europe and Asia, sedges, rushes, and grasses are used, including *Carex, Eriophorum, Festuca, Poa, Rhynchospora alba,* and *Stipa* (Higgins and Riley 1970; Tuzov 1997).

**Subspecies:** BC populations are subspecies *yukonensis* Holland, 1900 (TL: Dawson, YT). The only other North American subspecies are *kodiak* W.H. Edwards, 1869 (TL: Kodiak, AK), which occurs in southwestern Alaska, and *viluiensis* Ménétriés, 1859 (TL: Siberia), which occurs on the Seward Peninsula of western Alaska. Both subspecies are entirely shades of grey. Subspecies *mixturata* Alpheraky, 1897 (TL: Kamchatka, Siberia) has previously been incorrectly attributed to North America instead of *viluiensis*.

**Range and habitat:** Northern Ringlets occur in grass and alpine tundra habitats in northwestern BC.

**General distribution:** Northern Ringlets occur from Britain and Europe across Asia to eastern Siberia, AK, YT, and northwestern BC.

**Conservation status:** Not of concern (S4).

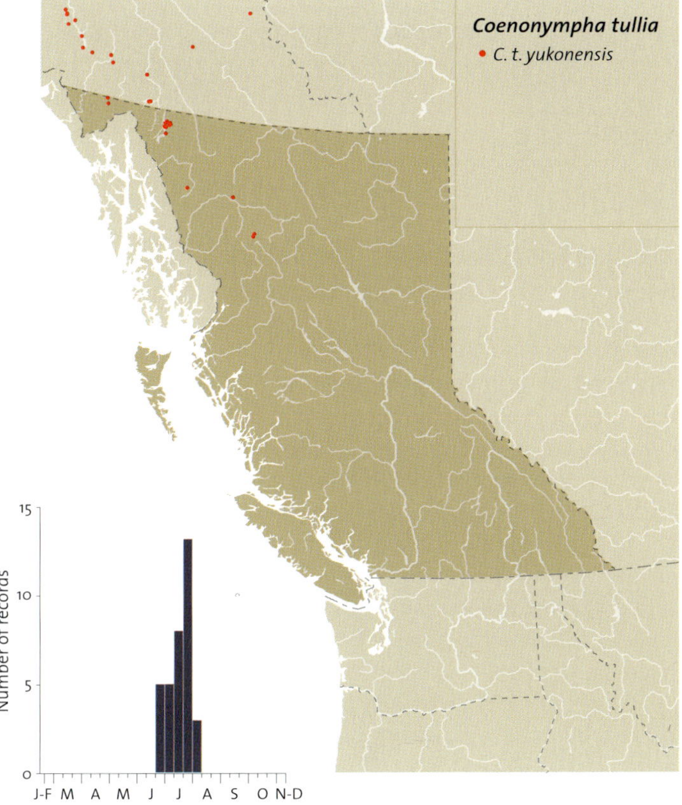

# Common Ringlet

*Coenonympha california* Westwood, [1851]

**ETYMOLOGY:** The species *california* was named for the type locality of California. Subspecies *benjamini* was named for the early entomologist F.H. Benjamin, who was one of Barnes's curators. The subspecies name *insulana* refers to the island nature of the Vancouver Island type locality. Subspecies *columbiana* was named for the British Columbia type locality. The common name was first used by Opler (1982) in reference to the species' wide distribution and frequent abundance.

**ADULT:** Common Ringlets are characterized by a predominantly white to orange brown or tan wing ground colour; lighter shades are present in females, and at least a ventral apical forewing ocellus is present in most subspecies. Male genitalia are long and slender, with the uncus at least 2.5 times longer than the tegumen. The valves and brachia are slender and strongly sinuous. The genitalia are golden brown, shading to black when heavily sclerotized (Fig. 73a). Electrophoretic data for Common Ringlet populations from California and adjacent areas demonstrate that subspecies *ampelos* is conspecific with subspecies *california* (Porter and Geiger 1988). Hence *ampelos* is a subspecies of *C. california*, and by extension *columbiana* and *insulana* must also be subspecies of *C. california* because of their clear phenotypic affinities to *ampelos*. We also place *benjamini* as a subspecies of *C. california* for lack of evidence to the contrary. The only similar butterfly in BC is the grey Northern Ringlet.

Ssp. *columbiana* ♂ D (3.6 cm)

Ssp. *columbiana* ♂ V (3.6 cm)

Ssp. *benjamini* ♂ D (3.6 cm)

Ssp. *benjamini* ♂ V (3.6 cm)

Ssp. *insulana* ♂ D (3.2 cm)

Ssp. *insulana* ♂ V (3.2 cm)

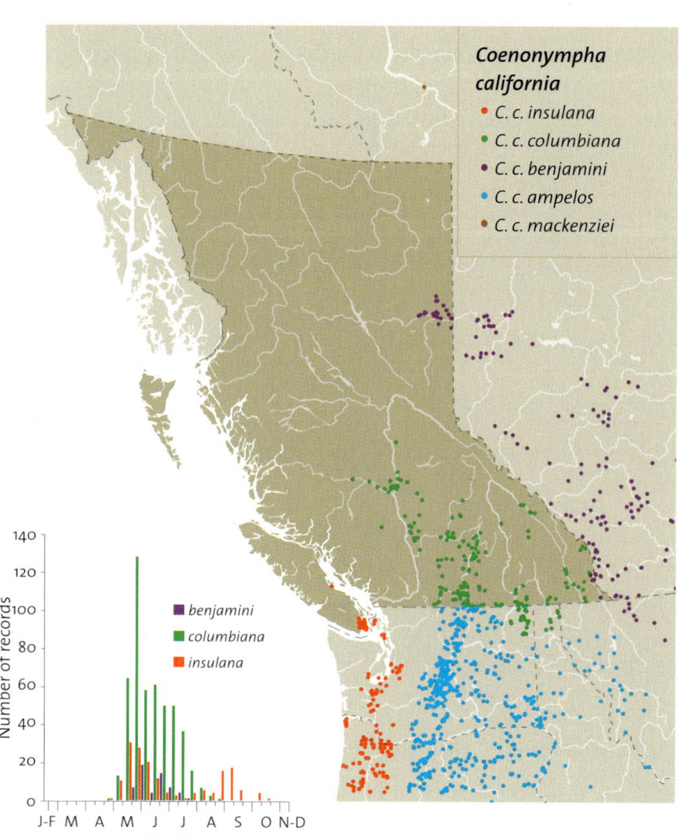



**IMMATURE STAGES:** The immature stages of subspecies *insulana* have been described from Vancouver Island. Eggs are barrel-shaped with 34 faint vertical ribs with faint cross-ribs. They are white to green yellow, with flecks and streaks of brown forming an irregular band around the middle. First instar larvae are pale, dull flesh-coloured or yellow green. The body tapers towards the back from the second thoracic segment, and has two short, fleshy conical tails. There is a mid-dorsal reddish line, and three similar ones on each side (subdorsal, lateral, and one in between). Mature larvae have a grass green head and body covered in white mushroomlike bodies, resulting in a blue grey bloom. The dorsal line and three subdorsal lines on each side are dark green. There is a yellow lateral line, which highlights a lateral fold. The spiracles appear as small black dots, and the anal processes are tinged with pink. Pupae are short and broad, smooth, and grass green, with a fuscous line on the costal and hind margins of the wing cases, and two short fuscous lines on the underside of the last abdominal segment, converging to form a "V" at the base of the cremaster. The pupa is suspended by its cremaster from a grass stem (Edwards 1887a; Hardy 1960).

**BIOLOGY:** Common Ringlets have one brood in May to July, and subspecies *insulana* has a second brood in August to

October. The two broods of *insulana* barely overlap, with extremely worn first-brood adults flying with fresh second-brood adults. Eggs of subspecies *insulana* are laid in May by first-brood adults; they produce some larvae that grow rapidly and mature into a second brood of adults in August and September. Other larvae grow slowly and hibernate as quite mature larvae, and produce first-brood adults the following year. Eggs laid by second-brood adults in the fall produce larvae that hibernate as young larvae, and produce more second-brood adults the following fall. The eggs hatch in about 12 days. Pupation in "fast-track" larvae occurs about eight weeks after hatching. Larvae of *insulana* that hibernate have four instars; those that do not go into diapause have three instars prior to pupation (Edwards 1886a; Hardy 1960). Adults lay eggs on grass, and the larvae require green grass for food. Hence on Vancouver Island, Common Ringlets can exist only in areas that are damp enough to maintain green grass throughout the driest period of the summer and yet do not flood excessively in the winter. Such areas became more common due to land clearing up to the 1950s, but are now becoming overgrown by brush and trees or destroyed through urbanization.

Common Ringlet (*Coenonympha california columbiana*)

Mature larva

Pupa

Larval foodplants are not firmly established, but are probably grasses and, less likely, sedges. Scott (1992) records various grasses and sedges as oviposition sites, but none are confirmed larval foodplants.

**SUBSPECIES:** Subspecies *insulana* McDunnough, 1928 (TL: Victoria, BC) occurs in BC on southern Vancouver Island. Adults are tan in colour and lack eyespots. Subspecies *columbiana* McDunnough, 1928 (TL: Aspen Grove, BC) occurs across central and southern BC east of the Coast Range. Adults are light orange brown and only very rarely have small eyespots. In the vicinity of Kilpoola Lake, south of Richter Pass, there is some intergradation with subspecies *ampelos* to the south, which has a greyer wing colour and small ocelli on the forewings and hindwings. Subspecies *benjamini* McDunnough, 1928 (TL: Waterton Lakes, AB) occurs in the Peace River area and in extreme southeastern BC. Adults are dark orange brown (ochre) and usually have an eyespot under the tip of each forewing.

**RANGE AND HABITAT:** Common Ringlets are found in meadows and grasslands at all elevations across southern BC and at low elevations in the Peace River area. Downes (1956) considered *insulana* as probably the most abundant butterfly on Vancouver Island. It was restricted to the Saanich Peninsula until about 1965, and then expanded northward to Chemainus (Guppy 1974; Shepard 1977). It is now uncommon, and increasing urbanization and invasion of habitat by Scotch broom are rapidly eliminating or isolating many populations.

**GENERAL DISTRIBUTION:** Common Ringlets occur from southern NT to CA west of the Great Plains, east across Canada from BC to NF, and the northeastern United States.

**CONSERVATION STATUS:** Subspecies *insulana* (S2S3) and *benjamini* (S3) are of Special Concern in BC, but subspecies *columbiana* is not of concern (S5).

## GENUS *CERCYONIS* SCUDDER, 1875  WOODNYMPHS

The name *Cercyonis* is derived from Cercyon, the son of Poseidon (Opler and Krizek 1984). The common name "woodnymphs" was first used by Holland (1898) in reference to the butterflies' bouncy flight in generally open forest habitat, similar to nymphs bounding through open forests.

Woodnymphs are medium-sized, dark brown butterflies with prominent eyespots on the forewings and some-times smaller eyespots on the hindwings. The eyespots are usually larger in females than in males, and are set in a lighter band. Woodnymphs have a peculiar bouncing flight, similar to that of ringlets.

Eggs are laid singly on grass blades. They are pale yellow, becoming tan and mottled with orange brown as they mature. They are cylindrical and squat, with a flat top and ridges down the side. First instar larvae are thinly covered with thick curved hairs, and are green with light and dark longitudinal stripes. Mature larvae are slender, green, or yellow green, with light and dark longitudinal stripes down the back and sides. They are thinly covered with hairs, and have two short red tails. Pupae are roughly cylindrical, rounded, and suspended from a cremaster. They are green to yellow green, and in most species have white or yellow markings (Emmel 1969).

Eggs hatch about 10 days after oviposition (at 25°C), and the first instar larvae immediately enter hibernation. By spring the larvae have shrunk to half their original length before they come out of hibernation. Once they commence feeding, they primarily feed at night. There are five (*oetus, sthenele*) or six (*pegala*) larval instars. It takes *oetus* about 2 months to pupate in the wild, and *sthenele* and *pegala* 2.5–3 months. Pupation occurs near the base of a grass clump, with the pupa hanging from grass blades. Adults emerge after about 20 days (Emmel 1969).

## COMMON WOODNYMPH
### *Cercyonis pegala* (Fabricius, 1775)

**ETYMOLOGY:** The species name *pegala* is from the Greek *pege* (well or fountain) and the suffix *al* (belonging to) (Bird et al. 1995), perhaps indicating that the first specimens were found near a spring. The subspecies name *ariane* perhaps refers to *Ariane,* a French play (1672). Subspecies *ino* is named

Common Woodnymph (*Cercyonis pegala ariane*)

after the Greek sea goddess Ino (Bird et al. 1995). The subspecies name *incana* is derived from the Latin *incanus* (grey). The common name was first used by Holland (1898).

**ADULT:** Common Woodnymphs are the largest species in the genus. They have two large well-defined eyespots on the forewings, with the rear eyespot at least slightly larger than the front eyespot (sometimes only slightly). The outer part of the ventral hindwings is a little lighter than, or the same shade as, the inner half of the wing.

**IMMATURE STAGES:** The immature stages are not known from BC, and the details of the coloration of larvae and pupae (see the description of the genus) are known to be geographically variable.

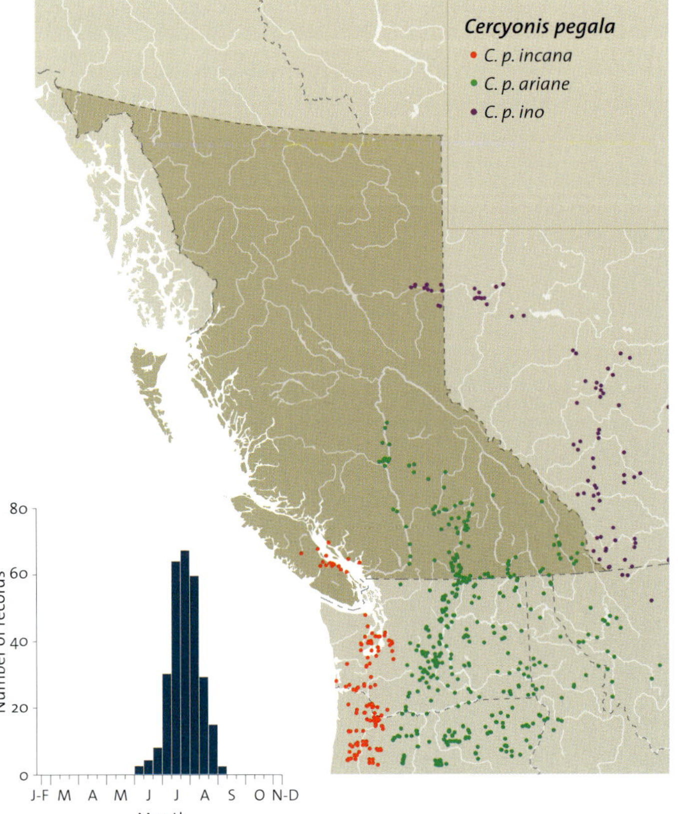

Ssp. *ariane* ♂ D (4.9 cm)

Ssp. *ariane* ♀ D (5.5 cm)

Ssp. *ariane* ♂ V (4.9 cm)

Ssp. *ino* ♂ D (4.4 cm)

Ssp. *incana* ♂ D (5.0 cm)

Ssp. *incana* ♀ D (5.0 cm)

Ssp. *incana* ♂ V (5.0 cm)

Ssp. *ino* ♀ D (4.9 cm)

**BIOLOGY:** Common Woodnymphs are univoltine, and fly from July to September. There are six larval instars, and a female lays 200–300 eggs (Emmel 1969). Adults feed on flowers and on willow and poplar sap.

Larval foodplants in BC are probably grasses. Outside BC recorded foodplants include grasses and sedges such as *Tridens flavus, Avena fatua, Stipa, Andropogon,* and *Carex* (Shapiro and Shapiro 1974; Layberry et al. 1998; Bird et al. 1995). Scott (1992) records many other oviposition sites; eggs are frequently laid without being attached to anything, and fall into the litter on the ground.

**SUBSPECIES:** Common Woodnymphs in southern BC, from Yale eastward, are subspecies *ariane* (Boisduval, 1852); TL: restricted to 2 miles south of Spanish Ranch, Plumas Co., CA (Emmel et al. 1998a). Subspecies *boopis* (Behr, 1864), into which BC populations have sometimes been placed, is restricted to California. Subspecies *ariane* is usually strongly marked on the underside of the wings, with the outer half of

the ventral hindwing paler than the basal half. They frequently have a ventral row of one or more small submarginal spots, and sometimes have dorsal hindwing spots. Subspecies *incana* (W.H. Edwards, 1880) (TL: Olympia, WA) occurs from Vancouver Island south

Ssp. *ino* ♂ V (4.4 cm)

to the Willamette Valley, OR. The underside of the wings is a uniform grey brown, and one spot is always present on each dorsal hindwing. Subspecies *ino* Hall, 1924 (TL: Calgary, AB) occurs in the Peace River grasslands, and may also occur in extreme southeastern BC near Flathead. The underside of both forewings and hindwings is evenly light grey brown, except for a pale outer band, and there are no submarginal spots.

**RANGE AND HABITAT:** Common Woodnymphs occur across southern BC in grassy forest openings, clearcuts, roadsides, meadows, and stream banks.

**GENERAL DISTRIBUTION:** Common Woodnymphs occur from southern BC south to central CA and AZ, and across the continent to the Atlantic.

**CONSERVATION STATUS:** Subspecies *incana* and *ino* are both of Special Concern in BC (S3). Subspecies *ariane* is not of concern (S5).

**Mature larva, ssp. *ariane***

**Pupa, ssp. *ariane***

# GREAT BASIN WOODNYMPH
*Cercyonis sthenele* (Boisduval, 1852)

**ETYMOLOGY:** The species name *sthenele* refers to Sthenelus, tragic poet and father of Eurystheus and/or Cycnus. The subspecies name *sineocellata* is derived from the Latin *sine* (without) and *ocellata* (eyes), in reference to the absence of a full complement of ocelli on the ventral hindwing (Austin and Emmel 1998a). The common name was first used by Pyle (1981) because the species' distribution is centred on the Great Basin area of the American Midwest.

**ADULT:** Great Basin Woodnymphs are similar to Common Woodnymphs in having two well-defined eyespots on the forewings, but the rear eyespot is smaller than the front eyespot. The difference in eyespot size is sometimes very small, and is most easily seen on the underside of the wings. The outer part of the ventral hindwings is much lighter than the inner half of the wings, with a strong dark line separating the two areas.

**IMMATURE STAGES:** Eggs are cream-coloured, and nearly spherical with light but regular sculpturing. Mature larvae are light green with a dark green dorsal stripe and whitish lateral stripes. The head and body are covered with fine white hairs. There are two reddish anal tails. Pupae are olive

♂ D  (4.3 CM)

♂ V  (4.3 CM)

♀ D  (4.5 CM)

♀ V  (4.5 CM)

green (Ferris and Brown 1981). The colour of the stripes is geographically variable, ranging from dark green to yellow or white (Emmel 1969).

**BIOLOGY:** Great Basin Woodnymphs are univoltine and fly in June and July at low elevations, and in August above timberline. There are five larval instars, and a female lays 100–150 eggs (Emmel 1969). Great Basin Woodnymphs are uncommon in BC. The larvae presumably feed on grasses, probably bunchgrass because it is the most abundant grass in their habitat. *Poa* species is reported as a larval foodplant by Ferris and Brown (1981).

**SUBSPECIES:** The subspecies found in BC is *C. s. sineocellata* Austin and Emmel, 1998 (TL: west side of Crump Lake, 8.2 rd. mi. north of Adel, Lake Co., OR).

**RANGE AND HABITAT:** Great Basin Woodnymphs occur in widely dispersed populations in dry low-elevation sagebrush, ponderosa pine, and Douglas-fir habitats in the southern and central interior. One population occurs above timberline in subalpine sagebrush grassland on Crater Mountain, near Keremeos. The three map records in Layberry et al. (1998) for the Kootenays are incorrect, and resulted from misidentifications of Small Woodnymphs.

**GENERAL DISTRIBUTION:** Great Basin Woodnymphs are found throughout most of the dry areas of western North America.

**CONSERVATION STATUS:** Not of concern (S4).

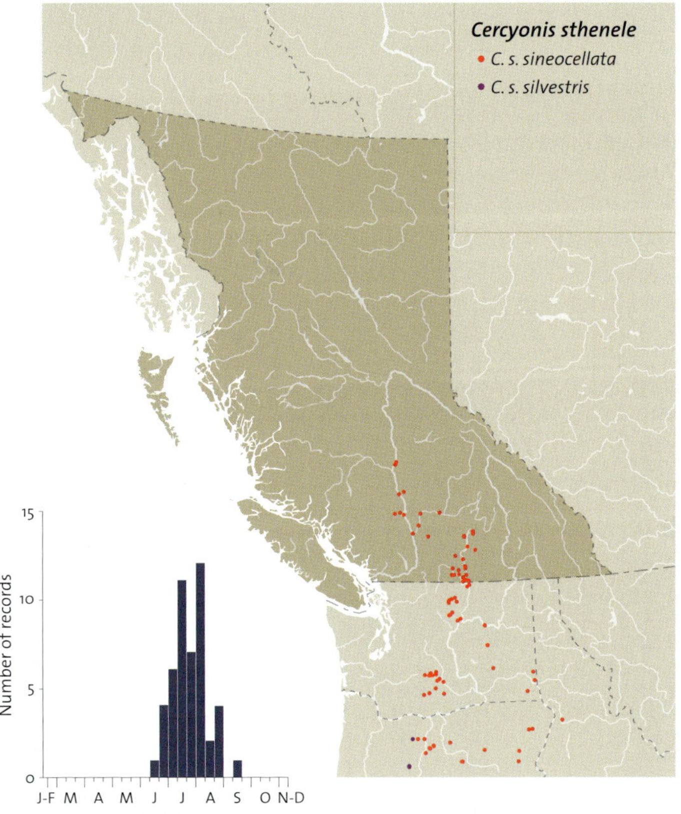

# Small Woodnymph

*Cercyonis oetus* (Boisduval, 1869)

**ETYMOLOGY:** The species name *oetus* is derived from the Greek *Oetes,* a mountain range in Thessaly where Hercules died. The subspecies name *phocus* refers to another part of Greece, Phocis, a region of central Greece. The common name was first used by Holland (1898) in reference to the fact that the adults are the smallest of the woodnymphs.

**ADULT:** Small Woodnymphs are small dark brown butterflies with small eyespots on the forewings. The front eyespot is distinctly larger than the rear one. The rear eyespot, and sometimes the front one as well, may be missing on the dorsal forewings. On the ventral forewings, there is no light band crossing the outer third of the wings, whereas in the other two woodnymphs there is always a definite but sometimes faint light band.

**IMMATURE STAGES:** For subspecies *charon* in Colorado, the eggs are lemon yellow, and barrel-shaped with 22 vertical ribs. First instar larvae are slender and taper towards the back, with two short "tails." The head is yellow brown, speckled with red brown, and the largest ocelli are emerald green. The body is pinkish yellow with red brown lines along the body, one down the middle of the back and four along each side. Mature larvae are cylindrical; they taper from the middle towards each end and have two short tails. The body is yellow green on the top half, more green on the bottom half. The tails are pale red, with yellow on the outer sides. Down the middle of the back is a dark green stripe bordered

Ssp. *phocus* ♂ D  (4.3 CM)

Ssp. *phocus* ♀ D  (4.2 CM)

Ssp. *phocus* ♂ V  (4.3 CM)

Ssp. *charon* ♂ V  (4.2 CM)

with yellow, and yellow lines on each side. The pupa is pale yellow green, thickly marked with whitish blotches on the underside. There is a whitish stripe down the middle of the back, and one on each side of the back. The wings have three streaks of darker green, and the edges of the wings, the top of the head, and the top of the thorax are lined with white. Pupal colour is highly variable in subspecies *charon;* others in the same batch were completely whitish green with no markings, greenish black with various markings, or dark brown with various markings (Edwards 1886b). The larvae and pupae are geographically variable in coloration (Emmel 1969).

**BIOLOGY:** Small Woodnymphs are univoltine and fly in June and July. There are five larval instars, and a female lays 100–150 eggs (Emmel 1969). Larvae hatch in the fall, and then hibernate without feeding. In spring they emerge from

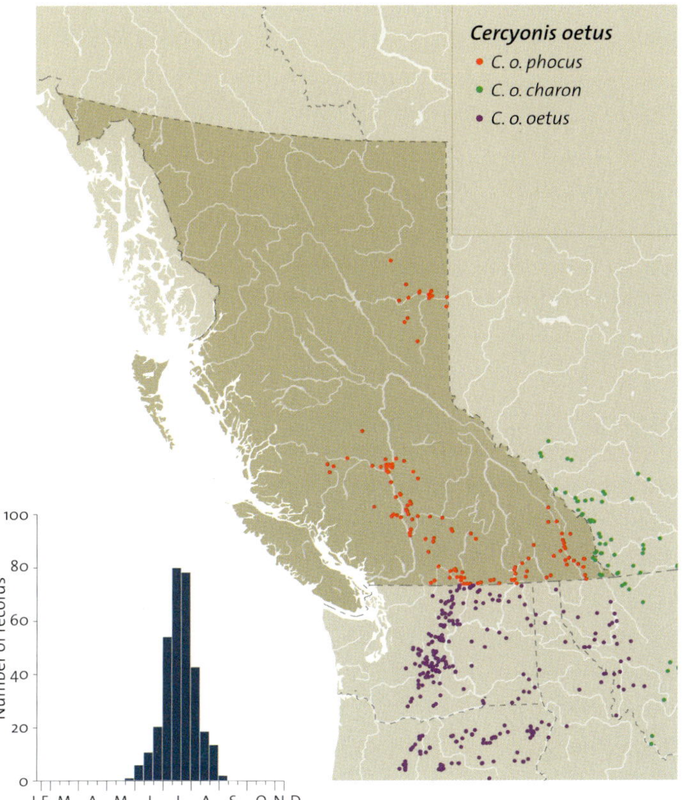

**Small Woodnymph (*Cercyones oetus phocus*)**

hibernation and begin feeding. Adults emerge about 10 weeks after the larvae come out of hibernation, and there is a prolonged emergence period spread over several weeks (Edwards 1886b).

Larvae feed on grasses, including bluegrass (*Poa*) species (Bird et al. 1995). Scott (1992) provides a list of oviposition substrates, but they have not been demonstrated to be larval foodplants.

**Subspecies:** Subspecies *phocus* (W.H. Edwards, 1874) (TL: Lac la Hache, BC) occurs in the Peace River, Chilcotin, northern Okanagan, Thompson, and Fraser drainages, and in the Kootenays and the Peace River. The ventral hindwings are smoothly dark brown, with the pattern present in subspecies *oetus;* TL: restricted to Mt. Judah, Placer Co., CA (Emmel et al. 1998a) only barely visible. There are a few small ventral submarginal spots. Intergrades with the nominate sub-

species occur in the Similkameen Valley and the southern Okanagan, with the ventral hindwing a more medium brown, with a contrasting darker band across the middle. There is a thin, dark brown line on each side of the dark band that is sometimes missing on the basal side. A few small submarginal spots are present. Subspecies *charon* (W.H. Edwards, 1872) (TL: near Twin Lakes, Lake Co., CO) occurs at Flathead, BC.

**Range and habitat:** Small Woodnymphs are found in dry grassland, sagebrush, and open woodland areas across southern and central BC east of the Coast Ranges. They also occur in the Peace River.

**General distribution:** Small Woodnymphs are found throughout the dry areas of western North America.

**Conservation status:** Subspecies *charon* is Endangered in BC (S1). Subspecies *phocus* is not of concern (S5).

---

## Genus *Erebia* Dalman, [1816] Alpines

The name *Erebia* is derived from the Greek Erebus, the region of darkness situated between earth and Hades (Reed 1871), in reference to the dark, dusky colour (Emmet 1991). The common name "alpines" was first used by Holland (1898) in reference to the alpine habitat of many species.

Alpines are medium-sized dark brown to black butterflies that have either submarginal eyespots or a red-flushed area on the forewings. In species with eyespots, there are usually orange-flushed areas around the spots. There are about 80 species worldwide, most of which are slow-flying.

The life histories of only some species are known. In these species, eggs are laid singly on leaves of grasses or sedges. They are white, cream, or yellow brown, and conical in shape with vertical ribs down the sides. First instar larvae are thinly covered with hairs, and are greenish with longitudinal stripes. Mature larvae are

slender, and yellow green with light and dark longitudinal stripes down the back and sides. They are thinly covered with hairs, and may have two short tails. Alpines hibernate as partly grown larvae, and there are five or six instars. Pupae are roughly cylindrical, rounded, and suspended from a cremaster. They are pale brown. All alpines have only one generation each year, and some may take two years to mature. *Erebia youngi* and *E. lafontainei* are occasionally difficult to separate reliably (worn specimens), in which case they can be distinguished by the shape of the valves of the male genitalia (Fig. 74).

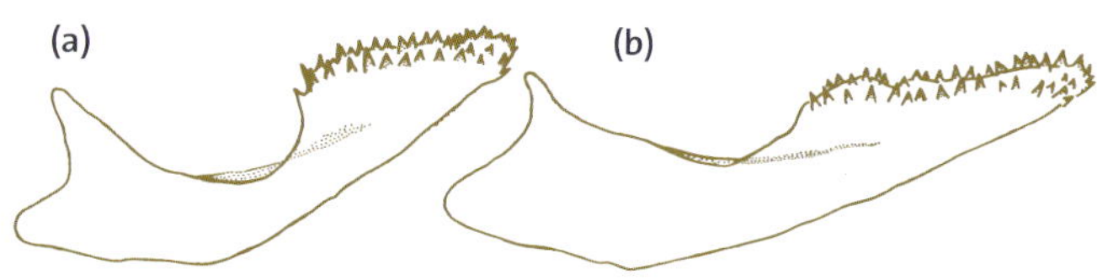


**74** Valves of *Erebia* species: (a) *E. youngi*, (b) *E. lafontainei*

## Vidler's Alpine
*Erebia vidleri* Elwes, 1898

**Etymology:** The species *vidleri* was named for Captain Vidler, who collected the type specimens. The common name was first used by Holland (1931).

**Adult:** Vidler's Alpines are easily identified by the sharply defined broad orange band surrounding the two or three eyespots on each forewing.

**Immature stages:** Undescribed.

**Biology:** Vidler's Alpines are univoltine, and are in flight in July and August. Larval foodplants have been reported by a number of authors as "probably grasses," but sedges are also possible. Pine grass

**Vidler's Alpine (*Erebia vidleri*)**

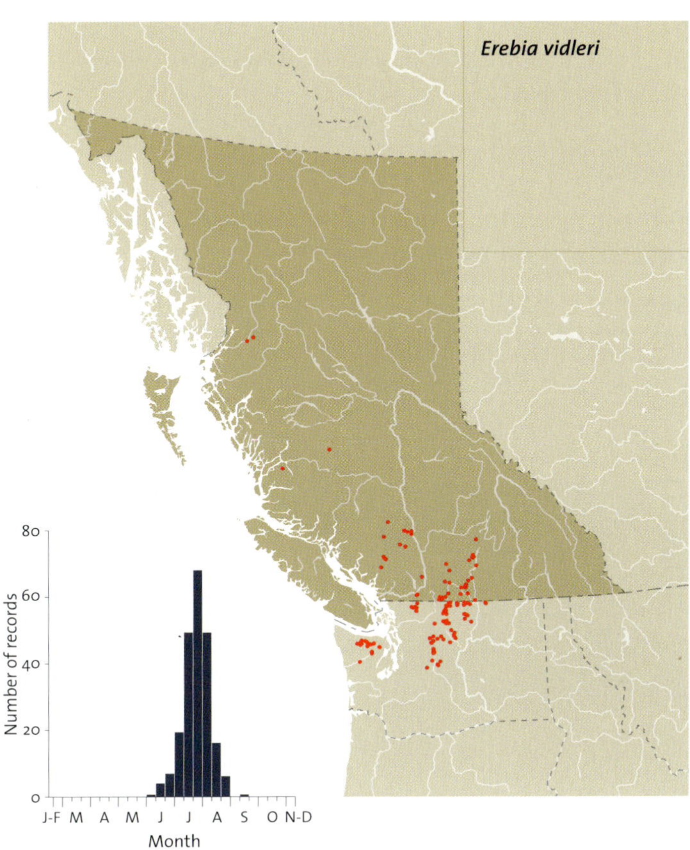

♂ D (4.6 cm)     ♂ V (4.6 cm)

(*Calamagrostis rubescens*) is generally associated with Vidler's Alpines, and may be the larval foodplant.

**SUBSPECIES:** None. The type locality of the species is the mountains above Seton Lake, near Lillooet, BC.

**RANGE AND HABITAT:** Vidler's Alpines inhabit open coniferous forests above 1,000 m elevation, as well as wet lower alpine meadows. They occur in the Cascades east to the western Okanagan, and in the Coast Range north to Mt. Hoadley, near New Aiyansh (Shepard and Shepard 1974).

**GENERAL DISTRIBUTION:** Vidler's Alpines are found only from the Olympic Peninsula and the northern Cascade Mountains of WA, north to New Aiyansh, BC, in the Coast Range. The entire world distribution is shown on the map.

**CONSERVATION STATUS:** Not of concern (S5).

## ROSS' ALPINE
*Erebia rossii* (Curtis, 1835)

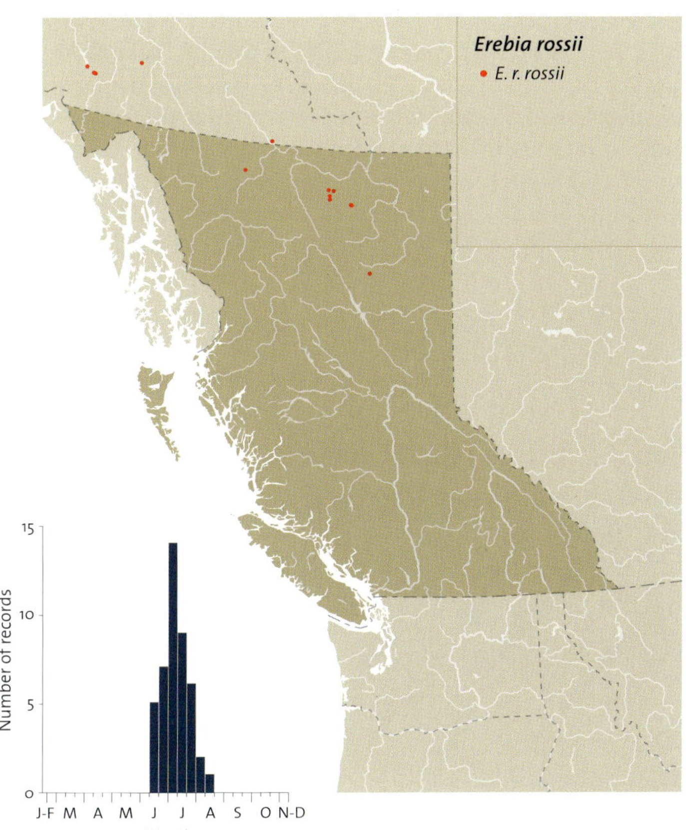

♂ D (4.4 cm)     ♂ V (4.4 cm)

**ETYMOLOGY:** The species *rossii* was named for arctic explorer Commander (later Captain) James Clark Ross (Curtis 1835). The types for Ross' Alpine were collected during the second voyage led by Captain John Ross to find the Northwest Passage.

**ADULT:** Ross' Alpines are entirely dark brown, appearing almost black. They have two (none or four occasionally) eyespots on each dorsal forewing, with orange surrounding the spots. The hindwings are without spots, and have faint bands on the underside.

**IMMATURE STAGES:** Undescribed.

**BIOLOGY:** Ross' Alpines are in flight in June and July. Freeman (1948) referred to them as "Bouncing Browns" because of their slow, bouncing flight across hummocky wet tundra. Oviposition on the sedges *Carex atrofusca* and *C. rariflora* has been observed at Churchill, MB (Scott 1986a). In BC these butterflies are most commonly seen along the Alaska Highway in Stone Mountain and Muncho Lake provincial parks, along road shoulders and rocky river banks.

**SUBSPECIES:** The nominate subspecies, *E. r. rossii* (Curtis, 1835) (TL: Boothia Peninsula, NU) occurs in BC. There is no difference between Ross' Alpines from the north coast of Canada and those in BC, although the subspecies name *gabrieli* dos Passos, 1949 (TL: Mt. McKinley National Park, AK) has been applied in the past.

**RANGE AND HABITAT:** Ross' Alpines inhabit wet subalpine or alpine tundra in northern BC.

**GENERAL DISTRIBUTION:** Ross' Alpines occur from Siberia across AK and northern CAN to Labrador.

**CONSERVATION STATUS:** Not of concern (S4).

## TAIGA ALPINE
*Erebia mancinus* Doubleday, [1849]

**ETYMOLOGY:** The name *mancinus* is derived from the Latin *mancus* (maimed, infirm, imperfect), perhaps referring to the butterfly's habit of dropping from flight as though injured (Bird et al. 1995). The common name was first used by Layberry et al. (1998), and refers to the butterfly's taiga and spruce bog habitat.

**ADULT:** Taiga Alpines are dark brown, appearing almost black. They have four eyespots on the dorsal forewing, surrounded by a diffuse orange area. There are no eyespots on the hindwings. The most characteristic feature is a small white spot near the centre of the ventral hindwing, with a second grey white spot usually present on the front edge.

Taiga Alpines have traditionally been considered a subspecies of the Disa Alpine (*Erebia disa*), which occurs from

♂ D  (4.6 CM)        ♂ V  (4.6 CM)

Scandinavia across the northern Asian and North American arctic as far east as Bathurst Inlet, NU. Taiga Alpines and Disa Alpines have different male genitalia, as described by Layberry et al. (1998) and shown in illustrations by Warren (1936). The saccus is a blunt triangle in *disa,* but is long, narrow and spikelike in *mancinus*. Taiga Alpines have larger submarginal spots, much more extensive red flushing around the spots, and a more even coloration of the ventral hindwing than Disa Alpines (Layberry et al. 1998). In addition, the third and fourth spots on the forewing of Disa Alpines are displaced outward relative to the front two spots, and usually are smaller than the front two spots. All four spots are usually roughly in line on Taiga Alpines, and the third and fourth spots are usually larger than the first two spots. The two species have been found flying together by Shepard at

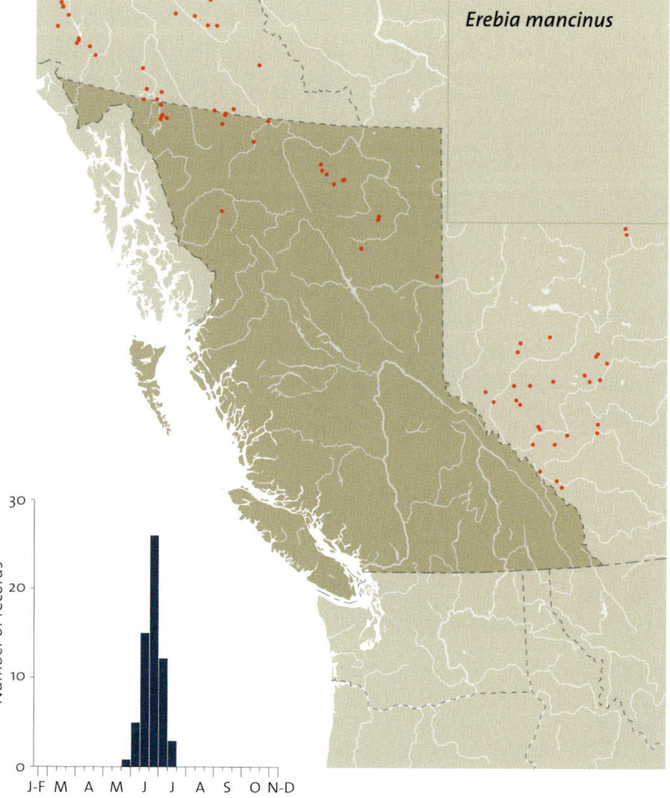

**Taiga Alpine (*Erebia mancinus*)**

several locations in the northern Northwest Territories, along the Dempster Highway, east of the Richardson Mountains (JHS).

**Immature stages:** Undescribed.

**Biology:** Taiga Alpines are in flight in June and July, and fly every year in BC. They appear to fly in alternate years in some areas, suggesting a two-year life cycle. Taiga Alpines in central Canada are usually restricted to bogs of black spruce and sphagnum, especially bogs with tall, dense stands of pure spruce. Strays occur in more open bogs or nearby road openings. There is a tendency to avoid bright sunlight, with most flight activity occurring before 11:00 AM and after 4:00 PM, or in the midday period on cloudy days. The butterflies fly slowly and steadily about 1 m off the ground straight through the spruce bog, and prefer to land in partial sunlight on low foliage near the base of a spruce tree. Jutta Arctics are usually found associated with Taiga Alpines. Each

population of Taiga Alpines is apparently biennial, but populations in a given area alternate randomly with each other in flying in odd or even years (Masters 1969). The biology of Taiga Alpines has not been studied in BC, but is apparently similar to that in central Canada. Larval foodplants are unknown, but are likely grasses or sedges.

**Subspecies:** None. The type locality of the species is Rock Lake, near Jasper, AB (Shepard 1984).

**Range and habitat:** Taiga Alpines inhabit spruce bogs across northern BC.

**General distribution:** Taiga Alpines occur from central AK and northern YT south to northern BC, east to Labrador and to MN in the USA. They occur in most of the boreal forests of North America, as well as wet tundra in the northern YT.

**Conservation status:** Not of concern (S5).

## Magdalena Alpine
*Erebia magdalena* Strecker, 1880

**Etymology:** The derivation of the name *magdalena* is unknown, with the association with Biblical Mary Magdalene suggested by Bird et al. (1995) being improbable. The subspecies name *saxicola* is derived from the Latin word *saxum* (rock) and suffix *icola* (inhabitant), in reference to the butterfly's habitation of alpine rock areas (Hilchie 1990). The common name was first used by Holland (1898).

♂ D Paratype (4.4 cm)

♀ D (4.8 cm)

♂ V Paratype (4.4 cm)

**Adult:** Magdalena Alpines are larger than any other alpine except the Mt. McKinley Alpine, and are a strikingly dark brown to black. There are no markings on either the dorsal or ventral wing surfaces. Males and females have scattered white hairs on the ventral wing surface, especially near the tip of the forewing. Antennae are ringed with grey white, with the club ferruginous above and black below. The uncus of the male genitalia has spines along the outer edge only (Hilchie 1990).

**Immature stages:** Eggs are cream-coloured, oval, and taller than wide, and have rounded ridges down the sides. First instar larvae are cream to light green with no dark maculations and only a few very fine hairs. Second instar larvae are green, with a brown head capsule. Third to fifth instar larvae have dark brown heads. The body is green, mottled with black maculations. They are sluglike in shape and covered with fine blunt hairs. The pupa has an olive green to brown

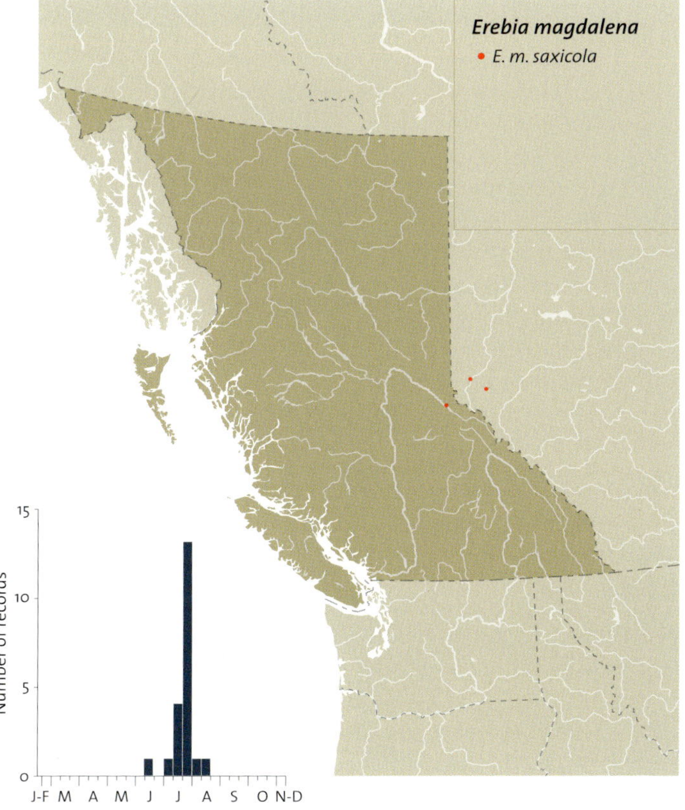

head and thorax; the abdomen is medium brown tinged with green (Hilchie 1990).

**BIOLOGY:** Magdalena Alpines are in flight in July, and have a one-year life cycle. They inhabit boulder fields on high alpine mountain summits. The boulders are covered with black lichens, with which the black adults blend when resting. The boulder fields include small patches of soil in the centre of polygon frost formations. These patches are thinly vegetated with alpine grasses, dwarf willows, and herbaceous flowering plants. Females have a wandering flight path over the rocks, and periodically land to bask or nectar. Males have a more direct flight pattern, and chase any dark butterflies flying near them. Oviposition occurs in areas of loose rock covered by black lichen, without reference to the presence of potential foodplants. Eggs are laid under the edges of boulders. Larvae probably hibernate in the second or third instar (Hilchie 1990).

Natural larval foodplants are unknown but are probably most grasses and sedges found within the species' habitat.

The nominate subspecies has been reared in captivity on the grass *Poa pratensis,* and oviposition has been observed in Colorado on the sedge *Carex* (Scott and Scott 1980). Hilchie (1990) reared subspecies *saxicola* in captivity on lawn grass, and the larvae also accepted barley.

**SUBSPECIES:** Magdalena Alpines of BC and AB belong to the subspecies *saxicola* Hilchie, 1990 (TL: Adams Lookout, Wilmore Wilderness Park, AB). The nominate subspecies occurs in the central Rockies in the United States.

**RANGE AND HABITAT:** Magdalena Alpines were found in BC near McBride by David Threatful in 1984. They inhabit barren rock rubble slopes, where their wings blend perfectly with the black lichen on the rocks.

**GENERAL DISTRIBUTION:** Magdalena Alpines occur in the Rocky Mountains of BC and AB, and also occur in scattered populations in the American Rocky Mountains, south to CO.

**CONSERVATION STATUS:** Magdalena Alpines are of Special Concern in BC (S3).

## MT. MCKINLEY ALPINE
*Erebia mackinleyensis* Gunder, 1932

**ETYMOLOGY:** The species *mackinleyensis* was named for the type locality of Mt. McKinley National Park, AK. "Mt. McKinley Alpine" was first used by Tilden and Smith (1986).

**ADULT:** Mt. McKinley Alpines are larger than any other alpine except the Magdalena Alpine, and are a strikingly dark

♂ D  (4.9 CM)

♀ D  (4.6 CM)

♂ V  (4.9 CM)

brown to black. There is usually at least a trace of a red flush on the dorsal wings, which is very prominent in the northern Yukon. Males and females lack white hairs on the ventral wing surface. Antennae clubs are testaceous to two-toned yellow. In the uncus of the male genitalia, the spination is expanded onto the inner surface (Hilchie 1990).

**IMMATURE STAGES:** Eggs are cream-coloured, oval, and taller than wide, and have rounded ridges down the sides. Larvae are indistinguishable from those of the Magdalena Alpine (Hilchie 1990).

**BIOLOGY:** Mt. McKinley Alpines are in flight in June and July, and have a one-year life cycle. They inhabit boulder fields on high alpine mountain summits and slopes. The boulders are covered with black lichens, with which the black adults blend when resting. Natural larval foodplants are unknown, but are

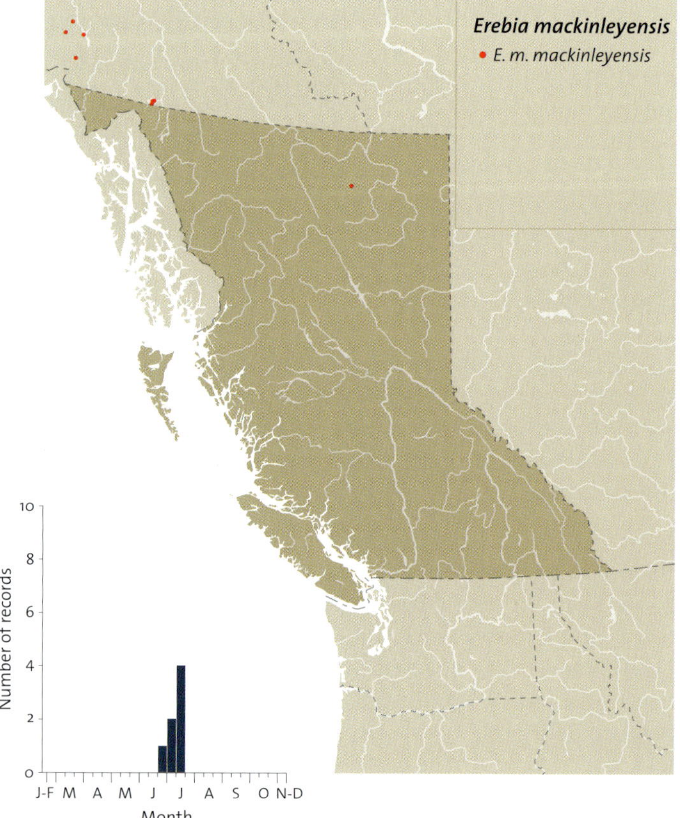

probably most grasses and sedges found within the species' habitat.

**Subspecies:** The type locality is Sable Pass, Mt. McKinley National Park, AK. The taxon *mackinleyensis* was described on 31 December 1932, but *erinnyn* has date priority by two weeks and thus apparently is the correct species name. However, as discussed by Ferris (1989), *erinnyn* is the accusative case of *erynnis* and therefore cannot be used (ICZN Article 11.9.1). The first available name is therefore *E. mackinleyensis*.

**Range and habitat:** Mt. McKinley Alpines are known only from Summit Lake in BC, but they also occur on Montana Mountain, YT, just north of the northwestern BC border. They are likely to occur in the mountains across northern BC, on barren rock rubble slopes and peaks.

Hatched egg

First instar larva

**General distribution:** Mt. McKinley Alpines occur from Siberia east across AK to the YT/NT border, and south to northern BC.

**Conservation status:** Mt. McKinley Alpines are of Special Concern in BC (S3).

## Banded Alpine
*Erebia fasciata* Butler, 1868

**Etymology:** The species name *fasciata* is derived from the Latin *fascia* (band), in reference to the banded pattern on the ventral hindwing. The common name was first used by Holland (1931) in reference to the ventral hindwing pattern.

**Adult:** Banded Alpines are similar to Mt. McKinley Alpines in having black dorsal wings with a broad red flush. The ventral wings are crossed by a series of well-defined bands, with two light and two dark bands alternating. There are no eyespots.

♂ D (4.2 cm)

♀ D (4.5 cm)

♂ V (4.2 cm)

**Immature stages:** Mature larvae are similar to those of Magdalena Alpines, but the pattern of mottling is different (Layberry et al. 1998).

**Biology:** Banded Alpines are univoltine, and are in flight in June and July. They fly with a slow, bouncing flight pattern. They are found in association with cotton grass (*Eriophorum*), which may be their larval foodplant. In Asia *Carex* sedges are used (Tuzov 1997).

**Subspecies:** The nominate subspecies (TL: "Arctic America") occurs in BC.

**Range and habitat:** Banded Alpines occur on Montana Mountain, YT, just north of the BC border. They should occur in the mountains near Atlin, west to the St. Elias Mountains, in wet cotton grass tundra.

**General distribution:** Banded Alpines occur from eastern Siberia across arctic and subarctic North America to Hudson Bay.

**Conservation status:** Not yet known from BC.

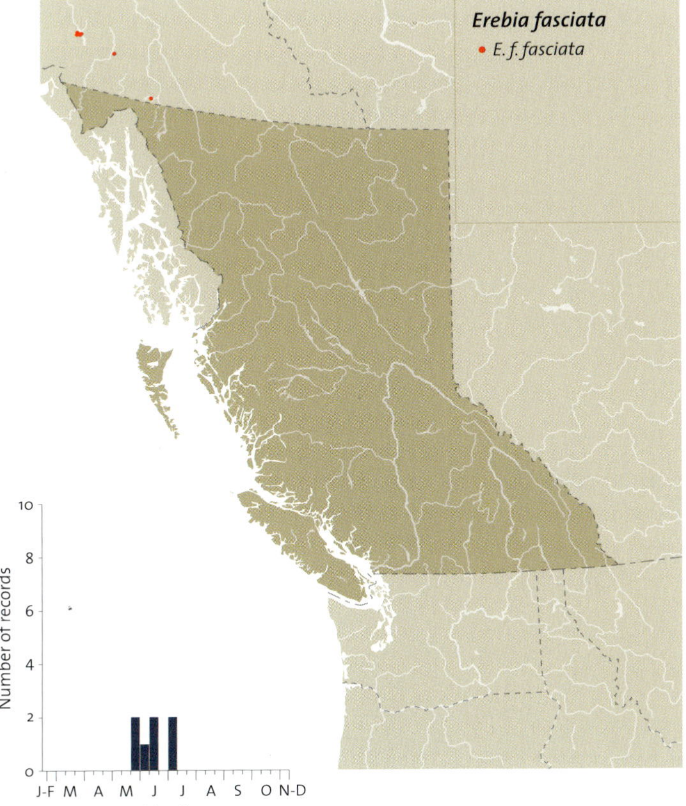

# Red-disked Alpine

*Erebia discoidalis* (W. Kirby, 1837)

**Etymology:** The species name *discoidalis* is derived from the Greek *diskos* (disc) and the suffix *oid* (resembling) (Bird et al. 1995), in reference to the red disc-shaped patch on the forewing. The common name Red-disked Alpine was first used by Holland (1931) in reference to the central area (disc) of the forewing being reddish. Holland, being American, used "disk" rather than "disc," which has been retained for the common name.

**Adult:** Red-disked Alpines are easily identified by the lack of eyespots combined with the red central area of the dorsal forewings. The ventral hindwings are mottled grey, especially in the outer area.

**Immature stages:** Eggs are barrel-shaped. Mature larvae are cream-coloured and covered with short hairs. A dark diagonal stripe is present on most segments, starting dorsolaterally on the front edge of the segment and running back and down to the back edge of the segment (Bird et al. 1995).

**Biology:** Red-disked Alpines are in flight in May and June, and are among the first butterflies in flight in spring. In the north-central USA, they prefer to fly either quite early or quite late in the day, but fly in the midday period in cool or partly cloudy weather. They fly slowly, weakly, and close to the ground. When attacked, they either rise into the air to be carried away by the wind or work their way down into the vegetation to hide. They appear to have a single-year life cycle (Masters 1971). Mature larvae spin flimsy cocoons before pupating (Kondla et al. 1994). Larval foodplants are probably grasses. At The Pas, MB, the non-native grass *Poa lucida* is used as the larval foodplant (Krivda 1968).

**Subspecies:** Only the nominate subspecies occurs in North America. Subspecies *mcdunnoughi* dos Passos, 1940 (TL: Whitehorse, YT) is a synonym of subspecies *discoidalis* W. Kirby, 1837 (TL: Cumberland House, SK). The amount of red on the forewing (the purported difference) and all other characters have the same range of variation for populations from the Canadian Prairies west to Atlin and the southern YT, and north to the Dempster Highway.

**Range and habitat:** Red-disked Alpines in BC are known only from a few localities in the north. They inhabit wet or dry grassy areas in forest openings, meadows, and alpine tundra in BC and adjacent areas, but in the north-central USA they inhabit bogs and occasionally other areas of damp acid soil (Masters 1971).

**General distribution:** Red-disked Alpines occur in Asia and from central AK across boreal Canada to PQ.

**Conservation status:** Red-disked Alpines are of Special Concern in BC because they may not occur between the known northwestern and northeastern populations (S3S4).

♂ D  (4.4 cm)

♀ D  (4.0 cm)

♂ V  (4.4 cm)

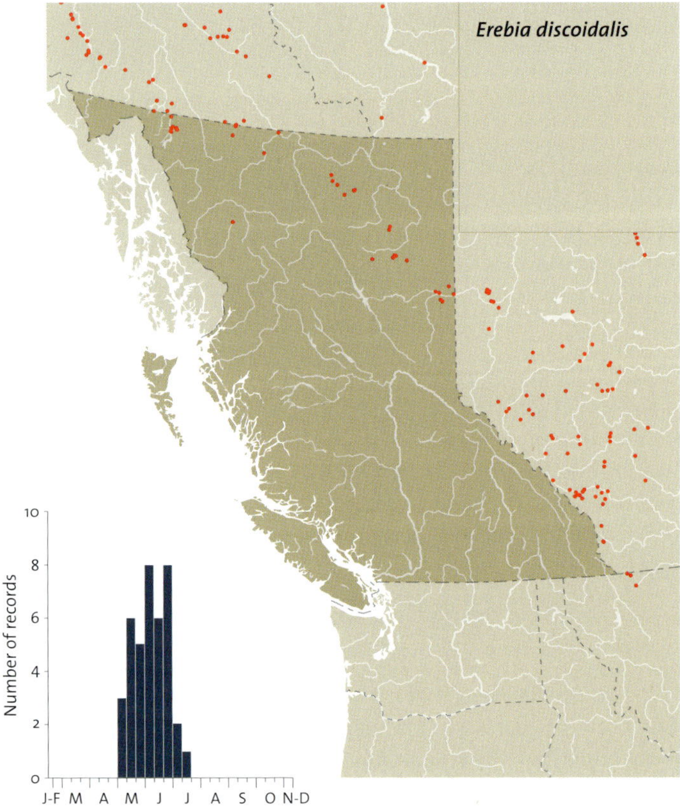

# Mountain Alpine

*Erebia pawlowskii* Ménétriés, 1859

**Etymology:** The species *pawlowskii* was named for M. Pavlovskiy, a Russian traveller who collected insects in Yakutia in the mid-19th century (Korshunov and Gorbunov 1995). The letters "v" and "w" are frequently interchanged when translating from Russian to English, hence the difference in spelling between the Latin name of the species and the name of the person. Subspecies *alaskensis* was named for the Alaskan type locality. The common name is used for the first time here, in reference to the species' mountain habitats.

**Adult:** Mountain Alpines are small butterflies with a row of spots across both the forewings and the hindwings that are red on the upperside and cream or pale yellow on the underside of the wings. This species has traditionally been called *Erebia theano* (Tauscher, 1806) in North America; however, *E. theano* is a more strongly marked species restricted to the mountains of eastern Siberia and Mongolia (Korshunov and Gorbunov 1995; Tuzov 1997).

**Immature stages:** In Colorado, subspecies *ethela* eggs are weakly ribbed and are cream-coloured with red brown spots (Bird et al. 1995). First instar larvae are cream with faint longitudinal lines (Scott 1992). Mature larvae are tan with a

♂ D (3.4 cm)   ♀ D (3.6 cm)

♂ V (3.4 cm)

dark brown dorsal longitudinal stripe, and three dark brown stripes on each side. There are two bumps at the hind end, covered in hairs (Bird et al. 1995).

**Biology:** Mountain Alpines are univoltine. They have a two-year life cycle and are in flight in July. In many Colorado populations, Mountain Alpine adults fly only every second year (Ferris and Brown 1981), mostly during even years but in some locations during odd years. Young larvae probably hibernate the first year, and mature larvae the second year (Scott 1992). Adults are very weak fliers, and are easily observed once found. The larval foodplants are unknown, but eggs are laid on dead grass or sedge leaves, and larvae eat *Poa pratensis* in the lab (Scott 1992).

**Subspecies:** The nominate subspecies was described from the Great Sibagly River basin, central Yakutia, Russia. The subspecies found in BC is *E. p. alaskensis* Holland, 1900 (TL: Eagle City and American Creek, AK).

**Range and habitat:** In BC Mountain Alpines are known only from Stone Mountain Provincial Park. They inhabit subalpine wet grassy meadows and bogs. Outside the arctic, their colonies are usually very small in area and widely scattered. This, combined with their flight in alternate years and their weak flight pattern, makes detection of populations very difficult.

**General distribution:** Mountain Alpines occur from the Sayan Mountains and northern Mongolia in Asia east across Siberia, AK, and YT to Churchill, MB, and in scattered colonies down the Rocky Mountains to CO.

**Conservation status:** Mountain Alpines are of Special Concern in BC (S3).

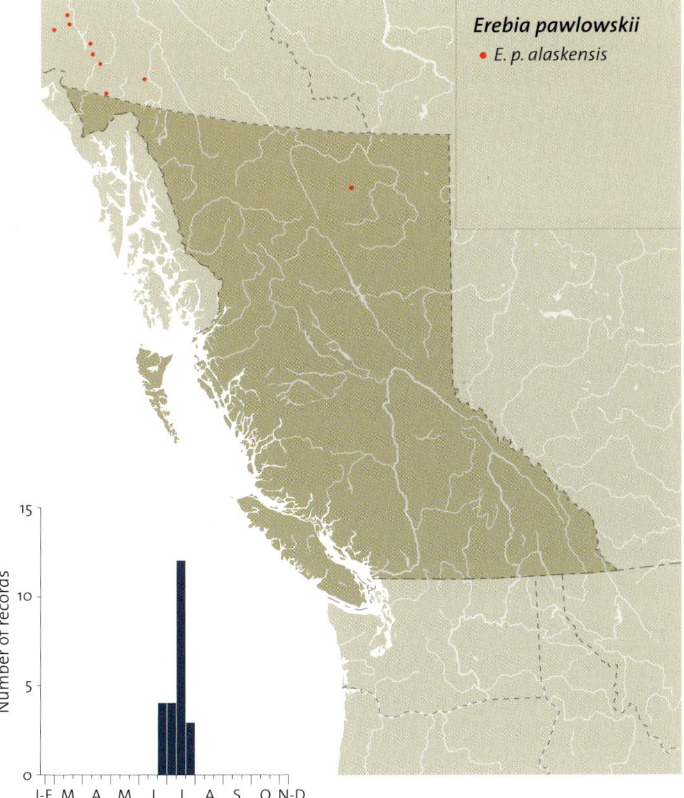

## YOUNG'S ALPINE

*Erebia youngi* Holland, 1900

**ETYMOLOGY:** The species *youngi* was named for the collector of the type specimens, Dr. S. Hall Young (Holland 1931). The common name was first used by Holland (1931).

**ADULT:** Young's Alpines are small dark brown butterflies with four (occasionally five) eyespots on the dorsal forewings that increase in size towards the back. The spots are found within diffuse, dull orange patches, the first two of which are usually fused. On the ventral forewing there is a broad, well-developed orange band surrounding comparatively large black spots. The dorsal hindwing has two to four small eyespots. The ventral hindwing is blackish brown with reddish hairs. A grey band across the ventral hindwing is speckled with black scales, producing a dark and uneven-appearing grey, and contains two to four small eyespots.

**IMMATURE STAGES:** Undescribed.

♂ D  (4.1 cm)

♀ D  (3.9 cm)

♂ V  (4.1 cm)

**BIOLOGY:** Young's Alpines are univoltine and fly in June and early July. The foodplants are unknown.

**SUBSPECIES:** The nominate subspecies, *E. y. youngi* Holland, 1900 (TL: mountains between Fortymile and Mission Creeks, AK), should occur in BC.

**RANGE AND HABITAT:** Young's Alpines are not yet known in BC, but they occur on Montana Mountain, YT, just north of the BC border. They should occur in the mountains near Atlin, west to the St. Elias Mountains, in dry short-grass alpine tundra and scree.

**GENERAL DISTRIBUTION:** Young's Alpines are found in the mountains of northern and eastern AK, western YT, and extreme northwestern NT.

**CONSERVATION STATUS:** Not yet known from BC.

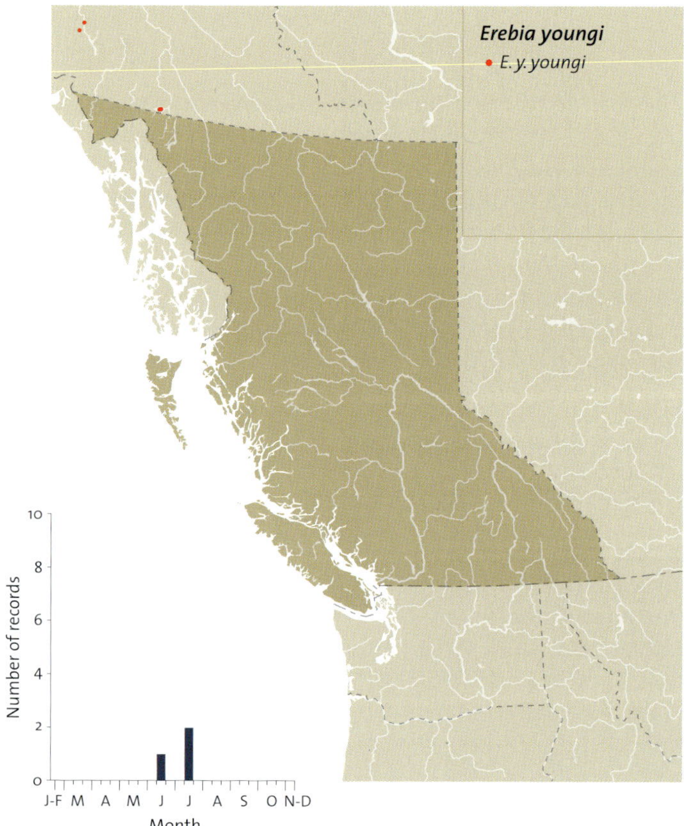

**Young's Alpine (*Erebia youngi*)**

## LAFONTAINE'S ALPINE

*Erebia lafontainei* Troubridge & Philip, 1983

**ETYMOLOGY:** The species *lafontainei* was named for Dr. J. Donald Lafontaine, a prominent lepidopterist currently doing research at the Canadian National Collection of Insects, Arachnids and Nematodes in Ottawa. The common name is used for the first time here.

**ADULT:** Lafontaine's Alpines are small dark brown butterflies with four eyespots on the dorsal forewings, usually increasing in size from front to back. The eyespots are found within well-defined and separated orange patches. The orange band on the ventral forewing is paler than that of Young's Alpines, and is broken into discrete spots. There are one to three (occasionally none or four) small orange spots on the dorsal hindwings that are always missing from the ventral hindwings. The ventral hindwing is reddish brown, with a grey brown band with a pinkish sheen crossing the wing.

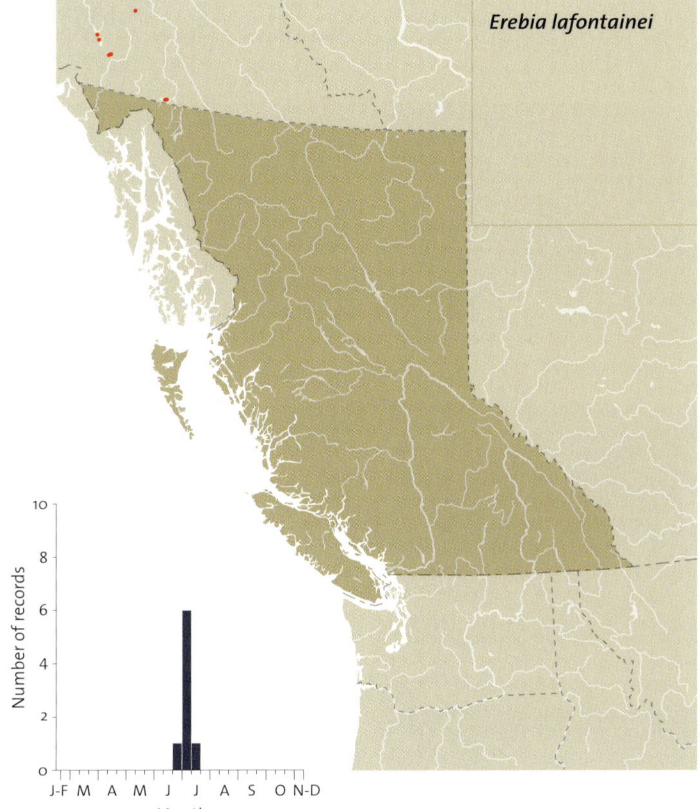

♂ D  (3.8 cm)

♀ D  (4.0 cm)

♂ V  (3.8 cm)

**IMMATURE STAGES:** Undescribed.

**BIOLOGY:** Lafontaine's Alpines are univoltine and fly in June and early July. The foodplants are unknown.

**SUBSPECIES:** None. The type locality of the species is Mt. Decoeli, St. Elias Mountains, YT.

**RANGE AND HABITAT:** Lafontaine's Alpines have not yet been found in BC, but have been found on adjacent Montana Mountain, YT, and should occur in extreme northwestern BC. They inhabit open shrub alpine tundra, generally at slightly lower elevations than the Young's Alpine.

**GENERAL DISTRIBUTION:** Lafontaine's Alpines are found in the mountains across northern AK and YT south to Montana Mountain, YT.

**CONSERVATION STATUS:** Not yet known from BC.

## COMMON ALPINE
*Erebia epipsodea* Butler, 1868

**ETYMOLOGY:** The name *epipsodea* is derived from the Greek "epi" meaning "towards" and *psodea*, the name widely used in 1868 for what is now called *Erebia medusa*, a European species closely related to *E. epipsodea* (Belik 2000). The subspecies name *sineocellata* is derived from the Latin *sine* (without) and *ocellata* (eyes), in reference to the absence of white spots in the centre of the black submarginal spots in the forewing of the type specimen, making it "without eyes" (Skinner 1889). Subspecies *remingtoni* was named for the prominent entomologist Dr. Charles Remington of Yale University, CT. The common name was first used by Holland (1898) in reference to the species' wide distribution and abundance.

**ADULT:** Common Alpines have two to four large orange patches on each dorsal forewing, which are frequently fused. Within the orange patches are up to four black spots, frequently with white centres. They are the only alpine with well-developed eyespots on the hindwings. In all populations there is a great deal of variation in the number and size of dorsal and ventral submarginal eyespots on both the forewings and hindwings. In all subspecies, specimens from alpine populations are smaller and the eyespots are reduced or absent.

**IMMATURE STAGES:** Immature stages of the nominate subspecies, *epipsodea*, were described by Edwards (1887–97), and of subspecies *sineocellata* by Lyman (1896). Eggs are round,

Egg

First instar larva

slightly taller than wide, with 22 vertical ribs, and white in colour. First instar larvae have large pale brownish heads, pitted with dark depressions and with brown spots. There is a brown stripe down the back, and the body is green white (*epipsodea*) to pale tan (*sineocellata*), with three (*epipsodea*) to five (*sineocellata*) brown stripes on each side. Dorsal, subdorsal, and lateral rows of blackish tubercles run along the body, and there are two short tails at right angles to each other. Mature larvae are greenish brown (*sineocellata*) to delicate yellow green with a greener underside (*epipsodea*). There is a dark brown stripe down the back bordered by a creamy yellow stripe on each side. Halfway down each side there is a creamy yellow stripe, with a brown line along the lower edge, and another creamy yellow line along the side just above the spiracles (*sineocellata*). The basal ridge is greenish (*epipsodea*). There are two short tails at right angles to each other. The head is pale yellow brown and covered with brown pits. Pupae are pale brown with dark brown markings and orange spiracles (*sineocellata*), or white brown with brown yellow and brown markings, 11–12 black longitudinal streaks on the wings, brown lines at the junction of the abdominal segments, and dorsal and lateral lines of black dots on the abdomen (*epipsodea*).

**BIOLOGY:** Common Alpines are in flight in June at low elevations and in July and August at high elevations. They have a one-year life cycle, and hibernate as third instar larvae (Lyman 1896). They are usually found in association with tall green grass, indicating moister sites. Eggs are laid on or near grass (Scott 1992). In captivity they sometimes produce

SSP. *epipsodea* ♂ D  (4.3 CM)

SSP. *epipsodea* ♀ D  (4.2 CM)

SSP. *epipsodea* ♂ V  (4.3 CM)

SSP. *sineocellata* ♂ V  (4.1 CM)

SSP. *sineocellata* ♂ D  (4.1 CM)

SSP. *sineocellata* ♀ D  (4.3 CM)

SSP. *remingtoni* ♂ D  (4.1 CM)

SSP. *remingtoni* ♀ D  (4.1 CM)

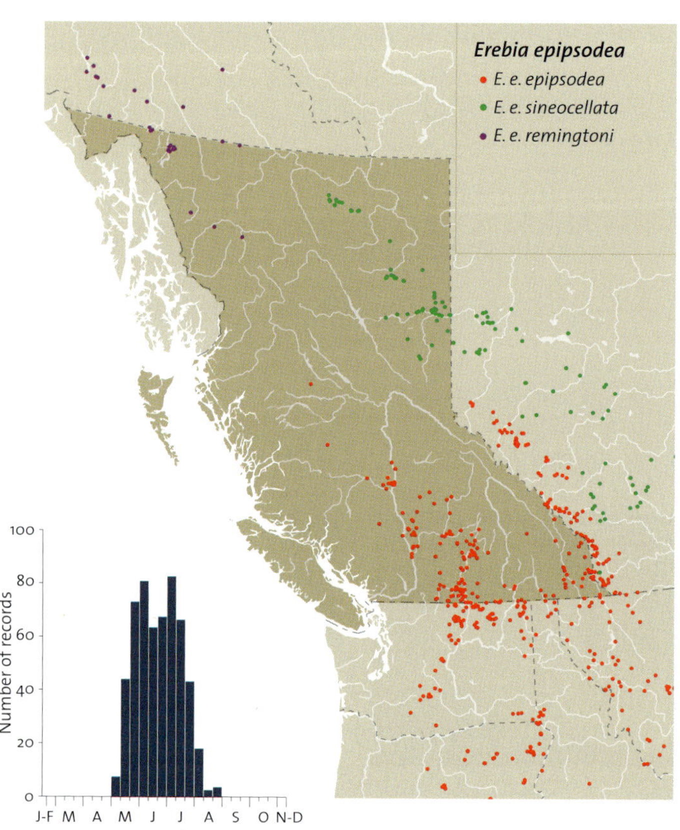

SSP. *remingtoni* ♂ V  (4.1 CM)

adults without hibernation, in which case there are only four instars. If they hibernate, there are five instars (Lyman 1896). Common Alpines are widespread and frequently numerous.

Oviposition on "grass" has been observed in the Southern Interior of BC by CSG. They feed on the grass *Poa pratensis* in captivity, and oviposit on many grasses, sedges, and other substrates (Scott 1992).

**SUBSPECIES:** The nominate subspecies, TL: Rock Lake, near Jasper, AB (Kondla 1996), occurs throughout southern and central BC. The dorsal orange patches are dull orange with poorly defined boundaries. The ventral hindwing of males is generally evenly dark brown, or with the outer third only slightly lighter than the rest; spots are rare. At high elevations the markings are reduced in size. Subspecies *sineocellata* Skinner, 1889 (TL: near Fort Qu'appelle, SK) is a prairie subspecies occurring in the Peace River of BC. The orange patches on the upperside of the wings are bright orange with well-defined boundaries. In the ventral hindwing of males, the outer third is much greyer than the rest, and spots are usually present. Subspecies *remingtoni* P. Ehrlich, 1952 (TL: Dawson, YT) occurs in northwestern BC. The dorsal orange patches are well developed, with moderately well defined boundaries. The spots within the orange patches are small, with the hindwing frequently missing all spots and the forewing normally with only two to three spots. The ventral ground colour is relatively even. Subspecies *freemani* Ehrlich, 1954; TL: restricted to the vicinity of Rivercourse, AB (Kondla 1996), is a synonym of subspecies *sineocellata*. Contrary to the opinion of Ehrlich (1955) and others, Skinner described *sineocellata* as a geographic variety, not as an aberration, and therefore Ehrlich's replacement name was unnecessary. Subspecies *hopfingeri* Ehrlich, 1954 (TL: Black Canyon, Okanogan Co., WA) is a synonym of the nominate subspecies. The *hopfingeri* phenotype is the low-elevation form of *E. e. epipsodea*.

**RANGE AND HABITAT:** Common Alpines occur in both southern and northern BC. They inhabit moist grassy fields, ditches, low-elevation grasslands, and subalpine and alpine meadows.

**GENERAL DISTRIBUTION:** Common Alpines occur from central AK south through BC, AB, and SK, and south to OR and NM.

**CONSERVATION STATUS:** Not of concern, with subspecies *remingtoni* and *sineocellata* both S4, and *epipsodea* S5.

## GENUS *OENEIS* HÜBNER, [1819]  ARCTICS

The name *Oeneis* refers to Oeneus, king of the ancient city of Calydon in western Greece, husband of Althaea and father of Meleagr and Tydeus. The name of the European genus *Melanargia* is derived from Meleagr, and another species of Satyrinae was derived from Tydeus. The common name "arctics" was first used by Holland (1898) in reference to the arctic and alpine distribution of many species.

Arctics are medium-sized brown or grey butterflies. They usually have eyespots on the wings. They fly rapidly and erratically over short distances, and then drop suddenly to the ground or onto a tree trunk. Arctics all have a two-year life cycle, with the young larvae hibernating the first winter and the almost mature larvae hibernating the second winter. The two-year life cycle results in many species having adults in flight only every second year, with butterflies in alternate years being greatly reduced in abundance or missing entirely in some or all areas.

Eggs are white or off-white in colour, and are conical in shape, with vertical ribs down the side. First instar larvae are thinly covered with hairs, and are tan or greenish. Mature larvae are slender and are tan or greenish with longitudinal stripes of various colours down the back and sides. They are thinly covered with hairs that are frequently reddish in colour. Pupae are roughly cylindrical and rounded, and have brown, yellow brown, and olive markings. Descriptions of the immature stages are all from outside BC, with the exception of the Great Arctic.

Larval foodplants are usually grasses and sedges. One species, the Jutta Arctic, also feeds on rushes. Eggs are laid singly on leaves of the foodplant, or nearby on dead leaves or debris. The foodplants naturally utilized in BC are not known for any species; the little information that is available is from Manitoba, Alberta, or the American Rocky Mountains.

Arctics fall into three basic ecological groups (Masters 1969): forest-dwelling species (*macounii, nevadensis, jutta*); prairie and steppe species (*uhleri, chryxus, alberta*); and arctic taiga-tundra/alpine summit species (*bore, melissa, polixenes*). *Oeneis bore* and *polixenes* can sometimes be difficult to identify by wing pattern alone, but the valves of the male genitalia are distinctly different (Fig. 75). *Oeneis rosovi* is also difficult to distinguish from *O. polixenes,* but there are no genitalic differences between the two species.

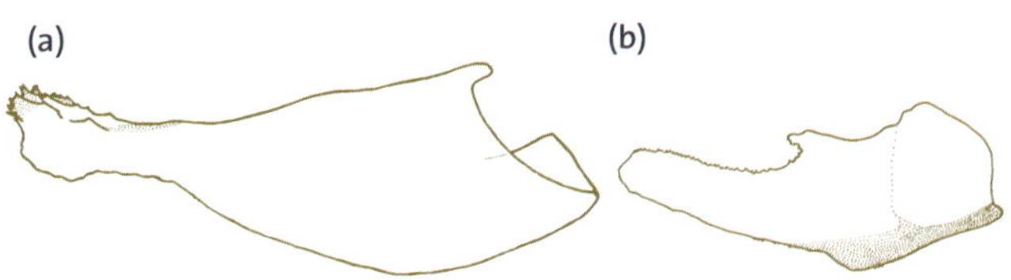


**75** Valves of *Oeneis* species: (a) *O. polixenes*, (b) *O. bore*

# Great Arctic

*Oeneis nevadensis* (C. & R. Felder, [1866])

**Etymology:** The species *nevadensis* was named for the type locality, Sierra Nevada, CA. The subspecies name *gigas* is Greek for giant, referring to the subspecies' large size. The common name "Great Arctic" was used for the first time by Holland (1989), in reference to the fact that this species is the largest arctic.

**Adult:** Great Arctics are large, golden brown butterflies with dark wing borders. They have two eyespots on the dorsal forewing and one on the dorsal hindwing. The ventral hindwings are mottled grey and brown, without a distinct band crossing the wing. Males have a large dark stigma, a patch of sex pheromone scales, covering the forewing discal cell. There are white patches along the margin of the hindwings, which gives a scalloped appearance to the wing edge. In females the dorsal forewing is heavily dusted with grey scales out to at least half, and frequently three-quarters, of the discal cell.

**Immature stages:** Eggs from Mt. Finlayson, near Goldstream on Vancouver Island, are subconical and have 18–19 vertical ribs. First instar larvae are initially pale red grey and then turn white green, and have two short tails. There is a narrow red brown dorsal and subdorsal line, and a broad red brown lateral band. The basal ridge is lighter than the ground colour, and there is a fine brown line under it. The head, underside, legs, and prolegs are green yellow with a brown tint (Edwards 1887–97). Mature larvae of the nominate subspecies in California are buff brown. The dorsal stripe is

♂ D  (5.7 cm)  ♀ D  (5.9 cm)

♂ V  (5.7 cm)

black, below which is a whitish stripe shading outward to brown and streaked longitudinally with dark brown and black. The subdorsal line is black, and below it the ground colour is buff with black speckles. The lateral band is deep black, the spiracular band green buff, the basal ridge yellow white, and the underside, legs, and prolegs brown buff (Edwards 1887–97).

**Biology:** Great Arctics fly in June and July at low to middle elevations, and in August in subalpine habitats. Masters (1974) stated that they fly only in even-numbered years throughout their distribution. On Vancouver Island, however, they fly in both odd and even years. At any one site on Vancouver Island, they tend to be much more abundant in alternate years, but the greater abundance may occur in either odd or even years. Great Arctics were common in the Goldstream area in both 1891 and 1893 (1892 was too rainy), when C. de Blois Green and W.G. Wright collected them (Edwards 1887–97). In Oregon several colonies of subspecies *nevadensis* fly in both odd and even years (Newcomer 1964d). Natural larval foodplants are unknown, but are probably grasses, on which the larvae feed in captivity.

**Subspecies:** The subspecies of Great Arctic that occurs in BC is *gigas* (Butler, 1868), which was described from southern Vancouver Island. The mainland populations are slightly different from *gigas*, but are closer to that subspecies than to any other named subspecies.

**Range and habitat:** Great Arctics occur on Vancouver Island and in the southern Coast Range. They inhabit forest openings and edges of meadows, from sea level to above timberline. The males are usually found on ridgetop clearings but also in forest openings lower down (Guppy 1962, 1970).

**General distribution:** Great Arctics occur only along the Pacific coast, from Vancouver Island and Lillooet, BC, south to northern CA.

**Conservation status:** Great Arctics are of Special Concern in BC (S3).

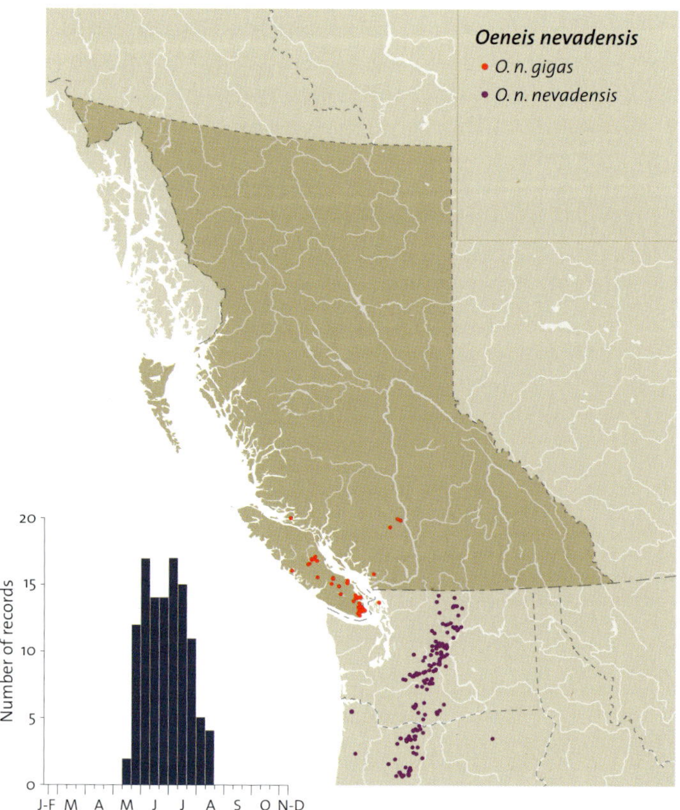

# Macoun's Arctic

*Oeneis macounii* (W.H. Edwards, 1885)

**Etymology:** The species *macounii* was named for John Macoun, who collected the type specimens. He was botanist to the Geological and Natural History Survey of Canada when he collected the types in 1884 (Edwards 1885a). The common name was first used by Holland (1898).

**Adult:** Macoun's Arctics are large, golden brown butterflies with dark wing borders. They have two eyespots on the dorsal forewing and one on the dorsal hindwing. The ventral hindwings are grey brown, and have a distinct band crossing the wing. Males do not have a large dark stigma covering the forewing discal cell. In females the dorsal forewing is only lightly dusted with grey scales near the wing base, usually less than one-quarter of the discal cell.

**Immature stages:** Eggs are round and white, slightly flattened on the top and bottom. There are 17–21 vertical ribs. First instar larvae are grey green with two short subconical tails. The longitudinal lines are red brown, the basal ridge is buff with a brown line below it, and the head, underside, legs, and prolegs are green yellow. Mature larvae are brown buff. The dorsal stripe is pale black, the lateral band is black with a green under colour, the spiracular band is green buff, the basal ridge is clear buff, and the underside, legs, and prolegs are grey green (Beutenmüller 1889; Edwards 1887–97).

**Biology:** Macoun's Arctics fly in June and July, usually in even years in BC but in odd years in the Peace River and near

♂ D  (5.2 cm)

♀ D  (5.9 cm)

♂ V  (5.2 cm)

Bridge Lake. They have a biennial life cycle that produces adults only once every two years. In the boreal forests of Alberta, they fly only in odd-numbered years (Masters and Sorensen 1969; Masters 1974). From Alberta to central Ontario they occur primarily in association with jackpine. In BC they occur in open lodgepole pine and Douglas-fir forests. The natural larval foodplants are unknown, but in captivity the larvae feed on grasses and sedges.

Females fly slowly and aimlessly through pine forests, while males actively defend territories in small forest clearings. A male perches on a tree branch or on top of a bush and patrols the clearing, darting at other butterflies when they come near, to drive away males and court females. When a male is captured, a new one frequently replaces him, using the same perch to defend the territory. Territories usually contain a few flowers used as nectar sources (Masters and Sorensen 1969). Near Kelowna a male used the top of a broken stump as a perch, in an opening in a lodgepole pine forest on a hilltop (CSG).

**Subspecies:** None. The type locality of the species is Lake Nipigon, ON.

**Range and habitat:** Macoun's Arctics occur in widely scattered localities from the Southern Interior of BC north to the Alaska Highway. They inhabit open pine and fir forests, especially thinly forested hilltops.

**General distribution:** Macoun's Arctics are found from BC east across the boreal forest regions of CAN to southern PQ, entering the USA only in MN and MI.

**Conservation status:** Not of concern (S5).

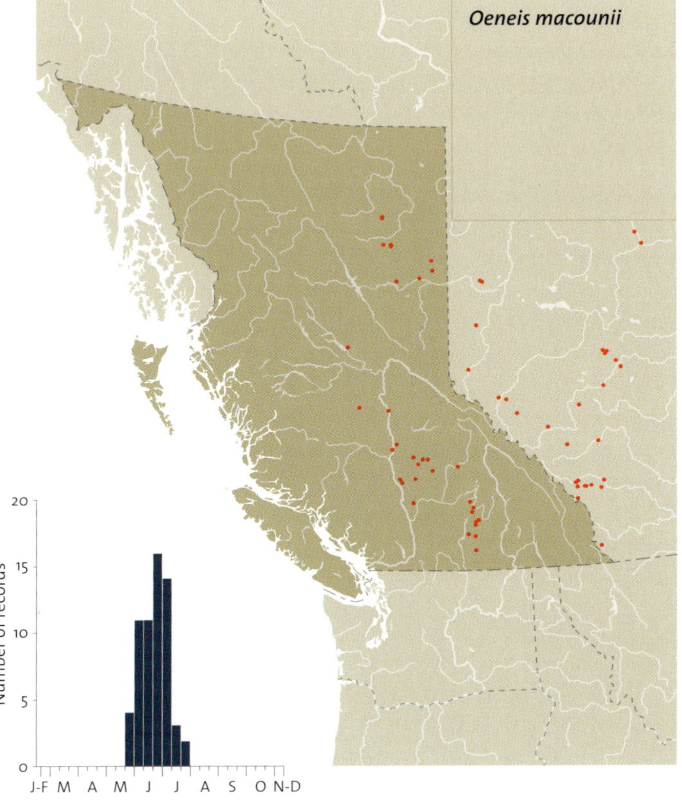

**Macoun's Arctic (*Oeneis macounii*)**

# Chryxus Arctic
*Oeneis chryxus* (Doubleday, [1849])

**ETYMOLOGY:** The species name *chryxus* is derived from the Greek *chryxos* (gold), in reference to the golden amber colour of the upperside of the wings (Bird et al. 1995). Subspecies *caryi* was named after Merritt Cary, who collected the type specimen in 1903 (Dyar 1904a). The common name was first used by Holland (1898).

**ADULT:** Chryxus Arctics are medium-sized golden brown to tan butterflies without dark wing borders. They have one to three eyespots on the dorsal forewing, and none on the hindwing. The ventral hindwing has a well-defined dark band across the middle. A dark line crossing the ventral forewing juts out sharply towards the wing margin about one-third of the way back from the front edge of the wing. Males have a large dark stigma covering the forewing discal cell.

**IMMATURE STAGES:** Eggs of the nominate subspecies from Banff, AB, are barrel-shaped, with the base roundly flattened. They are broadest just below the midline, with the upper part narrowing slightly and the top flattened. There are 19 vertical ribs. Mature larvae are stout. They are thickest in the middle and have an arched back and two short tails. The colour is various shades of buff, with longitudinal stripes. Down the middle of the back is a narrow black stripe bordered on both sides with a yellow buff line. There is a dorsolateral broad black stripe, bordered above with a brown buff stripe and below with a red buff stripe with a reddish line in the centre. In sequence below that is a yellow line, a brown buff lateral line, a yellow line along the ventrolateral ridge, and finally a brown buff line. The feet and prolegs are yellow brown, and the head is yellow brown with brown stripes. In Colorado the overall colour and the stripes are lighter brown, with narrower dorsal and lateral stripes. The pupae are brown with darker patches and stripes; the top of the thorax and abdomen is light yellow brown with black spots and streaks (Edwards 1887–97).

**BIOLOGY:** Chryxus Arctics have a two-year life cycle (Scott 1992), at least in some areas. In most areas of BC they fly every year. They are in flight in June and July at low elevations, into August at high elevations. They overwinter as partly grown larvae (Edwards 1887–97). Females lay eggs on tree bark and twigs on or near the ground (Scott 1992). In open meadows, males perch on open dirt cut banks along roads, and females move in short flights through the grassy areas. In forest areas, males perch on fallen logs and branches along the forest edge in sunny patches. The larval foodplant may be *Festuca idahoensis* in Washington (Pyle 1981), and *Carex* may be used in Colorado (Scott 1992). Larval foodplants in eastern Canada are grasses such as *Danthonia spicata, Oryzopsis pungens,* and *Phalaris arundinacea* (Klassen et al. 1989).

**SUBSPECIES:** The nominate subspecies, *O. c. chryxus* (Doubleday, [1849]); TL: Rock Lake, near Jasper, AB (Shepard 1984), occurs throughout most of mainland BC, north to the Yukon

Ssp. *chryxus* ♂ D (5.4 cm)

Ssp. *chryxus* ♀ D (5.2 cm)

Ssp. *chryxus* ♂ V (5.4 cm)

Ssp. *caryi* ♂ V (4.5 cm)

Ssp. *caryi* ♂ D (4.5 cm)

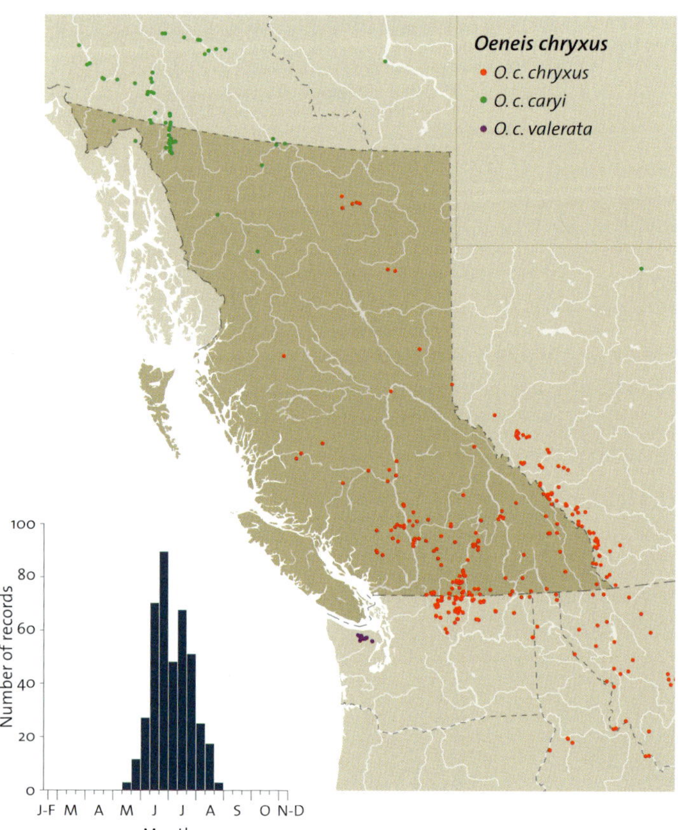

Chryxus arctic (*Oeneis chryxus*)

border. Subspecies *caryi* Dyar, 1904; TL: 59°54'N, 111°40'W near the west side of the Slave River, AB (Kondla 1995), occurs in northwestern BC, and is small and dull tan on the upperside of the wings. The underside of the hindwings is highly variable, ranging from similar to that of *chryxus* to light and finely flecked with grey all over. Females range from tan to brown on the upperside.

**RANGE AND HABITAT:** Chryxus Arctics are found throughout interior BC, east of the Coast Range. They inhabit dry grassland, forest openings, and dry alpine tundra.

**GENERAL DISTRIBUTION:** Chryxus Arctics are found from southeastern AK southeast across boreal Canada, and south through mountainous areas to NM and northern CA.

**CONSERVATION STATUS:** Not of concern, with both subspecies S5.

## UHLER'S ARCTIC
*Oeneis uhleri* (Reakirt, 1866)

**ETYMOLOGY:** The species *uhleri* was named for Reakirt's friend Dr. Philip Reese Uhler of Cambridge, MA (Reakirt 1866), a specialist in Hemiptera. The subspecies name *varuna* is derived from Varuna, king of the Vedic pantheon, who owned the West, the night, and the water. Thus the butterfly *varuna* is of the west, where the commercial collector H.K. Morrison was exploring for W.H. Edwards. Morrison would never tell Edwards exactly where in Dakota Territory (present-day eastern Montana, North Dakota, and South Dakota) the type series came from, and thus the ambiguous name meaning

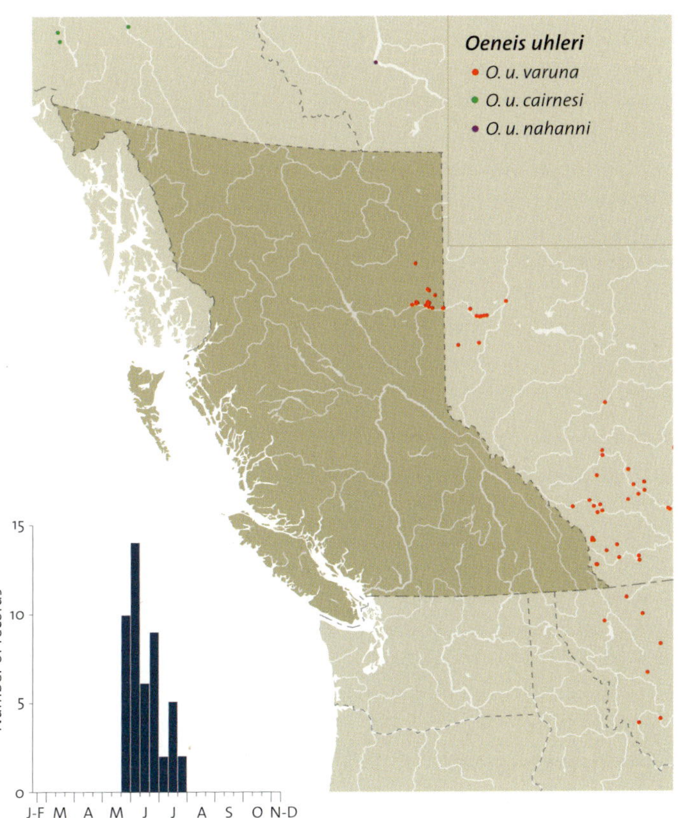

♂ D (4.0 CM)

♂ V (4.0 CM)

"from the west." The common name "Uhler's Arctic" was first used by Holland (1898).

**ADULT:** Uhler's Arctics are small golden brown butterflies. They usually have several eyespots on the dorsal forewing, and also 1–3 eyespots on the dorsal hindwing. The ventral hindwing is coloured with white and brown striations. The band across the middle of the ventral hindwing is pale or missing, and is not visible from the upperside. There is no dark angular line crossing the ventral forewing.

**IMMATURE STAGES:** Eggs of subspecies *varuna* are conical, with 20–24 vertical ribs; they are white. First instar larvae are grey green with two short subconical tails. The dorsal and subdorsal lines are pale brown; the lateral stripe is grey green on the thorax and pale brown on the abdomen; and the underside, legs, and prolegs are green white. Mature larvae have two short subconical tails. The colour is highly variable, with some being white buff with green stripes, others yellow buff with black and green stripes, and still others green with black stripes. All the colour forms have reddish streaks. Pupae are green yellow with tints of brown (Edwards 1887–97).

**BIOLOGY:** Uhler's Arctics are in flight in June and July, and in some areas fly only in alternate years. There are two flight periods during one summer at some sites in the Alberta

(CSG) and Colorado Rocky Mountains (Edwards 1887–97; Ferris and Brown 1981). Males establish territories along slope breaks in grasslands, and periodically fly up and flutter in fixed positions for a minute or so, apparently scanning the slope below for females (Masters and Sorensen 1969). Females are most active in the afternoon (Bird et al. 1995). Eggs are laid on or near the tip of grass blades. They hatch in about 20 days, and larvae hibernate in the third (Bird et al. 1995) or fourth instar (Ferris and Brown 1981). Pupation occurs in loose soil (Edwards 1887–97). Natural larval foodplants are unknown, but the larvae feed on grass in captivity. Grasses in the genera *Festuca, Koeleria,* and *Poa* may be larval foodplants (Scott 1992), but are unproven.

**SUBSPECIES:** The subspecies of Uhler's Arctic found in BC is *O. u. varuna* (W.H. Edwards, 1882) (TL: "plains of Dakotah Terr. on the way to Montana, elevation about 1200 feet").

**RANGE AND HABITAT:** Uhler's Arctics are known in BC only from the Peace River near the AB border. They inhabit dry bunchgrass hillsides.

**GENERAL DISTRIBUTION:** Uhler's Arctics are found in YT, and from the Peace River south and east to southern MB and south to NM and SD.

**CONSERVATION STATUS:** Uhler's Arctics are of Special Concern in BC (S3).

## ALBERTA ARCTIC
*Oeneis alberta* Elwes, 1893

**ETYMOLOGY:** The name *alberta* refers to the type locality being in Alberta. The common name was first used by Holland (1931).

**ADULT:** Alberta Arctics, the smallest of the arctics, are light brown or grey brown butterflies. They usually have several eyespots on the dorsal forewing, and none or one on the hindwing. The ventral hindwing has an irregular band with dark edges across the middle. A dark line crossing the ventral forewing juts out sharply towards the wing margin about one-third of the way back from the front edge of the wing. The wings are thin, with the underside markings visible from the upperside.

♂ D (4.3 CM)

♂ V (4.3 CM)

**IMMATURE STAGES:** Eggs from Alberta are barrel-shaped and grey white; they have 19–20 vertical ribs. First instar larvae have only two slightly conical projections for tails. They are grey white with pale brown longitudinal stripes; the basal ridge is white; the underside, legs, and prolegs are translucent white; and the head is yellow green with a tint of brown. Mature larvae have two short, blunt tails, and are dark brown. The dorsal stripe is black and edged with narrow yellow and white lines; the lateral band is black on a green ground colour and with a pale brown line along the lower edge; the spiracular band is green and speckled with black; and the basal ridge is brown. The underside, legs, and prolegs are white. The head is brown green. The pupa has a green grey thorax and head. The wings are dark olive green with lighter veins, and the abdomen is yellow brown with longitudinal rows of blackish dots and dashes (Edwards 1887–97).

**BIOLOGY:** Alberta Arctics are in flight in May and June, and probably fly every year. A two-year life cycle has not been confirmed for this species. A partial second flight may occur in some years in the American Rocky Mountains (Ferris and Brown 1981). The larval foodplant is unknown, but is likely the bunchgrass *Festuca idahoensis* in Colorado (Scott 1992).

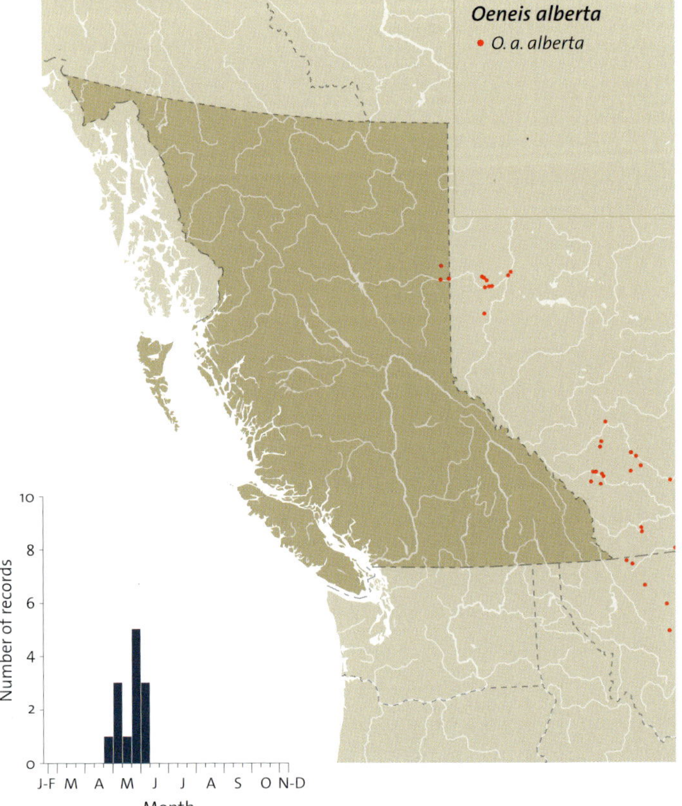

**Fourth instar larva**

**SUBSPECIES:** The nominate subspecies, *O. a. alberta* Elwes, 1893; TL: restricted to Fish Creek Provincial Park, Calgary, AB (Kondla 1996), occurs in BC.

**RANGE AND HABITAT:** Alberta Arctics are known in BC only from the Peace River near the AB border. They inhabit dry bunchgrass hillsides.

**GENERAL DISTRIBUTION:** Alberta Arctics are found from the Peace River in BC south along the dry areas of the Rockies to AZ and NM, and east to MB.

**CONSERVATION STATUS:** Alberta Arctics are Threatened in BC (S2S3).

## WHITE-VEINED ARCTIC
*Oeneis bore* (Schneider, 1792)

**ETYMOLOGY:** The name *bore* is Latin for north, in reference to the Lapland type locality. Subspecies *mckinleyensis* was named for the type locality of Mt. McKinley National Park in Alaska. Subspecies *edwardsi* was named for the prominent early lepidopterist W.H. Edwards. The common name "White-veined Arctic" for the North American subspecies was first used by Klots (1951) in reference to the white highlighting of the veins that is frequently present on the ventral hindwings. The European name for the species is the Arctic Grayling (Higgins and Riley 1970).

**ADULT:** White-veined Arctics are medium-sized grey brown arctics with slightly translucent wings. On the upperside, females are frequently orange tan, and males have a darker grey sex patch in the middle of the forewing. Often the upperside of the hindwings, and sometimes the forewings, have pale diffuse spots near the outer wing margin. The ventral hindwing has a dark brown band across the centre,

Ssp. *mckinleyensis* ♂ D (4.7 CM)

Ssp. *mckinleyensis* ♂ V (4.7 CM)

Ssp. *mckinleyensis* ♀ D (4.7 CM)

Ssp. *mckinleyensis* ♀ V (4.7 CM)

Ssp. *edwardsi* ♂ D (3.9 CM)

Ssp. *edwardsi* ♂ V (3.9 CM)

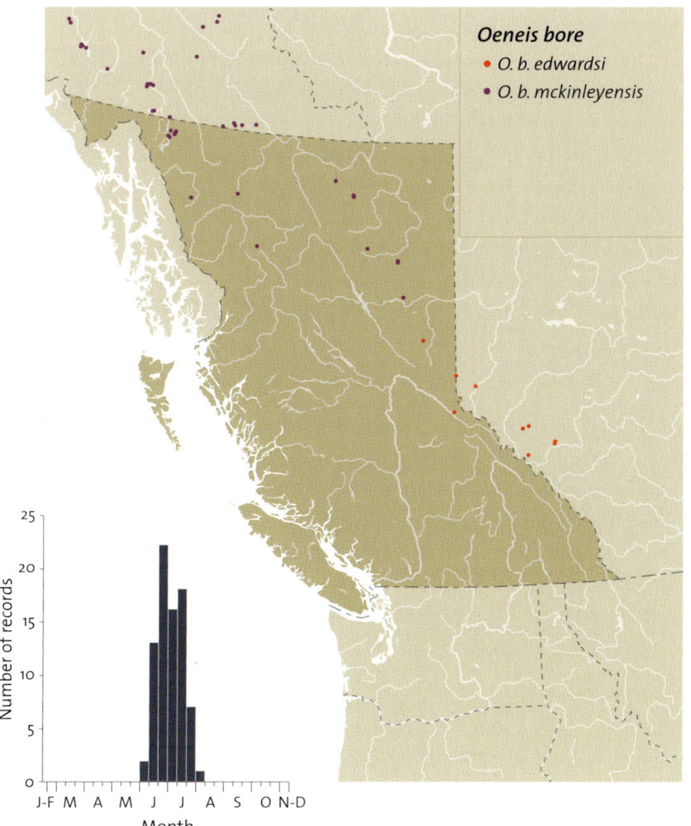


with whitish areas on both sides. The wings are thin and translucent, making this dark band visible from above. The veins on the ventral hindwings are frequently outlined in white.

**IMMATURE STAGES:** Eggs are white. Mature larvae from arctic Manitoba (subspecies *hanburyi*) are brown with a darker brown line down the back and with whitish brown, reddish white, and brown longitudinal stripes (Scott 1986b). In contrast, mature larvae of Norwegian *Oeneis bore* are bright brownish yellow, with a narrow black line down the back and one broader black line on each side (Holland 1891). The considerable difference in larval coloration between the nominate *bore* and the North American taxa suggests that more than one species may be involved.

**White-veined Arctic (*Oeneis bore mckinleyensis*)**

**BIOLOGY:** White-veined Arctics are in flight in June and July, and in most areas are present every year. In Norway, *O. bore* hibernate twice as larvae, once as a young larva and once as a nearly mature larva, and therefore have a two-year life cycle (Holland 1891). This also appears to be the case for arctic populations of White-veined Arctics in North America (Scott 1986b). White-veined Arctics in Colorado have a single-year life cycle (Edwards 1891). The length of the life cycle in BC is not known. Larval foodplants are probably sedges and grasses, *Carex* and *Festuca*. *Festuca ovina* is used in Europe (Higgins and Riley 1970).

**SUBSPECIES:** Subspecies *mckinleyensis* dos Passos, 1949 (TL: Mt. McKinley National Park, AK) occurs across northern BC. Adults are usually orange brown dorsally, but sometimes grey brown, and the ventral hindwings have pale bases and pale outer areas, with a prominent dark band across the middle. Subspecies *edwardsi* dos Passos, 1949 (TL: Rio Grande Pyramid, San Juan Mts., CO) is known in BC only from McBride Peak, but may occur further south in the BC Rockies to about Golden. Compared with *mckinleyensis*, *edwardsi* adults are smaller, the ventral hindwing is darker, and the ventral hindwing is relatively uniform in ground colour, with the darker middle band showing less contrast with the basal and distal areas.

**RANGE AND HABITAT:** White-veined Arctics occur across northern BC in boreal forest openings, moist alpine tundra, and rocky ridges, and in the Rocky Mountains along the AB border.

**GENERAL DISTRIBUTION:** White-veined Arctics are found from AK across the arctic to Greenland. There are scattered populations extending south through the Rocky Mountains to CO.

**CONSERVATION STATUS:** Subspecies *edwardsi* is of Special Concern in BC (S3), but subspecies *mckinleyensis* is not of concern (S5).

## JUTTA ARCTIC
*Oeneis jutta* (Hübner, 1805–1806)

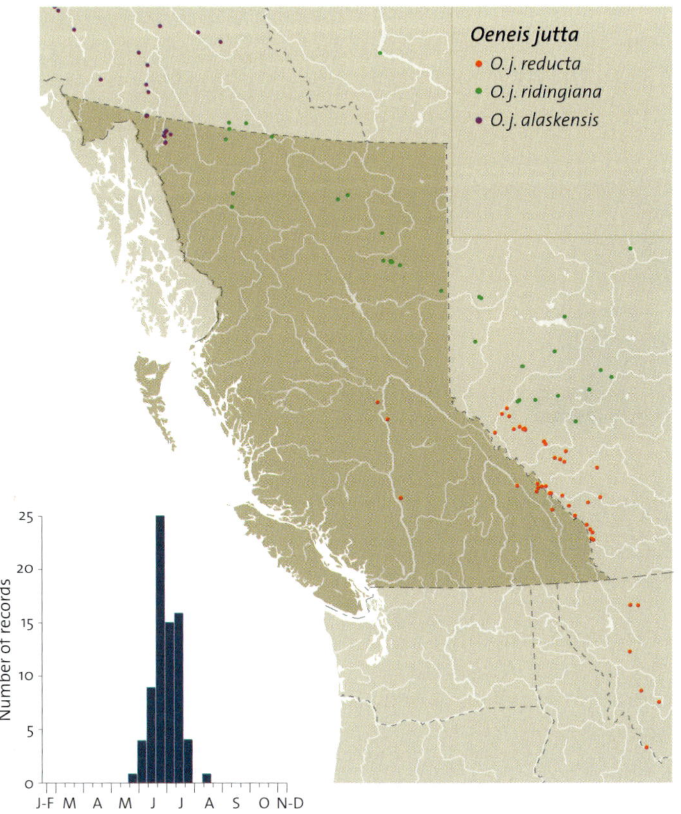


**ETYMOLOGY:** The species name *jutta* may refer to the old Danish state of Jutland, to which the Germanic tribe called the Jutes migrated. The Jutes later invaded Britain with the Saxons and destroyed the remnants of Roman civilization. The name may also just refer to wild forested habitat. The subspecies name *reducta* refers to the reduced size of the dorsal submarginal patches. Subspecies *ridingiana* was named for the type locality, the Riding Mountains of Manitoba. Subspecies *alaskensis* was named for the type locality, in Alaska. The common name "Jutta Arctic" for the North American subspecies was first used by Klots (1951). The European name is the Baltic Grayling (Higgins and Riley 1970).

**ADULT:** Jutta Arctics are dark grey brown above, and the ventral hindwings are striated with grey and brown, closely resembling bark. Near the margin of the dorsal forewings and hindwings is a row of orange brown patches, some of which contain eyespots. Males have a prominent dark sex patch on the dorsal forewing, and the orange patches are enlarged in females.

**IMMATURE STAGES:** In Quebec eggs are cream white, with 16–21 zigzag vertical ridges and with a small stalk at the base. The bodies of first instar larvae are pale amber to grey



FAMILY NYMPHALIDAE (BRUSHFOOTS)

white with a pink tinge; there are two tails and brown tubercles with light spines. The dorsal, subdorsal, and lateral stripes are yellow brown; the spiracular line is pale brown; and the basal ridge is dull white with a pale brown line below. The head, underside, legs, and prolegs are yellow green. Mature larvae are green buff, with two short subconical tails. The dorsal stripe is pale green, with blackish dots between segments; the lateral band is broad and pale green with a blackened upper edge; the basal ridge is yellowish; and the underside, legs, and prolegs are green buff. The head is green. The pupal head is amber, with a brown dash on either side. The wings are green with a brown outline and streaks. The abdomen is pale yellowish green, with a line of darker green down the back, and numerous rows of brown dots. There is a rosy tip to the abdomen (Fyles 1888; Edwards 1887–97).

**Biology:** Jutta Arctics are univoltine and fly in June and July. They take two years to mature, with larvae hibernating twice; adults are produced only every second year in some populations (Masters 1973; Tuzov 1997). In captivity in eastern Canada, however, larvae have six instars, hibernate in the sixth instar, and pupate on the ground or in moss (Fyles 1888, 1889), suggesting that some populations may have annual life cycles. Males establish territories in open areas, and have favourite perches from which they dart out to see whether something flying by is a female (Masters 1973). Females fly apparently aimlessly through the bog, sometimes flying to the top of a tree (Masters and Sorensen 1969), although Braun (Edwards 1887–97) specifically stated that near Bangor, Maine, females never fly to treetops but drop to the ground instead.

In Minnesota subspecies *ascerta* uses cotton grass (*Eriophorum vaginatum* var. *spissum*) as a larval foodplant (Masters 1973), and the sedge *Carex* is eaten in captivity (Fyles 1889). In Eurasia larval foodplants include *Carex*, *Eriophorum*, *Glyceria*, *Juncus*, *Molinia caerulea*, and *Scirpus caespitosus* (Tuzov 1997).

**Subspecies:** Subspecies *reducta* McDunnough, 1929 (TL: upper Gallatin Canyon, MT) occurs in southern BC from the Rockies west to near Prince George. The dorsal yellow brown submarginal patches are small and somewhat separated from each other. The ventral hindwing has a well-defined dark band across the middle, with pale highlighting along both the outer and inner edge of the band. Subspecies *chermocki* Wyatt, 1965 (TL: near Banff, AB) is a synonym of *reducta*. Subspecies *ridingiana* F. & R. Chermock, 1940 (TL: Riding Mountains, MB) occurs from the Peace River lowlands west across most of northern BC. The dorsal yellow brown submarginal patches are large and form a continuous

Ssp. *reducta* ♂ D (5.1 cm)

Ssp. *reducta* ♂ V (5.1 cm)

Ssp. *alaskensis* ♂ D (4.6 cm)

Ssp. *alaskensis* ♂ V (4.6 cm)

Ssp. *ridingiana* ♂ D (5.4 cm)

Ssp. *ridingiana* ♂ V (5.4 cm)

band broken only by brown veins in females. The ventral hindwing is a mottled grey brown, with only a weakly defined dark band across the middle. Subspecies *alaskensis* Holland, 1900 (TL: mountains between Fortymile and Mission Creeks, AK, and American Creek, AK) occurs in northwestern BC. The dorsal and ventral wings are a darker brown. The dorsal yellow orange submarginal patches are small and well separated, sometimes diffused and indistinct and sometimes sharply defined. The ventral hindwing is a strongly mottled grey brown; a dark band across the middle is generally without white highlighting on the basal side.

**Range and habitat:** Jutta Arctics occur across northern BC and in scattered locations through the Rockies and the Cariboo. They inhabit spruce bogs and open pine forests, and occasionally alpine tundra.

**General distribution:** Jutta Arctics are found from Europe and Asia across the boreal forests of CAN to NF. Scattered populations occur in the Rocky Mountains as far south as CO.

**Conservation status:** Subspecies *alaskensis* is of Special Concern in BC (S3). Subspecies *reducta* and *ridingiana* are not of concern (both S4).

# Melissa Arctic

*Oeneis melissa* (Fabricius, 1775)

**Etymology:** The species name *melissa* may refer to Melissa, the wife of Periander, tyrant of Corinth, who murdered her. Subspecies *beanii* was named for Thomas E. Bean, who collected for Edwards in the Lake Louise area of Alberta in the 1890s. Subspecies *atlinensis* is named for the type locality in the vicinity of Atlin, BC. The common name was first used by Macy and Shepard (1941).

**Adult:** Melissa Arctics are smoky grey to grey brown in colour, and are quite transparent. They lack eyespots on the dorsal surface, but the ventral hindwings usually have a row of small spots near the wing margin. The ventral hindwing is striated with dark grey, with at most a faint dark band across the middle in southern populations.

**Immature stages:** Eggs are subconical, and the base is flattened and rounded, with about 20 vertical ribs. First instar larvae are pale green white, with two very short, stubby tails. Dorsal, subdorsal, and lateral lines are pale brown, the underside is dull white, and feet and legs are translucent whitish. Mature larvae are shades of buff. The dorsal line is grey green, with dark spots edged with white lines between segments; below this, the ground colour is grey buff with blackish longitudinal streaks, and there are two brown dorsolateral lines below that. The lateral band is broad and deep black, and has a light buff line below. The spiracular band is dark grey. The underside, legs, and prolegs are grey buff. The head is green yellow with a brown tint. In pupae in Colorado (subspecies *lucilla* Barnes & McDunnough,

Ssp. *atlinensis* ♂ D Holotye (4.4 cm)

Ssp. *atlinensis* ♂ V Holotype (4.4 cm)

Ssp. *atlinensis* ♀ D (5.0 cm)

Ssp. *atlinensis* ♀ V (5.0 cm)

Ssp. *beanii* ♂ D (4.3 cm)

Ssp. *beanii* ♂ V (4.3 cm)

1918), the back of the head and the thorax are green yellow with a brown tint. The underside of the head and wings is black brown; the abdomen has alternating bands of yellow and brown grey, each of the grey bands finely edged in carmine; and the spiracular band is tinged with carmine (Edwards 1887–97).

**Biology:** Melissa Arctics are in flight in July and August in southern alpine areas, and in June and July in northern alpine areas. They are in flight every year, but have a two-year life cycle. They are challenging to closely observe or capture, as Phair (1919) discovered on the summit of Mt. McLean in Lillooet: "The underside is exactly the colour of the moss or lichen-covered rocks, and when they alight they turn over on their sides, but they are on the alert, and one does well to get within ten feet. When they start they rise high in the air and are off down the mountain several hundred feet." Larvae feed at night, and they pupate under or between rocks. They hibernate in the first instar, and then again as mature larvae (Scudder 1889a).

Larval foodplants are unknown for BC. Subspecies *semidea* uses *Carex bigelowii* on Mt. Washington, ME (Scudder 1889a). Larvae feed on the grass *Poa pratensis* in the lab (Scott 1992). Grasses and sedges, *Carex* and *Deschampsia caespitosa* ssp. *glauca*, are used in Asia (Tuzov 1997).

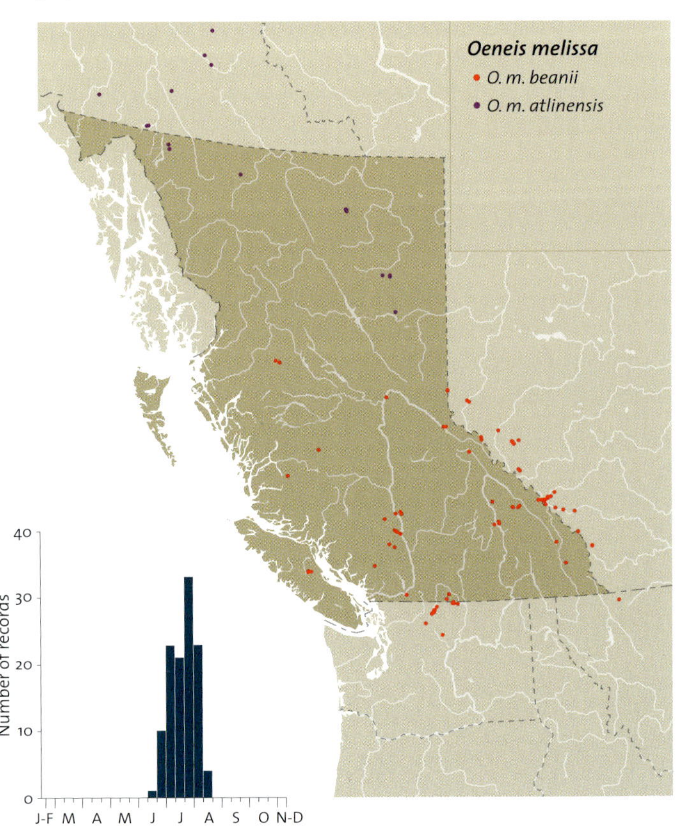

**SUBSPECIES:** Subspecies *beanii* Elwes, 1893 (TL: Laggan [vicinity Lake Louise], AB) occurs in southern and central BC from Vancouver Island to the Rockies, and south through the Rockies to Wyoming. The underside of the hindwings is an even mottled grey, with only an indistinct median band. The northwestern BC populations have traditionally been called subspecies *gibsoni* Holland, 1931 (TL: Bethel, Kuskoquim River, AK), from which they differ significantly. We provide a new subspecies name here.

**Oeneis melissa atlinensis Guppy & Shepard, new subspecies**. *Oeneis melissa atlinensi*s has a clearly defined median band, and grey and white checkered wing fringes and costal margin, that distinguish it from subspecies *beanii*. Males have little or no dark marginal ventral hindwing markings, which are characteristic of subspecies *gibsoni*. When dark marginal markings are present, they are much smaller and paler than in *gibsoni*. The ventral hindwing veins are greyish, rather than white as in *gibsoni*. In females the dark medial band on the ventral hindwing is more distinct than in subspecies *gibsoni*. **Types**. Holotype: male, BC, Atlin, Mt. Vaughan, el. 5,200+ ft., 28 July 1976, C.S. Guppy; a label "HOLOTYPE / *Oeneis melissa* / *atlinensis* Guppy & Shepard"

is attached. The holotype is deposited in the Royal British Columbia Museum, Victoria, BC, CAN. Paratypes: 8 males, 2 females, same data as holotype (CSG); 1 male, Atlin, Mt. Vaughan summit, el. 1,900 m, 25 June 1999, C.S. Guppy (CSG); 1 male, BC, Boulder Cr., 12 mi. E Atlin, 6,000′, 20 June 1973, Jon H. Shepard (JHS); 11 males, 8 females, YT, Montana Mt., near Carcross, el. 6,500′, 16 June 1989, J. and S. Shepard (JHS); 1 female, same locality, 17 June 1989 (JHS); 1 female, 7 July 1989 (JHS).

**RANGE AND HABITAT:** Melissa Arctics occur in northwestern BC and in alpine habitat in the coastal mountains and the Rocky Mountains of BC. There is one record from Strathcona Provincial Park on Vancouver Island. They inhabit dry alpine rock rubble slopes and rocky alpine tundra.

**GENERAL DISTRIBUTION:** Melissa Arctics are found from Siberia across arctic North America to Labrador. They occur in most of the mountainous areas of BC and AB; isolated colonies occur in the Rocky Mountains as far south as NM.

**CONSERVATION STATUS:** Not of concern, with both subspecies S5.

## POLIXENES ARCTIC
*Oeneis polixenes* (Fabricius, 1775)

**ETYMOLOGY:** The species name *polixenes* refers to Polyxena, the daughter of Priam and Hecuba of Troy, who was loved by Achilles and died shortly after the battle of Troy. Her death

Polixenes Arctic (*Oeneis polixenes brucei*)

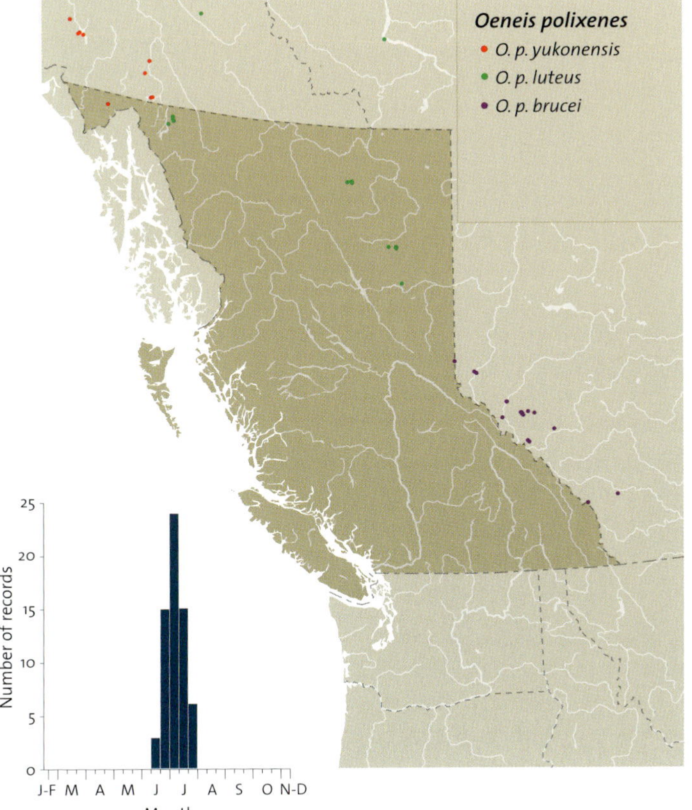


resulted either from battle wounds or from her being sacrificed to appease Achilles' ghost. Subspecies *brucei* was named for David Bruce, a mural painter who supplied the type specimens and many other butterflies to W.H. Edwards. The subspecies name *luteus* is Latin for yellow, in reference to the yellowish colour of the dorsal wings. Subspecies *yukonensis* was named for the type locality, in the Yukon. The common name was first used by Klots (1951).

**ADULT:** Polixenes Arctics in BC are grey brown, and males usually have a wide yellow brown submarginal band on the dorsal hindwings. The dorsal wings of females are usually mostly yellow brown. There are small pale yellowish dots along the margins of the dorsal hindwings. The ventral hindwings have a dark band across them, bordered by lighter striated grey areas. The wings are thinly scaled and translucent.

**IMMATURE STAGES:** Eggs of subspecies *brucei* from Colorado are subconical, broadest about one-third up from the base and narrowing towards the top. There are 20 vertical ribs, dull white in colour. First instar larvae are pale green white, turning green grey, with two short, stubby tails. The dorsal, subdorsal, and lateral stripes are pale brown, the underside is dull white, and the legs and prolegs are whitish and translucent. Mature larvae are buff, with two brown lines down the middle of the back, a broad deep black lateral band with a light buff line along the lower edge, and a dark grey spiracular band. The basal ridge is light buff with a grey line below it, and the underside, legs, and prolegs are grey buff. The head is yellow (Edwards 1887–97).

**BIOLOGY:** Polixenes Arctics are univoltine. They are in flight in July and August, and fly every year. Larvae may hibernate immediately after hatching or in the fourth instar (Ferris and Brown 1981). Larval foodplants are unknown in BC. In the lab the larvae feed on the grass *Poa pratensis,* and likely eat grasses and sedges in the wild (Scott 1992).

**SUBSPECIES:** Subspecies *luteus* Troubridge & Parshall, 1988 (TL: Pink Mountain, BC) occurs throughout most of northern BC. Males have orange brown submarginal areas, and the females have generally yellow brown dorsal wing surfaces. Subspecies *yukonensis* Gibson, 1920 (TL: Klutlan Glacier, YT, 8,200 feet) occurs along the Haines Highway in extreme northwestern BC. The dorsal wing surfaces of this subspecies are grey brown without any orange brown or yellow brown, and the wings are translucent. Subspecies *brucei* (W.H. Edwards, 1891) (TL: vic. Bullion and Hayden Mtns, Hall Valley, Park Co., CO) occurs in the Rocky Mountains of Alberta but

Ssp. *yukonensis* ♂ D (4.0 cm)

Ssp. *luteus* ♂ D (4.5 cm)

Ssp. *yukonensis* ♂ V (4.0 cm)

Ssp. *luteus* ♂ V (4.5 cm)

Ssp. *luteus* ♀ D (4.8 cm)

has yet to be found in BC. The dorsal wings are an even soft, translucent brown.

**RANGE AND HABITAT:** Polixenes Arctics occur in the mountains of northern BC. They inhabit moist alpine tundra and rocky ridgelines.

**GENERAL DISTRIBUTION:** Polixenes Arctics occur from AK across arctic Canada to Labrador. There are isolated populations scattered south through the Rocky Mountains to NM, and in the east south to Maine.

**CONSERVATION STATUS:** Subspecies *yukonensis* is of Special Concern in BC (S3). Subspecies *luteus* is not of concern (S4).

# ROSOV'S ARCTIC
*Oeneis rosovi* Kurentzov, 1960

**ETYMOLOGY:** The species *rosovi* was named for the Russian naturalist V.N. Rosov, hence the common name, which is used for the first time here. Subspecies *philipi* was named for Kenelm Philip, who for many years has been the only lepidopterist in Alaska (Troubridge and Parshall 1988). The common name "Philip's Arctic" was first used by Layberry et al. (1998) for the North American subspecies.

**ADULT:** Rosov's Arctics are highly variable in appearance. The North American subspecies *philipi* has dorsal wings that are an even grey and less translucent than in Polixenes Arctics. Rosov's Arctics are also larger than Polixenes Arctics. The ventral hindwings are heavily striated with blackish brown, and in the submarginal area there is more grey and less

brown than in Polixenes Arctics. The two species are very difficult to separate.

**IMMATURE STAGES:** Undescribed.

**BIOLOGY:** Rosov's Arctics are in flight in June, and fly at least in odd-numbered years and possibly every year. Adults fly in open spruce bogs, landing on tree trunks and flying up into the trees when startled (Layberry et al. 1998). The larvae feed on cotton grass (*Eriophorum*).

**SUBSPECIES:** Subspecies *philipi* Troubridge, 1988 (TL: Km 1 Dempster Hwy at the Klondike R., YT) occurs in North America. The taxonomy of this species is still subject to change. Layberry et al. (1998) determined *philipi* to be a subspecies of *O. rosovi* based on examination of the type specimen of *rosovi* in Vladivostok, Russia. Tuzov (1997), however, treated *O. philipi* as a synonym of *Oeneis oeno* Boisduval, 1932, which most authors consider to be a synonym of *O. melissa*.

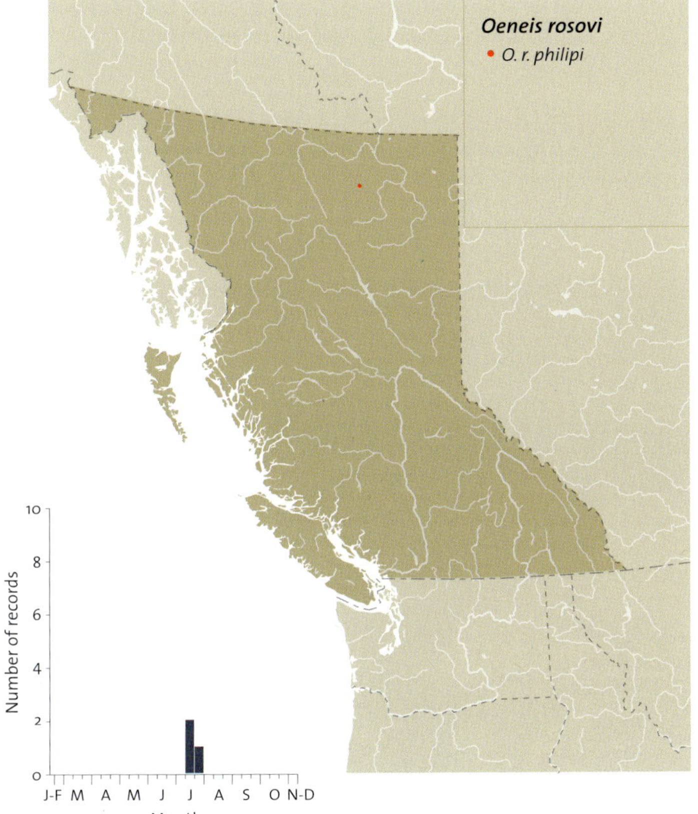

♂ D  (4.3 cm)

♀ D  (4.9 cm)

♂ V  (4.3 cm)

♀ V  (4.9 cm)

**RANGE AND HABITAT:** Rosov's Arctics are known in BC only from Stone Mountain Provincial Park, with doubtful records from Pink Mountain and near Muncho Lake. They inhabit open spruce forests and alpine tundra just above timberline.

**GENERAL DISTRIBUTION:** Rosov's Arctics occur across eastern Siberia, central AK, and northern YT. There is a disjunct population in Stone Mountain Provincial Park, in northern BC.

**CONSERVATION STATUS:** Rosov's Arctics are of Special Concern in BC (S3).

The subfamily name is derived from the type genus, *Danaus*. The common name "milkweed butterflies" refers to the use of milkweed as the larval foodplant of many of the species in the subfamily.

Milkweed butterflies are a small subfamily of large butterflies with a wide range of wing patterns. Adults are robust and brightly coloured, and usually contain distasteful or toxic chemicals acquired by the larvae from their foodplants. The larvae of many species feed on plants in the milkweed family (Asclepiadaceae), Solanaceae (Bird et al. 1995), and Apocynaceae (Layberry et al. 1998). The antennae do not have any scales on them, and the males have "hair pencils" on the abdomen that release pheromones attractive to the females. Males have dark scent patches on the dorsal hindwings.

Only two species of milkweed butterflies occur in temperate North America; another two are found in the extreme southern United States.

*Danaus similis*, a tropical member of the Danainae, escaped from a butterfly house in Victoria.

### GENUS *DANAUS* KLUK, 1780  ROYALS

The name *Danaus* is linked to the Danai, the name Linnaeus gave to a group of butterflies primarily composed of present-day Pieridae and Satyrinae. Danaus was the name of the king of Argos, after whom the Argives and often the Greeks as a whole were called Danai ("children of Danaus") by Homer (Emmet 1991). The common name "royals" is used here for the first time, because there is no existing generic common name.

Three royals are found in the USA and Canada: the Monarch, the Queen, and the Soldier. All are large orange brown butterflies with black markings. Only one species, the Monarch, occurs in BC.

## MONARCH
*Danaus plexippus* (Linnaeus, 1758)

**ETYMOLOGY:** The name *plexippus* refers to Plexippos, one of the 50 sons of Aegytus, the brother of Danaus in Greek mythology (Bird et al. 1995). The species and generic names reflect the relationship of the brothers. Plexippos took part in the boar hunt at Calydon and was killed by Meleager because he tried to take the prize for success from Atalanta (Emmet 1991). The common name "Monarch" was used for the first time by Scudder (1874b), because "it is one of the largest of our butterflies, and rules a vast domain."

**ADULT:** Monarchs are similar to only one other butterfly found in BC, the Viceroy, which is a mimic of the Monarch. Monarchs are larger than Viceroys, and lack a black line across the middle of the hindwings. Males have a black sex patch on each hindwing; females do not.

**IMMATURE STAGES:** Eggs are conical, with ridges down the side, and are green white or cream. Mature larvae are ringed with alternate black, yellow, and white stripes. On the 3rd and 12th segments are two long, black, fleshy tendril-like horns. The prolegs are black and there is a large white spot at the base of each one. The pupa is cylindrical and bright green, with an oval gold spot on each side of the antennae. A row of 11 gold spots circles the lower part of the pupa; there is a second row of gold spots above it, with a black line along the top of the row (Saunders 1869c). A line drawing of the pupa with terminology for spots is provided by Urquhart and Tang (1971).

**BIOLOGY:** Monarchs are multivoltine in southern North America, and migrate north into low-elevation areas of southern BC each summer. The Monarch's larval foodplant in BC, showy milkweed (*Asclepias speciosa*), is the only milkweed native to BC and occurs in the dry areas of the Southern Interior of BC. When fourth instar larvae collected on showy milkweed near Keremeos in late June were reared, adults emerged in mid-July (CSG). Many species of milkweed have been recorded as larval foodplants outside BC, as summarized by Malcolm and Brower (1987).

Monarchs lay eggs on the milkweed and at least one generation matures successfully each summer in BC. The number of adult Monarchs in BC varies from year to year, but the species is generally uncommon. In the late summer and fall, BC Monarchs presumably migrate south to California to hibernate. More than 200 hibernation sites have been recorded along the California coast and Baja California, south from San Francisco. Monarchs now hibernate in stands of introduced Australian eucalyptus trees, as a result of the cutting of the native stands of Monterey pine (*Pinus radiata*) and Monterey cypress (*Cupressus macrocarpa*). The hibernation sites in California have little legal protection against

♂ D  (10.0 CM)

♀ D  (9.7 CM)

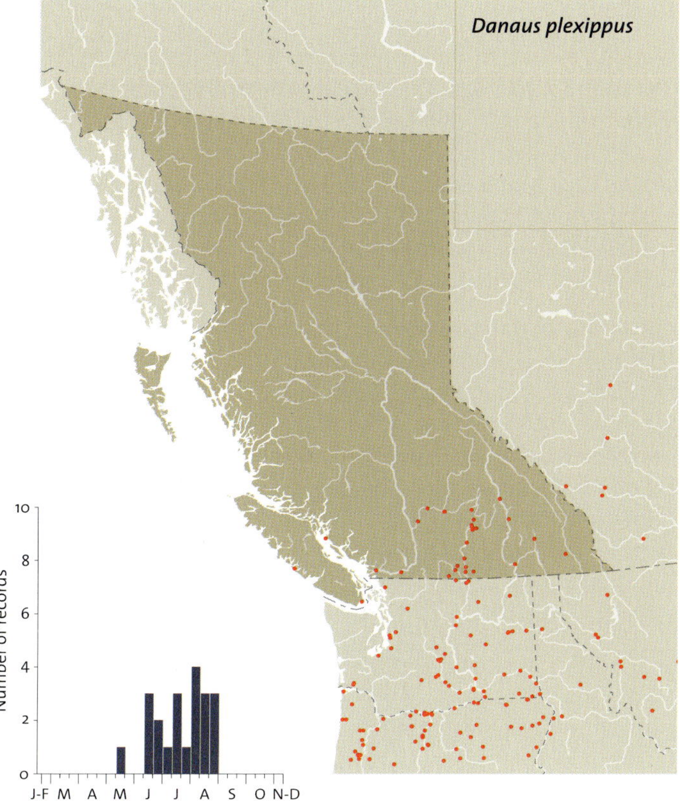

*Danaus plexippus*

destruction, and real estate developments threaten most of them (Crolla and Lafontaine 1996).

Some Monarchs fly further west than normal when migrating north, and end up on the west coast, with the most western record being from Tofino (AGG). Similarly, the eastern North American populations are regularly blown long distances out to sea by strong westerly winds during their southward migration in the fall (Urquhart and Urquhart 1979), resulting in migration from North America to Great Britain (Williams 1949b). Milkweed is not native to the west coast of BC, so normally migrants to that area cannot breed successfully. Monarchs, however, are good at finding very isolated patches of milkweed and using them as larval foodplants (Shapiro 1982d). They move constantly between areas, instead of staying in a patch of milkweed once they find it (Zalucki and Kitching 1985), which increases the probability of finding new patches of milkweed. This occurs regularly on Vancouver Island, where milkweed is grown in many gardens and where gardeners look forward to watching maturing Monarch larvae.

Monarch larvae accumulate and concentrate cardiac glycosides (heart poisons) in their bodies from the milkweed they eat. The poisons are retained in the bodies of the adults, making them toxic to most bird and mammalian predators. Some milkweeds do not contain cardiac glycosides, and the adults are therefore palatable. Some predators can also eat Monarchs without being affected by the cardiac glycosides. The Viceroy is also unpalatable to predators, and both butterflies benefit from its mimicry of the Monarch.

SUBSPECIES: None. The type locality of the species is "Pennsylvania."

RANGE AND HABITAT: Monarchs are frequently seen in the dry Southern Interior of BC, and infrequently in the Lower Fraser

Mature larva

Pupa

Valley, on Vancouver Island, and in the Rocky Mountain Trench. We have determined that the two map records for northeastern BC in Layberry et al. (1998) are data errors, and are actually Painted Lady records. Monarchs are frequently reared in captivity on potted or garden milkweed plants, and are then released into the wild in areas that normally have few Monarchs. This makes it difficult to determine whether a sighting or capture of a Monarch in an unusual area is that of a migrant or just of a reared butterfly. Monarchs in the Kootenays may be the eastern North American population; those in the rest of the province are part of the western North American population.

The Monarchs in BC are of Special Concern because of destruction of their hibernation sites in California as a result of urban development. In addition, milkweed is considered a noxious weed by the agricultural industry in many provinces, and is sometimes deliberately eliminated (Crolla and Lafontaine 1996). Milkweed is still abundant in the Southern Interior of BC, but may become less common in the future as more land is cleared or as range weed control programs eliminate it.

GENERAL DISTRIBUTION: Monarchs are found in many tropical and subtropical areas of the world. In North America the migratory populations hibernate in CA (western populations) and MEX (eastern populations). They migrate north to central CAN. Monarchs colonized New Zealand about 1840, Australia in 1870, and the Canary Islands in 1880, following the introduction of weedy asclepiads upon which the larvae could feed (Higgins and Riley 1970).

CONSERVATION STATUS: Monarchs have been designated as of Special Concern throughout Canada by the Committee on the Status of Endangered Wildlife in Canada (COSEWIC). In 1993 the winter roosts in Mexico and California were designated endangered phenomena by the International Union for the Conservation of Nature and Natural Resources (Crolla and Lafontaine 1996). The provincial status is S3BSZN, because the provincial breeding population requires monitoring and there are also non-breeding migrants in areas without milkweed.

Monarch (*Danaus plexippus*)

Photo on previous page: Rocky Mountain Apollo (*Parnassius smintheus magnus*)

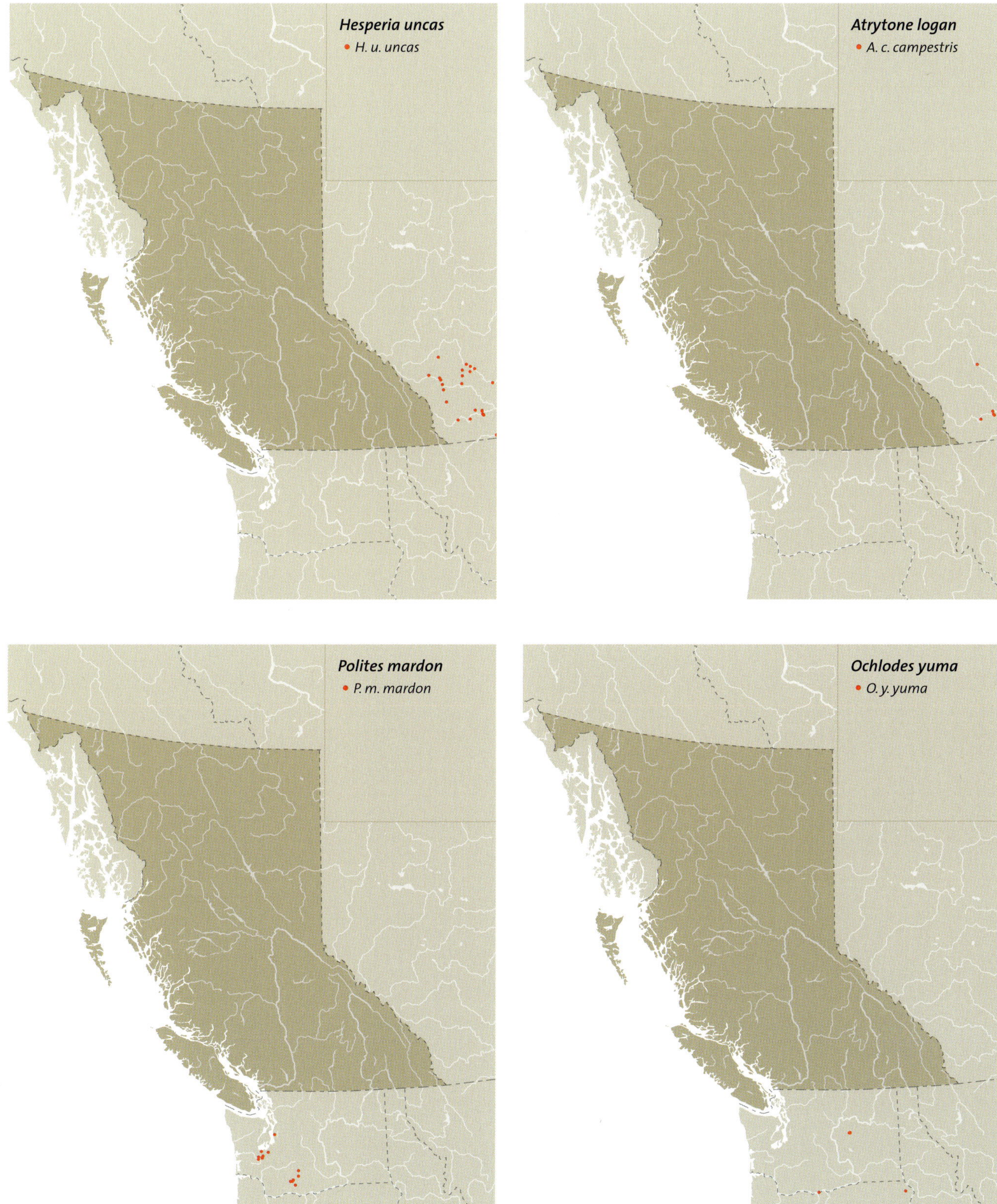


Hesperia uncas
H. u. uncas
Atrytone logan
A. c. campestris
Polites mardon
P. m. mardon
Ochlodes yuma
O. y. yuma

Lycaena editha
L. e. montana

Euchloe olympia

Habrodais grunus
H. g. herri

Satyrium acadicum
S. a. montanensis

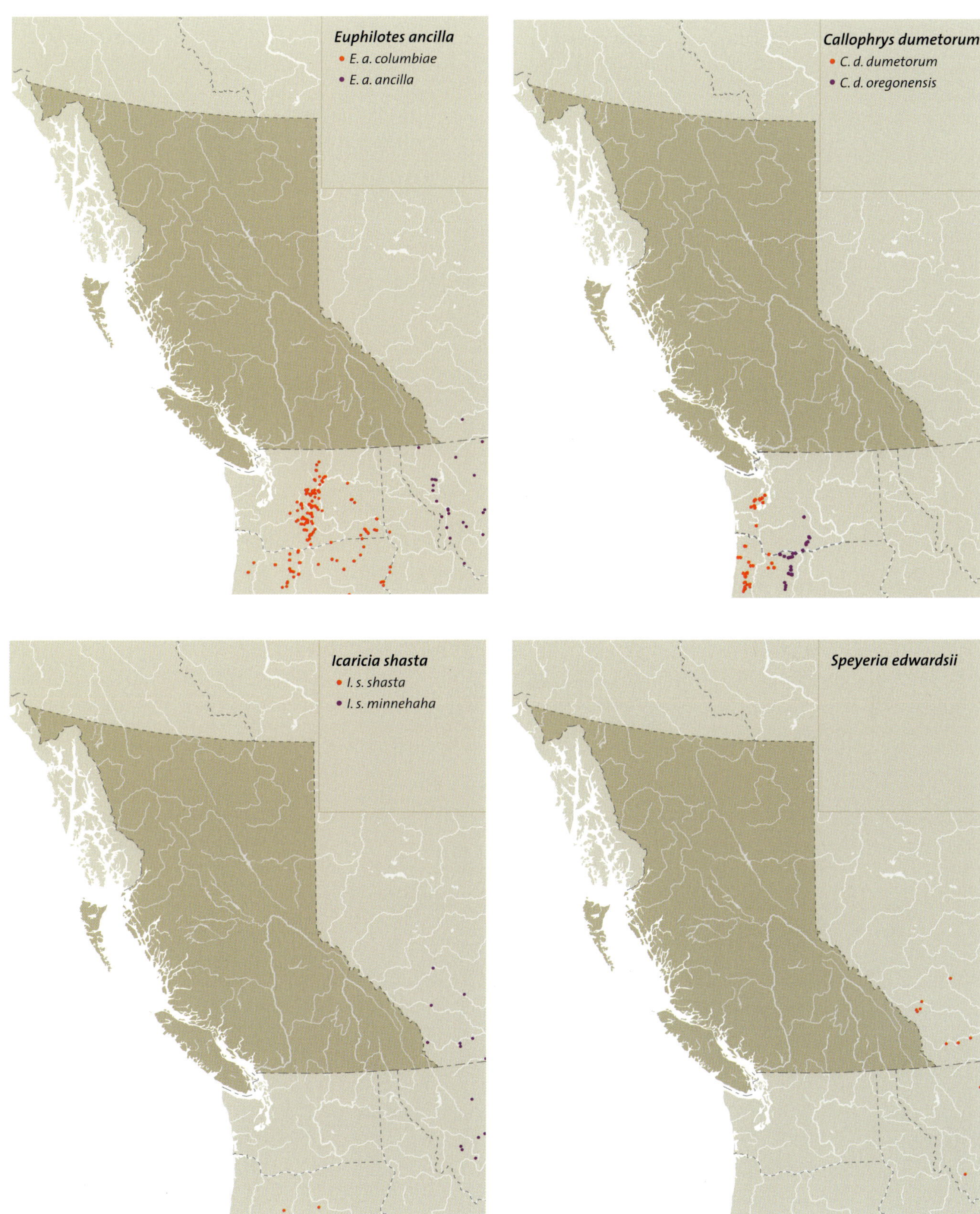

Euphilotes ancilla
E. a. columbiae
E. a. ancilla

Callophrys dumetorum
C. d. dumetorum
C. d. oregonensis

Icaricia shasta
I. s. shasta
I. s. minnehaha

Speyeria edwardsii

Speyeria coronis
S. c. simaetha

Charidryas gorgone
C. g. carlota


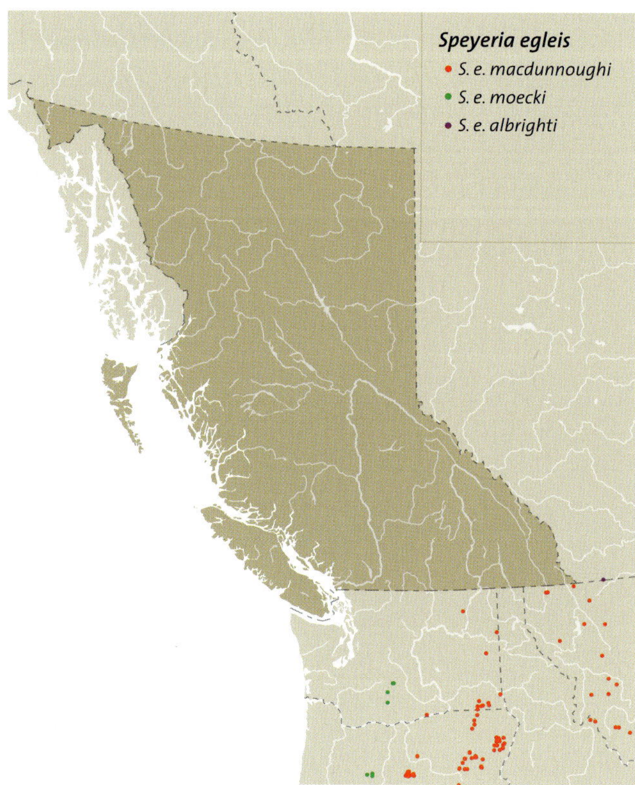

Speyeria egleis
S. e. macdunnoughi
S. e. moecki
S. e. albrighti

Charidryas acastus
C. a. acastus
C. a. dorothyae

Limentis
weidemeyerii
L. w. oberfoelli

Coenonympha
haydeni

Adelpha bredowii
A. b. californica

Neominois ridingsii
N. r. ridingsii
N. r. minimus

This checklist of the butterflies and skippers of British Columbia includes all known species and subspecies through 2000. The list includes hypothetical species and subspecies that may eventually be found in the province, indicated by a single asterisk (*). It also includes species and subspecies known from adjacent areas that are shown on the distribution maps but are not expected to occur in the province; these are indicated by a double asterisk (**). For these latter species new synonyms are listed in the checklist because the species are not discussed in the text. Otherwise new synonyms are listed in the text.

Superfamily **Hesperioidea** Latreille, 1809

Family **Hesperiidae** Latreille, 1809

Subfamily **Pyrginae** Burmeister, 1878

Genus *Epargyreus* Hübner, [1819]
1. *Epargyreus clarus* (Cramer, [1775])
   a. *E. c. californicus* MacNeill, 1975
   b. *E. c. clarus* (Cramer, [1775])

Genus *Thorybes* Scudder, 1872
2. *Thorybes pylades* (Scudder, 1870)
   a. *T. p. pylades* (Scudder, 1870)

Genus *Erynnis* Schrank, 1801
3. *Erynnis icelus* (Scudder & Burgess, 1870)
4. *Erynnis propertius* (Scudder & Burgess, 1870)
5. *Erynnis pacuvius* (Lintner, 1878)
   a. *E. p. lilius* (Dyar, 1904)
6. *Erynnis afranius* (Lintner, 1878)
7. *Erynnis persius* (Scudder, 1863)

Genus *Pyrgus* Hübner, [1819]
8. *Pyrgus centaureae* (Rambur, 1842)
   a. *P. c. loki* Evans, 1953
   b. *P. c. freija* (Warren, [1924])
9. *Pyrgus ruralis* (Boisduval, 1852)
   a. *P. r. ruralis* (Boisduval, 1852)
10. *Pyrgus communis* (Grote, 1872)
   a. *P. c. communis* (Grote, 1872)

Genus *Heliopetes* Billberg, 1820
* *Heliopetes ericetorum* (Boisduval, 1852)

Genus *Pholisora* Scudder, 1872
11. *Pholisora catullus* (Fabricius, 1793)

Subfamily **Hesperiinae** Latreille, 1809

Genus *Carterocephalus* Lederer, 1852
12. *Carterocephalus palaemon* (Pallas, 1771)
   a. *C. p. mandan* (Edwards, 1863)
   b. *C. p. skada* (Edwards, 1870)
   c. *C. p. magnus* Mattoon & Tilden, 1998

Genus *Oarisma* Scudder, 1872
13. *Oarisma garita* (Reakirt, 1866)

Genus *Thymelicus* Hübner, [1819]
14. *Thymelicus lineola* (Ochsenheimer, 1808)

Genus *Hesperia* Fabricius, 1793
** *Hesperia uncas* Edwards, 1863
15. *Hesperia juba* (Scudder, 1874)
16. *Hesperia comma* (Linnaeus, 1758)
   a. *H. c. oregonia* (Edwards, 1883)
   b. *H. c. harpalus* (Edwards, 1881)
   c. *H. c. manitoba* (Scudder, 1874)
   d. *H. c. assiniboia* (Lyman, 1892)
   ** *H. c. hulbirti* Lindsey, 1939
17. *Hesperia nevada* (Scudder, 1874)
   a. *H. n. nevada* (Scudder, 1874)

Genus *Polites* Scudder, 1872
18. *Polites peckius* (Kirby, 1837)
19. *Polites sabuleti* (Boisduval, 1852)
   a. *P. s. sabuleti* (Boisduval, 1852)
** *Polites mardon* (Edwards, 1881)
20. *Polites draco* (Edwards, 1871)
21. *Polites themistocles* (Latreille, [1824])
   a. *P. t. turneri* Freeman, 1944
   b. *P. t. themistocles* (Latreille, [1824])
22. *Polites mystic* (Edwards, 1863)
23. *Polites sonora* (Scudder, 1872)
   a. *P. s. sonora* (Scudder, 1872)
   ** *P. s. utahensis* (Skinner, 1911)
   ** *P. s. siris* (Edwards, 1881)

Genus *Atalopedes* Scudder, 1872
24. *Atalopedes campestris* (Boisduval, 1852)
   a. *A. c. campestris* (Boisduval, 1852)

Genus *Atrytone* Scudder, 1872
**  *Atrytone logan* (Edwards, 1863)

Genus *Ochlodes* Scudder, 1872
25. *Ochlodes sylvanoides* (Boisduval, 1852)
**  *Ochlodes yuma* (Edwards, 1873)
    **  *O. y. yuma* (Edwards, 1873)

Genus *Euphyes* Scudder, 1872
26. *Euphyes vestris* (Boisduval, 1852)
    a.  *E. v. vestris* (Boisduval, 1852)
    **  *E. v. kiowah* (Reakirt, 1866)

Genus *Amblyscirtes* Scudder, 1872
27. *Amblyscirtes vialis* (Edwards, 1862)

Superfamily **Papilionoidea** Latreille, [1802]

Family **Papilionidae** Latreille, [1802]

Subfamily **Parnassiinae** Duponchel, [1835]

Genus *Parnassius* Latreille, 1804
28. *Parnassius eversmanni* Ménétriés, [1851]
    a.  *P. e. thor* Hy. Edwards, 1881
29. *Parnassius clodius* Ménétriés, 1855
    a.  *P. c. claudianus* Stichel, 1907
    b.  *P. c. pseudogallatinus* Bryk, 1913
    c.  *P. c. altaurus* Dyar, 1903
    **  *P. c. shepardi* Eisner, 1966
30. *Parnassius smintheus* Doubleday, [1847]
    a.  *P. s. olympiannus* Burdick, 1941
    b.  *P. s. magnus* Wright, 1905
    c.  *P. s. smintheus* Doubleday, [1847]
    d.  *P. s. yukonensis* Eisner, 1969
31. *Parnassius phoebus* (Fabricius, 1793)
    a.  *P. p. apricatus* Stichel, 1906

Subfamily **Papilioninae** Latreille, [1802]

Genus *Papilio* Linnaeus, 1758
32. *Papilio bairdii* Edwards, 1869
    a.  *P. b. oregonius* Edwards, 1876
    b.  *P. b. pikei* Sperling, 1987
    c.  *P. b. dodi* McDunnough, 1939
33. *Papilio machaon* Linnaeus, 1758
    a.  *P. m. aliaska* Scudder, 1869
    **  *P. m. hudsonianus* A. H. Clarke, 1932
34. *Papilio zelicaon* Lucas, 1852
35. *Papilio indra* Reakirt, 1866
    a.  *P. i. indra* Reakirt, 1866
36. *Papilio canadensis* Rothschild & Jordan, 1906
37. *Papilio rutulus* Lucas, 1852
    a.  *P. r. rutulus* Lucas, 1852
38. *Papilio multicaudatus* Kirby, 1884
    a.  *P. m. pusillus* Austin & J. Emmel, 1998
39. *Papilio eurymedon* Lucas, 1852

Family **Pieridae** Duponchel, [1835]

Subfamily **Pierinae** Duponchel, [1835]

Genus *Neophasia* Behr, 1869
40. *Neophasia menapia* (C. & R. Felder, 1859)
    a.  *N. m. tau* (Scudder, 1861)

Genus *Pontia* Fabricius, 1807
41. *Pontia beckerii* (Edwards, 1871)
42. *Pontia sisymbrii* (Boisduval, 1852)
    a.  *P. s. flavitincta* (Comstock, 1924)
    b.  *P. s. beringiensis* Guppy & Kondla, 2001
43. *Pontia protodice* (Boisduval & Leconte, 1829)
44. *Pontia occidentalis* (Reakirt, 1866)
    a.  *P. o. occidentalis* (Reakirt, 1866)
    b.  *P. o. nelsoni* (Edwards, 1883)

Genus *Pieris* Schrank, 1801
45. *Pieris marginalis* Scudder, 1861
    a.  *P. m. marginalis* Scudder, 1861
    b.  *P. m. reicheli* Eitschberger, 1983
    c.  *P. m. guppyi* Eitschberger, 1983
    d.  *P. m. tremblayi* Eitschberger, 1983
46. *Pieris oleracea* Harris, 1829
    a.  *P. o. oleracea* Harris, 1829
47. *Pieris angelika* Eitschberger, 1983
48. *Pieris rapae* Linnaeus, 1758
    a.  *P. r. rapae* Linnaeus, 1758

Subfamily **Anthocharinae** Tutt, 1896

Genus *Euchloe* Hübner, [1819]
49. *Euchloe ausonides* (Lucas, 1852)
    a.  *E. a. mayi* F. & R. Chermock, 1940
    b.  *E. a. ogilvia* Back, 1990
    c.  *E. a. insulanus* Guppy & Shepard, 2001
50. *Euchloe creusa* (Doubleday, [1847])
51. *Euchloe naina* Kozhantshikov, 1923
52. Euchloe *lotta* Beutenmüller, 1898
**  *Euchloe olympia* (Edwards, 1871)

Genus *Anthocharis* Boisduval, Rambur, Duméril & Graslin, [1833]
53. *Anthocharis sara* Lucas, 1852
    a.  *A. s. flora* Wright, 1892
    b.  *A. s. alaskensis* Gunder, 1932
54. *Anthocharis stella* Edwards, 1879
    a.  *A. s. stella* Edwards, 1879

Subfamily **Coliadinae** Swainson, 1827

Genus *Colias* Fabricius, 1807
55. *Colias philodice* Godart, [1819]
    a.  *C. p. eriphyle* Edwards, 1876
    b.  *C. p. vitabunda* Hovanitz, 1943
56. *Colias eurytheme* Boisduval, 1852
57. *Colias alexandra* Edwards, 1863
    a.  *C. a. columbiensis* Ferris, 1973
    b.  *C. a. pseudocolumbiensis* Guppy & Shepard, 2001
    **  *C. a. alexandra* Edwards, 1863
    **  *C. a. edwardsii* Edwards, 1870

58. *Colias christina* Edwards, 1863
    a. *C. c. christina* Edwards, 1863
    * *C. c. kluanensis* Ferris, 1981
59. *Colias occidentalis* Scudder, 1862
    a. *C. o. occidentalis* Scudder, 1862
    ** *C. o. pseudochristina* Ferris, 1989
60. *Colias meadii* Edwards, 1871
    a. *C. m. elis* Strecker, 1885
    ** *C. m. lemhiensis* Curtis & Ferris, 1985
61. *Colias hecla* Lefèbvre, 1836
    a. *Colias h. hecla* Lefèbvre, 1836
62. *Colias canadensis* Ferris, 1982
63. *Colias nastes* Boisduval, [1834]
    a. *C. n. streckeri* Grum-Grschimailo, 1895
    b. *C. n. aliaska* Bang-Haas, 1927
64. *Colias chippewa* Edwards, 1872
    a. *C. c. chippewa* Edwards, 1872
65. *Colias interior* Scudder, 1862
66. *Colias pelidne* Boisduval & Leconte, 1829
    a. *C. p. minisni* Bean, 1895
    ** *C. p. skinneri* Barnes, 1897
67. *Colias gigantea* Strecker, 1900
    a. *C. g. gigantea* Strecker, 1900
    b. *C. g. mayi* F. & R. Chermock, 1940
    ** *C. g. harroweri* Klots, 1940

Genus *Eurema* Hübner, [1819]
    * *Eurema nicippe* (Cramer, [1779])

Genus *Nathalis* Boisduval, 1836
    * *Nathalis iole* Boisduval, 1836

## Family **Lycaenidae** Leach, 1815

### Subfamily **Lycaeninae** Leach, 1815

Genus *Lycaena* Fabricius, 1807
68. *Lycaena (Lycaena) phlaeas* (Linnaeus, 1761)
    a. *L. p. arethusa* (Wolley-Dod, 1907)
    ** *L. p. arctodon* Ferris, 1974
69. *Lycaena (Lycaena) cuprea* (Edwards, 1870)
    a. *L. c. henryae* (Cadbury, 1937)
    ** *L. c. cuprea* (Edwards, 1870)
70. *Lycaena (Hyllolycaena) hyllus* (Cramer, [1775])
71. *Lycaena (Chalceria) dione* (Scudder, [1870])
    ** *Lycaena (Chalceria) editha* (Mead, 1878)
    ** *L. e. montana* (Field, 1936)
    * *Lycaena (Chalceria) rubida* (Behr, 1866)
    * *L. r. perkinsorum* Johnson & Balogh, 1977
    ** *L. r. duofacies* Johnson & Balogh, 1977
    ** *L. r. sirius* (Edwards, 1871)
72. *Lycaena (Chalceria) heteronea* Boisduval, 1852
    a. *L. h. heteronea* Boisduval, 1852
73. *Lycaena (Epidemia) dorcas* Kirby, 1837
    a. *L. d. dorcas* Kirby, 1837
74. *Lycaena (Epidemia) helloides* (Boisduval, 1852)
75. *Lycaena (Epidemia) nivalis* (Boisduval, 1869)
    a. *L. n. browni* dos Passos, 1938

76. *Lycaena (Epidemia) mariposa* (Reakirt, 1866)
    a. *L. m. mariposa* (Reakirt, 1866)
    b. *L. m. charlottensis* (Holland, 1930)
    c. *L. m. penroseae* Field, 1938

## Subfamily **Theclinae** Swainson, 1831

Genus *Habrodais* Scudder, 1876
    ** *Habrodais grunus* (Boisduval, 1852)
    ** *H. g. herri* Field, 1938

Genus *Satyrium* Scudder, 1876
77. *Satyrium titus* (Fabricius, 1793)
    a. *S. t. immaculosus* (W.P. Comstock, 1913)
    b. *S. t. titus* (Fabricius, 1793)
78. *Satyrium behrii* (Edwards, 1870)
    a. *S. b. columbia* (McDunnough, 1944)
79. *Satyrium fuliginosum* (Edwards, 1861)
    a. *S. f. semiluna* Klots, 1930
80. *Satyrium californicum* (Edwards, 1862)
    ** *Satyrium acadicum* (Edwards, 1862)
    ** *S. a. montanensis* (Watson & Comstock, 1920)
    (= *coolinensis* (Watson & Comstock, 1920))
    (= *watrini* (Dufrane, 1939))
81. *Satyrium sylvinum* (Boisduval, 1852)
    a. *S. s. nootka* Fisher, 1998
    ** *S. s.* undescribed
82. *Satyrium liparops* (Leconte, 1833)
    a. *S. l. fletcheri* (Michener & dos Passos, 1942)
    ** *S. l. aliparops* (Michener & dos Passos, 1942)
83. *Satyrium saepium* (Boisduval, 1852)
    a. *S. s. okanaganum* (McDunnough, 1944)

Genus *Callophrys* Billberg, 1820
    ** *Callophrys perplexa* Barnes & Benjamin, 1923
    ** *C. p. oregonensis* Gorelick, 1969
84. *Callophrys affinis* (Edwards, 1862)
    a. *C. a. washingtonia* Clench, 1944
    ** *C. a. affinis* (Edwards, 1862)
85. *Callophrys sheridanii* (Edwards, 1877)
    a. *C. s. newcomeri* Clench, 1963

Genus *Loranthomitoura* Ballmer & Pratt, 1992
86. *Loranthomitoura spinetorum* (Hewitson, 1867)
87. *Loranthomitoura johnsoni* (Skinner, 1904)

Genus *Mitoura* Scudder, 1872
88. *Mitoura rosneri* Johnson, 1976
    a. *M. r. rosneri* Johnson, 1976
    b. *M. r. plicataria* Johnson, 1976
89. *Mitoura siva* (Edwards, 1874)
    a. *M. s. barryi* Johnson, 1976

Genus *Incisalia* Scudder, 1872
90. *Incisalia augustinus* (Westwood, [1852])
    a. *I. a. augustinus* (Westwood, [1852])
91. *Incisalia iroides* (Boisduval, 1852)
    a. *I. i. iroides* (Boisduval, 1852)

92. *Incisalia mossii* (Hy. Edwards, 1881)
    a. *I. m. mossii* (Hy. Edwards, 1881)
    b. *I. m. schryveri* Cross, 1937
93. *Incisalia polia* (Cook & Watson, 1907)
    a. *I. p. polia* (Cook & Watson, 1907)
    ** *I. p. maritima* Emmel, Emmel, & Matton, 1998
94. *Incisalia niphon* (Hübner, [1823])
    a. *I. n. clarki* T.N. Freeman, 1938
95. *Incisalia eryphon* (Boisduval, 1852)
    a. *I. e. eryphon* (Boisduval, 1852)
    b. *I. e. sheltonensis* F. Chermock & Frechin, 1948

Genus *Strymon* Hübner, [1818]
96. *Strymon melinus* Hübner, [1818]
    a. *S. m. atrofasciatus* McDunnough, 1921
    b. *S. m. setonia* McDunnough, 1927
    ** *S. m. franki* Field, 1938

## Subfamily **Polyommatinae** Swainson, 1827

### Tribe **Everini** Tutt, 1908

Genus *Everes* Hübner, [1819]
97. *Everes comyntas* (Godart, [1824])
    a. *E. c. comyntas* (Godart, [1824])
98. *Everes amyntula* (Boisduval, 1852)
    a. *E. a. amyntula* (Boisduval, 1852)

### Tribe **Celastrini** Tutt, 1908

Genus *Celastrina* Tutt, 1906
99. *Celastrina echo* (Edwards, 1864)
    a. *C. e. echo* (Edwards, 1864)
100. *Celastrina ladon* (Cramer, [1780])
    a. *C. l. lucia* (Kirby, 1837)

### Tribe **Scolitandini** Tutt, 1909

Genus *Euphilotes* Mattoni, [1978]
101. *Euphilotes battoides* (Behr, 1867)
    a. *E. b. glaucon* (Edwards, 1871)
  ** *Euphilotes ancilla* (Barnes & McDunnough, 1917)
    ** *E. a. ancilla* (Barnes & McDunnough, 1917)
    ** *E. a. columbiae* (Mattoni, [1955])

Genus *Glaucopsyche* Scudder, 1872
102. *Glaucopsyche piasus* (Boisduval, 1852)
    a. *G. p. toxeuma* Brown, 1971
103. *Glaucopsyche lygdamus* (Doubleday, 1841)
    a. *G. l. columbia* (Skinner, 1917)
    b. *G. l. couperi* Grote, 1873

### Tribe **Polyommatini** Swainson, 1827

Genus *Lycaeides* Hübner, [1819]
104. *Lycaeides idas* (Linnaeus, 1761)
    a. *L. i. atrapraetextus* (Field, 1939)
    b. *L. i. scudderi* (Edwards, 1861)
    c. *L. i. alaskensis* (F. Chermock, 1945)

105. *Lycaeides anna* (Edwards, 1861)
    a. *L. a. ricei* (Cross, 1937)
    b. *L. a. vancouverensis* Guppy & Shepard, 2001
106. *Lycaeides melissa* (Edwards, 1873)
    a. *L. m. melissa* (Edwards, 1873)

Genus *Plebeius* Kluk, 1780
107. *Plebeius saepiolus* (Boisduval, 1852)
    a. *P. s. insulanus* Blackmore, 1919
    b. *P. s. amica* (Edwards, 1863)

Genus *Icaricia* Nabokov, 1944
108. *Icaricia icarioides* (Boisduval, 1852)
    a. *I. i. blackmorei* (Barnes & McDunnough, 1919)
    b. *I. i. montis* (Blackmore, 1923)
    c. *I. i. pembina* (Edwards, 1862)
    ** *I. i. fenderi* (Macy, 1931)
  ** *Icaricia shasta* (Edwards, 1862)
    ** *I. s. shasta* (Edwards, 1862)
    ** *I. s. minnehaha* (Scudder, 1875)
109. *Icaricia acmon* (Westwood & Hewitson, [1852])
    a. *I. a. lutzi* (dos Passos, 1938)
    ** *I. a. spangelatus* (Burdick, 1942)

Genus *Vacciniina* Tutt, 1909
110. *Vacciniina optilete* (Knoch, 1781)
    a. *V. o. yukona* (Holland, 1900)

Genus *Agriades* Hübner, [1819]
111. *Agriades glandon* (de Prunner, 1798)
    a. *A. g. megalo* (McDunnough, 1927)
    b. *A. g. lacustris* (T. Freeman, 1939)

# Family **Riodinidae** Grote, 1895

## Subfamily **Riodininae** Grote, 1895

Genus *Apodemia* C. & R. Felder, [1865]
112. *Apodemia mormo* (C. & R. Felder, 1859)
    a. *A. m. mormo* (C. & R. Felder, 1859)

# Family **Nymphalidae** Swainson, 1827

## Subfamily **Nymphalinae** Swainson, 1827

Genus *Polygonia* Hübner, [1819]
113. *Polygonia satyrus* (Edwards, 1869)
114. *Polygonia faunus* (Edwards, 1862)
    a. *P. f. rusticus* (Edwards, 1874)
115. *Polygonia zephyrus* (Edwards, 1870)
116. *Polygonia gracilis* (Grote & Robinson, 1867)
117. *Polygonia oreas* (Edwards, 1869)
    a. *P. o. silenus* (Edwards, 1870)
    b. *P. o. threatfuli* Guppy & Shepard, 2001
118. *Polygonia progne* (Cramer, [1776])

Genus *Roddia* Korshunov, 1995
119. *Roddia l-album* (Esper, 1780)
    a. *R. l. watsoni* (Hall, 1924)

Genus *Nymphalis* Kluk, 1780
120.   *Nymphalis californica* (Boisduval, 1852)
121.   *Nymphalis antiopa* (Linnaeus, 1758)

Genus *Aglais* Dalman, 1816
122.   *Aglais milberti* (Godart, [1819])
        a.   *A. m. milberti* (Godart, [1819])

Genus *Vanessa* Fabricius, 1807
123.   *Vanessa virginiensis* (Drury, 1773)
124.   *Vanessa cardui* (Linnaeus, 1758)
125.   *Vanessa annabella* (Field, 1971)
126.   *Vanessa atalanta* (Linnaeus, 1758)
        a.   *V. a. rubria* (Fruhstorfer, 1909)

Subfamily **Argynninae** Duponchel, [1835]

Genus *Euptoieta* Doubleday, [1848]
127.   *Euptoieta claudia* (Cramer, [1775])

Genus *Speyeria* Scudder, 1872
128.   *Speyeria cybele* (Fabricius, 1775)
        a.   *S. c. leto* (Behr, 1862)
        b.   *S. c. pseudocarpenteri* (F. & R. Chermock, 1940)
       **   *S. c. pugetensis* (F. Chermock & Frechin, 1947)
  **    *Speyeria edwardsii* (Reakirt, 1866)
129.   *Speyeria aphrodite* (Fabricius, 1787)
        a.   *S. a. columbia* (Hy. Edwards, 1877)
        b.   *S. a. whitehousei* (Gunder, 1932)
        c.   *S. a. manitoba* (F. & R. Chermock, 1940)
   **   *Speyeria coronis* (Behr, 1864)
       **   *S. c. simaetha* (dos Passos & Grey, 1945)
130.   *Speyeria zerene* (Boisduval, 1852)
        a.   *S. z. bremnerii* (Edwards, 1872)
        b.   *S. z. picta* (McDunnough, 1924)
        c.   *S. z. garretti* (Gunder, 1932)
        *    *S. z. sitka* Hammond, Harry & McCorkle, 2001
       **   *S. z. hippolyta* (Edwards, 1879)
131.   *Speyeria callippe* (Boisduval, 1852)
        a.   *S. c. semivirida* (McDunnough, 1924)
        b.   *S. c. chilcotinensis* Guppy & Shepard, 2001
       **   *S. c. calgariana* (McDunnough, 1924)
       **   *S. c. elaine* dos Psssos & Grey, 1945
  **    *Speyeria egleis* (Behr, 1862)
       **   *S. e. macdunnoughi* (Gunder, 1932)
       **   *S. e. moecki* (Hammond & Dornfield, 1983)
       **   *S. e. albrighti* (Gunder, 1932)
132.   *Speyeria atlantis* (Edwards, 1862)
        a.   *S. a. hollandi* (F. & R. Chermock, 1940)
133.   *Speyeria hesperis* (Edwards, 1864)
        a.   *S. h. helena* dos Passos & Grey, 1957
        b.   *S. h. beani* (Barnes & Benjamin, 1926)
       **   *S. h. dodgei* (Gunder, 1931)
134.   *Speyeria hydaspe* (Boisduval, 1869)
        a.   *S. h. rhodope* (Edwards, 1874)
        b.   *S. h. minor* dos Passos & Grey, 1947
        c.   *S. h. sakuntala* (Skinner, 1911)

135.   *Speyeria mormonia* (Boisduval, 1869)
        a.   *S. m. bischoffii* (Edwards, 1870)
        b.   *S. m. opis* (Edwards, 1874)
        c.   *S. m. erinna* (Edwards, 1883)
       **   *S. m. eurynome* (W. H. Edwards, 1872)

Genus *Boloria* Moore, [1900]
136.   *Boloria napaea* (Hoffmansegg, 1804)
        a.   *B. n. alaskensis* (Holland, 1900)

Genus *Clossiana* Reuss, 1920
137.   *Clossiana eunomia* (Esper, 1799)
        a.   *C. e. dawsoni* (Barnes & McDunnough, 1916)
        *    *C. e. nichollae* (Barnes & Benjamin, 1926)
138.   *Clossiana selene* (Denis & Schiffermüller, [1775])
        a.   *C. s. atrocostalis* (Huard, 1927)
139.   *Clossiana bellona* (Fabricius, 1775)
        a.   *C. b. toddi* (Holland, 1928)
        b.   *C. b. jenistorum* Stallings & Turner, 1947
140.   *Clossiana frigga* (Thunberg, 1791)
        a.   *C. f. saga* (Staudinger, 1861)
141.   *Clossiana improba* (Butler, 1877)
        a.   *C. i. youngi* (Holland, 1900)
142.   *Clossiana epithore* (Edwards, [1864])
        a.   *C. e. chermocki* E. & S. Perkins, 1966
        b.   *C. e. sigridae* Shepard, 2001
143.   *Clossiana polaris* (Boisduval, [1828])
        a.   *C. p. polaris* (Boisduval, [1828])
144.   *Clossiana alberta* (Edwards, 1890)
145.   *Clossiana freija* (Thunberg, 1791)
        a.   *C. f. freija* (Thunberg, 1791)
146.   *Clossiana natazhati* (Gibson, 1920)
        a.   *C. n. nabokovi* (Stallings & Turner, 1947)
        *    *C. n. natazhati* (Gibson, 1920)
147.   *Clossiana tritonia* (Boeber, 1812)
        a.   *C. t. astarte* (Doubleday, [1847])
        b.   *C. t. distincta* (Gibson, 1920)
148.   *Clossiana chariclea* (Schneider, 1794)
        a.   *C. c. rainieri* (Barnes & McDunnough, 1913)
        b.   *C. c. grandis* (Barnes & McDunnough, 1916)
        c.   *C. c. butleri* (Edwards, 1883)

Subfamily **Melitaeinae** Grote, 1897

Genus *Phyciodes* Hübner, [1819]
149.   *Phyciodes tharos* (Drury, [1773])
        a.   *P. t. pascoensis* Wright, 1905
        b.   *P. t. cocyta* (Cramer, [1777])
       **   *P. t. tharos* (Drury, [1773])
150.   *Phyciodes batesii* (Reakirt, 1865)
        a.   *P. b. lakota* Scott, 1994
151.   *Phyciodes pratensis* (Behr, 1863)
        a.   *P. p. pratensis* (Behr, 1863)
152.   *Phyciodes pallidus* (Edwards, 1864)
        a.   *P. p. barnesi* Skinner, 1897
153.   *Phyciodes mylitta* (Edwards, 1861)
        a.   *P. m. mylitta* (Edwards, 1861)

Genus *Charidryas* Scudder, 1872
- ** *Charidryas gorgone* (Hübner, 1810)
  - ** *C. g. carlota* (Reakirt, 1866)
154. *Charidryas palla* (Boisduval, 1852)
  - a. *C. p. calydon* (Holland, 1931)
  - ** *C. p. sterope* (Edwards, 1870)
  - ** *C. p. palla* (Boisduval, 1852)
- ** *Charidryas acastus* (Edwards, 1874)
  - ** *C. a. acastus* (Edwards, 1874)
  - ** *C. a. dorothyae* Bauer, 1975
155. *Charidryas whitneyi* (Behr, 1863)
  - a. *C. w. damoetas* (Skinner, 1902)
156. *Charidryas hoffmanni* (Behr, 1863)
  - a. *C. h. manchada* (Bauer, 1960)
  - ** *C. h. segregata* (Barnes & McDunnough, 1918)

Genus *Euphydryas* Scudder, 1872
157. *Euphydryas (Euphydryas) chalcedona* (Doubleday, [1847])
  - a. *E. c. paradoxa* McDunnough, 1927
  - ** *E. c. wallacensis* (Gunder, 1928)
  - ** *E. c. colon* (Edwards, 1881)
158. *Euphydryas (Euphydryas) anicia* (Doubleday, [1847])
  - a. *E. a. anicia* (Doubleday, [1847])
  - b. *E. a. hopfingeri* Gunder, 1934
  - * *E. a. helvia* (Scudder, 1869)
  - ** *E. a. bakeri* D. Stallings & Turner, 1945
  - ** *E. a. veasieae* Fender & Jewett, 1953
159. *Euphydryas (Euphydryas) editha* (Boisduval, 1852)
  - a. *E. e. taylori* (Edwards, 1888)
  - b. *E. e. beani* (Skinner, 1897)
  - ** *E. e. colonia* (Wright, 1905)
  - ** *E. e. edithana* (Strand, 1914)
  - ** *E. e. hutchinsi* McDunnough, 1928
  - ** *E. e. lawrencei* Gunder, 1931
160. *Euphydryas (Hypodryas) gillettii* (Barnes, 1897)

Subfamily **Limenitidinae** Behr, 1864

Genus *Limenitis* Fabricius, 1807
161. *Limenitis arthemis* (Drury, [1773])
  - a. *L. a. rubrofasciata* (Barnes & McDunnough, 1916)
162. *Limenitis archippus* (Cramer, [1776])
  - a. *L. a. idaho* Austin, 1998
  - ** *L. a. archippus* (Cramer, [1776])
- ** *Limenitis weidemeyerii* (Edwards, 1861)
  - ** *L. w. oberfoelli* (Brown, 1960)
163. *Limenitis lorquini* Boisduval, 1852
  - a. *L. l. itelkae* Guppy, 2001
  - b. *L. l. ilgae* Guppy, 2001

Genus *Adelpha* Hübner, [1819]
- ** *Adelpha bredowii* (Geyer, 1827)
  - ** *A. b. californica* (Butler, 1865)

Subfamily **Satyrinae** Boisduval, 1833

Genus *Coenonympha* Hübner, [1819]
- ** *Coenonympha haydeni* (Edwards. 1972)
164. *Coenonympha tullia* Müller, 1764
  - a. *C. t. yukonensis* Holland, 1900
165. *Coenonympha california* Westwood, [1851]
  - a. *C. c. insulana* McDunnough, 1928
  - b. *C. c. columbiana* McDunnough, 1928
  - c. *C. c. benjamini* McDunnough, 1928
  - * *C. c. mackenziei* Davenport, 1936
  - ** *C. c. ampelos* Edwards, 1871

Genus *Cercyonis* Scudder, 1875
166. *Cercyonis pegala* (Fabricius, 1775)
  - a. *C. p. incana* (Edwards, 1880)
  - b. *C. p. ariane* (Boisduval, 1852)
  - c. *C. p. ino* Hall, 1924
167. *Cercyonis sthenele* (Boisduval, 1852)
  - a. *C. s. sineocellata* Austin & J. Emmel, 1998
  - ** *C. s. silvestris* (Edwards, 1861)
168. *Cercyonis oetus* (Boisduval, 1869)
  - a. *C. o. phocus* (Edwards, 1874)
  - b. *C. o. charon* (Edwards, 1872)
  - ** *C. o. oetus* (Boisduval, 1869)

Genus *Erebia* Dalman, 1816
169. *Erebia vidleri* Elwes, 1898
170. *Erebia rossii* (Curtis, 1835)
  - a. *E. r. rossii* (Curtis, 1835)
171. *Erebia mancinus* Doubleday, [1849]
172. *Erebia magdalena* Strecker, 1880
  - a. *E. m. saxicola* Hilchie, 1990
173. *Erebia mackinleyensis* Gunder, 1932
  - a. *E. m. mackinleyensis* Gunder, 1932
- * *Erebia fasciata* Butler, 1868
  - * *E. f. fasciata* Butler, 1868
174. *Erebia discoidalis* (Kirby, 1837)
175. *Erebia pawlowskii* Ménétriés, 1859
  - a. *E. p. alaskensis* Holland, 1900
- * *Erebia youngi* Holland, 1900
  - * *E. y. youngi* Holland, 1900
- * *Erebia lafontainei* Troubridge & Philip, 1983
176. *Erebia epipsodea* Butler, 1868
  - a. *E. e. epipsodea* Butler, 1868
  - b. *E. e. sineocellata* Skinner, 1889
  - c. *E. e. remingtoni* Ehrlich, 1952

Genus *Neominois* Scudder 1975
- ** *Neominois ridingsii* (Edwards, 1865)
  - ** *N. r. ridingsii* (Edwards, 1865)
  - ** *N. r. minimus* Austin, 1986

Genus *Oeneis* Hübner, [1819]
177. *Oeneis nevadensis* (C. & R. Felder, [1866])
  - a. *O. n. gigas* (Butler, 1868)
  - ** *O. n. nevadensis* (C. & R. Felder, [1866])
178. *Oeneis macounii* (Edwards, 1885)

179. *Oeneis chryxus* (Doubleday, [1849])
   a. *O. c. chryxus* (Doubleday, [1849])
   b. *O. c. caryi* Dyar, 1904
   ** *O. c. valerata* Burdick, 1958
180. *Oeneis uhleri* (Reakirt, 1866)
   a. *O. u. varuna* (Edwards, 1882)
   ** *O. u. cairnesi* Gibson, 1920
   ** *O. u. nahanni* Dyar, 1904
181. *Oeneis alberta* Elwes, 1893
   a. *O. a. alberta* Elwes, 1893
182. *Oeneis bore* (Schneider, 1792)
   a. *O. b. edwardsi* dos Passos, 1949
   b. *O. b. mckinleyensis* dos Passos, 1949
183. *Oeneis jutta* (Hübner, 1805–1806)
   a. *O. j. reducta* McDunnough, 1929
   b. *O. j. ridingiana* F. & R. Chermock, 1940
   c. *O. j. alaskensis* Holland, 1900
184. *Oeneis melissa* (Fabricius, 1775)
   a. *O. m. beanii* Elwes, 1893
   b. *O. m. atlinensis* Guppy & Shepard, 2001
185. *Oeneis polixenes* (Fabricius, 1775)
   a. *O. p. yukonensis* Gibson, 1920
   b. *O. p. luteus* Troubridge & Parshall, 1988
   c. *O. p. brucei* (Edwards, 1891)
186. *Oeneis rosovi* Kurentzov, 1960
   a. *O. r. philipi* Troubridge, 1988

## Subfamily **Danainae** Boisduval, 1833

Genus *Danaus* Kluk, 1780
187. *Danaus plexippus* (Linnaeus, 1758)

## DATA FOR BUTTERFLY PHOTOGRAPHS

*Aglais milberti* **D/V**: BC, Goldbridge, Taylor Cr., el. 2,100–2,200 m, 9 Aug. 1977, C.S. Guppy (CSG). **1st instar larva**: BC, Williams Lake, 12 May 1991, Anna Roberts. **Eggs**: BC, Williams Lake, 22 Apr. 1988, Anna Roberts. **Mature larva**: BC, Williams Lake, 23 May 1990, Anna Roberts. **Pupa**: BC, Williams Lake, 11 Sep. 1990, Anna Roberts. **Live adult**: BC, Lytton, Botanie Valley, 4 July 1969, C.S. Guppy.

*Agriades glandon lacustris* ♂ **V**: BC, N of Taylor, high breaks, 30 May 1997, J.H. Shepard (JHS). ♀ **D**: BC, Beaton R., Hwy 131, 5 June 1997, J.H. Shepard (JHS).

*Agriades glandon megalo* ♂ **D/V**: BC, Strathcona Prov. Park, Cream Lk., el. 1,400 m, 22 Aug. 1988, C.S. Guppy (CSG). ♀ **D**: BC, Forbidden Plateau, Mt. Becher summit, 14 Aug. 1954, G.A. Hardy (RBCM).

*Amblyscirtes vialis* ♂ **D/V**: BC, Peace R. at Clayhurst Bridge, N side, 23 May 1997, J.H. Shepard (JHS). ♀ **D**: BC, Pend-d'Oreille R. at Four Mile Cr., 14 May 1998, J. and S. Shepard (JHS).

*Anthocharis sara alaskensis* ♂ **D/V**: BC, 5 km S of Atlin Lk., el. 750 m, 27 June 1973, J.L. Gordon (UBC). ♀ **D**: BC, Kemano, 16 June 1979, G. Anweiler (UBC).

*Anthocharis sara flora* ♂ **D**, ♀ **D**: BC, Metchosin, Camas Hill, 19 Apr. 1987, C.S. Guppy (CSG). **Egg**: BC, Lions Bay, M Cr., 1 Apr. 1987, C.S. Guppy. **Mature larva**: BC, Metchosin, Camas Hill, 19 May 1987, A.G. Guppy. **Pupation sequence**: BC, Metchosin, Camas Hill, 11 May 1987, A.G. Guppy. **Pupa**: OR, Polk Co., D.V. McCorkle. ♂ **live adult V**: OR, Portland, 21 Apr. 1973, C.S. Guppy.

*Anthocharis stella* ♂ **D/V**: BC, West Crescent Valley, el. 900 m, 1 May 1967, J.H. Shepard (JHS). ♀ **D**: BC, Hedley, 9 May 1971, R.G. Bartman (UBC). **Live adult**: BC, Williams Lake, 28 Apr. 1998, Anna Roberts.

*Apodemia mormo mormo* ♂ **D/V**: BC, Keremeos, 9 Aug. 1998, J.H. Shepard (JHS). **Live adult, pupa**: CA, Victorville, ex larva coll. 12 Apr. 1974, C.S. Guppy. **Mature larva**: WA, Klickitat Co., D.V. McCorkle.

*Atalopedes campestris* ♂ **D/V**: TX, 19 km N of Alpine, 10 Sep. 1969, J.A. Scott (JHS). ♀ **D**: OR, Corvallis, ex egg on lawn grass, emerged 10 Apr. 1980, C.S. Guppy (CSG). **Live adult**: OR, Corvallis, ex egg, photo 4 Apr. 1980, C.S. Guppy. **Egg hatching, 1st instar larva**: OR, Corvallis, ex egg, photo 27 Nov. 1979, C.S. Guppy. **Mature larva, pupa**: OR, Corvallis, ex egg, photo Oct. 1979, C.S. Guppy.

*Boloria napaea alaskensis* ♂ **D/V**, ♀ **D/V**: BC, St. Elias Mtns., Tkope R. near Tatshenshini R., el. 1,359 m, 23 July 1992, C.S. Guppy (RBCM).

*Callophrys affinis* ♂ **D/V**: BC, Vernon, Goose Lk., el. 580 m, 17 May 1985, C.S. Guppy (CSG). ♀ **D**: BC, Oyama, 6 km N, 10 May 1977, C.S. Guppy (CSG). **Live adult**: BC, Oyama, 6 km N, 17 May 1975, C.S. Guppy. **Live adult**: BC, Williams Lake, 16 Aug. 1990, Anna Roberts.

*Callophrys sheridanii* ♂ **D/V**: BC, Grand Forks, 13 km E at Stubb Cr., 27 May 1976, J.H. Shepard (JHS). ♀ **D**: BC, Lytton, Botanie Mtn. LO rd., el. 1,500 m, 21 May 1976, J.H. Shepard (JHS). **Egg**: BC, Lytton, Botanie Lk., 5 June 1973, C.S. Guppy. **1st instar larva**: BC, Lytton, Botanie Lk., 13 June 1973, C.S. Guppy. **Live adult V**: BC, Lytton, Botanie Lk., 27 May 1973, C.S. Guppy.

*Carterocephalus palaemon magnus* ♂ **D/V**: BC, Nazko, Redwater Cr. rd. 3.8 km W of Nazko rd., 3 June 1995, C.S. Guppy (CSG). **Live adult**: BC, Kootenay Nat. Park, 1 July 1975, C.S. Guppy. **Live adult D/V**: BC, Williams Lake, 21 May 1998, Anna Roberts.

*Carterocephalus palaemon mandan* ♂ **V**: BC, Peace R. on N side of Clayhurst Bridge, 1 June 1997, J.H. Shepard (JHS).

*Carterocephalus palaemon skada* ♂ **V**: BC, Atlin, 5 km S, el. 750 m, 27 June 1973, J.H. Shepard (JHS).

*Celastrina echo* ♂ **D/V**: BC, Lions Bay, M Cr., 17 Apr. 1987, C.S. Guppy (CSG). ♀ **D/V**: BC, Ruby Cr., W of Hope on Hwy 7, 7 June 1987, C.S. Guppy (CSG). ♀ **summer form D/V**: BC, Osoyoos Lk. (N end), 4 Aug. 1981, J. and S. Shepard (JHS). **Mature larva**: CA, San Diego Co., Laguna Mtns., Boiling Springs, coll. 20 June 1982, G.E. Pratt. **Feeding on ashes**: OR, Troutdale (near Portland), 21 Apr. 1973, C.S. Guppy. ♂ **live adult**: BC, Metchosin, Camas Hill, 1 May 1990, C.S. Guppy.

*Celastrina ladon lucia* ♂ **D/V**: BC, Quesnel, Schiste Cr., 20 Apr. 1994, C.S. Guppy (CSG). ♀ **D/V**: BC, Bowron Lakes Prov. Park, 1 km S of boundary, 21 May 1995, C.S. Guppy (CSG). ♂ ♀ **Mating**: BC, Williams Lake, 1 May 1994, Anna Roberts.

*Cercyonis oetus phocus* ♂ **D/V**: BC, Lytton, 0.5 km S at Lytton Cr., 3 July 1988, G.E. Hutchings (RBCM). ♀ **D**: BC, Wasa Lk. Prov. Park, 13 July 1997, D.W. Knight (JHS).

*Cercyonis pegala ariane* ♂ **D/V**: BC, Soda Cr., 5 July 1997, C.S. Guppy (CSG). ♀ **D**: BC, Osoyoos, Mt. Kobau, el. 1,100 m, 9 July 1991, D.C.A. Blades and Carol Maier (RBCM). **Mature larva**: BC, Lytton, Botanie Lake, ex. egg from E coll. 26 August 1993, Richard Beard. **Pupa**: BC, Chilcotin, 10 July 1991, Anna Roberts. **Live adult**: BC, Chilcotin, 22 July 1991, Anna Roberts.

*Cercyonis pegala incana* ♂ **D/V**: BC, Hornby Island, W of Helliwell Prov. Park, 23 July 1995, J. and S. Shepard (JHS). ♀ **D**: BC, Vanc. Isl., Nanoose CDN, 27 July 1995, J.H. Shepard (JHS).

*Cercyonis pegala ino* ♂ **D/V**, ♀ **D**: BC, Beaton R., 1.5 km N of Hwy 131, 28 July 1997, J.H. Shepard (CSG).

*Cercyonis sthenele* ♂ **D/V**: BC, Osoyoos Lk., NE of Osoyoos, 30 June 1996, J. and S. Shepard (JHS).

*Charidryas hoffmanni* ♂ **D/V**: BC, Manning Prov. Park, June 1968, R. Carcasson (RBCM). ♀ **D**: BC, Manning Prov. Park, Gibsons Pass, 23 July 1978, J.H. Shepard (JHS).

*Charidryas palla* ♂ **D/V**: BC, Toby Cr. 13 km W of Athalmer, 19 June 1969, J. and S. Shepard (JHS). ♀ **D**: BC, 3 km E of Glenemma, 20 June 1985, J. and S. Shepard (JHS). **Eggs**: BC, Lytton, Botanie Lk., June 1973, C.S. Guppy. **1st instar larvae**: BC, 10 km S of Logan Lake, el. 1,400 m, ♀ laid about 140 eggs in wild on underside of *Rhinanthus crysta-galli* leaf 9 July 1991, Richard Beard. **Prepupal larva**: BC, Lytton, Botanie Lk., 13 June 1973, C.S. Guppy. **Pupa**: BC, Lytton, Botanie Lk., 20 June 1973, C.S. Guppy. **Live adult D**: BC, Williams Lake, 5 June 1998, Anna Roberts. **Live adult V**: BC, Williams Lake, 10 June 1989, Anna Roberts.

*Charidryas whitneyi damoetas* ♂ **D/V**: BC, Ilgachuz Mtns., "Cross Mtn.," el. 2,000–2,300 m, 3 Aug. 1993, Dick Beard (CSG). ♀ **D**: BC, Mt. McLean, el. 2,250 m, 1–2 Aug. 1974, J. and S. Shepard (JHS).

*Clossiana alberta* ♂ **D/V**, ♀ **D**: AB, Plateau Mtn., el. 2,500 m, 18 July 1979, C.S. Guppy (CSG).

*Clossiana bellona jenistorum* ♂ **D/V**: BC, Peace R. at Clayhurst Bridge, 7 June 1997, J.H. Shepard (JHS).

*Clossiana bellona toddi* ♂ **D/V**: BC, Bridge Lk., 17 June 1977, J. and S. Shepard (JHS).

*Clossiana chariclea butleri* ♂ **D/V**: BC, Atlin, Boulder Cr., el. 1,200 m, 20 July 1977, J.H. Shepard (JHS). ♀ **D**: BC, St. Elias Mtns., Tkope R. near Tatshenshini R., el. 1,350 m, 23 July 1992, C.S. Guppy (RBCM).

*Clossiana chariclea grandis* ♂ **D/V**: BC, Ashcroft, Cornwall Hills, el. 2,000 m, 10 Aug. 1989, C.S. Guppy (RBCM). ♀ **D**: BC, Ilgachuz Mtns., Pan Cr., el. 1,700 m, 28 July 1987, R.A. Cannings (RBCM). **Live adult D**: BC, Tatla Lk., Little Meadow Mtn, 17 June 1989, Anna Roberts.

*Clossiana chariclea rainieri* ♂ **D/V**, ♀ **D**: BC, Mt. Cheam, el. 1,800 m, 9 Aug. 1981, G.B. Straley (RBCM). ♂ ♀ **Mating**: WA, Olympic Nat. Park, Hurricane Ridge, 7 Aug. 1979, C.S. Guppy.

*Clossiana epithore chermocki* ♂ **D/V**: BC, Trail, Violin Lk., 15 June 1952, A.C. Jenkins (JHS). **Live adult V**: BC, Lytton, Botanie Valley, 13 June 1970, A.G. Guppy.

*Clossiana epithore sigridae* ♂ **holotype D/V**, ♀ **D/V**: BC, St. Elias Mtns., Tats Lk., el. 770 m, 22 July 1992, C.S. Guppy (RBCM).

*Clossiana eunomia dawsoni* ♂ **D/V**: BC, Summit Lake, Alaska Hwy Mile 392.5, el. 1,350 m, 15 July 1977, J. and S. Shepard (JHS).

*Clossiana eunomia nichollae* ♂ **D/V**: AB, Wilcox Pass, el. 2,300 m, 29 July 1980, D.L. Threatful (JHS).

*Clossiana freija freija* ♂ **D/V**: BC, Keremeos, Crater Mtn., el. 2,100 m, 13 June 1976, C.S. Guppy (CSG). ♀ **D**: BC, Atlin, "Monarch Mtn.," el. 1,350 m, 25 June 1979, C.S. Guppy (CSG). **Live adult**: BC, McIntosh Lk., 20 May 1991, Anna Roberts.

*Clossiana frigga saga* ♂ **D/V**: BC, Atlin District, 26 June 1914 (RBCM).

*Clossiana improba youngi* ♂ **D/V**: BC, Pink Mtn. L.O., 15 July 1984, J. and S. Shepard (JHS). ♀ **D/V**: BC, Sweeney Mtn., el. 1,800 m, 12 July 1987, J. and S. Shepard (JHS). **Live adult D**: AK, Mt. McKinley Park, Highway Pass, 18 July 1976, C.S. Guppy.

*Clossiana natazhati nabokovi* ♂ **D/V**, ♀ **D**: BC, Mt. St. Paul, S ridge, el. 2,100 m, 2 Aug. 1999, J.H. Shepard (JHS).

*Clossiana natazhati natazhati* ♂ **D/V**, ♀ **D**: YT, Montana Mtn. near Carcross, el. 2,000 m, 17 June 1989, J. and S. Shepard (JHS).

*Clossiana polaris polaris* ♂ **D/V**: BC, Summit Lake, el. 1,450–1,650 m, 21 June 1976, C.S. Guppy (CSG). ♀ **D/V**: BC, Pink Mtn. L.O., 15 July 1984, J. and S. Shepard (JHS). **Live adult D**: BC, Summit Lake, Alaska Hwy Mile 392, el. 1,550-1,650 m, 21 June 1976, C.S. Guppy.

*Clossiana selene* ♂ **D/V**: BC, Galloway, 1.5 km SE, 17 June 1968, J. and S. Shepard (JHS). **Mature larva, pupae**: WA, Yakima Co., D.V. McCorkle. **Live adult**: BC, Williams Lake, 6 June 1990, Anna Roberts.

*Clossiana tritonia astarte* ♂ **D/V**: BC, Dore R. near McBride, el. 2,380 m, 17 July 1985, D.L. Threatful (JHS).

*Clossiana tritonia distincta* ♂ **D/V**: BC, "Monarch Mtn.," 5 km S of Atlin, el. 1,350 m, 4 July 1973, J.H. Shepard (JHS). **Live adult D**: BC, Atlin, "Monarch Mtn.," 28 June 1975, C.S. Guppy.

*Coenonympha california benjamini* ♂ **D/V**: BC, Beaton R., Hwy 131, 5 June 1997, J.H. Shepard (JHS).

*Coenonympha california columbiana* ♂ **D/V**: BC, Bull R. highway crossing, 12 May 1990, C.S. Guppy (CSG). **Mature larva**: BC, Williams Lake, 1 June 1990, Anna Roberts. **Pupa**: BC, Williams Lake, 18 June 1990, Anna Roberts. **Live adult**: BC, Oyama, 6 km N, 17 May 1975, C.S. Guppy.

*Coenonympha california insulana* ♂ **D/V**: BC, Shawnigan Lk.–Mill Bay rd. powerline, 24 May 1995, J.H. Shepard (JHS).

*Coenonympha tullia yukonensis* ♂ **D/V**: BC, Atlin, 28 June 1914 (RBCM). ♀ **D**: BC, Atlin, "Monarch Mtn.," el. 1,350 m, 19 July 1977, J. and S. Shepard (JHS).

*Colias alexandra columbiensis* ♀ **D/V**: BC. Lillooet Mtn., el. 1,500 m, 8 July 1968, Helmut Kimmich (UBC). ♂ **D/V**: BC, Riske Creek, McIntyre Lk., el. 930 m, 21 June 1998, C.S. Guppy (CSG).

*Colias alexandra pseudocolumbiensis* ♀ **D/V**: BC, White Lk., Kearns Cr., el. 550 m, 29 June 1996, C.S. Guppy (CSG). ♂ **holotype D/V**: BC, Hall Cr. at Hwy 33, 19 July 1986, J. and S. Shepard (JHS).

*Colias canadensis* ♂ **D/V**: BC, Summit Lake, 24 km E, 21 June 1976, C.S. Guppy (CSG). ♀ **orange form D/V**: BC, Atlin, Pine Cr., el. 900 m, 29 June 1973, J.H. Shepard (JHS). ♀ **white form D**: BC, Summit Lake, 23 km E, 21 June 1976, C.S. Guppy (CSG). **Mature larva**: BC, Atlin, Ruffner Mine rd., el. 930 m, ex ♀ 24 June 1999, photo 7 Aug. 1999, C.S. Guppy.

*Colias chippewa* ♂ **D/V**: BC, Atlin, "Monarch Mtn.," el. 1,350 m, 25 July 1975, C.S. Guppy (CSG). ♀ **yellow form D/V**: BC, Summit Lake, el. 1,650 m, 15 July 1977, J. and S. Shepard (JHS). ♀ **white form D**: BC, 16 km N of Stikine R. crossing, Hwy 37, 10 July 1989, J. and S. Shepard (JHS).

*Colias christina christina* ♂ **D/V**, ♀ **D/V**: BC, Pink Mtn., el. 1,200–1,500 m, 17 July 1998, C.S. Guppy (CSG).

*Colias christina kluanensis* ♂ ♀ **mating**: YT, Sandpete Cr., 29 July 1976, C.S. Guppy.

*Colias eurytheme* ♂ **D/V**: BC, Lk. Windemere, 22 Aug. 1918, W.B. Anderson (UBC). ♀ **orange form D/V**, ♀ **white form D**: BC, Manning Prov. Park, 19 July 1961, Helmut Kimmich (UBC).

*Colias gigantea gigantea* ♂ **D/V**, ♀ **D**: BC, Atlin, Ruffner Mine rd. km 6, 16 July 1991, N.G. Kondla (NGK).

*Colias gigantea mayi* ♂ **D/V**: BC, 20 km S of Wonowon on Alaska Hwy, 19 July 1984, J. and S. Shepard (JHS). ♀ **D**: BC, Pink Mtn., 2 km E of Alaska Hwy, el. 1,000 m, 17 July 1998, C.S. Guppy (CSG).

*Colias hecla* ♂ **D/V**: BC, Haines Hwy km 122, 24 July 1972, C.S. Guppy (CSG). ♀ **orange form D/V**: BC, Nadahini Mtn., Haines Hwy km 110 68, el. 1,100–1,400 m, C.S. Guppy (CSG). ♀ **white form D**: YT, Destruction Bay, Kluane Lk., el. 1,200 m, 10 July 1979, G. Anweiler (UBC).

*Colias interior* ♂ **D/V**, ♀ **D/V**: BC, Quesnel, Puntataekut Cr., el. 960 m, 18 July 1997, C.S. Guppy (CSG). **Live adult**: BC, Moffat Cr., 8 July 1990, Anna Roberts.

*Colias meadii* ♂ **D/V**, ♀ **D/V**: BC, Watch Peak, NE ridge, el. 2,400 m, 9 Aug. 1981, J. and S. Shepard (JHS).

*Colias nastes aliaska* ♂ **D/V**: YT, Montana Mtn. near Carcross, el. 2,000 m, 7 July 1989, J. and S. Shepard (JHS).

*Colias nastes ssp.* **1st instar larva**: AK, Seward Penninsula, 1 July 1980, C.S. Guppy.

*Colias nastes streckeri* ♀ **D**: BC, Camelsfoot Range, Poison Mtn., el. 2,100 m, 1 Aug. 1978, C.S. Guppy (CSG). ♂ **D/V**: BC, Mt. Cartier, el. 2,400 m, 4 Aug. 1983, D.L. Threatful (JHS). **Mature larvae in diapause**: WA, Okanogan Co., Windy Peak, D.V. McCorkle. **Live adult V**: AB, Parker Ridge, 6 July 1975, C.S. Guppy.

*Colias occidentalis* ♂ **D/V**: BC, Saanich, Royal Oak, 11 June 1955, G.A. Hardy (RBCM). ♀ **D/V**: BC, Cowichan Lk., 18 June 1913 (RBCM).

*Colias pelidne* ♂ **D/V**: BC, NE ridge of Watch Peak, el. 2,400 m, 9 Aug. 1981, J. and S. Shepard (JHS). ♀ **yellow form D/V**: AB, N of Dolomite Peak, 19 July 1969, Helmut Kimmich (UBC). ♀ **white form D**: AB, Bow Pass, el. 2,100 m, 22 July 1969, Helmut Kimmich (UBC).

*Colias philodice eriphyle* ♂ **spring form D/V**, ♀ **spring form D**: BC, Penticton, Mt. Campbell, 25 May 1963, Helmut Kimmich (UBC).

♂ **summer form V**: BC, Vernon, Swan Lk., 29 July 1984, C.S. Guppy (CSG). ♀ **white form D**: BC, 3 km N of Fort Steele, 13 May 1973, J.L. Gordon (UBC). **Egg**: BC, Quesnel, Fishpot Lk., female coll. 27 June 1997 oviposited in captivity on red clover, photo 28 June 1997, C.S. Guppy. **Mature larva**: BC, Dog Creek, 19 July 1991, on *Hedysarum boreale*, Anna Roberts. **Pupa**: BC, Quesnel, Fishpot Lk., female coll. 27 June 1997 oviposited in captivity on red clover, photo 6 Aug. 1997, C.S. Guppy. **Live adult**: BC, Williams Lake, 22 Sep. 1989, Anna Roberts.

*Colias philodice vitabunda* ♀ **D/V**: BC, Pink Mtn., el. 1,200–1,500 m, 17 July 1998, C.S. Guppy (CSG). ♂ **V**: BC, 16 km N of Stikine R. crossing, Hwy 37, 10 July 1989, J. and S. Shepard (JHS).

*Danaus plexippus* ♂ **D**: OH, Lake Co., Maddison, 3 Aug. 1972, L.C. Koehn (CSG). ♀ **D**: BC, Ashnola R. at Similkameen R., ex larva on milkweed, emerged 8 Aug. 1972, C.S. Guppy (CSG). **Mature larva, pupa**: BC, Ashnola R. at Similkameen R., larva found on milkweed 10 July 1972, photo 11 July 1972, A.G. Guppy. ♀ **live adult**: BC, Ashnola R. at Similkameen R., larva found on milkweed 10 July 1972, photo 8 Aug. 1972, C.S. Guppy.

*Danaus similis* **Live adult**: BC, Victoria, 2 June 1993, C.S. Guppy.

*Epargyreus clarus californicus* ♂ **V**: BC, Vancouver, 27 May 1934, T.P.O. Menzies (RBCM).

*Epargyreus clarus clarus* ♂ **V**: BC, Wasa Lk. Prov. Park, E of head-quarters, 13 July 1997, D.W. Knight (JHS). ♀ **D**: BC, Pend-d'Oreille R. 4 km W of 7 Mile Dam, 13 July 1999, N.G. Kondla (NGK).

*Erebia discoidalis* ♂ **D/V**: YT, Dempster Hwy at Ogilvie R., el. 600 m, 19 June 1979, J.H. Shepard (JHS). ♀ **D**: BC, 5 km N of Pink Mtn. on Alaska Hwy, 8 June 1979, J.H. Shepard (JHS).

*Erebia epipsodea epipsodea* ♂ **D/V**: BC, Princeton, Whipsaw Cr., 11 June 1988, C.S. Guppy (CSG). ♀ **D**: BC, Pavilion Mtn., el. 2,200 m, 11 Aug. 1989, C.S. Guppy (RBCM). **Egg**: BC, Falkland, 16 km N, 26 June 1974, C.S. Guppy. **1st instar larva**: BC, Falkland, 16 km N, 23 July 1974, C.S. Guppy.

*Erebia epipsodea remingtoni* ♂ **D/V**: BC, Boulder Cr., 19 km E of Atlin, el. 1,200 m, 30 June 1973, J.L. Gordon (UBC). ♀ **D**: BC, Mt. Klappan rd. km 13, el. 860 m, 11 July 1989, J. and S. Shepard (JHS).

*Erebia epipsodea sineocellata* ♂ **D/V**: BC, Beaton R., 1.5 km N of Hwy 131, 3 June 1997, J.H. Shepard (JHS). ♀ **D**: BC, Beaton R. mouth, 2 June 1997, J.H. Shepard (JHS).

*Erebia fasciata* ♂ **D/V**: YT, Montana Mtn. near Carcross, el. 2,000 m, 17 June 1989, J. and S. Shepard (JHS). ♀ **D**: NWT, Dempster Hwy 22 km E of YT border, el. 800 m, 21 June 1979, J.H. Shepard (JHS).

*Erebia lafontainei* ♂ **D/V**: YT, Montana Mtn. near Carcross, el. 2,000 m, 17 June 1989, J. and S. Shepard (JHS). ♀ **D**: AK, Atigun Gorge, N ridge, 13 July 1979, J.H. Shepard (JHS).

*Erebia mackinleyensis mackinleyensis* ♂ **D/V**: YT, Montana Mtn. near Carcross, el. 2,000 m, 17 June 1989, J. and S. Shepard (JHS). ♀ **D**: BC, Summit Lake, 15 June 1992, N.G. Kondla (NGK). **Hatched egg**: YT, Dempster Hwy km 155, 22 June 1979, C.S. Guppy. **1st instar larva**: YT, Dempster Hwy km 155, 8 July 1979, C.S. Guppy.

*Erebia magdalena* ♂ **paratype D/V**: AB, Adams LO, 24 July 1982, G.J. Hilchie (JHS). ♀ **D**: BC, Dore R. near McBride, el. 2,200 m, 13 July 1985, D.L. Threatful (JHS).

*Erebia mancinus* ♂ **D/V**: BC, Racing R., 7 km N of junction with Delano Cr., 22 June 1976, C.S. Guppy (CSG). **Live adult**: BC, Racing R., 7 km N of junction with Delano Cr., 22 June 1976, A.G. Guppy.

*Erebia pawlowskii* ♂ **D/V**: YT, Donjek R., Alaska Hwy km 1823, 3 July 1979, G. Anweiler (RBCM). ♀ **D**: YT, White R., Miles Ridge, 30 June 1979, G. Anweiler (UBC).

*Erebia rossii rossii* ♂ **D/V**: BC, Summit Lake, 21 June 1976, C.S. Guppy (JHS).

*Erebia vidleri* ♂ **D/V**: BC, Lillooet, Mission Ridge, el. 1,900 m, 23 July 1988, C.S. Guppy (CSG). **Live adult**: BC, Manning Prov. Park, 10 July 1975, C.S. Guppy.

*Erebia youngi* ♂ **D/V**: YT, Richardson Mtns., Dempster Hwy km 479, 3–4 July 1979, C.S. Guppy (CSG). ♀ **D**: YT, Richardson Mtns., Dempster Hwy km 406, 19 July 1981, J. Troubridge (CSG). ♂ **Live adult D**: AK, Mt. Fairplay, el. 1,200 m, 4 July 1976, C.S. Guppy.

*Erynnis afranius* ♂ **D/V**: BC, New Aiyansh, 1940–1950, H. and L. Hughan (JHS).

*Erynnis icelus* ♂ **D/V**: BC, West Vancouver, 15 May 1960, A.C. Jenkins (JHS). ♀ **D**: BC, Peace R. at Clayhurst Bridge (N side), 23 May 1997, J.H. Shepard (JHS).

*Erynnis pacuvius* ♂ **D/V**: BC, Lytton, 0.5 km S on Hwy 1, 21 May 1983, C.S. Guppy (CSG).

*Erynnis persius* ♂ **D**: BC, Nanaimo, Green Mtn., el. 1,000 m, 25 June 1989, CSG (CSG). ♀ **D/V**: BC, Wasa Lk. Prov. Park, E of headquarters, 29 May 1997, D.W. Knight (JHS).

*Erynnis propertius* ♂ **D/V**: BC, Hornby Island, W of Helliwell Prov. Park, 27 Apr. 1995, J.H. Shepard (JHS). ♀ **D**: BC, Vanc. Isl., Rocky Point CDN, 23 Apr. 1995, J.H. Shepard (JHS). ♂ ♀ **mating**: BC, Royal Oak, 19 May 1976, C.S. Guppy.

*Euchloe ausonides insulanus* ♂ **D/V**: BC, Victoria, Beacon Hill Park, 28 May 1999, E. Anderson (RBCM). ♀ **holotype D/V**: BC, Wellington, 3 June 1904, G.W. Taylor (CNC).

*Euchloe ausonides mayi* ♂ **D/V**: BC, Pavilion, 15 June 1989, R.A. Cannings (RBCM). ♀ **D**: BC, Wasa Lk. Prov. Park, E of headquarters, 14 May 1997, D.W. Knight (JHS). **Egg**: BC, Quesnel, Baezaeko R. (3900 rd.), eggs coll. on *Arabis* sp. 26 June 1997, photo 28 June 1997, C.S. Guppy. **Mature larva, prepupal larva**: BC, Quesnel, Baker Cr. ("Hangman Springs"), larvae coll. on *Arabis* sp. 8 July 1997, photo 13 July 1997, C.S. Guppy. **Pupa**: BC, Quesnel, Baezaeko R. (3900 rd.), eggs coll. on *Arabis* sp. 26 June 1997, photo 15 July 1997, C.S. Guppy. **Caught by crab spider**: BC, Vernon Hill near Becker Lk., June 1982, C.S. Guppy. **Live adult**: BC, Chilcotin, 18 May 1991, Anna Roberts.

*Euchloe ausonides ogilvia* ♀ **V**: BC, Atlin, Warm Springs, el. 750 m, 27 June 1975, J.H. Shepard (JHS). ♂ **V**: YT, Jakes Corner, 18 km S on Atlin rd., 29 June 1976, C.S. Guppy (CSG).

*Euchloe creusa* ♂ **D/V**: BC, Atlin rd., 48 km S of Jakes Corner, YT, 25 June 1979, C.S. Guppy (CSG).

*Euchloe lotta* ♂ **D/V**: BC, Spences Bridge, Hwy 1, in town, 4 Apr. 1998, C.S. Guppy (CSG). ♀ **D/V**: BC, Vaseux Lk., 22 Apr., 1976, C.S. Guppy (CSG). **Live adult**: BC, Spences Bridge, May 1998, C.S. Guppy.

*Euchloe naina* ♀ **D/V**: BC, Mt. Racine, 26 June 1999, N.G. Kondla (NGK). ♂ **D/V**: BC, Atlin, 2.9 km N of Hitchcock Cr., 24 June 1999, C.S. Guppy (CSG).

*Euphilotes battoides glaucon* ♂ **D/V**: BC, Olalla, 20 May 1979, G.B. Straley (RBCM). ♀ **D**: BC, Vernon Hill, Becker Lk., 19 June 1983, C.S. Guppy (CSG). ♂ ♀ **mating**: BC, Chilcotin, 4 July 1993, Anna Roberts.

*Euphilotes battoides oregonensis* **mature larva**: OR, Harney Co., near Alvord Hot Springs, coll. 22 June 1998, G.E. Pratt.

*Euphilotes battoides* ssp. **pupa**: CA, Riverside, coll. July 1982, G.E. Pratt.

*Euphydryas anicia anicia* ♂ **D/V**: BC, Wasa Lk. Prov. Park, E of head-quarters, 15 June 1997, D.W. Knight (JHS). ♀ **D**: BC, 4 km S of Hwy3 on Corbin rd., 17 July 1990, J. and S. Shepard (JHS).

*Euphydryas anicia hopfingeri* ♂ **D/V**, ♀ **D**: BC, Lytton, Botanie Mtn. LO. rd., el. 1,500 m, 5 July 1978, C.S. Guppy (CSG). **Mature larva**: BC, Williams Lake, 28 May 1998, Anna Roberts. **Pupa**: BC, Williams Lake,

4 June 1998, Anna Roberts. **Live adult**: BC, Williams Lake, 19 June 1998, Anna Roberts. ♂ **live aberrant** D: BC, Oyama, 6 km N, 17 May 1975, C.S. Guppy.

*Euphydryas chalcedona* ♂ **D/V**: BC, Goldstream, 24 May 1904 (RBCM). ♀ **D**: BC, Goldstream, 24 May 1903 (RBCM).

*Euphydryas editha beani* ♂ **D/V**: BC, Manning Prov. Park, 16 June 1962, A.C. Jenkins (JHS). ♀ **D**: BC, Mt. Cheam, el. 2,000 m, 8 Aug. 1981, G.B. Straley (RBCM).

*Euphydryas editha taylori* ♂ **D/V**: BC, Hornby Island W of Helliwell Prov. Park, 27 Apr. 1995, J.H. Shepard (JHS). ♀ **D**: BC, Hornby Island W of Helliwell Prov. Park, ex larva, emerged 25–26 Apr. 1977, J. and S. Shepard (JHS). **Egg, live adult**: BC, Mill Bay, Shawnigan Lk.–Mill Bay rd. powerline, 14 June 1987, C.S. Guppy.

*Euphydryas gillettii* ♂ **D/V**: AB, Kananaskis rd. near Livingston Cmpgd, 22 July 1977, L. Paul Grey (CSG). ♀ **D**: BC, Proctor Lk., 21 July 1990, J. and S. Shepard (JHS).

*Euphyes vestris* ♂ **D/V**: BC, Nanaimo Lakes rd., 16 km W of Hwy 1, 25 June 1988, C.S. Guppy (CSG). ♂ **live adult**: BC, Cobble Hill, June 1995, Derrick Marvin.

*Euptoieta claudia* ♂ **D/V**: AB, Parker Ridge, el. 2,300 m, 18 July 1976, Helmut Kimmich (UBC).

*Eurema nicippe* ♂ **D/V**: AZ, Apache Junction, 6 Sep. 1957, R.W. Dawson (JHS). ♀ **D**: AZ, Apache Junction, 29 Aug. 1957, R.W. Dawson (JHS).

*Everes amyntula* ♀ **D/V**: BC, S of Hope, 26 Apr. 1981, G.B. Straley (RBCM). ♂ **D/V**: BC, Nazko, Redwater Cr. rd., 2.8 km W Nazko Hwy, 3 June 1995, C.S. Guppy (CSG). **Mature larva**: CA, Santa Rosa Island, coll. May 1989 by J. and T. Emmel, photo G.E. Pratt. **Pupa**: CA, San Bernardino Co., Big Bear City, ex egg on *Astragalus douglasii*, 30 May 1987, G.E. Pratt. **Live adult**: BC, Williams Lake, 1 June 1990, Anna Roberts.

*Everes comyntas* ♂ **D/V**: BC, Pend-d'Oreille R., 5 km E of 7-Mile Dam, 12 July 1997, N.G. Kondla (NGK). ♀ **D**: BC, Waneta Dam, 4 July 1998, N.G. Kondla (NGK). ♀ **V**: BC, Proctor Lk. near Flathead rd., 7 July 1991, J. and S. Shepard (JHS). **Mature larva**: CA, Kern Co., reared ex egg by Ken Davenport, photo G.E. Pratt.

*Glaucopsyche lygdamus columbia* ♂ **D/V**: BC, Princeton, 6 km E, 9 June 1986, C.S. Guppy (CSG). ♀ **D**: BC, Ladysmith, 13 June 1981, G. Anweiler (RBCM). **Mature larva**: BC, Dog Cr., 19 July 1991, on *Hedysarum boreale*, Anna Roberts. **Pupa**: BC, Dog Cr., 1 Aug. 1991, Anna Roberts. **Live adult**: BC, Agassiz, 8 May 1975, C.S. Guppy.

*Glaucopsyche lygdamus couperi* ♂ **D/V**: BC, Dease Lake, 5 km SW on Telegraph Cr. rd., el. 800 m, 25 June 1973, J.H. Shepard (JHS). ♀ **D**: BC, Jesmond, 8 km S, 29 June 1972, K. Howard, D. Paull, and S. Shepard (JHS).

*Glaucopsyche piasus piasus* **mature larva**: CA, San Bernardino Co., Forest Home, coll. on *Lupinus*, 13 June 1982, G.E. Pratt.

*Glaucopsyche piasus* ssp. **pupa**: CA, G.E. Pratt.

*Glaucopsyche piasus toxeuma* ♂ **D/V**: BC, Osoyoos, July 1970, R. Carcasson (RBCM). ♀ **D**: BC, Douglas Lk. rd., 40 km E of Merritt, 22 May 1976, J.H. Shepard (JHS). **Live adult**: BC, Oyama, 6 km N, 17 May 1975, C.S. Guppy.

*Heliopetes ericetorum* ♀ **D**: WA, Benton Co., Kennewick, el. 130 m, 1 Oct. 1960, J.H. Shepard (JHS). ♂ **D/V**: WA, Chelan Co., Swakane Canyon, el. 400 m, 17 June 1994, J. and S. Shepard (JHS).

*Hesperia comma assiniboia* ♂ **V**: BC, Beaton R., NE of Fort St. John, 7 Aug. 1993, N.G. Kondla (JHS).

*Hesperia comma harpalus* ♂ **V**: BC, Kilpoola Lk., 1 July 1996, J. and S. Shepard (JHS).

*Hesperia comma manitoba* ♂ **V**: BC, Skagway-Carcross rd., 0.5–3 km S of BC/YT border, 8 July 1989, J. and S. Shepard (JHS).

*Hesperia comma oregonia* ♂ **D/V**: BC, Royal Oak, Saanich, 29 July 1956, G.A. Hardy (RBCM). ♀ **D**: BC, Hudson Bay Woods, Saanich, 23 Aug. 1953, G.A. Hardy (RBCM). ♂ **live adult**: WA, Olympic Nat. Park, Hurricane Ridge, 1 Aug. 1979, C.S. Guppy.

*Hesperia juba* ♂ **D/V**: BC, Lytton, 0.5 km S on Hwy 1, 21 May 1983, C.S. Guppy (CSG). ♀ **D**: BC, Canford, above Hwy 8, 2 July 1988, C.S. Guppy (CSG). ♂ **Live adult**: BC, Keremeos, Crater Mtn., low elevation, 17 Aug. 1974, C.S. Guppy.

*Hesperia nevada* ♂ **D/V**: BC, Keremeos, Fairview Trail, 1950 ?, J.K. Cooper (RBCM). ♀ **D**: BC, Bella Vista near Vernon, el. 800 m, 14 May 1985, D.L. Threatful (JHS).

*Icaricia acmon acmon* **mature larva**: CA, San Bernardino Co., Mountain Home Village, coll. on *Eriogonum elongatum* 23 Jan. 1988, G.E. Pratt. **Pupa**: MEX, Baja California Norte, Sierra Juarez, Guadalupe Canyon, larva coll. 23 Mar. 1986, G.E. Pratt.

*Icaricia acmon lutzi* ♂ **D/V**: BC, Olalla, 20 May 1979, G.B. Straley (RBCM). ♀ **D**: BC, Osoyoos, Anarchist Pass, el. 975 m, 22 June 1975, C.S. Guppy (CSG). ♂ **live adult**: WA, SE of Wenatchee, 31 May 1970, A.G. Guppy.

*Icaricia icarioides blackmorei* ♂ **D/V**: BC, Duncan, 3 June 1921, A.W. Hanham (RBCM). ♀ **D**: BC, Vanc. Isl., Spectacle Lk., 11 June 1953, G.A. Hardy (RBCM).

*Icaricia icarioides montis* ♂ **D/V**: BC, Botanie Valley 20 km NE of Lytton, 2 July 1978, J.H. Shepard (JHS). ♀ **D**: BC, Lillooet, Mt. McLean, el. 1,800 m, 1–2 Aug. 1974, J.H. Shepard (JHS). **Mature larva tended by ant, mature larva**: BC, Keremeos, Crater Mtn., el. 2,100 m, 1 Sep. 1974, C.S. Guppy.

*Icaricia icarioides pembina* ♂ **D/V**: BC, Mt. Kobau summit, 1 July 1996, C.S. Guppy (CSG). ♀ **D**: BC, Kitchener, Goat R. FSR km 8–16, 15 June 1976, C.S. Guppy (CSG). **Live adult**: BC, Oyama, 6 km N, 17 May 1975, C.S. Guppy.

*Incisalia augustinus* ♂ **D/V**: BC, Quesnel, 3 km N of Baker Cr. Falls, 13 May 1995, C.S. Guppy (CSG). ♀ **D**: BC, Quesnel, Baker Cr. at "Hangman Springs," 15 May 1999, C.S. Guppy (CSG).

*Incisalia eryphon eryphon* ♂ **D/V**: BC, Nelson, Sproule Cr., el. 700 m, 19 Apr. 1992, J.H. Shepard (JHS). ♀ **D**: BC, Quesnel, Dragon Lk., 8 May 1994, C.S. Guppy (CSG). **Live adult**: BC, Williams Lake, 11 May 1994, Anna Roberts.

*Incisalia eryphon sheltonensis* ♂ **V**: BC, Metchosin, Camas Hill, 15 Apr. 1987, A.G. Guppy (CSG). **Mature larva**: BC, Wolf Cr., Nanaimo Lakes rd., C.S. Guppy.

*Incisalia iroides iroides* ♂ **D/V**, ♀ **D**: BC, Osoyoos, E on Hwy 3, el. 330 m, 8 May 1976, C.S. Guppy (CSG). **Egg**: BC, Metchosin, Camas Hill, ex egg laid on *Gaultheria shallon* flowers, 30 Apr. 1983, A.G. Guppy. **Mature larva D**: BC, Burnaby, Burns Bog, photographed 30 May 1993, Richard Beard. **Mature larva lateral**: BC, Metchosin, Camas Hill, ex egg laid on *Gaultheria shallon* flowers, 3 June 1983, A.G. Guppy. ♀ **Live adult V**: BC, Duncan, Mt. Prevost, May 1997, Derrick Marvin.

*Incisalia mossi mossi* ♂ **D/V**: BC, Metchosin, Camas Hill, 9 Apr. 1989, C.S. Guppy (CSG). ♀ **D**: BC, Spectacle Lk. powerline, 30 Apr. 1989, C.S. Guppy (CSG). **Live adult**: BC, Metchosin, Camas Hill, 4 Apr. 1996, A.G. Guppy.

*Incisalia mossi schryveri* ♂ **V**: BC, Creston, 24 Mar. 1987, J.H. Shepard (JHS). **Mature larva**: WA, Bear Canyon, 28 May 1971, D.V. McCorkle.

*Incisalia niphon clarki* ♀ **D/V**: BC, W of Liard Hwy at Nelson R., mixed jack pine and lodgepole pine stand, 16 June 1999, C.S. Guppy (CSG). ♂ **D**: ME, Passadunkeng, 11 May 1976, L.P. Grey (CSG). **Mature larva**: NJ, Chatsworth, ex ♀ coll. 28 Apr. 1990, G.E. Pratt.

*Incisalia polia maritima* **mature larva**: OR, Driftwood Park, coll. on *Arctostaphylos uva-ursi*, D.V. McCorkle.

*Incisalia polia polia* ♂ **D/V**: BC, Creston, 5 km E at Arrow Cr., 13 May 1973, J. and S. Shepard (JHS). ♀ **D**: BC, Ft. Steele, 3 km N, 13 May 1973, J. and S. Shepard (JHS).

*Limenitis archippus* ♂ **D/V**: BC, Lillooet, 6 July 1930, A.W.A. Phair (RBCM).

*Limenitis arthemis* ♂ **5 larval instars D/V**: BC, Titetown Lk., 30 June 1997, C.S. Guppy (CSG). ♀ **D**, ♂ **4 larval instars D**: BC, Titetown Lk., ex ova emerged Aug. 1997, C.S. Guppy (CSG). **Eggs**: BC, Quesnel, Titetown Lk., ♀ coll. 30 June 1997 oviposited in captivity on trembling aspen, photo 4 July 1997, C.S. Guppy. **1st instar larva**: BC, Quesnel, Titetown Lk., ♀ coll. 30 June 1997 oviposited in captivity on trembling aspen, photo 9 July 1997, C.S. Guppy.
**2nd instar larva hibernaculum**: BC, Quesnel, Titetown Lk., ♀ coll. 30 June 1997 oviposited in captivity on trembling aspen, photo 24 July 1997, C.S. Guppy. **Mature larva, pupa**: BC, Quesnel, Titetown Lk., ♀ coll. 30 June 1997 oviposited in captivity on trembling aspen, photo 31 July 1997, C.S. Guppy. **Live adult hanging from pupa**: BC, Titetown Lk., ex ova emerged Aug. 1997, C.S. Guppy (CSG). **Live adult**: BC, Williams Lake, 17 July 1991, Anna Roberts.

*Limenitis lorquini* × *arthemis* hybrid ♂ **D/V**: BC, Salmon Arm, 6 km SE on E side of Hwy 97B, rocky ridge, 12 June 1983, C.S. Guppy (CSG).

*Limenitis lorquini ilgae* ♂ **holotype D/V**: BC, Bamberton, Jones Cr. rd. to Oliphant Lk., 9 July 1988, C.S. Guppy (CSG). ♀ **D/V**: BC, Shawnigan Lk.–Mill Bay rd. powerline, 24 May 1995, J.H. Shepard (JHS). **Egg**: BC, Vancouver, C.S. Guppy. **3rd instar larva hibernaculum**: BC, Vancouver, June 1981, C.S. Guppy. **Mature larva**: BC, Vancouver, 1 June 1979, C.S. Guppy. **Pupa pre-emergence**: BC, Surrey, 28 June 1972, C.S. Guppy. **Live adult**: BC, Vancouver, 22 June 1979, A.G. Guppy. **Live adult**: Richard Beard.

*Limenitis lorquini itelkae* ♂ **holotype D/V**: BC, Keremeos, 1.5 km N of Hwy 3A on Mt. Apex rd., 8 June 1982, C.S. Guppy (CSG). ♀ **D/V**: BC, Creston, 8 km W on Hwy 3, 16 June 1967, J.H. Shepard (JHS).

*Loranthomitoura johnsoni* ♀ **D/V**: BC, West Vancouver, el. 150 m, 25 May 1963, A.C. Jenkins (JHS). **Mature larva, pupa**: WA, Pierce Co., Greenwater, 1968, D.V. McCorkle.

*Loranthomitoura spinetorum* ♂ **D/V**: BC, Summerland, N boundary on Hwy 97, 2 June 1976, C.S. Guppy (CSG). ♀ **D**: BC, Keremeos, Hwy 3A, 1 km E of Mt. Apex rd., 7 May 1980, C.S. Guppy (CSG). **Mature larva**: OR, Klamath Co., Beaver Marsh Hwy 97, larva found on mistletoe on *Pinus contorta* 20 July 1969, D.V. McCorkle. **Pupa**: CA, San Diego Co., Laguna Mtns., Boiling Springs tract, ex egg coll. 7 Aug. 1985, G.E. Pratt. ♀ **live adult**: BC, Kootenay Nat. Park, 2 July 1975, C.S. Guppy.

*Lycaeides anna ricei* ♂ **D/V**, ♀ **D/V**: BC, Manning Prov. Park, Gibsons Pass, 23 July 1978, J.H. Shepard (JHS). ♂ ♀ **Mating**: Richard Beard.

*Lycaeides anna vancouverensis* ♂ **holotype D/V**, ♀ **D/V**: BC, Strathcona Prov. Park, Cream Lk., el. 1,400 m, 22 Aug. 1988, C.S. Guppy (CSG). ♀ **live adult**: BC, Strathcona Prov. Park, Flower Ridge, el. 1,500 m, 12 Aug. 1974, C.S. Guppy.

*Lycaeides idas alaskensis* ♂ **D/V**, ♀ **D**: BC, Atlin, 22 July 1975, C.S. Guppy (CSG).

*Lycaeides idas atrapraetextus* ♂ **D/V**: BC, Hall Cr. at Hwy 33, 19 July 1986, J. and S. Shepard (JHS). ♀ **D**: BC, Athalmer, 13 km W, 1 July 1975, C.S. Guppy (CSG).

*Lycaeides idas scudderi* ♂ **D/V**, ♀ **D**: BC, Vernon, Aberdeen Lk. rd., el. 1,250 m, 14 Aug. 1997, D.L. Threatful (JHS). **Mature larva**: BC, Chilcotin, 15 June 1992, Anna Roberts. **Live adult**: BC, French Bar Cr., 13 Sep. 1983, Anna Roberts.

*Lycaeides melissa melissa* ♂ **D/V**: BC, Osoyoos, Mt. Kobau, el. 990 m, 24 Aug. 1991, D.C.A. Blades and Carol Maier (RBCM). ♀ **D**: BC, Vernon, Goose Lk., 22 May 1989, C.S. Guppy (RBCM). **Mature larva**:

CA, Inyo Co., Big Pine, Owens R., coll. 16 Aug. 1994, with *Formica* ants, G.E. Pratt. **Live adult**: BC, Osoyoos, Richter Pass, 14 May 1975, C.S. Guppy.

*Lycaena cuprea cuprea* **pupa**: CA, Mono Co., White Mtns., vicinity Mt. Barcroft Research Station, coll. 16 Aug. 1987, G.E. Pratt.

*Lycaena cuprea henryae* ♀ **D**: BC, Goldbridge, "Cinnibar Mtn.," 24 July 1977, Helmut Kimmich (UBC). ♂ **D/V**: BC, Mt. Cartier, 15 km S of Revelstoke, el. 2,377 m, 5 Aug. 1984, D.L. Threatful (JHS). **Mature larva**: BC, Pemberton, Blowdown Pass, ex egg, larva photographed 12 May 1994, Richard Beard.

*Lycaena dione* ♂ **D/V**, ♀ **D**: BC, Cranbrook, Elizabeth Lk., 16 July 1990, J. and S. Shepard (JHS).

*Lycaena dorcas* ♂ **D/V**: BC, Valemont, "Cranberry Bog," 7 July 1987, J. and S. Shepard (JHS). ♀ **D**: YT, Kluane Lk., 19 July 1980, G. Anweiler (RBCM).

*Lycaena helloides* ♂ **D/V**: BC, Saanich, Rithet's Bog, 19 June 1988, C.S. Guppy (CSG). ♀ **D**: BC, Saanich, Rithet's Bog, 30 Aug. 1988, C.S. Guppy (CSG). **Egg**: BC, Riske Creek, oviposition on *Polygonum amphibium* stem and dead leaf 6 Sep. 1997, C.S. Guppy. **Mature larva**: BC, Tsawwassen, larvae found 27 June 1992, Richard Beard. **Live adult**: BC, Williams Lake, 26 July 1990, Anna Roberts.

*Lycaena heteronea heteronea* ♂ **D/V**: BC, Kilpoola Lk., el. 800 m, 1 July 1996, C.S. Guppy (CSG). ♀ **D**: WA, Pend Oreille Co., 6 km S of Ruby on Hwy 131, 14 July 1976, J.H. Shepard (JHS). **Mature larva**: CA, Ventura Co., Frazier Park, coll. on *Eriogonum fasciculatum* May 1983, G.E. Pratt. **Pupa**: CA, Siskyou Co., Ball Mtn., el. 2,100 m, ex ♀ coll. 24 July 1987 by A.M. Shapiro, photo G.E. Pratt. **Live adult V**: BC, Osoyoos, Anarchist Pass, 1 July 1975, C.S. Guppy.

*Lycaena hyllus* ♂ **D/V**: MD, Columbia, 2 July (CSG). ♀ **D**: ON, Toronto, Don Valley, 1 July 1959, P. Watkins (JHS). **Mature larva**: G.E. Pratt.

*Lycaena mariposa charlottensis* ♀ **D/V**: BC, Graham Island, Drizzle Lk., 27 July 1980, J.H. Shepard (JHS).

*Lycaena mariposa mariposa* ♂ **D**, ♀ **D/V**: BC, Mt. Cokely Ski Hill rd., 3 Aug. 1995, J. and S. Shepard (JHS). **Mature larva, pupa**: BC, Burnaby, Burns Bog, 4 eggs wild laid on *Vaccinium uliginosum* 22 July 1991, 5 eggs wild laid on *Vaccinium oxycoccus* and 2 eggs on *Andromeda* 23 July 1991, larva photographed 6 May 1992, pupation 7 May 1992, Richard Beard.

*Lycaena mariposa penroseae* ♀ **D/V**: BC, Hall Cr. at Hwy 33, 19 July 1986, J. and S. Shepard (JHS).

*Lycaena nivalis browni* ♀ **D**: BC, Vernon, Terrace Mtn., el. 1,650 m, 17 Aug. 1983, C.S. Guppy (CSG). ♂ **D/V**: BC, Westbank, Last Mtn. rd., el. 1,200 m, C.S. Guppy (CSG).

*Lycaena nivalis nivalis* **mature larva**: CA, Modoc Co., Warner Mtns., Mt. Bidwell, ex ♀ coll. 15 July 1985, G.E. Pratt.

*Lycaena phlaeas arethusa* ♂ **D/V**: AB, Plateau Mtn., 5 km E Kananaskis rd., el. 2,000 m, 20 July 1977, C.S. Guppy (CSG). ♀ **D**: AB, Plateau Mtn., circ on E side, 23 July 1977, C.S. Guppy (CSG).

*Lycaena phlaeas* ssp. **mature larva**: YT, Ogilvie Mtns., Dempster Hwy km 155, 26 June 1979, C.S. Guppy.

*Lycaena rubida perkinsorum* ♂ **D/V**, ♀ **D**: WA, Kittitas Co., Umtanum Cr. at Yakima R., 14 June 1969, J.H. Shepard (JHS).

*Mitoura rosneri plicataria* ♂ **V**: BC, Saturna Island, Winter Cove Park, 16 May 1995, J.H. Shepard (JHS). **Pupa**, ♂ **live adult**: BC, Goldstream Prov. Park, ex egg in captivity, photo 9 Feb. 1990, C.S. Guppy.

*Mitoura rosneri rosneri* ♂ **D/V**, ♀ **D**: BC, Sicamous, Shuswap Lk., 8 May 1990, C.S. Guppy (CSG). **Mature larva**: BC, Sicamous, Shushap Lk., ex ♀ coll. 8 May 1990, C.S. Guppy.

*Mitoura siva* ♂ **D/V**: BC, Keremeos, 1.5 km N of junction of Hwy 3A and Mt. Apex rd., 31 May 1975, C.S. Guppy (CSG). ♀ **D**: BC, Keremeos Cr., Green Mtn., el. 1,200 m, 18 June 1981, C.S. Guppy

(CSG). **Pupa**: BC, Vernon, Kalamalka Lk., ex egg in captivity, photo 9 Feb. 1990, C.S. Guppy. **Mature larva**: BC, Vernon, Kalamalka Lk., 17 June 1990, C.S. Guppy. **Live adult**: BC, Williams Lake, 19 May 1991, Anna Roberts.

*Nathalis iole* ♂ **D/V**: TX, Uralde Co., Blewett, 21 Apr. 1975, J. Vernon (CSG).

*Neophasia menapia* ♂ **D/V**: BC, Vernon, 13 July 1963, Helmut Kimmich (UBC). ♀ **V**: BC, Okanagan Landing, 19 July 1964, Helmut Kimmich (UBC). **Live adult**: BC, North Vancouver, 1 Aug. 1979, C.S. Guppy.

*Nymphalis antiopa* **D/V**: BC, Vancouver, East End, emerged 17 July 1986 (ex larva), R.A. Ashton (RBCM). **Pupa**: BC, North Vancouver, el. 300 m, 16 July 1968, A.G. Guppy. **Eggs**: BC, Victoria, 15 May 1998, Steve Ansell. **Mature larva**: BC, Williams Lake, 28 June 1988, Anna Roberts. **Feeding on sap from birch stump**: BC, Quesnel, Gravelle Ferry, 16 Apr. 1999, C.S. Guppy (CSG). **Ovipositing on** *Salix exigua*: BC, Williams Lake, 19 June 1990, Anna Roberts. **Live adult**: BC, Tofino, August 1968, A.G. Guppy.

*Nymphalis californica* **D/V**: BC, Shawnigan District, 21 Aug. 1945, J.R.L. Jones (UBC). **Egg**: BC, North Vancouver, el. 300 m, 13 May 1973, C.S. Guppy. **3rd instar larva**: BC, Lytton, Botanie Valley, 7 July 1974, C.S. Guppy. **Mature larva, pupa**: WA, Satus Pass, D.V. McCorkle.

*Oarisma garita* ♂ **D/V**: BC, Beaton R., 1.5 km N of Hwy 131, 30 June 1997, J.H. Shepard (JHS). ♀ **D**: BC, West Crescent Valley, 18 June 1993, J.H. Shepard (JHS).

*Ochlodes sylvanoides* ♂ **D/V**: BC, Mt. McLean, el. 1,800 m, 1–2 Aug. 1974, J. and S. Shepard (JHS). ♀ **D**: BC, Clearwater, 3.7 km N of Spahats Cr., 23 Aug. 1989, C.S. Guppy (CSG). **Live adult**: BC, Vancouver, UBC, 15 Aug. 1973, A.G. Guppy.

*Oeneis alberta* ♂ **D/V**: BC, Peace R. at Clayhurst Bridge (N side), N.G. Kondla (JHS). **4th instar larva**: AB, Edmonton, ex eggs laid 9 May 1992, Richard Beard.

*Oeneis bore edwardsi* ♂ **D/V**: BC, McBride Peak, el. 2,200 m, 21 July 1986, J. and S. Shepard (JHS).

*Oeneis bore mckinleyensis* ♂ **D/V**, ♀ **D/V**: BC, Telegraph Creek rd., 5 km SW of Dease Lake, el. 750 m, 23 June 1973, J.H. Shepard (JHS). **Live adult**: BC, Summit Lake, Alaska Hwy Mile 392, el. 1,450-1,650 m, 21 June 1976, C.S. Guppy.

*Oeneis chryxus caryi* ♂ **D/V**: BC, Atlin rd., 50 km S of Jakes Corner, YT, 25 June 1979, C.S. Guppy (CSG).

*Oeneis chryxus chryxus* ♂ **D/V**: BC, Osoyoos, Anarchist Pass (McGuire rd.), 11 June 1988, C.S. Guppy (RBCM). ♀ **D**: BC, Hat Cr., Finney Cr. sage flats, 15 June 1989, C.S. Guppy (RBCM). **Live adult**: BC, Sheep Cr. Bridge E of Riske Creek, 14 June 1990, Anna Roberts.

*Oeneis jutta alaskensis* ♂ **D/V**: YT, Minto, 9 km N, el. 600 m, 13 June 1979, J.H. Shepard (JHS).

*Oeneis jutta reducta* ♂ **D/V**: AB, The Wistlers, el. 2,300 m, 13 July 1966, C.S. Guppy (CSG).

*Oeneis jutta ridingiana* ♂ **D/V**: AB, Seven Mile Cr., Forest Trunk rd., 23 June 1970, J. and S. Shepard (JHS).

*Oeneis macounii* ♂ **D/V**: BC, Beaton R., 16 km N of Rose Prairie, 10 June 1997, J.H. Shepard (JHS). ♀ **D**: BC, Riske Creek, Meldrum Creek rd., 850 m, 24 June 1997, A.I. Fischer (RBCM). **Live adult**: BC, Williams Lake, 23 May 1991, Anna Roberts.

*Oeneis melissa atlinensis* ♂ **holotype D/V**: BC, Atlin, Mt. Vaughan, el. 1,550+ m, 28 July 1976, C.S. Guppy (CSG). ♀ **D/V**: BC, Summit Lake, Alaska Hwy Mile 392.5, el. 1,650 m, 15 July 1997, J. and S. Shepard (JHS).

*Oeneis melissa beanii* ♂ **D/V**: BC, Cathedral Lakes Prov. Park, Cathedral Rim, 6 July 1986, S.G. Cannings (UBC).

*Oeneis nevadensis* ♂ **D/V**: BC, Quamichan District, 1–7 June 1920, A.W. Hanham (RBCM). ♀ **D**: BC, Mt. Finlayson summit, 27 May 1954, G.A. Hardy (RBCM).

*Oeneis polixenes brucei* **Live adult**: AB, Parker Ridge, el. 2,200 m, 6 July 1975, C.S. Guppy.

*Oeneis polixenes luteus* ♂ **D/V**, ♀ **D**: BC, Pink Mtn. LO, 15 July 1984, J. and S. Shepard (JHS).

*Oeneis polixenes yukonensis* ♂ **D/V**: YT, Montana Mtn. near Carcross, el. 2,000 m, 16 July 1989, J. and S. Shepard (JHS).

*Oeneis rosovi* ♂ **D/V**: BC, Alaska Hwy Mile 392.5, el. 1,650 m, 14 July 1977, J. and S. Shepard (JHS). ♀ **D/V**: YT, Richardson Mtns., Dempster Hwy km 413, 22 June 1989, J. and S. Shepard (JHS).

*Oeneis uhleri* ♂ **D/V**: BC, Peace R. at Clayhurst Bridge (N side), 31 May 1997, J.H. Shepard (JHS).

*Papilio bairdii dodi* ♀ **V**: BC, Cranbrook, Bummer's Flats, 9 May 1998, C.S. Guppy (CSG).

*Papilio bairdii oregonius* ♂ **D/V**: BC, Vernon, Goose Lk., 6 June 1983, C.S. Guppy (CSG). **Eggs, 1st instar larva**: BC, Okanagan Valley, July 1982, C.S. Guppy. **Mature larva**: BC, Sheep Cr. Bridge E of Riske Creek, 20 Aug. 1991, Anna Roberts. **Pupa**: BC, Sheep Cr. Bridge E of Riske Creek, 15 Oct. 1991, Anna Roberts. **Live adult D**: BC, Sheep Cr. Bridge E of Riske Creek, 23 May 1988, Anna Roberts.

*Papilio bairdii pikei* ♂ **D**: BC, Beaton R. mouth, 9 June 1997, J.H. Shepard (JHS).

*Papilio canadensis* ♂ **D/V**: BC, Quesnel, Hallis Lk., 1 June 1995, C.S. Guppy (CSG). **Egg**: BC, Quesnel, Long Bar rd., ex ♀ coll. 14 June 1997, oviposited in captivity on aspen, photo 21 June 1997, C.S. Guppy. **Prepupal larva**: BC, Quesnel, Long Bar rd., ♀ coll. 14 June 1997, oviposited in captivity on aspen, photo 2 Aug. 1997, C.S. Guppy. **Mature larva**: BC, Quesnel, Schiste Cr., 5th instar larva coll. on *Salix* species 22 Aug. 1997, photo 22 Aug. 1997, C.S. Guppy. **Pupa**: BC, Williams Lake, 15 Sep. 1998, Anna Roberts. **Mud-puddling adults**: BC, Teeter Cr., about 8 km W of Liard R., 17 June 1999, C.S. Guppy. ♀ **Live adult V**: BC, Horsefly, July 1998, Derrick Marvin. ♂ **Live adult V**: BC, Shuswap Lk., W end, 24 June 1974, C.S. Guppy.

*Papilio eurymedon × canadensis* ♂ **D/V**: BC, 10 km S of Galloway on Hwy 3/93, 19 May 1983, J. and S. Shepard (JHS).

*Papilio eurymedon* ♂ **D**: BC, Metchosin, Camas Hill, ex egg on red alder (*Alnus rubra*), emerged June 1988, C.S. Guppy (CSG). **Egg**: BC, Metchosin, Camas Hill, ex egg on red alder, photo July 1987, C.S. Guppy. **Mature larva**: BC, Metchosin, Camas Hill, ex egg on red alder, photo Aug. 1987, C.S. Guppy. **Prepupal larva**: BC, Metchosin, Camas Hill, ex egg on red alder, photo Aug. 1988, C.S. Guppy. **Pupa**: BC, Lytton, ex egg from ♀ coll. 15 May 1991, pupation 25 July 1991, Richard Beard.

*Papilio eurymedon and P. rutulus* **Mudpuddling**: 3 adults: BC, Tatlayoko Lk., SE end at Cheshi Cr. (on sand at end of creek), 3 June 2000, S.D. Walker.

*Papilio indra* ♂ **D**: WA, Chelan Co., Leavenworth, Tumwater Canyon, 20 May 1979, J.H. Shepard (JHS).

*Papilio machaon* ♂ **D**: BC, Boulder Cr., 19 km E of Atlin, el. 1,500 m, 1 July 1973, J.L. Gordon (UBC).

*Papilio multicaudatus* ♂ **D/V**: BC, Richter Pass, 2.2 km S of Old Richter Pass rd., 1 July 1996, C.S. Guppy (CSG). **Live adult D**: BC, Okanagan Valley, 27 June 1974, C.S. Guppy. **Live adult D**: BC, Richard Beard.

*Papilio rutulus × multicaudatus* ♂ **D/V**: BC, Vernon, Commonage, 6 June 1971, R.P. Nelson (RBCM).

*Papilio rutulus* ♂ **D/V**: BC, Sumas Mtn., 21 June 1978, C.S. Guppy (CSG). **Egg**: BC, Okanagan Valley, July 1982, C.S. Guppy. **Mature larva**: BC, North Vancouver, el. 300 m, 27 Sep. 1976, A.G. Guppy. **Live adult**: BC, North Vancouver, el. 300 m, July 1980, C.S. Guppy.

*Papilio zelicaon* ♂ **D**: BC, Quesnel, Schiste Cr., 14 May 1995, C.S. Guppy (CSG). **Mature larva**: BC, Quesnel, Lavington rd. 4th instar larva found on cow parsnip (*Heracleum maximum*) leaf 13 Aug. 1997, photo 18 Aug. 1997, C.S. Guppy. **2nd to 5th instar larvae**: BC, Tofino, May 1972, C.S. Guppy. **Green pupa**: BC, Quesnel, Lavington rd. 4th instar larva found on cow parsnip leaf Aug. 13, 1997, photo 22 Aug. 1997, C.S. Guppy. **Brown pupa**: BC, Tofino, ex larvae, 5 Aug. 1968, C.S. Guppy.

*Parnassius clodius altaurus* ♂ **D**: BC, Idaho Peak near New Denver, el. 2,200 m, 31 July 1974, J.H. Shepard (JHS). ♀ **D**: ID, Moscow Mtn., el. 1,400 m, 6 July 1964, J.H. Shepard (JHS).

*Parnassius clodius claudianus* ♀ **D**: BC, Mt. Cheam, el. 1,500–1,800 m, 27 July 1977, C.S. Guppy (CSG). ♂ **D**: BC, Squamish, 15 June 1966, A.C. Jenkins (JHS). **Eggs**: BC, Strathcona Prov. Park, Buttle Lk., 31 July 1974, C.S. Guppy. **Mature larva**: BC, Strathcona Prov. Park, Buttle Lk., 2 May 1975, C.S. Guppy. **Live adult**: BC, Duncan, Mt. Prevost, June 1997, Derrick Marvin.

*Parnassius clodius pseudogallatinus* ♂ **D**: BC, Yale, Inkawatha Lk., el. 1,200 m, 19 July 1978, C.S. Guppy (CSG). ♀ **D**: BC, Yale, 10 km S at Emory Cr., el. 300 m, 19 July 1978, C.S. Guppy (CSG).

*Parnassius eversmanni thor* ♂ **D**, ♀ **D**: BC, Boulder Cr., 19 km E of Atlin, el. 1,500 m, 1 July 1973, J.L. Gordon (UBC).

*Parnassius phoebus apricatus* ♂ **D**, ♀ **D**: YT, North Fork Pass, Demspter Hwy km 81, el. 1,500 m, 4 July 1989, J. and S. Shepard (JHS).

*Parnassius phoebus phoebus* **larva yellow spots**: Russia, West Siberia, Altai Republic, upper Dzhazator River basin, Yuzhno-Chuiskii Mtns., el. 3,000 m, 15 July 1998, Oleg Kosterin. **Larva orange spots**: Russia, West Siberia, Central Altai Mtns., Kosh-Agach District, Katunskiy Mtns., 15 km W of Dzhazator, el. 2,200 m, 19 July 1998, Oleg Kosterin.

*Parnassius smintheus magnus* ♂ **D**, ♀ **D**: BC, Enderby, Brash Cr., el. 700 m, 14 June 1986, C.S. Guppy (CSG). **Egg**: BC, Lytton, 12 June 1973, C.S. Guppy. **Mature larva**: BC, Chilcotin, 11 July 1991, Anna Roberts. **Pupa**: WA, Yakima Co., D.V. McCorkle. ♂ **live adult**: BC, Vernon, Silver Star Mt. Summit, Aug. 1982, C.S. Guppy. ♀ **live adult**, ♂♀ **mating**: BC, Vernon Hill near Becker Lk., July 1982, C.S. Guppy. ♀ **ovipositing**, ♂ **with bird bite**, ♂ **with small mammal damage**: BC, Vernon Hill near Becker Lk., June 1983, C.S. Guppy. ♂ **caught by orb-weaving spider**: BC, Vernon Hill near Becker Lk., June 1982, C.S. Guppy.

*Parnassius smintheus olympiannus* ♂ **D**, ♀ **D**: BC, Strathcona Prov. Park, Cream Lk., 1,400 m, 22 Aug. 1988, C.S. Guppy (RBCM).

*Parnassius smintheus smintheus* ♂ **D**: BC, 30 km SE of Morrisey on rd. to Flathead, el. 900 m, 27 July 1989, C.S. Guppy (RBCM). ♀ **D**: AB, Cat Cr., 18 July 1977, C.S. Guppy (CSG).

*Parnassius smintheus yukonensis* ♂ **D**, ♀ **D**: BC, Tuya R. at Telegraph Cr. rd., 10 July 1989, J. and S. Shepard (JHS).

*Pholisora catullus* ♂ **D/V**, ♀ **D**: BC, Similkameen R., 8 km N of BC/WA border, 2 Aug. 1987, J. and S. Shepard (JHS).

*Phyciodes batesii* ♂ **D/V**: BC, Peace R. at Clayhurst Bridge, 2 July 1997, J.H. Shepard (JHS). ♀ **D**: BC, Attachie, 4 July 1981, N.G. Kondla (JHS).

*Phyciodes mylitta mylitta* ♂ **D/V**: BC, Caulfield, 11 May 1958, A.C. Jenkins (JHS). ♀ **D**: BC, Saltspring Island, rd. to Mt. Maxwell, 20 July 1995, J. and S. Shepard (JHS). **Mature larva, pupa**: BC, Metchosin, Camas Hill, 30 July 1983, A.G. Guppy.

*Phyciodes pallidus* ♂ **D/V**: BC, Spences Bridge, el. 250 m, 19 May 1989, C.S. Guppy (RBCM). ♀ **D**: BC, Nicola Valley, 26 km E of Spences Bridge, 5 June 1964, A.C. Jenkins (JHS). **Eggs, mature larva, pupa**: BC, 21 km N of Lillooet, eggs laid 11 June 1991, larva photographed April 1992, pupation 2 May 1992, Richard Beard. ♀ **live adult**: BC,

Fraser R. at Sheep Cr. Bridge E of Riske Creek, ex larva on *Cirsium undulatum*, 2 July 1999, Anna Roberts.

*Phyciodes pratensis pratensis* ♂ **D/V**: BC, Pavilion, 2 km N, 18 July 1981, J. and S. Shepard (JHS). ♀ **D**: BC, Victoria, 15 June 1916, E.H. Blackmore (RBCM). **Mature larva**: WA, D.V. McCorkle. **Live adult D**: BC, Williams Lake, 19 July 1987, Anna Roberts.

*Phyciodes tharos cocyta* ♂ **D/V**, ♀ **D**: BC, Peace R. at Clayhurst Bridge, 14 July 1984, J.H. Shepard (JHS).

*Phyciodes tharos pascoensis* ♂ **D/V**: BC, Quesnel, Schiste Cr., 15 June 1995, C.S. Guppy (CSG). ♀ **D**: BC, Elko, 16 km E on Hwy 3, 11 June 1988, C.S. Guppy (CSG). **Live adult**: BC, Moffat Cr., 8 July 1998, Anna Roberts.

*Pieris angelika* ♂ **D/V**: BC, Warm Springs, 24 km S of Atlin, 27 June 1973, J.L. Gordon (UBC). ♀ **D/V**: YT, Dempster Hwy, km 84, 19 June 1973, Helmut Kimmich (UBC). **Egg**: YT, Klondike R., S at Dempster Hwy km 68, 23 June 1981, C.S. Guppy. **Mature larva**: YT, Klondike R., Dempster Hwy km 68, June 1981, C.S. Guppy.

*Pieris marginalis guppyi* ♂ **D/V**: BC, Tats Lk. near Tatshenshini R., el. 770 m, 22 July 1992, C.S. Guppy (RBCM). ♂ **V**: AK, Skagway, 5 June 1981, C.S. Guppy (UBC).

*Pieris marginalis marginalis* ♂ **spring form D/V**: BC, Sumas Mtn., 24 Apr. 1977, G.B. Straley (RBCM). ♀ **spring form D**: BC, Delta, Boundary Bay Dike, 12 May 1993, C.S. Guppy (CSG). ♀ **summer form D/V**: BC, Sumas Mtn., 21 July 1978, C.S. Guppy (CSG). ♂ **live adult**: BC, Vedder Crossing, 8 May 1975, C.S. Guppy.

*Pieris marginalis reicheli* ♀ **spring form D/V**, ♂ **spring form V**: BC, Revelstoke, 3 May 1980, D.L. Threatful (RBCM).

*Pieris marginalis tremblayi* ♂ **V**: BC, McBride, Holmes R. FSR km 7.5, el. 800 m, 9 June 1998, C.S. Guppy (CSG). ♀ **D/V**: BC, Atlin rd., 48 km S of Jakes Corner, YT, 25 June 1979, C.S. Guppy (CSG). **Mature larva, pupa, pupa with yellow ♀ about to emerge**: BC, McBride, Holmes R. FSR km 16.2, ex ova in captivity, summer 1998, C.S. Guppy.

*Pieris oleracea* ♀ **D/V**: BC, Quesnel, Dragon Lk. powerline, el. 700 m, 8 May 1994, C.S. Guppy (CSG). ♂ **D/V**: BC, McBride, Holmes R. FSR km 16.2, el. 800 m, 9 June 1998, C.S. Guppy (CSG). **Eggs**: BC, Quesnel, Coglistako R., eggs coll. on *Arabis* sp. 27 June 1997, photo 28 June 1997, C.S. Guppy. **Mature larva**: BC, Atlin, 2 km S of Hitchcock Cr., ex ♀ coll. 23 June 1999, photo 12 July 1999, C.S. Guppy. **Pupa with yellow ♀ about to emerge**: BC, Prince George, Moores Meadow, ♀ coll. 21 June 1997 oviposited in captivity on annual crucifer, photo 1 Aug. 1997, C.S. Guppy. **Live adult**: BC, Williams Lake, 4 May 1998, Anna Roberts.

*Pieris rapae* ♂ **summer form D/V**: BC, Duncan, Skutz Falls, 12 July 1988, C.S. Guppy (CSG). ♀ **summer form D**: BC, Burnaby Lk., 8 Aug. 1975, C.S. Guppy (CSG). **Live adult**: BC, Metchosin, Camas Hill, Aug. 1998, A.G. Guppy. **Mature larva**: BC, Quesnel, eggs coll. 30 Aug. 1997 on weedy mustard, photo 21 Sep. 1997, C.S. Guppy.

*Plebeius saepiolus amica* ♀ **D**: BC, Vernon Hill near Becker Lk., 13 June 1985, D.L. Threatful (JHS). ♂ **V**: BC, Atlin, "Monarch Mtn.," el. 1,150 m, 25 June 1979, C.S. Guppy (CSG). **Live adult**: BC, Dog Cr., 24 June 1990, Anna Roberts.

*Plebeius saepiolus insulanus* ♂ **D/V**: BC, Vanc. Isl., Spectacle Lk., 6 June 1960, G.A. Hardy (RBCM). ♀ **D/V**: BC, Vanc. Isl., Spectacle Lk., 31 May 1957, G.A. Hardy (RBCM).

*Plebeius saepiolus saepiolus* **mature larva**: CA, San Bernardino Co., San Bernardino Mtns., Wildhorse Meadow, reared ex ♀ coll. 8 July 1984, G.E. Pratt.

*Polites draco* ♂ **D/V**: AB, Ram R. campground, Forest Trunk rd., 7 July 1969, J. and S. Shepard (JHS). ♀ **D**: YT, rd. E of Hwy 2 and 25 km S of Lac Laberge campgrd, 18 June 1989, J. and S. Shepard (JHS).

*Polites mystic* ♂ D, ♀ D/V: BC, West Crescent Valley, el. 900 m, 18 June 1993, J.H. Shepard (JHS).

*Polites peckius* ♂ D/V, ♀ D: BC, West Crescent Valley, 24 June 1981, J.H. Shepard (JHS).

*Polites sabuleti sabuleti* ♂ D/V: OR, Harney Co., Wildhorse Cr. rd., 3 June 1992, J. and S. Shepard (JHS). ♀ D: BC, Silver Star Mtn. at lodge, el. 1,650 m, 4 Aug. 1997, D.L. Threatful (JHS).

*Polites sonora* ♂ D/V: BC, Crater Mtn. near Keremeos, el. 1,150 m, 21 June 1975, J.L. Gordon (JHS).

*Polites themistocles* ♂ D, ♀ D/V: BC, Vernon, W of Goose Lk., el. 800 m, 6 June 1983, C.S. Guppy (CSG). **Egg:** BC, Quesnel, Long Bar rd., ex ♀ coll. 29 June 1997, ovipos. on grass, 4 July 1997, C.S. Guppy.

*Polygonia faunus* ♂ D/V: BC, Alison Cr., N of Princeton, 28 Sep. 1987, C.S. Guppy (RBCM). ♀ V: BC, Peace R. at Clayhurst Bridge, 20 Aug. 1989, C.S. Guppy (CSG). **Mature larva, pupa:** OR, Lane Co., D.V. McCorkle. ♂ **adult feeding on moose bones:** BC, Quesnel, Gravelle Ferry, 21 Apr. 1999, C.S. Guppy (CSG). **Live adult:** BC, Williams Lake, 29 Mar. 1995, Anna Roberts.

*Polygonia gracilis* ♂ D/V: BC, Quesnel, Schiste Cr., el. 800 m, 30 July 1998, C.S. Guppy (CSG).

*Polygonia oreas silenus* ♂ D/V: BC, Quesnel, Schiste Cr., 8 Apr. 1996, C.S. Guppy (CSG). **Mature larva, pupa:** OR, Polk Co., D.V. McCorkle.

*Polygonia oreas threatfuli* ♀ **holotype** D/V: BC, Vernon, Kalamalka Lk. Prov. Park, el. 550 m, 21 Sep. 1982, D.L. Threatful (CSG).

*Polygonia progne* ♂ D/V: BC, Quesnel, forestry building, el. 1,100 m, 10 Aug. 1995, C.S. Guppy (CSG). ♀ D/V: BC, Quesnel, Schiste Cr., 9 Sep. 1995, C.S. Guppy (CSG). **Live adult:** BC, Williams Lake, 4 Aug. 1995 Anna Roberts. **Feeding on rotten snowberry fruit:** BC, Williams Lake, 20 Apr. 1988, Anna Roberts.

*Polygonia satyrus* ♂ D/V: BC, University of BC bush (on sap), 29 July 1972, R.A. Ashton (RBCM). ♀ V: BC, Stanley Park, emerg. 15 July 1988 (ex larva), R.A. Ashton (RBCM). **Egg:** BC, Vancouver, University of BC, 14 June 1987, C.S. Guppy. **Mature larva:** BC, Williams Lake, 4 June 1988, Anna Roberts. **Pupa:** BC, Liard R. Hotsprings, 1 Aug. 1972, ex Larva, C.S. Guppy. **Live adult:** BC, Sheep Cr. Bridge E of Riske Creek, 14 Apr. 1991, Anna Roberts.

*Polygonia zephyrus* ♂ D/V: BC, Hunters Range, 26 km E of Salmon Arm, 15 Aug. 1983, C.S. Guppy (CSG). ♀ V: BC, Vanc. Isl., Mt. Cokely ski hill rd., 3 Aug. 1995, J. and S. Shepard (JHS). **Mature larva, pupa:** BC, Big White Mtn. rd., low elevation, ex larva coll. on *Ribes* 23 July 1974, C.S. Guppy.

*Polygonia zephyrus* and *P. faunus* **feeding on dung:** BC, Mt. Arrowsmith, 27 Aug. 1998, Steve Ansell.

*Pontia beckerii* ♂ **summer form** D: BC, Lillooet, 24 June 1962, Helmut Kimmich (UBC). ♀ **summer form** D/V: BC, Princeton, 26 Aug. 1962, Helmut Kimmich (UBC). ♂ **spring form** V: BC, Oliver, 13 Apr. 1963, Helmut Kimmich (UBC). ♀ **spring form** V: BC, Oliver, 20 Apr. 1963, Helmut Kimmich (UBC). **Pupa:** BC, Okanagan Valley, 30 July 1969, C.S. Guppy. ♂ **live adult:** BC, Princeton, 13 May 1975, C.S. Guppy.

*Pontia occidentalis nelsoni* ♂ D/V: BC, Atlin, "Monarch Mtn.," el. 1,350 m, 25 June 1979, C.S. Guppy (CSG). ♀ D: BC, Alaska Hwy km 309, 31 July 1973, A.G. Guppy (CSG).

*Pontia occidentalis occidentalis* ♀ **summer form** D/V: BC, Okanagan Falls, 20 July 1964, Helmut Kimmich (UBC). ♂ **spring form** D/V: BC, Penticton, Mt. Campbell, 19 May 1964, Helmut Kimmich (UBC). ♀ **spring form** V: BC, Oliver, 22 Apr. 1962, Helmut Kimmich (UBC). **Eggs, mature larva, pupa:** BC, Spences Bridge, ex egg, spring 1998, C.S. Guppy. ♂♀ **Live adults mating:** BC, Spences Bridge, spring 1998, C.S. Guppy.

*Pontia protodice* ♂ D/V: OR, Harney Co., Wild Horse Cr. rd., 3 June 1992, J. and S. Shepard (JHS). ♀ D/V: ID, Bear Lake Co., Emmigration Cr. Cmpgd, 14 July 1993, J. and S. Shepard (JHS).

*Pontia sisymbrii beringiensis* ♂ **holotype** D/V: BC, Atlin, 9 km N at Ruffner Mine rd., 21 June 1999, C.S. Guppy (CSG). ♀ D/V: YT, Atlin rd. at Tarfu Lk. rd., 15 June 1999, N.G. Kondla (NGK). **4th instar larva:** BC, Atlin, 2.9 km N of Hitchcock Cr. (rock cliffs), ex eggs wild coll. 24 June 1999, C.S. Guppy.

*Pontia sisymbrii flavitincta* ♀ D/V: BC, Kamloops, 17 May 1954, J.K. Cooper (RBCM). ♂ D/V: BC, Okanagan Falls, White Lk., 20 Apr. 1988, R.A. Cannings (RBCM). ♂ **Live adult** V: BC, Hedley, 8 km E, 12 May 1975, C.S. Guppy.

*Pyrgus centaureae freija* ♂ D/V: BC, Boulder Cr., 19 km E of Atlin, el. 1,200 m, 1 July 1973, J.H. Shepard (JHS).

*Pyrgus centaureae loki* ♂ D/V: AB, Plateau Mtn., 10 km E, 15 June 1988, C.S. Guppy (CSG).

*Pyrgus communis* ♂ D/V, ♀ D: BC, Esling Cr. near Trail, 27 June 1952, A.C. Jenkins (JHS).

*Pyrgus ruralis* ♂ D/V: BC, Metchosin, Camas Hill, 21 Apr. 1985, C.S. Guppy (CSG). **Live adult:** BC, Skagit Valley, Ross Lk., 10 May 1975, C.S. Guppy.

*Roddia l-album watsoni* ♂ V: BC, Quesnel, Schiste Cr., 2 Sep. 1995, C.S. Guppy (CSG). ♀ **grey form** D/V: BC, Keremeos, 1 km E, 2 Aug. 1991, C.S. Guppy (CSG). ♀ **brown form** V: BC, Slocan Lk., 25 Aug. 1970, C.S. Guppy (CSG). **Egg and 1st instar larva:** BC, Okanagan Falls, Sleeping Waters Lk., ex eggs oviposited on birch, 1 Apr. 1990, C.S. Guppy. **Mature larva, prepupal larva:** BC, Okanagan Falls, Sleeping Waters Lk., ex eggs oviposited on birch, 1 June 1990, C.S. Guppy. **Pupa, live adult:** BC, Okanagan Falls, Sleeping Waters Lk., ex eggs oviposited on birch, 17 June 1990, C.S. Guppy. **Feeding on mountain alder sap:** BC, Horsefly Lk. rd., 5 Aug. 1989, Anna Roberts.

*Satyrium behrii behrii* **mature larva:** CA, Mono Co., Scherwin Summit, el. 2,000 m, coll. on *Purshia tridentata,* 24 May 1985, G.E. Pratt. **Pupa:** CA, San Bernardino Co., Mojave R. Forks, coll. on *Purshia tridentata,* May 1982, G.E. Pratt.

*Satyrium behrii columbia* ♂ D/V, ♀ D: BC, Osoyoos Lk., E side of S end, el. 300 m, 30 June 1996, C.S. Guppy (CSG). **Live adult:** BC, Oliver, 30 June 1996, Anna Roberts.

*Satyrium californicum* ♂ D/V, ♀ D: BC, Ashnola R. rd., 11 June 1983, G.B. Straley (RBCM). **Mature larva:** CA, San Bernardino Co., Mountain Home Village, coll. on *Ceanothus leucodermis,* 24 May 1998, G.E. Pratt. **Live adult:** BC, Oliver, 30 June 1996, Anna Roberts.

*Satyrium fuliginosum fuliginosum* **mature larva:** CA, Modoc Co., Warner Mtns., Horse Ridge, ex ♀ coll. 14 July 1985, G.E. Pratt. **Pupa:** CA, Mono Co., 5 km N of Lee Vining, coll. 24 May 1985, G.E. Pratt.

*Satyrium fuliginosum semiluna* ♀ D: BC, Osoyoos, 3 km SE on Hwy 3, el. 300 m, 4 July 1976, J. and S. Shepard (JHS). ♂ D/V: WA, Chelan Co., 5 km NE of Wenatchee, el. 865 m, 2 June 1994, D. Rolfs (JHS).

*Satyrium liparops fletcheri* ♂ D/V, ♀ D: BC, Peace R. at Clayhurst Ferry, 14 July 1984, J. and S. Shepard (JHS).

*Satyrium liparops* ssp. **mature larva:** G.E. Pratt.

*Satyrium saepium* ♂ D/V: BC, Lytton, Botanie Valley, 10 June 1982, G.B. Straley (RBCM). ♀ D: BC, Lytton, 1.5 km S, 15 July 1974, C.S. Guppy (CSG). **Mature larva, prepupal larva, pupa:** BC, Lytton, 1.5 km S, June 1974, C.S. Guppy.

*Satyrium sylvinum nootka* ♂ D: BC, Rossland, 29 July 1952, A.C. Jenkins (JHS). ♂ V: BC, Vanc. Isl., Spectacle Lk., 22 June 1961 (ex ova), G.A. Hardy (RBCM). ♀ D: BC, Revelstoke vicinity, 19 July 1970, D.L. Threatful (UBC).

*Satyrium sylvinum sylvinum* **mature larva**: CA, near Seven Oaks, ex ♀ coll. Aug. 1982, G.E. Pratt.

*Satyrium titus immaculosus* ♀ **D/V**: BC, Nicola R. at Petit Cr., 2 July 1988, C.S. Guppy (CSG). ♂ **D**: BC, Fraser R. at Hwy 20, 20 July 1981, J.H. Shepard (JHS). **Mature larva**: CA, Modoc Co., Fort Bidwell, ex ♀ coll. 15 July 1985, G.E. Pratt.

*Satyrium titus titus* ♀ **V**: BC, Beaton R., Hwy 131, E bluff above river, 28 July 1997, J.H. Shepard (JHS).

*Speyeria aphrodite columbia* ♂ **D/V**: BC, Riske Creek, 20 July 1997, A.I. Fischer (CSG).

*Speyeria aphrodite manitoba* ♂ **D/V**: BC, near mouth of Farwell Cr., 3 Aug. 1997, J.H. Shepard (JHS).

*Speyeria aphrodite whitehousei* ♂ **D/V**: BC, Windemere Lk., 23 July 1985, J. and S. Shepard (JHS). **Mature larva**: BC, Brisco, D.V. McCorkle.

*Speyeria atlantis* ♂ **D/V**: BC, 3 km W of Brisco, 24 July 1985, J. and S. Shepard (JHS). ♀ **V**: BC, 3 km W of Brisco, 24 July 1985, J. and S. Shepard (JHS).

*Speyeria callippe chilcotinensis* ♂ **holotype D/V**: BC, Harper Lk. W of Hanceville, 12 July 1997, C.S. Guppy and A.I. Fischer (CSG). ♀ **D/V**: BC, Riske Creek, 23 July 1981, J. and S. Shepard (JHS).

*Speyeria callippe semivirida* ♂ **D/V**: BC, Crater Mtn. near Keremeos, el. 900 m, 22 July 1978, J. and S. Shepard (JHS). **Mature larva**, **pupa**: WA, Okanogan Co., D.V. McCorkle.

*Speyeria coronis simaetha* ♂ **D**: WA, Yakima Co., Oak Cr., 1 June 1959, J.H. Shepard (JHS). ♀ **V**: WA, Yakima Co., Ahtanum Cr. near Guard Station, 31 May 1965, J. and S. Shepard (JHS).

*Speyeria cybele leto* ♂ **D/V**: BC, Vernon Hill near Becker Lk., 28 June 1983, C.S. Guppy (CSG). ♀ **D**: BC, Vernon Hill near Becker Lk., 29 July 1984, C.S. Guppy (CSG). **Gyandromorph D/V**: BC, Sproule Ck., el. 700 m, 1 Aug. 1994, J. and S. Shepard (JHS).

*Speyeria cybele pseudocarpenteri* ♀ **D/V**: BC, Peace R. at Clayhurst Ferry, 14 July 1984, J. and S. Shepard (JHS).

*Speyeria cybele pugetensis* **mature larva**: OR, Benton Co., D.V. McCorkle. **Pupa**: OR, Polk Co., D.V. McCorkle.

*Speyeria egleis macdunnoughi* ♂ **D/V**: WA, Dayton, el. 1,500 m, 2 July 1961, R.E. Miller (JHS).

*Speyeria hesperis beani* ♂ **D/V**: BC, 1.5 km S of Valemont, 25 June 1970, J. and S. Shepard (JHS). **Live adult**: BC, Williams Lake, 7 June 1998, Anna Roberts.

*Speyeria hesperis helena* ♂ **D/V**: BC, N side of Peace R. at Clayhurst Ferry, 14 July 1984, J. and S. Shepard (JHS).

*Speyeria hydaspe minor* ♂ **V**: BC, Mt. McLean, el. 1,350 m, 3 Aug. 1974, J. and S. Shepard (JHS).

*Speyeria hydaspe rhodope* ♂ **D/V**: BC, Duncan, Mt. Prevost summit, 12 July 1987, C.S. Guppy (CSG). **Mature larva**: BC, Squamish, Diamond Head trail, ex egg from mating pair coll. 10 August 1991, photograph April 1992, Richard Beard. ♂ ♀ **mating**: BC, Duncan, Mt. Prevost, July 1998, Derrick Marven.

*Speyeria hydaspe sakuntala* ♂ **V**: BC, 16 km SE of Windemere, 21 July 1977, G.B. Straley (RBCM).

*Speyeria mormonia bischoffii* ♂ **V**: BC, Warm Springs S of Atlin, el. 730 m, 19 July 1977, J. and S. Shepard (JHS).

*Speyeria mormonia erinna* ♂ **V**: BC, 3 km N of Hwy 3 on Conkle Lk. rd., 3 Aug. 1998, N.G. Kondla (NGK).

*Speyeria mormonia opis* ♂ **silvered form D/V**, ♂ **unsilvered form V**: BC, Botanie Valley, 20 km NE of Lytton, el. 700 m, 2 July 1978, J. and S. Shepard (JHS). **4th instar larva**: BC, near Tweedsmuir Park, D.V. McCorkle. ♂ ♀ **mating**: BC, Keremeos, Crater Mtn., el. 2,250 m, 24 Aug. 1974, C.S. Guppy.

*Speyeria zerene bremnerii* ♂ **D/V**: BC, Mt. Tuam, Saltspring Island, 18 July 1995, J. and S. Shepard (JHS).

*Speyeria zerene garretti* ♀ **D/V**: BC, Rossland, 15 Aug. 1963, A.C. Jenkins (JHS).

*Speyeria zerene picta* ♂ **D/V**: BC, Kilpoola Lk., el. 800 m, 1 July 1998, C.S. Guppy (CSG). **Live adult V**: BC, Kamloops, Lac La Jeune rd., 15 Aug. 1972, A.G. Guppy.

*Speyeria zerene silka* ♂ **holotype D/V**, ♀ **D/V**: AK, Haines, 11 Aug. 1949, G.E. Pollard (CAS). **Mature larva**, **pupa**: AK, N of Haines, D.V. McCorkle.

*Strymon melinus atrofasciatus* ♂ **D/V**: BC, Bamberton, 30 Apr. 1989, C.S. Guppy (CSG). **Mature larva**: BC, Metchosin, Camas Hill, 1 Sep. 1997, A.G. Guppy. **Pupa**: BC, Metchosin, Camas Hill, 9 Sep. 1997, A.G. Guppy. **Live adult**: BC, Metchosin, Camas Hill, 4 July 1987, A.G. Guppy.

*Strymon melinus setonia* ♂ **V**: BC, Mowhokan Cr., 26 km S of Lytton, 3 July 1988, C.S. Guppy (CSG).

*Thorybes pylades* ♂ **D/V**: BC, Peace R. at Clayhurst Bridge (N side), 1 June 1997, J.H. Shepard (JHS). **Live adult**: BC, Williams Lake, 1 June 1990, Anna Roberts.

*Thymelicus lineola* ♂ **D/V**: BC, Sicamous, 19 June 1980, D.L. Threatful (JHS). ♀ **D**: BC, Sicamous, 12 June 1980, D.L. Threatful (JHS).

*Vacciniina optilete* ♂ **D/V**: YT, Whitehorse, 18 July 1981, G. Anweiler (RBCM). ♀ **D**: BC, Atlin, 5 km S at "Monarch Mtn.," el. 1,350 m, 19 July 1977, J. and S. Shepard (JHS).

*Vanessa annabella* **D/V**: BC, Vancouver, Musqueam I.R., 18 Aug. 1973, R.A. Ashton (RBCM). **Mature larva**, **pupa**: BC, Tsawwassen, egg hatched 18 May 1992, pupation 1 June 1992, Richard Beard. **Live adult**: BC, Victoria, Beacon Hill Park, 19 Sep. 1997, Steve Ansell.

*Vanessa atalanta* **D/V**: BC, Vancouver, Wreck Beach, ex larva, emerged 3 Aug. 1987, C.S. Guppy (CSG). **Egg**: BC, Lytton, Botanie Valley, 12 June 1973, C.S. Guppy. **Mature larva**, **pupa**: BC, West Vancouver, instar IV larva photographed 22 June 1990, mature larva photographed 26 June 1990, pupation 27 June 1990, Richard Beard. **Larval leaf shelter**: BC, West Vancouver, 22 June 1990, Richard Beard. **Live adult**: BC, North Vancouver, el. 300 m, 27 Aug. 1971, C.S. Guppy.

*Vanessa cardui* **D/V**: BC, Vancouver, Musqueam I.R., 8 Aug. 1973, R.A. Ashton (RBCM). **Eggs**: BC, Lytton, Botanie Valley, 3 June 1973, C.S. Guppy. **Mature larva**: on *Cirsium edule*: BC, Chilcotin, 15 June 1991, Anna Roberts. **Pupa**: BC, Chilcotin, 4 July 1991, Anna Roberts. **Live adult tagged**: BC, West Vancouver, June 1994, Richard Beard. **Live adult**: BC, Metchosin, Camas Hill, 15 July 1988, A.G. Guppy.

*Vanessa virginiensis* **D/V**: BC, Victoria, 6 Sep. 1977, R. Carcasson (RBCM).

**Figure 28**. Lateral view of the male genitalia of *Oeneis bore* – BC, McBride Peak, el. 2,460 m, 29 July 1985, J. and S. Shepard

**Figure 56**. Left valve of *Erynnis* species:
(a)  *E. icelus* – BC, Trail, 18 July 1952, A.C. Jenkins
(b)  *E. afranius* – BC, New Aiyansh, 1940–50, H. and L. Hughan
(c)  *E. persius* – BC, Sage Cr. Campground, 7 July 1991, J. and S. Shepard
(d)  *E. pacuvius* – BC, Spinekop Ridge, near Winfield, el. 628 m, 22 June 1984, D. Threatful
(e)  *E. propertius* – BC, Royal Oak Res., Victoria, 19 May 1976, J. Shepard

**Figure 57**. Left valve of *Hesperia* species:
(a)  *H. uncas* – AB, Dorothy, 24 June 1979, N.G. Kondla
(b)  *H. nevada* – BC, Crater Mtn., near Keremeos, 29 June 1980, J. and S. Shepard
(c)  *H. juba* – WA, Franklin Co., Byer's Landing, 5 May 1960, J. Shepard
(d)  *H. comma* – BC, Sage Cr. Campground, 7 July 1991, J. and S. Shepard

**Figure 58**. Lateral view of the uncus and tegumen of *Hesperia* species:
(a)  *H. nevada* – BC, Terrace Mtn., near Vernon, 30–31 May 1983, D. Threatful
(b)  *H. comma* – BC, Sage Cr. Campground, 7 July 1991, J. and S. Shepard

**Figure 59**. Ventral view of the uncus of *Hesperia* species:
(a)  *H. comma* – BC, Sage Cr. Campground, 7 July 1991, J. and S. Shepard
(b)  *H. juba* – WA, Walla Walla, Kooskooskie, el. 500 m, 14 May 1960, J. Shepard

**Figure 66**. Dorsal view of the uncus of *Everes* species:
(a)  *E. comyntas* – BC, Pend-d'Oreille R. at Four Mile Cr., 14 May 1998, J. and S. Shepard
(b)  *E. amyntula* – BC, Sage Cr. Campground, 7 July 1991, J. and S. Shepard

**Figure 67**. Valves of *Euphilotes* species:
(a)  *E. ancilla* – WA, Yakima Co., Oak Cr., 30 June 1960, J. Shepard
(b)  *E. battoides* – WA, Yakima Co., Blue Slide LO, el. 2,100 m, 14 July 1959, J. Shepard

**Figure 68**. Gnathos of *Lycaeides* species:
(a)  *L. melissa* – WA, Kennewick, 6 June 1960, J. Shepard
(b)  *L. anna* – WA, Okanogan Co., Pasayten Airport, 27 June 1961, J. Shepard

**Figure 71**. Harpe, located on the inner surface of valves, *Euphydryas* species:
(a)  *E. chalcedona* – BC, Mt. McLean, el. 1,900 m, 1–2 August 1974, J. and S. Shepard
(b)  *E. anicia* – BC, Kettle R., 13 mi. S Christian Valley, 22 June 1966, J. and S. Shepard
(c)  *E. editha* – BC, Manning Provincial Park, 15 June 1962, A.C. Jenkins
(d)  *E. gillettii* – AB, Savana Cr., Kananaskis Rd., 27 July 1969, J. and S. Shepard

**Figure 73**. Male genitalia of North American *Coenonympha* species:
(a)  *C. california* – BC, Princeton, Whipsaw Cr., Hwy. 3, el. 850 m, 11 June 1988, C.S. Guppy
(b)  *C. tullia* – BC, Atlin, 5 mi. N, 22 July 1975, C.S. Guppy

**Figure 74**. Valves of *Erebia* species:
(a)  *E. youngi* – YT, Dempster Hwy., mi. 254, el. 800 m, 20 June 1979, J. Shepard
(b)  *E. lafontainei* – AK, 148°40′, 69°25′, near Sagwon, el. 200 m, 9 July 1979, J. Shepard

**Figure 75**. Valves of *Oeneis* species:
(a)  *O. polixenes* – BC, Alaska Hwy., mi. 392.5, el. 1,700 m, 15 July 1977, J. and S. Shepard
(b)  *O. bore* – BC, McBride Peak, el. 2,300 m, 29 July 1985, J. and S. Shepard

(ADOPTED BY THE EXECUTIVE COUNCIL ON 13 JUNE 1996, HOUSTON, TEXAS)

The Lepidopterists' Society affirms that collecting Lepidoptera is one of many legitimate activities enabling professional and avocational lepidopterists to further the scientifically sound and progressive study of Lepidoptera and education about Lepidoptera as well as encouraging interaction between professional and avocational lepidopterists.

The foregoing Statement of The Lepidopterists' Society is accompanied by the following Collecting Guidelines. The Guidelines elucidate the manner in which collecting should be conducted. Practitioners are encouraged to adopt these Guidelines and to use the Guidelines for the instruction of others.

## COLLECTING GUIDELINES

### PREAMBLE

Our responsibility to assess and preserve natural resources, for the increase of knowledge and for the maintenance of biological diversity in perpetuity, requires that lepidopterists examine the practices of collecting Lepidoptera for the purpose of governing their own activities.

To this end, the following guidelines are outlined, based on these premises:

I. Lepidoptera is one of the largest orders of insects. Lepidopterans are an important component of biological diversity.

II. Lepidoptera are conspicuous and scientifically well known, thus they are frequently used as indicator groups for conservation programs.

III. The collection of Lepidoptera

A. is a means of introducing children and adults to awareness and study of their natural environment;

B. has an essential role in the elucidation of scientific information, both for its own sake and as a basis from which to develop rational means for protecting the environment, its resources, human health, and the world food supply;

C. is an educational activity which generally can be pursued in a manner not detrimental to the resource involved.

### GUIDELINES

I. **PURPOSES OF COLLECTING (consistent with the above):**

A. To create a reference collection for study and appreciation.

B. To document regional diversity, frequency, and variability of species, and as voucher material for published records.

C. To document faunal representation in environments undergoing or threatened with alteration by humans or natural forces.

D. To participate in development of regional checklists and institutional reference collections.

E. To complement a planned research endeavor.

F. To aid in dissemination of educational information.

G. To augment understanding of taxonomic and ecologic relationships for medical and economic purposes.

II. **COLLECTING METHODS:**

A. Collecting adults or immature stages should be limited to sampling, not depleting, the population concerned. Numbers collected should be consistent with the purposes outlined in sections 1.1 through 1.7.

B. Where the extent and/or the fragility of the population is unknown, caution and restraint should be exercised.

III. **DATA SHARING:**

A. All data should be recorded, and the data should be made available to appropriate interested parties.

IV. **LIVE MATERIAL:**

A. Rearing to elucidate life histories and to obtain series of immature stages and adults is to be encouraged, provided that collection of the rearing stock is in keeping with these guidelines.

B. Reared material in excess of need should be released only in the region where it originated, and in suitable habitat.

V. **ENVIRONMENTAL:**

A. Protection of the supporting habitat must be recognized as the sine qua non of protection of a species.

B. Collecting should be performed in a manner such as to minimize trampling or other damage to the habitat or to specific foodplants.

C. Property rights and sensibilities of others must be respected.

D. Collectors must comply with regulations relating to publicly controlled areas, to individual species, and to habitats.

VI. **RESPONSIBILITY FOR COLLECTED MATERIAL:**

A. All material should be preserved with all known data attached.

B. All material should be protected from physical damage and deterioration, e.g., light, molds, and museum pests.

C. Collections should be made available for examination by qualified researchers.

D. Collections or specimens, and their associated written and photographic records, should be willed or offered to the care of an appropriate scientific institution, if the collector lacks space or loses interest, or in anticipation of death.

E. Type specimens, especially holotype or allotype, should be deposited in appropriate scientific institutions.

VII. **RELATED ACTIVITIES OF COLLECTORS**:

A. Collecting should include permanently recorded field notes regarding habitat, conditions, and other pertinent information.

B. Recording of observations of behavior and of biological interactions should receive as high priority as collecting.

C. Photographic records, with full data, are to be encouraged.

D. Education of the public regarding collecting and conservation, as reciprocally beneficial activities, should be undertaken whenever possible.

E. All known data should be recorded with the specimens, e.g., date, location, collector, habitat, caterpillar host plant data, and parentage of immatures, when known.

VIII. **TRAFFIC IN LEPIDOPTERAN SPECIMENS**:

A. Collection of specimens for exchange or sale should be performed in accordance with these guidelines.

B. Rearing of specimens for exchange or sale should be from stock obtained in a manner consistent with these guidelines, and so documented.

C. Mass collecting of Lepidoptera for commercial purposes and collection of specimens for creation of saleable artifacts are not included among the purposes of the Society.

IX. **LEGAL CONSIDERATIONS**:

A. Collectors should comply with local, state or provincial, federal and national, and international laws and regulations that govern collecting and possession, commerce and exchange, import and export, and protection of species. Collectors should comply with additional local, state or provincial, federal and national, and international laws and regulations governing live material.

# Glossary

**aedeagus** — The male genital structure through which the sperm passes to the female.

**aestivate** — To rest in an inactive state during the summer (*see also* hibernate).

**alkaline** — Non-acidic, with pH > 7.0, in reference to soil or the fluids in an insect's intestine.

**alkaloids** — A group of chemicals found in plants that are toxic to many insects.

**allelochemicals** — The term used for all chemicals found in plants that are toxic to insects and other animals.

**allopatric** — Living in separate geographic areas (*see also* sympatric).

**allotype** — A term for a designated specimen of the opposite sex to the holotype.

**androconial scales** — Specialized scales on a butterfly's wing that produce scent (pheromones) for sex and species recognition during courtship and mating.

**annual** — Yearly, in reference to a plant that germinates from a seed, grows, flowers, sets seed, and dies within a period of one year.

**anterior** — The front part of a butterfly or caterpillar.

**Beringia** — The area of Siberia, Alaska, and Yukon that was a single, continuous ice-free area during the last ice age.

**biennial** — Occurring in two-year life cycles. Usually refers to a plant that germinates from a seed, grows, flowers, sets seed, and dies within a period of two years. The plant only flowers and sets seed in the second year. Some butterflies have a biennial life cycle.

**bivoltine** — Two full generations (life cycles) of a butterfly per year, with hibernation only occurring in one of the generations.

**boreal** — This term is used generally for northern forested areas, including both taiga and boreal forests. More specifically, boreal forest is the north temperate coniferous forest characterized by white spruce mixed with black spruce, trembling aspen, balsam poplar, larch, lodgepole pine, paper birch, and subalpine fir. Boreal forest occurs south of the taiga (*see* taiga).

**butterfly** — A butterfly is an insect in the Order Lepidoptera, which is made up of butterflies, skippers, and moths. The Lepidoptera are characterized by having scaled ("lepido" in Latin) wings ("ptera"). Butterflies and skippers both have simple antennae that gradually or abruptly expand near the tip, and their wings have no frenulum and retinaculum (defined below). The differences between skippers (Superfamily Hesperioidea) and butterflies (Superfamily Papilionoidea) are described in those superfamily treatments in the Species Accounts.

**chemoreception** — The process of sensing chemicals. Smell is chemoreception of air-borne chemicals; taste is chemoreception of liquid or solid chemicals.

**cline** — A gradual change over a wide geographic area in the appearance of a butterfly species, usually with respect to butterfly colour, markings, or size.

**cocoon** — A bag-like structure made of silk by the caterpillar that surrounds the pupa (chrysalis) and protects it. More common with moths than butterflies.

**conspecific** — The same species.

**COSEWIC** — Committee on the Status of Endangered Wildlife in Canada. This group determines the national conservation status of wildlife in Canada.

**cremaster** — A structure at the anal (rear) end of a pupa, usually covered with many hooks that hold the pupa to a silk pad attached to the substrate.

**chrysalis** — *See* pupa.

**cryptic** — Coloured and/or shaped to blend into the surroundings; camouflaged.

**det.** — Abbreviation for "determined by," meaning "identified by."

**diapause** — A resting stage, including hibernation and aestivation (*see also* those terms).

**dicotyledon** — The majority of flowering plants, in which the leaf veins are normally branched. A monocotydon, in contrast, has parallel unbranched leaf veins.

**dimorphic** — Occurring in two forms; for butterflies, either one sex having two colour forms or the two sexes having different colour forms.

**disjunct** — The range of a species that is formed of two or more geographically separate areas.

**dorsal** — The upper surface of a butterfly or caterpillar.

dorsolateral — The upper area of the side of a butterfly or caterpillar.

electrophoresis — The process of comparing some of the physical and electrical characteristics of the same protein; for butterflies, usually comparing species or subspecies to obtain an estimate of how similar they are.

emerge — When the adult butterfly sheds the pupal skin.

endemic — Occurring only in a small geographic area.

entomology — The study of insects.

etymology — An account of the origin and development of a word.

extant — Still existing.

fauna — The species of animals present in an area or time period.

feral — Domestic animals that are living in the wild.

FIS — Forest Insect Survey. A long-term project by the federal government to survey the insects of the forests of Canada.

flora — The species of plants present in an area or time period.

frenulum — A spine on the base of the hindwing that fits into the retinaculum on the forewing.

gynandromorph — A butterfly with portions of the body being of both sexes. A perfect bilateral gynandromorph has one side completely male and the other side completely female. Others may be predominantly of one sex, with only a small patch of wing of the other sex.

hemolymph — The "blood" of an insect. Unlike the blood of vertebrates, hemolymph does not contain red blood cells and therefore does not carry oxygen through the body.

hibernaculum — A silk and rolled leaf structure within which some partly grown caterpillars hibernate.

hibernate — To rest in an inactive state during the winter (*see also* aestivate).

holarctic — Occurring throughout the arctic and temperate regions of both Eurasia and North America.

Holocene — The period of time from about 10,000 years ago to the present.

holotype — The single specimen designated during the original description to represent the identity of a species or subspecies.

homonym — Each of two or more available species or subspecies names that have the same spelling and that denote different taxa in the same genus.

hybrid — A butterfly that is the offspring of a male and a female from different species.

hyperparasitoid — For butterflies, a parasitoid whose host is the parasitoid of a caterpillar.

hypsithermal — The climatically warm period after the end of the last glacial period, from about 9,000 to 2,500 years ago, with maximum temperatures occurring about 7,000 years ago.

ICZN — International Commission on Zoological Nomenclature and its publication the *International Code of Zoological Nomenclature*.

intersegmental — Between the body segments of a caterpillar.

lateral — Referring to the side of a butterfly or caterpillar.

latex — A sticky, milky, thick fluid in some plants, such as dandelions.

lectotype — The single specimen designated to represent the identity of a nominal species or subspecies, selected (at a date after the original description) from the series of specimens (syntypes) used in the original description.

median — The middle area, lengthwise, down the centre of the body, or the middle area of a butterfly's wing.

mesic — The average or normal environment of an area.

migrant — A butterfly species that makes regular long-distance flights from one area to another, or an individual butterfly that has migrated into an area.

monophagous — Feeding on only one species or genus of plants (*see also* polyphagous).

monotypic — A genus or family that includes only one species, or a species that has no subspecies.

moth — A moth is an insect in the Order Lepidoptera, which is made up of butterflies, skippers, and moths. Moths usually have either branched or feathery antennae, or simple thread-like antennae. The few moths that have simple antennae gradually expanding near the tip, as in many butterflies and skippers, also have wings with a frenulum and retinaculum.

moult — When a caterpillar sheds its skin.

MRR — Mark-Release-Recapture. A method used to determine butterfly population size.

multivoltine — Many full generations (life cycles) of a butterfly per year, with hibernation occurring in none or only one of the generations.

myrmecophilous — Living in association with ants, with the ants usually protecting the caterpillar.

Nearctic — The arctic, temperate, and subtropical area that includes Canada, the USA, Greenland, and Mexico.

nec — Latin for "not."

Neotropical — The tropical areas of North, South, and Central America.

nomen nudum — Latin for "naked name," a name for a butterfly that was not accompanied by a definition, description, or reference as required by the ICZN.

| | |
|---|---|
| nomenclature | A system of names, and the rules for their formation and use. |
| nominate | The subspecies that is the one that was originally described as a species, as opposed to other subspecies of a species. |
| oligophagous | Feeding on a few genera, and sometimes a few closely related families, of plants. |
| oviposit | A female laying eggs. |
| paratype | A term for one or more specimens designated at the time of the original description of a species or subspecies as being part of the series upon which the description is based. |
| patronym | A name of a butterfly that refers to a person. |
| perennial | A plant that lives many years and flowers and sets seed more than once. |
| pH | The measure of alkalinity or acidity. A pH of 7 is neutral, a pH less than 7 is acidic, and a pH greater than 7 is alkaline. |
| phenology | The seasonal pattern of the life cycle of a butterfly. |
| phenotype | The visible characteristics of a butterfly or caterpillar, as opposed to the genotype, which includes many characteristics that have not been expressed in the phenotype. |
| pheromone | Air-borne chemicals produced by butterflies that act as sex attractants or for species recognition. |
| Pliocene | The final epoch of the Tertiary period. |
| pluvial | Water-deposited soils, usually in reference to areas that were once lakes or large rivers. |
| PNW | Pacific Northwest, including Oregon, Washington, Idaho, and southern BC. |
| polyphagous | Feeding on many genera, and usually several families, of plants. |
| postglacial | After the retreat of the continental glaciers of the last ice age. |
| post-hypsithermal | The cooler period after the hypsithermal period (*see also* hypsithermal). |
| prepupal | The stage of a mature caterpillar in which it stops feeding and prepares to pupate. There are often obvious changes in behaviour and colour pattern during this stage. |
| pupa | The stage of a butterfly's life cycle in which it transforms from a caterpillar into an adult butterfly. |
| refugium | An area, north of the southern limit of glaciation during the last ice age, that remained ice-free and maintained populations of species through the ice age. |
| relict | A population left behind, far from the main range of a species, when the species has been extirpated from the area between, especially due to glaciation during a glacial period. |
| retinaculum | A hook on the forewing into which fits the hindwing frenulum, to hold the wings together. |
| riparian | A habitat associated with streams, rivers, lakes, and some wetlands that is characterized by a unique vegetation community that is dependant on the water table maintained by the adjacent body of water. |
| scree | Loose, angular gravels and rocks flaked from bedrock that accumulates at the base of cliffs. |
| sex brand | *See* stigma. |
| skeletonize | The process of a caterpillar eating all of a leaf except the veins. |
| skipper | A skipper is an insect in the Order Lepidoptera, which is made up of butterflies, skippers, and moths. Skippers, like butterflies, have simple antennae that gradually or abruptly expand near the tip, and their wings have no frenulum and retinaculum. The one exception is an Australian skipper in the subfamily Pyrginae, *Euschemon rufflesia* (Macl.), whose wings have a frenulum and retinaculum (Common 1970). Skippers are frequently grouped with butterflies in casual conversation because they are day-flying. The differences between skippers (Superfamily Hesperioidea) and butterflies (Superfamily Papilionoidea) are described in those superfamily treatments in the Species Accounts. |
| solifluction | The process of the soils above permafrost slowly moving down a slope in a wave-like pattern during successive spring thaws, forming alternating flat terraces and steep slopes (each a few metres wide). Also known as gelifluction. |
| sp. | The abbreviation for "species," with the plural abbreviation being "spp." |
| ssp. | The abbreviation for "subspecies," with the plural abbreviation being "sspp." |
| stigma | Also called a sex brand. A dark patch of androconial scales on the forewing of some male butterflies, especially in the Hesperiidae and some hairstreaks (*see also* androconial scales). |
| subspiracular | Below the spiracles of a caterpillar or adult butterfly. |
| sympatric | Living in the same geographic area (*see also* allopatric). |
| taiga | The subarctic coniferous forest characterized by open black spruce, larch, and wetland vegetation. Taiga is between the boreal forest (*see* boreal) and the arctic tundra. |
| taxon | The general term used for a subspecies and species. The plural is taxa. |
| teneral | Freshly emerged from the pupa, with the wings and body not fully hardened. |

| | |
|---|---|
| **TL** | The abbreviation used for "type locality," the place where the holotype or lectotype specimen was collected. |
| **trivoltine** | Three full generations (life cycles) of a butterfly per year, with hibernation occurring in only one of the generations. |
| **tundra** | A grass- and sedge-dominated treeless environment at elevations too high, or at latitudes too far north, to permit tree growth. |
| **type** | The original specimen(s) on which the description of a species was based. If the description was based on only a single specimen then it is the holotype (*see* holotype), even if it was not designated as such. If there was more than one specimen in the original series and no holotype was designated, then all the specimens are syntypes. If a holotype was designated, then the remaining specimens in the original series are paratypes, with the allotype being a designated paratype of the opposite sex to the holotype. |
| **univoltine** | Having one full generation per year. |
| **ventral** | The under surface of a butterfly or caterpillar. |
| **ventrolateral** | The lower area of the side of a butterfly or caterpillar. |
| **vestigial** | Remaining as a structure that no longer has the function for which it evolved. |
| **xeric** | Very dry, desert-like. |

# Bibliography


Abbot, John, and J.E. Smith. 1797. *The natural history of the rarer lepidopterous insects of Georgia.* Vol. 1. London: J. Edwards, Cadell and Davies, J. White. xv + 102 pp. + pls. 1–51.

Adler, P.H. 1982. Soil- and puddle-visiting habits of moths. *J. Lep. Soc.* 36(3):161–73.

Adler, P.H. and D.L. Pearson. 1982. Why do male butterflies visit mud puddles? *Can. J. Zool.* 60(3):322–25.

Ae, S.A. 1958a. Comparative studies of developmental rates, hibernation, and food plants in North American *Colias* (Lepidoptera: Pieridae). *Amer. Midl. Nat.* 60(1):84–96.

–. 1958b. Effects of photoperiod on *Colias eurytheme. Lep. News* 11(6):207–14.

Allen, W.W., and R.F. Smith. 1958. Some factors influencing the efficiency of *Apanteles medicaginis* Muesebeck (Hymenoptera: Braconidae) as a parasite of the alfalfa caterpillar, *Colias philodice eurytheme* Boisduval. *Hilgardia* 28(1):1–42.

Anderson, E.M. 1904. *Catalogue of British Columbia Lepidoptera.* Victoria, BC: Brit. Columbia Provincial Museum. 56 pp.

Anderson, W.B. 1923. The relation of botany to entomology. *Proc. Ent. Soc. Brit. Columbia* 17/19:172–74.

Anonymous. 1875. Proceedings of the club. Psyche 1(19):120.

–. 1906. Checklist of British Columbia Lepidoptera. Victoria, BC: Brit. Columbia Dept. Agric. 49 pp.

–. 1929. Obituary Ernest Henry Blackmore. *Can. Ent.* 61(9):218.

Arms, K., P. Feeny, and R.C. Lederhouse. 1974. Sodium: Stimulus for puddling behavior by Tiger Swallowtail butterflies, *Papilio glaucus. Science* 185(4148):372–74.

Arthur, A.P. 1966. The present status of the introduced skipper, *Thymelicus lineola* (Ochs.) (Lepidoptera: Hesperiidae), in North America and possible methods of control. *Can. Ent.* 98(6):622–26.

Atkinson, R.N. 1978. *Butterflies and moths.* Penticton, BC: Penticton Mus. Archive File 6-3634. 13 pp.

Austin, G.T. 1998a. New subspecies of Pieridae (Lepidoptera) from Nevada. Pp. 533–38 in *Systematics of western North American butterflies,* edited by T.C. Emmel. Gainesville, FL: Mariposa Press. xxviii + 878 pp.

–. 1998b. *Limenitis archippus* (Cramer) (Lepidoptera: Nymphalidae) in western United States with special reference to its biogeography in the Great Basin. Pp. 751–62 in *Systematics of western North American butterflies,* edited by T.C. Emmel. Gainesville, FL: Mariposa Press. xxviii + 878 pp.

Austin, G.T., and J.F. Emmel. 1998a. New subspecies of butterflies (Lepidoptera) from Nevada and California. Pp. 501–22 in *Systematics of western North American butterflies,* edited by T.C. Emmel. Gainesville, FL: Mariposa Press. xxviii + 878 pp.

–. 1998b. A review of *Papilio multicaudatus* Kirby (Lepidoptera: Papilionidae). Pp. 691–700 in *Systematics of western North American butterflies,* edited by T.C. Emmel. Gainesville, FL: Mariposa Press. xxviii + 878 pp.

Ballmer, G.R. and G.F. Pratt. 1989. A survey of the last instar larvae of the Lycaenidae of California. *J. Res. Lep.* 27(1):1–81.

–. 1992a. Quantification of ant attendance (myrmecophily) of lycaenid larvae. *J. Res. Lep.* 30(1–2):95–112.

–. 1992b. *Loranthomitoura,* a new genus of Eumaeini (Lepidoptera: Lycaenidae: Theclinae). *Tropical Lep.* 3(1):37–46.

Barbosa, Pedro. 1993. Lepidopteran foraging on plants in agroecosystems: Constraints and consequences. Pp. 523–66 in *Caterpillars: Ecological and evolutionary constraints on foraging,* edited by N.E. Stamp and T.M. Casey. New York: Chapman and Hall. xiii + 587 pp.

Barnes, William, and J.H. McDunnough. 1916. Some new races and species of North American Lepidoptera. *Can. Ent.* 48(7):221–26.

–. 1919. A new race of *Plebeius icarioides* from Vancouver Island. *Can. Ent.* 51(4):92–93.

Barrows, E.M. 1984. *Cypripedium* flowers entrap adult *Thymelicus* (Lepidoptera: Hesperiidae) in northern Michigan. *J. Lep. Soc.* 37(4):265–68.

Bauer, D.L. 1960. A new geographical subspecies of *Chlosyne hoffmanni* (Nymphalidae) from Washington state. *J. Lep. Soc.* 13(4):207–11.

Bean, T.E. 1890. The butterflies of Laggan, NWT; account of certain species inhabiting the Rocky Mountains in latitude 51°25′. *Can. Ent.* 22(5):94–99, 126–32.

Beck, A.F., and W.J. Garnett. 1984. Distribution and notes on the Great Dismal Swamp population of *Mitoura hesseli* Rawson and Ziegler (Lycaenidae). *J. Lep. Soc.* 37(4):289–300.

Behrens, James. 1874. Vernacular names for butterflies. *Psyche* 1(3): 9–10.

Belik, A.G. 2000. On the correct placement of *Erebia epipsodea* Butler, 1868 within the genus *Erebia* Dalman, 1816 (Lepidoptera: Satyridae). *J. Res. Lep.* 36:16-23.

Bell, E.A., and J.J. Dayton. 1986. Predation on monarch butterflies (*Danaus plexippus*) by chestnut-backed chickadees (*Parus rufescens*) at a California overwintering site. P. 5 in *Abstracts of the Second International Conference on the Monarch Butterfly "MONCON-2," 2–5 September 1986,* edited by J.P. Donahue. Los Angeles: Natural History Museum of Los Angeles County. vii + 23 pp.

Berger, T.A. and R.C. Lederhouse. 1986. Puddling by single male and female Tiger Swallowtails, *Papilio glaucus* L. (Papilionidae). *J. Lep. Soc.* 39(4):339–40.

Berkhousen, A.E. and A.M. Shapiro. 1994. Persistent pollen as a tracer for hibernating butterflies: The case of *Hesperia juba* (Lepidoptera: Hesperiidae). *Great Basin Nat.* 54(1):71–78.

Bethune, C.J.S. 1873a. On some of our common insects. III. Cabbage Whites. *Can. Ent.* 5(3):41–43.

–. 1873b. On some of our common insects. 10. The Clouded Sulphur Butterfly. *Can. Ent.* 5(12):221–23.

–. 1874. On some of our common insects. 13. The Disippus Butterfly – *Limenitis disippus,* Godt. *Can. Ent.* 6(3):46–49.

Beutenmüller, William. 1889. On early stages of some Lepidoptera. *Can. Ent.* 21(8):160.

Bird, C.D., G.J. Hilchie, N.G. Kondla, E.M. Pike, and F.A.H. Sperling. 1995. *Alberta butterflies.* Edmonton, AB: Provincial Museum of Alberta. viii + 349 pp.

Blackmore, E.H. 1927. *Check-list of the macrolepidoptera of British Columbia (butterflies and moths).* Victoria, BC: Brit. Columbia Provincial Museum. 47 pp.

Blau, W.S. 1981. Notes on the natural history of *Papilio polyxenes stabilis* (Papilionidae) in Costa Rica. *J. Lep. Soc.* 34(3):321–24.

Boggs, C.L. 1998. The Y files: Salt, sex and the single butterfly. *Amer. Butterflies* 6(1):4–9.

Boggs, C.L., and L.E. Gilbert. 1979. Male nutrient contribution to egg production in butterflies: evidence for transfer of nutrients at mating. *Science* 206(4414):83–84.

Boisduval, J.B.A. 1852. Lepidoptères de la Californie. *Annls. Soc. Ent. Fr.* (2)10:275–324.

Bower, H.M. 1911. Early stages of *Lycaena lygdamus* Doubleday (Lepid.). *Ent. News* 22(8):359–63.

Bowers, M.D., I.L. Brown, and D. Wheye. 1985. Bird predation as a selective agent in a butterfly population. *Evolution* 39(1):93–103.

Bowles, G.J. 1872. Notes on *Pieris rapae. Can. Ent.* 4(6):102–5.

Brako, Lois, A.Y. Rossman, and D.F. Farr. 1995. Scientific and common names of 7,000 vascular plants in the United States. St. Paul, MN: APS Press. v + 295 pp.

Breedlove, D.E. and P.R. Ehrlich. 1972. Coevolution: Patterns of legume predation by a lycaenid butterfly. *Oecologia* 10(1):99–104.

Bretherton, R.F. 1990a. [*Parnassius apollo* (Linnaeus). The Apollo or The Crimson-ringed Butterfly]. Pp. 73–75 in *The moths and butterflies of Great Britain and Ireland.* Vol. 7, Pt. I, *Hesperiidae to Nymphalidae,* edited by A.M. Emmet and John Heath. Colchester, UK: Harley Books. ix + 370 pp.

–. 1990b. [Subsp. *gorganus* Fruhstorfer. The Continental Swallowtail]. Pp. 78–81 in *The moths and butterflies of Great Britain and Ireland.* Vol. 7, Pt. I, *Hesperiidae to Nymphalidae,* edited by A.M. Emmet and John Heath. Colchester, UK: Harley Books. ix + 370 pp.

Bretherton, R.F. and A.M. Emmet. 1990. *Vanessa atalanta* (Linnaeus). The Red Admiral. Pp. 190–93 in *The moths and butterflies of Great Britain and Ireland.* Vol. 7, Pt. I, *Hesperiidae to Nymphalidae,* edited by A.M. Emmet and John Heath. Colchester, UK: Harley Books. ix + 370 pp.

Bridges, C.A. 1993. *Bibliography (Lepidoptera: Rhopalocera),* 2nd ed. Urbana, IL: C.A. Bridges. v + ii + 464 + ii + 167 + ii + 52 pp.

Britten, H.B., P.F. Brussard, and D.D. Murphy. 1994. The pending extinction of the Uncompahgre Fritillary butterfly. *Conser. Biol.* 8(1):86–94.

Britton, D.R., T.R. New, and A. Jelinek. 1995. Rare Lepidoptera at Mount Piper, Victoria – the role of a threatened butterfly community in advancing understanding of insect conservation. *J. Lep. Soc.* 49(2):97–113.

Brodie, William, and J.E. White. 1883. *Check list of insects of the Dominion of Canada.* Toronto, ON: C. Blackett. 67 pp.

Brower, L.P. 1959. Larval foodplant specificity in butterflies of the *Papilio glaucus* group. *Lep. News* 12(3–4):103–114.

–. 1969. Ecological chemistry. *Sci. Amer.* 220(2):22–29.

Brower, L.P., and W.H. Calvert. 1985. Foraging dynamics of bird predators on overwintering monarch butterflies in Mexico. *Evolution* 39(4):852–68.

Brower, L.P., L.S. Fink, A.V.Z. Brower, K. Leong, K. Oberhauser, S. Altizer, O. Taylor, D. Vickerman, W.H. Calvert, T. van Hook, A. Alonso-Mejia, S.B. Malcolm, D.F. Owen, and M.P. Zalucki. 1995. On the dangers of interpopulational transfers of monarch butterflies. *BioScience* 45(8):540–44.

Brown, F.M. 1965. Itineraries of the Wheeler Survey naturalists, 1871. *Lep. News* 9(6):185–90.

–. 1966. The types of nymphalid butterflies described by William Henry Edwards – Part II, Melitaeinae. *Trans. Amer. Ent. Soc.* 92(4):357–468.

–. 1971. The "Arrowhead Blue," *Glaucopsyche piasus* Boisduval (Lycaenidae: Plebejinae). *J. Lep. Soc.* 25(4):240–46.

Brown, F.M., Donald Eff, and Bernard Rotger. 1957. *Colorado butterflies.* Denver, CO: Denver Museum of Natural History. vii + 368 + [I] pp. + backpiece.

Brown, F.M. and L.D. Miller. 1980. The types of the hesperiid butterflies named by William Henry Edwards. Part II, Hesperiidae: Hesperiinae, section II. *Trans. Amer. Ent. Soc.* 106(1):43–88.

Brown, W.D. and J. Alcock. 1991. Hilltopping by the Red Admiral butterfly: Mate searching alongside congeners. *J. Res. Lep.* 29(1–2):1–10.

Bruce, David. 1891. Attracting butterflies in Colorado. *Can. Ent.* 23(5):110.

Bruggemann, P.F. 1948. Lepidoptera hibernation. *Lep. News* 2(2):13.

Buckell, E.R. 1947. A list of the Lepidoptera collected in the Shuswap Lake district of British Columbia by Dr. W.R. Buckell. *Proc. Ent. Soc. Brit. Columbia* 43:11–21.

Burns, J.M. 1964. *Evolution in skipper butterflies of the genus* Erynnis. Univ. Calif. Publ. Ent. 37: i–iv, 1–214, pl. 1.

–. 1966. Expanding distribution and evolutionary potential of *Thymelicus lineola* (Lepidoptera: Hesperiidae), an introduced skipper, with special reference to its appearance in British Columbia. *Can. Ent.* 98(8):859–66.

Byers, J.R. and K.W. Richards. 1986. Spiny elm caterpillars, *Nymphalis antiopa* (Nymphalidae: Lepidoptera), feeding on sainfoin, *Onobrychis viciaefolia* (Leguminosae). *Can. Ent.* 118(9):941–42.

Byers, J.R., B.T. Roth, R.D. Thomson, and A.K. Topinka. 1984. Contamination of mustard and canola seed by frass of painted lady caterpillars, *Vanessa cardui* (Lepidoptera: Nymphalidae). *Can. Ent.* 116(10):1431–32.

Cadbury, J.W. 1937. Lepidoptera collected in northern British Columbia by Miss Josephine de N. Henry. Part I – Rhopalocera. *Proc. Acad. Nat. Sci. Philad.* 89:387–413.

Caldas, Astrid. 1995. Mortality of *Anaea ryphea* (Lepidoptera: Nymphalidae) immatures in Panama. *J. Res. Lep.* 31(3–4):195–204.

Calvert, W.H. 1994. Behavioral response of monarch butterflies (Nymphalidae) to disturbances in their habitat – a group startle response? *J. Lep. Soc.* 48(2):157–65.

Calvert, W.H., L.E. Hedrick, and L.P. Brower. 1979. Mortality of the monarch butterfly (*Danaus plexippus* L.): Avian predation at five overwintering sites in Mexico. *Science* 204(4395):847–51.

Campbell, R.W., N.K. Dawe, Ian McTaggart-Cowan, J.M. Cooper, G.W. Kaiser, and M.C.E. McNall. 1990. *The birds of British Columbia, Vol. 1. Nonpasserines.* Victoria, BC: Royal British Columbia Museum and Environment Canada. xvii + 514 pp.

Carl, G.C. 1966. In Memoriam George Austin Hardy (1888–1966). *J. Ent. Soc. Brit. Columbia* 63:43–44.

Casey, T.M. 1976. Activity patterns, body temperature and thermal ecology in two desert caterpillars (Lepidoptera: Sphingidae). *Ecology* 57(3):485–97.

–. 1993. Effects of temperature on foraging of caterpillars. Pp. 5–28 in *Caterpillars: Ecological and evolutionary constraints on foraging,* edited by N.E. Stamp and T.M. Casey. New York: Chapman and Hall. xiii + 587 pp.

Cassie, B., Jeff Glassberg, Paul Opler, Robert Robbins, and G. Tudor. 1995. *North American Butterfly Association (NABA) checklist of*

*English names of North American butterflies.* Morristown, NJ: North American Butterfly Association. 43 pp.

Caulfield, F.B. 1875. Notes on the larva of *Grapta faunus* Edwards. *Can. Ent.* 7(3):49–50.

Chambers, D.S. 1963. A preliminary study of foodplant preference in the *Lycaena helloides* complex (Lycaenidae) in Colorado. *J. Lep. Soc.* 17(1):24–26.

Chang, V.C.S. 1963. Quantitative analysis of certain wing and genitalia characters of *Pieris* in western North America. *J. Res. Lep.* 2(2):97–125.

Chermock, F.H. and R.L. Chermock. 1940. Some new diurnal Lepidoptera from the Riding Mountains and the Sand Ridge, Manitoba. *Can. Ent.* 72(4):81–83.

Chew, F.S. 1980. Natural interspecific pairing between *Pieris virginiensis* and *P. napi oleracea* (Pieridae). *J. Lep. Soc.* 34(2):259–61.

Chew, F.S. and R.K. Robbins. 1984. Egg laying in butterflies. Pp. 65–79 in *The biology of butterflies. Symposium of the Royal Entomological Society of London number 11,* edited by R.I. Vane-Wright and P.R. Ackery. London: Academic Press. xxiv + 429 pp.

Clague, J.J. 1985. Deglaciation of the Prince Rupert–Kitimat area, British Columbia. *Can. J. Earth Sci.* 22(2):256–65.

–. 1987. Quaternary stratigraphy and history, Williams Lake, British Columbia. *Can. J. Earth Sci.* 24(1):147–58.

–. 1988. Quaternary stratigraphy and history, Quesnel, British Columbia. *Geog. Phys. Quat.* 429(3):279–88.

Clark, A.H. 1932. The butterflies of the District of Columbia and vicinity. *Bull. US Natn.Mus.* 157:i–x, 1–337.

Clench, H.K. 1958. The "pumping" of certain moths at water. *Lep. News* 11(1–3):18–21.

–. 1961. Tribe Theclini. Pp. 177–220 in *The butterflies,* edited by P.R. Ehrlich and A.H. Ehrlich. Dubuque, IA: Wm. C. Brown. v + 262 pp.

–. 1963. *Callophrys* (Lycaenidae) from the Pacific Northwest. *J. Res. Lep.* 2(2):151–60.

–. 1966. Behavioral thermoregulation in butterflies. *Ecology* 4766: 1021–34.

Cockle, J.W. 1910. Notes on a few butterflies found at Kaslo and in northern British Columbia. *Can. Ent.* 42(6):203–4.

–. 1915. Notes of the habits of some Lepidoptera. *Proc. Ent. Soc. Brit. Columbia* 5:91–94.

–. 1920. A swarm of *Vanessa californica* and some notes on a swarm of *Plusia californica. Proc. Ent. Soc. Brit. Columbia* 14:20–21.

Cole, J.M. 1979. *Exile in the wilderness.* Don Mills, ON: Burns and MacEachern. xviii + 268 pp.

Cole, L.R. 1959. On the defences of lepidopterous pupae in relation to the oviposition behaviour of certain Ichneumonidae. *J. Lep. Soc.* 13(1):1–10.

Coles, H.J. 1948. Spring flight of *Nymphalis californica* near Nelson, BC (Lepidoptera: Nymphalidae). *Proc. Ent. Soc. Brit. Columbia* 44:34.

Common, I.F.B. 1970. Lepidoptera (moths and butterflies). In *The insects of Australia,* edited by CSIRO. Carlton, Australia: Melbourne University Press. xiii + 1029 pp. + 8 pls.

Comstock, J.A. 1927. *The butterflies of California.* Los Angeles: Dr. John Adams Comstock. 334 pp. + 63 pls.

–. 1928. Studies in Pacific coast Lepidoptera. *Bull. S. Calif. Acad. Sci.* 27(2):63–66.

–. 1929. Studies in Pacific coast Lepidoptera (life history studies) (continued). *Bull. S. Calif. Acad. Sci.* 28(2):22–32.

–. 1933. Studies in Pacific coast Lepidoptera (continued). *Bull. S. Calif. Acad. Sci.* 32(3):113–20.

Comstock, J.A., and C.M. Dammers. 1932. Metamorphoses of five Californian diurnals (Lepidoptera). *Bull. S. Calif. Acad. Sci.* 31(2): 33–45.

–. 1933. Notes on the life histories of four Californian lepidopterous insects. *Bull. S. Calif. Acad. Sci.* 32(2):77–83.

–. 1934. Additional notes on the early stages of Californian Lepidoptera. *Bull. S. Calif. Acad. Sci.* 33(1):25–34.

–. 1935. Notes on the early stages of two butterflies and one moth. *Bull. S. Calif. Acad. Sci.* 34(1):81–87.

–. 1936. Notes on the life histories of three butterflies and three moths from California. *Bull. S. Calif. Acad. Sci.* 34(3):211–25.

–. 1938. Notes on the metamorphosis of *Mitoura spinetorum* Hew. (Lepidoptera, Theclinae). *Bull. S. Calif. Acad. Sci.* 37(1):30–32.

Comstock, J.H., and A.B. Comstock. 1915. *How to know the butterflies. A manual of the butterflies of the eastern United States.* New York: D. Appleton. xii + 311 pp.

Cook, J.H. 1906. Studies in the genus *Incisalia.* II. *Incisalia augustus. Can. Ent.* 38(7):214–17.

–. 1908. Studies in the genus *Incisalia.* V. *Incisalia polios. Can. Ent.* 40(2):37–43.

Cook, J.H., and H. Cook. 1904. Notes on *Incisalia augustus. Can. Ent.* 36(5):137.

Coolidge, K.R. 1909. Further notes on the Rhopalocera of Santa Clara County, California. *Can. Ent.* 41(6):187–88.

–. 1923. The life history of *Hesperia ericetorum* Boisd. *Ent. News* 34(5): 140–46.

Courant, A.V., A.E. Holbrook, E.D. Van der Reijden, and F.S. Chew. 1994. Native pierine butterfly (Pieridae) adapting to naturalized crucifer? *J. Lep. Soc.* 48(2):168–70.

Crolla, J.P., and J.D. Lafontaine. 1996. Status report on the Monarch butterfly (*Danaus plexippus*) in Canada. Ottawa, ON: Can. Wildlife Ser., Water and Habitat Conser. Br., Habitat Conser. Div. 23 pp. + map.

Curtis, John. 1835. Order Lepidoptera. Pp. lxv-lxxv, plate A, in *Appendix of narrative of a second voyage in search of a Northwest Passage, and of a residence in the Arctic Regions during the years 1829, 1830, 1831, 1832, 1833,* Sir John Ross. London: Webster. xii + 120 + cii pp., 20 pls.

Damman, Hans. 1993. Patterns of interaction among herbivore species. Pp. 132–69 in *Caterpillars: Ecological and evolutionary constraints on foraging,* edited by N.E. Stamp and T.M. Casey. New York: Chapman and Hall. xiii + 587 pp.

Danby, W.H. 1890. Food plant of *Melitaea taylori,* Edw. *Can. Ent.* 22(6):121–22.

–. 1891. *Vanessa californica* on Vancouver Island. *Can. Ent.* 23(5):113.

–. 1894. Notes on Lepidoptera found on Vancouver Island. *J. New York Ent. Soc.* 29(1):31–36.

Danby, W.H., and C. de B. Green. 1893. Report on the entomology of British Columbia. *Bull. Nat. Hist. Soc. BC* 1:11–18.

Davenport, Dermorest. 1941. The butterflies of the satyrid genus *Coenonympha. Bull. Harvard Mus. Comp. Zool.* 87(4):215–349.

Dawson, P.M. 1889. *Grapta j-album. Can. Ent.* 21(9):179–80.

Dempster, J.P. 1967. The control of *Pieris rapae* with DDT. I. The natural mortality of the young stages of *Pieris. J. Appl. Ecol.* 4(2):485–500.

–. 1968. The control of *Pieris rapae* with DDT. II. Survival of the young stages of *Pieris* after spraying. *J. Appl. Ecol.* 5(2):451–62.

–. 1984. Natural enemies of butterflies. Pp. 65–79 in *The biology of butterflies. Symposium of the Royal Entomological Society of London number 11,* edited by R.I. Vane-Wright and P.R. Ackery. London: Academic Press. xxiv + 429 pp.

Dempster, J. P. and A. M. Emmet. 1990. [*Pieris rapae* (Linnaeus), The Small White]. Pp. 105–7 in *The moths and butterflies of Great Britain and Ireland*. Vol. 7, Pt. I, *Hesperiidae to Nymphalidae,* edited by A.M. Emmet and John Heath. Colchester, UK: Harley Books. ix + 370 pp.

Denton, S.W. 1889. Early stages of *Grapta j-album. Can. Ent.* 21(9): 164–65.

Dethier, V.G. 1940. The life history of *Polites peckius* Kby. *Bull. S. Calif. Acad. Sci.* 38(3):188–90.

–. 1944. The life history of *Polites sabuleti* Bdv. *Bull. S. Calif. Acad. Sci.* 42(3):128–31.

Dimock, T.E. 1973. Three natural hybrids of *Vanessa atalanta rubria* × *Cynthia annabella* (Nymphalidae). *J. Lep. Soc.* 27(4):274–78.

–. 1978. Notes on the life cycle and natural history of *Vanessa annabella* (Nymphalidae). *J. Lep. Soc.* 32(2):88–96.

Dodge, G.M. 1882. *Pieris rapae* to Nebraska. *Can. Ent.* 14(2):39–40.

Dolinger, P.M., P.R. Ehrlich, W.L. Fitch, and D.E. Breedlove. 1973. Alkaloid and predation patterns in Colorado lupine populations. *Oecologia* 13(2):191–204.

Dornfield, E.J. 1980. *The butterflies of Oregon.* Forest Grove, OR: Timber Press. xiv + 276 pp.

dos Passos, C.F. 1964. A synonymic list of the Nearctic Rhopalocera. *Lep. Soc. Mem.* 1:i–v, 1–145.

–. 1977. A taxonomic note on *Polygonia faunus arcticus* Leussler (Lepidoptera: Nymphalidae). *Pan-Pac. Ent.* 53(3):179–80.

dos Passos, C.F., and L.P. Grey. 1945. A genitalic survey of Argynninae (Lepidoptera, Nymphalidae). *Amer. Mus. Novit.* 1296:1–29.

–. 1947. Systematic catalogue of *Speyeria* (Lepidoptera: Nymphalidae) with designations of type and fixations of type localities. *Amer. Mus. Novit.* 1370:1–30.

Douglas, G.W., G.B. Straley, Del Meidinger, and Jim Pojar. 1998a. *Illustrated flora of British Columbia.* Vol. 1, *Gymnosperms and Dicotyledons (Aceraceae through Asteraceae).* Victoria, BC: Brit. Columbia Ministry of Environment, Lands and Parks and Ministry of Forests. v + 436 pp.

–. 1998b. *Illustrated flora of British Columbia.* Vol. 2. *Dicotyledons (Balsaminaceae through Cuscutaceae).* Victoria, BC: Brit. Columbia Ministry of Environment, Lands and Parks and Ministry of Forests. v + 400 pp.

Douglas, G.W., G.B. Straley, Del Meidinger, and Jim Pojar. 1999. *Illustrated flora of British Columbia.* Vol. 3. *Dicotyledons (Diapensiaceae through Onagraceae).* Victoria, BC: Brit. Columbia Ministry of Environment, Lands and Parks and Ministry of Forests. v + 423 pp.

Douglas, M.M. 1986. *The lives of butterflies.* Ann Arbor, MI: University of Michigan Press. xvii + 241 pp.

Douwes, Per. 1977. An area census method for estimating butterfly population numbers. *J. Res. Lep.* 15(3):146–52.

Downes, J.A. 1973. Lepidoptera feeding at puddle-margins, dung and carrion. *J. Lep. Soc.* 27(2):89–99.

Downes, William. 1943. In Memoriam George O. Day, FES. *Proc. Ent. Soc. Brit. Columbia* 40:34–35.

–. 1945. In Memoriam Abdiel William Hanham, 1857–1944. *Proc. Ent. Soc. Brit. Columbia* 42:27–28.

–. 1948. The hibernation of *Nymphalis californica* (Bdv.), the California Tortoiseshell butterfly; a query. *Proc. Ent. Soc. Brit. Columbia* 44:34.

–. 1956. Observations on the effect of drought in insect populations with especial reference to Heteroptera, Homoptera, and Lepidoptera. *Proc. Ent. Soc. Brit. Columbia* 52:12–16.

Downey, J.C. 1962a. Variation in *Plebejus icarioides* (Lepidoptera, Lycaenidae) II. Parasites of the immature stages. *Ann. Ent. Soc. Amer.* 55(4):367–73.

–. 1962b. Myrmecophily in *Plebejus* (*Icaricia*) *icarioides* (Lepid.: Lycaenidae). *Ent. News* 73(3):57–66.

Downey, J.C., and A.C. Allyn. 1973. Butterfly ultrastructure. 1. Sound production and associated abdominal structures in pupae of Lycaenidae and Riodinidae. *Bull. Allyn Mus.* 14:1–47.

Downey, J.C., and D.B. Dunn. 1964. Variation in the lycaenid butterfly *Plebejus icarioides*. III. Additional data on food-plant specificity. *Ecology* 45(1):172–78.

Downey, J.C., and W.C. Fuller. 1961. Variation in *Plebejus icarioides* (Lycaenidae). I. Foodplant specificity. *J. Lep. Soc.* 15(1):34–42.

Duchesne, R.M., and J.N. McNeil. 1978. Transport of the European skipper, *Thymelicus lineola* (Lepidoptera: Hesperiidae), associated with the production of certified timothy seed. *Can. Ent.* 110(3): 245–47.

Durden, C.J. 1965. *Speyeria callippe* and *Artemisia,* a possible foodplant. *J. Lep. Soc.* 19(3):186–87.

Dussourd, D.E. 1993. Foraging with finesse: Caterpillar adaptations for circumventing plant defenses. Pp. 92–131 in *Caterpillars: Ecological and evolutionary constraints on foraging,* edited by N.E. Stamp and T.M. Casey. New York: Chapman and Hall. xiii + 587 pp.

Dyar, H.G. 1891. *Limenitis lorquini. Can. Ent.* 23(9):201.

–. 1904a. Two new forms of *Oeneis* Hübner. *Proc. Ent. Soc. Wash.* 6(3):142.

–. 1904b. The Lepidoptera of the Kootenai [sic] district of British Columbia. *Proc. US Natn. Mus.* 27(1376):779–938.

Edwards, W.H. 1863. Description of certain species of diurnal Lepidoptera found within the limits of the United States and British America. No. 2. *Proc. Ent. Soc. Philad.* 2(2):78–82.

–. 1868–72. *The butterflies of North America.* Vol. 1. Philadelphia, PA: Amer. Ent. Soc. [iii] + ii + [163] + v + 52 pp. + [50] pls.

–. 1874–84. *The butterflies of North America.* Vol. 2. Boston: Houghton Mifflin. [v] + [358] [N.B. pp. 316–30 are numbered 2–16] pp. + [51] pls.

–. 1878. Notes on *Lycaena pseudargiolus* and its larval history. *Can. Ent.* 10(1):1–14.

–. 1880. Description of the preparatory stages of *Grapta progne,* Cramer. *Can. Ent.* 12(1):9–14.

–. 1881. Description of the preparatory stages of *Terias nicippe. Can. Ent.* 13(4):61–63.

–. 1882a. Notes on certain butterflies, their habits, etc. No. 2. *Can. Ent.* 14(3):49–56.

–. 1882b. Description of the preparatory stages of *Pyrameis atalanta,* Linn. *Can. Ent.* 14(12):229–34.

–. 1883. Description of the preparatory stages of *Pyrameis atalanta,* Linn. (continued). *Can. Ent.* 15(1):14–20.

–. 1884. Notes on butterflies, with directions for breeding them from the egg. *Can. Ent.* 16(5):81–89.

–. 1885a. Description of a new species of *Chionobas* from British America. *Can. Ent.* 17(5):74–75.

–. 1885b. Note on habit of larva of *P. atalanta. Can. Ent.* 17(9):179.

–. 1885c. History of the preparatory stages of *Vanessa milberti,* Godart. *Can. Ent.* 17(10):181–88.

–. 1885d. Description of the preparatory stages of *Pholisora catullus,* Fabricius. *Can. Ent.* 17(12):245–48.

–. 1886a. Miscellaneous notes on butterflies, their larvae, etc. *Can. Ent.* 18(1):14–18.

–. 1886b. Description of the preparatory stages of *Satyrus charon,* Edw. *Can. Ent.* 18(5):88–92.

–. 1887–97. *The butterflies of North America.* Vol. 3. Boston: Houghton Mifflin. [viii] + [432] pp. + [51] pls.

–. 1887a. Description of the preparatory stages of *Coenonympha ampelos*. *Can. Ent.* 19(3):41–44.

–. 1887b. History of the preparatory stages of *Colias alexandra* Edw. *Can. Ent.* 19(12):226–30.

–. 1889. Description of the preparatory stages of *Colias meadii*, Edwards. *Can. Ent.* 21(3):41–43.

–. 1891. *Chionobas bore. Can. Ent.* 23(1):16.

–. 1892. Miscellaneous notes on butterflies, larvae, etc. *Can. Ent.* 24(3):50–56, 24(5):105–11.

Ehrlich, P.R. 1955. The distribution and subspeciation of *Erebia epipsodea* Butler (Lepidoptera: Satyridae). *Univ. Kan. Sci. Bull.* 37(1):175–94.

–. 1962. A biting midge ectoparasitic on Arizona Lycaenids. *J. Lep. Soc.* 16(1):20–22.

Ehrlich, P.R., D.E. Breedlove, P.F. Brussard, M.A. Sharp. 1972. Weather and the regulation of subalpine populations. *Ecology* 53(2): 243–47.

Ehrlich, P.R., and S.E. Davidson. 1961. Techniques for capture-recapture studies of Lepidoptera populations. *J. Lep. Soc.* 14(4):227–29.

Ehrlich, P.R., and D.D. Murphy. 1982. Butterfly nomenclature: a critique. *J. Res. Lep.* 20(1):1–11.

Eitschberger, Ulf. 1981. Die nordamerikanischen Arten aus der *Pieris napi-bryoniae*-Gruppe (Lep., Pieridae) [The North American species of the *Pieris napi-bryoniae* group (Lep., Pieridae)]. *Atalanta* 11(5):366–71.

–. 1983. Systematische untersuchungen am *Pieris napi-bryoniae*-komplex (s. l.) (Lepidoptera, Pieridae). Marktleuthen, Ger.: *Herbipoliana* 1(1):i–xxii, 1–504; 1(2):1–601.

–. 1993. Die struktur der eihullen einiger *Papilio*-arten im vergleich unter dem REM/SEM (Lepidoptera, Papilionidae). *Atalanta* 24 (1/2):15–32.

Eliot, J.N., and A. Kawazoe. 1983. *Blue butterflies of the* Lycaenopsis *group*. London: British Museum of Natural History. 309 pp.

Ellis, S.L. 1974. Field observations on *Colias alexandra* Edwards (Pieridae). *J. Lep. Soc.* 28(2):114–25.

Elrod, M.J. 1906. *The butterflies of Montana*. Missoula, MT: University of Montana. xvi + 173 pp. + xiii pls.

Elwes, H.J. 1889. A revision of the genus *Argynnis*. *Trans. Ent. Soc. Lond.* 37(4):535–75.

Emmel, J.F. 1975. Subfamily Papilioninae. Pp. 390–402 in *The butterflies of North America*, edited by W.H. Howe. Garden City, NY: Doubleday. xiii + 633 pp. + 97 pls.

–. 1982. Two new subspecies of the *Papilio indra* complex from California (Papilionidae). *J. Lep. Soc.* 35(4):297–302.

Emmel, J.F., and T.C. Emmel. 1963. Larval food-plant records for six western papilios. *J. Res. Lep.* 1(3):191–93.

–. 1964. The life history of *Papilio indra minori*. *J. Lep. Soc.* 18(2):65–73.

–. 1998. A new species of *Agriades* (Lepidoptera: Lycaenidae) from the Sierra Nevada and Trinity Alps of California, and the biology and geographic variation of *Agriades podarce* in California. Pp. 287–302 in *Systematics of western North American butterflies*, edited by T.C. Emmel. Gainesville, FL: Mariposa Press. xxviii + 878 pp.

Emmel, J.F., T.C. Emmel, and S.O. Mattoon. 1998a. The types of California butterflies named by Jean Alphonse Boisduval: Designation of lectotypes and a neotype, and fixation of type localities. Pp. 4–76 in *Systematics of western North American butterflies*, edited by T.C. Emmel. Gainesville, FL: Mariposa Press. xxviii + 878 pp.

–. 1998b. The types of California butterflies named by Pierre Hippolyte Lucas: Designation of lectotypes and fixation of type localities. Pp. 77–82 in *Systematics of western North American butterflies*, edited by T.C. Emmel. Gainesville, FL: Mariposa Press. xxviii + 878 pp.

–. 1998c. Fixation of a type for *Thecla spinetorum* Hewitson (Lepidoptera: Lycaenidae). Pp. 83–86 in *Systematics of western North American butterflies*, edited by T.C. Emmel. Gainesville, FL: Mariposa Press. xxviii + 878 pp.

–. 1998d. The types of California and Nevada butterflies named by Cajetan and Rudolph Felder: Designation of lectotypes and fixation of type localities. Pp. 87–94 in *Systematics of western North American butterflies*, edited by T.C. Emmel. Gainesville, FL: Mariposa Press. xxviii + 878 pp.

–. 1998e. The types of California butterflies named by Herman Behr: Designation of neotypes and fixation of type localities. Pp. 95–113 in *Systematics of western North American butterflies*, edited by T.C. Emmel. Gainesville, FL: Mariposa Press. xxviii + 878 pp.

Emmel, J.F., Oakley Shields, and D.E. Breedlove. 1971. Larval foodplant records for North American Rhopalocera. Pt. 2. *J. Res. Lep.* 9(4): 233–42.

Emmel, T.C. 1969. Taxonomy, distribution and biology of the genus *Cercyonis* (Satyridae). I. Characteristics of the genus. *J. Lep. Soc.* 23(3):165–75.

Emmel, T.C., and J.F. Emmel. 1962. Ecological studies of Rhopalocera at Donner Pass, California. I. *J. Lep. Soc.* 16(1):23–44.

–. 1967. The biology of *Papilio indra kaibabensis* in the Grand Canyon. *J. Lep. Soc.* 21(1):41–48.

–. 1973. *The butterflies of southern California*. Sci. Ser. Nat. Hist. Mus. Los Angeles 26:i–xi, 1–148.

–. 1974. Ecological studies of Rhopalocera in a Sierra Nevadan Community – Donner Pass, California. V. Faunal additions and foodplant records since 1962. *J. Lep. Soc.* 28(4):344–48.

Emmet, A.M. 1990a. [*Anthocharis cardamines* (Linnaeus). The Orange-tip]. Pp. 114–17 in *The moths and butterflies of Great Britain and Ireland*. Vol. 7, Pt. I, *Hesperiidae to Nymphalidae*, edited by A.M. Emmet and John Heath. Colchester, UK: Harley Books. ix + 370 pp.

–. 1990b. The vernacular names and early history of British butterflies. Pp. 7–21 in *The moths and butterflies of Great Britain and Ireland*. Vol. 7, Pt. I, *Hesperiidae to Nymphalidae*, edited by A.M. Emmet and John Heath. Colchester, UK: Harley Books. ix + 370 pp.

–. 1991. *The scientific names of the British Lepidoptera, their history and meaning*. Colchester, UK: Harley Books. 288 pp.

Emmons, Ebenezer. 1854. Butterflies. Pp. 198–216 in *Natural history of New York*. Vol. 5, *Agriculture of New York*. Albany, NY: C. van Benthysen. viii + 272 pp. + pls. A-C + 47.

Evans, W.H. 1951. *A catalogue of the American Hesperiidae in the British Museum*. Pt. I, *Pyrrhopyginae*. London: British Museum of Natural History. x + 92 pp. + 9 pls.

–. 1952. *A catalogue of the American Hesperiidae in the British Museum*. Pt. II, *Pyrginae*, section 1. London: British Museum of Natural History. v + 178 pp. + 16 pls.

–. 1953. *A catalogue of the American Hesperiidae in the British Museum*. Pt. III, *Pyrginae*, section 2. London: British Museum of Natural History. v + 246 pp. + 28 pls.

–. 1955. *A catalogue of the American Hesperiidae in the British Museum*. Pt. IV, *Hesperiinae and Megathyminae*. London: British Museum of Natural History. v + 499 pp. + 35 pls.

–. 1975. Seasonal forms of *Anthocharis sara* (Pieridae). *J. Lep. Soc.* 29(1):52–55.

F., T.W. 1912. The Rev. George W. Taylor, FRSC, FZS. *Can. Ent.* 44(10): 285–87.

Fales, J.H. 1976. More records of butterflies as prey for ambush bugs (Heteroptera). *J. Lep. Soc.* 30(2):147–48.

Fales, J.H., and D.T. Jennings. 1977. Butterflies as prey for crab spiders (Thomisidae). *J. Lep. Soc.* 31(4):280–82.

Ferguson, D.C. 1954. The Lepidoptera of Nova Scotia. *Proc. Nova Scotia Inst. Sci.* 23:1–375.

Fernald, C.H. 1884. *The butterflies of Maine.* Augusta, ME: Ann. Rep. State College Agriculture and Mechanical Arts 1883:3–106.

Fernald, Mrs. C.H. 1886. Notes on *Papilio turnus* and *Pyrameis cardui. Can. Ent.* 18(3):50–51.

Ferris, C.D. 1972. Notes on certain species of *Colias* (Lepidoptera: Pieridae) found in Wyoming and associated regions. *Bull. Allyn Mus.* 5:1–23.

–. 1973a. A revision of the *Colias alexandra* complex (Pieridae) aided by ultraviolet reflectance photography with designation of a new subspecies. *J. Lep. Soc.* 27(1):57–73.

–. 1973b. Polymorphism in two species of Alaskan *Boloria. J. Res. Lep.* 11(4):255–59.

–. 1974. Distribution of arctic-alpine *Lycaena phlaeas* L. (Lycaenidae) in North America with designation of a new subspecies. *Bull. Allyn Mus.* 18:1–13.

–. 1982. On *Colias hecla* Lefèbvre re a recent paper by Oosting and Parshall (Lepidoptera: Pieridae). *J. Res. Lep.* 20(1):52–53.

–. 1988. Revision of the North American Ericaciae-feeding [sic] *Colias* species (Pieridae: Coliadinae). *Bull. Allyn Mus.* 122:1–34.

–, ed. 1989. Supplement to: A catalogue/checklist of the butterflies of America North of Mexico. *Lep. Soc. Mem.* 3:i–vii, 1–103.

Ferris, C.D., and F.M. Brown. 1981. *Butterflies of the Rocky Mountain states.* Norman, OK: University of Oklahoma Press. xix + 442 pp.

Ferris, C.D., C.F. dos Passos, J.A. Ebner, and J.D. Lafontaine. 1983. An annotated list of the butterflies (Lepidoptera) of the Yukon Territory, Canada. *Can. Ent.* 115(7):823–40.

Fiedler, Konrad, and Dorthe Hagemann. 1995. The influence of larval age and ant number on myrmecophilous interactions of the African Grass Blue butterfly, *Zizeria knysna* (Lepidoptera: Lycaenidae). *J. Res. Lep.* 31(3–4):213–32.

Fiedler, Konrad, Peter Seufert, N.E. Pierce, John G. Pearson, and H.-T. Baumgarten. 1995. Exploitation of lycaenid-ant mutualisms by braconid parasitoids. *J. Res. Lep.* 31(3–4):153–68.

Field, W.D. 1936. New North American Rhopalocera. *J. Ent. Zool.* 28(2):17–26.

–. 1971. Butterflies of the genus *Vanessa* and of the resurrected genera *Bassaris* and *Cynthia* (Lepidoptera: Nymphalidae). *Smith. Cont. Zool.* 84:10105.

Fletcher, James. 1888. List of diurnal Lepidoptera. *Geol. Nat. His. Sur. Canada. Ann. Rep. n. s.* 3(1)(Report B):229B–31B.

–. 1889a. A trip to Nepigon. *Ann. Rep. Ent. Soc. Ontario* 19:74–91.

–. 1889b. Notes on the preparatory stages of *Carterocephalus mandan. Can. Ent.* 21(6):113–16.

–. 1889c. Popular and economic entomology – No. 5. The Tiger-Swallowtail (*Papilio turnus*, L.). *Can. Ent.* 21(11): 201–4.

–. 1899. *Papilio ajax,* var. *marcellus,* in British Columbia. *Can. Ent.* 31(1):8.

–. 1902. Entomological record, 1901. *Ann. Rep. Ent. Soc. Ontario* 32: 99–107.

–. 1904. The Rev. George William Taylor, FRSC, FES, FZS. *Can. Ent.* 36(1):1–2, pl. I.

–. 1905. Entomological record, 1904. *Ann. Rep. Ent. Soc. Ontario* 35: 56–75.

–. 1908. Entomological record, 1907. *Ann. Rep. Ent. Soc. Ontario* 38: 113–27.

Foxlee, H.R. 1945. A preliminary list of the Heterocera of the Nelson-Robson-Trail district of British Columbia (Insecta: Lepidoptera). *Proc. Ent. Soc. Brit. Columbia* 42:9–14.

–. 1947. Harry Cane, 1860–1935. *Proc. Ent. Soc. Brit. Columbia* 43: 45–46.

Freeman, T.N. 1937. Notes on the specific grouping of the genus *Lycaena* (Lepidoptera). *Can. Ent.* 68(12):277–79.

–. 1948. The Arctic Lepidoptera of Baker Lake, North West Territories. *Lep. News* 2(6):63–65.

Friedlander, T.P. 1984. Insect parasites and predators of hackberry butterflies (Nymphalidae: *Asterocampa*). *J. Lep. Soc.* 38(1):60–61.

Funk, R.S. 1975. Association of ants with ovipositing *Lycaena rubidus* (Lycaenidae). *J. Lep. Soc.* 29(4):261–62.

Fyles, T.W. 1888. Description of the preparatory stages of *Chionobas jutta. Can. Ent.* 20(7):131–33.

Fyles, T.W. 1889. Further notes on *Chionobas jutta. Can. Ent.* 21(1): 12–13.

Gall, L.F. 1983. The "white male" variant of *Colias* (Pieridae): Two new records from Colorado. *J. Lep. Soc.* 37(2):177–79.

–. 1985. Measuring the size of lepidopteran populations. *J. Res. Lep.* 24(2):97–116.

Gall, L.F., and F.A.H. Sperling. 1980. A new high altitude species of *Boloria* from southwestern Colorado (Nymphalidae), with a discussion of phenetics and hierarchical decisions. *J. Lep. Soc.* 34(2):230–52.

Gardiner, B.O.C. 1974. *Pieris brassicae* L. established in Chile; another Palearctic pest crosses the Atlantic (Pieridae). *J. Lep. Soc.* 28(3): 269–77.

Garraway, Eric, and A.J.A. Bailey. 1992. Parasitoid induced mortality in the eggs of the endangered giant swallowtail butterfly *Papilio homerus* (Papilionidae). *J. Lep. Soc.* 46(3):233–34.

Garth, J.S., and J.W. Tilden. 1986. *California butterflies.* Berkeley: University of California Press. xvi + 246 pp.

Geddes, Gamble. 1884. List of diurnal Lepidoptera collected in the north-west territory and the Rocky Mountains. *Can. Ent.* 15(12): 221–23.

Geiger, H.J. 1990. Enzyme electrophoretic methods in studies of systematics and evolutionary biology of butterflies. Chapter 10, pp. 397–436, in *Butterflies of Europe,* Vol. 2, edited by Otakar Kudrna. Wiesbaden, Germany: AULA-Verlag. 559 pp.

Geiger, H.J., and A. Scholl. 1985. Systematics and evolution of Holarctic Pierinae (Lepidoptera): An enzyme electrophoretic approach. *Experientia* 41(1):24–29.

Geiger, H.J., and A.M. Shapiro. 1986. Electrophoretic evidence for speciation within the nominal species *Anthocharis sara* Lucas (Pieridae). *J. Res. Lep.* 25(1):15–24.

–. 1992. Genetics, systematics and evolution of Holarctic *Pieris napi* species group populations (Lepidoptera, Pieridae). *Zeit. Zool. Syst. Evol.* 30(2):100–22.

Gibo, D.L. 1981. Some observations on soaring flight in the Mourning Cloak butterfly (*Nymphalis antiopa* L.) in southern Ontario. *J. New York Ent. Soc.* 89(2):98–101.

Gibson, Arthur. 1910. Notes on the larva of *Thymelicus garita* Reakirt. *Can. Ent.* 42(4):145–47.

–. 1915. The Entomological record, 1914. *Ann. Rep. Ent. Soc. Ontario* 45:123–36.

Gilbert, L.E. 1972. Pollen feeding and reproductive biology of *Heliconius* butterflies. *Proc. Nat. Acad. Sci.* 69(6):1403–7.

Gilbert, L.E., and M.C. Singer. 1973. Dispersal and gene flow in a butterfly species. *Amer. Nat.* 107(953):58–73.

Goodpasture, Carll. 1973. Biology and systematics of the *Plebejus* (*Icaricia*) *acmon* group (Lepidoptera: Lycaenidae). I. Review of the group. *J. Kan. Ent. Soc.* 46(4):468–85.

Gosse, P.H. 1840. *The Canadian naturalist. A series of conversations on the natural history of Lower Canada.* London: John Van Voorst. xii + 372 pp.

–. 1841a. Analytical notice of "The Canadian naturalist." A series of conversations on the natural history of Lower Canada. *Entomologist* 1(6):81–88.

–. 1841b. List of butterflies taken at Compton, in Lower Canada. *Entomologist* 1(9):137–39.

–. 1844. Notes on a voyage up the Alabama River. *Zoologist* 2:703–9.

–. 1859. *Letters from Alabama (US) chiefly relating to natural history.* London: Morgan and Chase. xii + 306 pp.

–. 1883. Notes on butterflies obtained at Carbonear Island, Newfoundland, 1832–1835. *Can. Ent.* 15(3):44–51.

Graham, Kenneth. 1966. In Memoriam George Johnston Spencer Jan. 16, 1888; July 24, 1966. *J. Ent. Soc. Brit. Columbia* 63:41–43.

Grant, Jim. 1963. The eversible glands of *Papilio multicaudatus* Kby. *Proc. Ent. Soc. Brit. Columbia* 60:52.

Green, G.M. 1947. In Memoriam William Arthur Dashwood-Jones, 1858–1928. *Proc. Ent. Soc. Brit. Columbia* 43:41–43.

Grey, L.P. 1989. Sundry Argynnine concepts revisited (Nymphalidae). *J. Lep. Soc.* 43(1):1–10.

Grey, L.P., and A.H. Moeck. 1962. Notes on overlapping subspecies. I. An example in *Speyeria zerene. J. Lep. Soc.* 16(2):81–97.

Grossmueller, D.W., and R.C. Lederhouse. 1987. The role of nectar source distribution in habitat use and oviposition by the Tiger Swallowtail Butterfly. *J. Lep. Soc.* 41(3):159–65.

Gunder, J.D. 1929a. North American institutions featuring Lepidoptera. IV. The Provincial Museum, Victoria, BC, Canada. *Ent. News* 40(5):135–36, pl. 6.

–. 1929b. The genus *Euphydryas* Scud. of boreal America (Lepidoptera, Nymphalidae). *Pan-Pac.* 6(1):1–8, pls. 1–16.

–. 1934. A check list revision of the genus *Basilarchia* Scudder. *Can. Ent.* 66(2):39–48.

Guppy, C.S. 1986a. Geographic variation in wing melanism of the butterfly *Parnassius phoebus* F. (Lepidoptera: Papilionidae). *Can. J. Zool.* 64(4):956–62.

–. 1986b. The adaptive significance of alpine melanism in the butterfly *Parnassius phoebus* F. (Lepidoptera: Papilionidae). *Oecologia* 70(2):205–13.

–. 1989. Evidence for genetic determination of variation in adult size and wing melanism of *Parnassius phoebus* F. *J. Lep. Soc.* 43(2):148–51.

–. 1998. Notes on *Parnassius smintheus* Doubleday (Papilionidae) on Vancouver Island. *J. Lep. Soc.* 52(1):115–18.

Guppy, C.S., J.H. Shepard, and N.G. Kondla. 1994. Butterflies and skippers of conservation concern in British Columbia. *Can. Field-Nat.* 108(1):31–40.

Guppy, Richard. 1951. Mortality of *Nymphalis milberti* larvae. *Lep. News* 5(6–7):69.

–. 1954. Some host plant records derived from rearing experiments. *Lep. News* 8(3–4):101.

–. 1955. Further remarks on the habits of *Nymphalis milberti. Lep. News* 9(1):15–16.

–. 1956. Distribution of butterflies on Vancouver Island. *Lep. News* 10(5):169–71.

–. 1959. Host plants of *Strymon melinus atrofasciata. J. Lep. Soc.* 13(3):170.

–. 1962. Collecting *Oeneis nevadensis* (Satyridae) and other genera on Vancouver Island, with a theory to account for hilltopping. *J. Lep. Soc.* 16(1):64–66.

–. 1970. Further observations on "hilltopping" in *Papilio zelicaon. J. Res. Lep.* 8(3):105–17.

–. 1974. Butterflies attracted to amber glass. *J. Lep. Soc.* 28(3):248.

Hagen, H.A. 1884. Contributions from the northern transcontinental survey. Notes on the genus *Pieris. Proc. Boston Soc. Nat. Hist.* 22(2):134–41.

Hagen, R.H., R.C. Lederhouse, J.L. Bossart, and J.M. Scriber. 1992. *Papilio canadensis* and *P. glaucus* (Papilionidae) are distinct species. *J. Lep. Soc.* 45(4):245–58.

Hagen, R.H., and J.M. Scriber. 1991. Systematics of the *Papilio glaucus* and *P. troilus* species groups (Lepidoptera: Papilionidae): Inferences from allonyms. *Ann. Ent. Soc. Amer.* 84(4):380–95.

Hancock, D.L. 1983. Classification of the Papilionidae (Lepidoptera): A phylogenetic approach. *Smithersia* 2:1–48.

Hanham, A.W., and F.W. Fyles. 1913. Rev. George W. Taylor, FRSC, FZS. *Proc. Ent. Soc. Brit. Columbia* 2:1–4.

Harcombe, Andrew, and J.E. Underhill. 1970. *Moths and butterflies of Manning Park.* Victoria, BC: Brit. Columbia Parks Branch. i + 9 pp.

Harcourt, D.G. 1966. Major factors in survival of the immature stages of *Pieris rapae* (L.). *Can. Ent.* 98(6):653–62.

Hardy, G.A. 1947. California Tortoise-shell butterfly in British Columbia in 1945 (Lepidoptera). *Proc. Ent. Soc. Brit. Columbia* 43:36.

–. 1953. Notes on the occurrence of the Painted Lady, *Vanessa cardui* L. on Vancouver and the Queen Charlotte Islands in 1952. *Proc. Ent. Soc. Brit. Columbia* 50:37.

–. 1954. Notes on the life-history of *Hesperia comma* L. *manitoba* Scud. (Lepidoptera, Rhopalocera) on Vancouver Island. *Proc. Ent. Soc. Brit. Columbia* 51:21–22.

–. 1957. Notes on the life histories of five species of Lepidoptera from southern Vancouver Island, British Columbia. *Proc. Ent. Soc. Brit. Columbia* 54:40–43.

–. 1958a. Notes on the life histories of three species of Lepidoptera from southern Vancouver Island, British Columbia. *Proc. Ent. Soc. Brit. Columbia* 55:27–30.

–. 1958b. Notes on the life-histories of five species of Lepidoptera occurring on Vancouver Island. *Rep. Prov. Mus. Nat. Hist. Anthro.* 1957:30–36.

–. 1959a. On the life history of *Incisalia eryphon* (Lycaenidae) on southern Vancouver Island. *J. Lep. Soc.* 13(2):70.

–. 1959b. Painted Lady, *Vanessa cardui,* on Vancouver Island. *Proc. Ent. Soc. Brit. Columbia* 56:39.

–. 1960. Notes on the life histories of two butterflies and one moth from Vancouver Island. *Proc. Ent. Soc. Brit. Columbia* 57:27–29.

–. 1961. The California Tortoise-shell, *Nymphalis californica* Bdv., on Vancouver Island. *Proc. Ent. Soc. Brit. Columbia* 58:32.

–. 1962a. *Phyciodes mytilla* Edw. on Vancouver Island. *Proc. Ent. Soc. Brit. Columbia* 59:14.

–. 1962b. Additional notes on *Nymphalis californica* Bdv. *Proc. Ent. Soc. Brit. Columbia* 59:34.

–. 1962c. Notes on the life histories of one butterfly and three moths from Vancouver Island (Lepidoptera: Lycaenidae, Phalaenidae and Geometridae). *Proc. Ent. Soc. Brit. Columbia* 59:35–38.

–. 1963. Note on the life histories of four moths and one butterfly from Vancouver Island (Lepidoptera: Phalaenidae, Lasiocampidae and Lycaenidae). *Proc. Ent. Soc. Brit. Columbia* 60:35–40.

–. 1964. Notes on the life histories of one butterfly and three moths from southern Vancouver Island (Lepidoptera: Nymphalidae and Phalaenidae). *Proc. Ent. Soc. Brit. Columbia* 61:31–36.

Harris, Moses. 1766. *The Aurelian or natural history of English insects namely moths and butterflies.* London: M Harris. 77 pp. + 41 pls.

Harris, T.W. 1829. American turnip butterfly. *New England Farmer* 7:402.

–. 1862. *A treatise on some of the insects injurious to vegetation.* 3rd ed. Boston: William White. xi + 604 pp. + 8 pls.

Harrison, Susan. 1989. Long-distance dispersal and colonization in the Bay Checkerspot butterfly, *Euphydryas editha bayensis. Ecology* 70(5):1236–43.

Harvey, R.V. 1904. *A guide to some of the butterflies of British Columbia.* Victoria, BC: Thompson Stationery. 16 pp.

–, ed. 1908. Food plants of British Columbian Lepidoptera. First list of butterflies. *Proc. Ent. Soc. Brit. Columbia* 10:4.

Hayes, J.L. 1981. Some aspects of the biology of the developmental stages of *Colias alexandra* (Pieridae). *J. Lep. Soc.* 34(4):345–52.

Hayes, J.L. 1984. *Colias alexandra:* A model for the study of natural populations of butterflies. *J. Res. Lep.* 23(2):113–24.

Hebda, R.J., and Fran Aitkens, eds. 1993. *Garry oak-meadow colloquium, Victoria 1993, proceedings.* Victoria: Garry Oak Meadow Society. ix + 93 pp.

Heinrich, Bernd. 1993. How avian predators constrain caterpillar foraging. Pp. 224–47 in *Caterpillars: Ecological and evolutionary constraints on foraging,* edited by N.E. Stamp and T.M. Casey. New York: Chapman and Hall. xiii + 587 pp.

Heinrich, Bernd, and S.L. Collins. 1983. Caterpillar leaf damage, and the game of hide-and-seek with birds. *Ecology* 64(3):592–602.

Heitzman, J.R. 1965. The early stages of *Euphyes vestris. J. Res. Lep.* 3(4):151–53.

Henriksen, H.J., and Ib Kreutzer. 1982. *The butterflies of Scandinavia in nature.* Odense, Denmark: Skandinavisk Bogforlag. 215 pp.

Heusser, L.E. 1983. Palynology and paleoecology of postglacial sediments in an anoxic basin, Saanich Inlet, British Columbia. *Can. J. Earth Sci.* 20(5):873–85.

Higgins, L.G. 1975. *The classification of European butterflies.* London: Collins. 320 pp.

Higgins, L.G., and N.D. Riley. 1970. *A field guide to the butterflies of Britain and Europe.* London: Collins. 380 pp.

Hilchie, G.J. 1990. Classification, relationships, life history, and evolution of *Erebia magdalena* Stecker (Lepidoptera: Satyridae). *Quaestiones Ent.* 26(4):665–93.

Hinchliff, John. 1996. *The distribution of the butterflies of Washington.* Corvallis, OR: The Evergreen Aurelians. vi + 162 pp. + map.

Hiruma, Kiyoshi, J.P. Pelham, and Herve Bouhin. 1997. Termination of pupal diapause in *Callophrys sheridanii* (Lycaenidae). *J. Lep. Soc.* 51(1):75–82.

Hoegh-Guldberg, Ove. 1972. Pupal sound production of some Lycaenidae. *J. Res. Lep.* 10(2):127–47.

Hoffman, R.J. 1978. Environmental uncertainty and evolution of physiological adaptation in *Colias* butterflies. *Amer. Nat.* 112(988):999–1015.

Holland, W.J. 1891. [Translation of Sandberg's 1885 article "Observations upon the metamorphosis of arctic Lepidoptera"]. *Can. Ent.* 23(1):16–19.

–. 1898. *The butterfly book.* New York: Doubleday, Page. xx + 382 pp. + 48 pls.

–. 1915. *The butterfly guide.* New York: Doubleday, Page. 237 pp.

–. 1931. *The butterfly book.* Rev. ed. Garden City, NY: Doubleday. xii + 424 pp. + 77 pls.

Hooper, R.R. 1973. *The butterflies of Saskatchewan.* Regina: Saskatchewan Department of Natural Resources. v + 215 pp.

–. 1986. Revised checklist of Saskatchewan butterflies. *Blue Jay* 44(3):154–63.

Hopfinger, J.C. 1947. Field summary of Lepidoptera – 1947 Season. 2. Northwest – Oregon, Washington, Idaho, and British Columbia. *Lep. News* 1(8):90.

–. 1960. Season's summary for 1959. 2. Northwest – Oregon, Washington, Idaho, Montana, British Columbia. *News Lep. Soc.* 1960(3):4–6.

–. 1961. Season's summary, 1960. 2. Northwest – Oregon, Washington, Idaho, Montana, British Columbia. *News Lep. Soc.* 1961(4):2–5.

Hornblower, Simon, and Antony Spawforth. 1996. *The Oxford classical dictionary.* 3rd ed. Oxford: Oxford University Press. lvii + 1,640 pp.

Hovanitz, William. 1948. Differences in the field activity of two female color phases of *Colias* butterflies at various times of the day. *Contr. Lab. Vertebr. Biol. Univ. Mich.* 41:1–37

–. 1950. The biology of *Colias* butterflies. II. Parallel geographic variation of dimorphic color phases in North American species. *Wasmann J. Biol.* 8(2):197–219.

–. 1962. The distribution of the species of the genus *Pieris* in North America. *J. Res. Lep.* 1(1):73–83.

–. 1969. Habitat: *Pieris beckeri. J. Res. Lep.* 7(1):56.

Hovanitz, William, and V.C.S. Chang. 1962. The effect of various foodplants on survival and growth rate of *Pieris. J. Res. Lep.* 1(1): 21–42.

Howe, W.H., ed. 1975. *The butterflies of North America.* Garden City, NY: Doubleday. xiii + 633 pp. + 97 pls.

Irwin, R.R. 1968. *Thymelicus lineola* (Hesperiidae) in Illinois. *J. Lep. Soc.* 22(1):21–26.

Jennings, D.T., and M.E. Toliver. 1976. Crab spider preys on *Neophasia menapia* (Pieridae). *J. Lep. Soc.* 30(3):236–37.

Johnson, Kurt. 1976. Three new Nearctic species of *Callophrys (Mitoura)*, with a diagnosis of all Nearctic consubgeners (Lepidoptera: Lycaenidae). *Bull. Allyn Mus.* 38:1–30.

Johnson, Kurt, and George Balogh. 1977. Studies in the Lycaeninae (Lycaenidae). 2. Taxonomy and evolution of the Nearctic *Lycaena rubidus* complex, with description of a new species. *Bull. Allyn Mus.* 43:1–62.

Jones, J.R.J. Llewellyn. 1933. Some food plants of lepidopterous larvae. *Proc. Ent. Soc. Brit. Columbia* 30:25–26.

–. 1935. Some food plants of lepidopterous larvae list 2. *Proc. Ent. Soc. Brit. Columbia* 31:28–32.

–. 1936. Some food plants of lepidopterous larvae list 3. *Proc. Ent. Soc. Brit. Columbia* 32:29–31.

–. 1938. Some food plants of lepidopterous larvae list 5. *Proc. Ent. Soc. Brit. Columbia* 34:20–21.

–. 1939. Some food plants of lepidopterous larvae list 6. *Proc. Ent. Soc. Brit. Columbia* 35:24–27.

–. 1940. Some food plants of lepidopterous larvae list 7. *Proc. Ent. Soc. Brit. Columbia* 36:13–14.

–. 1942. Some food plants of lepidopterous larvae list 8. *Proc. Ent. Soc. Brit. Columbia* 38:18–19.

–. 1943. Some food plants of lepidopterous larvae list 9. *Proc. Ent. Soc. Brit. Columbia* 40:27.

–. 1951. *An annotated check list of the macrolepidoptera of British Columbia.* Occ. Pap. Ent. Soc. Brit. Columbia No. 1:i–v, i–ii, 1–148.

Jones, R.E. 1977. Movement patterns and egg distribution in cabbage butterflies. *J. Anim. Ecol.* 46(1):195–212.

Karban, Richard, and Steven Courtney. 1987. Intraspecific host plant choice: Lack of consequences for *Streptanthus tortuosus* (Cruciferae) and *Euchloe hyantis* (Lepidoptera: Pieridae). *Oikos* 48(3):243–48.

Karsholt, Ole, and Jozef Razowski. 1996. *The Lepidoptera of Europe. A distributional checklist.* Stenstrup, Denmark: Apollo Books. 380 pp. + CD.

Kartesz, J.T. 1994a. *A synonymized checklist of the vascular flora of the United States, Canada, and Greenland.* 2nd ed. Vol. 1, *Checklist.* Portland, OR: Timber Press. lxi + 622 pp.

–. 1994b. *A synonymized checklist of the vascular flora of the United States, Canada, and Greenland.* 2nd ed. Vol. 2, *Thesaurus.* Portland, OR: Timber Press. vii + 816 pp.

Keji, J.A. 1951. Oviposition observations. *Lep. News* 5(6–7):69.

Kellogg, T.A. 1986. Egg dispersion patterns and egg avoidance behavior in the butterfly *Pieris sisymbrii. J. Lep. Soc.* 39(4):268–75.

Kendall, R.O. 1964. Larval foodplants for twenty-six species of Rhopalocera (Papilionidae) from Texas. *J. Lep. Soc.* 18(3):129–57.

Kingsolver, J.G. 1985. Butterfly thermoregulation: Organismic mechanisms and population consequences. *J. Res. Lep.* 24(1):1–20.

Kirby, William. 1837. [Lepidoptera: Diurna.] Pp. 286-300, pls. 3,4, in Part 4, Insects, in *Fauna Boreali-Americana or the zoology of the northern parts of British America,* John Richardson. Norwich, England: Josiah Fletcher. v-xxxix + 325 pp. + cpl 1-8 + page i (errata).

Klassen, Paul, A.R. Westwood, W.B. Preston, and W.B. McKillop. 1989. *The butterflies of Manitoba.* Winnipeg, MB: Manitoba Museum of Man and Nature. vi + 290 pp.

Klots, A.B. 1951. *A field guide to the butterflies.* Boston: Houghton Mifflin. xvi + 349 pp. + 40 pls.

–. 1975. Genus *Colias* Fabricius. Pp. 354–67 in *The butterflies of North America,* edited by W.H. Howe. Garden City, NY: Doubleday. xiii + 633 pp. + 97 pls.

Koçak, A.Ö. 1981. Critical check-list of European Papilionoidea (Lepidoptera). *Priamus* 1(2):46–92.

Kondla, N.G. 1986. A discussion on the correct status of *Colias eurytheme alberta. Utahensis* 6(4):44–45.

–. 1995. Type localities of the butterflies of Cary's Arctic (*Oeneis caryi* Dyar) and Christina Sulphur (*Colias christina* Edw.). *Alberta Nat.* 25(4):75–76.

–. 1996. Clarification of some Alberta butterfly type localities. *Alberta Nat.* 26(2):39–41.

Kondla, N.G., E.M. Pike, and F.A.H. Sperling. 1994. Butterflies of the Peace River region of Alberta and British Columbia. *Blue Jay* 52(2):71–90.

Kondla, N.G., and J.P. Pelham. 1995. First record of the butterfly *Euchloe naina* (Lepidoptera: Pieridae) from North America. *Can. Field-Nat.* 109(2): 259.

Korshunov, Yu. P., and P.Y. Gorbunov. 1995. [Butterflies of the Asiatic part of Russia]. Yekaterinbyerg, Russia: Izdatyelbstbo uralbskoto Gosudarsbyenogc Unibyersityeta. [in Russian] 202 pp.

Krivda, W.V. 1968. Notes on *Erebia discoidalis* (Kirby) and its foodplant at The Pas, Manitoba (Lepidoptera: Satyridae). *Bull. Assoc. Minnesota Ent.* 2(3):47–49.

Kudrna, Otakar. 1986. *Butterflies of Europe.* Vol. 8, *Aspects of the conservation of butterflies in Europe.* Wiesbaden, Germany: AULA-Verlag GmbH. 323 pp.

Kudrna, Otakar, and Hansjuerg Geiger. 1985. A critical review of "Systematische Untersuchungen am *Pieris napi-bryoniae*-Komplex (s. l.)" (Lepidoptera: Pieridae) by Ulf Eitschberger. *J. Res. Lep.* 24(1):47–60.

Lane, R.P. 1984. Host specificity of ectoparasitic midges on butterflies. Pp. 105–8 in *The biology of butterflies. Symposium of the Royal Entomological Society of London number 11,* edited by R.I. Vane-Wright and P.R. Ackery. London: Academic Press. xxiv + 429 pp.

Langston, R.L. 1975. Genus *Celastrina* Tutt. Pp. 335–37 in *The butterflies of North America,* edited by W.H. Howe. Garden City, NY: Doubleday. xiii + 633 pp. + 97 pls.

Larsen, E.M., E. Rodrick, and R. Milner, eds. 1995. *Management recommendations for Washington's priority species.* Vol. 1, *Invertebrates.* Olympia, WA: Washington Fish and Wildlife. x + 97 pp.

Latheef, M.A., and R.D. Irwin. 1979. The effect of companionate planting on lepidopteran pests of cabbage. *Can. Ent.* 111(7):863–64.

Latheef, M.A., and J.H. Ortiz. 1983a. The influence of companion herbs on egg distribution of the imported cabbageworm, *Pieris rapae* (Lepidoptera: Pieridae), on collard plants. *Can. Ent.* 115(8):1031–38.

–. 1983b. The influence of companion herbs on oviposition of imported cabbageworm, *Pieris rapae* (Lepidoptera: Pieridae), and cabbage looper, *Trichoplusia ni* (Lepidoptera: Noctuidae) on collard plants. *Can. Ent.* 115(11):1529–31.

Lawrence, D.A., and J.C. Downey. 1967. Morphology of the immature stages of *Everes comyntas* Godart (Lycaenidae). *J. Res. Lep.* 5(2): 61–96.

Layberry, R.A., P.W. Hall, and J.D. Lafontaine. 1998. *The butterflies of Canada.* Toronto: University of Toronto Press. viii + 280 pp.

Lederhouse, R.C. 1993. Territoriality along flyways as mate-locating behavior in male *Limenitis arthemis* (Nymphalidae). *J. Lep. Soc.* 47(1):22–31.

Leighton, B.V. 1946. The butterflies of Washington. Univ. Wash. Publ. Bio. 9(2):47–63.

Leong, K.L.H., D. Frey, and C. Nagano. 1990. Wasp predation on overwintering monarch butterflies (Lepidoptera: Danaidae) in central California. *Pan-Pac. Ent.* 66(4):326–28.

Leong, K.L.H., M.A. Yoshimura, and H.K. Kaya. 1992. Low susceptibility of overwintering monarch butterflies to *Bacillus thuringiensis* Berliner. *Pan-Pac. Ent.* 68(1):66–68.

Lester, R.T., and J.P. Myers. 1990. Global warming, climate disruption, and biological diversity. *Audubon Wildlife Report* 1989/1990:176–221.

Lindsey, A.W. 1923. The egg and larva of *Hesperia juba* Bdv. *Denison Univ. Bull., J. Sci. Lab.* 20:121–25, pl. 16.

–. 1927. Notes on phylogeny in *Erynnis* Schr. (*Thanaos* Auct.). *Denison Univ. Bull., J. Sci. Lab.* 22:109–15.

Lindsey, A.W., E.L. Bell, and R.C. Williams. 1931. The Hesperioidea of North America. *Denison Univ. Bull., J. Sci. Lab.* 26(1):1–142.

Lord, John Keast. 1866. *The naturalist in Vancouver Island and British Columbia.* Vol. 2. London: Richard Bentley. 1 + x + 375 pp.

Losey, J.E., L.S. Rayor, and M.R. Carter. 1999. Transgenic pollen harms monarch larvae. *Nature* 399(6733):214.

Luternauer, J.L., J.J. Clague, K.W. Conway, J.V. Barrie, B. Blaise, and R.W. Mathewes. 1989. Late Pleistocene terrestrial deposits on the continental shelf of western Canada: Evidence for rapid sea-level change at the end of the last glaciation. *Geology* 17(4):357–60.

Lyall, David. 1863. Account of the botanical collections made by David Lyall, surgeon and naturalist of the North American Boundary Commission. *Proc. Linn. Soc. – Botany* 7:124–41.

Lyman, H.H. 1896. Notes on the preparatory stages of *Erebia epipsodea* (Butler). *Can. Ent.* 28(11):274–78.

–. 1897. Notes on the life history of *Colias interior,* Scud. *Can. Ent.* 29(11):249–58.

MacDonald, S.O., and J.A. Cook. 1996. The land mammal fauna of southeast Alaska. *Can. Field-Nat.* 110(4):571–98.

MacNeill, C.D. 1964. The skippers of the genus *Hesperia* in western North America with special reference to California (Lepidoptera: Hesperiidae). Univ. Calif. Publ. Ent. 35:i–iv, 1–221, pls. 1–8.

–. 1975. Family Hesperiidae The Skippers. In *The butterflies of North America,* edited by W.H. Howe. Garden City, NY: Doubleday. xiii + 633 pp. + 97 pls.

–. 1993. Comments on the genus *Polites,* with the description of a new species of the *themistocles* group from Mexico (Hesperiidae: Hesperiinae). *J. Lep. Soc.* 47(3):177–98.

Macy, R.W., and H.H. Shepard. 1941. *Butterflies. A handbook of the butterflies of the United States, complete for the region north of the Potomac and Ohio Rivers and east of the Dakotas.* Minneapolis, MN: University of Minnesota Press. vii + 247 pp.

Malcolm, S.B., and L.P. Brower. 1987. Selective oviposition by monarch butterflies (*Danaus plexippus* L.) in a mixed stand of *Asclepias curassavica* L. and *A. incarnata* L. in southern Florida. *J. Lep. Soc.* 40(4):255–63.

Masters, J.H. 1969. Ecological and distributional notes on *Erebia disa* in central Canada. *J. Res. Lep.* 7(1):19–22.

–. 1971. Ecological and distributional notes on *Erebia discoidalis* (Satyridae) in the north central states. *J. Res. Lep.* 9(1):11–16.

–. 1973. Habitat: *Oeneis jutta ascerta* Masters and Sorensen. *J. Res. Lep.* 11(2):94.

–. 1974. Biennialism in *Oeneis macounii* (Satyridae). *J. Lep. Soc.* 28(3):237–42.

Masters, J.H., and J.T. Sorensen. 1969. Field observations on forest *Oeneis* (Satyridae). *J. Lep. Soc.* 23(3):155–61.

Mathews, R.W., and G.E. Rouse. 1974. Palynology and paleoecology of postglacial sediments from the lower Fraser River canyon of British Columbia. *Can. J. Earth Sci.* 12(5):745–56.

Mattoni, R.H.T., and M.S.B. Seiger. 1963. Techniques in the study of population structure in *Philotes sonorensis. J. Res. Lep.* 1(4):237–44.

Mattoon, S.O., and J.W. Tilden. 1998. Re-evaluation of North American *Carterocephalus palaemon* (Pallas) (Lepidoptera: Hesperiidae) and description of a new subspecies. Pp. 641–60 in *Systematics of western North American butterflies,* edited by T.C. Emmel. Gainesville, FL: Mariposa Press. xxviii + 878 pp.

Maynard, C.J. 1886. *The butterflies of New England; with original descriptions of one hundred and six species, accompanied by eight lithographic plates, in which are given at least two hand-colored figures of each species.* Boston: Bradlee Whidden. iv + 65 + 3 pp. + 8 pls.

–. 1891. *A manual of North American butterflies.* Boston: De Wolfe, Fiske. iv + 226 pp. + 10 pls.

McCorkle, D.V., and P.C. Hammond. 1986. Observations on the biology of *Parnassius clodius* (Papilionidae) in the Pacific Northwest. *J. Lep. Soc.* 39(3):156–62.

–. 1988. Biology of *Speyeria zerene hippolyta* (Nymphalidae) in a marine-modified environment. *J. Lep. Soc.* 42(3):184–92.

–. 1989. Genetic experiments with a *calverleyi*-like mutation isolated from *Papilio bairdii oregonius* (Papilionidae). *J. Res. Lepid.* 27(3–4): 186–91.

McDunnough, J.H. 1922. Notes on the Lepidoptera of Alberta. *Can. Ent.* 54(6):134–41.

–. 1927. The Lepidoptera of the Seton Lake region, British Columbia. *Can. Ent.* 59(7):152–62; (8):193–99; (9):207–14; (10):239–46; (11):266–77.

McGugan, B.M. 1958. Forest Lepidoptera of Canada recorded by the Forest Insect Survey. Vol. 1, Papilionidae to Arctiidae. Can. Dept. Agric. Forest Biol. Div. Publ. 1034:[i]–[iv], 1–76.

Mead, T.L. 1869. Extension of habitat of *Pieris rapae,* Linn. *Can. Ent.* 2(3):36.

–. 1878. Notes on certain Californian diurnals. *Psyche* 2(53–56): 179–84.

Middleton, M.S. 1913. Report from the Kootenay. *Proc. Ent. Soc. Brit. Columbia* 2:17–19.

Miller, J.C. 1990. Field assessment of the effects of a microbial pest control agent on nontarget Lepidoptera. *Amer. Ent.* 36(2):135–39.

–. 1992. Effects of microbial insecticide, *Bacillus thuringiensis kurstaki,* on nontarget Lepidoptera in a spruce budworm–infested forest. *J. Res. Lep.* 29(4):267–76.

Miller, J.C., and K.J. West. 1987. Efficiency of *Bacillus thuringiensis* and diflubenzeron on Douglas-fir and oak for gypsy moth control in Oregon. *J. Arboriculture* 13(10):240–42.

Miller, J.Y., ed. 1992. *The common names of North American butterflies.* Washington, DC: Smithsonian Institution Press. ix + 177 pp.

Miller, L.D., and F.M. Brown. 1979. Studies in the Lycaeninae (Lycaenidae). 4. The higher classification of the American coppers. *Bull. Allyn Mus.* 51:1–30.

–. 1981. A catalogue/checklist of the butterflies of America north of Mexico. *Lepid. Soc. Mem.* 2:i–vii, 1–280.

Moilliet, T.K. 1947. Theodore Albert Moilliet, 1883–1935. *Proc. Ent. Soc. Brit. Columbia* 43:43.

Morin, N.R., ed. 1993. *Flora of North America north of Mexico.* Vol. 2, *Pteridophytes and Gymnospermae.* Oxford: Oxford University Press. xvi + 475 pp.

–, ed. 1997. *Flora of North America north of Mexico.* Vol. 3, *Magnoliophyta: Magnoliidae and Hamamelidae.* Oxford: Oxford University Press. xxiii + 590 pp.

Morton, A.C. 1982. The effects of marking and capture on recapture frequencies of butterflies. *Oecologia* 53(1):105–10.

Munroe, Eugene. 1948. The field season summary of North American Lepidoptera for 1948. 7. Northeast – Maryland north to southern Quebec. *Lep. News* 2(Suppl.):9.

Murphy, D.D., A.E. Launer, and P.R. Ehrlich. 1983. The role of adult feeding in egg production and population dynamics of the checkerspot butterfly *Euphydryas editha. Oecologia* 56(2):257–63.

Murphy, D.D., M.S. Menninger, and P.R. Ehrlich. 1984. Nectar distribution as a determinant of oviposition host species in *Euphydryas chalcedona. Oecologia* 62(2):269–71.

Nabokov, Vladimir. 1949. The Nearctic members of the genus *Lycaeides* Hübner (Lycaenidae, Lepidoptera). *Bull. Harvard Mus. Comp. Zool.* 101(4):479–541, pls. 1–9.

Neck, R.W. 1977. Bizarre capture of a butterfly by an ambush bug. *J. Lep. Soc.* 31(1):22.

–. 1980. Utilization of grass inflorescences as adult resources by Rhopalocera. *J. Lep. Soc.* 34(2):261–62.

Newcomb, W.W. 1910. *Chrysophanus dorcas* Kirby, and related species in the Upper Peninsula of Michigan. *Can. Ent.* 42(5):153–57, pls. 4–5.

Newcomer, E.J. 1911. The life histories of two lycaenid butterflies. *Can. Ent.* 43(3):83–88.

–. 1964a. Butterflies of Yakima County, Washington. *J. Lep. Soc.* 18(4):217–28.

–. 1964b. Life histories of *Papilio indra* and *P. oregonius. J. Res. Lep.* 3(1):49–62.

–. 1964c. The synonymy, variability and biology of *Lycaena nivalis. J. Res. Lep.* 2(4):271–80.

–. 1964d. North American season summary for 1963. Zone 2: Pacific Northwest – British Columbia, Washington, Oregon, Idaho, Montana. *News Lep. Soc.* 1964(4):4–5.

–. 1967a. Early stages of *Chlosyne hoffmanni manchada* (Nymphalidae). *J. Lep. Soc.* 21(1):71–73.

–. 1967b. Life histories of three western species of *Polites. J. Res. Lep.* 5(4):243–47.

Nielsen, M.C. 1977. *Sphinx luscitiosa* (Sphingidae) feeding on decayed fish. *J. Lep. Soc.* 31(4):275.

–. 1985. Notes on the habitat and foodplant of *Incisalia henrici* (Lycaenidae) and *Pygrus centaureae* (Hesperiidae) in Michigan. *J. Lep. Soc.* 39(1):62–63.

Nielsen, M.C., and L.A. Ferge. 1982. Observations of *Lycaeides argyrognomon nabokovi* in the Great Lakes region (Lycaenidae). *J. Lep. Soc.* 36(3):233–34.

Nowak, C.L., R.S. Nowak, R.J. Tausch, and P.E. Wigand. 1994. A 30,000 year record of vegetation dynamics at a semi-arid locale in the Great Basin. *J. Vegetation Sci.* 5(4):579–90.

Oosting, D.P., and D.K. Parshall. 1980. Ecological notes on the butterflies of the Churchill region of northern Manitoba. *J. Res. Lep.* 17(3):188–203.

Opler, P.A. 1967a. New host plant records for *Anthocharis* (Pieridae). *J. Lep. Soc.* 21(3):212.

–. 1967b. Studies on the Nearctic *Euchloe*. Part 1. Introduction. *J. Res. Lep.* 5(1):39–40.

–. 1967c. Studies on the Nearctic *Euchloe*. Part 2. Chronological review of the literature and bibliography. *J. Res. Lep.* 5(1):41–50.

–. 1967d. Studies on the Nearctic *Euchloe*. Part 3. Complete synonymical treatment. *J. Res. Lep.* 5(3):185–90.

–. 1967e. Studies on the Nearctic *Euchloe*. Part 4. Type data and type locality restrictions. *J. Res. Lep.* 5(3):190–95.

–. 1970. Studies on Nearctic *Euchloe*. Part 5. Distribution. *J. Res. Lep.* 7(2):65–86.

–. 1971. Studies on Nearctic *Euchloe*. Part 6. Systematics of adults. *J. Res. Lep.* 8(4):153–68.

–. 1975. Studies on Nearctic *Euchloe*. Part 7. Comparative life histories, hosts and the morphology of immature stages. *J. Res. Lep.* 13(1):1–20.

–. 1999. *A field guide to western butterflies*. Boston: Houghton Mifflin. xiv + 540 pp.

Opler, P.A., and G.O. Krizek. 1984. *Butterflies east of the Great Plains*. Baltimore, MD: Johns Hopkins University Press. xvii + 294 pp.

Orive, M.E., and J.F. Baughman. 1989. Effects of handling on *Euphydryas editha* (Nymphalidae). *J. Lep. Soc.* 43(3):244–47.

Osburn, Gerald, and B.H. Luckman. 1988. Holocene glacier fluctuations in the Canadian Cordillera (Alberta and British Columbia). *Quat. Sci. Rev.* 7(2):115–28.

Otvos, I.S., and Sandy Vanderveen. 1993. *Environmental report and current status of* Bacillus thuringiensis *var.* kurstaki. *Use for control of forest and agricultural insect pests*. Victoria, BC: Brit. Columbia Ministry of Forests. vii + 81 pp.

Paclt, Jiri von. 1955. Die Gattungsnamen von Kluk 1780. *Danaus, Heliconius, Nymphalis* und *Plebejus. Beitrage Ent.* 5:428-31.

Parmesan, Camille. 1996. Climate and species' range. *Nature* 382(6594):765–66.

Payne, J.A., and E.W. King. 1969. Lepidoptera associated with pig carrion. *J. Lep. Soc.* 23(3):191–95.

Peck, W.D. 1796. The description and history of the cankerworm. *Mass. Mag.* 1795(7):323–27.

Pellmyr, Olle. 1983. Plebeian courtship revisited: Studies on the female-produced male behavior-eliciting signals in *Lycaeides idas* courtship (Lycaenidae). *J. Res. Lep.* 21(3):147–57.

Perkins, E.M., and E.V. Gage. 1971. On the occurrence of *Limenitis archippus* × *L. lorquini* hybrids (Nymphalidae). *J. Res. Lep.* 9(4):223–26.

Perkins, E.M., and S.F. Perkins. 1966. A review of the *Limenitis lorquini* complex (Nymphalidae). *J. Lep. Soc.* 20(3):172–76.

Perkins, S.F., and E.M. Perkins Jr. 1975. Genus *Limenitis* Fabricius. Pp. 130–36 in *The butterflies of North America,* edited by W.H. Howe. Garden City, NY: Doubleday. xiii + 633 pp. + 97 pls.

Petersen, Björn. 1967. Comparative speciation in two butterfly families, Pieridae and Nymphalidae. *J. Res. Lep.* 5(2):113–26.

Petiver, James. 1702–05. *Gazophylacium naturae et artis decades 10.* London: Bateman. 12 pp. + 100 pls.

Phair, A.W.A. 1919. Three years collecting in the Lillooet district. *Proc. Ent. Soc. Brit. Columbia* 12:34–36.

Pielou, E.C. 1991. *After the Ice Age: The return of life to glaciated North America.* Chicago: University of Chicago Press. ix + 366 pp.

Pierce, N.E., and S. Easteal. 1986. The selective advantage of attendant ants for the larvae of a lycaenid butterfly, *Glaucopsyche lygdamus. J. Anim. Ecol.* 55(2):451–62.

Pierce, N.E., and P.S. Mead. 1981. Parasitoids as selective agents in the symbiosis between lycaenid butterfly larvae and ants. *Science* 211(4487):1185–87.

Pike, E.M. 1980. Origins of tundra butterflies in Alberta. *Quaestiones Ent.* 16(3):555–96.

Pinheiro, C.E.G. 1991. Territorial hilltopping behavior of three swallowtail butterflies (Lepidoptera, Papilionidae) in western Brazil. *J. Res. Lep.* 29(1–2):134–42.

Pivnick, K.A., and J.N. McNeil. 1985. Effects of nectar concentration on butterfly feeding: Measured feeding rates for *Thymelicus lineola* (Lepidoptera: Hesperiidae) and a general feeding model for adult Lepidoptera. *Oecologia* 66(2):226–37.

Platt, A.P. 1979. Editor's note to Scott, J.A. 1979. Mid-valval flexion in the left valva of asymmetric genitalia of *Erynnis* (Hesperiidae). *J. Lep. Soc.* 32(4):305.

Pliske, T.E. 1973. Factors determining mating frequency in some New World butterflies and skippers. *Ann. Ent. Soc. Amer.* 66(1):164–69.

Pollard, Ernest. 1977. A method for assessing changes in the abundance of butterflies. *Biol. Conser.* 12(2):115–34.

–. 1984. Synoptic studies of butterfly abundance. Pp. 59–61 in *The biology of butterflies. Symposium of the Royal Entomological Society of London number 11,* edited by R.I. Vane-Wright and P.R. Ackery. London: Academic Press. xxiv + 429 pp.

Pollard, Ernest, and A.M. Emmet. 1990. [*Ladoga camilla* (Linnaeus). The White Admiral]. Pp. 183–85 in *The moths and butterflies of Great Britain and Ireland*. Vol. 7, Pt. I, *Hesperiidae to Nymphalidae,* edited by A.M. Emmet and John Heath. Colchester, UK: Harley Books. ix + 370 pp.

Porter, A.H., and Hansjuerg Geiger. 1988. Genetic and phenotypic population structure of the *Coenonympha tullia* complex (Lepidoptera: Nymphalidae: Satyrinae) in California: No evidence for species boundaries. *Can. J. Zool.* 66(12):2751–65.

Porter, A.H., and J.C. Mueller. 1998. Partial genetic isolation between *Phyciodes tharos* and *P. cocyta* (Nymphalidae). *J. Lep. Soc.* 52(2): 182–205.

Porter, Keith. 1982. Basking behaviour in larvae of the butterfly *Euphydryas aurinia. Oikos* 38(3):308–12.

–. 1983. Multivoltinism in *Apanteles bignellii* and the influence of weather on synchronization with its host *Euphydryas aurinia. Ent. Exp. Appl.* 34(2):155–62.

Powell, J.A. 1968. Foodplants of *Callophrys* (*Incisalia*) *iroides* (Lycaenidae). *J. Lep. Soc.* 22(4):225–26.

Pratt, G.F., and G.R. Ballmer. 1986. Clarification of the larval host plant of *Epidemia mariposa* (Lycaenidae) in northern California. *J. Lep. Soc.* 40(2):127.

Pratt, G.F., D.M. Wright, and Harry Pavulaan. 1994. The various taxa and hosts of the North American *Celastrina* (Lepidoptera: Lycaenidae). *Proc. Ent. Soc. Wash.* 96(3):566–78.

Prest, V.K. 1969. Retreat of Wisconsin and recent ice in North America. Geol. Sur. Can. Map 1257A.

Price, H.F. 1962. Lepidoptera as prey of other insects. *J. Lep. Soc.* 15(2):93–94.

Pronin, G.F. 1955. Notes on the life-history and methods of rearing the Giant Tiger Swallowtail, *Papilio multicaudatus. Lep. News* 9(4–5):137–40.

Pyle, R.M. 1974. *Watching Washington butterflies.* Seattle, WA: Seattle Audubon Society. viii + 109 pp.

–. 1981. *The Audubon Society field guide to North American butterflies.* New York: Alfred A. Knopf. 924 pp.

Rausher, M.D. 1978. Search image for leaf shape in a butterfly. *Science* 200(4345):1071–73.

Rawlins, J.E. 1980. Thermoregulation by the Black Swallowtail butterfly, *Papilio polyxenes* (Lepidoptera: Papilionidae). *Ecology* 61(2):345–57.

Reakirt, Tryon. 1865. Descriptions of some new species of *Eresia. Proc. Ent. Soc. Philad.* 5(2):224–27.

Reed, C.C. 1997. Diurnal Lepidoptera of native and reconstructed prairies in eastern Minnesota. *J. Lep. Soc.* 51(2):179–84.

Reed, E.B. 1870. Accentuated list of Canadian Lepidoptera. *Can. Ent.* 2(10–11):149–51.

–. 1871. Accentuated list of Canadian Lepidoptera. *Can. Ent.* 3(5):95–96.

Reed, H.B. Jr. 1958. A study of dog carcass communities in Tennessee, with special reference to the insects. *Amer. Midl. Nat.* 59(1):213–45.

Reinthal, W.J. 1963. About the "pumping action" of a *Papilio* at water. *J. Lep. Soc.* 17(1):35–36.

Remington, C.L. 1952. The biology of Nearctic Lepidoptera. I. Food-plants and life-histories of Colorado Papilionoidea. *Psyche* 59(2):61–70.

–. 1958. New records of larval host plants of *Mitoura spinetorum* (Lycaenidae). *Lep. News* 12(1–2):14.

Renwick, J.A.A., and F.S. Chew. 1994. Oviposition behaviour in Lepidoptera. *Ann. Rev. Ent.* 39:377–400.

Reuss, Theodore. 1922. Eine Androconialform von "*Argynnis*" *niobe* L., f.n., und durch entsprechende gekennzeichnete ostasiatische Formen oder Arten, die bisher zu "*adippe*" L. (rect. *cydippe* L.) gerechnet wurden, sich aber nunmehr durch Art und Verteilung der Androconien abtrennen lassen. Mit einer Revision des "Genus *Argynnis* F." *Arch. Naturgesch.* (A)87(11):180–230.

Riotte, J.C.E. 1992. *Annotated list of Ontario Lepidoptera.* Life Sciences Miscellaneous Publications. Toronto: Royal Ontario Museum. viii + 208 pp.

Ritland, D.B. 1995. Comparative unpalatability of mimetic Viceroy butterflies (*Limenitis archippus*) from four south-eastern United States populations. *Oecologia* 103(3):327–36.

Ritland, D.B., and L.P. Brower. 1991. The Viceroy butterfly is not a Batesian mimic. *Nature* 350(6318):497–98.

Robbins, R.K. 1980. The lycaenid "false head" hypothesis: Historical review and quantitative analysis. *J. Lep. Soc.* 34(2):194–208.

–. 1986. Independent evolution of "false head" behavior in the Riodinidae. *J. Lep. Soc.* 39(3):224–25.

Robbins, R.K., and P.M. Hensen. 1986. Why *Pieris rapae* is a better name than *Artogeia rapae* (Pieridae). *J. Lep. Soc.* 40(2):79–92.

Roland, Jens. 1982. Melanism and diel activity of alpine *Colias* (Lepidoptera: Pieridae). *Oecologia* 53:214–21.

Room, Adrian. 1983. *Room's classical dictionary.* London: Routledge and Kegan Paul. ii + 343 pp.

Ross, A.H. 1913. Letter to R.C. Treherne, in Notes and observations on the season. *Proc. Ent. Soc. Brit. Columbia* 2:72–73.

Ross, D.A. 1966. In Memoriam Edmund Peter Venables. *J. Ent. Soc. Brit. Columbia* 63:45.

Ruhmann, M.H. 1935. The imported cabbage-worm. Victoria, BC: Brit Columbia Dept. Agric. Hort. Circ. 37:1–4.

Rutowski, R.L. 1984. Sexual selection and the evolution of butterfly mating behavior. *J. Res. Lep.* 23(2):125–42.

Sakai, W.H. 1994. Avian predation on the Monarch butterfly, *Danaus plexippus* (Nymphalidae: Danainae), at a California overwintering site. *J. Lep. Soc.* 48(2):148–56.

Sargent, T.D. 1973. Studies on the *Catocala* (Noctuidae) of southern New England. IV. A preliminary analysis of beak-damaged specimens, with discussion of anomaly as a potential anti-predator function of hindwing diversity. *J. Lep. Soc.* 27(3):175–92.

–. 1995. On the relative acceptabilities of local butterflies and moths to local birds. *J. Lep. Soc.* 49(2):148–62.

Sattler, Klaus. 1989. Anmerkungen zu zwei artiklen v. Mentzers über die Genera bei Denis and Schiffermüller. *Nota Lep.* 12(2):170–71.

Sattler, Klaus, and W.G. Tremewan. 1984. The Lepidoptera names of Denis and Schiffermüller – a case for stability. *Nota Lep.* 7(3):282–85.

Saunders, William. 1868. Entomological notes during a trip to Saguenay. *Can. Ent.* 1(2):11–13.

–. 1869a. Entomological notes. *Can. Ent.* 1(7):53–57.

–. 1869b. Entomological notes. *Can. Ent.* 1(8):65–67.

–. 1869c. Entomological notes. *Can. Ent.* 1(9):73–77.

–. 1869d. Entomological notes. *Can. Ent.* 1(11):93–101.

–. 1886. Annual address of the president of the Entomological Society of Ontario. *Can. Ent.* 18(10):184–88.

–. 1916. European butterfly found at London, Ont. *Ottawa Nat.* 30:116.

Schurian, K.G. 1993. Parasitoids exploit secretions of myrmecophilous lycaenid butterfly caterpillars (Lycaenidae). *J. Lep. Soc.* 47(2):150–54.

Scoble, M.J. 1995. *The Lepidoptera: Form, function and diversity.* Oxford: Oxford University Press. xi + 404 pp. + 4 pls.

Scott, J.A. 1973a. Down-valley flight of adult Theclini (Lycaenidae) in search of nourishment. *J. Lep. Soc.* 27(4):283–87.

–. 1973b. Mating of butterflies. *J. Res. Lep.* 11(2):99–127.

–. 1974. Mate-locating behavior of butterflies. *Amer. Midl. Nat.* 91(1):103–17.

–. 1980. A survey of the valvae of *Euphydryas chalcedona, E. c. colon,* and *E. c. anicia. J. Res. Lep.* 17(4):245–52.

–. 1981. New Papilionoidea and Hesperioidea from North America. *Papilio* (n.s.) 1:1–12.

–. 1982. The life history and biology of an alpine relict, *Boloria improba acrocnema* (Lepidoptera: Nymphalidae), illustrating a new mathematical population census method. *Papilio* (n.s.) 2:1–12.

–. 1983. Mate-locating behavior of western North American butter-flies. II. New observations and morphological adaptations. *J. Res. Lep.* 21(3):177–87.

–. 1986a. Larval hostplant records for butterflies and skippers (mainly from western US), with notes on their natural history. *Papilio* (n.s.) 4:1–37.

–. 1986b. *The butterflies of North America: A natural history and field guide.* Stanford, CA: Stanford University Press. xiii + 581 pp.

–. 1988. Biology of *Polygonia progne nigrozephyrus* and related taxa (Nymphalidae). *J. Lep. Soc.* 42(1):46–56.

–. 1992. Hostplant records for butterflies and skippers (mostly from Colorado) 1959–1992, with new life histories and notes on oviposition, immatures, and ecology. *Papilio* (n.s.). 6:1–185.

–. 1994. Biology and systematics of *Phyciodes* (*Phyciodes*). *Papilio* (n.s.) 7:1–120.

Scott, J.M., and G.R. Scott. 1980. Ecology and distribution of the butterflies of southern central Colorado. *J. Res. Lep.* 17(2):73–128.

Scriber, J.M. 1995. Overview of swallowtail butterflies: Taxonomic and distributional latitude. Pp. 3–8 in *Swallowtail butterflies: Their ecology and evolutionary biology,* edited by J.M. Scriber et al. Gainesville, FL: Scientific Publishers. vii + 459 pp.

Scriber, J.M., and S.H. Gage. 1995. Pollution and global climate change: Plant ecotones, butterfly hybrid zones and changes in biodiversity. Pp. 319–44 in *Swallowtail butterflies: Their ecology and evolutionary biology,* edited by J.M. Scriber et al. Gainesville, FL: Scientific Publishers. vii + 459 pp.

Scudder, G.G.E. 1996. *Terrestrial and freshwater invertebrates of British Columbia: Priorities for inventory and descriptive research.* Victoria, BC: Brit. Columbia Ministry of Forests. vi + 206 pp.

Scudder, S.H. 1861. Notice of some North American species of *Pieris. Proc. Boston Soc. Nat. Hist.* 8:178–85.

–. 1862. On the genus *Colias* in North America. *Proc. Boston Soc. Nat. Hist.* 9:103–10.

–. 1872a. A variety of *Pieris rapae* unknown in Europe. *Can. Ent.* 4(4):79.

–. 1872b. A systematic revision of some of the American butterflies with brief notes on those known to occur in Essex County, Mass. *Rep. Peabody Acad. Sci.* 4(1871):24–82.

–. 1874a. The species of the lepidopterous genus *Pamphila. Mem. Bos. Soc. Nat. His.* 2(3):341–53, pls. 10, 11.

–. 1874b. English names for butterflies. *Psyche* 1(1):2–3; 1(3):10–11; 1(8):31

–. 1875. English names for butterflies. *Psyche* 1(9):40; 1(10):43–44; 1(11):56.

–. 1887. The introduction and spread of *Pieris rapae* in North America, 1860–1886, with a map. *Mem. Boston Soc. Nat. Hist.* 4(3):53–69, pl. 8.

–. 1889a. *The butterflies of the eastern United States and Canada, with special reference to New England.* Vol. 1, *Introduction, Nymphalidae.* Cambridge, MA: S.H. Scudder. xxiv + 766 pp. + frontispiece.

–. 1889b. *The butterflies of the eastern United States and Canada with special reference to New England.* Vol. 2, *Lycaenidae, Papilionidae, Hesperidae.* Cambridge, MA: S.H. Scudder. xii + 767-1774 pp. + frontispiece + map.

–. 1893. *Brief guide to the commoner butterflies of the northern United States and Canada.* New York: Henry Holt. xi + 206 pp. + 22 pls.

Sevastopulo, D.G. 1974. Lepidoptera feeding at puddle-margins, dung and carrion. *J. Lep. Soc.* 28(2):167–68.

Shapiro, A.M. 1966. *Butterflies of the Delaware Valley.* Philadelphia, PA: Amer. Ent. Soc. Spec. Publ. i–vi, 1–79.

–. 1974a. Altitudinal migration of butterflies in the central Sierra Nevada. *J. Res. Lep.* 12(4):231–35.

–. 1974b. Butterflies and skippers of New York state. *Search* 4(3):1–60.

–. 1975a. Altitudinal migration of central California butterflies. *J. Res. Lep.* 13(3):157–61.

–. 1975b. Butterflies of the Suisun Marsh, California. *J. Res. Lep.* 13(3):191–206.

–. 1976a. The biological status of Nearctic taxa in the *Pieris protodice-occidentalis* group (Pieridae). *J. Lep. Soc.* 30(4):289–300.

–. 1976b. The genetics of subspecific phenotype differences in *Pieris occidentalis* Reakirt and of variation in *P. o. nelsoni* W.H. Edwards (Pieridae). *J. Res. Lep.* 14(2):61–83.

–. 1976c. Why do California Tortoiseshells migrate? *J. Res. Lep.* 14(2):93–97.

–. 1976d. The role of watercress, *Nasturtium officinale,* as a host of native and introduced pierid butterflies in California. *J. Res. Lep.* 14:(3)158–68.

–. 1977. Habitat: *Pieris occidentalis* (Pieridae). *J. Res. Lep.* 15(3):182–83.

–. 1978. Photoperiod and temperature in phenotype determination of Pacific slope Pierini: Biosystematics implications. *J. Res. Lep.* 16(4):193–200.

–. 1979. *Nymphalis milberti* (Nymphalidae) near sea level in California. *J. Lep. Soc.* 33(3):200–1.

–. 1980a. Genetic incompatibility between *Pieris callidice* and *Pieris occidentalis nelsoni:* Differentiation within a periglacial relict complex (Lepidoptera: Pieridae). *Can. Ent.* 112(5):463–68.

–. 1980b. Evidence for a return migration of *Vanessa cardui* in northern California (Lepidoptera: Nymphalidae). *Pan-Pac. Ent.* 56(4):319–22.

–. 1981a. The pierid red-egg syndrome. *Amer. Nat.* 117(3):276–94.

–. 1981b. Egg-load assessment and carryover diapause in *Anthocharis* (Pieridae). *J. Lep. Soc.* 34(3):307–15.

–. 1981c. Egg-mimics of *Streptanthus* (Cruciferae) deter oviposition by *Pieris sisymbrii* (Lepidoptera: Pieridae). *Oecologia* 48(1):142–43.

–. 1982a. Susceptibility of *Pieris napi microstriata* (Pieridae) to *Apanteles glomeratus* (Hymenoptera, Braconidae). *J. Lep. Soc.* 35(3):256.

–. 1982b. The pierid fauna of jewel flower at a mid-elevation Sierran locality. *J. Lep. Soc.* 35(4):322–24.

–. 1982c. Redundancy in pierid polyphenism: Pupal chilling induces vernal phenotype in *Pieris occidentalis* (Pieridae). *J. Lep. Soc.* 36(3):174–77.

–. 1982d. A recondite breeding site for the monarch (*Danaus plexippus,* Danaidae) in the montane Sierra Nevada. *J. Res. Lep.* 20(1):50–51.

–. 1982e. Survival of refrigerated *Tatochila* butterflies (Lepidoptera: Pieridae) as an indicator of male nutrient investment in reproduction. *Oecologia* 53(1):139–40.

–. 1985a. "Edge effect" in oviposition behavior: A natural experiment with *Euchloe ausonides* (Pieridae). *J. Lep. Soc.* 38(3):242–45.

–. 1985b. Book review. *J. Lep. Soc.* 38(4):324–27.

–. 1985c. The impact of pierid feeding on seed production by a native California crucifer. *J. Res. Lep.* 24(2):191–94.

–. 1993. Extirpation and recolonization of the Buckeye, *Junonia coenia* (Nymphalidae), following the northern California freeze of December 1990. *J. Res. Lep.* 30(3–4):209–20.

Shapiro, A.M., and J.D. Biggs. 1970. A hybrid *Limenitis* from New York. *J. Res. Lep.* 7(3):149–52.

Shapiro, A.M., and S.P. Courtney. 1985. Loss of the "pierid mate-refusal posture" in *Phulia* and allied high-Andean genera (Lepidoptera: Pieridae). *Anim. Behav.* 33(4):1388–90.

Shapiro, A.M., and J.E. DeVay. 1987. Hypersensitivity reaction of *Brassica nigra* L. (Cruciferae) kills eggs of *Pieris* butterflies (Lepidoptera: Pieridae). *Oecologia* 71(4):631–32.

Shapiro, A.M., and Hansjuerg Geiger. 1986. Electrophoretic confirmation of the species status of *Pontia protodice* and *P. occidentalis* (Pieridae). *J. Res. Lep.* 25(1):39–47.

–. 1989. Electrophoretic comparisons of vicariant *Vanessa:* Genetic differentiation between *V. annabella* and *V. carye* (Nymphalidae) since the Great American Interchange. *J. Lep. Soc.* 43(2):81–92.

Shapiro, A.M., C.A. Palm, and K.L. Weislo. 1981. The ecology and biogeography of the butterflies of the Trinity Alps and Mount Eddy, northern California. *J. Res. Lep.* 18(2):69–151.

Shapiro, A.M., and A.R. Shapiro. 1974. The ecological associations of the butterflies of Staten Island (Richmond County, New York). *J. Res. Lep.* 12(2):65–128.

Sharp, M.A., and D.R. Parks. 1973. Habitat selection and population structure in *Plebejus saepiolus* Boisduval (Lycaenidae). *J. Lep. Soc.* 27(1):17–22.

Shepard, J.H. 1964. The genus *Lycaeides* in the Pacific Northwest. *J. Res. Lep.* 3(1):25–36.

–. 1966. A study of the hilltopping behavior of *Pieris occidentalis* Reakirt (Lepidoptera: Pieridae). *Pan-Pac. Ent.* 42(4):287–94.

–. 1975. Genus *Boloria* Reuss. Pp. 243–52 in *The butterflies of North America,* edited by W.H. Howe. Garden City, NY: Doubleday. xiii + 633 pp. + 97 pls.

–. 1977. Immigration of *Phyciodes mylitta* to Vancouver Island, British Columbia (Lepidoptera: Nymphalidae). *Pan-Pac. Ent.* 53(3):167–68.

–. 1983. Additional records of *Thymelicus lineola* (Oshsenheimer) in British Columbia (Lepidoptera: Hesperiidae). *Pan-Pac. Ent.* 58(3):260.

–. 1984. Type locality restrictions and lectotype designations for the "Rocky Mountain" butterflies described by Edward Doubleday in "The Genera of Diurnal Lepidoptera" 1847–1849. *Quaestiones Ent.* 20(1):35–44.

–. 1995. The status of butterflies of conservation concern on southeastern Vancouver Island and the adjacent Gulf Islands. Unpublished report. Victoria, BC: Brit. Columbia Ministry of Environment, Lands and Parks and BC Conservation Data Centre.

–. 1998. The correct name for the *Boloria chariclea/titania* complex in North America (Lepidoptera: Nymphalidae). Pp. 727–30 in *Systematics of western North American butterflies,* edited by T.C. Emmel. Gainesville, FL: Mariposa Press. xxviii + 878 pp.

–. 2000a. Status of five butterfles and skippers in British Columbia. Victoria, BC: Brit. Columbia Ministry of Environment, Lands and Parks, Wildlife Branch and Resources Inventory Branch. Wildlife Working Rep. No. WR-101. 7 + 27.

–. 2000b. Final report for the 1997 and 1999 survey of macrolepidoptera of the Peace River canyon. Unpublished report, Brit. Columbia Ministry of Environment, Lands and Parks, Environment Branch, Fort St. John, BC. 15 pp. + detailed appendices.

Shepard, J.H., Lars Crabo, and J.P. Pelham. 1998. Revision of *Boloria natazhati* (Gibson) (Lepidoptera: Papilionidae). Pp. 731–36 in *Systematics of western North American butterflies,* edited by T.C. Emmel. Gainesville, FL: Mariposa Press. xxviii + 878 pp.

Shepard, J.H., and N. Kondla. 1993. The type locality of *Boloria freija nabokovi* Stallings & Turner (Lepidoptera: Nymphalidae). *Pan-Pac. Ent.* 69(3):273–74.

Shepard, J.H., and T.R. Manley. 1998. A species revision of the *Parnassius phoebus* complex in North America (Lepidoptera: Papilionidae). Pp. 717–26 in *Systematics of western North American butterflies,* edited by T.C. Emmel. Gainesville, FL: Mariposa Press. xxviii + 878 pp.

Shepard, J.H., and S.M. Shepard. 1974. One new species and two range extensions for British Columbia butterflies. *J. Lep. Soc.* 28(4):348.

Shepard, J.H., and S.S. [sic] Shepard. 1975. Subfamily Parnassiinae. Pp. 403–9 in *The butterflies of North America,* edited by W.H. Howe. Garden City, NY: Doubleday. xiii + 633 pp. + 97 pls.

Sherman, P.W., and W.B. Watt. 1973. The thermal ecology of some *Colias* butterfly larvae. *J. Comp. Physiol.* 83(1):25–40.

Sherman, R.S. 1918. In Memoriam Captain R.V. Harvey. *Proc. Ent. Soc. Brit. Columbia* 8:29–30, plate.

Shields, Oakley. 1966 *Callophrys (Mitoura) spinetorum* and *C. (M.) johnsoni:* their known range, habits, variation, and history. *J. Res. Lep.* 4(4):233–50.

–. 1968a. Fixation of the type locality of *Lycaena phlaeas hypophlaeas* (Boisduval) and a foodplant correction. *J. Res. Lep.* 6(1):22.

–. 1968b. Hilltopping, an ecological study of summit congregation behavior of butterflies on a southern California hill. *J. Res. Lep.* 6(2):69–178.

–. 1973. A review of carrying pair behavior and mating times in butterflies. *J. Res. Lep.* 12(1):25–64.

–. 1974. Resistance in butterfly foodplants. *J. Lep. Soc.* 28(3):288.

–. 1977. Fossil butterflies and the evolution of Lepidoptera. *J. Res. Lep.* 15(3):132–43.

Shields, Oakley, J.F. Emmel, and D.E. Breedlove. 1970. Butterfly larval foodplant records and a procedure for reporting foodplants. *J. Res. Lep.* 8(1):21–36.

Shields, Oakley, and J.C. Montgomery. 1967. The distribution and bionomics of the arctic-alpine *Lycaena phlaeas* subspecies in North America. *J. Res. Lep.* 5(4):231–42, 265–66.

Shirôzu, Takashi. 1956. A generic revision of the phylogeny of the tribe Theclini (Lepidoptera: Lycaenidae). *Sieboldia* 1(4):329–423, pls. 35–85.

Shirôzu, Takashi, and Hideho Yamamoto. 1953. Morphology of the male genital organ of *Argyronome laodice japonica* Ménétriés (Lepidoptera, Nymphalidae). *Sieboldia* 1(2):161–68, pls. 20–23.

Shuey, J.A. 1994. Phylogeny and biogeography of *Euphyes* Scudder (Hesperiidae). *J. Lep. Soc.* 47(4):261–78.

Shull, E.M. 1977. Colony of *Pieris napi oleracea* (Pieridae) in Indiana. *J. Lep. Soc.* 31(1):68–70.

Sibatani, Atuhiro. 1974. A new genus for two new species of Lycaeninae (S. Str.) (Lepidoptera: Lycaenidae) from Papua New Guinea. *J. Aust. Ent. Soc.* 13(2):95–110.

Silberglied, R.E. 1984. Visual communication and sexual selection among butterflies. Pp. 207–23 in *The biology of butterflies. Symposium of the Royal Entomological Society of London number 11,* edited by R.I. Vane-Wright and P.R. Ackery. London: Academic Press. xxiv + 429 pp.

Singer, M.C., D. Ng, and C.D. Thomas. 1988. Heritability of oviposition preference and its relationship to offspring performance within a single insect population. *Evolution* 42(5):977–85.

Singer, M.C., and P. Wedlake. 1981. Capture does affect probability of recapture in a butterfly species. *Ecol. Ent.* 6(2):215–16.

Skinner, Henry. 1889. A list of the butterflies of Philadelphia, PA. *Can. Ent.* 21(7):126–31.

–. 1893. The larva and chrysalis of *Chrysophanus dione. Can. Ent.* 25(1):22.

–. 1904. A new *Thecla* from the Northwest. *Ent. News* 15(9):298–99.

Slansky, Frank Jr. 1993. Nutritional ecology: The fundamental quest for nutrients. Pp. 29–91 in *Caterpillars: Ecological and evolutionary constraints on foraging,* edited by N.E. Stamp and T.M. Casey. New York: Chapman and Hall. xiii + 587 pp.

Smith, B.E., ed. 1902a. *The century dictionary and cyclopedia.* Vol. 9, *Proper names.* New York: The Century Co. vii + 1085 pp.

–, ed. 1902b. *The century dictionary and cyclopedia.* Vol. 10, *The century atlas of the world.* New York: The Century Co. xxx + 401 pp. + 118 maps.

Smith, L.V. 1982. Longevity estimates of four individual butterflies. *J. Lep. Soc.* 35(3):172.

–. 1984. A twelve-year count of three California butterflies. *J. Lep. Soc.* 37(4):275–80.

Spencer, G.J. 1956. In Memoriam James Rushton John Llewllyn-Jones, 1894–1953. *Proc. Ent. Soc. Brit. Columbia* 52:47.

–. 1957. The collections of Lepidoptera in the Department of Entomology, University of British Columbia. *Proc. Ent. Soc. Brit. Columbia* 54:51.

Sperling, F.A.H. 1987. Evolution of the *Papilio machaon* species group in western Canada (Lepidoptera: Papilionidae). *Quaestiones Ent.* 23(2):198–315.

–. 1993. Mitochondrial DNA phylogeny of the *Papilio machaon* species group (Lepidoptera: Papilionidae). *Mem. Ent. Soc. Can.* 165:233–42.

Sprague, P.S. 1868. Capture of *Pieris rapae* in the U. States. *Can. Ent.* 1(3):21.

–. 1871. Parasite on *Pieris rapae*. *Can. Ent.* 3(12):235.

Stallings, D.B., and J.R. Turner. 1947. New American butterflies. *Can. Ent.* 78(7–8):134–37.

Stamp, N.E. 1984. Interactions of parasitoids and checkerspot caterpillars *Euphydryas* spp. (Nymphalidae). *J. Res. Lep.* 23(1):2–18.

–. 1987. Physical constraints of defense and response to invertebrate predators by pipevine caterpillars (*Battus philenor:* Papilionidae). *J. Lep. Soc.* 40(3):191–205.

Stamp, N.E., and M.D. Bowers. 1990. Body temperature, behavior, and growth of early-spring caterpillars (*Hemileuca lucina:* Saturniidae). *J. Lep. Soc.* 44(3):143–55.

Stanford, R.E., and P.A. Opler. 1993. *Atlas of western USA butterflies.* Denver and Fort Collins, CO: Stanford and Opler. x + 275 pp. + map.

Strecker, Herman. 1885. Description of a new *Colias* from the Rocky Mountains and of an example of polymelanism in *Samia cercropia. Proc. Acad. Nat. Sci. Phil.* 37:24–27.

–. 1892. On *Argynnis astarte* Doubl. – Hew. and other matters. *Ent. News* 3(9):218–20.

Struzik, Ed. 1998. Butterfly blips. *Equinox* (Dec./Jan.):48–55.

Sugden, B.A. 1970. Annotated list of forest insects of British Columbia, Pt. 14, *Polygonia, Nymphalis,* and *Limenitis* (Nymphalidae). *J. Ent. Soc. Brit. Columbia* 67:30–31.

Sugden, B.A., and D.A. Ross. 1963. Annotated list of forest insects of British Columbia. Pt. 11, *Papilio* spp. (Papilionidae). *Proc. Ent. Soc. Brit. Columbia* 60:17–18.

Taylor, E.D. 1990. A very gentle man. The Reverend George William Taylor, MA, FRSC, FZS. 1854–1912. Unpublished manuscript.

Taylor, G.W. 1884. Notes on the entomology of Vancouver Island. *Can. Ent.* 16(4):61–62.

–. 1888. Visit to the home of *Chionobas gigas,* Butler. *Rep. Ent. Soc. Ontario* 18:24–25.

Taylor, R.L., and Bruce MacBride. 1977. Vascular plants of British Columbia, a descriptive resource inventory. Univ. Brit. Columbia Bot. Garden Tech. Bull. 4:i–xxiv, 1–754.

Thomas, J.A. 1983. A quick method for estimating butterfly numbers during surveys. *Biol. Conser.* 27(2):195–211.

–. 1984. Conservation of butterflies in temperate countries: Past efforts and lessons for the future. Pp. 333–53 in *The biology of butterflies. Symposium of the Royal Entomological Society of London number 11,* edited by R.I. Vane-Wright and P.R. Ackery. London: Academic Press. xxiv + 429 pp.

Thompson, J.N., and O. Pellmyr. 1991. Evolution of oviposition behavior and host preference in Lepidoptera. *Ann. Rev. Ent.* 36:65–89.

Threatful, D.L. 1989. A list of the butterflies and skippers on Mount Revelstoke and Glacier National Parks, British Columbia, Canada (Lepidoptera). *J. Res. Lep.* 27(3–4):213–21.

Tietz, H.M. 1972. *An index to the described life histories, early stages and hosts of the macrolepidoptera of the continental United States and Canada.* Sarasota, FL: Allyn Museum of Entomology. iv + 1041 pp.

Tilden, J.W. 1947. An occurrence of the pupa of *Glaucopsyche lygdamus behrii* (Edwards) in an ant nest (Lepidoptera, Lycaenidae). *Pan-Pac. Ent.* 23(1):42–43.

–. 1962. General characteristics of the movements of *Vanessa cardui* (L.). *J. Res. Lep.* 1(1):43–49.

–. 1963. An analysis of the North American species of the genus *Callophrys. J. Res. Lep.* 1(4):281–300.

–. 1975. An analysis of the W.G. Wright butterfly and skipper plesiotypes in the collection of the California Academy of Sciences. Occ. Pap. Calif. Acad. Sci. 118:1–44.

Tilden, J.W., and A.C. Smith. 1986. *A field guide to western butterflies.* Boston: Houghton Mifflin. xiv + 373 pp.

Tong, M.L., and A.M. Shapiro. 1989. Genetic differentiation among California populations of the Anise Swallowtail butterfly, *Papilio zelicaon* Lucas. *J. Lep. Soc.* 43(3):217–28.

Treherne, R.C. 1914. Report from Vancouver District: Insects economically important in the Lower Fraser Valley. *Proc. Ent. Soc. Brit. Columbia* 6:19–33.

–. 1915. Shade-tree and ornamental insects of British Columbia. *Proc. Ent. Soc. Brit. Columbia* 7:35–41.

Troubridge, J.T., and D.K. Parshall. 1988. A review of the *Oeneis polixenes* (Fabricius) (Lepidoptera: Satyrinae) complex in North America. *Can. Ent.* 120(7): 679–96.

Tschudi-Rein, Kathrin, and George Benz. 1990. Mechanisms of sperm transfer in female *Pieris brassicae* (Lepidoptera: Pieridae). *Ann. Ent. Soc. Amer.* 83(6):1158–64.

Turner, N.J. 1994. Burning mountainsides for better crops: Aboriginal landscape burning in British Columbia. *Inter. J. Ecoforestry* 10(3):116–22.

Tuzov, V.K., ed. 1997. *Guide to the butterflies of Russia and adjacent territories (Lepidoptera, Rhopalocera).* Vol. 1. Sofia – Moscow: Pensoft. 480 pp.

Urquhart, F.A. 1960. *The Monarch butterfly.* Toronto: University of Toronto Press. 361 pp.

–. 1976. Found at last: The Monarch's winter home. *National Geographic* 150(2):160–73.

–. 1987. *The Monarch butterfly: International traveler.* Chicago: Nelson Hall. 232 pp.

Urquhart, F.A., and A.P.S. Tang. 1971. The effect of cauterizing the PPM ["gold spots" of authors] of the pupa of the Monarch butterfly (*D. plexippus*). *J. Res. Lep.* 9(3):157–67.

Urquhart, F.A., and N.R. Urquhart. 1979. Aberrant autumnal migration of the eastern population of the Monarch butterfly, *Danaus p. plexippus* (Lepidoptera: Danaidae) as it relates to the occurrence of strong westerly winds. *Can. Ent.* 111(11):1281–86.

Venables, B.A.B., and E.M. Barrows. 1986. Skippers: Pollinators or nectar thieves? *J. Lep. Soc.* 39(4):299–312.

Venables, E.P. 1912. Report from Okanagan district. *Proc. Ent. Soc. Brit. Columbia* 1:8–11.

–. 1913. Report from Okanagan District. *Proc. Ent. Soc. Brit. Columbia* 2:11–13.

Voss, E.G. 1961. Season's summary, 1960. 5. Central Region: Missouri to West Virginia, north to Ontario. *News Lep. Soc.* 1961(4):7–9.

Wagner, Diane. 1993. Species-specific effects of tending ants on the development of lycaenid butterfly larvae. *Oecologia* 96(2):276–81.

Warren, B.C.S. 1936. *Monograph of the genus* Erebia. London: British Museum (Nat. His.).

–. 1944. Review of the classification of the Argynnidi:with a systematic revision of the genus *Boloria* (Lepidoptera; Nymphalidae). *Trans. R. Ent. Soc. Lond.* 94(1):1–153.

Warren, B.C.S., C.F. dos Passos, and L.P. Grey. 1946. Supplementary notes on the classification of Argynninae (Lepidoptera, Nymphalidae). *Proc. R. Ent. Soc. Lond.* (B)15(5–6):71–73.

Warren, M.S. 1990. [*Leptidea sinapsis* (Linnaeus). The Wood White]. Pp. 84–86 in *The moths and butterflies of Great Britain and Ireland.* Vol. 7, Pt. I, *Hesperiidae to Nymphalidae,* edited by A.M. Emmet and John Heath. Colchester, UK: Harley Books. ix + 370 pp.

Watanabe, Manotue. 1976. A preliminary study on population dynamics of the swallowtail butterfly, *Papilio xuthus* L. in a deforested area. *Res. Pop. Ecol.* 17(2):200–10.

–. 1981. Population dynamics of the swallowtail butterfly, *Papilio xuthus* L. in a deforested area. *Res. Pop. Ecol.* 23(1):74–93.

Watt, W.B., F.S. Chew, L.R.G. Snyder, A.G. Watt, and D.E. Rothschild. 1977. Population structure of pierid butterflies. I. Numbers and movements of some montane *Colias* species. *Oecologia* 27(1):1–22.

Watt, W.B., P.C. Hoch, and S.G. Mills. 1974. Nectar resource use by *Colias* butterflies. *Oecologia* 14(3/4):353–74.

Webb, C.J., and K.S. Bawa. 1983. Pollen dispersal by hummingbirds and butterflies: A comparative study of two lowland tropical plants. *Evolution* 37(6):1258–70.

Webster, F.M. 1894. Butterflies common to Norway and arctic North America. *Can. Ent.* 26(5):117–20.

Wehling, W.F. 1994. Geography of host use, oviposition preference, and gene flow in the Anise Swallowtail butterfly (*Papilio zelicaon*). Ph.D. thesis, Washington State University (Pullman). xi + 118 pp.

Weiss, H.B. 1998. The pioneer century of American entomology. Chapter 3, The early years of the 19th century. *Lep. News* 1998(3):6–18.

Weller, S.J., D.P. Pashley, and J.A. Martin. 1996. Reassessment of butterfly family relationships using independent genes and morphology. *Ann Ent. Soc. Amer.* 89(2):184–92.

Welling, E.C. 1959. More observations of the "pumping" action of moths at water. *Lep. News* 12(5–6):170–72.

Whitney, W.D., ed. 1902. *The century dictionary and cyclopedia*. Vols. 1–8, *The century dictionary*. New York: The Century Co. ii + xviii + 7,046 + 30 pp.

Wickman, P.O. 1986. Male determined mating duration in butterflies? *J. Lep. Soc.* 39(4):341–42.

Wiklund, Christer. 1984. Egg-laying patterns in butterflies in relation to their phenology and the visual apparency and abundance of their host plants. *Oecologia* 63(1):23–29.

Williams, C.B. 1949a. Migrant butterflies in North America. *Lep. News* 3(2):17–18.

–. 1949b. Migrant butterflies outside North America. *Lep. News* 3(4–5):39.

Williams, E.H. 1988. Habitat and range of *Euphydryas gillettii* (Nymphalidae). *J. Lep. Soc.* 42(1):37–45.

–. 1995. Fire-burned habitat and reintroductions of the butterfly *Euphydryas gillettii* (Nymphalidae). *J. Lep. Soc.* 49(3):183–91.

Williams, E.H., C.E. Holdren, and P.R. Ehrlich. 1984. The life history and ecology of *Euphydryas gillettii* Barnes (Nymphalidae). *J. Lep. Soc.* 38(1):1–12.

Williams, F.X. 1908. The life-history of *Lycaena antiacis* Bdv., with other notes on other species. *Ent. News* 19(10):476–83.

–. 1910. The butterflies of San Francisco, California. *Ent. News* 21(1):30–41.

Williams, K.S., and L.E. Gilbert. 1981. Insects as selective agents on plant vegetative morphology: Egg mimicry reduces egg laying by butterflies. *Science* 212(4493):467–69.

Williams, R.T. 1889. *Williams' British Columbia directory 1889. Containing general information and directories of the various cities and settlements in the province, with a classified business directory*. Victoria: R.T. Williams. xiv + 592 pp. + 1 map.

–. 1892. *Williams' British Columbia directory 1892. Containing general information and directories of the cities and settlements in British Columbia, with a classified business directory*. Victoria: R.T. Williams. xx + 1,253 pp. + 2 maps.

Winn, A.F. 1914. A protected butterfly. *Can. Ent.* 46(3):109.

Wolff, N.L. 1964. The Lepidoptera of Greenland. *Meddelelser om Grönland* 159(11):1–74, pls. i–xxi.

Woods, R.S. 1944. *The naturalist's lexicon. A list of classical Greek and Latin words used or suitable for use in biological nomenclature*. Pasadena, CA: Abbey Garden Press. xvii + 282 pp.

Wourms, M.K., and F.K. Wasserman. 1986. Bird predation on Lepidoptera and the reliability of beak-marks in determining predation pressure. *J. Lep. Soc.* 39(4):239–61.

Wright, W.G. 1884. Notes on the preparatory stages of *Lycaena amyntula*. *Papilio* 4(7–8):126–28.

–. 1891. *Vanessa californica*. *Can. Ent.* 23(2):27–28.

–. 1892a. Description of a new *Anthocharis*. *Can. Ent.* 24(7):154–55.

–. 1892b. In Alaska. *Ent. News* 3(4):74–77.

–. 1896. *Pieris rapae* and *Agraulis vanillae*. *Can. Ent.* 28(4):102.

–. 1905. *The butterflies of the west coast of the United States*. San Bernardino, CA: W.G. Wright. 257 + vii pp. + 32 pls.

Young, A.M. 1977. Butterflies associated with an army ant swarm raid in Honduras: The "feeding hypothesis" as an alternate explanation. *J. Lep. Soc.* 31(3):190.

Young, R.M. 1987. Mass emergences of the Pine White, *Neophasia menapia menapia* (Felder and Felder), in Colorado (Pieridae). *J. Lep. Soc.* 40(4):314.

Zalucki, M.P., and R.L. Kitching. 1985. The dynamics of adult *Danaus plexippus* L. (Danaidae) within patches of its food plant, *Asclepias* spp. *J. Lep. Soc.* 38(3):209–19.

Ziegler, J.B. 1953. Notes on the life history of *Incisalia augustinus* and a new host plant record (Lycaenidae). *Lep. News* 7(2):33–35.

Zimmermann, Marie, Niklas Wahlberg, and Henri Descimon. 2000. Phylogeny of *Euphydryas* checkerspot butterflies (Lepidoptera: Nymphalidae) based on mitochondrial DNA sequence data. *Ann. Ent. Soc. Amer.* 93(3):347–55.

# CREDITS

## PHOTOGRAPHS

Crispin Guppy took all of the museum specimen photographs and many of the live photographs for this book. A number of other photographers also contributed live photographs. They are listed below, with the full data given in Appendix 3.

| PAGE | DESCRIPTION | PHOTOGRAPHER |
|---|---|---|
| 1 | Northwestern Fritillary | Anna Roberts |
| 2 | Mylitta Crescent | Arthur Guppy |
| 4-5 | Western Sulphur | Derrick Marven |
| 8 | Northern Checkerspot | Anna Roberts |
| 39 | Painted Lady | Richard Beard |
| 44 | Cabbage White | Arthur Guppy |
| 46 | Sara's Orangetip pupation sequence | Arthur Guppy |
| 47 | Egg and micropyle | Jon Shepard |
| 55 | Mourning Cloak | Anna Roberts |
| 68 | Compton Tortoiseshell | Anna Roberts |
| 69 | Anglewings feeding on dung | Steve Ansell |
|  | Grey Anglewing | Anna Roberts |
| 72 | Red Admiral | Richard Beard |
| 81 | Common Branded Skipper | Anna Roberts |
| 85 | Arctic Skipper | Anna Roberts |
| 88 | Northern Cloudywing | Anna Roberts |
| 101 | Arctic Skipper | Anna Roberts |
| 116 | Woodland Skipper | Arthur Guppy |
|  | Dun Skipper | Derrick Marven |
| 119 | Painted Lady | Anna Roberts |
| 120 | Two-tailed Swallowtail | Richard Beard |
| 124 | Clodius Apollo | Derrick Marven |
| 126 | Mature larva, Rocky Mountain Apollo | Anna Roberts |
|  | Pupa, Rocky Mountain Apollo | David McCorkle |
| 127 | Mature larva, Phoebus Apollo (2 photos) | Oleg Kosterin |
| 130 | Mature larva and pupa, Baird's Swallowtail | Anna Roberts |
|  | Baird's Swallowtail | Anna Roberts |
| 135 | Pupa, Canadian Tiger Swallowtail | Anna Roberts |
| 136 | Mature larva, Western Tiger Swallowtail | Arthur Guppy |
| 138 | Pupa, Pale Swallowtail | Richard Beard |
| 139 | Western Tiger Swallowtail and Pale Swallowtails | Stephen Walker |
| 153 | Mustard White | Anna Roberts |
| 159 | Large Marble | Anna Roberts |
| 164 | Pupa, Sara's Orangetip | David McCorkle |
| 166 | Stella's Orangetip | Anna Roberts |
| 168 | Clouded Sulphur | Anna Roberts |
| 169 | Mature larva, Clouded Sulphur | Anna Roberts |
| 180 | Mature larvae, Arctic Sulphur | David McCorkle |
| 182 | Pink-edged Sulphur | Anna Roberts |
| 187 | Purplish Copper | Anna Roberts |
| 190 | Pupa, Lustrous Copper | Gordon Pratt |
|  | Mature larva, Lustrous Copper | Richard Beard |
| 191 | Mature larva, Bronze Copper | Gordon Pratt |

| PAGE | DESCRIPTION | PHOTOGRAPHER |
|---|---|---|
| 194 | Mature larva and pupa, Blue Copper | Gordon Pratt |
| 196 | Mature larva, Purplish Copper | Richard Beard |
| 197 | Mature larva, Lilac-bordered Copper | Gordon Pratt |
| 198 | Mature larva and pupa, Reakirt's Copper | Richard Beard |
| 200 | Mature larva, Coral Hairstreak | Gordon Pratt |
|  | Behr's Hairstreak | Anna Roberts |
| 201 | Mature larva and pupa, Behr's Hairstreak | Gordon Pratt |
| 202 | Mature larva and pupa, Sooty Hairstreak | Gordon Pratt |
| 203 | California Hairstreak | Anna Roberts |
|  | Mature larva, California Hairstreak | Gordon Pratt |
| 204 | Mature larva, Sylvan Hairstreak | Gordon Pratt |
| 205 | Mature larva, Striped Hairstreak | Gordon Pratt |
| 210 | Pupa, Thicket Hairstreak | Gordon Pratt |
| 211 | Mature larva and pupa, Johnson's Hairstreak | David McCorkle |
| 213 | Juniper Hairstreak | Anna Roberts |
| 216 | Western Elfin | Derrick Marven |
|  | Egg and mature larva (left), Western Elfin | Arthur Guppy |
|  | Mature larva (right), Western Elfin | Richard Beard |
| 217 | Moss' Elfin | Arthur Guppy |
|  | Mature larva, Moss' Elfin | David McCorkle |
| 218 | Mature larva, Hoary Elfin | David McCorkle |
| 219 | Mature larva, Eastern Pine Elfin | David McCorkle |
| 220 | Western Pine Elfin | Anna Roberts |
| 221 | Mature larva and pupa, Grey Hairstreak | Arthur Guppy |
| 222 | Grey Hairstreak | Arthur Guppy |
| 224 | Mature larva, Eastern Tailed Blue | Gordon Pratt |
| 225 | Western Tailed Blue | Anna Roberts |
|  | Mature larva and pupa, Western Tailed Blue | Gordon Pratt |
| 227 | Mature larva, Western Spring Azure | Gordon Pratt |
| 228 | Boreal Spring Azure | Anna Roberts |
| 230 | Square-spotted Blue | Anna Roberts |
|  | Mature larva and pupa, Square-spotted Blue | Gordon Pratt |
| 231 | Pupa and mature larva, Arrowhead Blue | Gordon Pratt |
| 232 | Pupa and mature larva, Silvery Blue | Anna Roberts |
| 234 | Northern Blue | Anna Roberts |
| 235 | Mature larva, Northern Blue | Anna Roberts |
| 237 | Anna's Blue | Richard Beard |
| 238 | Mature larva, Melissa's Blue | Gordon Pratt |
|  | Greenish Blue | Anna Roberts |
| 239 | Mature larva, Greenish Blue | Gordon Pratt |
| 242 | Acmon Blue | Arthur Guppy |
|  | Mature larva and pupa, Acmon Blue | Gordon Pratt |
| 247 | Mature larva, Mormon Metalmark | David McCorkle |

## DRAWINGS

*Note:* Page numbers in bold indicate the main discussion of a topic.

Ericaceae (family), 167, 188, 198, 243
*ericetorum, Heliopetes*, 97, **98**
*Eriogonum* (genus), 188, 194, 207, 208, 209, 229, 230, 242
*Eriogonum niveum*, 247
*Eriogonum umbellatum*, 194
*Eriophorum*, 333, 350
*Eriophorum* (genus), 321, 347
*Eriophorum vaginatum* var. *spissum*, 347
*Eriophyllum lanatum*, 266
Ermatinger, Edward, 17
*Eruca vesicaria* ssp. *virginicum*, 157
*Erynnis* (genus), 87, **89**, 94, 99
*Erynnis afranius*, 33, **92–3**
*Erynnis icelus*, **90**
*Erynnis pacuvius*, **92**
*Erynnis persius*, 92, **93–4**
*Erynnis propertius*, 33, 42, **91**
*eryphon, Incisalia*, 25, 42, **220**
*Erysimum asperum*, 160
eucalyptus, 353
*Euchloe* (genus), 33, **158**, 158–63
*Euchloe ausonia*, **159**
*Euchloe ausonides*, 29, 32, 33, 56, **159–61**
*Euchloe creusa*, **161**
*Euchloe hyantis*, 163
*Euchloe lotta*, **162–3**
*Euchloe naina*, 33, **162**
Eulophidae (family), 64
*eunomia, Clossiana*, 23, **284–5**
*Euphilotes* (genus), **229**
*Euphilotes ancilla*, 229
*Euphilotes battoides*, **229–30**
*Euphilotes enoptes*, 229
*Euphydryas* (genus), **307**
*Euphydryas anicia*, 25, **308–10**
*Euphydryas chalcedona*, 62, 64, 68, **307–8**
*Euphydryas editha*, 26, 34, 56, 64, 68, **310–11**
*Euphydryas gillettii*, 34, 43, **311–12**
*euphydryadis, Apanteles*, 64
*Euphyes* (genus), **116**
*Euphyes vestris*, 33, 111, **116–17**
*Euptoieta* (genus), **270**
*Euptoieta claudia*, 114, **271**
*Eurema* (genus), **184**
*Eurema nicippe*, 167, **184–5**
European Skipper (*Thymelicus lineola*), 27, 68, 75, 76, 102, **103**
*eurymedon, Papilio*, **138–9**
*eurytheme, Colias*, 41, 58, 75, 167, **170–1**, 174
*Everes* (genus), **223**
*Everes amyntula*, 78, **224–5**
*Everes comyntas*, 33, **223–4**
*Everini* (tribe), 223
*eversmanni, Parnassius*, 122
Eversmann's Apollo (*Parnassius eversmanni*), **122**
extirpated, definition, 31
Eyed Hawkmoths, 42

**F**abaceae (family), 87, 89, 167, 172, 174, 176, 177, 178, 199, 230
'false heads,' 63
*fasciata, Erebia*, **333**
*faunus, Polygonia*, **251–2**
*Feniseca traquinius*, 46
*Festuca* (genus), 104, 106, 321, 326, 346
*Festuca idahoensis*, 113, 342, 344
*Festuca ovina*, 346
Field Crescent (*Phyciodes pratensis*), 41, **301**
field mustard (*Brassica campestris*), 156
Findlay, G.H. (Rev.), 19
Fischer, Aud Itelke, 317
Fischer, Britta Ilge, 317
Fletcher, James, 18, 205
*Foeniculum* (genus), 132
foodplants. *See also* names of specific plants
  adult butterflies, **68–71**
  defences against oviposition, **56–7**
  larval, **39–40**, 41, **71–5**
  monophagous, 55
  nectar, 41, **68–9**
  oligophagous, 55
  polyphagous, 55
  recognition, **55–6**
*Forcipomyia* (genus), 60
Forcipomyiinae (subfamily), 60
Forest Insect Survey (FIS), 21
*Formica* (genus), 241
*Formica altipetens*, 193
*Formica pilicornis*, 67
Foxlee, Harold, 21
*Fragaria* (genus), 96
*Fragaria vesca* ssp. *bracteata*, 222
*Fragaria virginiana*, 95
*Frangula californica*, 139
*frass*, 47
*Fraxinus* (genus), 134
*freija, Clossiana*, **293**
Freija Fritillary (*Clossiana freija*), **293**
*frigga, Clossiana*, **287–8**
Frigga Fritillary (*Clossiana frigga*), **287–8**
fritillaries
  Argynninae, **270**, 270–97
  *Boloria*, 282–3
  *Clossiana*, 284–97
  *Euptoieta*, 270–1
  *Speyeria*, 75, 271–82
*fuliginosum, Satyrium*, 33, **202**
Fumariaceae (family), 122, 124

**g**ardens, butterfly, **41–4**
*garita, Oarisma*, 27, **102**, 286
Garita Skipperling (*Oarisma garita*), 27, **102**, 286
garlic mustard, 155
Garrett, A.B., 21
Garrett, C.B., 274
Garrett's Fritillary (*Speyeria zerene garretti*), **274, 275**

Garry oak (*Quercus garryana*), 23, 25, 26, 28, 29, 37, 42
*Gaultheria humifusa*, 183
*Gaultheria shallon*, 216, 222
Giant Sulphur (*Colias gigantea*), 33, **183–4**
*gigantea, Colias*, 33, **183–4**
Gillett, Clarence P., 311
*gillettii, Euphydryas*, 34, 43, **311–12**
Gillett's Checkerspot (*Euphydryas gillettii*), 34, 43, **311–12**
glaciers, and butterflies, 23–5
*glandon, Agriades*, 34, **244–5**
*Glaucopsyche* (genus), 229, **230**
*Glaucopsyche alexis*, **231**
*Glaucopsyche lygdamus*, 59, 67, **231–3**
*Glaucopsyche piasus*, **230–1**
*Glaucopsyche xerxes*, 230
*glaucus, Papilio*, 37, 68, 70, 71, **134**
global warming, effect on butterflies, 28
*Gluphisia septentrionis*, 70
*Glyceria* (genus), 347
*Glycyrrhiza lepidota*, 88, 238
*Gnaphalium* (genus), 266
*Gnaphalium bicolor*, 265
*Gnaphalium obtusifolium*, 265
*Gnaphalium palustre*, 265
*Gnaphalium polycephalum*, 265
*Gnaphalium purpureum*, 265
gooseberry (*Ribes rotundifolium*), 256
goosegrass (*Eleusine indica*), 114
gossamer wings (Lycaenidae), 33, 34, **187**, 187–245
Gosse, Phillip, 11, 12
*gracilis, Polygonia*, 69, **253**
grass skippers (Hesperiinae), **100**, 100–18
grasses, 41, 109, 118, 320, 321, 323, 324, 325, 326, 331, 332, 333, 334, 339, 342, 344, 346, 350
grazing, effect on butterfly habitat, 28, 29
Great Arctic (*Oeneis nevadensis*), 34, 52, **340**
Great Basin Woodnymph (*Cercyonis sthenele*), **326**
Great Copper (*Lycaena xanthoides*), 192
Great Spangled Fritillary (*Speyeria cybele*), 34, **272–3**
greater fritillaries (*Speyeria*), 270, **271**, 272–82
Green, Charles de Blois, 18, 22
green alder (*Alnus viridis*), 136
Green Comma (*Polygonia faunus*), 69, **251–2**
Green Hairstreaks (*Callophrys*), **207**, 214, 290
Green Marble (*Euchloe naina*), 33, **162**
Greenish Blue (*Plebeius saepiolus*), 32, 34, **238–9**
Gregson, John (Jack) D., 21
Grey Comma (*Polygonia progne*), **255–6**
Grey Hairstreak (*Strymon melinus*), 221–2
Grizzled Skipper (*Pyrgus centaureae*), 94, **95**, 96, 97
ground squirrel, 141
Gunder, Jean, 20, 21
Guppy, Richard, 21, 22
Gypsy Moth (*Lymantria dispar*), **36–7**

Uncompahgre Fritillary (*Clossiana improba acrocnema*), 28, **288**
Uncus Skipper (*Hesperia uncas*), **104,** 109
Underhill, James (Ted) Edward, 21
University of British Columbia, 13, 15, 16. *See also* Spencer Entomological Museum (UBC)
*Urtica dioica*, 41, 74, 250, 263, 264, 268, 269
*urticae, Aglais*, 64, 262

*Vacciniina* (genus), **243**
*Vacciniina optilete*, 34, **243–4**
*Vaccinium* (genus), 167, 182, 183, 188, 215, 228, 283
*Vaccinium caespitosum*, 181, 182, 198, 234–5, 293
*Vaccinium myrtilloides*, 182
*Vaccinium oxycoccus*, 198
*Vaccinium uliginosum*, 181, 198, 243
Vancouver, butterfly habitat disappearance, 28–9
Vancouver Island
    butterfly habitat, disappearance of, 28
    during Wisconsin glaciation, 12
*Vanessa* (genus), **264**
*Vanessa annabella*, 13, 41, 58, 74, 78, 263, **267–8**
*Vanessa atalanta*, 52, 58, 68, 74, **268–9**
*Vanessa cardui*, 39, 40, 41, 58, 64, 78, **265–7**
*Vanessa carye*, **268**
*Vanessa virginiensis*, 69, **264–5**
Variegated Fritillary (*Euptoieta claudia*), 114, **271**
Veined White (*Pieris marginalis venosa*), 151
Venables, E.P., 19
*Venusia cambrica*, 70
*vestris, Euphyes*, 33, 111, **116–17**
vetch (*Vicia*), 41, 173
*vialis, Amblyscirtes*, 99, **117–18**
*Viburnum* (genus), 228
Viceroy (*Limenitis archippus*), 27, 32, 34, 61, **315–16**
*Vicia* (genus), 41, 173
*Vicia americana*, 169, 171, 225
*Vicia sativa*, 171, 175
Vidler, Captain, 328

*vidleri, Erebia*, **328–9**
Vidler's Alpine (*Erebia vidleri*), **328–9**
*Viola* (genus), 271, 272, 274, 278, 279, 280, 282, 285, 287, 290
*Viola canadensis*, 287, 290
violet. *See Viola*
*virginiensis, Vanessa*, 69, **264–5**

**W**alker, Francis, 17
wasps, 60, 63–4
Watson, Frank Edward, 257
weather
    and butterfly movement, 65–6
    as source of butterfly mortality, 59
weeds, non-native, 26
weeping willow (*Salix*), 262
West Coast Lady (*Vanessa annabella*), 13, 41, 58, 74, 78, 263, **267–8**
Western Checkerspot (*Euphydryas chalcedona*), **308**
Western Elfin (*Incisalia iroides*), 71, 78, **215–16**
western hemlock (*Tsuga heterophylum*), 23, 26, 142, 211
Western Meadow Fritillary (*Clossiana epithore*), 34, 101, **289–90**
Western Pine Elfin (*Incisalia eryphon*), 42, **220**
western red cedar (*Thuja plicata*), 23, 212, 213, 214
Western Spring Azure (*Celastrina echo*), 70, 78, **226–7**
Western Sulphur (*Colias occidentalis*), 17, 33, **174–5**
Western Tailed Blue (*Everes amyntula*), 78, **224–5**
Western Tiger Swallowtail (*Papilio rutulus*), 42, 53, **135–6,** 138
Western White (*Pontia occidentalis*), 53, 57, 78, 79, **148–9**
western white pine (*Pinus monticola*), 142
White Admiral (*Limenitis arthemis*), 45, 53, 68, 69, 70, 78, **314–15**
white clover (*Trifolium repens*), 169, 171
white rhododendron (*Rhododendron albiflorum*), 253
White Skippers (*Heliopetes*), 94, **97,** 99

white sweet clover (*Melilotus albus*), 71
Whitehouse's Fritillary (*Speyeria aphrodite whitehousei*), 34, **274**
White-lined Sphinx (*Hyles lineata*), 72
whites
    Pieridae, 33, 42, **150**
    Pierinae, 141–57
    Pontia, **143**
White-spotted Skipper (*Epargyreus clarus*), 32, 33, 69, **87–8**
White-veined Arctic (*Oeneis bore*), 34, **345–6**
*whitneyi, Charidryas*, **305**
Whitney's Checkerspot (*Charidryas whitneyi*), **305**
willow (*Salix*), 42, 87, 89, 90, 93, 134, 167, 184, 203, 204, 251, 252, 258, 261, 264, 287, 291, 315, 316, 318, 325
winter butterflies, 77
Wisconsin glaciation, 23–5
*Wislizenia refracta*, 147
Woodland Skipper (*Ochlodes sylvanoides*), 78, 112, **115–16**
woodnymphs (*Cercyonis*), 9, 26, 68, 78
Wright, Henry William Greenwood, 12, 18, 19

*x*anthoides, Lycaena, 192
Xerces Society, 31, 38
*xerxes, Glaucopsyche*, 230
*xuthus, Papilio*, 59

**y**ellow sweet clover (*Melilotus officinalis*), 71, 171, 203
Young, S. Hall (Rev.), 288, 336
*youngi, Erebia*, **336**
Young's Alpine (*Erebia youngi*), **336**
*yreka, Pieris*, 75

**z**elicaon, Papilio, 28, 42, 45, 52, 56, 59, 74, 78, **131–3**
Zephyr Anglewing (*Polygonia zephyrus*), 77, **252–3**
*zephyrus, Polygonia*, 77, **252–3**
*zerene, Speyeria*, 34, 66, **274–6**
Zerene Fritillary (*Speyeria zerene*), 34, 66, **274–6**
Zerynthiini (tribe), **121**

Set in Thesis The Sans by Artegraphica Design Co.

Printed and bound in Canada by Friesens

Text design: Blakeley Design

Copy editor: Francis J. Chow

Cartographer: Eric Leinberger

Indexer: Annette Lorek